SECOND EDITION

Hands-On Machine Learning with Scikit-Learn, Keras, and TensorFlow

Concepts, Tools, and Techniques to Build Intelligent Systems

Aurélien Géron

Beijing · Boston · Farnham · Sebastopol · Tokyo

Hands-On Machine Learning with Scikit-Learn, Keras, and TensorFlow

by Aurélien Géron

Printed in Canada.

Published by O'Reilly Media, Inc., 1005 Gravenstein Highway North, Sebastopol, CA 95472.

O'Reilly books may be purchased for educational, business, or sales promotional use. Online editions are also available for most titles (*http://oreilly.com*). For more information, contact our corporate/institutional sales department: 800-998-9938 or *corporate@oreilly.com*.

Editors: Rachel Roumeliotis and Nicole Tache
Production Editor: Kristen Brown
Copyeditor: Amanda Kersey
Proofreader: Rachel Head

Indexer: Judith McConville
Interior Designer: David Futato
Cover Designer: Karen Montgomery
Illustrator: Rebecca Demarest

September 2019:   Second Edition

Revision History for the Second Edition
2019-09-05:   First Release
2019-10-11:   Second Release

See *http://oreilly.com/catalog/errata.csp?isbn=9781492032649* for release details.

978-1-492-03264-9

[TI]

Table of Contents

Preface. xv

Part I.   The Fundamentals of Machine Learning

1. The Machine Learning Landscape. 1
    What Is Machine Learning?   2
    Why Use Machine Learning?   2
    Examples of Applications   5
    Types of Machine Learning Systems   7
        Supervised/Unsupervised Learning   7
        Batch and Online Learning   14
        Instance-Based Versus Model-Based Learning   17
    Main Challenges of Machine Learning   23
        Insufficient Quantity of Training Data   23
        Nonrepresentative Training Data   25
        Poor-Quality Data   26
        Irrelevant Features   27
        Overfitting the Training Data   27
        Underfitting the Training Data   29
        Stepping Back   30
    Testing and Validating   30
        Hyperparameter Tuning and Model Selection   31
        Data Mismatch   32
    Exercises   33

2. End-to-End Machine Learning Project. 35
    Working with Real Data   35

Look at the Big Picture	37
Frame the Problem	37
Select a Performance Measure	39
Check the Assumptions	42
Get the Data	42
Create the Workspace	42
Download the Data	46
Take a Quick Look at the Data Structure	47
Create a Test Set	51
Discover and Visualize the Data to Gain Insights	56
Visualizing Geographical Data	56
Looking for Correlations	58
Experimenting with Attribute Combinations	61
Prepare the Data for Machine Learning Algorithms	62
Data Cleaning	63
Handling Text and Categorical Attributes	65
Custom Transformers	68
Feature Scaling	69
Transformation Pipelines	70
Select and Train a Model	72
Training and Evaluating on the Training Set	72
Better Evaluation Using Cross-Validation	73
Fine-Tune Your Model	75
Grid Search	76
Randomized Search	78
Ensemble Methods	78
Analyze the Best Models and Their Errors	78
Evaluate Your System on the Test Set	79
Launch, Monitor, and Maintain Your System	80
Try It Out!	83
Exercises	84
3. Classification.	**85**
MNIST	85
Training a Binary Classifier	88
Performance Measures	88
Measuring Accuracy Using Cross-Validation	89
Confusion Matrix	90
Precision and Recall	92
Precision/Recall Trade-off	93
The ROC Curve	97
Multiclass Classification	100

Error Analysis   102
Multilabel Classification   106
Multioutput Classification   107
Exercises   108

4. Training Models 111
Linear Regression   112
The Normal Equation   114
Computational Complexity   117
Gradient Descent   118
Batch Gradient Descent   121
Stochastic Gradient Descent   124
Mini-batch Gradient Descent   127
Polynomial Regression   128
Learning Curves   130
Regularized Linear Models   134
Ridge Regression   135
Lasso Regression   137
Elastic Net   140
Early Stopping   141
Logistic Regression   142
Estimating Probabilities   143
Training and Cost Function   144
Decision Boundaries   145
Softmax Regression   148
Exercises   151

5. Support Vector Machines 153
Linear SVM Classification   153
Soft Margin Classification   154
Nonlinear SVM Classification   157
Polynomial Kernel   158
Similarity Features   159
Gaussian RBF Kernel   160
Computational Complexity   162
SVM Regression   162
Under the Hood   164
Decision Function and Predictions   165
Training Objective   166
Quadratic Programming   167
The Dual Problem   168
Kernelized SVMs   169

       Online SVMs   172

Exercises   174

6. Decision Trees. **175**

Training and Visualizing a Decision Tree 175

Making Predictions 176

Estimating Class Probabilities 178

The CART Training Algorithm 179

Computational Complexity 180

Gini Impurity or Entropy? 180

Regularization Hyperparameters 181

Regression 183

Instability 185

Exercises 186

7. Ensemble Learning and Random Forests. **189**

Voting Classifiers 189

Bagging and Pasting 192

    Bagging and Pasting in Scikit-Learn 194

    Out-of-Bag Evaluation 195

Random Patches and Random Subspaces 196

Random Forests 197

    Extra-Trees 198

    Feature Importance 198

Boosting 199

    AdaBoost 200

    Gradient Boosting 203

Stacking 208

Exercises 211

8. Dimensionality Reduction. **213**

The Curse of Dimensionality 214

Main Approaches for Dimensionality Reduction 215

    Projection 215

    Manifold Learning 218

PCA 219

    Preserving the Variance 219

    Principal Components 220

    Projecting Down to d Dimensions 221

    Using Scikit-Learn 222

    Explained Variance Ratio 222

    Choosing the Right Number of Dimensions 223

PCA for Compression   224
Randomized PCA   225
Incremental PCA   225
Kernel PCA   226
Selecting a Kernel and Tuning Hyperparameters   227
LLE   230
Other Dimensionality Reduction Techniques   232
Exercises   233

9. Unsupervised Learning Techniques.................................... 235
Clustering   236
K-Means   238
Limits of K-Means   248
Using Clustering for Image Segmentation   249
Using Clustering for Preprocessing   251
Using Clustering for Semi-Supervised Learning   253
DBSCAN   255
Other Clustering Algorithms   258
Gaussian Mixtures   260
Anomaly Detection Using Gaussian Mixtures   266
Selecting the Number of Clusters   267
Bayesian Gaussian Mixture Models   270
Other Algorithms for Anomaly and Novelty Detection   274
Exercises   275

Part II.   Neural Networks and Deep Learning

10. Introduction to Artificial Neural Networks with Keras......................... 279
From Biological to Artificial Neurons   280
Biological Neurons   281
Logical Computations with Neurons   283
The Perceptron   284
The Multilayer Perceptron and Backpropagation   289
Regression MLPs   292
Classification MLPs   294
Implementing MLPs with Keras   295
Installing TensorFlow 2   296
Building an Image Classifier Using the Sequential API   297
Building a Regression MLP Using the Sequential API   307
Building Complex Models Using the Functional API   308
Using the Subclassing API to Build Dynamic Models   313

     Saving and Restoring a Model   314
     Using Callbacks   315
     Using TensorBoard for Visualization   317
   Fine-Tuning Neural Network Hyperparameters   320
     Number of Hidden Layers   323
     Number of Neurons per Hidden Layer   325
     Learning Rate, Batch Size, and Other Hyperparameters   325
   Exercises   327

11. Training Deep Neural Networks. **331**
   The Vanishing/Exploding Gradients Problems   332
     Glorot and He Initialization   333
     Nonsaturating Activation Functions   335
     Batch Normalization   338
     Gradient Clipping   345
   Reusing Pretrained Layers   345
     Transfer Learning with Keras   347
     Unsupervised Pretraining   349
     Pretraining on an Auxiliary Task   350
   Faster Optimizers   351
     Momentum Optimization   351
     Nesterov Accelerated Gradient   353
     AdaGrad   354
     RMSProp   355
     Adam and Nadam Optimization   356
     Learning Rate Scheduling   359
   Avoiding Overfitting Through Regularization   364
     ℓ_1 and ℓ_2 Regularization   364
     Dropout   365
     Monte Carlo (MC) Dropout   368
     Max-Norm Regularization   370
   Summary and Practical Guidelines   371
   Exercises   373

12. Custom Models and Training with TensorFlow. **375**
   A Quick Tour of TensorFlow   376
   Using TensorFlow like NumPy   379
     Tensors and Operations   379
     Tensors and NumPy   381
     Type Conversions   381
     Variables   382
     Other Data Structures   383

Customizing Models and Training Algorithms   384
   Custom Loss Functions   384
   Saving and Loading Models That Contain Custom Components   385
   Custom Activation Functions, Initializers, Regularizers, and Constraints   387
   Custom Metrics   388
   Custom Layers   391
   Custom Models   394
   Losses and Metrics Based on Model Internals   397
   Computing Gradients Using Autodiff   399
   Custom Training Loops   402
TensorFlow Functions and Graphs   405
   AutoGraph and Tracing   407
   TF Function Rules   409
Exercises   410

13. Loading and Preprocessing Data with TensorFlow. 413
The Data API   414
   Chaining Transformations   415
   Shuffling the Data   416
   Preprocessing the Data   419
   Putting Everything Together   420
   Prefetching   421
   Using the Dataset with tf.keras   423
The TFRecord Format   424
   Compressed TFRecord Files   425
   A Brief Introduction to Protocol Buffers   425
   TensorFlow Protobufs   427
   Loading and Parsing Examples   428
   Handling Lists of Lists Using the SequenceExample Protobuf   429
Preprocessing the Input Features   430
   Encoding Categorical Features Using One-Hot Vectors   431
   Encoding Categorical Features Using Embeddings   433
   Keras Preprocessing Layers   437
TF Transform   439
The TensorFlow Datasets (TFDS) Project   441
Exercises   442

14. Deep Computer Vision Using Convolutional Neural Networks. 445
The Architecture of the Visual Cortex   446
Convolutional Layers   448
   Filters   450
   Stacking Multiple Feature Maps   451

TensorFlow Implementation   453
Memory Requirements   456
Pooling Layers   456
TensorFlow Implementation   458
CNN Architectures   460
LeNet-5   463
AlexNet   464
GoogLeNet   466
VGGNet   470
ResNet   471
Xception   474
SENet   476
Implementing a ResNet-34 CNN Using Keras   478
Using Pretrained Models from Keras   479
Pretrained Models for Transfer Learning   481
Classification and Localization   483
Object Detection   485
Fully Convolutional Networks   487
You Only Look Once (YOLO)   489
Semantic Segmentation   492
Exercises   496

15. Processing Sequences Using RNNs and CNNs.   497
Recurrent Neurons and Layers   498
Memory Cells   500
Input and Output Sequences   501
Training RNNs   502
Forecasting a Time Series   503
Baseline Metrics   505
Implementing a Simple RNN   505
Deep RNNs   506
Forecasting Several Time Steps Ahead   508
Handling Long Sequences   511
Fighting the Unstable Gradients Problem   512
Tackling the Short-Term Memory Problem   514
Exercises   523

16. Natural Language Processing with RNNs and Attention.   525
Generating Shakespearean Text Using a Character RNN   526
Creating the Training Dataset   527
How to Split a Sequential Dataset   527
Chopping the Sequential Dataset into Multiple Windows   528

Building and Training the Char-RNN Model   530
Using the Char-RNN Model   531
Generating Fake Shakespearean Text   531
Stateful RNN   532
Sentiment Analysis   534
Masking   538
Reusing Pretrained Embeddings   540
An Encoder–Decoder Network for Neural Machine Translation   542
Bidirectional RNNs   546
Beam Search   547
Attention Mechanisms   549
Visual Attention   552
Attention Is All You Need: The Transformer Architecture   554
Recent Innovations in Language Models   563
Exercises   565

17. Representation Learning and Generative Learning Using Autoencoders and GANs. 567
Efficient Data Representations   569
Performing PCA with an Undercomplete Linear Autoencoder   570
Stacked Autoencoders   572
Implementing a Stacked Autoencoder Using Keras   572
Visualizing the Reconstructions   574
Visualizing the Fashion MNIST Dataset   574
Unsupervised Pretraining Using Stacked Autoencoders   576
Tying Weights   577
Training One Autoencoder at a Time   578
Convolutional Autoencoders   579
Recurrent Autoencoders   580
Denoising Autoencoders   581
Sparse Autoencoders   582
Variational Autoencoders   586
Generating Fashion MNIST Images   590
Generative Adversarial Networks   592
The Difficulties of Training GANs   596
Deep Convolutional GANs   598
Progressive Growing of GANs   601
StyleGANs   604
Exercises   607

18. Reinforcement Learning. 609
Learning to Optimize Rewards   610
Policy Search   612

Introduction to OpenAI Gym   613
Neural Network Policies   617
Evaluating Actions: The Credit Assignment Problem   619
Policy Gradients   620
Markov Decision Processes   625
Temporal Difference Learning   629
Q-Learning   630
    Exploration Policies   632
    Approximate Q-Learning and Deep Q-Learning   633
Implementing Deep Q-Learning   634
Deep Q-Learning Variants   639
    Fixed Q-Value Targets   639
    Double DQN   640
    Prioritized Experience Replay   640
    Dueling DQN   641
The TF-Agents Library   642
    Installing TF-Agents   643
    TF-Agents Environments   643
    Environment Specifications   644
    Environment Wrappers and Atari Preprocessing   645
    Training Architecture   649
    Creating the Deep Q-Network   650
    Creating the DQN Agent   652
    Creating the Replay Buffer and the Corresponding Observer   654
    Creating Training Metrics   655
    Creating the Collect Driver   656
    Creating the Dataset   658
    Creating the Training Loop   661
Overview of Some Popular RL Algorithms   662
Exercises   664

19. Training and Deploying TensorFlow Models at Scale......................... 667
Serving a TensorFlow Model   668
    Using TensorFlow Serving   668
    Creating a Prediction Service on GCP AI Platform   677
    Using the Prediction Service   682
Deploying a Model to a Mobile or Embedded Device   685
Using GPUs to Speed Up Computations   689
    Getting Your Own GPU   690
    Using a GPU-Equipped Virtual Machine   692
    Colaboratory   693
    Managing the GPU RAM   694

Placing Operations and Variables on Devices   697
Parallel Execution Across Multiple Devices   699
Training Models Across Multiple Devices   701
Model Parallelism   701
Data Parallelism   704
Training at Scale Using the Distribution Strategies API   709
Training a Model on a TensorFlow Cluster   711
Running Large Training Jobs on Google Cloud AI Platform   714
Black Box Hyperparameter Tuning on AI Platform   716
Exercises   717
Thank You!   718

A. Exercise Solutions.   719

B. Machine Learning Project Checklist.   755

C. SVM Dual Problem.   761

D. Autodiff.   765

E. Other Popular ANN Architectures.   773

F. Special Data Structures.   783

G. TensorFlow Graphs.   791

Index.   801

Preface

The Machine Learning Tsunami

In 2006, Geoffrey Hinton et al. published a paper (*https://homl.info/136*)[1] showing how to train a deep neural network capable of recognizing handwritten digits with state-of-the-art precision (>98%). They branded this technique "Deep Learning." A deep neural network is a (very) simplified model of our cerebral cortex, composed of a stack of layers of artificial neurons. Training a deep neural net was widely considered impossible at the time,[2] and most researchers had abandoned the idea in the late 1990s. This paper revived the interest of the scientific community, and before long many new papers demonstrated that Deep Learning was not only possible, but capable of mind-blowing achievements that no other Machine Learning (ML) technique could hope to match (with the help of tremendous computing power and great amounts of data). This enthusiasm soon extended to many other areas of Machine Learning.

A decade or so later, Machine Learning has conquered the industry: it is at the heart of much of the magic in today's high-tech products, ranking your web search results, powering your smartphone's speech recognition, recommending videos, and beating the world champion at the game of Go. Before you know it, it will be driving your car.

Machine Learning in Your Projects

So, naturally you are excited about Machine Learning and would love to join the party!

---

1 Geoffrey E. Hinton et al., "A Fast Learning Algorithm for Deep Belief Nets," *Neural Computation* 18 (2006): 1527–1554.

2 Despite the fact that Yann LeCun's deep convolutional neural networks had worked well for image recognition since the 1990s, although they were not as general-purpose.

Perhaps you would like to give your homemade robot a brain of its own? Make it recognize faces? Or learn to walk around?

Or maybe your company has tons of data (user logs, financial data, production data, machine sensor data, hotline stats, HR reports, etc.), and more than likely you could unearth some hidden gems if you just knew where to look. With Machine Learning, you could accomplish the following and more (*https://homl.info/usecases*):

- Segment customers and find the best marketing strategy for each group.
- Recommend products for each client based on what similar clients bought.
- Detect which transactions are likely to be fraudulent.
- Forecast next year's revenue.

Whatever the reason, you have decided to learn Machine Learning and implement it in your projects. Great idea!

Objective and Approach

This book assumes that you know close to nothing about Machine Learning. Its goal is to give you the concepts, tools, and intuition you need to implement programs capable of *learning from data*.

We will cover a large number of techniques, from the simplest and most commonly used (such as Linear Regression) to some of the Deep Learning techniques that regularly win competitions.

Rather than implementing our own toy versions of each algorithm, we will be using production-ready Python frameworks:

- Scikit-Learn (*http://scikit-learn.org/*) is very easy to use, yet it implements many Machine Learning algorithms efficiently, so it makes for a great entry point to learning Machine Learning. It was created by David Cournapeau in 2007, and is now led by a team of researchers at the French Institute for Research in Computer Science and Automation (Inria).

- TensorFlow (*https://tensorflow.org/*) is a more complex library for distributed numerical computation. It makes it possible to train and run very large neural networks efficiently by distributing the computations across potentially hundreds of multi-GPU (graphics processing unit) servers. TensorFlow (TF) was created at Google and supports many of its large-scale Machine Learning applications. It was open sourced in November 2015, and version 2.0 was released in September 2019.

- Keras (*https://keras.io/*) is a high-level Deep Learning API that makes it very simple to train and run neural networks. It can run on top of either TensorFlow,

Theano, or Microsoft Cognitive Toolkit (formerly known as CNTK). TensorFlow comes with its own implementation of this API, called *tf.keras*, which provides support for some advanced TensorFlow features (e.g., the ability to efficiently load data).

The book favors a hands-on approach, growing an intuitive understanding of Machine Learning through concrete working examples and just a little bit of theory. While you can read this book without picking up your laptop, I highly recommend you experiment with the code examples available online as Jupyter notebooks at *https://github.com/ageron/handson-ml2*.

Prerequisites

This book assumes that you have some Python programming experience and that you are familiar with Python's main scientific libraries—in particular, NumPy (*http://numpy.org/*), pandas (*http://pandas.pydata.org/*), and Matplotlib (*http://matplotlib.org/*).

Also, if you care about what's under the hood, you should have a reasonable understanding of college-level math as well (calculus, linear algebra, probabilities, and statistics).

If you don't know Python yet, *http://learnpython.org/* is a great place to start. The official tutorial on Python.org (*https://docs.python.org/3/tutorial/*) is also quite good.

If you have never used Jupyter, Chapter 2 will guide you through installation and the basics: it is a powerful tool to have in your toolbox.

If you are not familiar with Python's scientific libraries, the provided Jupyter notebooks include a few tutorials. There is also a quick math tutorial for linear algebra.

Roadmap

This book is organized in two parts. Part I, *The Fundamentals of Machine Learning*, covers the following topics:

- What Machine Learning is, what problems it tries to solve, and the main categories and fundamental concepts of its systems
- The steps in a typical Machine Learning project
- Learning by fitting a model to data
- Optimizing a cost function
- Handling, cleaning, and preparing data
- Selecting and engineering features

- Selecting a model and tuning hyperparameters using cross-validation
- The challenges of Machine Learning, in particular underfitting and overfitting (the bias/variance trade-off)
- The most common learning algorithms: Linear and Polynomial Regression, Logistic Regression, k-Nearest Neighbors, Support Vector Machines, Decision Trees, Random Forests, and Ensemble methods
- Reducing the dimensionality of the training data to fight the "curse of dimensionality"
- Other unsupervised learning techniques, including clustering, density estimation, and anomaly detection

Part II, *Neural Networks and Deep Learning*, covers the following topics:

- What neural nets are and what they're good for
- Building and training neural nets using TensorFlow and Keras
- The most important neural net architectures: feedforward neural nets for tabular data, convolutional nets for computer vision, recurrent nets and long short-term memory (LSTM) nets for sequence processing, encoder/decoders and Transformers for natural language processing, autoencoders and generative adversarial networks (GANs) for generative learning
- Techniques for training deep neural nets
- How to build an agent (e.g., a bot in a game) that can learn good strategies through trial and error, using Reinforcement Learning
- Loading and preprocessing large amounts of data efficiently
- Training and deploying TensorFlow models at scale

The first part is based mostly on Scikit-Learn, while the second part uses TensorFlow and Keras.

 Don't jump into deep waters too hastily: while Deep Learning is no doubt one of the most exciting areas in Machine Learning, you should master the fundamentals first. Moreover, most problems can be solved quite well using simpler techniques such as Random Forests and Ensemble methods (discussed in Part I). Deep Learning is best suited for complex problems such as image recognition, speech recognition, or natural language processing, provided you have enough data, computing power, and patience.

Changes in the Second Edition

This second edition has six main objectives:

1. Cover additional ML topics: more unsupervised learning techniques (including clustering, anomaly detection, density estimation, and mixture models); more techniques for training deep nets (including self-normalized networks); additional computer vision techniques (including Xception, SENet, object detection with YOLO, and semantic segmentation using R-CNN); handling sequences using covolutional neural networks (CNNs, including WaveNet); natural language processing using recurrent neural networks (RNNs), CNNs, and Transformers; and GANs.

2. Cover additional libraries and APIs (Keras, the Data API, TF-Agents for Reinforcement Learning) and training and deploying TF models at scale using the Distribution Strategies API, TF-Serving, and Google Cloud AI Platform. Also briefly introduce TF Transform, TFLite, TF Addons/Seq2Seq, and TensorFlow.js.

3. Discuss some of the latest important results from Deep Learning research.

4. Migrate all TensorFlow chapters to TensorFlow 2, and use TensorFlow's implementation of the Keras API (tf.keras) whenever possible.

5. Update the code examples to use the latest versions of Scikit-Learn, NumPy, pandas, Matplotlib, and other libraries.

6. Clarify some sections and fix some errors, thanks to plenty of great feedback from readers.

Some chapters were added, others were rewritten, and a few were reordered. See *https://homl.info/changes2* for more details on what changed in the second edition.

Other Resources

Many excellent resources are available to learn about Machine Learning. For example, Andrew Ng's ML course on Coursera (*https://homl.info/ngcourse*) is amazing, although it requires a significant time investment (think months).

There are also many interesting websites about Machine Learning, including of course Scikit-Learn's exceptional User Guide (*https://homl.info/skdoc*). You may also enjoy Dataquest (*https://www.dataquest.io/*), which provides very nice interactive tutorials, and ML blogs such as those listed on Quora (*https://homl.info/1*). Finally, the Deep Learning website (*http://deeplearning.net/*) has a good list of resources to check out to learn more.

There are many other introductory books about Machine Learning. In particular:

- Joel Grus's *Data Science from Scratch* (O'Reilly) presents the fundamentals of Machine Learning and implements some of the main algorithms in pure Python (from scratch, as the name suggests).

- Stephen Marsland's *Machine Learning: An Algorithmic Perspective* (Chapman & Hall) is a great introduction to Machine Learning, covering a wide range of topics in depth with code examples in Python (also from scratch, but using NumPy).

- Sebastian Raschka's *Python Machine Learning* (Packt Publishing) is also a great introduction to Machine Learning and leverages Python open source libraries (Pylearn 2 and Theano).

- François Chollet's *Deep Learning with Python* (Manning) is a very practical book that covers a large range of topics in a clear and concise way, as you might expect from the author of the excellent Keras library. It favors code examples over mathematical theory.

- Andriy Burkov's *The Hundred-Page Machine Learning Book* is very short and covers an impressive range of topics, introducing them in approachable terms without shying away from the math equations.

- Yaser S. Abu-Mostafa, Malik Magdon-Ismail, and Hsuan-Tien Lin's *Learning from Data* (AMLBook) is a rather theoretical approach to ML that provides deep insights, in particular on the bias/variance trade-off (see Chapter 4).

- Stuart Russell and Peter Norvig's *Artificial Intelligence: A Modern Approach*, 3rd Edition (Pearson), is a great (and huge) book covering an incredible amount of topics, including Machine Learning. It helps put ML into perspective.

Finally, joining ML competition websites such as Kaggle.com will allow you to practice your skills on real-world problems, with help and insights from some of the best ML professionals out there.

Conventions Used in This Book

The following typographical conventions are used in this book:

Italic
Indicates new terms, URLs, email addresses, filenames, and file extensions.

`Constant width`
Used for program listings, as well as within paragraphs to refer to program elements such as variable or function names, databases, data types, environment variables, statements and keywords.

`Constant width bold`
Shows commands or other text that should be typed literally by the user.

Constant width italic

Shows text that should be replaced with user-supplied values or by values determined by context.

This element signifies a tip or suggestion.

This element signifies a general note.

This element indicates a warning or caution.

Code Examples

There is a series of Jupyter notebooks full of supplemental material, such as code examples and exercises, available for download at *https://github.com/ageron/handson-ml2*.

Some of the code examples in the book leave out repetitive sections or details that are obvious or unrelated to Machine Learning. This keeps the focus on the important parts of the code and saves space to cover more topics. If you want the full code examples, they are all available in the Jupyter notebooks.

Note that when the code examples display some outputs, these code examples are shown with Python prompts (>>> and ...), as in a Python shell, to clearly distinguish the code from the outputs. For example, this code defines the square() function, then it computes and displays the square of 3:

```
>>> def square(x):
...     return x ** 2
...
>>> result = square(3)
>>> result
9
```

When code does not display anything, prompts are not used. However, the result may sometimes be shown as a comment, like this:

```
def square(x):
    return x ** 2

result = square(3)  # result is 9
```

Using Code Examples

This book is here to help you get your job done. In general, if example code is offered with this book, you may use it in your programs and documentation. You do not need to contact us for permission unless you're reproducing a significant portion of the code. For example, writing a program that uses several chunks of code from this book does not require permission. Selling or distributing a CD-ROM of examples from O'Reilly books does require permission. Answering a question by citing this book and quoting example code does not require permission. Incorporating a significant amount of example code from this book into your product's documentation does require permission.

We appreciate, but do not require, attribution. An attribution usually includes the title, author, publisher, and ISBN. For example: "*Hands-On Machine Learning with Scikit-Learn, Keras, and TensorFlow*, 2nd Edition, by Aurélien Géron (O'Reilly). Copyright 2019 Kiwisoft S.A.S., 978-1-492-03264-9." If you feel your use of code examples falls outside fair use or the permission given above, feel free to contact us at *permissions@oreilly.com*.

O'Reilly Online Learning

For almost 40 years, *O'Reilly Media* has provided technology and business training, knowledge, and insight to help companies succeed.

Our unique network of experts and innovators share their knowledge and expertise through books, articles, conferences, and our online learning platform. O'Reilly's online learning platform gives you on-demand access to live training courses, in-depth learning paths, interactive coding environments, and a vast collection of text and video from O'Reilly and 200+ other publishers. For more information, please visit *http://oreilly.com*.

How to Contact Us

Please address comments and questions concerning this book to the publisher:

O'Reilly Media, Inc.
1005 Gravenstein Highway North
Sebastopol, CA 95472
800-998-9938 (in the United States or Canada)
707-829-0515 (international or local)
707-829-0104 (fax)

We have a web page for this book, where we list errata, examples, and any additional information. You can access this page at *https://homl.info/oreilly2*.

To comment or ask technical questions about this book, send email to *bookquestions@oreilly.com*.

For more information about our books, courses, conferences, and news, see our website at *http://www.oreilly.com*.

Find us on Facebook: *http://facebook.com/oreilly*

Follow us on Twitter: *http://twitter.com/oreillymedia*

Watch us on YouTube: *http://www.youtube.com/oreillymedia*

Acknowledgments

Never in my wildest dreams did I imagine that the first edition of this book would get such a large audience. I received so many messages from readers, many asking questions, some kindly pointing out errata, and most sending me encouraging words. I cannot express how grateful I am to all these readers for their tremendous support. Thank you all so very much! Please do not hesitate to file issues on GitHub (*https://homl.info/issues2*) if you find errors in the code examples (or just to ask questions), or to submit errata (*https://homl.info/errata2*) if you find errors in the text. Some readers also shared how this book helped them get their first job, or how it helped them solve a concrete problem they were working on. I find such feedback incredibly motivating. If you find this book helpful, I would love it if you could share your story with me, either privately (e.g., via LinkedIn (*https://www.linkedin.com/in/aurelien-geron/*)) or publicly (e.g., in a tweet or through an Amazon review (*https://homl.info/amazon2*)).

I am also incredibly thankful to all the amazing people who took time out of their busy lives to review my book with such care. In particular, I would like to thank François Chollet for reviewing all the chapters based on Keras and TensorFlow and giving me some great in-depth feedback. Since Keras is one of the main additions to this second edition, having its author review the book was invaluable. I highly recommend François's book *Deep Learning with Python* (*https://homl.info/cholletbook*) (Manning): it has the conciseness, clarity, and depth of the Keras library itself. Special thanks as well to Ankur Patel, who reviewed every chapter of this second edition and gave me excellent feedback, in particular on Chapter 9, which covers unsupervised learning techniques. He could write a whole book on the topic… oh, wait, he did! Do check out *Hands-On Unsupervised Learning Using Python: How to Build Applied Machine Learning Solutions from Unlabeled Data* (*https://homl.info/patel*) (O'Reilly). Huge thanks as well to Olzhas Akpambetov, who reviewed all the chapters in the second part of the book, tested much of the code, and offered many great suggestions. I'm grateful to Mark Daoust, Jon Krohn, Dominic Monn, and Josh Patterson for reviewing the second part of this book so thoroughly and offering their expertise. They left no stone unturned and provided amazingly useful feedback.

While writing this second edition, I was fortunate enough to get plenty of help from members of the TensorFlow team—in particular Martin Wicke, who tirelessly answered dozens of my questions and dispatched the rest to the right people, including Karmel Allison, Paige Bailey, Eugene Brevdo, William Chargin, Daniel "Wolff" Dobson, Nick Felt, Bruce Fontaine, Goldie Gadde, Sandeep Gupta, Priya Gupta, Kevin Haas, Konstantinos Katsiapis ,Viacheslav Kovalevskyi, Allen Lavoie, Clemens Mewald, Dan Moldovan, Sean Morgan, Tom O'Malley, Alexandre Passos, André Susano Pinto, Anthony Platanios, Oscar Ramirez, Anna Revinskaya, Saurabh Saxena, Ryan Sepassi, Jiri Simsa, Xiaodan Song, Christina Sorokin, Dustin Tran, Todd Wang, Pete Warden (who also reviewed the first edition) Edd Wilder-James, and Yuefeng Zhou, all of whom were tremendously helpful. Huge thanks to all of you, and to all other members of the TensorFlow team, not just for your help, but also for making such a great library! Special thanks to Irene Giannoumis and Robert Crowe of the TFX team for reviewing Chapters 13 and 19 in depth.

Many thanks as well to O'Reilly's fantastic staff, in particular Nicole Taché, who gave me insightful feedback and was always cheerful, encouraging, and helpful: I could not dream of a better editor. Big thanks to Michele Cronin as well, who was very helpful (and patient) at the start of this second edition, and to Kristen Brown, the production editor for the second edition, who saw it through all the steps (she also coordinated fixes and updates for each reprint of the first edition). Thanks as well to Rachel Monaghan and Amanda Kersey for their thorough copyediting (respectively for the first and second edition), and to Johnny O'Toole who managed the relationship with Amazon and answered many of my questions. Thanks to Marie Beaugureau, Ben Lorica, Mike Loukides, and Laurel Ruma for believing in this project and helping me define its scope. Thanks to Matt Hacker and all of the Atlas team for answering all my technical questions regarding formatting, AsciiDoc, and LaTeX, and thanks to Nick Adams, Rebecca Demarest, Rachel Head, Judith McConville, Helen Monroe, Karen

Montgomery, Rachel Roumeliotis, and everyone else at O'Reilly who contributed to this book.

I would also like to thank my former Google colleagues, in particular the YouTube video classification team, for teaching me so much about Machine Learning. I could never have started the first edition without them. Special thanks to my personal ML gurus: Clément Courbet, Julien Dubois, Mathias Kende, Daniel Kitachewsky, James Pack, Alexander Pak, Anosh Raj, Vitor Sessak, Wiktor Tomczak, Ingrid von Glehn, and Rich Washington. And thanks to everyone else I worked with at YouTube and in the amazing Google research teams in Mountain View. Many thanks as well to Martin Andrews, Sam Witteveen, and Jason Zaman for welcoming me into their Google Developer Experts group in Singapore, with the kind support of Soonson Kwon, and for all the great discussions we had about Deep Learning and TensorFlow. Anyone interested in Deep Learning in Singapore should definitely join their Deep Learning Singapore meetup (*https://homl.info/meetupsg*). Jason deserves special thanks for sharing some of his TFLite expertise for Chapter 19!

I will never forget the kind people who reviewed the first edition of this book, including David Andrzejewski, Lukas Biewald, Justin Francis, Vincent Guilbeau, Eddy Hung, Karim Matrah, Grégoire Mesnil, Salim Sémaoune, Iain Smears, Michel Tessier, Ingrid von Glehn, Pete Warden, and of course my dear brother Sylvain. Special thanks to Haesun Park, who gave me plenty of excellent feedback and caught several errors while he was writing the Korean translation of the first edition of this book. He also translated the Jupyter notebooks into Korean, not to mention TensorFlow's documentation. I do not speak Korean, but judging by the quality of his feedback, all his translations must be truly excellent! Haesun also kindly contributed some of the solutions to the exercises in this second edition.

Last but not least, I am infinitely grateful to my beloved wife, Emmanuelle, and to our three wonderful children, Alexandre, Rémi, and Gabrielle, for encouraging me to work hard on this book. I'm also thankful to them for their insatiable curiosity: explaining some of the most difficult concepts in this book to my wife and children helped me clarify my thoughts and directly improved many parts of it. And they keep bringing me cookies and coffee! What more can one dream of?

## PART I

The Fundamentals of Machine Learning

CHAPTER 1

The Machine Learning Landscape

When most people hear "Machine Learning," they picture a robot: a dependable butler or a deadly Terminator, depending on who you ask. But Machine Learning is not just a futuristic fantasy; it's already here. In fact, it has been around for decades in some specialized applications, such as Optical Character Recognition (OCR). But the first ML application that really became mainstream, improving the lives of hundreds of millions of people, took over the world back in the 1990s: the *spam filter*. It's not exactly a self-aware Skynet, but it does technically qualify as Machine Learning (it has actually learned so well that you seldom need to flag an email as spam anymore). It was followed by hundreds of ML applications that now quietly power hundreds of products and features that you use regularly, from better recommendations to voice search.

Where does Machine Learning start and where does it end? What exactly does it mean for a machine to *learn* something? If I download a copy of Wikipedia, has my computer really learned something? Is it suddenly smarter? In this chapter we will start by clarifying what Machine Learning is and why you may want to use it.

Then, before we set out to explore the Machine Learning continent, we will take a look at the map and learn about the main regions and the most notable landmarks: supervised versus unsupervised learning, online versus batch learning, instance-based versus model-based learning. Then we will look at the workflow of a typical ML project, discuss the main challenges you may face, and cover how to evaluate and fine-tune a Machine Learning system.

This chapter introduces a lot of fundamental concepts (and jargon) that every data scientist should know by heart. It will be a high-level overview (it's the only chapter without much code), all rather simple, but you should make sure everything is crystal clear to you before continuing on to the rest of the book. So grab a coffee and let's get started!

If you already know all the Machine Learning basics, you may want to skip directly to Chapter 2. If you are not sure, try to answer all the questions listed at the end of the chapter before moving on.

What Is Machine Learning?

Machine Learning is the science (and art) of programming computers so they can *learn from data*.

Here is a slightly more general definition:

> [Machine Learning is the] field of study that gives computers the ability to learn without being explicitly programmed.
>
> —Arthur Samuel, 1959

And a more engineering-oriented one:

> A computer program is said to learn from experience E with respect to some task T and some performance measure P, if its performance on T, as measured by P, improves with experience E.
>
> —Tom Mitchell, 1997

Your spam filter is a Machine Learning program that, given examples of spam emails (e.g., flagged by users) and examples of regular (nonspam, also called "ham") emails, can learn to flag spam. The examples that the system uses to learn are called the *training set*. Each training example is called a *training instance* (or *sample*). In this case, the task T is to flag spam for new emails, the experience E is the *training data*, and the performance measure P needs to be defined; for example, you can use the ratio of correctly classified emails. This particular performance measure is called *accuracy*, and it is often used in classification tasks.

If you just download a copy of Wikipedia, your computer has a lot more data, but it is not suddenly better at any task. Thus, downloading a copy of Wikipedia is not Machine Learning.

Why Use Machine Learning?

Consider how you would write a spam filter using traditional programming techniques (Figure 1-1):

1. First you would consider what spam typically looks like. You might notice that some words or phrases (such as "4U," "credit card," "free," and "amazing") tend to come up a lot in the subject line. Perhaps you would also notice a few other patterns in the sender's name, the email's body, and other parts of the email.

2. You would write a detection algorithm for each of the patterns that you noticed, and your program would flag emails as spam if a number of these patterns were detected.

3. You would test your program and repeat steps 1 and 2 until it was good enough to launch.

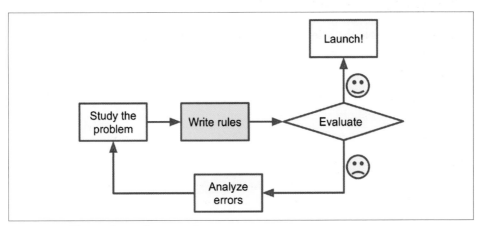

Figure 1-1. The traditional approach

Since the problem is difficult, your program will likely become a long list of complex rules—pretty hard to maintain.

In contrast, a spam filter based on Machine Learning techniques automatically learns which words and phrases are good predictors of spam by detecting unusually frequent patterns of words in the spam examples compared to the ham examples (Figure 1-2). The program is much shorter, easier to maintain, and most likely more accurate.

What if spammers notice that all their emails containing "4U" are blocked? They might start writing "For U" instead. A spam filter using traditional programming techniques would need to be updated to flag "For U" emails. If spammers keep working around your spam filter, you will need to keep writing new rules forever.

In contrast, a spam filter based on Machine Learning techniques automatically notices that "For U" has become unusually frequent in spam flagged by users, and it starts flagging them without your intervention (Figure 1-3).

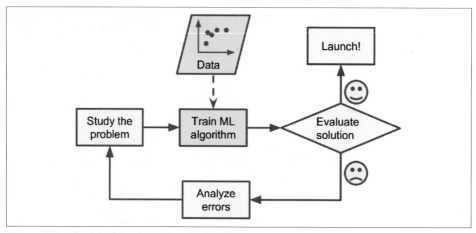

Figure 1-2. The Machine Learning approach

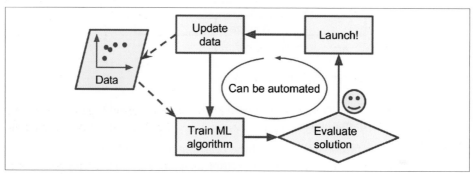

Figure 1-3. Automatically adapting to change

Another area where Machine Learning shines is for problems that either are too com-plex for traditional approaches or have no known algorithm. For example, consider speech recognition. Say you want to start simple and write a program capable of dis-tinguishing the words "one" and "two." You might notice that the word "two" starts with a high-pitch sound ("T"), so you could hardcode an algorithm that measures high-pitch sound intensity and use that to distinguish ones and twos—but obviously this technique will not scale to thousands of words spoken by millions of very differ-ent people in noisy environments and in dozens of languages. The best solution (at least today) is to write an algorithm that learns by itself, given many example record-ings for each word.

Finally, Machine Learning can help humans learn (Figure 1-4). ML algorithms can be inspected to see what they have learned (although for some algorithms this can be tricky). For instance, once a spam filter has been trained on enough spam, it can easily be inspected to reveal the list of words and combinations of words that it believes are the best predictors of spam. Sometimes this will reveal unsuspected correlations or new trends, and thereby lead to a better understanding of the problem. Applying ML techniques to dig into large amounts of data can help discover patterns that were not immediately apparent. This is called *data mining*.

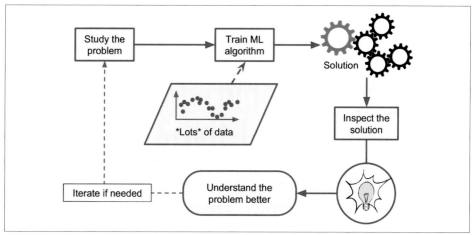

Figure 1-4. Machine Learning can help humans learn

To summarize, Machine Learning is great for:

- Problems for which existing solutions require a lot of fine-tuning or long lists of rules: one Machine Learning algorithm can often simplify code and perform better than the traditional approach.
- Complex problems for which using a traditional approach yields no good solution: the best Machine Learning techniques can perhaps find a solution.
- Fluctuating environments: a Machine Learning system can adapt to new data.
- Getting insights about complex problems and large amounts of data.

Examples of Applications

Let's look at some concrete examples of Machine Learning tasks, along with the techniques that can tackle them:

Analyzing images of products on a production line to automatically classify them
    This is image classification, typically performed using convolutional neural networks (CNNs; see Chapter 14).

Detecting tumors in brain scans
> This is semantic segmentation, where each pixel in the image is classified (as we want to determine the exact location and shape of tumors), typically using CNNs as well.

Automatically classifying news articles
> This is natural language processing (NLP), and more specifically text classification, which can be tackled using recurrent neural networks (RNNs), CNNs, or Transformers (see Chapter 16).

Automatically flagging offensive comments on discussion forums
> This is also text classification, using the same NLP tools.

Summarizing long documents automatically
> This is a branch of NLP called text summarization, again using the same tools.

Creating a chatbot or a personal assistant
> This involves many NLP components, including natural language understanding (NLU) and question-answering modules.

Forecasting your company's revenue next year, based on many performance metrics
> This is a regression task (i.e., predicting values) that may be tackled using any regression model, such as a Linear Regression or Polynomial Regression model (see Chapter 4), a regression SVM (see Chapter 5), a regression Random Forest (see Chapter 7), or an artificial neural network (see Chapter 10). If you want to take into account sequences of past performance metrics, you may want to use RNNs, CNNs, or Transformers (see Chapters 15 and 16).

Making your app react to voice commands
> This is speech recognition, which requires processing audio samples: since they are long and complex sequences, they are typically processed using RNNs, CNNs, or Transformers (see Chapters 15 and 16).

Detecting credit card fraud
> This is anomaly detection (see Chapter 9).

Segmenting clients based on their purchases so that you can design a different marketing strategy for each segment
> This is clustering (see Chapter 9).

Representing a complex, high-dimensional dataset in a clear and insightful diagram
> This is data visualization, often involving dimensionality reduction techniques (see Chapter 8).

Recommending a product that a client may be interested in, based on past purchases
> This is a recommender system. One approach is to feed past purchases (and other information about the client) to an artificial neural network (see Chap-

ter 10), and get it to output the most likely next purchase. This neural net would typically be trained on past sequences of purchases across all clients.

Building an intelligent bot for a game

This is often tackled using Reinforcement Learning (RL; see Chapter 18), which is a branch of Machine Learning that trains agents (such as bots) to pick the actions that will maximize their rewards over time (e.g., a bot may get a reward every time the player loses some life points), within a given environment (such as the game). The famous AlphaGo program that beat the world champion at the game of Go was built using RL.

This list could go on and on, but hopefully it gives you a sense of the incredible breadth and complexity of the tasks that Machine Learning can tackle, and the types of techniques that you would use for each task.

Types of Machine Learning Systems

There are so many different types of Machine Learning systems that it is useful to classify them in broad categories, based on the following criteria:

- Whether or not they are trained with human supervision (supervised, unsupervised, semisupervised, and Reinforcement Learning)
- Whether or not they can learn incrementally on the fly (online versus batch learning)
- Whether they work by simply comparing new data points to known data points, or instead by detecting patterns in the training data and building a predictive model, much like scientists do (instance-based versus model-based learning)

These criteria are not exclusive; you can combine them in any way you like. For example, a state-of-the-art spam filter may learn on the fly using a deep neural network model trained using examples of spam and ham; this makes it an online, model-based, supervised learning system.

Let's look at each of these criteria a bit more closely.

Supervised/Unsupervised Learning

Machine Learning systems can be classified according to the amount and type of supervision they get during training. There are four major categories: supervised learning, unsupervised learning, semisupervised learning, and Reinforcement Learning.

Supervised learning

In *supervised learning*, the training set you feed to the algorithm includes the desired solutions, called *labels* (Figure 1-5).

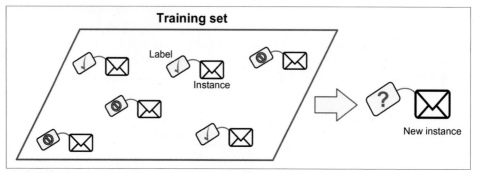

Figure 1-5. A labeled training set for spam classification (an example of supervised learning)

A typical supervised learning task is *classification*. The spam filter is a good example of this: it is trained with many example emails along with their *class* (spam or ham), and it must learn how to classify new emails.

Another typical task is to predict a *target* numeric value, such as the price of a car, given a set of *features* (mileage, age, brand, etc.) called *predictors*. This sort of task is called *regression* (Figure 1-6).[1] To train the system, you need to give it many examples of cars, including both their predictors and their labels (i.e., their prices).

 In Machine Learning an *attribute* is a data type (e.g., "mileage"), while a *feature* has several meanings, depending on the context, but generally means an attribute plus its value (e.g., "mileage = 15,000"). Many people use the words *attribute* and *feature* interchangeably.

Note that some regression algorithms can be used for classification as well, and vice versa. For example, *Logistic Regression* is commonly used for classification, as it can output a value that corresponds to the probability of belonging to a given class (e.g., 20% chance of being spam).

---

1 Fun fact: this odd-sounding name is a statistics term introduced by Francis Galton while he was studying the fact that the children of tall people tend to be shorter than their parents. Since the children were shorter, he called this *regression to the mean*. This name was then applied to the methods he used to analyze correlations between variables.

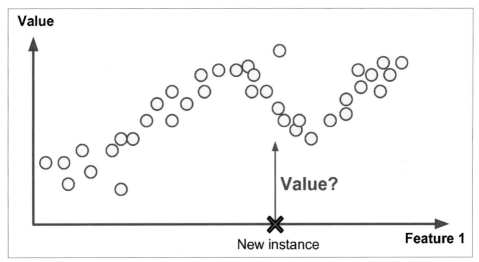

Figure 1-6. A regression problem: predict a value, given an input feature (there are usually multiple input features, and sometimes multiple output values)

Here are some of the most important supervised learning algorithms (covered in this book):

- k-Nearest Neighbors
- Linear Regression
- Logistic Regression
- Support Vector Machines (SVMs)
- Decision Trees and Random Forests
- Neural networks[2]

Unsupervised learning

In *unsupervised learning*, as you might guess, the training data is unlabeled (Figure 1-7). The system tries to learn without a teacher.

---

2 Some neural network architectures can be unsupervised, such as autoencoders and restricted Boltzmann machines. They can also be semisupervised, such as in deep belief networks and unsupervised pretraining.

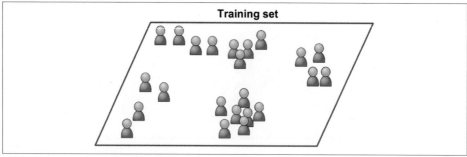

Figure 1-7. An unlabeled training set for unsupervised learning

Here are some of the most important unsupervised learning algorithms (most of these are covered in Chapters 8 and 9):

- Clustering
  - K-Means
  - DBSCAN
  - Hierarchical Cluster Analysis (HCA)
- Anomaly detection and novelty detection
  - One-class SVM
  - Isolation Forest
- Visualization and dimensionality reduction
  - Principal Component Analysis (PCA)
  - Kernel PCA
  - Locally Linear Embedding (LLE)
  - t-Distributed Stochastic Neighbor Embedding (t-SNE)
- Association rule learning
  - Apriori
  - Eclat

For example, say you have a lot of data about your blog's visitors. You may want to run a *clustering* algorithm to try to detect groups of similar visitors (Figure 1-8). At no point do you tell the algorithm which group a visitor belongs to: it finds those connections without your help. For example, it might notice that 40% of your visitors are males who love comic books and generally read your blog in the evening, while 20% are young sci-fi lovers who visit during the weekends. If you use a *hierarchical clustering* algorithm, it may also subdivide each group into smaller groups. This may help you target your posts for each group.

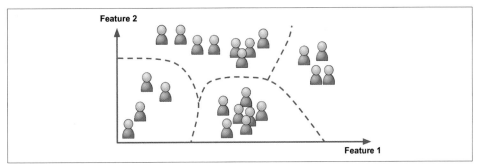

Figure 1-8. Clustering

Visualization algorithms are also good examples of unsupervised learning algorithms: you feed them a lot of complex and unlabeled data, and they output a 2D or 3D representation of your data that can easily be plotted (Figure 1-9). These algorithms try to preserve as much structure as they can (e.g., trying to keep separate clusters in the input space from overlapping in the visualization) so that you can understand how the data is organized and perhaps identify unsuspected patterns.

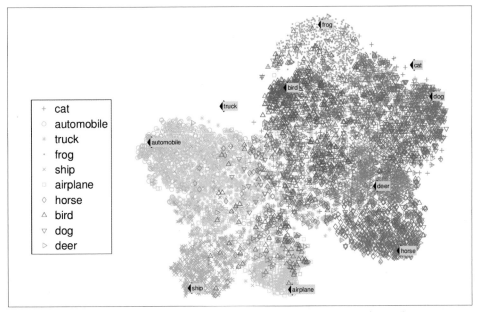

Figure 1-9. Example of a t-SNE visualization highlighting semantic clusters[3]

---

3 Notice how animals are rather well separated from vehicles and how horses are close to deer but far from birds. Figure reproduced with permission from Richard Socher et al., "Zero-Shot Learning Through Cross-Modal Transfer," *Proceedings of the 26th International Conference on Neural Information Processing Systems* 1 (2013): 935–943.

A related task is *dimensionality reduction*, in which the goal is to simplify the data without losing too much information. One way to do this is to merge several correlated features into one. For example, a car's mileage may be strongly correlated with its age, so the dimensionality reduction algorithm will merge them into one feature that represents the car's wear and tear. This is called *feature extraction*.

> It is often a good idea to try to reduce the dimension of your training data using a dimensionality reduction algorithm before you feed it to another Machine Learning algorithm (such as a supervised learning algorithm). It will run much faster, the data will take up less disk and memory space, and in some cases it may also perform better.

Yet another important unsupervised task is *anomaly detection*—for example, detecting unusual credit card transactions to prevent fraud, catching manufacturing defects, or automatically removing outliers from a dataset before feeding it to another learning algorithm. The system is shown mostly normal instances during training, so it learns to recognize them; then, when it sees a new instance, it can tell whether it looks like a normal one or whether it is likely an anomaly (see Figure 1-10). A very similar task is *novelty detection*: it aims to detect new instances that look different from all instances in the training set. This requires having a very "clean" training set, devoid of any instance that you would like the algorithm to detect. For example, if you have thousands of pictures of dogs, and 1% of these pictures represent Chihuahuas, then a novelty detection algorithm should not treat new pictures of Chihuahuas as novelties. On the other hand, anomaly detection algorithms may consider these dogs as so rare and so different from other dogs that they would likely classify them as anomalies (no offense to Chihuahuas).

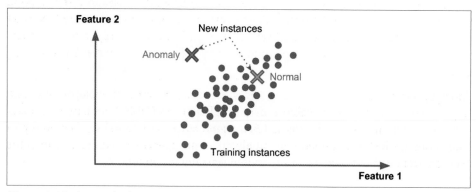

Figure 1-10. Anomaly detection

Finally, another common unsupervised task is *association rule learning*, in which the goal is to dig into large amounts of data and discover interesting relations between attributes. For example, suppose you own a supermarket. Running an association rule on your sales logs may reveal that people who purchase barbecue sauce and potato chips also tend to buy steak. Thus, you may want to place these items close to one another.

Semisupervised learning

Since labeling data is usually time-consuming and costly, you will often have plenty of unlabeled instances, and few labeled instances. Some algorithms can deal with data that's partially labeled. This is called *semisupervised learning* (Figure 1-11).

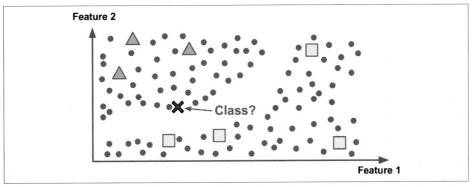

Figure 1-11. Semisupervised learning with two classes (triangles and squares): the unlabeled examples (circles) help classify a new instance (the cross) into the triangle class rather than the square class, even though it is closer to the labeled squares

Some photo-hosting services, such as Google Photos, are good examples of this. Once you upload all your family photos to the service, it automatically recognizes that the same person A shows up in photos 1, 5, and 11, while another person B shows up in photos 2, 5, and 7. This is the unsupervised part of the algorithm (clustering). Now all the system needs is for you to tell it who these people are. Just add one label per person[4] and it is able to name everyone in every photo, which is useful for searching photos.

Most semisupervised learning algorithms are combinations of unsupervised and supervised algorithms. For example, *deep belief networks* (DBNs) are based on unsupervised components called *restricted Boltzmann machines* (RBMs) stacked on top of one another. RBMs are trained sequentially in an unsupervised manner, and then the whole system is fine-tuned using supervised learning techniques.

---

4 That's when the system works perfectly. In practice it often creates a few clusters per person, and sometimes mixes up two people who look alike, so you may need to provide a few labels per person and manually clean up some clusters.

Reinforcement Learning

Reinforcement Learning is a very different beast. The learning system, called an *agent* in this context, can observe the environment, select and perform actions, and get *rewards* in return (or *penalties* in the form of negative rewards, as shown in Figure 1-12). It must then learn by itself what is the best strategy, called a *policy*, to get the most reward over time. A policy defines what action the agent should choose when it is in a given situation.

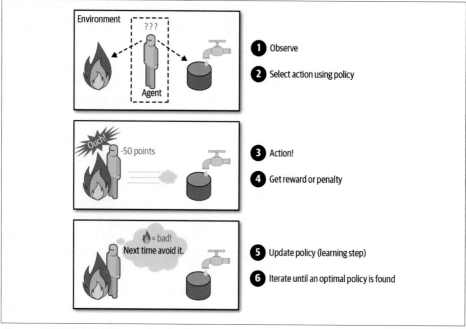

Figure 1-12. Reinforcement Learning

For example, many robots implement Reinforcement Learning algorithms to learn how to walk. DeepMind's AlphaGo program is also a good example of Reinforcement Learning: it made the headlines in May 2017 when it beat the world champion Ke Jie at the game of Go. It learned its winning policy by analyzing millions of games, and then playing many games against itself. Note that learning was turned off during the games against the champion; AlphaGo was just applying the policy it had learned.

Batch and Online Learning

Another criterion used to classify Machine Learning systems is whether or not the system can learn incrementally from a stream of incoming data.

Batch learning

In *batch learning*, the system is incapable of learning incrementally: it must be trained using all the available data. This will generally take a lot of time and computing resources, so it is typically done offline. First the system is trained, and then it is launched into production and runs without learning anymore; it just applies what it has learned. This is called *offline learning*.

If you want a batch learning system to know about new data (such as a new type of spam), you need to train a new version of the system from scratch on the full dataset (not just the new data, but also the old data), then stop the old system and replace it with the new one.

Fortunately, the whole process of training, evaluating, and launching a Machine Learning system can be automated fairly easily (as shown in Figure 1-3), so even a batch learning system can adapt to change. Simply update the data and train a new version of the system from scratch as often as needed.

This solution is simple and often works fine, but training using the full set of data can take many hours, so you would typically train a new system only every 24 hours or even just weekly. If your system needs to adapt to rapidly changing data (e.g., to predict stock prices), then you need a more reactive solution.

Also, training on the full set of data requires a lot of computing resources (CPU, memory space, disk space, disk I/O, network I/O, etc.). If you have a lot of data and you automate your system to train from scratch every day, it will end up costing you a lot of money. If the amount of data is huge, it may even be impossible to use a batch learning algorithm.

Finally, if your system needs to be able to learn autonomously and it has limited resources (e.g., a smartphone application or a rover on Mars), then carrying around large amounts of training data and taking up a lot of resources to train for hours every day is a showstopper.

Fortunately, a better option in all these cases is to use algorithms that are capable of learning incrementally.

Online learning

In *online learning*, you train the system incrementally by feeding it data instances sequentially, either individually or in small groups called *mini-batches*. Each learning step is fast and cheap, so the system can learn about new data on the fly, as it arrives (see Figure 1-13).

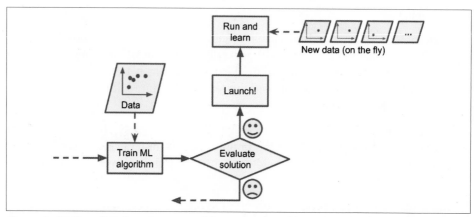

Figure 1-13. In online learning, a model is trained and launched into production, and then it keeps learning as new data comes in

Online learning is great for systems that receive data as a continuous flow (e.g., stock prices) and need to adapt to change rapidly or autonomously. It is also a good option if you have limited computing resources: once an online learning system has learned about new data instances, it does not need them anymore, so you can discard them (unless you want to be able to roll back to a previous state and "replay" the data). This can save a huge amount of space.

Online learning algorithms can also be used to train systems on huge datasets that cannot fit in one machine's main memory (this is called *out-of-core* learning). The algorithm loads part of the data, runs a training step on that data, and repeats the process until it has run on all of the data (see Figure 1-14).

Out-of-core learning is usually done offline (i.e., not on the live system), so *online learning* can be a confusing name. Think of it as *incremental learning*.

One important parameter of online learning systems is how fast they should adapt to changing data: this is called the *learning rate*. If you set a high learning rate, then your system will rapidly adapt to new data, but it will also tend to quickly forget the old data (you don't want a spam filter to flag only the latest kinds of spam it was shown). Conversely, if you set a low learning rate, the system will have more inertia; that is, it will learn more slowly, but it will also be less sensitive to noise in the new data or to sequences of nonrepresentative data points (outliers).

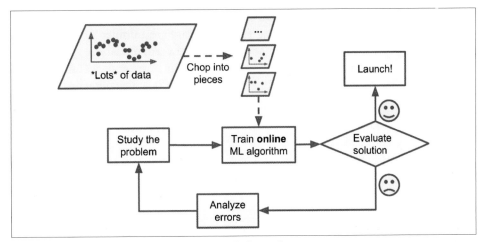

Figure 1-14. Using online learning to handle huge datasets

A big challenge with online learning is that if bad data is fed to the system, the system's performance will gradually decline. If it's a live system, your clients will notice. For example, bad data could come from a malfunctioning sensor on a robot, or from someone spamming a search engine to try to rank high in search results. To reduce this risk, you need to monitor your system closely and promptly switch learning off (and possibly revert to a previously working state) if you detect a drop in performance. You may also want to monitor the input data and react to abnormal data (e.g., using an anomaly detection algorithm).

Instance-Based Versus Model-Based Learning

One more way to categorize Machine Learning systems is by how they *generalize*. Most Machine Learning tasks are about making predictions. This means that given a number of training examples, the system needs to be able to make good predictions for (generalize to) examples it has never seen before. Having a good performance measure on the training data is good, but insufficient; the true goal is to perform well on new instances.

There are two main approaches to generalization: instance-based learning and model-based learning.

Instance-based learning

Possibly the most trivial form of learning is simply to learn by heart. If you were to create a spam filter this way, it would just flag all emails that are identical to emails that have already been flagged by users—not the worst solution, but certainly not the best.

Instead of just flagging emails that are identical to known spam emails, your spam filter could be programmed to also flag emails that are very similar to known spam emails. This requires a *measure of similarity* between two emails. A (very basic) similarity measure between two emails could be to count the number of words they have in common. The system would flag an email as spam if it has many words in common with a known spam email.

This is called *instance-based learning*: the system learns the examples by heart, then generalizes to new cases by using a similarity measure to compare them to the learned examples (or a subset of them). For example, in Figure 1-15 the new instance would be classified as a triangle because the majority of the most similar instances belong to that class.

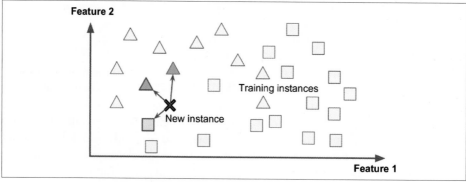

Figure 1-15. Instance-based learning

Model-based learning

Another way to generalize from a set of examples is to build a model of these examples and then use that model to make *predictions*. This is called *model-based learning* (Figure 1-16).

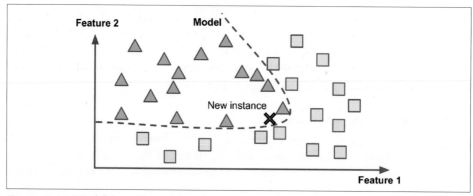

Figure 1-16. Model-based learning

For example, suppose you want to know if money makes people happy, so you download the Better Life Index data from the OECD's website (*https://homl.info/4*) and stats about gross domestic product (GDP) per capita from the IMF's website (*https://homl.info/5*). Then you join the tables and sort by GDP per capita. Table 1-1 shows an excerpt of what you get.

Table 1-1. Does money make people happier?

Country	GDP per capita (USD)	Life satisfaction
Hungary	12,240	4.9
Korea	27,195	5.8
France	37,675	6.5
Australia	50,962	7.3
United States	55,805	7.2

Let's plot the data for these countries (Figure 1-17).

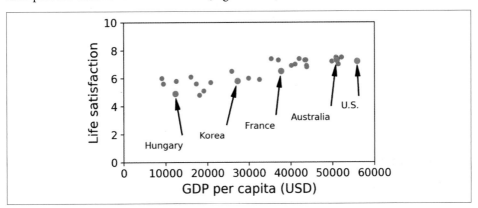

Figure 1-17. Do you see a trend here?

There does seem to be a trend here! Although the data is *noisy* (i.e., partly random), it looks like life satisfaction goes up more or less linearly as the country's GDP per capita increases. So you decide to model life satisfaction as a linear function of GDP per capita. This step is called *model selection*: you selected a *linear model* of life satisfaction with just one attribute, GDP per capita (Equation 1-1).

Equation 1-1. A simple linear model

$$\text{life\_satisfaction} = \theta_0 + \theta_1 \times \text{GDP\_per\_capita}$$

This model has two *model parameters*, θ_0 and θ_1.[5] By tweaking these parameters, you can make your model represent any linear function, as shown in Figure 1-18.

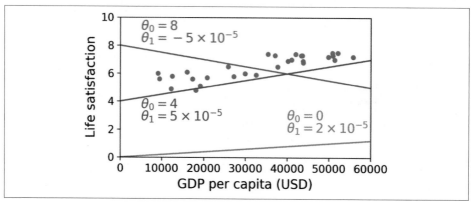

Figure 1-18. A few possible linear models

Before you can use your model, you need to define the parameter values θ_0 and θ_1. How can you know which values will make your model perform best? To answer this question, you need to specify a performance measure. You can either define a *utility function* (or *fitness function*) that measures how *good* your model is, or you can define a *cost function* that measures how *bad* it is. For Linear Regression problems, people typically use a cost function that measures the distance between the linear model's predictions and the training examples; the objective is to minimize this distance.

This is where the Linear Regression algorithm comes in: you feed it your training examples, and it finds the parameters that make the linear model fit best to your data. This is called *training* the model. In our case, the algorithm finds that the optimal parameter values are $\theta_0 = 4.85$ and $\theta_1 = 4.91 \times 10^{-5}$.

Confusingly, the same word "model" can refer to a *type of model* (e.g., Linear Regression), to a *fully specified model architecture* (e.g., Linear Regression with one input and one output), or to the *final trained model* ready to be used for predictions (e.g., Linear Regression with one input and one output, using $\theta_0 = 4.85$ and $\theta_1 = 4.91 \times 10^{-5}$). Model selection consists in choosing the type of model and fully specifying its architecture. Training a model means running an algorithm to find the model parameters that will make it best fit the training data (and hopefully make good predictions on new data).

---

5 By convention, the Greek letter θ (theta) is frequently used to represent model parameters.

Now the model fits the training data as closely as possible (for a linear model), as you can see in Figure 1-19.

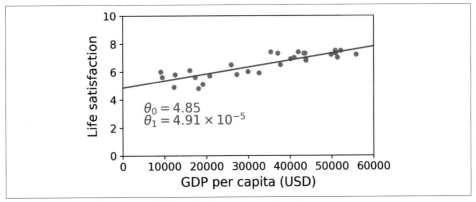

Figure 1-19. The linear model that fits the training data best

You are finally ready to run the model to make predictions. For example, say you want to know how happy Cypriots are, and the OECD data does not have the answer. Fortunately, you can use your model to make a good prediction: you look up Cyprus's GDP per capita, find \$22,587, and then apply your model and find that life satisfaction is likely to be somewhere around $4.85 + 22{,}587 \times 4.91 \times 10^{-5} = 5.96$.

To whet your appetite, Example 1-1 shows the Python code that loads the data, prepares it,[6] creates a scatterplot for visualization, and then trains a linear model and makes a prediction.[7]

Example 1-1. Training and running a linear model using Scikit-Learn

```
import matplotlib.pyplot as plt
import numpy as np
import pandas as pd
import sklearn.linear_model

# Load the data
oecd_bli = pd.read_csv("oecd_bli_2015.csv", thousands=',')
gdp_per_capita = pd.read_csv("gdp_per_capita.csv",thousands=',',delimiter='\t',
                             encoding='latin1', na_values="n/a")
```

---

6 The `prepare_country_stats()` function's definition is not shown here (see this chapter's Jupyter notebook if you want all the gory details). It's just boring pandas code that joins the life satisfaction data from the OECD with the GDP per capita data from the IMF.

7 It's OK if you don't understand all the code yet; we will present Scikit-Learn in the following chapters.

```
# Prepare the data
country_stats = prepare_country_stats(oecd_bli, gdp_per_capita)
X = np.c_[country_stats["GDP per capita"]]
y = np.c_[country_stats["Life satisfaction"]]

# Visualize the data
country_stats.plot(kind='scatter', x="GDP per capita", y='Life satisfaction')
plt.show()

# Select a linear model
model = sklearn.linear_model.LinearRegression()

# Train the model
model.fit(X, y)

# Make a prediction for Cyprus
X_new = [[22587]]  # Cyprus's GDP per capita
print(model.predict(X_new)) # outputs [[ 5.96242338]]
```

If you had used an instance-based learning algorithm instead, you would have found that Slovenia has the closest GDP per capita to that of Cyprus ($20,732), and since the OECD data tells us that Slovenians' life satisfaction is 5.7, you would have predicted a life satisfaction of 5.7 for Cyprus. If you zoom out a bit and look at the two next-closest countries, you will find Portugal and Spain with life satisfactions of 5.1 and 6.5, respectively. Averaging these three values, you get 5.77, which is pretty close to your model-based prediction. This simple algorithm is called *k-Nearest Neighbors* regression (in this example, $k = 3$).

Replacing the Linear Regression model with k-Nearest Neighbors regression in the previous code is as simple as replacing these two lines:

```
import sklearn.linear_model
model = sklearn.linear_model.LinearRegression()
```

with these two:

```
import sklearn.neighbors
model = sklearn.neighbors.KNeighborsRegressor(
    n_neighbors=3)
```

If all went well, your model will make good predictions. If not, you may need to use more attributes (employment rate, health, air pollution, etc.), get more or better-quality training data, or perhaps select a more powerful model (e.g., a Polynomial Regression model).

In summary:

- You studied the data.
- You selected a model.
- You trained it on the training data (i.e., the learning algorithm searched for the model parameter values that minimize a cost function).
- Finally, you applied the model to make predictions on new cases (this is called *inference*), hoping that this model will generalize well.

This is what a typical Machine Learning project looks like. In Chapter 2 you will experience this firsthand by going through a project end to end.

We have covered a lot of ground so far: you now know what Machine Learning is really about, why it is useful, what some of the most common categories of ML systems are, and what a typical project workflow looks like. Now let's look at what can go wrong in learning and prevent you from making accurate predictions.

Main Challenges of Machine Learning

In short, since your main task is to select a learning algorithm and train it on some data, the two things that can go wrong are "bad algorithm" and "bad data." Let's start with examples of bad data.

Insufficient Quantity of Training Data

For a toddler to learn what an apple is, all it takes is for you to point to an apple and say "apple" (possibly repeating this procedure a few times). Now the child is able to recognize apples in all sorts of colors and shapes. Genius.

Machine Learning is not quite there yet; it takes a lot of data for most Machine Learning algorithms to work properly. Even for very simple problems you typically need thousands of examples, and for complex problems such as image or speech recognition you may need millions of examples (unless you can reuse parts of an existing model).

The Unreasonable Effectiveness of Data

In a famous paper (*https://homl.info/6*) published in 2001, Microsoft researchers Michele Banko and Eric Brill showed that very different Machine Learning algorithms, including fairly simple ones, performed almost identically well on a complex problem of natural language disambiguation[8] once they were given enough data (as you can see in Figure 1-20).

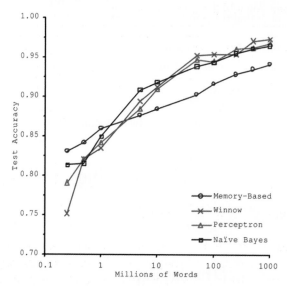

Figure 1-20. The importance of data versus algorithms[9]

As the authors put it, "these results suggest that we may want to reconsider the trade-off between spending time and money on algorithm development versus spending it on corpus development."

The idea that data matters more than algorithms for complex problems was further popularized by Peter Norvig et al. in a paper titled "The Unreasonable Effectiveness of Data" (*https://homl.info/7*), published in 2009.[10] It should be noted, however, that small- and medium-sized datasets are still very common, and it is not always easy or cheap to get extra training data—so don't abandon algorithms just yet.

---

8 For example, knowing whether to write "to," "two," or "too," depending on the context.

9 Figure reproduced with permission from Michele Banko and Eric Brill, "Scaling to Very Very Large Corpora for Natural Language Disambiguation," *Proceedings of the 39th Annual Meeting of the Association for Computational Linguistics* (2001): 26–33.

10 Peter Norvig et al., "The Unreasonable Effectiveness of Data," *IEEE Intelligent Systems* 24, no. 2 (2009): 8–12.

Nonrepresentative Training Data

In order to generalize well, it is crucial that your training data be representative of the new cases you want to generalize to. This is true whether you use instance-based learning or model-based learning.

For example, the set of countries we used earlier for training the linear model was not perfectly representative; a few countries were missing. Figure 1-21 shows what the data looks like when you add the missing countries.

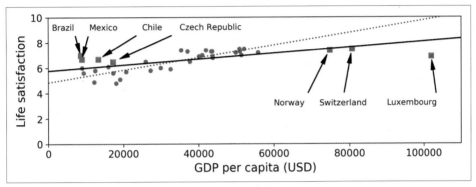

Figure 1-21. A more representative training sample

If you train a linear model on this data, you get the solid line, while the old model is represented by the dotted line. As you can see, not only does adding a few missing countries significantly alter the model, but it makes it clear that such a simple linear model is probably never going to work well. It seems that very rich countries are not happier than moderately rich countries (in fact, they seem unhappier), and conversely some poor countries seem happier than many rich countries.

By using a nonrepresentative training set, we trained a model that is unlikely to make accurate predictions, especially for very poor and very rich countries.

It is crucial to use a training set that is representative of the cases you want to generalize to. This is often harder than it sounds: if the sample is too small, you will have *sampling noise* (i.e., nonrepresentative data as a result of chance), but even very large samples can be nonrepresentative if the sampling method is flawed. This is called *sampling bias*.

Examples of Sampling Bias

Perhaps the most famous example of sampling bias happened during the US presidential election in 1936, which pitted Landon against Roosevelt: the *Literary Digest* conducted a very large poll, sending mail to about 10 million people. It got 2.4 million answers, and predicted with high confidence that Landon would get 57% of the votes. Instead, Roosevelt won with 62% of the votes. The flaw was in the *Literary Digest's* sampling method:

- First, to obtain the addresses to send the polls to, the *Literary Digest* used telephone directories, lists of magazine subscribers, club membership lists, and the like. All of these lists tended to favor wealthier people, who were more likely to vote Republican (hence Landon).

- Second, less than 25% of the people who were polled answered. Again this introduced a sampling bias, by potentially ruling out people who didn't care much about politics, people who didn't like the *Literary Digest*, and other key groups. This is a special type of sampling bias called *nonresponse bias*.

Here is another example: say you want to build a system to recognize funk music videos. One way to build your training set is to search for "funk music" on YouTube and use the resulting videos. But this assumes that YouTube's search engine returns a set of videos that are representative of all the funk music videos on YouTube. In reality, the search results are likely to be biased toward popular artists (and if you live in Brazil you will get a lot of "funk carioca" videos, which sound nothing like James Brown). On the other hand, how else can you get a large training set?

Poor-Quality Data

Obviously, if your training data is full of errors, outliers, and noise (e.g., due to poor-quality measurements), it will make it harder for the system to detect the underlying patterns, so your system is less likely to perform well. It is often well worth the effort to spend time cleaning up your training data. The truth is, most data scientists spend a significant part of their time doing just that. The following are a couple of examples of when you'd want to clean up training data:

- If some instances are clearly outliers, it may help to simply discard them or try to fix the errors manually.

- If some instances are missing a few features (e.g., 5% of your customers did not specify their age), you must decide whether you want to ignore this attribute altogether, ignore these instances, fill in the missing values (e.g., with the median age), or train one model with the feature and one model without it.

Irrelevant Features

As the saying goes: garbage in, garbage out. Your system will only be capable of learning if the training data contains enough relevant features and not too many irrelevant ones. A critical part of the success of a Machine Learning project is coming up with a good set of features to train on. This process, called *feature engineering*, involves the following steps:

- *Feature selection* (selecting the most useful features to train on among existing features)
- *Feature extraction* (combining existing features to produce a more useful one—as we saw earlier, dimensionality reduction algorithms can help)
- Creating new features by gathering new data

Now that we have looked at many examples of bad data, let's look at a couple of examples of bad algorithms.

Overfitting the Training Data

Say you are visiting a foreign country and the taxi driver rips you off. You might be tempted to say that *all* taxi drivers in that country are thieves. Overgeneralizing is something that we humans do all too often, and unfortunately machines can fall into the same trap if we are not careful. In Machine Learning this is called *overfitting*: it means that the model performs well on the training data, but it does not generalize well.

Figure 1-22 shows an example of a high-degree polynomial life satisfaction model that strongly overfits the training data. Even though it performs much better on the training data than the simple linear model, would you really trust its predictions?

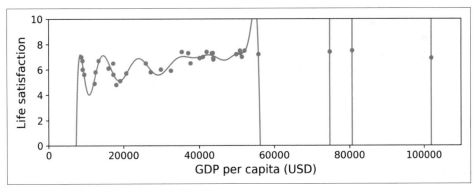

Figure 1-22. Overfitting the training data

Complex models such as deep neural networks can detect subtle patterns in the data, but if the training set is noisy, or if it is too small (which introduces sampling noise), then the model is likely to detect patterns in the noise itself. Obviously these patterns will not generalize to new instances. For example, say you feed your life satisfaction model many more attributes, including uninformative ones such as the country's name. In that case, a complex model may detect patterns like the fact that all countries in the training data with a *w* in their name have a life satisfaction greater than 7: New Zealand (7.3), Norway (7.4), Sweden (7.2), and Switzerland (7.5). How confident are you that the *w*-satisfaction rule generalizes to Rwanda or Zimbabwe? Obviously this pattern occurred in the training data by pure chance, but the model has no way to tell whether a pattern is real or simply the result of noise in the data.

> Overfitting happens when the model is too complex relative to the amount and noisiness of the training data. Here are possible solutions:
>
> - Simplify the model by selecting one with fewer parameters (e.g., a linear model rather than a high-degree polynomial model), by reducing the number of attributes in the training data, or by constraining the model.
> - Gather more training data.
> - Reduce the noise in the training data (e.g., fix data errors and remove outliers).

Constraining a model to make it simpler and reduce the risk of overfitting is called *regularization*. For example, the linear model we defined earlier has two parameters, θ_0 and θ_1. This gives the learning algorithm two *degrees of freedom* to adapt the model to the training data: it can tweak both the height (θ_0) and the slope (θ_1) of the line. If we forced $\theta_1 = 0$, the algorithm would have only one degree of freedom and would have a much harder time fitting the data properly: all it could do is move the line up or down to get as close as possible to the training instances, so it would end up around the mean. A very simple model indeed! If we allow the algorithm to modify θ_1 but we force it to keep it small, then the learning algorithm will effectively have somewhere in between one and two degrees of freedom. It will produce a model that's simpler than one with two degrees of freedom, but more complex than one with just one. You want to find the right balance between fitting the training data perfectly and keeping the model simple enough to ensure that it will generalize well.

Figure 1-23 shows three models. The dotted line represents the original model that was trained on the countries represented as circles (without the countries represented as squares), the dashed line is our second model trained with all countries (circles and squares), and the solid line is a model trained with the same data as the first model but with a regularization constraint. You can see that regularization forced the model to have a smaller slope: this model does not fit the training data (circles) as well as the first model, but it actually generalizes better to new examples that it did not see during training (squares).

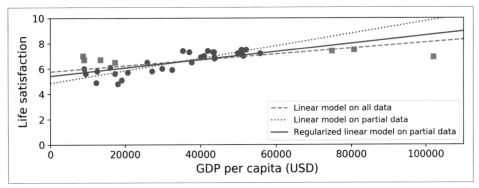

Figure 1-23. Regularization reduces the risk of overfitting

The amount of regularization to apply during learning can be controlled by a *hyperparameter*. A hyperparameter is a parameter of a learning algorithm (not of the model). As such, it is not affected by the learning algorithm itself; it must be set prior to training and remains constant during training. If you set the regularization hyperparameter to a very large value, you will get an almost flat model (a slope close to zero); the learning algorithm will almost certainly not overfit the training data, but it will be less likely to find a good solution. Tuning hyperparameters is an important part of building a Machine Learning system (you will see a detailed example in the next chapter).

Underfitting the Training Data

As you might guess, *underfitting* is the opposite of overfitting: it occurs when your model is too simple to learn the underlying structure of the data. For example, a linear model of life satisfaction is prone to underfit; reality is just more complex than the model, so its predictions are bound to be inaccurate, even on the training examples.

Here are the main options for fixing this problem:

- Select a more powerful model, with more parameters.
- Feed better features to the learning algorithm (feature engineering).
- Reduce the constraints on the model (e.g., reduce the regularization hyperparameter).

Stepping Back

By now you know a lot about Machine Learning. However, we went through so many concepts that you may be feeling a little lost, so let's step back and look at the big picture:

- Machine Learning is about making machines get better at some task by learning from data, instead of having to explicitly code rules.

- There are many different types of ML systems: supervised or not, batch or online, instance-based or model-based.

- In an ML project you gather data in a training set, and you feed the training set to a learning algorithm. If the algorithm is model-based, it tunes some parameters to fit the model to the training set (i.e., to make good predictions on the training set itself), and then hopefully it will be able to make good predictions on new cases as well. If the algorithm is instance-based, it just learns the examples by heart and generalizes to new instances by using a similarity measure to compare them to the learned instances.

- The system will not perform well if your training set is too small, or if the data is not representative, is noisy, or is polluted with irrelevant features (garbage in, garbage out). Lastly, your model needs to be neither too simple (in which case it will underfit) nor too complex (in which case it will overfit).

There's just one last important topic to cover: once you have trained a model, you don't want to just "hope" it generalizes to new cases. You want to evaluate it and fine-tune it if necessary. Let's see how to do that.

Testing and Validating

The only way to know how well a model will generalize to new cases is to actually try it out on new cases. One way to do that is to put your model in production and monitor how well it performs. This works well, but if your model is horribly bad, your users will complain—not the best idea.

A better option is to split your data into two sets: the *training set* and the *test set*. As these names imply, you train your model using the training set, and you test it using the test set. The error rate on new cases is called the *generalization error* (or *out-of-sample error*), and by evaluating your model on the test set, you get an estimate of this error. This value tells you how well your model will perform on instances it has never seen before.

If the training error is low (i.e., your model makes few mistakes on the training set) but the generalization error is high, it means that your model is overfitting the training data.

It is common to use 80% of the data for training and *hold out* 20% for testing. However, this depends on the size of the dataset: if it contains 10 million instances, then holding out 1% means your test set will contain 100,000 instances, probably more than enough to get a good estimate of the generalization error.

Hyperparameter Tuning and Model Selection

Evaluating a model is simple enough: just use a test set. But suppose you are hesitating between two types of models (say, a linear model and a polynomial model): how can you decide between them? One option is to train both and compare how well they generalize using the test set.

Now suppose that the linear model generalizes better, but you want to apply some regularization to avoid overfitting. The question is, how do you choose the value of the regularization hyperparameter? One option is to train 100 different models using 100 different values for this hyperparameter. Suppose you find the best hyperparameter value that produces a model with the lowest generalization error—say, just 5% error. You launch this model into production, but unfortunately it does not perform as well as expected and produces 15% errors. What just happened?

The problem is that you measured the generalization error multiple times on the test set, and you adapted the model and hyperparameters to produce the best model *for that particular set*. This means that the model is unlikely to perform as well on new data.

A common solution to this problem is called *holdout validation*: you simply hold out part of the training set to evaluate several candidate models and select the best one. The new held-out set is called the *validation set* (or sometimes the *development set*, or *dev set*). More specifically, you train multiple models with various hyperparameters on the reduced training set (i.e., the full training set minus the validation set), and you select the model that performs best on the validation set. After this holdout validation process, you train the best model on the full training set (including the validation set), and this gives you the final model. Lastly, you evaluate this final model on the test set to get an estimate of the generalization error.

This solution usually works quite well. However, if the validation set is too small, then model evaluations will be imprecise: you may end up selecting a suboptimal model by mistake. Conversely, if the validation set is too large, then the remaining training set will be much smaller than the full training set. Why is this bad? Well, since the final model will be trained on the full training set, it is not ideal to compare candidate models trained on a much smaller training set. It would be like selecting the fastest sprinter to participate in a marathon. One way to solve this problem is to perform repeated *cross-validation*, using many small validation sets. Each model is evaluated once per validation set after it is trained on the rest of the data. By averaging out all the evaluations of a model, you get a much more accurate measure of its performance. There is a drawback, however: the training time is multiplied by the number of validation sets.

Data Mismatch

In some cases, it's easy to get a large amount of data for training, but this data probably won't be perfectly representative of the data that will be used in production. For example, suppose you want to create a mobile app to take pictures of flowers and automatically determine their species. You can easily download millions of pictures of flowers on the web, but they won't be perfectly representative of the pictures that will actually be taken using the app on a mobile device. Perhaps you only have 10,000 representative pictures (i.e., actually taken with the app). In this case, the most important rule to remember is that the validation set and the test set must be as representative as possible of the data you expect to use in production, so they should be composed exclusively of representative pictures: you can shuffle them and put half in the validation set and half in the test set (making sure that no duplicates or near-duplicates end up in both sets). But after training your model on the web pictures, if you observe that the performance of the model on the validation set is disappointing, you will not know whether this is because your model has overfit the training set, or whether this is just due to the mismatch between the web pictures and the mobile app pictures. One solution is to hold out some of the training pictures (from the web) in yet another set that Andrew Ng calls the *train-dev set*. After the model is trained (on the training set, *not* on the train-dev set), you can evaluate it on the train-dev set. If it performs well, then the model is not overfitting the training set. If it performs poorly on the validation set, the problem must be coming from the data mismatch. You can try to tackle this problem by preprocessing the web images to make them look more like the pictures that will be taken by the mobile app, and then retraining the model. Conversely, if the model performs poorly on the train-dev set, then it must have overfit the training set, so you should try to simplify or regularize the model, get more training data, and clean up the training data.

No Free Lunch Theorem

A model is a simplified version of the observations. The simplifications are meant to discard the superfluous details that are unlikely to generalize to new instances. To decide what data to discard and what data to keep, you must make *assumptions*. For example, a linear model makes the assumption that the data is fundamentally linear and that the distance between the instances and the straight line is just noise, which can safely be ignored.

In a famous 1996 paper (*https://homl.info/8*),[11] David Wolpert demonstrated that if you make absolutely no assumption about the data, then there is no reason to prefer one model over any other. This is called the *No Free Lunch* (NFL) theorem. For some datasets the best model is a linear model, while for other datasets it is a neural network. There is no model that is *a priori* guaranteed to work better (hence the name of the theorem). The only way to know for sure which model is best is to evaluate them all. Since this is not possible, in practice you make some reasonable assumptions about the data and evaluate only a few reasonable models. For example, for simple tasks you may evaluate linear models with various levels of regularization, and for a complex problem you may evaluate various neural networks.

Exercises

In this chapter we have covered some of the most important concepts in Machine Learning. In the next chapters we will dive deeper and write more code, but before we do, make sure you know how to answer the following questions:

1. How would you define Machine Learning?
2. Can you name four types of problems where it shines?
3. What is a labeled training set?
4. What are the two most common supervised tasks?
5. Can you name four common unsupervised tasks?
6. What type of Machine Learning algorithm would you use to allow a robot to walk in various unknown terrains?
7. What type of algorithm would you use to segment your customers into multiple groups?
8. Would you frame the problem of spam detection as a supervised learning problem or an unsupervised learning problem?

---

11 David Wolpert, "The Lack of A Priori Distinctions Between Learning Algorithms," *Neural Computation* 8, no. 7 (1996): 1341–1390.

9. What is an online learning system?

10. What is out-of-core learning?

11. What type of learning algorithm relies on a similarity measure to make predictions?

12. What is the difference between a model parameter and a learning algorithm's hyperparameter?

13. What do model-based learning algorithms search for? What is the most common strategy they use to succeed? How do they make predictions?

14. Can you name four of the main challenges in Machine Learning?

15. If your model performs great on the training data but generalizes poorly to new instances, what is happening? Can you name three possible solutions?

16. What is a test set, and why would you want to use it?

17. What is the purpose of a validation set?

18. What is the train-dev set, when do you need it, and how do you use it?

19. What can go wrong if you tune hyperparameters using the test set?

Solutions to these exercises are available in Appendix A.

## CHAPTER 2

End-to-End Machine Learning Project

In this chapter you will work through an example project end to end, pretending to be a recently hired data scientist at a real estate company.[1] Here are the main steps you will go through:

1. Look at the big picture.
2. Get the data.
3. Discover and visualize the data to gain insights.
4. Prepare the data for Machine Learning algorithms.
5. Select a model and train it.
6. Fine-tune your model.
7. Present your solution.
8. Launch, monitor, and maintain your system.

Working with Real Data

When you are learning about Machine Learning, it is best to experiment with real-world data, not artificial datasets. Fortunately, there are thousands of open datasets to choose from, ranging across all sorts of domains. Here are a few places you can look to get data:

---

1 The example project is fictitious; the goal is to illustrate the main steps of a Machine Learning project, not to learn anything about the real estate business.

- Popular open data repositories
  - UC Irvine Machine Learning Repository (*http://archive.ics.uci.edu/ml/*)
  - Kaggle datasets (*https://www.kaggle.com/datasets*)
  - Amazon's AWS datasets (*https://registry.opendata.aws/*)
- Meta portals (they list open data repositories)
  - Data Portals (*http://dataportals.org/*)
  - OpenDataMonitor (*http://opendatamonitor.eu/*)
  - Quandl (*http://quandl.com/*)
- Other pages listing many popular open data repositories
  - Wikipedia's list of Machine Learning datasets (*https://homl.info/9*)
  - Quora.com (*https://homl.info/10*)
  - The datasets subreddit (*https://www.reddit.com/r/datasets*)

In this chapter we'll use the California Housing Prices dataset from the StatLib repository[2] (see Figure 2-1). This dataset is based on data from the 1990 California census. It is not exactly recent (a nice house in the Bay Area was still affordable at the time), but it has many qualities for learning, so we will pretend it is recent data. For teaching purposes I've added a categorical attribute and removed a few features.

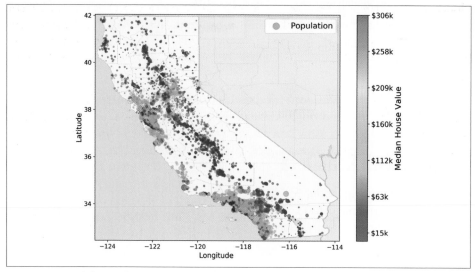

Figure 2-1. California housing prices

---

2 The original dataset appeared in R. Kelley Pace and Ronald Barry, "Sparse Spatial Autoregressions," *Statistics & Probability Letters* 33, no. 3 (1997): 291–297.

Look at the Big Picture

Welcome to the Machine Learning Housing Corporation! Your first task is to use California census data to build a model of housing prices in the state. This data includes metrics such as the population, median income, and median housing price for each block group in California. Block groups are the smallest geographical unit for which the US Census Bureau publishes sample data (a block group typically has a population of 600 to 3,000 people). We will call them "districts" for short.

Your model should learn from this data and be able to predict the median housing price in any district, given all the other metrics.

 Since you are a well-organized data scientist, the first thing you should do is pull out your Machine Learning project checklist. You can start with the one in Appendix B; it should work reasonably well for most Machine Learning projects, but make sure to adapt it to your needs. In this chapter we will go through many checklist items, but we will also skip a few, either because they are self-explanatory or because they will be discussed in later chapters.

Frame the Problem

The first question to ask your boss is what exactly the business objective is. Building a model is probably not the end goal. How does the company expect to use and benefit from this model? Knowing the objective is important because it will determine how you frame the problem, which algorithms you will select, which performance measure you will use to evaluate your model, and how much effort you will spend tweaking it.

Your boss answers that your model's output (a prediction of a district's median housing price) will be fed to another Machine Learning system (see Figure 2-2), along with many other signals.[3] This downstream system will determine whether it is worth investing in a given area or not. Getting this right is critical, as it directly affects revenue.

---

3 A piece of information fed to a Machine Learning system is often called a *signal*, in reference to Claude Shannon's information theory, which he developed at Bell Labs to improve telecommunications. His theory: you want a high signal-to-noise ratio.

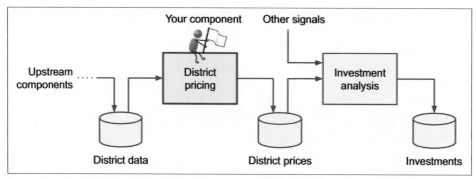

Figure 2-2. A Machine Learning pipeline for real estate investments

> # Pipelines
>
> A sequence of data processing components is called a data *pipeline*. Pipelines are very common in Machine Learning systems, since there is a lot of data to manipulate and many data transformations to apply.
>
> Components typically run asynchronously. Each component pulls in a large amount of data, processes it, and spits out the result in another data store. Then, some time later, the next component in the pipeline pulls this data and spits out its own output. Each component is fairly self-contained: the interface between components is simply the data store. This makes the system simple to grasp (with the help of a data flow graph), and different teams can focus on different components. Moreover, if a component breaks down, the downstream components can often continue to run normally (at least for a while) by just using the last output from the broken component. This makes the architecture quite robust.
>
> On the other hand, a broken component can go unnoticed for some time if proper monitoring is not implemented. The data gets stale and the overall system's performance drops.

The next question to ask your boss is what the current solution looks like (if any). The current situation will often give you a reference for performance, as well as insights on how to solve the problem. Your boss answers that the district housing prices are currently estimated manually by experts: a team gathers up-to-date information about a district, and when they cannot get the median housing price, they estimate it using complex rules.

This is costly and time-consuming, and their estimates are not great; in cases where they manage to find out the actual median housing price, they often realize that their estimates were off by more than 20%. This is why the company thinks that it would be useful to train a model to predict a district's median housing price, given other data about that district. The census data looks like a great dataset to exploit for this purpose, since it includes the median housing prices of thousands of districts, as well as other data.

With all this information, you are now ready to start designing your system. First, you need to frame the problem: is it supervised, unsupervised, or Reinforcement Learning? Is it a classification task, a regression task, or something else? Should you use batch learning or online learning techniques? Before you read on, pause and try to answer these questions for yourself.

Have you found the answers? Let's see: it is clearly a typical supervised learning task, since you are given *labeled* training examples (each instance comes with the expected output, i.e., the district's median housing price). It is also a typical regression task, since you are asked to predict a value. More specifically, this is a *multiple regression* problem, since the system will use multiple features to make a prediction (it will use the district's population, the median income, etc.). It is also a *univariate regression* problem, since we are only trying to predict a single value for each district. If we were trying to predict multiple values per district, it would be a *multivariate regression* problem. Finally, there is no continuous flow of data coming into the system, there is no particular need to adjust to changing data rapidly, and the data is small enough to fit in memory, so plain batch learning should do just fine.

 If the data were huge, you could either split your batch learning work across multiple servers (using the MapReduce technique) or use an online learning technique.

Select a Performance Measure

Your next step is to select a performance measure. A typical performance measure for regression problems is the Root Mean Square Error (RMSE). It gives an idea of how much error the system typically makes in its predictions, with a higher weight for large errors. Equation 2-1 shows the mathematical formula to compute the RMSE.

Equation 2-1. Root Mean Square Error (RMSE)

$$\text{RMSE}(\mathbf{X}, h) = \sqrt{\frac{1}{m} \sum_{i=1}^{m} \left(h\left(\mathbf{x}^{(i)}\right) - y^{(i)} \right)^2}$$

Notations

This equation introduces several very common Machine Learning notations that we will use throughout this book:

- m is the number of instances in the dataset you are measuring the RMSE on.
  - For example, if you are evaluating the RMSE on a validation set of 2,000 districts, then $m = 2,000$.
- $\mathbf{x}^{(i)}$ is a vector of all the feature values (excluding the label) of the ith instance in the dataset, and $y^{(i)}$ is its label (the desired output value for that instance).
  - For example, if the first district in the dataset is located at longitude –118.29°, latitude 33.91°, and it has 1,416 inhabitants with a median income of $38,372, and the median house value is $156,400 (ignoring the other features for now), then:

$$\mathbf{x}^{(1)} = \begin{pmatrix} -118.29 \\ 33.91 \\ 1,416 \\ 38,372 \end{pmatrix}$$

  and:

$$y^{(1)} = 156,400$$

- $\mathbf{X}$ is a matrix containing all the feature values (excluding labels) of all instances in the dataset. There is one row per instance, and the ith row is equal to the transpose of $\mathbf{x}^{(i)}$, noted $(\mathbf{x}^{(i)})^{\mathsf{T}}$.[4]
  - For example, if the first district is as just described, then the matrix $\mathbf{X}$ looks like this:

$$\mathbf{X} = \begin{pmatrix} \left(\mathbf{x}^{(1)}\right)^{\mathsf{T}} \\ \left(\mathbf{x}^{(2)}\right)^{\mathsf{T}} \\ \vdots \\ \left(\mathbf{x}^{(1999)}\right)^{\mathsf{T}} \\ \left(\mathbf{x}^{(2000)}\right)^{\mathsf{T}} \end{pmatrix} = \begin{pmatrix} -118.29 & 33.91 & 1,416 & 38,372 \\ \vdots & \vdots & \vdots & \vdots \end{pmatrix}$$

---

4 Recall that the transpose operator flips a column vector into a row vector (and vice versa).

- *h* is your system's prediction function, also called a *hypothesis*. When your system is given an instance's feature vector $\mathbf{x}^{(i)}$, it outputs a predicted value $\hat{y}^{(i)} = h(\mathbf{x}^{(i)})$ for that instance ($\hat{y}$ is pronounced "y-hat").

  — For example, if your system predicts that the median housing price in the first district is \$158,400, then $\hat{y}^{(1)} = h(\mathbf{x}^{(1)}) = 158{,}400$. The prediction error for this district is $\hat{y}^{(1)} - y^{(1)} = 2{,}000$.

- RMSE($\mathbf{X}$,*h*) is the cost function measured on the set of examples using your hypothesis *h*.

We use lowercase italic font for scalar values (such as *m* or $y^{(i)}$) and function names (such as *h*), lowercase bold font for vectors (such as $\mathbf{x}^{(i)}$), and uppercase bold font for matrices (such as $\mathbf{X}$).

Even though the RMSE is generally the preferred performance measure for regression tasks, in some contexts you may prefer to use another function. For example, suppose that there are many outlier districts. In that case, you may consider using the *mean absolute error* (MAE, also called the average absolute deviation; see Equation 2-2):

Equation 2-2. Mean absolute error (MAE)

$$\text{MAE}(\mathbf{X}, h) = \frac{1}{m} \sum_{i=1}^{m} \left| h(\mathbf{x}^{(i)}) - y^{(i)} \right|$$

Both the RMSE and the MAE are ways to measure the distance between two vectors: the vector of predictions and the vector of target values. Various distance measures, or *norms*, are possible:

- Computing the root of a sum of squares (RMSE) corresponds to the *Euclidean norm*: this is the notion of distance you are familiar with. It is also called the ℓ_2 *norm*, noted $\|\cdot\|_2$ (or just $\|\cdot\|$).

- Computing the sum of absolutes (MAE) corresponds to the ℓ_1 *norm*, noted $\|\cdot\|_1$. This is sometimes called the *Manhattan norm* because it measures the distance between two points in a city if you can only travel along orthogonal city blocks.

- More generally, the ℓ_k *norm* of a vector $\mathbf{v}$ containing *n* elements is defined as $\|\mathbf{v}\|_k = (|v_0|^k + |v_1|^k + \dots + |v_n|^k)^{1/k}$. ℓ_0 gives the number of nonzero elements in the vector, and ℓ_∞ gives the maximum absolute value in the vector.

- The higher the norm index, the more it focuses on large values and neglects small ones. This is why the RMSE is more sensitive to outliers than the MAE. But when outliers are exponentially rare (like in a bell-shaped curve), the RMSE performs very well and is generally preferred.

Check the Assumptions

Lastly, it is good practice to list and verify the assumptions that have been made so far (by you or others); this can help you catch serious issues early on. For example, the district prices that your system outputs are going to be fed into a downstream Machine Learning system, and you assume that these prices are going to be used as such. But what if the downstream system converts the prices into categories (e.g., "cheap," "medium," or "expensive") and then uses those categories instead of the prices themselves? In this case, getting the price perfectly right is not important at all; your system just needs to get the category right. If that's so, then the problem should have been framed as a classification task, not a regression task. You don't want to find this out after working on a regression system for months.

Fortunately, after talking with the team in charge of the downstream system, you are confident that they do indeed need the actual prices, not just categories. Great! You're all set, the lights are green, and you can start coding now!

Get the Data

It's time to get your hands dirty. Don't hesitate to pick up your laptop and walk through the following code examples in a Jupyter notebook. The full Jupyter notebook is available at *https://github.com/ageron/handson-ml2*.

Create the Workspace

First you will need to have Python installed. It is probably already installed on your system. If not, you can get it at *https://www.python.org/*.[5]

Next you need to create a workspace directory for your Machine Learning code and datasets. Open a terminal and type the following commands (after the $ prompts):

```
$ export ML_PATH="$HOME/ml"      # You can change the path if you prefer
$ mkdir -p $ML_PATH
```

You will need a number of Python modules: Jupyter, NumPy, pandas, Matplotlib, and Scikit-Learn. If you already have Jupyter running with all these modules installed, you can safely skip to "Download the Data" on page 46. If you don't have them yet, there are many ways to install them (and their dependencies). You can use your system's packaging system (e.g., apt-get on Ubuntu, or MacPorts or Homebrew on macOS), install a Scientific Python distribution such as Anaconda and use its packaging system, or just use Python's own packaging system, pip, which is included by

---

5 The latest version of Python 3 is recommended. Python 2.7+ may work too, but now that it's deprecated, all major scientific libraries are dropping support for it, so you should migrate to Python 3 as soon as possible.

default with the Python binary installers (since Python 2.7.9).[6] You can check to see if pip is installed by typing the following command:

```
$ python3 -m pip --version
pip 19.0.2 from [...]/lib/python3.6/site-packages (python 3.6)
```

You should make sure you have a recent version of pip installed. To upgrade the pip module, type the following (the exact version may differ):[7]

```
$ python3 -m pip install --user -U pip
Collecting pip
[...]
Successfully installed pip-19.0.2
```

Creating an Isolated Environment

If you would like to work in an isolated environment (which is strongly recommended so that you can work on different projects without having conflicting library versions), install virtualenv[8] by running the following pip command (again, if you want virtualenv to be installed for all users on your machine, remove --user and run this command with administrator rights):

```
$ python3 -m pip install --user -U virtualenv
Collecting virtualenv
[...]
Successfully installed virtualenv
```

Now you can create an isolated Python environment by typing this:

```
$ cd $ML_PATH
$ virtualenv my_env
Using base prefix '[...]'
New python executable in [...]/ml/my_env/bin/python3.6
Also creating executable in [...]/ml/my_env/bin/python
Installing setuptools, pip, wheel...done.
```

Now every time you want to activate this environment, just open a terminal and type the following:

---

6 I'll show the installation steps using pip in a bash shell on a Linux or macOS system. You may need to adapt these commands to your own system. On Windows, I recommend installing Anaconda instead.

7 If you want to upgrade pip for all users on your machine rather than just your own user, you should remove the --user option and make sure you have administrator rights (e.g., by adding sudo before the whole command on Linux or macOS).

8 Alternative tools include venv (very similar to virtualenv and included in the standard library), virtualenv-wrapper (provides extra functionalities on top of virtualenv), pyenv (allows easy switching between Python versions), and pipenv (a great packaging tool by the same author as the popular requests library, built on top of pip and virtualenv).

```
$ cd $ML_PATH
$ source my_env/bin/activate # on Linux or macOS
$ .\my_env\Scripts\activate  # on Windows
```

To deactivate this environment, type **deactivate**. While the environment is active, any package you install using pip will be installed in this isolated environment, and Python will only have access to these packages (if you also want access to the system's packages, you should create the environment using virtualenv's `--system-site-packages` option). Check out virtualenv's documentation for more information.

Now you can install all the required modules and their dependencies using this simple pip command (if you are not using a virtualenv, you will need the `--user` option or administrator rights):

```
$ python3 -m pip install -U jupyter matplotlib numpy pandas scipy scikit-learn
Collecting jupyter
  Downloading jupyter-1.0.0-py2.py3-none-any.whl
Collecting matplotlib
  [...]
```

To check your installation, try to import every module like this:

```
$ python3 -c "import jupyter, matplotlib, numpy, pandas, scipy, sklearn"
```

There should be no output and no error. Now you can fire up Jupyter by typing the following:

```
$ jupyter notebook
[I 15:24 NotebookApp] Serving notebooks from local directory: [...]/ml
[I 15:24 NotebookApp] 0 active kernels
[I 15:24 NotebookApp] The Jupyter Notebook is running at: http://localhost:8888/
[I 15:24 NotebookApp] Use Control-C to stop this server and shut down all
kernels (twice to skip confirmation).
```

A Jupyter server is now running in your terminal, listening to port 8888. You can visit this server by opening your web browser to *http://localhost:8888/* (this usually happens automatically when the server starts). You should see your empty workspace directory (containing only the *env* directory if you followed the preceding virtualenv instructions).

Now create a new Python notebook by clicking the New button and selecting the appropriate Python version[9] (see Figure 2-3). Doing that will create a new notebook file called *Untitled.ipynb* in your workspace, start a Jupyter Python kernel to run the notebook, and open this notebook in a new tab. You should start by renaming this

---

9 Note that Jupyter can handle multiple versions of Python, and even many other languages such as R or Octave.

notebook to "Housing" (this will automatically rename the file to *Housing.ipynb*) by clicking Untitled and typing the new name.

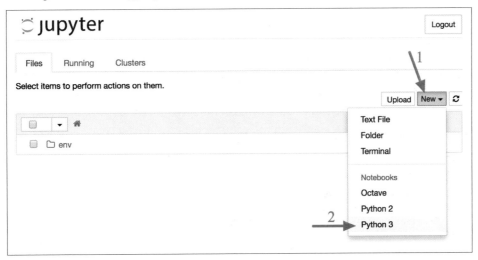

Figure 2-3. Your workspace in Jupyter

A notebook contains a list of cells. Each cell can contain executable code or formatted text. Right now the notebook contains only one empty code cell, labeled "In [1]:". Try typing **print("Hello world!")** in the cell and clicking the play button (see Figure 2-4) or pressing Shift-Enter. This sends the current cell to this notebook's Python kernel, which runs it and returns the output. The result is displayed below the cell, and since you've reached the end of the notebook, a new cell is automatically created. Go through the User Interface Tour from Jupyter's Help menu to learn the basics.

Figure 2-4. Hello world Python notebook

Download the Data

In typical environments your data would be available in a relational database (or some other common data store) and spread across multiple tables/documents/files. To access it, you would first need to get your credentials and access authorizations[10] and familiarize yourself with the data schema. In this project, however, things are much simpler: you will just download a single compressed file, *housing.tgz*, which contains a comma-separated values (CSV) file called *housing.csv* with all the data.

You could use your web browser to download the file and run `tar xzf housing.tgz` to decompress it and extract the CSV file, but it is preferable to create a small function to do that. Having a function that downloads the data is useful in particular if the data changes regularly: you can write a small script that uses the function to fetch the latest data (or you can set up a scheduled job to do that automatically at regular intervals). Automating the process of fetching the data is also useful if you need to install the dataset on multiple machines.

Here is the function to fetch the data:[11]

```
import os
import tarfile
import urllib

DOWNLOAD_ROOT = "https://raw.githubusercontent.com/ageron/handson-ml2/master/"
HOUSING_PATH = os.path.join("datasets", "housing")
HOUSING_URL = DOWNLOAD_ROOT + "datasets/housing/housing.tgz"

def fetch_housing_data(housing_url=HOUSING_URL, housing_path=HOUSING_PATH):
    os.makedirs(housing_path, exist_ok=True)
    tgz_path = os.path.join(housing_path, "housing.tgz")
    urllib.request.urlretrieve(housing_url, tgz_path)
    housing_tgz = tarfile.open(tgz_path)
    housing_tgz.extractall(path=housing_path)
    housing_tgz.close()
```

Now when you call `fetch_housing_data()`, it creates a *datasets/housing* directory in your workspace, downloads the *housing.tgz* file, and extracts the *housing.csv* file from it in this directory.

---

10 You might also need to check legal constraints, such as private fields that should never be copied to unsafe data stores.

11 In a real project you would save this code in a Python file, but for now you can just write it in your Jupyter notebook.

Now let's load the data using pandas. Once again, you should write a small function to load the data:

```
import pandas as pd

def load_housing_data(housing_path=HOUSING_PATH):
    csv_path = os.path.join(housing_path, "housing.csv")
    return pd.read_csv(csv_path)
```

This function returns a pandas DataFrame object containing all the data.

Take a Quick Look at the Data Structure

Let's take a look at the top five rows using the DataFrame's head() method (see Figure 2-5).

In [5]:	housing = load_housing_data() housing.head()						
Out[5]:		longitude	latitude	housing_median_age	total_rooms	total_bedrooms	population
	0	-122.23	37.88	41.0	880.0	129.0	322.0
	1	-122.22	37.86	21.0	7099.0	1106.0	2401.0
	2	-122.24	37.85	52.0	1467.0	190.0	496.0
	3	-122.25	37.85	52.0	1274.0	235.0	558.0
	4	-122.25	37.85	52.0	1627.0	280.0	565.0

Figure 2-5. Top five rows in the dataset

Each row represents one district. There are 10 attributes (you can see the first 6 in the screenshot): longitude, latitude, housing_median_age, total_rooms, total_bedrooms, population, households, median_income, median_house_value, and ocean_proximity.

The info() method is useful to get a quick description of the data, in particular the total number of rows, each attribute's type, and the number of nonnull values (see Figure 2-6).

```
In [6]: housing.info()

        <class 'pandas.core.frame.DataFrame'>
        RangeIndex: 20640 entries, 0 to 20639
        Data columns (total 10 columns):
        longitude             20640 non-null float64
        latitude              20640 non-null float64
        housing_median_age    20640 non-null float64
        total_rooms           20640 non-null float64
        total_bedrooms        20433 non-null float64
        population            20640 non-null float64
        households            20640 non-null float64
        median_income         20640 non-null float64
        median_house_value    20640 non-null float64
        ocean_proximity       20640 non-null object
        dtypes: float64(9), object(1)
        memory usage: 1.6+ MB
```

Figure 2-6. Housing info

There are 20,640 instances in the dataset, which means that it is fairly small by Machine Learning standards, but it's perfect to get started. Notice that the total_bedrooms attribute has only 20,433 nonnull values, meaning that 207 districts are missing this feature. We will need to take care of this later.

All attributes are numerical, except the ocean_proximity field. Its type is object, so it could hold any kind of Python object. But since you loaded this data from a CSV file, you know that it must be a text attribute. When you looked at the top five rows, you probably noticed that the values in the ocean_proximity column were repetitive, which means that it is probably a categorical attribute. You can find out what categories exist and how many districts belong to each category by using the value_counts() method:

```
>>> housing["ocean_proximity"].value_counts()
<1H OCEAN     9136
INLAND        6551
NEAR OCEAN    2658
NEAR BAY      2290
ISLAND           5
Name: ocean_proximity, dtype: int64
```

Let's look at the other fields. The describe() method shows a summary of the numerical attributes (Figure 2-7).

```
In [8]:  housing.describe()
```

Out[8]:

	longitude	latitude	housing_median_age	total_rooms	total_bedr
count	20640.000000	20640.000000	20640.000000	20640.000000	20433.0000
mean	-119.569704	35.631861	28.639486	2635.763081	537.870553
std	2.003532	2.135952	12.585558	2181.615252	421.385070
min	-124.350000	32.540000	1.000000	2.000000	1.000000
25%	-121.800000	33.930000	18.000000	1447.750000	296.000000
50%	-118.490000	34.260000	29.000000	2127.000000	435.000000
75%	-118.010000	37.710000	37.000000	3148.000000	647.000000
max	-114.310000	41.950000	52.000000	39320.000000	6445.0000

Figure 2-7. Summary of each numerical attribute

The count, mean, min, and max rows are self-explanatory. Note that the null values are ignored (so, for example, the count of total_bedrooms is 20,433, not 20,640). The std row shows the *standard deviation*, which measures how dispersed the values are.[12] The 25%, 50%, and 75% rows show the corresponding *percentiles*: a percentile indicates the value below which a given percentage of observations in a group of observations fall. For example, 25% of the districts have a housing_median_age lower than 18, while 50% are lower than 29 and 75% are lower than 37. These are often called the 25th percentile (or first *quartile*), the median, and the 75th percentile (or third quartile).

Another quick way to get a feel of the type of data you are dealing with is to plot a histogram for each numerical attribute. A histogram shows the number of instances (on the vertical axis) that have a given value range (on the horizontal axis). You can either plot this one attribute at a time, or you can call the hist() method on the whole dataset (as shown in the following code example), and it will plot a histogram for each numerical attribute (see Figure 2-8):

```
%matplotlib inline    # only in a Jupyter notebook
import matplotlib.pyplot as plt
housing.hist(bins=50, figsize=(20,15))
plt.show()
```

---

12 The standard deviation is generally denoted σ (the Greek letter sigma), and it is the square root of the *variance*, which is the average of the squared deviation from the mean. When a feature has a bell-shaped *normal distribution* (also called a *Gaussian distribution*), which is very common, the "68-95-99.7" rule applies: about 68% of the values fall within 1σ of the mean, 95% within 2σ, and 99.7% within 3σ.

The hist() method relies on Matplotlib, which in turn relies on a user-specified graphical backend to draw on your screen. So before you can plot anything, you need to specify which backend Matplotlib should use. The simplest option is to use Jupyter's magic command %matplotlib inline. This tells Jupyter to set up Matplotlib so it uses Jupyter's own backend. Plots are then rendered within the notebook itself. Note that calling show() is optional in a Jupyter notebook, as Jupyter will automatically display plots when a cell is executed.

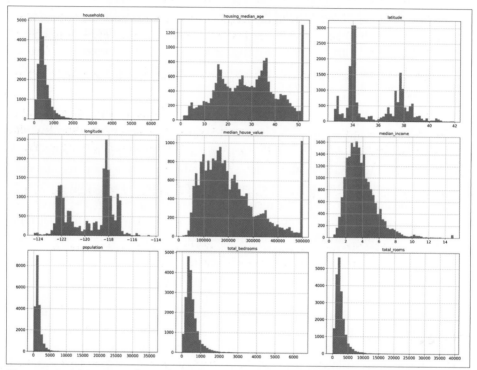

Figure 2-8. A histogram for each numerical attribute

There are a few things you might notice in these histograms:

1. First, the median income attribute does not look like it is expressed in US dollars (USD). After checking with the team that collected the data, you are told that the data has been scaled and capped at 15 (actually, 15.0001) for higher median incomes, and at 0.5 (actually, 0.4999) for lower median incomes. The numbers represent roughly tens of thousands of dollars (e.g., 3 actually means about $30,000). Working with preprocessed attributes is common in Machine Learning,

and it is not necessarily a problem, but you should try to understand how the data was computed.

2. The housing median age and the median house value were also capped. The latter may be a serious problem since it is your target attribute (your labels). Your Machine Learning algorithms may learn that prices never go beyond that limit. You need to check with your client team (the team that will use your system's output) to see if this is a problem or not. If they tell you that they need precise predictions even beyond $500,000, then you have two options:

   a. Collect proper labels for the districts whose labels were capped.

   b. Remove those districts from the training set (and also from the test set, since your system should not be evaluated poorly if it predicts values beyond $500,000).

3. These attributes have very different scales. We will discuss this later in this chapter, when we explore feature scaling.

4. Finally, many histograms are *tail-heavy*: they extend much farther to the right of the median than to the left. This may make it a bit harder for some Machine Learning algorithms to detect patterns. We will try transforming these attributes later on to have more bell-shaped distributions.

Hopefully you now have a better understanding of the kind of data you are dealing with.

Wait! Before you look at the data any further, you need to create a test set, put it aside, and never look at it.

Create a Test Set

It may sound strange to voluntarily set aside part of the data at this stage. After all, you have only taken a quick glance at the data, and surely you should learn a whole lot more about it before you decide what algorithms to use, right? This is true, but your brain is an amazing pattern detection system, which means that it is highly prone to overfitting: if you look at the test set, you may stumble upon some seemingly interesting pattern in the test data that leads you to select a particular kind of Machine Learning model. When you estimate the generalization error using the test set, your estimate will be too optimistic, and you will launch a system that will not perform as well as expected. This is called *data snooping* bias.

Creating a test set is theoretically simple: pick some instances randomly, typically 20% of the dataset (or less if your dataset is very large), and set them aside:

```
import numpy as np

def split_train_test(data, test_ratio):
    shuffled_indices = np.random.permutation(len(data))
    test_set_size = int(len(data) * test_ratio)
    test_indices = shuffled_indices[:test_set_size]
    train_indices = shuffled_indices[test_set_size:]
    return data.iloc[train_indices], data.iloc[test_indices]
```

You can then use this function like this:[13]

```
>>> train_set, test_set = split_train_test(housing, 0.2)
>>> len(train_set)
16512
>>> len(test_set)
4128
```

Well, this works, but it is not perfect: if you run the program again, it will generate a different test set! Over time, you (or your Machine Learning algorithms) will get to see the whole dataset, which is what you want to avoid.

One solution is to save the test set on the first run and then load it in subsequent runs. Another option is to set the random number generator's seed (e.g., with `np.random.seed(42)`)[14] before calling `np.random.permutation()` so that it always generates the same shuffled indices.

But both these solutions will break the next time you fetch an updated dataset. To have a stable train/test split even after updating the dataset, a common solution is to use each instance's identifier to decide whether or not it should go in the test set (assuming instances have a unique and immutable identifier). For example, you could compute a hash of each instance's identifier and put that instance in the test set if the hash is lower than or equal to 20% of the maximum hash value. This ensures that the test set will remain consistent across multiple runs, even if you refresh the dataset. The new test set will contain 20% of the new instances, but it will not contain any instance that was previously in the training set.

Here is a possible implementation:

```
from zlib import crc32

def test_set_check(identifier, test_ratio):
    return crc32(np.int64(identifier)) & 0xffffffff < test_ratio * 2**32
```

---

13 In this book, when a code example contains a mix of code and outputs, as is the case here, it is formatted like in the Python interpreter, for better readability: the code lines are prefixed with >>> (or ... for indented blocks), and the outputs have no prefix.

14 You will often see people set the random seed to 42. This number has no special property, other than to be the Answer to the Ultimate Question of Life, the Universe, and Everything.

```
def split_train_test_by_id(data, test_ratio, id_column):
    ids = data[id_column]
    in_test_set = ids.apply(lambda id_: test_set_check(id_, test_ratio))
    return data.loc[~in_test_set], data.loc[in_test_set]
```

Unfortunately, the housing dataset does not have an identifier column. The simplest solution is to use the row index as the ID:

```
housing_with_id = housing.reset_index()   # adds an `index` column
train_set, test_set = split_train_test_by_id(housing_with_id, 0.2, "index")
```

If you use the row index as a unique identifier, you need to make sure that new data gets appended to the end of the dataset and that no row ever gets deleted. If this is not possible, then you can try to use the most stable features to build a unique identifier. For example, a district's latitude and longitude are guaranteed to be stable for a few million years, so you could combine them into an ID like so:[15]

```
housing_with_id["id"] = housing["longitude"] * 1000 + housing["latitude"]
train_set, test_set = split_train_test_by_id(housing_with_id, 0.2, "id")
```

Scikit-Learn provides a few functions to split datasets into multiple subsets in various ways. The simplest function is `train_test_split()`, which does pretty much the same thing as the function `split_train_test()`, with a couple of additional features. First, there is a `random_state` parameter that allows you to set the random generator seed. Second, you can pass it multiple datasets with an identical number of rows, and it will split them on the same indices (this is very useful, for example, if you have a separate DataFrame for labels):

```
from sklearn.model_selection import train_test_split

train_set, test_set = train_test_split(housing, test_size=0.2, random_state=42)
```

So far we have considered purely random sampling methods. This is generally fine if your dataset is large enough (especially relative to the number of attributes), but if it is not, you run the risk of introducing a significant sampling bias. When a survey company decides to call 1,000 people to ask them a few questions, they don't just pick 1,000 people randomly in a phone book. They try to ensure that these 1,000 people are representative of the whole population. For example, the US population is 51.3% females and 48.7% males, so a well-conducted survey in the US would try to maintain this ratio in the sample: 513 female and 487 male. This is called *stratified sampling*: the population is divided into homogeneous subgroups called *strata*, and the right number of instances are sampled from each stratum to guarantee that the test set is representative of the overall population. If the people running the survey used purely random sampling, there would be about a 12% chance of sampling a skewed test set

---

15 The location information is actually quite coarse, and as a result many districts will have the exact same ID, so they will end up in the same set (test or train). This introduces some unfortunate sampling bias.

that was either less than 49% female or more than 54% female. Either way, the survey results would be significantly biased.

Suppose you chatted with experts who told you that the median income is a very important attribute to predict median housing prices. You may want to ensure that the test set is representative of the various categories of incomes in the whole dataset. Since the median income is a continuous numerical attribute, you first need to create an income category attribute. Let's look at the median income histogram more closely (back in Figure 2-8): most median income values are clustered around 1.5 to 6 (i.e., $15,000–$60,000), but some median incomes go far beyond 6. It is important to have a sufficient number of instances in your dataset for each stratum, or else the estimate of a stratum's importance may be biased. This means that you should not have too many strata, and each stratum should be large enough. The following code uses the `pd.cut()` function to create an income category attribute with five categories (labeled from 1 to 5): category 1 ranges from 0 to 1.5 (i.e., less than $15,000), category 2 from 1.5 to 3, and so on:

```
housing["income_cat"] = pd.cut(housing["median_income"],
                               bins=[0., 1.5, 3.0, 4.5, 6., np.inf],
                               labels=[1, 2, 3, 4, 5])
```

These income categories are represented in Figure 2-9:

```
housing["income_cat"].hist()
```

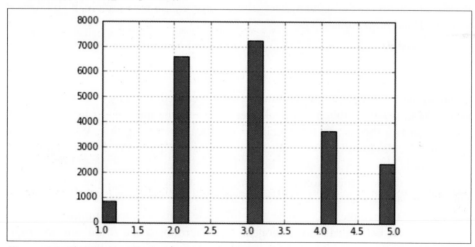

Figure 2-9. Histogram of income categories

Now you are ready to do stratified sampling based on the income category. For this you can use Scikit-Learn's `StratifiedShuffleSplit` class:

```
from sklearn.model_selection import StratifiedShuffleSplit

split = StratifiedShuffleSplit(n_splits=1, test_size=0.2, random_state=42)
for train_index, test_index in split.split(housing, housing["income_cat"]):
    strat_train_set = housing.loc[train_index]
    strat_test_set = housing.loc[test_index]
```

Let's see if this worked as expected. You can start by looking at the income category proportions in the test set:

```
>>> strat_test_set["income_cat"].value_counts() / len(strat_test_set)
3    0.350533
2    0.318798
4    0.176357
5    0.114583
1    0.039729
Name: income_cat, dtype: float64
```

With similar code you can measure the income category proportions in the full dataset. Figure 2-10 compares the income category proportions in the overall dataset, in the test set generated with stratified sampling, and in a test set generated using purely random sampling. As you can see, the test set generated using stratified sampling has income category proportions almost identical to those in the full dataset, whereas the test set generated using purely random sampling is skewed.

	Overall	Stratified	Random	Rand. %error	Strat. %error
1	0.039826	0.039729	0.040213	0.973236	-0.243309
2	0.318847	0.318798	0.324370	1.732260	-0.015195
3	0.350581	0.350533	0.358527	2.266446	-0.013820
4	0.176308	0.176357	0.167393	-5.056334	0.027480
5	0.114438	0.114583	0.109496	-4.318374	0.127011

Figure 2-10. Sampling bias comparison of stratified versus purely random sampling

Now you should remove the income_cat attribute so the data is back to its original state:

```
for set_ in (strat_train_set, strat_test_set):
    set_.drop("income_cat", axis=1, inplace=True)
```

We spent quite a bit of time on test set generation for a good reason: this is an often neglected but critical part of a Machine Learning project. Moreover, many of these ideas will be useful later when we discuss cross-validation. Now it's time to move on to the next stage: exploring the data.

Discover and Visualize the Data to Gain Insights

So far you have only taken a quick glance at the data to get a general understanding of the kind of data you are manipulating. Now the goal is to go into a little more depth.

First, make sure you have put the test set aside and you are only exploring the training set. Also, if the training set is very large, you may want to sample an exploration set, to make manipulations easy and fast. In our case, the set is quite small, so you can just work directly on the full set. Let's create a copy so that you can play with it without harming the training set:

```
housing = strat_train_set.copy()
```

Visualizing Geographical Data

Since there is geographical information (latitude and longitude), it is a good idea to create a scatterplot of all districts to visualize the data (Figure 2-11):

```
housing.plot(kind="scatter", x="longitude", y="latitude")
```

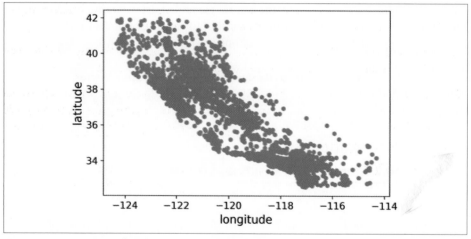

Figure 2-11. A geographical scatterplot of the data

This looks like California all right, but other than that it is hard to see any particular pattern. Setting the alpha option to 0.1 makes it much easier to visualize the places where there is a high density of data points (Figure 2-12):

```
housing.plot(kind="scatter", x="longitude", y="latitude", alpha=0.1)
```

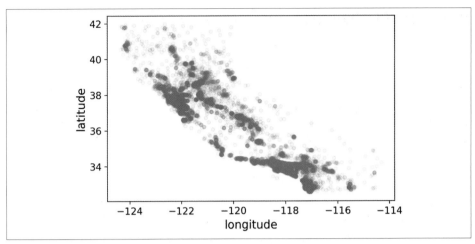

Figure 2-12. A better visualization that highlights high-density areas

Now that's much better: you can clearly see the high-density areas, namely the Bay Area and around Los Angeles and San Diego, plus a long line of fairly high density in the Central Valley, in particular around Sacramento and Fresno.

Our brains are very good at spotting patterns in pictures, but you may need to play around with visualization parameters to make the patterns stand out.

Now let's look at the housing prices (Figure 2-13). The radius of each circle represents the district's population (option s), and the color represents the price (option c). We will use a predefined color map (option cmap) called jet, which ranges from blue (low values) to red (high prices):[16]

```
housing.plot(kind="scatter", x="longitude", y="latitude", alpha=0.4,
    s=housing["population"]/100, label="population", figsize=(10,7),
    c="median_house_value", cmap=plt.get_cmap("jet"), colorbar=True,
)
plt.legend()
```

---

16 If you are reading this in grayscale, grab a red pen and scribble over most of the coastline from the Bay Area down to San Diego (as you might expect). You can add a patch of yellow around Sacramento as well.

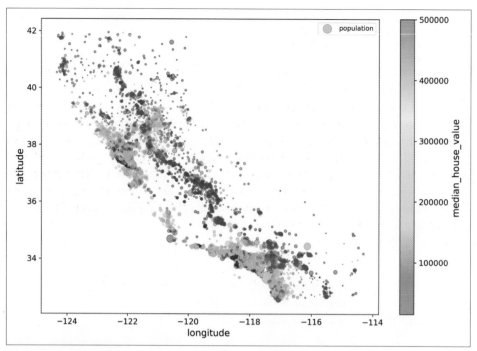

Figure 2-13. California housing prices: red is expensive, blue is cheap, larger circles indicate areas with a larger population

This image tells you that the housing prices are very much related to the location (e.g., close to the ocean) and to the population density, as you probably knew already. A clustering algorithm should be useful for detecting the main cluster and for adding new features that measure the proximity to the cluster centers. The ocean proximity attribute may be useful as well, although in Northern California the housing prices in coastal districts are not too high, so it is not a simple rule.

Looking for Correlations

Since the dataset is not too large, you can easily compute the *standard correlation coefficient* (also called *Pearson's r*) between every pair of attributes using the corr() method:

```
corr_matrix = housing.corr()
```

Now let's look at how much each attribute correlates with the median house value:

```
>>> corr_matrix["median_house_value"].sort_values(ascending=False)
median_house_value    1.000000
median_income         0.687170
total_rooms           0.135231
housing_median_age    0.114220
```

```
households          0.064702
total_bedrooms      0.047865
population         -0.026699
longitude          -0.047279
latitude           -0.142826
Name: median_house_value, dtype: float64
```

The correlation coefficient ranges from –1 to 1. When it is close to 1, it means that there is a strong positive correlation; for example, the median house value tends to go up when the median income goes up. When the coefficient is close to –1, it means that there is a strong negative correlation; you can see a small negative correlation between the latitude and the median house value (i.e., prices have a slight tendency to go down when you go north). Finally, coefficients close to 0 mean that there is no linear correlation. Figure 2-14 shows various plots along with the correlation coefficient between their horizontal and vertical axes.

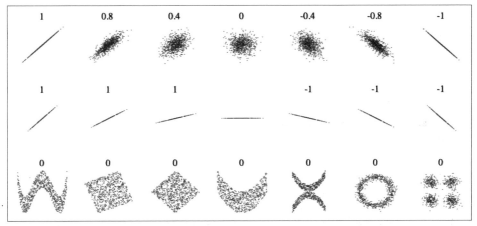

Figure 2-14. Standard correlation coefficient of various datasets (source: Wikipedia; public domain image)

 The correlation coefficient only measures linear correlations ("if x goes up, then y generally goes up/down"). It may completely miss out on nonlinear relationships (e.g., "if x is close to 0, then y generally goes up"). Note how all the plots of the bottom row have a correlation coefficient equal to 0, despite the fact that their axes are clearly not independent: these are examples of nonlinear relationships. Also, the second row shows examples where the correlation coefficient is equal to 1 or –1; notice that this has nothing to do with the slope. For example, your height in inches has a correlation coefficient of 1 with your height in feet or in nanometers.

Another way to check for correlation between attributes is to use the pandas `scatter_matrix()` function, which plots every numerical attribute against every other numerical attribute. Since there are now 11 numerical attributes, you would get $11^2 = 121$ plots, which would not fit on a page—so let's just focus on a few promising attributes that seem most correlated with the median housing value (Figure 2-15):

```python
from pandas.plotting import scatter_matrix

attributes = ["median_house_value", "median_income", "total_rooms",
              "housing_median_age"]
scatter_matrix(housing[attributes], figsize=(12, 8))
```

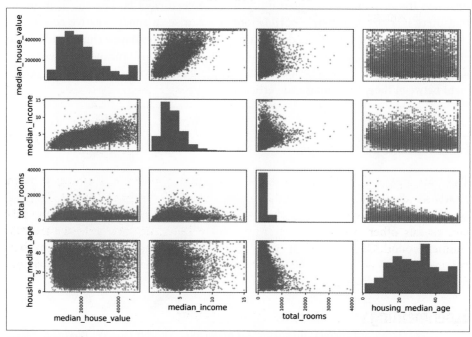

Figure 2-15. This scatter matrix plots every numerical attribute against every other numerical attribute, plus a histogram of each numerical attribute

The main diagonal (top left to bottom right) would be full of straight lines if pandas plotted each variable against itself, which would not be very useful. So instead pandas displays a histogram of each attribute (other options are available; see the pandas documentation for more details).

The most promising attribute to predict the median house value is the median income, so let's zoom in on their correlation scatterplot (Figure 2-16):

```python
housing.plot(kind="scatter", x="median_income", y="median_house_value",
             alpha=0.1)
```

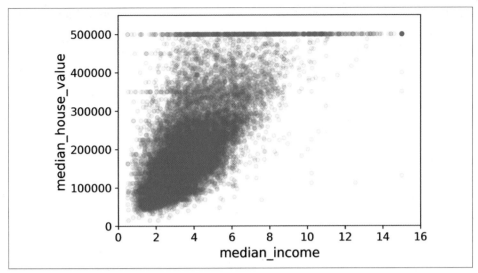

Figure 2-16. Median income versus median house value

This plot reveals a few things. First, the correlation is indeed very strong; you can clearly see the upward trend, and the points are not too dispersed. Second, the price cap that we noticed earlier is clearly visible as a horizontal line at $500,000. But this plot reveals other less obvious straight lines: a horizontal line around $450,000, another around $350,000, perhaps one around $280,000, and a few more below that. You may want to try removing the corresponding districts to prevent your algorithms from learning to reproduce these data quirks.

Experimenting with Attribute Combinations

Hopefully the previous sections gave you an idea of a few ways you can explore the data and gain insights. You identified a few data quirks that you may want to clean up before feeding the data to a Machine Learning algorithm, and you found interesting correlations between attributes, in particular with the target attribute. You also noticed that some attributes have a tail-heavy distribution, so you may want to transform them (e.g., by computing their logarithm). Of course, your mileage will vary considerably with each project, but the general ideas are similar.

One last thing you may want to do before preparing the data for Machine Learning algorithms is to try out various attribute combinations. For example, the total number of rooms in a district is not very useful if you don't know how many households there are. What you really want is the number of rooms per household. Similarly, the total number of bedrooms by itself is not very useful: you probably want to compare it to the number of rooms. And the population per household also seems like an interesting attribute combination to look at. Let's create these new attributes:

```
housing["rooms_per_household"] = housing["total_rooms"]/housing["households"]
housing["bedrooms_per_room"] = housing["total_bedrooms"]/housing["total_rooms"]
housing["population_per_household"]=housing["population"]/housing["households"]
```

And now let's look at the correlation matrix again:

```
>>> corr_matrix = housing.corr()
>>> corr_matrix["median_house_value"].sort_values(ascending=False)
median_house_value          1.000000
median_income               0.687160
rooms_per_household         0.146285
total_rooms                 0.135097
housing_median_age          0.114110
households                  0.064506
total_bedrooms              0.047689
population_per_household    -0.021985
population                  -0.026920
longitude                   -0.047432
latitude                    -0.142724
bedrooms_per_room           -0.259984
Name: median_house_value, dtype: float64
```

Hey, not bad! The new bedrooms_per_room attribute is much more correlated with the median house value than the total number of rooms or bedrooms. Apparently houses with a lower bedroom/room ratio tend to be more expensive. The number of rooms per household is also more informative than the total number of rooms in a district—obviously the larger the houses, the more expensive they are.

This round of exploration does not have to be absolutely thorough; the point is to start off on the right foot and quickly gain insights that will help you get a first rea‐sonably good prototype. But this is an iterative process: once you get a prototype up and running, you can analyze its output to gain more insights and come back to this exploration step.

Prepare the Data for Machine Learning Algorithms

It's time to prepare the data for your Machine Learning algorithms. Instead of doing this manually, you should write functions for this purpose, for several good reasons:

- This will allow you to reproduce these transformations easily on any dataset (e.g., the next time you get a fresh dataset).
- You will gradually build a library of transformation functions that you can reuse in future projects.
- You can use these functions in your live system to transform the new data before feeding it to your algorithms.

- This will make it possible for you to easily try various transformations and see which combination of transformations works best.

But first let's revert to a clean training set (by copying strat_train_set once again). Let's also separate the predictors and the labels, since we don't necessarily want to apply the same transformations to the predictors and the target values (note that drop() creates a copy of the data and does not affect strat_train_set):

```
housing = strat_train_set.drop("median_house_value", axis=1)
housing_labels = strat_train_set["median_house_value"].copy()
```

Data Cleaning

Most Machine Learning algorithms cannot work with missing features, so let's create a few functions to take care of them. We saw earlier that the total_bedrooms attribute has some missing values, so let's fix this. You have three options:

1. Get rid of the corresponding districts.
2. Get rid of the whole attribute.
3. Set the values to some value (zero, the mean, the median, etc.).

You can accomplish these easily using DataFrame's dropna(), drop(), and fillna() methods:

```
housing.dropna(subset=["total_bedrooms"])     # option 1
housing.drop("total_bedrooms", axis=1)         # option 2
median = housing["total_bedrooms"].median()   # option 3
housing["total_bedrooms"].fillna(median, inplace=True)
```

If you choose option 3, you should compute the median value on the training set and use it to fill the missing values in the training set. Don't forget to save the median value that you have computed. You will need it later to replace missing values in the test set when you want to evaluate your system, and also once the system goes live to replace missing values in new data.

Scikit-Learn provides a handy class to take care of missing values: SimpleImputer. Here is how to use it. First, you need to create a SimpleImputer instance, specifying that you want to replace each attribute's missing values with the median of that attribute:

```
from sklearn.impute import SimpleImputer

imputer = SimpleImputer(strategy="median")
```

Since the median can only be computed on numerical attributes, you need to create a copy of the data without the text attribute ocean_proximity:

```
housing_num = housing.drop("ocean_proximity", axis=1)
```

Now you can fit the `imputer` instance to the training data using the `fit()` method:

```
imputer.fit(housing_num)
```

The `imputer` has simply computed the median of each attribute and stored the result in its `statistics_` instance variable. Only the `total_bedrooms` attribute had missing values, but we cannot be sure that there won't be any missing values in new data after the system goes live, so it is safer to apply the `imputer` to all the numerical attributes:

```
>>> imputer.statistics_
array([ -118.51 , 34.26 , 29. , 2119.5 , 433. , 1164. , 408. , 3.5409])
>>> housing_num.median().values
array([ -118.51 , 34.26 , 29. , 2119.5 , 433. , 1164. , 408. , 3.5409])
```

Now you can use this "trained" `imputer` to transform the training set by replacing missing values with the learned medians:

```
X = imputer.transform(housing_num)
```

The result is a plain NumPy array containing the transformed features. If you want to put it back into a pandas DataFrame, it's simple:

```
housing_tr = pd.DataFrame(X, columns=housing_num.columns,
                          index=housing_num.index)
```

Scikit-Learn Design

Scikit-Learn's API is remarkably well designed. These are the main design principles (*https://homl.info/11*):[17]

Consistency
All objects share a consistent and simple interface:

Estimators
Any object that can estimate some parameters based on a dataset is called an *estimator* (e.g., an `imputer` is an estimator). The estimation itself is performed by the `fit()` method, and it takes only a dataset as a parameter (or two for supervised learning algorithms; the second dataset contains the labels). Any other parameter needed to guide the estimation process is considered a hyperparameter (such as an `imputer`'s `strategy`), and it must be set as an instance variable (generally via a constructor parameter).

Transformers
Some estimators (such as an `imputer`) can also transform a dataset; these are called *transformers*. Once again, the API is simple: the transformation is

---

17 For more details on the design principles, see Lars Buitinck et al., "API Design for Machine Learning Software: Experiences from the Scikit-Learn Project" ," arXiv preprint arXiv:1309.0238 (2013).

performed by the transform() method with the dataset to transform as a parameter. It returns the transformed dataset. This transformation generally relies on the learned parameters, as is the case for an imputer. All transformers also have a convenience method called fit_transform() that is equivalent to calling fit() and then transform() (but sometimes fit_transform() is optimized and runs much faster).

Predictors
Finally, some estimators, given a dataset, are capable of making predictions; they are called *predictors*. For example, the LinearRegression model in the previous chapter was a predictor: given a country's GDP per capita, it predicted life satisfaction. A predictor has a predict() method that takes a dataset of new instances and returns a dataset of corresponding predictions. It also has a score() method that measures the quality of the predictions, given a test set (and the corresponding labels, in the case of supervised learning algorithms).[18]

Inspection
All the estimator's hyperparameters are accessible directly via public instance variables (e.g., imputer.strategy), and all the estimator's learned parameters are accessible via public instance variables with an underscore suffix (e.g., imputer.statistics_).

Nonproliferation of classes
Datasets are represented as NumPy arrays or SciPy sparse matrices, instead of homemade classes. Hyperparameters are just regular Python strings or numbers.

Composition
Existing building blocks are reused as much as possible. For example, it is easy to create a Pipeline estimator from an arbitrary sequence of transformers followed by a final estimator, as we will see.

Sensible defaults
Scikit-Learn provides reasonable default values for most parameters, making it easy to quickly create a baseline working system.

Handling Text and Categorical Attributes

So far we have only dealt with numerical attributes, but now let's look at text attributes. In this dataset, there is just one: the ocean_proximity attribute. Let's look at its value for the first 10 instances:

---

18 Some predictors also provide methods to measure the confidence of their predictions.

```
>>> housing_cat = housing[["ocean_proximity"]]
>>> housing_cat.head(10)
      ocean_proximity
17606        <1H OCEAN
18632        <1H OCEAN
14650       NEAR OCEAN
3230            INLAND
3555         <1H OCEAN
19480           INLAND
8879         <1H OCEAN
13685           INLAND
4937         <1H OCEAN
4861         <1H OCEAN
```

It's not arbitrary text: there are a limited number of possible values, each of which represents a category. So this attribute is a categorical attribute. Most Machine Learning algorithms prefer to work with numbers, so let's convert these categories from text to numbers. For this, we can use Scikit-Learn's OrdinalEncoder class:[19]

```
>>> from sklearn.preprocessing import OrdinalEncoder
>>> ordinal_encoder = OrdinalEncoder()
>>> housing_cat_encoded = ordinal_encoder.fit_transform(housing_cat)
>>> housing_cat_encoded[:10]
array([[0.],
       [0.],
       [4.],
       [1.],
       [0.],
       [1.],
       [0.],
       [1.],
       [0.],
       [0.]])
```

You can get the list of categories using the categories_ instance variable. It is a list containing a 1D array of categories for each categorical attribute (in this case, a list containing a single array since there is just one categorical attribute):

```
>>> ordinal_encoder.categories_
[array(['<1H OCEAN', 'INLAND', 'ISLAND', 'NEAR BAY', 'NEAR OCEAN'],
      dtype=object)]
```

One issue with this representation is that ML algorithms will assume that two nearby values are more similar than two distant values. This may be fine in some cases (e.g., for ordered categories such as "bad," "average," "good," and "excellent"), but it is obviously not the case for the ocean_proximity column (for example, categories 0 and 4 are clearly more similar than categories 0 and 1). To fix this issue, a common solution

---

19 This class is available in Scikit-Learn 0.20 and later. If you use an earlier version, please consider upgrading, or use the pandas Series.factorize() method.

is to create one binary attribute per category: one attribute equal to 1 when the category is "<1H OCEAN" (and 0 otherwise), another attribute equal to 1 when the category is "INLAND" (and 0 otherwise), and so on. This is called *one-hot encoding*, because only one attribute will be equal to 1 (hot), while the others will be 0 (cold). The new attributes are sometimes called *dummy* attributes. Scikit-Learn provides a OneHotEncoder class to convert categorical values into one-hot vectors:[20]

```
>>> from sklearn.preprocessing import OneHotEncoder
>>> cat_encoder = OneHotEncoder()
>>> housing_cat_1hot = cat_encoder.fit_transform(housing_cat)
>>> housing_cat_1hot
<16512x5 sparse matrix of type '<class 'numpy.float64'>'
   with 16512 stored elements in Compressed Sparse Row format>
```

Notice that the output is a SciPy *sparse matrix*, instead of a NumPy array. This is very useful when you have categorical attributes with thousands of categories. After one-hot encoding, we get a matrix with thousands of columns, and the matrix is full of 0s except for a single 1 per row. Using up tons of memory mostly to store zeros would be very wasteful, so instead a sparse matrix only stores the location of the nonzero elements. You can use it mostly like a normal 2D array,[21] but if you really want to convert it to a (dense) NumPy array, just call the toarray() method:

```
>>> housing_cat_1hot.toarray()
array([[1., 0., 0., 0., 0.],
       [1., 0., 0., 0., 0.],
       [0., 0., 0., 0., 1.],
       ...,
       [0., 1., 0., 0., 0.],
       [1., 0., 0., 0., 0.],
       [0., 0., 0., 1., 0.]])
```

Once again, you can get the list of categories using the encoder's categories_ instance variable:

```
>>> cat_encoder.categories_
[array(['<1H OCEAN', 'INLAND', 'ISLAND', 'NEAR BAY', 'NEAR OCEAN'],
       dtype=object)]
```

---

20 Before Scikit-Learn 0.20, the method could only encode integer categorical values, but since 0.20 it can also handle other types of inputs, including text categorical inputs.

21 See SciPy's documentation for more details.

If a categorical attribute has a large number of possible categories (e.g., country code, profession, species), then one-hot encoding will result in a large number of input features. This may slow down training and degrade performance. If this happens, you may want to replace the categorical input with useful numerical features related to the categories: for example, you could replace the ocean_proximity feature with the distance to the ocean (similarly, a country code could be replaced with the country's population and GDP per capita). Alternatively, you could replace each category with a learnable, low-dimensional vector called an *embedding*. Each category's representation would be learned during training. This is an example of *representation learning* (see Chapters 13 and 17 for more details).

Custom Transformers

Although Scikit-Learn provides many useful transformers, you will need to write your own for tasks such as custom cleanup operations or combining specific attributes. You will want your transformer to work seamlessly with Scikit-Learn functionalities (such as pipelines), and since Scikit-Learn relies on duck typing (not inheritance), all you need to do is create a class and implement three methods: fit() (returning self), transform(), and fit_transform().

You can get the last one for free by simply adding TransformerMixin as a base class. If you add BaseEstimator as a base class (and avoid *args and **kargs in your constructor), you will also get two extra methods (get_params() and set_params()) that will be useful for automatic hyperparameter tuning.

For example, here is a small transformer class that adds the combined attributes we discussed earlier:

```python
from sklearn.base import BaseEstimator, TransformerMixin

rooms_ix, bedrooms_ix, population_ix, households_ix = 3, 4, 5, 6

class CombinedAttributesAdder(BaseEstimator, TransformerMixin):
    def __init__(self, add_bedrooms_per_room = True): # no *args or **kargs
        self.add_bedrooms_per_room = add_bedrooms_per_room
    def fit(self, X, y=None):
        return self  # nothing else to do
    def transform(self, X, y=None):
        rooms_per_household = X[:, rooms_ix] / X[:, households_ix]
        population_per_household = X[:, population_ix] / X[:, households_ix]
        if self.add_bedrooms_per_room:
            bedrooms_per_room = X[:, bedrooms_ix] / X[:, rooms_ix]
            return np.c_[X, rooms_per_household, population_per_household,
                         bedrooms_per_room]
```

```
        else:
            return np.c_[X, rooms_per_household, population_per_household]

attr_adder = CombinedAttributesAdder(add_bedrooms_per_room=False)
housing_extra_attribs = attr_adder.transform(housing.values)
```

In this example the transformer has one hyperparameter, add_bedrooms_per_room, set to True by default (it is often helpful to provide sensible defaults). This hyperparameter will allow you to easily find out whether adding this attribute helps the Machine Learning algorithms or not. More generally, you can add a hyperparameter to gate any data preparation step that you are not 100% sure about. The more you automate these data preparation steps, the more combinations you can automatically try out, making it much more likely that you will find a great combination (and saving you a lot of time).

Feature Scaling

One of the most important transformations you need to apply to your data is *feature scaling*. With few exceptions, Machine Learning algorithms don't perform well when the input numerical attributes have very different scales. This is the case for the housing data: the total number of rooms ranges from about 6 to 39,320, while the median incomes only range from 0 to 15. Note that scaling the target values is generally not required.

There are two common ways to get all attributes to have the same scale: *min-max scaling* and *standardization*.

Min-max scaling (many people call this *normalization*) is the simplest: values are shifted and rescaled so that they end up ranging from 0 to 1. We do this by subtracting the min value and dividing by the max minus the min. Scikit-Learn provides a transformer called MinMaxScaler for this. It has a feature_range hyperparameter that lets you change the range if, for some reason, you don't want 0–1.

Standardization is different: first it subtracts the mean value (so standardized values always have a zero mean), and then it divides by the standard deviation so that the resulting distribution has unit variance. Unlike min-max scaling, standardization does not bound values to a specific range, which may be a problem for some algorithms (e.g., neural networks often expect an input value ranging from 0 to 1). However, standardization is much less affected by outliers. For example, suppose a district had a median income equal to 100 (by mistake). Min-max scaling would then crush all the other values from 0–15 down to 0–0.15, whereas standardization would not be much affected. Scikit-Learn provides a transformer called StandardScaler for standardization.

As with all the transformations, it is important to fit the scalers to the training data only, not to the full dataset (including the test set). Only then can you use them to transform the training set and the test set (and new data).

Transformation Pipelines

As you can see, there are many data transformation steps that need to be executed in the right order. Fortunately, Scikit-Learn provides the `Pipeline` class to help with such sequences of transformations. Here is a small pipeline for the numerical attributes:

```
from sklearn.pipeline import Pipeline
from sklearn.preprocessing import StandardScaler

num_pipeline = Pipeline([
        ('imputer', SimpleImputer(strategy="median")),
        ('attribs_adder', CombinedAttributesAdder()),
        ('std_scaler', StandardScaler()),
    ])

housing_num_tr = num_pipeline.fit_transform(housing_num)
```

The `Pipeline` constructor takes a list of name/estimator pairs defining a sequence of steps. All but the last estimator must be transformers (i.e., they must have a `fit_transform()` method). The names can be anything you like (as long as they are unique and don't contain double underscores, __); they will come in handy later for hyperparameter tuning.

When you call the pipeline's `fit()` method, it calls `fit_transform()` sequentially on all transformers, passing the output of each call as the parameter to the next call until it reaches the final estimator, for which it calls the `fit()` method.

The pipeline exposes the same methods as the final estimator. In this example, the last estimator is a `StandardScaler`, which is a transformer, so the pipeline has a `trans form()` method that applies all the transforms to the data in sequence (and of course also a `fit_transform()` method, which is the one we used).

So far, we have handled the categorical columns and the numerical columns separately. It would be more convenient to have a single transformer able to handle all columns, applying the appropriate transformations to each column. In version 0.20, Scikit-Learn introduced the `ColumnTransformer` for this purpose, and the good news is that it works great with pandas DataFrames. Let's use it to apply all the transformations to the housing data:

```
from sklearn.compose import ColumnTransformer

num_attribs = list(housing_num)
cat_attribs = ["ocean_proximity"]

full_pipeline = ColumnTransformer([
        ("num", num_pipeline, num_attribs),
        ("cat", OneHotEncoder(), cat_attribs),
    ])

housing_prepared = full_pipeline.fit_transform(housing)
```

First we import the ColumnTransformer class, next we get the list of numerical column names and the list of categorical column names, and then we construct a ColumnTransformer. The constructor requires a list of tuples, where each tuple contains a name,[22] a transformer, and a list of names (or indices) of columns that the transformer should be applied to. In this example, we specify that the numerical columns should be transformed using the num_pipeline that we defined earlier, and the categorical columns should be transformed using a OneHotEncoder. Finally, we apply this ColumnTransformer to the housing data: it applies each transformer to the appropriate columns and concatenates the outputs along the second axis (the transformers must return the same number of rows).

Note that the OneHotEncoder returns a sparse matrix, while the num_pipeline returns a dense matrix. When there is such a mix of sparse and dense matrices, the ColumnTransformer estimates the density of the final matrix (i.e., the ratio of nonzero cells), and it returns a sparse matrix if the density is lower than a given threshold (by default, sparse_threshold=0.3). In this example, it returns a dense matrix. And that's it! We have a preprocessing pipeline that takes the full housing data and applies the appropriate transformations to each column.

 Instead of using a transformer, you can specify the string "drop" if you want the columns to be dropped, or you can specify "passthrough" if you want the columns to be left untouched. By default, the remaining columns (i.e., the ones that were not listed) will be dropped, but you can set the remainder hyperparameter to any transformer (or to "passthrough") if you want these columns to be handled differently.

If you are using Scikit-Learn 0.19 or earlier, you can use a third-party library such as sklearn-pandas, or you can roll out your own custom transformer to get the same functionality as the ColumnTransformer. Alternatively, you can use the FeatureUnion

---

22 Just like for pipelines, the name can be anything as long as it does not contain double underscores.

class, which can apply different transformers and concatenate their outputs. But you cannot specify different columns for each transformer; they all apply to the whole data. It is possible to work around this limitation using a custom transformer for column selection (see the Jupyter notebook for an example).

Select and Train a Model

At last! You framed the problem, you got the data and explored it, you sampled a training set and a test set, and you wrote transformation pipelines to clean up and prepare your data for Machine Learning algorithms automatically. You are now ready to select and train a Machine Learning model.

Training and Evaluating on the Training Set

The good news is that thanks to all these previous steps, things are now going to be much simpler than you might think. Let's first train a Linear Regression model, like we did in the previous chapter:

```
from sklearn.linear_model import LinearRegression

lin_reg = LinearRegression()
lin_reg.fit(housing_prepared, housing_labels)
```

Done! You now have a working Linear Regression model. Let's try it out on a few instances from the training set:

```
>>> some_data = housing.iloc[:5]
>>> some_labels = housing_labels.iloc[:5]
>>> some_data_prepared = full_pipeline.transform(some_data)
>>> print("Predictions:", lin_reg.predict(some_data_prepared))
Predictions: [ 210644.6045  317768.8069  210956.4333  59218.9888  189747.5584]
>>> print("Labels:", list(some_labels))
Labels: [286600.0, 340600.0, 196900.0, 46300.0, 254500.0]
```

It works, although the predictions are not exactly accurate (e.g., the first prediction is off by close to 40%!). Let's measure this regression model's RMSE on the whole training set using Scikit-Learn's `mean_squared_error()` function:

```
>>> from sklearn.metrics import mean_squared_error
>>> housing_predictions = lin_reg.predict(housing_prepared)
>>> lin_mse = mean_squared_error(housing_labels, housing_predictions)
>>> lin_rmse = np.sqrt(lin_mse)
>>> lin_rmse
68628.19819848922
```

This is better than nothing, but clearly not a great score: most districts' `median_hous ing_values` range between $120,000 and $265,000, so a typical prediction error of $68,628 is not very satisfying. This is an example of a model underfitting the training data. When this happens it can mean that the features do not provide enough information to make good predictions, or that the model is not powerful enough. As we saw in the previous chapter, the main ways to fix underfitting are to select a more powerful model, to feed the training algorithm with better features, or to reduce the constraints on the model. This model is not regularized, which rules out the last option. You could try to add more features (e.g., the log of the population), but first let's try a more complex model to see how it does.

Let's train a `DecisionTreeRegressor`. This is a powerful model, capable of finding complex nonlinear relationships in the data (Decision Trees are presented in more detail in Chapter 6). The code should look familiar by now:

```
from sklearn.tree import DecisionTreeRegressor

tree_reg = DecisionTreeRegressor()
tree_reg.fit(housing_prepared, housing_labels)
```

Now that the model is trained, let's evaluate it on the training set:

```
>>> housing_predictions = tree_reg.predict(housing_prepared)
>>> tree_mse = mean_squared_error(housing_labels, housing_predictions)
>>> tree_rmse = np.sqrt(tree_mse)
>>> tree_rmse
0.0
```

Wait, what!? No error at all? Could this model really be absolutely perfect? Of course, it is much more likely that the model has badly overfit the data. How can you be sure? As we saw earlier, you don't want to touch the test set until you are ready to launch a model you are confident about, so you need to use part of the training set for training and part of it for model validation.

Better Evaluation Using Cross-Validation

One way to evaluate the Decision Tree model would be to use the `train_test_split()` function to split the training set into a smaller training set and a validation set, then train your models against the smaller training set and evaluate them against the validation set. It's a bit of work, but nothing too difficult, and it would work fairly well.

A great alternative is to use Scikit-Learn's *K-fold cross-validation* feature. The following code randomly splits the training set into 10 distinct subsets called *folds*, then it trains and evaluates the Decision Tree model 10 times, picking a different fold for evaluation every time and training on the other 9 folds. The result is an array containing the 10 evaluation scores:

```
from sklearn.model_selection import cross_val_score
scores = cross_val_score(tree_reg, housing_prepared, housing_labels,
                         scoring="neg_mean_squared_error", cv=10)
tree_rmse_scores = np.sqrt(-scores)
```

Scikit-Learn's cross-validation features expect a utility function (greater is better) rather than a cost function (lower is better), so the scoring function is actually the opposite of the MSE (i.e., a negative value), which is why the preceding code computes `-scores` before calculating the square root.

Let's look at the results:

```
>>> def display_scores(scores):
...     print("Scores:", scores)
...     print("Mean:", scores.mean())
...     print("Standard deviation:", scores.std())
...
>>> display_scores(tree_rmse_scores)
Scores: [70194.33680785 66855.16363941 72432.58244769 70758.73896782
 71115.88230639 75585.14172901 70262.86139133 70273.6325285
 75366.87952553 71231.65726027]
Mean: 71407.68766037929
Standard deviation: 2439.4345041191004
```

Now the Decision Tree doesn't look as good as it did earlier. In fact, it seems to perform worse than the Linear Regression model! Notice that cross-validation allows you to get not only an estimate of the performance of your model, but also a measure of how precise this estimate is (i.e., its standard deviation). The Decision Tree has a score of approximately 71,407, generally ±2,439. You would not have this information if you just used one validation set. But cross-validation comes at the cost of training the model several times, so it is not always possible.

Let's compute the same scores for the Linear Regression model just to be sure:

```
>>> lin_scores = cross_val_score(lin_reg, housing_prepared, housing_labels,
...                              scoring="neg_mean_squared_error", cv=10)
...
>>> lin_rmse_scores = np.sqrt(-lin_scores)
>>> display_scores(lin_rmse_scores)
Scores: [66782.73843989 66960.118071   70347.95244419 74739.57052552
 68031.13388938 71193.84183426 64969.63056405 68281.61137997
 71552.91566558 67665.10082067]
Mean: 69052.46136345083
Standard deviation: 2731.674001798348
```

That's right: the Decision Tree model is overfitting so badly that it performs worse than the Linear Regression model.

Let's try one last model now: the `RandomForestRegressor`. As we will see in Chapter 7, Random Forests work by training many Decision Trees on random subsets of the features, then averaging out their predictions. Building a model on top of many other models is called *Ensemble Learning*, and it is often a great way to push ML algorithms even further. We will skip most of the code since it is essentially the same as for the other models:

```
>>> from sklearn.ensemble import RandomForestRegressor
>>> forest_reg = RandomForestRegressor()
>>> forest_reg.fit(housing_prepared, housing_labels)
>>> [...]
>>> forest_rmse
18603.515021376355
>>> display_scores(forest_rmse_scores)
Scores: [49519.80364233 47461.9115823  50029.02762854 52325.28068953
 49308.39426421 53446.37892622 48634.8036574  47585.73832311
 53490.10699751 50021.5852922 ]
Mean: 50182.303100336096
Standard deviation: 2097.0810550985693
```

Wow, this is much better: Random Forests look very promising. However, note that the score on the training set is still much lower than on the validation sets, meaning that the model is still overfitting the training set. Possible solutions for overfitting are to simplify the model, constrain it (i.e., regularize it), or get a lot more training data. Before you dive much deeper into Random Forests, however, you should try out many other models from various categories of Machine Learning algorithms (e.g., several Support Vector Machines with different kernels, and possibly a neural network), without spending too much time tweaking the hyperparameters. The goal is to shortlist a few (two to five) promising models.

 You should save every model you experiment with so that you can come back easily to any model you want. Make sure you save both the hyperparameters and the trained parameters, as well as the cross-validation scores and perhaps the actual predictions as well. This will allow you to easily compare scores across model types, and compare the types of errors they make. You can easily save Scikit-Learn models by using Python's pickle module or by using the joblib library, which is more efficient at serializing large NumPy arrays (you can install this library using pip):

```
import joblib

joblib.dump(my_model, "my_model.pkl")
# and later...
my_model_loaded = joblib.load("my_model.pkl")
```

Fine-Tune Your Model

Let's assume that you now have a shortlist of promising models. You now need to fine-tune them. Let's look at a few ways you can do that.

Grid Search

One option would be to fiddle with the hyperparameters manually, until you find a great combination of hyperparameter values. This would be very tedious work, and you may not have time to explore many combinations.

Instead, you should get Scikit-Learn's `GridSearchCV` to search for you. All you need to do is tell it which hyperparameters you want it to experiment with and what values to try out, and it will use cross-validation to evaluate all the possible combinations of hyperparameter values. For example, the following code searches for the best combination of hyperparameter values for the `RandomForestRegressor`:

```
from sklearn.model_selection import GridSearchCV

param_grid = [
    {'n_estimators': [3, 10, 30], 'max_features': [2, 4, 6, 8]},
    {'bootstrap': [False], 'n_estimators': [3, 10], 'max_features': [2, 3, 4]},
  ]

forest_reg = RandomForestRegressor()

grid_search = GridSearchCV(forest_reg, param_grid, cv=5,
                           scoring='neg_mean_squared_error',
                           return_train_score=True)

grid_search.fit(housing_prepared, housing_labels)
```

 When you have no idea what value a hyperparameter should have, a simple approach is to try out consecutive powers of 10 (or a smaller number if you want a more fine-grained search, as shown in this example with the n_estimators hyperparameter).

This `param_grid` tells Scikit-Learn to first evaluate all 3 × 4 = 12 combinations of `n_estimators` and `max_features` hyperparameter values specified in the first `dict` (don't worry about what these hyperparameters mean for now; they will be explained in Chapter 7), then try all 2 × 3 = 6 combinations of hyperparameter values in the second `dict`, but this time with the `bootstrap` hyperparameter set to `False` instead of `True` (which is the default value for this hyperparameter).

The grid search will explore 12 + 6 = 18 combinations of `RandomForestRegressor` hyperparameter values, and it will train each model 5 times (since we are using five-fold cross validation). In other words, all in all, there will be 18 × 5 = 90 rounds of training! It may take quite a long time, but when it is done you can get the best combination of parameters like this:

```
>>> grid_search.best_params_
{'max_features': 8, 'n_estimators': 30}
```

Since 8 and 30 are the maximum values that were evaluated, you should probably try searching again with higher values; the score may continue to improve.

You can also get the best estimator directly:

```
>>> grid_search.best_estimator_
RandomForestRegressor(bootstrap=True, criterion='mse', max_depth=None,
        max_features=8, max_leaf_nodes=None, min_impurity_decrease=0.0,
        min_impurity_split=None, min_samples_leaf=1,
        min_samples_split=2, min_weight_fraction_leaf=0.0,
        n_estimators=30, n_jobs=None, oob_score=False, random_state=None,
        verbose=0, warm_start=False)
```

If GridSearchCV is initialized with refit=True (which is the default), then once it finds the best estimator using cross-validation, it retrains it on the whole training set. This is usually a good idea, since feeding it more data will likely improve its performance.

And of course the evaluation scores are also available:

```
>>> cvres = grid_search.cv_results_
>>> for mean_score, params in zip(cvres["mean_test_score"], cvres["params"]):
...     print(np.sqrt(-mean_score), params)
...
63669.05791727153 {'max_features': 2, 'n_estimators': 3}
55627.16171305252 {'max_features': 2, 'n_estimators': 10}
53384.57867637289 {'max_features': 2, 'n_estimators': 30}
60965.99185930139 {'max_features': 4, 'n_estimators': 3}
52740.98248528835 {'max_features': 4, 'n_estimators': 10}
50377.344409590376 {'max_features': 4, 'n_estimators': 30}
58663.84733372485 {'max_features': 6, 'n_estimators': 3}
52006.15355973719 {'max_features': 6, 'n_estimators': 10}
50146.465964159885 {'max_features': 6, 'n_estimators': 30}
57869.25504027614 {'max_features': 8, 'n_estimators': 3}
51711.09443660957 {'max_features': 8, 'n_estimators': 10}
49682.25345942335 {'max_features': 8, 'n_estimators': 30}
62895.088889905004 {'bootstrap': False, 'max_features': 2, 'n_estimators': 3}
54658.14484390074 {'bootstrap': False, 'max_features': 2, 'n_estimators': 10}
59470.399594730654 {'bootstrap': False, 'max_features': 3, 'n_estimators': 3}
52725.01091081235 {'bootstrap': False, 'max_features': 3, 'n_estimators': 10}
57490.612956065226 {'bootstrap': False, 'max_features': 4, 'n_estimators': 3}
51009.51445842374 {'bootstrap': False, 'max_features': 4, 'n_estimators': 10}
```

In this example, we obtain the best solution by setting the max_features hyperparameter to 8 and the n_estimators hyperparameter to 30. The RMSE score for this combination is 49,682, which is slightly better than the score you got earlier using the default hyperparameter values (which was 50,182). Congratulations, you have successfully fine-tuned your best model!

 Don't forget that you can treat some of the data preparation steps as hyperparameters. For example, the grid search will automatically find out whether or not to add a feature you were not sure about (e.g., using the `add_bedrooms_per_room` hyperparameter of your `CombinedAttributesAdder` transformer). It may similarly be used to automatically find the best way to handle outliers, missing features, feature selection, and more.

Randomized Search

The grid search approach is fine when you are exploring relatively few combinations, like in the previous example, but when the hyperparameter search space is large, it is often preferable to use `RandomizedSearchCV` instead. This class can be used in much the same way as the `GridSearchCV` class, but instead of trying out all possible combinations, it evaluates a given number of random combinations by selecting a random value for each hyperparameter at every iteration. This approach has two main benefits:

- If you let the randomized search run for, say, 1,000 iterations, this approach will explore 1,000 different values for each hyperparameter (instead of just a few values per hyperparameter with the grid search approach).

- Simply by setting the number of iterations, you have more control over the computing budget you want to allocate to hyperparameter search.

Ensemble Methods

Another way to fine-tune your system is to try to combine the models that perform best. The group (or "ensemble") will often perform better than the best individual model (just like Random Forests perform better than the individual Decision Trees they rely on), especially if the individual models make very different types of errors. We will cover this topic in more detail in Chapter 7.

Analyze the Best Models and Their Errors

You will often gain good insights on the problem by inspecting the best models. For example, the `RandomForestRegressor` can indicate the relative importance of each attribute for making accurate predictions:

```
>>> feature_importances = grid_search.best_estimator_.feature_importances_
>>> feature_importances
array([7.33442355e-02, 6.29090705e-02, 4.11437985e-02, 1.46726854e-02,
```

```
        1.41064835e-02, 1.48742809e-02, 1.42575993e-02, 3.66158981e-01,
        5.64191792e-02, 1.08792957e-01, 5.33510773e-02, 1.03114883e-02,
        1.64780994e-01, 6.02803867e-05, 1.96041560e-03, 2.85647464e-03])
```

Let's display these importance scores next to their corresponding attribute names:

```
>>> extra_attribs = ["rooms_per_hhold", "pop_per_hhold", "bedrooms_per_room"]
>>> cat_encoder = full_pipeline.named_transformers_["cat"]
>>> cat_one_hot_attribs = list(cat_encoder.categories_[0])
>>> attributes = num_attribs + extra_attribs + cat_one_hot_attribs
>>> sorted(zip(feature_importances, attributes), reverse=True)
[(0.3661589806181342, 'median_income'),
 (0.1647809935615905, 'INLAND'),
 (0.10879295677551573, 'pop_per_hhold'),
 (0.07334423551601242, 'longitude'),
 (0.0629090704826203, 'latitude'),
 (0.05641917918195401, 'rooms_per_hhold'),
 (0.05335107734767581, 'bedrooms_per_room'),
 (0.041143798478729635, 'housing_median_age'),
 (0.014874280890402767, 'population'),
 (0.014672685420543237, 'total_rooms'),
 (0.014257599323407807, 'households'),
 (0.014106483453584102, 'total_bedrooms'),
 (0.010311488326303787, '<1H OCEAN'),
 (0.002856474637320158, 'NEAR OCEAN'),
 (0.00196041559947807, 'NEAR BAY'),
 (6.028038672736599e-05, 'ISLAND')]
```

With this information, you may want to try dropping some of the less useful features (e.g., apparently only one ocean_proximity category is really useful, so you could try dropping the others).

You should also look at the specific errors that your system makes, then try to understand why it makes them and what could fix the problem (adding extra features or getting rid of uninformative ones, cleaning up outliers, etc.).

Evaluate Your System on the Test Set

After tweaking your models for a while, you eventually have a system that performs sufficiently well. Now is the time to evaluate the final model on the test set. There is nothing special about this process; just get the predictors and the labels from your test set, run your full_pipeline to transform the data (call transform(), *not* fit_transform()—you do not want to fit the test set!), and evaluate the final model on the test set:

```
final_model = grid_search.best_estimator_

X_test = strat_test_set.drop("median_house_value", axis=1)
y_test = strat_test_set["median_house_value"].copy()

X_test_prepared = full_pipeline.transform(X_test)
```

```
final_predictions = final_model.predict(X_test_prepared)

final_mse = mean_squared_error(y_test, final_predictions)
final_rmse = np.sqrt(final_mse)   # => evaluates to 47,730.2
```

In some cases, such a point estimate of the generalization error will not be quite enough to convince you to launch: what if it is just 0.1% better than the model currently in production? You might want to have an idea of how precise this estimate is. For this, you can compute a 95% *confidence interval* for the generalization error using `scipy.stats.t.interval()`:

```
>>> from scipy import stats
>>> confidence = 0.95
>>> squared_errors = (final_predictions - y_test) ** 2
>>> np.sqrt(stats.t.interval(confidence, len(squared_errors) - 1,
...                          loc=squared_errors.mean(),
...                          scale=stats.sem(squared_errors)))
...
array([45685.10470776, 49691.25001878])
```

If you did a lot of hyperparameter tuning, the performance will usually be slightly worse than what you measured using cross-validation (because your system ends up fine-tuned to perform well on the validation data and will likely not perform as well on unknown datasets). It is not the case in this example, but when this happens you must resist the temptation to tweak the hyperparameters to make the numbers look good on the test set; the improvements would be unlikely to generalize to new data.

Now comes the project prelaunch phase: you need to present your solution (highlighting what you have learned, what worked and what did not, what assumptions were made, and what your system's limitations are), document everything, and create nice presentations with clear visualizations and easy-to-remember statements (e.g., "the median income is the number one predictor of housing prices"). In this California housing example, the final performance of the system is not better than the experts' price estimates, which were often off by about 20%, but it may still be a good idea to launch it, especially if this frees up some time for the experts so they can work on more interesting and productive tasks.

Launch, Monitor, and Maintain Your System

Perfect, you got approval to launch! You now need to get your solution ready for production (e.g., polish the code, write documentation and tests, and so on). Then you can deploy your model to your production environment. One way to do this is to save the trained Scikit-Learn model (e.g., using `joblib`), including the full preprocessing and prediction pipeline, then load this trained model within your production environment and use it to make predictions by calling its `predict()` method. For example, perhaps the model will be used within a website: the user will type in some data about a new district and click the Estimate Price button. This will send a query containing the data to the web server, which will forward it to your web application, and finally your code will simply call the model's predict() method (you want to load the model upon server startup, rather than every time the model is used). Alternatively, you can wrap the model within a dedicated web service that your web application can query through a REST API[23] (see Figure 2-17). This makes it easier to upgrade your model to new versions without interrupting the main application. It also simplifies scaling, since you can start as many web services as needed and load-balance the requests coming from your web application across these web services. Moreover, it allows your web application to use any language, not just Python.

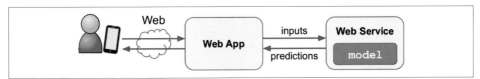

Figure 2-17. A model deployed as a web service and used by a web application

Another popular strategy is to deploy your model on the cloud, for example on Google Cloud AI Platform (formerly known as Google Cloud ML Engine): just save your model using joblib and upload it to Google Cloud Storage (GCS), then head over to Google Cloud AI Platform and create a new model version, pointing it to the GCS file. That's it! This gives you a simple web service that takes care of load balancing and scaling for you. It take JSON requests containing the input data (e.g., of a district) and returns JSON responses containing the predictions. You can then use this web service in your website (or whatever production environment you are using). As we will see in Chapter 19, deploying TensorFlow models on AI Platform is not much different from deploying Scikit-Learn models.

But deployment is not the end of the story. You also need to write monitoring code to check your system's live performance at regular intervals and trigger alerts when it drops. This could be a steep drop, likely due to a broken component in your infrastructure, but be aware that it could also be a gentle decay that could easily go unnoticed for a long time. This is quite common because models tend to "rot" over time: indeed, the world changes, so if the model was trained with last year's data, it may not be adapted to today's data.

---

23 In a nutshell, a REST (or RESTful) API is an HTTP-based API that follows some conventions, such as using standard HTTP verbs to read, update, create, or delete resources (GET, POST, PUT, and DELETE) and using JSON for the inputs and outputs.

Even a model trained to classify pictures of cats and dogs may need to be retrained regularly, not because cats and dogs will mutate overnight, but because cameras keep changing, along with image formats, sharpness, brightness, and size ratios. Moreover, people may love different breeds next year, or they may decide to dress their pets with tiny hats—who knows?

So you need to monitor your model's live performance. But how do you that? Well, it depends. In some cases, the model's performance can be inferred from downstream metrics. For example, if your model is part of a recommender system and it suggests products that the users may be interested in, then it's easy to monitor the number of recommended products sold each day. If this number drops (compared to non-recommended products), then the prime suspect is the model. This may be because the data pipeline is broken, or perhaps the model needs to be retrained on fresh data (as we will discuss shortly).

However, it's not always possible to determine the model's performance without any human analysis. For example, suppose you trained an image classification model (see Chapter 3) to detect several product defects on a production line. How can you get an alert if the model's performance drops, before thousands of defective products get shipped to your clients? One solution is to send to human raters a sample of all the pictures that the model classified (especially pictures that the model wasn't so sure about). Depending on the task, the raters may need to be experts, or they could be nonspecialists, such as workers on a crowdsourcing platform (e.g., Amazon Mechanical Turk). In some applications they could even be the users themselves, responding for example via surveys or repurposed captchas.[24]

Either way, you need to put in place a monitoring system (with or without human raters to evaluate the live model), as well as all the relevant processes to define what to do in case of failures and how to prepare for them. Unfortunately, this can be a lot of work. In fact, it is often much more work than building and training a model.

If the data keeps evolving, you will need to update your datasets and retrain your model regularly. You should probably automate the whole process as much as possible. Here are a few things you can automate:

- Collect fresh data regularly and label it (e.g., using human raters).
- Write a script to train the model and fine-tune the hyperparameters automatically. This script could run automatically, for example every day or every week, depending on your needs.

---

24 A captcha is a test to ensure a user is not a robot. These tests have often been used as a cheap way to label training data.

- Write another script that will evaluate both the new model and the previous model on the updated test set, and deploy the model to production if the performance has not decreased (if it did, make sure you investigate why).

You should also make sure you evaluate the model's input data quality. Sometimes performance will degrade slightly because of a poor-quality signal (e.g., a malfunctioning sensor sending random values, or another team's output becoming stale), but it may take a while before your system's performance degrades enough to trigger an alert. If you monitor your model's inputs, you may catch this earlier. For example, you could trigger an alert if more and more inputs are missing a feature, or if its mean or standard deviation drifts too far from the training set, or a categorical feature starts containing new categories.

Finally, make sure you keep backups of every model you create and have the process and tools in place to roll back to a previous model quickly, in case the new model starts failing badly for some reason. Having backups also makes it possible to easily compare new models with previous ones. Similarly, you should keep backups of every version of your datasets so that you can roll back to a previous dataset if the new one ever gets corrupted (e.g., if the fresh data that gets added to it turns out to be full of outliers). Having backups of your datasets also allows you to evaluate any model against any previous dataset.

You may want to create several subsets of the test set in order to evaluate how well your model performs on specific parts of the data. For example, you may want to have a subset containing only the most recent data, or a test set for specific kinds of inputs (e.g., districts located inland versus districts located near the ocean). This will give you a deeper understanding of your model's strengths and weaknesses.

As you can see, Machine Learning involves quite a lot of infrastructure, so don't be surprised if your first ML project takes a lot of effort and time to build and deploy to production. Fortunately, once all the infrastructure is in place, going from idea to production will be much faster.

Try It Out!

Hopefully this chapter gave you a good idea of what a Machine Learning project looks like as well as showing you some of the tools you can use to train a great system. As you can see, much of the work is in the data preparation step: building monitoring tools, setting up human evaluation pipelines, and automating regular model training. The Machine Learning algorithms are important, of course, but it is probably preferable to be comfortable with the overall process and know three or four algorithms well rather than to spend all your time exploring advanced algorithms.

So, if you have not already done so, now is a good time to pick up a laptop, select a dataset that you are interested in, and try to go through the whole process from A to Z. A good place to start is on a competition website such as *http://kaggle.com/*: you will have a dataset to play with, a clear goal, and people to share the experience with. Have fun!

Exercises

The following exercises are all based on this chapter's housing dataset:

1. Try a Support Vector Machine regressor (`sklearn.svm.SVR`) with various hyper-parameters, such as `kernel="linear"` (with various values for the `C` hyperparameter) or `kernel="rbf"` (with various values for the `C` and `gamma` hyperparameters). Don't worry about what these hyperparameters mean for now. How does the best SVR predictor perform?

2. Try replacing `GridSearchCV` with `RandomizedSearchCV`.

3. Try adding a transformer in the preparation pipeline to select only the most important attributes.

4. Try creating a single pipeline that does the full data preparation plus the final prediction.

5. Automatically explore some preparation options using `GridSearchCV`.

Solutions to these exercises can be found in the Jupyter notebooks available at *https://github.com/ageron/handson-ml2*.

## CHAPTER 3

Classification

In Chapter 1 I mentioned that the most common supervised learning tasks are regression (predicting values) and classification (predicting classes). In Chapter 2 we explored a regression task, predicting housing values, using various algorithms such as Linear Regression, Decision Trees, and Random Forests (which will be explained in further detail in later chapters). Now we will turn our attention to classification systems.

MNIST

In this chapter we will be using the MNIST dataset, which is a set of 70,000 small images of digits handwritten by high school students and employees of the US Census Bureau. Each image is labeled with the digit it represents. This set has been studied so much that it is often called the "hello world" of Machine Learning: whenever people come up with a new classification algorithm they are curious to see how it will perform on MNIST, and anyone who learns Machine Learning tackles this dataset sooner or later.

Scikit-Learn provides many helper functions to download popular datasets. MNIST is one of them. The following code fetches the MNIST dataset:[1]

```
>>> from sklearn.datasets import fetch_openml
>>> mnist = fetch_openml('mnist_784', version=1)
>>> mnist.keys()
dict_keys(['data', 'target', 'feature_names', 'DESCR', 'details',
            'categories', 'url'])
```

---

1 By default Scikit-Learn caches downloaded datasets in a directory called *$HOME/scikit_learn_data*.

Datasets loaded by Scikit-Learn generally have a similar dictionary structure, including the following:

- A DESCR key describing the dataset
- A data key containing an array with one row per instance and one column per feature
- A target key containing an array with the labels

Let's look at these arrays:

```
>>> X, y = mnist["data"], mnist["target"]
>>> X.shape
(70000, 784)
>>> y.shape
(70000,)
```

There are 70,000 images, and each image has 784 features. This is because each image is 28 × 28 pixels, and each feature simply represents one pixel's intensity, from 0 (white) to 255 (black). Let's take a peek at one digit from the dataset. All you need to do is grab an instance's feature vector, reshape it to a 28 × 28 array, and display it using Matplotlib's imshow() function:

```
import matplotlib as mpl
import matplotlib.pyplot as plt

some_digit = X[0]
some_digit_image = some_digit.reshape(28, 28)

plt.imshow(some_digit_image, cmap="binary")
plt.axis("off")
plt.show()
```

This looks like a 5, and indeed that's what the label tells us:

```
>>> y[0]
'5'
```

Note that the label is a string. Most ML algorithms expect numbers, so let's cast y to integer:

```
>>> y = y.astype(np.uint8)
```

To give you a feel for the complexity of the classification task, Figure 3-1 shows a few more images from the MNIST dataset.

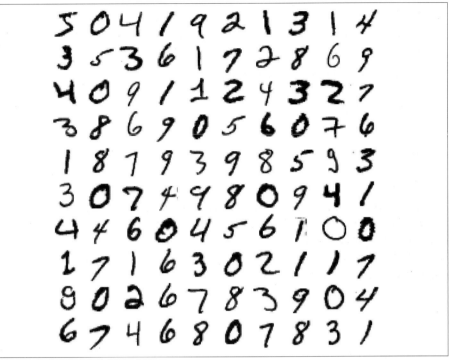

Figure 3-1. Digits from the MNIST dataset

But wait! You should always create a test set and set it aside before inspecting the data closely. The MNIST dataset is actually already split into a training set (the first 60,000 images) and a test set (the last 10,000 images):

```
X_train, X_test, y_train, y_test = X[:60000], X[60000:], y[:60000], y[60000:]
```

The training set is already shuffled for us, which is good because this guarantees that all cross-validation folds will be similar (you don't want one fold to be missing some digits). Moreover, some learning algorithms are sensitive to the order of the training instances, and they perform poorly if they get many similar instances in a row. Shuffling the dataset ensures that this won't happen.[2]

---

2 Shuffling may be a bad idea in some contexts—for example, if you are working on time series data (such as stock market prices or weather conditions). We will explore this in the next chapters.

Training a Binary Classifier

Let's simplify the problem for now and only try to identify one digit—for example, the number 5. This "5-detector" will be an example of a *binary classifier*, capable of distinguishing between just two classes, 5 and not-5. Let's create the target vectors for this classification task:

```
y_train_5 = (y_train == 5)  # True for all 5s, False for all other digits
y_test_5 = (y_test == 5)
```

Now let's pick a classifier and train it. A good place to start is with a *Stochastic Gradient Descent* (SGD) classifier, using Scikit-Learn's `SGDClassifier` class. This classifier has the advantage of being capable of handling very large datasets efficiently. This is in part because SGD deals with training instances independently, one at a time (which also makes SGD well suited for online learning), as we will see later. Let's create an `SGDClassifier` and train it on the whole training set:

```
from sklearn.linear_model import SGDClassifier

sgd_clf = SGDClassifier(random_state=42)
sgd_clf.fit(X_train, y_train_5)
```

 The `SGDClassifier` relies on randomness during training (hence the name "stochastic"). If you want reproducible results, you should set the `random_state` parameter.

Now we can use it to detect images of the number 5:

```
>>> sgd_clf.predict([some_digit])
array([ True])
```

The classifier guesses that this image represents a 5 (`True`). Looks like it guessed right in this particular case! Now, let's evaluate this model's performance.

Performance Measures

Evaluating a classifier is often significantly trickier than evaluating a regressor, so we will spend a large part of this chapter on this topic. There are many performance measures available, so grab another coffee and get ready to learn many new concepts and acronyms!

Measuring Accuracy Using Cross-Validation

A good way to evaluate a model is to use cross-validation, just as you did in Chapter 2.

Implementing Cross-Validation

Occasionally you will need more control over the cross-validation process than what Scikit-Learn provides off the shelf. In these cases, you can implement cross-validation yourself. The following code does roughly the same thing as Scikit-Learn's `cross_val_score()` function, and it prints the same result:

```python
from sklearn.model_selection import StratifiedKFold
from sklearn.base import clone

skfolds = StratifiedKFold(n_splits=3, random_state=42)

for train_index, test_index in skfolds.split(X_train, y_train_5):
    clone_clf = clone(sgd_clf)
    X_train_folds = X_train[train_index]
    y_train_folds = y_train_5[train_index]
    X_test_fold = X_train[test_index]
    y_test_fold = y_train_5[test_index]

    clone_clf.fit(X_train_folds, y_train_folds)
    y_pred = clone_clf.predict(X_test_fold)
    n_correct = sum(y_pred == y_test_fold)
    print(n_correct / len(y_pred))  # prints 0.9502, 0.96565, and 0.96495
```

The `StratifiedKFold` class performs stratified sampling (as explained in Chapter 2) to produce folds that contain a representative ratio of each class. At each iteration the code creates a clone of the classifier, trains that clone on the training folds, and makes predictions on the test fold. Then it counts the number of correct predictions and outputs the ratio of correct predictions.

Let's use the `cross_val_score()` function to evaluate our `SGDClassifier` model, using K-fold cross-validation with three folds. Remember that K-fold cross-validation means splitting the training set into K folds (in this case, three), then making predictions and evaluating them on each fold using a model trained on the remaining folds (see Chapter 2):

```python
>>> from sklearn.model_selection import cross_val_score
>>> cross_val_score(sgd_clf, X_train, y_train_5, cv=3, scoring="accuracy")
array([0.96355, 0.93795, 0.95615])
```

Wow! Above 93% accuracy (ratio of correct predictions) on all cross-validation folds? This looks amazing, doesn't it? Well, before you get too excited, let's look at a very dumb classifier that just classifies every single image in the "not-5" class:

```
from sklearn.base import BaseEstimator

class Never5Classifier(BaseEstimator):
    def fit(self, X, y=None):
        pass
    def predict(self, X):
        return np.zeros((len(X), 1), dtype=bool)
```

Can you guess this model's accuracy? Let's find out:

```
>>> never_5_clf = Never5Classifier()
>>> cross_val_score(never_5_clf, X_train, y_train_5, cv=3, scoring="accuracy")
array([0.91125, 0.90855, 0.90915])
```

That's right, it has over 90% accuracy! This is simply because only about 10% of the images are 5s, so if you always guess that an image is *not* a 5, you will be right about 90% of the time. Beats Nostradamus.

This demonstrates why accuracy is generally not the preferred performance measure for classifiers, especially when you are dealing with *skewed datasets* (i.e., when some classes are much more frequent than others).

Confusion Matrix

A much better way to evaluate the performance of a classifier is to look at the *confusion matrix*. The general idea is to count the number of times instances of class A are classified as class B. For example, to know the number of times the classifier confused images of 5s with 3s, you would look in the fifth row and third column of the confusion matrix.

To compute the confusion matrix, you first need to have a set of predictions so that they can be compared to the actual targets. You could make predictions on the test set, but let's keep it untouched for now (remember that you want to use the test set only at the very end of your project, once you have a classifier that you are ready to launch). Instead, you can use the cross_val_predict() function:

```
from sklearn.model_selection import cross_val_predict

y_train_pred = cross_val_predict(sgd_clf, X_train, y_train_5, cv=3)
```

Just like the cross_val_score() function, cross_val_predict() performs K-fold cross-validation, but instead of returning the evaluation scores, it returns the predictions made on each test fold. This means that you get a clean prediction for each instance in the training set ("clean" meaning that the prediction is made by a model that never saw the data during training).

Now you are ready to get the confusion matrix using the confusion_matrix() function. Just pass it the target classes (y_train_5) and the predicted classes (y_train_pred):

```
>>> from sklearn.metrics import confusion_matrix
>>> confusion_matrix(y_train_5, y_train_pred)
array([[53057,  1522],
       [ 1325,  4096]])
```

Each row in a confusion matrix represents an *actual class*, while each column represents a *predicted class*. The first row of this matrix considers non-5 images (the *negative class*): 53,057 of them were correctly classified as non-5s (they are called *true negatives*), while the remaining 1,522 were wrongly classified as 5s (*false positives*). The second row considers the images of 5s (the *positive class*): 1,325 were wrongly classified as non-5s (*false negatives*), while the remaining 4,096 were correctly classified as 5s (*true positives*). A perfect classifier would have only true positives and true negatives, so its confusion matrix would have nonzero values only on its main diagonal (top left to bottom right):

```
>>> y_train_perfect_predictions = y_train_5  # pretend we reached perfection
>>> confusion_matrix(y_train_5, y_train_perfect_predictions)
array([[54579,     0],
       [    0,  5421]])
```

The confusion matrix gives you a lot of information, but sometimes you may prefer a more concise metric. An interesting one to look at is the accuracy of the positive predictions; this is called the *precision* of the classifier (Equation 3-1).

Equation 3-1. Precision

$$\text{precision} = \frac{TP}{TP + FP}$$

TP is the number of true positives, and *FP* is the number of false positives.

A trivial way to have perfect precision is to make one single positive prediction and ensure it is correct (precision = 1/1 = 100%). But this would not be very useful, since the classifier would ignore all but one positive instance. So precision is typically used along with another metric named *recall*, also called *sensitivity* or the *true positive rate* (TPR): this is the ratio of positive instances that are correctly detected by the classifier (Equation 3-2).

Equation 3-2. Recall

$$\text{recall} = \frac{TP}{TP + FN}$$

FN is, of course, the number of false negatives.

If you are confused about the confusion matrix, Figure 3-2 may help.

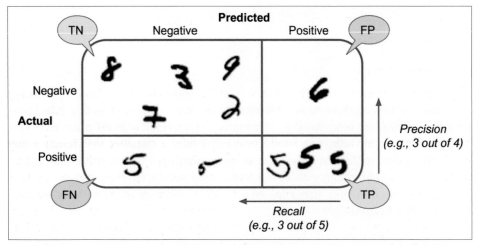

Figure 3-2. An illustrated confusion matrix shows examples of true negatives (top left), false positives (top right), false negatives (lower left), and true positives (lower right)

Precision and Recall

Scikit-Learn provides several functions to compute classifier metrics, including precision and recall:

```
>>> from sklearn.metrics import precision_score, recall_score
>>> precision_score(y_train_5, y_train_pred) # == 4096 / (4096 + 1522)
0.7290850836596654
>>> recall_score(y_train_5, y_train_pred) # == 4096 / (4096 + 1325)
0.7555801512636044
```

Now your 5-detector does not look as shiny as it did when you looked at its accuracy. When it claims an image represents a 5, it is correct only 72.9% of the time. Moreover, it only detects 75.6% of the 5s.

It is often convenient to combine precision and recall into a single metric called the F_1 *score*, in particular if you need a simple way to compare two classifiers. The F_1 score is the *harmonic mean* of precision and recall (Equation 3-3). Whereas the regular mean treats all values equally, the harmonic mean gives much more weight to low values. As a result, the classifier will only get a high F_1 score if both recall and precision are high.

Equation 3-3. F_1

$$F_1 = \frac{2}{\frac{1}{precision} + \frac{1}{recall}} = 2 \times \frac{precision \times recall}{precision + recall} = \frac{TP}{TP + \frac{FN + FP}{2}}$$

To compute the F_1 score, simply call the `f1_score()` function:

```
>>> from sklearn.metrics import f1_score
>>> f1_score(y_train_5, y_train_pred)
0.7420962043663375
```

The F_1 score favors classifiers that have similar precision and recall. This is not always what you want: in some contexts you mostly care about precision, and in other contexts you really care about recall. For example, if you trained a classifier to detect videos that are safe for kids, you would probably prefer a classifier that rejects many good videos (low recall) but keeps only safe ones (high precision), rather than a classifier that has a much higher recall but lets a few really bad videos show up in your product (in such cases, you may even want to add a human pipeline to check the classifier's video selection). On the other hand, suppose you train a classifier to detect shoplifters in surveillance images: it is probably fine if your classifier has only 30% precision as long as it has 99% recall (sure, the security guards will get a few false alerts, but almost all shoplifters will get caught).

Unfortunately, you can't have it both ways: increasing precision reduces recall, and vice versa. This is called the *precision/recall trade-off*.

Precision/Recall Trade-off

To understand this trade-off, let's look at how the SGDClassifier makes its classification decisions. For each instance, it computes a score based on a *decision function*. If that score is greater than a threshold, it assigns the instance to the positive class; otherwise it assigns it to the negative class. Figure 3-3 shows a few digits positioned from the lowest score on the left to the highest score on the right. Suppose the *decision threshold* is positioned at the central arrow (between the two 5s): you will find 4 true positives (actual 5s) on the right of that threshold, and 1 false positive (actually a 6). Therefore, with that threshold, the precision is 80% (4 out of 5). But out of 6 actual 5s, the classifier only detects 4, so the recall is 67% (4 out of 6). If you raise the threshold (move it to the arrow on the right), the false positive (the 6) becomes a true negative, thereby increasing the precision (up to 100% in this case), but one true positive becomes a false negative, decreasing recall down to 50%. Conversely, lowering the threshold increases recall and reduces precision.

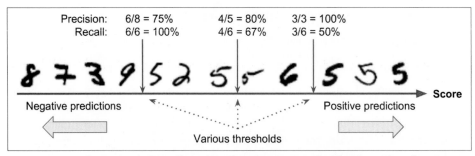

Figure 3-3. In this precision/recall trade-off, images are ranked by their classifier score, and those above the chosen decision threshold are considered positive; the higher the threshold, the lower the recall, but (in general) the higher the precision

Scikit-Learn does not let you set the threshold directly, but it does give you access to the decision scores that it uses to make predictions. Instead of calling the classifier's `predict()` method, you can call its `decision_function()` method, which returns a score for each instance, and then use any threshold you want to make predictions based on those scores:

```
>>> y_scores = sgd_clf.decision_function([some_digit])
>>> y_scores
array([2412.53175101])
>>> threshold = 0
>>> y_some_digit_pred = (y_scores > threshold)
array([ True])
```

The `SGDClassifier` uses a threshold equal to 0, so the previous code returns the same result as the `predict()` method (i.e., `True`). Let's raise the threshold:

```
>>> threshold = 8000
>>> y_some_digit_pred = (y_scores > threshold)
>>> y_some_digit_pred
array([False])
```

This confirms that raising the threshold decreases recall. The image actually represents a 5, and the classifier detects it when the threshold is 0, but it misses it when the threshold is increased to 8,000.

How do you decide which threshold to use? First, use the `cross_val_predict()` function to get the scores of all instances in the training set, but this time specify that you want to return decision scores instead of predictions:

```
y_scores = cross_val_predict(sgd_clf, X_train, y_train_5, cv=3,
                             method="decision_function")
```

With these scores, use the `precision_recall_curve()` function to compute precision and recall for all possible thresholds:

```
from sklearn.metrics import precision_recall_curve

precisions, recalls, thresholds = precision_recall_curve(y_train_5, y_scores)
```

Finally, use Matplotlib to plot precision and recall as functions of the threshold value (Figure 3-4):

```
def plot_precision_recall_vs_threshold(precisions, recalls, thresholds):
    plt.plot(thresholds, precisions[:-1], "b--", label="Precision")
    plt.plot(thresholds, recalls[:-1], "g-", label="Recall")
    [...] # highlight the threshold and add the legend, axis label, and grid

plot_precision_recall_vs_threshold(precisions, recalls, thresholds)
plt.show()
```

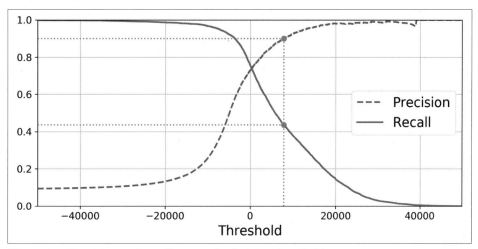

Figure 3-4. Precision and recall versus the decision threshold

You may wonder why the precision curve is bumpier than the recall curve in Figure 3-4. The reason is that precision may sometimes go down when you raise the threshold (although in general it will go up). To understand why, look back at Figure 3-3 and notice what happens when you start from the central threshold and move it just one digit to the right: precision goes from 4/5 (80%) down to 3/4 (75%). On the other hand, recall can only go down when the threshold is increased, which explains why its curve looks smooth.

Another way to select a good precision/recall trade-off is to plot precision directly against recall, as shown in Figure 3-5 (the same threshold as earlier is highlighted).

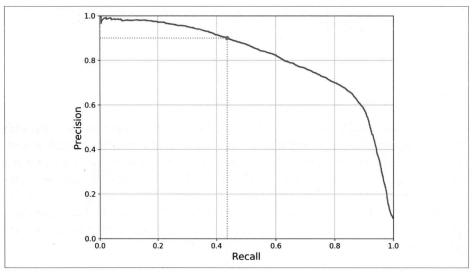

Figure 3-5. Precision versus recall

You can see that precision really starts to fall sharply around 80% recall. You will probably want to select a precision/recall trade-off just before that drop—for example, at around 60% recall. But of course, the choice depends on your project.

Suppose you decide to aim for 90% precision. You look up the first plot and find that you need to use a threshold of about 8,000. To be more precise you can search for the lowest threshold that gives you at least 90% precision (`np.argmax()` will give you the first index of the maximum value, which in this case means the first `True` value):

```
threshold_90_precision = thresholds[np.argmax(precisions >= 0.90)] # ~7816
```

To make predictions (on the training set for now), instead of calling the classifier's `predict()` method, you can run this code:

```
y_train_pred_90 = (y_scores >= threshold_90_precision)
```

Let's check these predictions' precision and recall:

```
>>> precision_score(y_train_5, y_train_pred_90)
0.9000380083618396
>>> recall_score(y_train_5, y_train_pred_90)
0.4368197749492714
```

Great, you have a 90% precision classifier! As you can see, it is fairly easy to create a classifier with virtually any precision you want: just set a high enough threshold, and you're done. But wait, not so fast. A high-precision classifier is not very useful if its recall is too low!

If someone says, "Let's reach 99% precision," you should ask, "At what recall?"

The ROC Curve

The *receiver operating characteristic* (ROC) curve is another common tool used with binary classifiers. It is very similar to the precision/recall curve, but instead of plotting precision versus recall, the ROC curve plots the *true positive rate* (another name for recall) against the *false positive rate* (FPR). The FPR is the ratio of negative instances that are incorrectly classified as positive. It is equal to 1 – the *true negative rate* (TNR), which is the ratio of negative instances that are correctly classified as negative. The TNR is also called *specificity*. Hence, the ROC curve plots *sensitivity* (recall) versus 1 – *specificity*.

To plot the ROC curve, you first use the roc_curve() function to compute the TPR and FPR for various threshold values:

```
from sklearn.metrics import roc_curve

fpr, tpr, thresholds = roc_curve(y_train_5, y_scores)
```

Then you can plot the FPR against the TPR using Matplotlib. This code produces the plot in Figure 3-6:

```
def plot_roc_curve(fpr, tpr, label=None):
    plt.plot(fpr, tpr, linewidth=2, label=label)
    plt.plot([0, 1], [0, 1], 'k--') # Dashed diagonal
    [...] # Add axis labels and grid

plot_roc_curve(fpr, tpr)
plt.show()
```

Once again there is a trade-off: the higher the recall (TPR), the more false positives (FPR) the classifier produces. The dotted line represents the ROC curve of a purely random classifier; a good classifier stays as far away from that line as possible (toward the top-left corner).

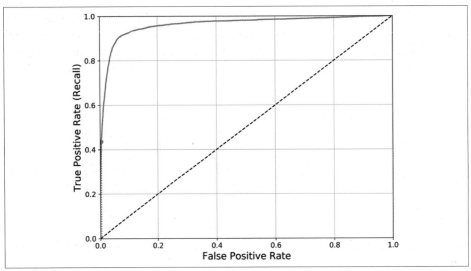

Figure 3-6. This ROC curve plots the false positive rate against the true positive rate for all possible thresholds; the red circle highlights the chosen ratio (at 43.68% recall)

One way to compare classifiers is to measure the *area under the curve* (AUC). A perfect classifier will have a ROC AUC equal to 1, whereas a purely random classifier will have a ROC AUC equal to 0.5. Scikit-Learn provides a function to compute the ROC AUC:

```
>>> from sklearn.metrics import roc_auc_score
>>> roc_auc_score(y_train_5, y_scores)
0.9611778893101814
```

Since the ROC curve is so similar to the precision/recall (PR) curve, you may wonder how to decide which one to use. As a rule of thumb, you should prefer the PR curve whenever the positive class is rare or when you care more about the false positives than the false negatives. Otherwise, use the ROC curve. For example, looking at the previous ROC curve (and the ROC AUC score), you may think that the classifier is really good. But this is mostly because there are few positives (5s) compared to the negatives (non-5s). In contrast, the PR curve makes it clear that the classifier has room for improvement (the curve could be closer to the top-left corner).

Let's now train a RandomForestClassifier and compare its ROC curve and ROC AUC score to those of the SGDClassifier. First, you need to get scores for each instance in the training set. But due to the way it works (see Chapter 7), the Random ForestClassifier class does not have a decision_function() method. Instead, it has a predict_proba() method. Scikit-Learn classifiers generally have one or the other, or both. The predict_proba() method returns an array containing a row per instance and a column per class, each containing the probability that the given instance belongs to the given class (e.g., 70% chance that the image represents a 5):

```
from sklearn.ensemble import RandomForestClassifier

forest_clf = RandomForestClassifier(random_state=42)
y_probas_forest = cross_val_predict(forest_clf, X_train, y_train_5, cv=3,
                                    method="predict_proba")
```

The roc_curve() function expects labels and scores, but instead of scores you can give it class probabilities. Let's use the positive class's probability as the score:

```
y_scores_forest = y_probas_forest[:, 1]   # score = proba of positive class
fpr_forest, tpr_forest, thresholds_forest = roc_curve(y_train_5,y_scores_forest)
```

Now you are ready to plot the ROC curve. It is useful to plot the first ROC curve as well to see how they compare (Figure 3-7):

```
plt.plot(fpr, tpr, "b:", label="SGD")
plot_roc_curve(fpr_forest, tpr_forest, "Random Forest")
plt.legend(loc="lower right")
plt.show()
```

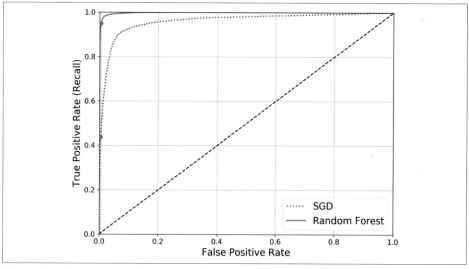

Figure 3-7. Comparing ROC curves: the Random Forest classifier is superior to the SGD classifier because its ROC curve is much closer to the top-left corner, and it has a greater AUC

As you can see in Figure 3-7, the `RandomForestClassifier`'s ROC curve looks much better than the `SGDClassifier`'s: it comes much closer to the top-left corner. As a result, its ROC AUC score is also significantly better:

```
>>> roc_auc_score(y_train_5, y_scores_forest)
0.9983436731328145
```

Try measuring the precision and recall scores: you should find 99.0% precision and 86.6% recall. Not too bad!

You now know how to train binary classifiers, choose the appropriate metric for your task, evaluate your classifiers using cross-validation, select the precision/recall trade-off that fits your needs, and use ROC curves and ROC AUC scores to compare various models. Now let's try to detect more than just the 5s.

Multiclass Classification

Whereas binary classifiers distinguish between two classes, *multiclass classifiers* (also called *multinomial classifiers*) can distinguish between more than two classes.

Some algorithms (such as SGD classifiers, Random Forest classifiers, and naive Bayes classifiers) are capable of handling multiple classes natively. Others (such as Logistic Regression or Support Vector Machine classifiers) are strictly binary classifiers. However, there are various strategies that you can use to perform multiclass classification with multiple binary classifiers.

One way to create a system that can classify the digit images into 10 classes (from 0 to 9) is to train 10 binary classifiers, one for each digit (a 0-detector, a 1-detector, a 2-detector, and so on). Then when you want to classify an image, you get the decision score from each classifier for that image and you select the class whose classifier outputs the highest score. This is called the *one-versus-the-rest* (OvR) strategy (also called *one-versus-all*).

Another strategy is to train a binary classifier for every pair of digits: one to distinguish 0s and 1s, another to distinguish 0s and 2s, another for 1s and 2s, and so on. This is called the *one-versus-one* (OvO) strategy. If there are N classes, you need to train $N \times (N - 1) / 2$ classifiers. For the MNIST problem, this means training 45 binary classifiers! When you want to classify an image, you have to run the image through all 45 classifiers and see which class wins the most duels. The main advantage of OvO is that each classifier only needs to be trained on the part of the training set for the two classes that it must distinguish.

Some algorithms (such as Support Vector Machine classifiers) scale poorly with the size of the training set. For these algorithms OvO is preferred because it is faster to train many classifiers on small training sets than to train few classifiers on large training sets. For most binary classification algorithms, however, OvR is preferred.

Scikit-Learn detects when you try to use a binary classification algorithm for a multi-class classification task, and it automatically runs OvR or OvO, depending on the algorithm. Let's try this with a Support Vector Machine classifier (see Chapter 5), using the `sklearn.svm.SVC` class:

```
>>> from sklearn.svm import SVC
>>> svm_clf = SVC()
>>> svm_clf.fit(X_train, y_train) # y_train, not y_train_5
>>> svm_clf.predict([some_digit])
array([5], dtype=uint8)
```

That was easy! This code trains the SVC on the training set using the original target classes from 0 to 9 (y_train), instead of the 5-versus-the-rest target classes (y_train_5). Then it makes a prediction (a correct one in this case). Under the hood, Scikit-Learn actually used the OvO strategy: it trained 45 binary classifiers, got their decision scores for the image, and selected the class that won the most duels.

If you call the `decision_function()` method, you will see that it returns 10 scores per instance (instead of just 1). That's one score per class:

```
>>> some_digit_scores = svm_clf.decision_function([some_digit])
>>> some_digit_scores
array([[ 2.92492871,  7.02307409,  3.93648529,  0.90117363,  5.96945908,
         9.5       ,  1.90718593,  8.02755089, -0.13202708,  4.94216947]])
```

The highest score is indeed the one corresponding to class 5:

```
>>> np.argmax(some_digit_scores)
5
>>> svm_clf.classes_
array([0, 1, 2, 3, 4, 5, 6, 7, 8, 9], dtype=uint8)
>>> svm_clf.classes_[5]
5
```

When a classifier is trained, it stores the list of target classes in its `classes_` attribute, ordered by value. In this case, the index of each class in the `classes_` array conveniently matches the class itself (e.g., the class at index 5 happens to be class 5), but in general you won't be so lucky.

If you want to force Scikit-Learn to use one-versus-one or one-versus-the-rest, you can use the `OneVsOneClassifier` or `OneVsRestClassifier` classes. Simply create an instance and pass a classifier to its constructor (it does not even have to be a binary classifier). For example, this code creates a multiclass classifier using the OvR strategy, based on an SVC:

```
>>> from sklearn.multiclass import OneVsRestClassifier
>>> ovr_clf = OneVsRestClassifier(SVC())
>>> ovr_clf.fit(X_train, y_train)
```

```
>>> ovr_clf.predict([some_digit])
array([5], dtype=uint8)
>>> len(ovr_clf.estimators_)
10
```

Training an SGDClassifier (or a RandomForestClassifier) is just as easy:

```
>>> sgd_clf.fit(X_train, y_train)
>>> sgd_clf.predict([some_digit])
array([5], dtype=uint8)
```

This time Scikit-Learn did not have to run OvR or OvO because SGD classifiers can directly classify instances into multiple classes. The decision_function() method now returns one value per class. Let's look at the score that the SGD classifier assigned to each class:

```
>>> sgd_clf.decision_function([some_digit])
array([[-15955.22628, -38080.96296, -13326.66695,   573.52692, -17680.68466,
          2412.53175, -25526.86498, -12290.15705, -7946.05205, -10631.35889]])
```

You can see that the classifier is fairly confident about its prediction: almost all scores are largely negative, while class 5 has a score of 2412.5. The model has a slight doubt regarding class 3, which gets a score of 573.5. Now of course you want to evaluate this classifier. As usual, you can use cross-validation. Use the cross_val_score() function to evaluate the SGDClassifier's accuracy:

```
>>> cross_val_score(sgd_clf, X_train, y_train, cv=3, scoring="accuracy")
array([0.8489802 , 0.87129356, 0.86988048])
```

It gets over 84% on all test folds. If you used a random classifier, you would get 10% accuracy, so this is not such a bad score, but you can still do much better. Simply scaling the inputs (as discussed in Chapter 2) increases accuracy above 89%:

```
>>> from sklearn.preprocessing import StandardScaler
>>> scaler = StandardScaler()
>>> X_train_scaled = scaler.fit_transform(X_train.astype(np.float64))
>>> cross_val_score(sgd_clf, X_train_scaled, y_train, cv=3, scoring="accuracy")
array([0.89707059, 0.8960948 , 0.90693604])
```

Error Analysis

If this were a real project, you would now follow the steps in your Machine Learning project checklist (see Appendix B). You'd explore data preparation options, try out multiple models (shortlisting the best ones and fine-tuning their hyperparameters using GridSearchCV), and automate as much as possible. Here, we will assume that you have found a promising model and you want to find ways to improve it. One way to do this is to analyze the types of errors it makes.

First, look at the confusion matrix. You need to make predictions using the `cross_val_predict()` function, then call the `confusion_matrix()` function, just like you did earlier:

```
>>> y_train_pred = cross_val_predict(sgd_clf, X_train_scaled, y_train, cv=3)
>>> conf_mx = confusion_matrix(y_train, y_train_pred)
>>> conf_mx
array([[5578,    0,   22,    7,    8,   45,   35,    5,  222,    1],
       [   0, 6410,   35,   26,    4,   44,    4,    8,  198,   13],
       [  28,   27, 5232,  100,   74,   27,   68,   37,  354,   11],
       [  23,   18,  115, 5254,    2,  209,   26,   38,  373,   73],
       [  11,   14,   45,   12, 5219,   11,   33,   26,  299,  172],
       [  26,   16,   31,  173,   54, 4484,   76,   14,  482,   65],
       [  31,   17,   45,    2,   42,   98, 5556,    3,  123,    1],
       [  20,   10,   53,   27,   50,   13,    3, 5696,  173,  220],
       [  17,   64,   47,   91,    3,  125,   24,   11, 5421,   48],
       [  24,   18,   29,   67,  116,   39,    1,  174,  329, 5152]])
```

That's a lot of numbers. It's often more convenient to look at an image representation of the confusion matrix, using Matplotlib's `matshow()` function:

```
plt.matshow(conf_mx, cmap=plt.cm.gray)
plt.show()
```

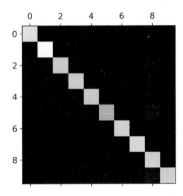

This confusion matrix looks pretty good, since most images are on the main diagonal, which means that they were classified correctly. The 5s look slightly darker than the other digits, which could mean that there are fewer images of 5s in the dataset or that the classifier does not perform as well on 5s as on other digits. In fact, you can verify that both are the case.

Let's focus the plot on the errors. First, you need to divide each value in the confusion matrix by the number of images in the corresponding class so that you can compare error rates instead of absolute numbers of errors (which would make abundant classes look unfairly bad):

```
row_sums = conf_mx.sum(axis=1, keepdims=True)
norm_conf_mx = conf_mx / row_sums
```

Fill the diagonal with zeros to keep only the errors, and plot the result:

```
np.fill_diagonal(norm_conf_mx, 0)
plt.matshow(norm_conf_mx, cmap=plt.cm.gray)
plt.show()
```

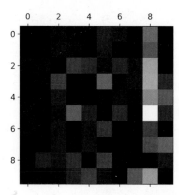

You can clearly see the kinds of errors the classifier makes. Remember that rows represent actual classes, while columns represent predicted classes. The column for class 8 is quite bright, which tells you that many images get misclassified as 8s. However, the row for class 8 is not that bad, telling you that actual 8s in general get properly classified as 8s. As you can see, the confusion matrix is not necessarily symmetrical. You can also see that 3s and 5s often get confused (in both directions).

Analyzing the confusion matrix often gives you insights into ways to improve your classifier. Looking at this plot, it seems that your efforts should be spent on reducing the false 8s. For example, you could try to gather more training data for digits that look like 8s (but are not) so that the classifier can learn to distinguish them from real 8s. Or you could engineer new features that would help the classifier—for example, writing an algorithm to count the number of closed loops (e.g., 8 has two, 6 has one, 5 has none). Or you could preprocess the images (e.g., using Scikit-Image, Pillow, or OpenCV) to make some patterns, such as closed loops, stand out more.

Analyzing individual errors can also be a good way to gain insights on what your classifier is doing and why it is failing, but it is more difficult and time-consuming. For example, let's plot examples of 3s and 5s (the `plot_digits()` function just uses Matplotlib's `imshow()` function; see this chapter's Jupyter notebook for details):

```
cl_a, cl_b = 3, 5
X_aa = X_train[(y_train == cl_a) & (y_train_pred == cl_a)]
X_ab = X_train[(y_train == cl_a) & (y_train_pred == cl_b)]
X_ba = X_train[(y_train == cl_b) & (y_train_pred == cl_a)]
X_bb = X_train[(y_train == cl_b) & (y_train_pred == cl_b)]
```

```
plt.figure(figsize=(8,8))
plt.subplot(221); plot_digits(X_aa[:25], images_per_row=5)
plt.subplot(222); plot_digits(X_ab[:25], images_per_row=5)
plt.subplot(223); plot_digits(X_ba[:25], images_per_row=5)
plt.subplot(224); plot_digits(X_bb[:25], images_per_row=5)
plt.show()
```

The two 5 × 5 blocks on the left show digits classified as 3s, and the two 5 × 5 blocks on the right show images classified as 5s. Some of the digits that the classifier gets wrong (i.e., in the bottom-left and top-right blocks) are so badly written that even a human would have trouble classifying them (e.g., the 5 in the first row and second column truly looks like a badly written 3). However, most misclassified images seem like obvious errors to us, and it's hard to understand why the classifier made the mistakes it did.[3] The reason is that we used a simple SGDClassifier, which is a linear model. All it does is assign a weight per class to each pixel, and when it sees a new image it just sums up the weighted pixel intensities to get a score for each class. So since 3s and 5s differ only by a few pixels, this model will easily confuse them.

The main difference between 3s and 5s is the position of the small line that joins the top line to the bottom arc. If you draw a 3 with the junction slightly shifted to the left, the classifier might classify it as a 5, and vice versa. In other words, this classifier is quite sensitive to image shifting and rotation. So one way to reduce the 3/5 confusion would be to preprocess the images to ensure that they are well centered and not too rotated. This will probably help reduce other errors as well.

---

3 But remember that our brain is a fantastic pattern recognition system, and our visual system does a lot of complex preprocessing before any information reaches our consciousness, so the fact that it feels simple does not mean that it is.

Multilabel Classification

Until now each instance has always been assigned to just one class. In some cases you may want your classifier to output multiple classes for each instance. Consider a face-recognition classifier: what should it do if it recognizes several people in the same picture? It should attach one tag per person it recognizes. Say the classifier has been trained to recognize three faces, Alice, Bob, and Charlie. Then when the classifier is shown a picture of Alice and Charlie, it should output [1, 0, 1] (meaning "Alice yes, Bob no, Charlie yes"). Such a classification system that outputs multiple binary tags is called a *multilabel classification* system.

We won't go into face recognition just yet, but let's look at a simpler example, just for illustration purposes:

```
from sklearn.neighbors import KNeighborsClassifier

y_train_large = (y_train >= 7)
y_train_odd = (y_train % 2 == 1)
y_multilabel = np.c_[y_train_large, y_train_odd]

knn_clf = KNeighborsClassifier()
knn_clf.fit(X_train, y_multilabel)
```

This code creates a `y_multilabel` array containing two target labels for each digit image: the first indicates whether or not the digit is large (7, 8, or 9), and the second indicates whether or not it is odd. The next lines create a `KNeighborsClassifier` instance (which supports multilabel classification, though not all classifiers do), and we train it using the multiple targets array. Now you can make a prediction, and notice that it outputs two labels:

```
>>> knn_clf.predict([some_digit])
array([[False,  True]])
```

And it gets it right! The digit 5 is indeed not large (`False`) and odd (`True`).

There are many ways to evaluate a multilabel classifier, and selecting the right metric really depends on your project. One approach is to measure the F_1 score for each individual label (or any other binary classifier metric discussed earlier), then simply compute the average score. This code computes the average F_1 score across all labels:

```
>>> y_train_knn_pred = cross_val_predict(knn_clf, X_train, y_multilabel, cv=3)
>>> f1_score(y_multilabel, y_train_knn_pred, average="macro")
0.976410265560605
```

This assumes that all labels are equally important, however, which may not be the case. In particular, if you have many more pictures of Alice than of Bob or Charlie, you may want to give more weight to the classifier's score on pictures of Alice. One simple option is to give each label a weight equal to its *support* (i.e., the number of instances with that target label). To do this, simply set `average="weighted"` in the preceding code.[4]

Multioutput Classification

The last type of classification task we are going to discuss here is called *multioutput–multiclass classification* (or simply *multioutput classification*). It is simply a generalization of multilabel classification where each label can be multiclass (i.e., it can have more than two possible values).

To illustrate this, let's build a system that removes noise from images. It will take as input a noisy digit image, and it will (hopefully) output a clean digit image, represented as an array of pixel intensities, just like the MNIST images. Notice that the classifier's output is multilabel (one label per pixel) and each label can have multiple values (pixel intensity ranges from 0 to 255). It is thus an example of a multioutput classification system.

The line between classification and regression is sometimes blurry, such as in this example. Arguably, predicting pixel intensity is more akin to regression than to classification. Moreover, multioutput systems are not limited to classification tasks; you could even have a system that outputs multiple labels per instance, including both class labels and value labels.

Let's start by creating the training and test sets by taking the MNIST images and adding noise to their pixel intensities with NumPy's `randint()` function. The target images will be the original images:

```
noise = np.random.randint(0, 100, (len(X_train), 784))
X_train_mod = X_train + noise
noise = np.random.randint(0, 100, (len(X_test), 784))
X_test_mod = X_test + noise
y_train_mod = X_train
y_test_mod = X_test
```

Let's take a peek at an image from the test set (yes, we're snooping on the test data, so you should be frowning right now):

---

4 Scikit-Learn offers a few other averaging options and multilabel classifier metrics; see the documentation for more details.

On the left is the noisy input image, and on the right is the clean target image. Now let's train the classifier and make it clean this image:

```
knn_clf.fit(X_train_mod, y_train_mod)
clean_digit = knn_clf.predict([X_test_mod[some_index]])
plot_digit(clean_digit)
```

Looks close enough to the target! This concludes our tour of classification. You should now know how to select good metrics for classification tasks, pick the appropriate precision/recall trade-off, compare classifiers, and more generally build good classification systems for a variety of tasks.

Exercises

1. Try to build a classifier for the MNIST dataset that achieves over 97% accuracy on the test set. Hint: the KNeighborsClassifier works quite well for this task; you just need to find good hyperparameter values (try a grid search on the weights and n_neighbors hyperparameters).

2. Write a function that can shift an MNIST image in any direction (left, right, up, or down) by one pixel.[5] Then, for each image in the training set, create four shifted copies (one per direction) and add them to the training set. Finally, train your best model on this expanded training set and measure its accuracy on the test set. You should observe that your model performs even better now! This technique of artificially growing the training set is called *data augmentation* or *training set expansion*.

---

5 You can use the shift() function from the scipy.ndimage.interpolation module. For example, shift(image, [2, 1], cval=0) shifts the image two pixels down and one pixel to the right.

3. Tackle the Titanic dataset. A great place to start is on Kaggle (*https://www.kaggle.com/c/titanic*).

4. Build a spam classifier (a more challenging exercise):

   • Download examples of spam and ham from Apache SpamAssassin's public datasets (*https://homl.info/spamassassin*).

   • Unzip the datasets and familiarize yourself with the data format.

   • Split the datasets into a training set and a test set.

   • Write a data preparation pipeline to convert each email into a feature vector. Your preparation pipeline should transform an email into a (sparse) vector that indicates the presence or absence of each possible word. For example, if all emails only ever contain four words, "Hello," "how," "are," "you," then the email "Hello you Hello Hello you" would be converted into a vector [1, 0, 0, 1] (meaning ["Hello" is present, "how" is absent, "are" is absent, "you" is present]), or [3, 0, 0, 2] if you prefer to count the number of occurrences of each word.

   You may want to add hyperparameters to your preparation pipeline to control whether or not to strip off email headers, convert each email to lowercase, remove punctuation, replace all URLs with "URL," replace all numbers with "NUMBER," or even perform *stemming* (i.e., trim off word endings; there are Python libraries available to do this).

   Finally, try out several classifiers and see if you can build a great spam classifier, with both high recall and high precision.

Solutions to these exercises can be found in the Jupyter notebooks available at *https://github.com/ageron/handson-ml2*.

CHAPTER 4
Training Models

So far we have treated Machine Learning models and their training algorithms mostly like black boxes. If you went through some of the exercises in the previous chapters, you may have been surprised by how much you can get done without knowing anything about what's under the hood: you optimized a regression system, you improved a digit image classifier, and you even built a spam classifier from scratch, all this without knowing how they actually work. Indeed, in many situations you don't really need to know the implementation details.

However, having a good understanding of how things work can help you quickly home in on the appropriate model, the right training algorithm to use, and a good set of hyperparameters for your task. Understanding what's under the hood will also help you debug issues and perform error analysis more efficiently. Lastly, most of the topics discussed in this chapter will be essential in understanding, building, and training neural networks (discussed in Part II of this book).

In this chapter we will start by looking at the Linear Regression model, one of the simplest models there is. We will discuss two very different ways to train it:

- Using a direct "closed-form" equation that directly computes the model parameters that best fit the model to the training set (i.e., the model parameters that minimize the cost function over the training set).

- Using an iterative optimization approach called Gradient Descent (GD) that gradually tweaks the model parameters to minimize the cost function over the training set, eventually converging to the same set of parameters as the first method. We will look at a few variants of Gradient Descent that we will use again and again when we study neural networks in Part II: Batch GD, Mini-batch GD, and Stochastic GD.

Next we will look at Polynomial Regression, a more complex model that can fit non-linear datasets. Since this model has more parameters than Linear Regression, it is more prone to overfitting the training data, so we will look at how to detect whether or not this is the case using learning curves, and then we will look at several regularization techniques that can reduce the risk of overfitting the training set.

Finally, we will look at two more models that are commonly used for classification tasks: Logistic Regression and Softmax Regression.

 There will be quite a few math equations in this chapter, using basic notions of linear algebra and calculus. To understand these equations, you will need to know what vectors and matrices are; how to transpose them, multiply them, and inverse them; and what partial derivatives are. If you are unfamiliar with these concepts, please go through the linear algebra and calculus introductory tutorials available as Jupyter notebooks in the online supplemental material (*https://github.com/ageron/handson-ml2*). For those who are truly allergic to mathematics, you should still go through this chapter and simply skip the equations; hopefully, the text will be sufficient to help you understand most of the concepts.

Linear Regression

In Chapter 1 we looked at a simple regression model of life satisfaction: *life_satisfaction = $\theta_0 + \theta_1 \times GDP\_per\_capita$*.

This model is just a linear function of the input feature `GDP_per_capita`. θ_0 and θ_1 are the model's parameters.

More generally, a linear model makes a prediction by simply computing a weighted sum of the input features, plus a constant called the *bias term* (also called the *intercept term*), as shown in Equation 4-1.

> *Equation 4-1. Linear Regression model prediction*
>
> $$\hat{y} = \theta_0 + \theta_1 x_1 + \theta_2 x_2 + \cdots + \theta_n x_n$$

In this equation:

- $\hat{y}$ is the predicted value.
- n is the number of features.
- x_i is the ith feature value.

- θ_j is the jth model parameter (including the bias term θ_0 and the feature weights $\theta_1, \theta_2, \cdots, \theta_n$).

This can be written much more concisely using a vectorized form, as shown in Equation 4-2.

Equation 4-2. Linear Regression model prediction (vectorized form)

$$\hat{y} = h_\theta(\mathbf{x}) = \mathbf{\theta} \cdot \mathbf{x}$$

In this equation:

- $\mathbf{\theta}$ is the model's *parameter vector*, containing the bias term θ_0 and the feature weights θ_1 to θ_n.
- $\mathbf{x}$ is the instance's *feature vector*, containing x_0 to x_n, with x_0 always equal to 1.
- $\mathbf{\theta} \cdot \mathbf{x}$ is the dot product of the vectors $\mathbf{\theta}$ and $\mathbf{x}$, which is of course equal to $\theta_0 x_0 + \theta_1 x_1 + \theta_2 x_2 + ... + \theta_n x_n$.
- h_θ is the hypothesis function, using the model parameters $\mathbf{\theta}$.

> In Machine Learning, vectors are often represented as *column vectors*, which are 2D arrays with a single column. If $\mathbf{\theta}$ and $\mathbf{x}$ are column vectors, then the prediction is $\hat{y} = \mathbf{\theta}^T \mathbf{x}$, where $\mathbf{\theta}^T$ is the *transpose* of $\mathbf{\theta}$ (a row vector instead of a column vector) and $\mathbf{\theta}^T \mathbf{x}$ is the matrix multiplication of $\mathbf{\theta}^T$ and $\mathbf{x}$. It is of course the same prediction, except that it is now represented as a single-cell matrix rather than a scalar value. In this book I will use this notation to avoid switching between dot products and matrix multiplications.

OK, that's the Linear Regression model—but how do we train it? Well, recall that training a model means setting its parameters so that the model best fits the training set. For this purpose, we first need a measure of how well (or poorly) the model fits the training data. In Chapter 2 we saw that the most common performance measure of a regression model is the Root Mean Square Error (RMSE) (Equation 2-1). Therefore, to train a Linear Regression model, we need to find the value of $\mathbf{\theta}$ that minimizes the RMSE. In practice, it is simpler to minimize the mean squared error (MSE) than the RMSE, and it leads to the same result (because the value that minimizes a function also minimizes its square root).[1]

The MSE of a Linear Regression hypothesis h_θ on a training set $\mathbf{X}$ is calculated using Equation 4-3.

Equation 4-3. MSE cost function for a Linear Regression model

$$\text{MSE}(\mathbf{X}, h_\theta) = \frac{1}{m} \sum_{i=1}^{m} \left(\theta^\mathsf{T} \mathbf{x}^{(i)} - y^{(i)} \right)^2$$

Most of these notations were presented in Chapter 2 (see "Notations" on page 40). The only difference is that we write h_θ instead of just h to make it clear that the model is parametrized by the vector θ. To simplify notations, we will just write $\text{MSE}(\theta)$ instead of $\text{MSE}(\mathbf{X}, h_\theta)$.

The Normal Equation

To find the value of θ that minimizes the cost function, there is a *closed-form solution* —in other words, a mathematical equation that gives the result directly. This is called the *Normal Equation* (Equation 4-4).

Equation 4-4. Normal Equation

$$\hat{\theta} = \left(\mathbf{X}^\mathsf{T} \mathbf{X} \right)^{-1} \mathbf{X}^\mathsf{T} \, \mathbf{y}$$

In this equation:

- $\hat{\theta}$ is the value of θ that minimizes the cost function.
- $\mathbf{y}$ is the vector of target values containing $y^{(1)}$ to $y^{(m)}$.

Let's generate some linear-looking data to test this equation on (Figure 4-1):

```
import numpy as np

X = 2 * np.random.rand(100, 1)
y = 4 + 3 * X + np.random.randn(100, 1)
```

---

[1] It is often the case that a learning algorithm will try to optimize a different function than the performance measure used to evaluate the final model. This is generally because that function is easier to compute, because it has useful differentiation properties that the performance measure lacks, or because we want to constrain the model during training, as you will see when we discuss regularization.

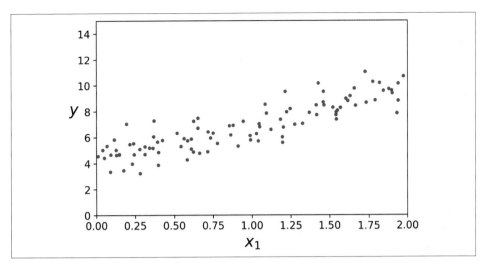

Figure 4-1. Randomly generated linear dataset

Now let's compute $\widehat{\theta}$ using the Normal Equation. We will use the inv() function from NumPy's linear algebra module (np.linalg) to compute the inverse of a matrix, and the dot() method for matrix multiplication:

```
X_b = np.c_[np.ones((100, 1)), X]  # add x0 = 1 to each instance
theta_best = np.linalg.inv(X_b.T.dot(X_b)).dot(X_b.T).dot(y)
```

The function that we used to generate the data is $y = 4 + 3x_1 +$ Gaussian noise. Let's see what the equation found:

```
>>> theta_best
array([[4.21509616],
       [2.77011339]])
```

We would have hoped for $\theta_0 = 4$ and $\theta_1 = 3$ instead of $\theta_0 = 4.215$ and $\theta_1 = 2.770$. Close enough, but the noise made it impossible to recover the exact parameters of the original function.

Now we can make predictions using $\widehat{\theta}$:

```
>>> X_new = np.array([[0], [2]])
>>> X_new_b = np.c_[np.ones((2, 1)), X_new]  # add x0 = 1 to each instance
>>> y_predict = X_new_b.dot(theta_best)
>>> y_predict
array([[4.21509616],
       [9.75532293]])
```

Let's plot this model's predictions (Figure 4-2):

```
plt.plot(X_new, y_predict, "r-")
plt.plot(X, y, "b.")
plt.axis([0, 2, 0, 15])
plt.show()
```

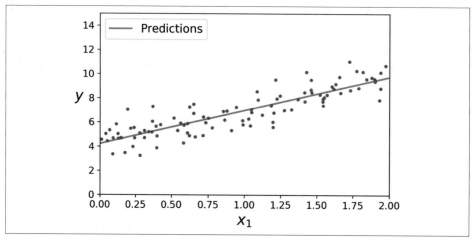

Figure 4-2. Linear Regression model predictions

Performing Linear Regression using Scikit-Learn is simple:[2]

```
>>> from sklearn.linear_model import LinearRegression
>>> lin_reg = LinearRegression()
>>> lin_reg.fit(X, y)
>>> lin_reg.intercept_, lin_reg.coef_
(array([4.21509616]), array([[2.77011339]]))
>>> lin_reg.predict(X_new)
array([[4.21509616],
       [9.75532293]])
```

The LinearRegression class is based on the scipy.linalg.lstsq() function (the name stands for "least squares"), which you could call directly:

```
>>> theta_best_svd, residuals, rank, s = np.linalg.lstsq(X_b, y, rcond=1e-6)
>>> theta_best_svd
array([[4.21509616],
       [2.77011339]])
```

This function computes $\hat{\theta} = \mathbf{X}^+\mathbf{y}$, where $\mathbf{X}^+$ is the *pseudoinverse* of $\mathbf{X}$ (specifically, the Moore-Penrose inverse). You can use np.linalg.pinv() to compute the pseudoinverse directly:

---

2 Note that Scikit-Learn separates the bias term (intercept_) from the feature weights (coef_).

```
>>> np.linalg.pinv(X_b).dot(y)
array([[4.21509616],
       [2.77011339]])
```

The pseudoinverse itself is computed using a standard matrix factorization technique called *Singular Value Decomposition* (SVD) that can decompose the training set matrix **X** into the matrix multiplication of three matrices **U** Σ **V**$^\top$ (see `numpy.linalg.svd()`). The pseudoinverse is computed as $\mathbf{X}^+ = \mathbf{V}\Sigma^+\mathbf{U}^\top$. To compute the matrix Σ^+, the algorithm takes Σ and sets to zero all values smaller than a tiny threshold value, then it replaces all the nonzero values with their inverse, and finally it transposes the resulting matrix. This approach is more efficient than computing the Normal Equation, plus it handles edge cases nicely: indeed, the Normal Equation may not work if the matrix $\mathbf{X}^\top\mathbf{X}$ is not invertible (i.e., singular), such as if $m < n$ or if some features are redundant, but the pseudoinverse is always defined.

Computational Complexity

The Normal Equation computes the inverse of $\mathbf{X}^\top\,\mathbf{X}$, which is an $(n + 1) \times (n + 1)$ matrix (where n is the number of features). The *computational complexity* of inverting such a matrix is typically about $O(n^{2.4})$ to $O(n^3)$, depending on the implementation. In other words, if you double the number of features, you multiply the computation time by roughly $2^{2.4} = 5.3$ to $2^3 = 8$.

The SVD approach used by Scikit-Learn's `LinearRegression` class is about $O(n^2)$. If you double the number of features, you multiply the computation time by roughly 4.

Both the Normal Equation and the SVD approach get very slow when the number of features grows large (e.g., 100,000). On the positive side, both are linear with regard to the number of instances in the training set (they are $O(m)$), so they handle large training sets efficiently, provided they can fit in memory.

Also, once you have trained your Linear Regression model (using the Normal Equation or any other algorithm), predictions are very fast: the computational complexity is linear with regard to both the number of instances you want to make predictions on and the number of features. In other words, making predictions on twice as many instances (or twice as many features) will take roughly twice as much time.

Now we will look at a very different way to train a Linear Regression model, which is better suited for cases where there are a large number of features or too many training instances to fit in memory.

Gradient Descent

Gradient Descent is a generic optimization algorithm capable of finding optimal solutions to a wide range of problems. The general idea of Gradient Descent is to tweak parameters iteratively in order to minimize a cost function.

Suppose you are lost in the mountains in a dense fog, and you can only feel the slope of the ground below your feet. A good strategy to get to the bottom of the valley quickly is to go downhill in the direction of the steepest slope. This is exactly what Gradient Descent does: it measures the local gradient of the error function with regard to the parameter vector θ, and it goes in the direction of descending gradient. Once the gradient is zero, you have reached a minimum!

Concretely, you start by filling θ with random values (this is called *random initialization*). Then you improve it gradually, taking one baby step at a time, each step attempting to decrease the cost function (e.g., the MSE), until the algorithm *converges* to a minimum (see Figure 4-3).

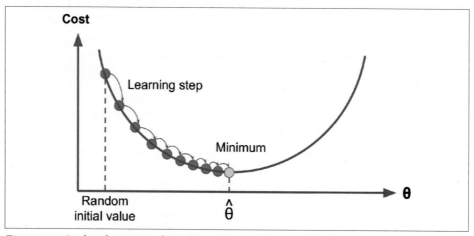

Figure 4-3. In this depiction of Gradient Descent, the model parameters are initialized randomly and get tweaked repeatedly to minimize the cost function; the learning step size is proportional to the slope of the cost function, so the steps gradually get smaller as the parameters approach the minimum

An important parameter in Gradient Descent is the size of the steps, determined by the *learning rate* hyperparameter. If the learning rate is too small, then the algorithm will have to go through many iterations to converge, which will take a long time (see Figure 4-4).

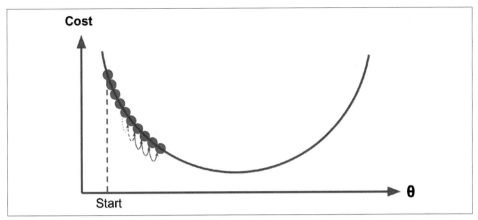

Figure 4-4. The learning rate is too small

On the other hand, if the learning rate is too high, you might jump across the valley and end up on the other side, possibly even higher up than you were before. This might make the algorithm diverge, with larger and larger values, failing to find a good solution (see Figure 4-5).

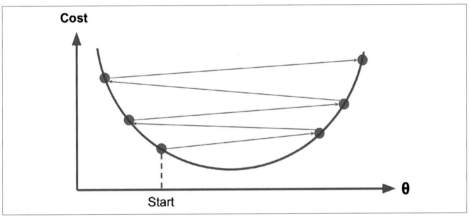

Figure 4-5. The learning rate is too large

Finally, not all cost functions look like nice, regular bowls. There may be holes, ridges, plateaus, and all sorts of irregular terrains, making convergence to the minimum difficult. Figure 4-6 shows the two main challenges with Gradient Descent. If the random initialization starts the algorithm on the left, then it will converge to a *local minimum*, which is not as good as the *global minimum*. If it starts on the right, then it will take a very long time to cross the plateau. And if you stop too early, you will never reach the global minimum.

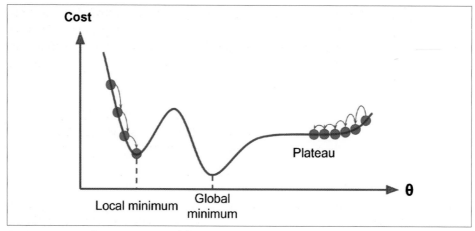

Figure 4-6. Gradient Descent pitfalls

Fortunately, the MSE cost function for a Linear Regression model happens to be a *convex function*, which means that if you pick any two points on the curve, the line segment joining them never crosses the curve. This implies that there are no local minima, just one global minimum. It is also a continuous function with a slope that never changes abruptly.[3] These two facts have a great consequence: Gradient Descent is guaranteed to approach arbitrarily close the global minimum (if you wait long enough and if the learning rate is not too high).

In fact, the cost function has the shape of a bowl, but it can be an elongated bowl if the features have very different scales. Figure 4-7 shows Gradient Descent on a training set where features 1 and 2 have the same scale (on the left), and on a training set where feature 1 has much smaller values than feature 2 (on the right).[4]

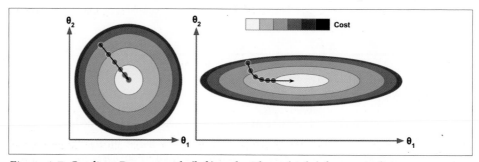

Figure 4-7. Gradient Descent with (left) and without (right) feature scaling

---

3 Technically speaking, its derivative is *Lipschitz continuous*.

4 Since feature 1 is smaller, it takes a larger change in θ_1 to affect the cost function, which is why the bowl is elongated along the θ_1 axis.

As you can see, on the left the Gradient Descent algorithm goes straight toward the minimum, thereby reaching it quickly, whereas on the right it first goes in a direction almost orthogonal to the direction of the global minimum, and it ends with a long march down an almost flat valley. It will eventually reach the minimum, but it will take a long time.

> When using Gradient Descent, you should ensure that all features have a similar scale (e.g., using Scikit-Learn's StandardScaler class), or else it will take much longer to converge.

This diagram also illustrates the fact that training a model means searching for a combination of model parameters that minimizes a cost function (over the training set). It is a search in the model's *parameter space*: the more parameters a model has, the more dimensions this space has, and the harder the search is: searching for a needle in a 300-dimensional haystack is much trickier than in 3 dimensions. Fortunately, since the cost function is convex in the case of Linear Regression, the needle is simply at the bottom of the bowl.

Batch Gradient Descent

To implement Gradient Descent, you need to compute the gradient of the cost function with regard to each model parameter θ_j. In other words, you need to calculate how much the cost function will change if you change θ_j just a little bit. This is called a *partial derivative*. It is like asking "What is the slope of the mountain under my feet if I face east?" and then asking the same question facing north (and so on for all other dimensions, if you can imagine a universe with more than three dimensions). Equation 4-5 computes the partial derivative of the cost function with regard to parameter θ_j, noted $\partial \text{MSE}(\theta) / \partial \theta_j$.

Equation 4-5. Partial derivatives of the cost function

$$\frac{\partial}{\partial \theta_j} \text{MSE}(\theta) = \frac{2}{m} \sum_{i=1}^{m} \left(\theta^{\mathsf{T}} x^{(i)} - y^{(i)} \right) x_j^{(i)}$$

Instead of computing these partial derivatives individually, you can use Equation 4-6 to compute them all in one go. The gradient vector, noted $\nabla_\theta \text{MSE}(\theta)$, contains all the partial derivatives of the cost function (one for each model parameter).

Equation 4-6. Gradient vector of the cost function

$$\nabla_{\boldsymbol{\theta}} \, \text{MSE}(\boldsymbol{\theta}) = \begin{pmatrix} \frac{\partial}{\partial \theta_0} \text{MSE}(\boldsymbol{\theta}) \\ \frac{\partial}{\partial \theta_1} \text{MSE}(\boldsymbol{\theta}) \\ \vdots \\ \frac{\partial}{\partial \theta_n} \text{MSE}(\boldsymbol{\theta}) \end{pmatrix} = \frac{2}{m} \mathbf{X}^{\mathsf{T}} (\mathbf{X}\boldsymbol{\theta} - \mathbf{y})$$

 Notice that this formula involves calculations over the full training set **X**, at each Gradient Descent step! This is why the algorithm is called *Batch Gradient Descent*: it uses the whole batch of training data at every step (actually, *Full Gradient Descent* would probably be a better name). As a result it is terribly slow on very large training sets (but we will see much faster Gradient Descent algorithms shortly). However, Gradient Descent scales well with the number of features; training a Linear Regression model when there are hundreds of thousands of features is much faster using Gradient Descent than using the Normal Equation or SVD decomposition.

Once you have the gradient vector, which points uphill, just go in the opposite direction to go downhill. This means subtracting $\nabla_{\boldsymbol{\theta}}\text{MSE}(\boldsymbol{\theta})$ from $\boldsymbol{\theta}$. This is where the learning rate η comes into play:[5] multiply the gradient vector by η to determine the size of the downhill step (Equation 4-7).

Equation 4-7. Gradient Descent step

$$\boldsymbol{\theta}^{(\text{next step})} = \boldsymbol{\theta} - \eta \, \nabla_{\boldsymbol{\theta}} \, \text{MSE}\Big(\boldsymbol{\theta}\Big)$$

Let's look at a quick implementation of this algorithm:

```
eta = 0.1  # learning rate
n_iterations = 1000
m = 100

theta = np.random.randn(2,1)  # random initialization

for iteration in range(n_iterations):
    gradients = 2/m * X_b.T.dot(X_b.dot(theta) - y)
    theta = theta - eta * gradients
```

---

5 Eta (η) is the seventh letter of the Greek alphabet.

That wasn't too hard! Let's look at the resulting `theta`:

```
>>> theta
array([[4.21509616],
       [2.77011339]])
```

Hey, that's exactly what the Normal Equation found! Gradient Descent worked perfectly. But what if you had used a different learning rate `eta`? Figure 4-8 shows the first 10 steps of Gradient Descent using three different learning rates (the dashed line represents the starting point).

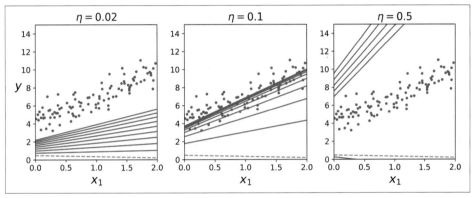

Figure 4-8. Gradient Descent with various learning rates

On the left, the learning rate is too low: the algorithm will eventually reach the solution, but it will take a long time. In the middle, the learning rate looks pretty good: in just a few iterations, it has already converged to the solution. On the right, the learning rate is too high: the algorithm diverges, jumping all over the place and actually getting further and further away from the solution at every step.

To find a good learning rate, you can use grid search (see Chapter 2). However, you may want to limit the number of iterations so that grid search can eliminate models that take too long to converge.

You may wonder how to set the number of iterations. If it is too low, you will still be far away from the optimal solution when the algorithm stops; but if it is too high, you will waste time while the model parameters do not change anymore. A simple solution is to set a very large number of iterations but to interrupt the algorithm when the gradient vector becomes tiny—that is, when its norm becomes smaller than a tiny number ϵ (called the *tolerance*)—because this happens when Gradient Descent has (almost) reached the minimum.

Stochastic Gradient Descent

The main problem with Batch Gradient Descent is the fact that it uses the whole training set to compute the gradients at every step, which makes it very slow when the training set is large. At the opposite extreme, *Stochastic Gradient Descent* picks a random instance in the training set at every step and computes the gradients based only on that single instance. Obviously, working on a single instance at a time makes the algorithm much faster because it has very little data to manipulate at every iteration. It also makes it possible to train on huge training sets, since only one instance needs to be in memory at each iteration (Stochastic GD can be implemented as an out-of-core algorithm; see Chapter 1).

On the other hand, due to its stochastic (i.e., random) nature, this algorithm is much less regular than Batch Gradient Descent: instead of gently decreasing until it reaches the minimum, the cost function will bounce up and down, decreasing only on average. Over time it will end up very close to the minimum, but once it gets there it will continue to bounce around, never settling down (see Figure 4-9). So once the algorithm stops, the final parameter values are good, but not optimal.

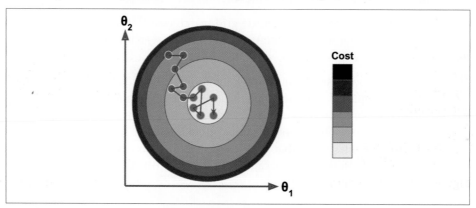

Figure 4-9. With Stochastic Gradient Descent, each training step is much faster but also much more stochastic than when using Batch Gradient Descent

When the cost function is very irregular (as in Figure 4-6), this can actually help the algorithm jump out of local minima, so Stochastic Gradient Descent has a better chance of finding the global minimum than Batch Gradient Descent does.

Therefore, randomness is good to escape from local optima, but bad because it means that the algorithm can never settle at the minimum. One solution to this dilemma is to gradually reduce the learning rate. The steps start out large (which helps make quick progress and escape local minima), then get smaller and smaller, allowing the algorithm to settle at the global minimum. This process is akin to *simulated annealing*, an algorithm inspired from the process in metallurgy of annealing, where molten metal is slowly cooled down. The function that determines the learning rate at each iteration is called the *learning schedule*. If the learning rate is reduced too quickly, you may get stuck in a local minimum, or even end up frozen halfway to the minimum. If the learning rate is reduced too slowly, you may jump around the minimum for a long time and end up with a suboptimal solution if you halt training too early.

This code implements Stochastic Gradient Descent using a simple learning schedule:

```
n_epochs = 50
t0, t1 = 5, 50  # learning schedule hyperparameters

def learning_schedule(t):
    return t0 / (t + t1)

theta = np.random.randn(2,1)  # random initialization

for epoch in range(n_epochs):
    for i in range(m):
        random_index = np.random.randint(m)
        xi = X_b[random_index:random_index+1]
        yi = y[random_index:random_index+1]
        gradients = 2 * xi.T.dot(xi.dot(theta) - yi)
        eta = learning_schedule(epoch * m + i)
        theta = theta - eta * gradients
```

By convention we iterate by rounds of *m* iterations; each round is called an *epoch*. While the Batch Gradient Descent code iterated 1,000 times through the whole training set, this code goes through the training set only 50 times and reaches a pretty good solution:

```
>>> theta
array([[4.21076011],
       [2.74856079]])
```

Figure 4-10 shows the first 20 steps of training (notice how irregular the steps are).

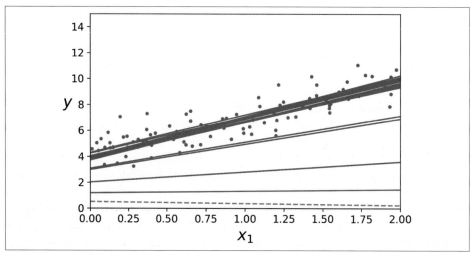

Figure 4-10. The first 20 steps of Stochastic Gradient Descent

Note that since instances are picked randomly, some instances may be picked several times per epoch, while others may not be picked at all. If you want to be sure that the algorithm goes through every instance at each epoch, another approach is to shuffle the training set (making sure to shuffle the input features and the labels jointly), then go through it instance by instance, then shuffle it again, and so on. However, this approach generally converges more slowly.

> When using Stochastic Gradient Descent, the training instances must be independent and identically distributed (IID) to ensure that the parameters get pulled toward the global optimum, on average. A simple way to ensure this is to shuffle the instances during training (e.g., pick each instance randomly, or shuffle the training set at the beginning of each epoch). If you do not shuffle the instances—for example, if the instances are sorted by label—then SGD will start by optimizing for one label, then the next, and so on, and it will not settle close to the global minimum.

To perform Linear Regression using Stochastic GD with Scikit-Learn, you can use the `SGDRegressor` class, which defaults to optimizing the squared error cost function. The following code runs for maximum 1,000 epochs or until the loss drops by less than 0.001 during one epoch (`max_iter=1000`, `tol=1e-3`). It starts with a learning rate of 0.1 (`eta0=0.1`), using the default learning schedule (different from the preceding one). Lastly, it does not use any regularization (`penalty=None`; more details on this shortly):

```
from sklearn.linear_model import SGDRegressor
sgd_reg = SGDRegressor(max_iter=1000, tol=1e-3, penalty=None, eta0=0.1)
sgd_reg.fit(X, y.ravel())
```

Once again, you find a solution quite close to the one returned by the Normal Equation:

```
>>> sgd_reg.intercept_, sgd_reg.coef_
(array([4.24365286]), array([2.8250878]))
```

Mini-batch Gradient Descent

The last Gradient Descent algorithm we will look at is called *Mini-batch Gradient Descent*. It is simple to understand once you know Batch and Stochastic Gradient Descent: at each step, instead of computing the gradients based on the full training set (as in Batch GD) or based on just one instance (as in Stochastic GD), Mini-batch GD computes the gradients on small random sets of instances called *mini-batches*. The main advantage of Mini-batch GD over Stochastic GD is that you can get a performance boost from hardware optimization of matrix operations, especially when using GPUs.

The algorithm's progress in parameter space is less erratic than with Stochastic GD, especially with fairly large mini-batches. As a result, Mini-batch GD will end up walking around a bit closer to the minimum than Stochastic GD—but it may be harder for it to escape from local minima (in the case of problems that suffer from local minima, unlike Linear Regression). Figure 4-11 shows the paths taken by the three Gradient Descent algorithms in parameter space during training. They all end up near the minimum, but Batch GD's path actually stops at the minimum, while both Stochastic GD and Mini-batch GD continue to walk around. However, don't forget that Batch GD takes a lot of time to take each step, and Stochastic GD and Mini-batch GD would also reach the minimum if you used a good learning schedule.

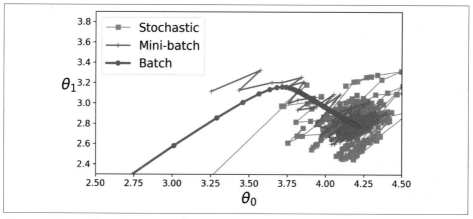

Figure 4-11. Gradient Descent paths in parameter space

Let's compare the algorithms we've discussed so far for Linear Regression[6] (recall that m is the number of training instances and n is the number of features); see Table 4-1.

Table 4-1. Comparison of algorithms for Linear Regression

Algorithm	Large m	Out-of-core support	Large n	Hyperparams	Scaling required	Scikit-Learn
Normal Equation	Fast	No	Slow	0	No	N/A
SVD	Fast	No	Slow	0	No	LinearRegression
Batch GD	Slow	No	Fast	2	Yes	SGDRegressor
Stochastic GD	Fast	Yes	Fast	≥2	Yes	SGDRegressor
Mini-batch GD	Fast	Yes	Fast	≥2	Yes	SGDRegressor

 There is almost no difference after training: all these algorithms end up with very similar models and make predictions in exactly the same way.

Polynomial Regression

What if your data is more complex than a straight line? Surprisingly, you can use a linear model to fit nonlinear data. A simple way to do this is to add powers of each feature as new features, then train a linear model on this extended set of features. This technique is called *Polynomial Regression.*

Let's look at an example. First, let's generate some nonlinear data, based on a simple *quadratic equation*[7] (plus some noise; see Figure 4-12):

```
m = 100
X = 6 * np.random.rand(m, 1) - 3
y = 0.5 * X**2 + X + 2 + np.random.randn(m, 1)
```

---

6 While the Normal Equation can only perform Linear Regression, the Gradient Descent algorithms can be used to train many other models, as we will see.

7 A quadratic equation is of the form $y = ax^2 + bx + c$.

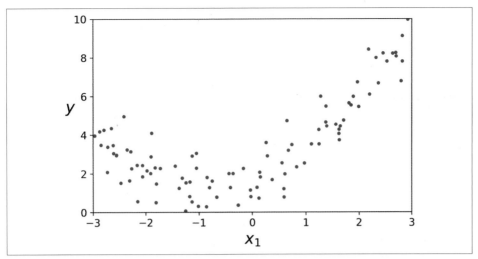

Figure 4-12. Generated nonlinear and noisy dataset

Clearly, a straight line will never fit this data properly. So let's use Scikit-Learn's `Poly nomialFeatures` class to transform our training data, adding the square (second-degree polynomial) of each feature in the training set as a new feature (in this case there is just one feature):

```
>>> from sklearn.preprocessing import PolynomialFeatures
>>> poly_features = PolynomialFeatures(degree=2, include_bias=False)
>>> X_poly = poly_features.fit_transform(X)
>>> X[0]
array([-0.75275929])
>>> X_poly[0]
array([-0.75275929, 0.56664654])
```

`X_poly` now contains the original feature of X plus the square of this feature. Now you can fit a `LinearRegression` model to this extended training data (Figure 4-13):

```
>>> lin_reg = LinearRegression()
>>> lin_reg.fit(X_poly, y)
>>> lin_reg.intercept_, lin_reg.coef_
(array([1.78134581]), array([[0.93366893, 0.56456263]]))
```

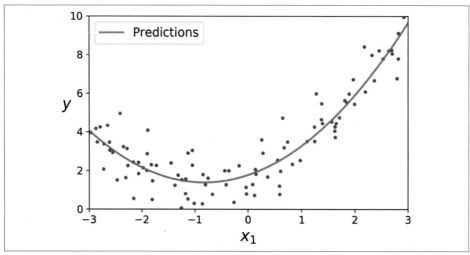

Figure 4-13. Polynomial Regression model predictions

Not bad: the model estimates $\hat{y} = 0.56x_1^2 + 0.93x_1 + 1.78$ when in fact the original function was $y = 0.5x_1^2 + 1.0x_1 + 2.0 +$ Gaussian noise.

Note that when there are multiple features, Polynomial Regression is capable of finding relationships between features (which is something a plain Linear Regression model cannot do). This is made possible by the fact that `PolynomialFeatures` also adds all combinations of features up to the given degree. For example, if there were two features *a* and *b*, `PolynomialFeatures` with `degree=3` would not only add the features a^2, a^3, b^2, and b^3, but also the combinations ab, a^2b, and ab^2.

> `PolynomialFeatures(degree=d)` transforms an array containing *n* features into an array containing $(n + d)! / d!n!$ features, where *n*! is the *factorial* of *n*, equal to $1 \times 2 \times 3 \times \cdots \times n$. Beware of the combinatorial explosion of the number of features!

Learning Curves

If you perform high-degree Polynomial Regression, you will likely fit the training data much better than with plain Linear Regression. For example, Figure 4-14 applies a 300-degree polynomial model to the preceding training data, and compares the result with a pure linear model and a quadratic model (second-degree polynomial). Notice how the 300-degree polynomial model wiggles around to get as close as possible to the training instances.

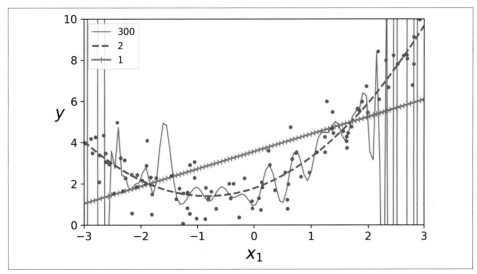

Figure 4-14. High-degree Polynomial Regression

This high-degree Polynomial Regression model is severely overfitting the training data, while the linear model is underfitting it. The model that will generalize best in this case is the quadratic model, which makes sense because the data was generated using a quadratic model. But in general you won't know what function generated the data, so how can you decide how complex your model should be? How can you tell that your model is overfitting or underfitting the data?

In Chapter 2 you used cross-validation to get an estimate of a model's generalization performance. If a model performs well on the training data but generalizes poorly according to the cross-validation metrics, then your model is overfitting. If it performs poorly on both, then it is underfitting. This is one way to tell when a model is too simple or too complex.

Another way to tell is to look at the *learning curves*: these are plots of the model's performance on the training set and the validation set as a function of the training set size (or the training iteration). To generate the plots, train the model several times on different sized subsets of the training set. The following code defines a function that, given some training data, plots the learning curves of a model:

```
from sklearn.metrics import mean_squared_error
from sklearn.model_selection import train_test_split

def plot_learning_curves(model, X, y):
    X_train, X_val, y_train, y_val = train_test_split(X, y, test_size=0.2)
    train_errors, val_errors = [], []
    for m in range(1, len(X_train)):
        model.fit(X_train[:m], y_train[:m])
        y_train_predict = model.predict(X_train[:m])
        y_val_predict = model.predict(X_val)
        train_errors.append(mean_squared_error(y_train[:m], y_train_predict))
        val_errors.append(mean_squared_error(y_val, y_val_predict))
    plt.plot(np.sqrt(train_errors), "r-+", linewidth=2, label="train")
    plt.plot(np.sqrt(val_errors), "b-", linewidth=3, label="val")
```

Let's look at the learning curves of the plain Linear Regression model (a straight line; see Figure 4-15):

```
lin_reg = LinearRegression()
plot_learning_curves(lin_reg, X, y)
```

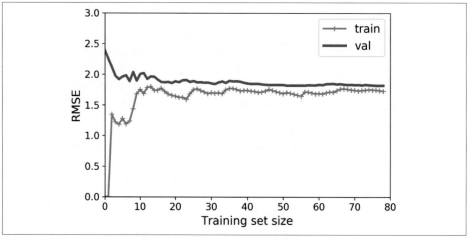

Figure 4-15. Learning curves

This model that's underfitting deserves a bit of explanation. First, let's look at the performance on the training data: when there are just one or two instances in the training set, the model can fit them perfectly, which is why the curve starts at zero. But as new instances are added to the training set, it becomes impossible for the model to fit the training data perfectly, both because the data is noisy and because it is not linear at all. So the error on the training data goes up until it reaches a plateau, at which point adding new instances to the training set doesn't make the average error much better or worse. Now let's look at the performance of the model on the validation data. When the model is trained on very few training instances, it is incapable of generalizing properly, which is why the validation error is initially quite big. Then, as the model is shown more training examples, it learns, and thus the validation error slowly goes down. However, once again a straight line cannot do a good job modeling the data, so the error ends up at a plateau, very close to the other curve.

These learning curves are typical of a model that's underfitting. Both curves have reached a plateau; they are close and fairly high.

If your model is underfitting the training data, adding more training examples will not help. You need to use a more complex model or come up with better features.

Now let's look at the learning curves of a 10th-degree polynomial model on the same data (Figure 4-16):

```
from sklearn.pipeline import Pipeline

polynomial_regression = Pipeline([
        ("poly_features", PolynomialFeatures(degree=10, include_bias=False)),
        ("lin_reg", LinearRegression()),
    ])

plot_learning_curves(polynomial_regression, X, y)
```

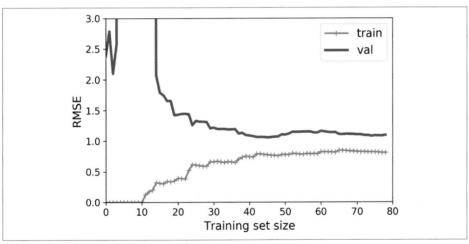

Figure 4-16. Learning curves for the 10th-degree polynomial model

These learning curves look a bit like the previous ones, but there are two very important differences:

- The error on the training data is much lower than with the Linear Regression model.

- There is a gap between the curves. This means that the model performs significantly better on the training data than on the validation data, which is the hallmark of an overfitting model. If you used a much larger training set, however, the two curves would continue to get closer.

One way to improve an overfitting model is to feed it more training data until the validation error reaches the training error.

The Bias/Variance Trade-off

An important theoretical result of statistics and Machine Learning is the fact that a model's generalization error can be expressed as the sum of three very different errors:

Bias
> This part of the generalization error is due to wrong assumptions, such as assuming that the data is linear when it is actually quadratic. A high-bias model is most likely to underfit the training data.[8]

Variance
> This part is due to the model's excessive sensitivity to small variations in the training data. A model with many degrees of freedom (such as a high-degree polynomial model) is likely to have high variance and thus overfit the training data.

Irreducible error
> This part is due to the noisiness of the data itself. The only way to reduce this part of the error is to clean up the data (e.g., fix the data sources, such as broken sensors, or detect and remove outliers).

Increasing a model's complexity will typically increase its variance and reduce its bias. Conversely, reducing a model's complexity increases its bias and reduces its variance. This is why it is called a trade-off.

Regularized Linear Models

As we saw in Chapters 1 and 2, a good way to reduce overfitting is to regularize the model (i.e., to constrain it): the fewer degrees of freedom it has, the harder it will be

---

8 This notion of bias is not to be confused with the bias term of linear models.

for it to overfit the data. A simple way to regularize a polynomial model is to reduce the number of polynomial degrees.

For a linear model, regularization is typically achieved by constraining the weights of the model. We will now look at Ridge Regression, Lasso Regression, and Elastic Net, which implement three different ways to constrain the weights.

Ridge Regression

Ridge Regression (also called *Tikhonov regularization*) is a regularized version of Linear Regression: a *regularization term* equal to $\alpha\sum_{i=1}^{n} \theta_i^2$ is added to the cost function. This forces the learning algorithm to not only fit the data but also keep the model weights as small as possible. Note that the regularization term should only be added to the cost function during training. Once the model is trained, you want to use the unregularized performance measure to evaluate the model's performance.

> It is quite common for the cost function used during training to be different from the performance measure used for testing. Apart from regularization, another reason they might be different is that a good training cost function should have optimization-friendly derivatives, while the performance measure used for testing should be as close as possible to the final objective. For example, classifiers are often trained using a cost function such as the log loss (discussed in a moment) but evaluated using precision/recall.

The hyperparameter α controls how much you want to regularize the model. If $\alpha = 0$, then Ridge Regression is just Linear Regression. If α is very large, then all weights end up very close to zero and the result is a flat line going through the data's mean. Equation 4-8 presents the Ridge Regression cost function.[9]

Equation 4-8. Ridge Regression cost function

$$J(\boldsymbol{\theta}) = \text{MSE}(\boldsymbol{\theta}) + \alpha\frac{1}{2}\sum_{i=1}^{n} \theta_i^2$$

Note that the bias term θ_0 is not regularized (the sum starts at $i = 1$, not 0). If we define **w** as the vector of feature weights (θ_1 to θ_n), then the regularization term is

---

9 It is common to use the notation $J(\boldsymbol{\theta})$ for cost functions that don't have a short name; we will often use this notation throughout the rest of this book. The context will make it clear which cost function is being discussed.

equal to $\frac{1}{2}(\| \mathbf{w} \|_2)^2$, where $\| \mathbf{w} \|_2$ represents the ℓ_2 norm of the weight vector.[10] For Gradient Descent, just add $\alpha\mathbf{w}$ to the MSE gradient vector (Equation 4-6).

 It is important to scale the data (e.g., using a StandardScaler) before performing Ridge Regression, as it is sensitive to the scale of the input features. This is true of most regularized models.

Figure 4-17 shows several Ridge models trained on some linear data using different α values. On the left, plain Ridge models are used, leading to linear predictions. On the right, the data is first expanded using PolynomialFeatures(degree=10), then it is scaled using a StandardScaler, and finally the Ridge models are applied to the resulting features: this is Polynomial Regression with Ridge regularization. Note how increasing α leads to flatter (i.e., less extreme, more reasonable) predictions, thus reducing the model's variance but increasing its bias.

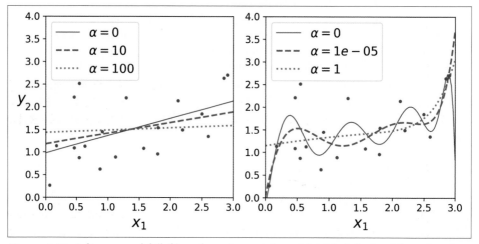

Figure 4-17. A linear model (left) and a polynomial model (right), both with various levels of Ridge regularization

As with Linear Regression, we can perform Ridge Regression either by computing a closed-form equation or by performing Gradient Descent. The pros and cons are the

---

10 Norms are discussed in Chapter 2.

same. Equation 4-9 shows the closed-form solution, where $\mathbf{A}$ is the $(n + 1) \times (n + 1)$ *identity matrix*,[11] except with a 0 in the top-left cell, corresponding to the bias term.

Equation 4-9. Ridge Regression closed-form solution

$$\hat{\boldsymbol{\theta}} = \left(\mathbf{X}^{\mathsf{T}}\mathbf{X} + \alpha\mathbf{A}\right)^{-1} \ \mathbf{X}^{\mathsf{T}} \ \mathbf{y}$$

Here is how to perform Ridge Regression with Scikit-Learn using a closed-form solution (a variant of Equation 4-9 that uses a matrix factorization technique by André-Louis Cholesky):

```
>>> from sklearn.linear_model import Ridge
>>> ridge_reg = Ridge(alpha=1, solver="cholesky")
>>> ridge_reg.fit(X, y)
>>> ridge_reg.predict([[1.5]])
array([[1.55071465]])
```

And using Stochastic Gradient Descent:[12]

```
>>> sgd_reg = SGDRegressor(penalty="l2")
>>> sgd_reg.fit(X, y.ravel())
>>> sgd_reg.predict([[1.5]])
array([1.47012588])
```

The `penalty` hyperparameter sets the type of regularization term to use. Specifying `"l2"` indicates that you want SGD to add a regularization term to the cost function equal to half the square of the ℓ_2 norm of the weight vector: this is simply Ridge Regression.

Lasso Regression

Least Absolute Shrinkage and Selection Operator Regression (usually simply called *Lasso Regression*) is another regularized version of Linear Regression: just like Ridge Regression, it adds a regularization term to the cost function, but it uses the ℓ_1 norm of the weight vector instead of half the square of the ℓ_2 norm (see Equation 4-10).

Equation 4-10. Lasso Regression cost function

$$J(\boldsymbol{\theta}) = \mathrm{MSE}(\boldsymbol{\theta}) + \alpha\sum_{i=1}^{n}|\theta_i|$$

---

11 A square matrix full of 0s except for 1s on the main diagonal (top left to bottom right).

12 Alternatively you can use the `Ridge` class with the `"sag"` solver. Stochastic Average GD is a variant of Stochastic GD. For more details, see the presentation "Minimizing Finite Sums with the Stochastic Average Gradient Algorithm" (*https://homl.info/12*) by Mark Schmidt et al. from the University of British Columbia.

Figure 4-18 shows the same thing as Figure 4-17 but replaces Ridge models with Lasso models and uses smaller α values.

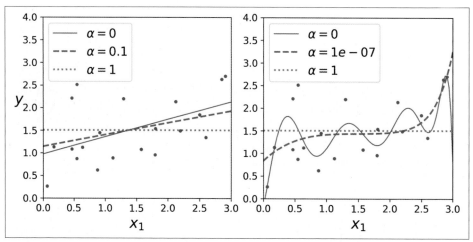

Figure 4-18. A linear model (left) and a polynomial model (right), both using various levels of Lasso regularization

An important characteristic of Lasso Regression is that it tends to eliminate the weights of the least important features (i.e., set them to zero). For example, the dashed line in the righthand plot in Figure 4-18 (with $\alpha = 10^{-7}$) looks quadratic, almost linear: all the weights for the high-degree polynomial features are equal to zero. In other words, Lasso Regression automatically performs feature selection and outputs a *sparse model* (i.e., with few nonzero feature weights).

You can get a sense of why this is the case by looking at Figure 4-19: the axes represent two model parameters, and the background contours represent different loss functions. In the top-left plot, the contours represent the ℓ_1 loss ($|\theta_1| + |\theta_2|$), which drops linearly as you get closer to any axis. For example, if you initialize the model parameters to $\theta_1 = 2$ and $\theta_2 = 0.5$, running Gradient Descent will decrement both parameters equally (as represented by the dashed yellow line); therefore θ_2 will reach 0 first (since it was closer to 0 to begin with). After that, Gradient Descent will roll down the gutter until it reaches $\theta_1 = 0$ (with a bit of bouncing around, since the gradients of ℓ_1 never get close to 0: they are either –1 or 1 for each parameter). In the top-right plot, the contours represent Lasso's cost function (i.e., an MSE cost function plus an ℓ_1 loss). The small white circles show the path that Gradient Descent takes to optimize some model parameters that were initialized around $\theta_1 = 0.25$ and $\theta_2 = -1$: notice once again how the path quickly reaches $\theta_2 = 0$, then rolls down the gutter and ends up bouncing around the global optimum (represented by the red square). If we increased α, the global optimum would move left along the dashed yellow line, while if we decreased α, the global optimum would move right (in this example, the optimal parameters for the unregularized MSE are $\theta_1 = 2$ and $\theta_2 = 0.5$).

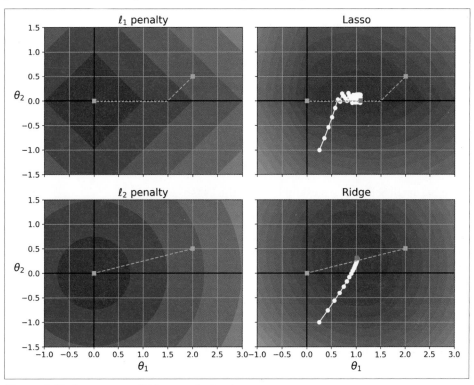

Figure 4-19. Lasso versus Ridge regularization

The two bottom plots show the same thing but with an ℓ_2 penalty instead. In the bottom-left plot, you can see that the ℓ_2 loss decreases with the distance to the origin, so Gradient Descent just takes a straight path toward that point. In the bottom-right plot, the contours represent Ridge Regression's cost function (i.e., an MSE cost function plus an ℓ_2 loss). There are two main differences with Lasso. First, the gradients get smaller as the parameters approach the global optimum, so Gradient Descent naturally slows down, which helps convergence (as there is no bouncing around). Second, the optimal parameters (represented by the red square) get closer and closer to the origin when you increase α, but they never get eliminated entirely.

> To avoid Gradient Descent from bouncing around the optimum at the end when using Lasso, you need to gradually reduce the learning rate during training (it will still bounce around the optimum, but the steps will get smaller and smaller, so it will converge).

The Lasso cost function is not differentiable at $\theta_i = 0$ (for $i = 1, 2, \cdots, n$), but Gradient Descent still works fine if you use a *subgradient vector* $\mathbf{g}$[13] instead when any $\theta_i = 0$. Equation 4-11 shows a subgradient vector equation you can use for Gradient Descent with the Lasso cost function.

Equation 4-11. Lasso Regression subgradient vector

$$g(\boldsymbol{\theta}, J) = \nabla_{\boldsymbol{\theta}} \text{MSE}(\boldsymbol{\theta}) + \alpha \begin{pmatrix} \text{sign}(\theta_1) \\ \text{sign}(\theta_2) \\ \vdots \\ \text{sign}(\theta_n) \end{pmatrix} \quad \text{where } \text{sign}(\theta_i) = \begin{cases} -1 & \text{if } \theta_i < 0 \\ 0 & \text{if } \theta_i = 0 \\ +1 & \text{if } \theta_i > 0 \end{cases}$$

Here is a small Scikit-Learn example using the `Lasso` class:

```
>>> from sklearn.linear_model import Lasso
>>> lasso_reg = Lasso(alpha=0.1)
>>> lasso_reg.fit(X, y)
>>> lasso_reg.predict([[1.5]])
array([1.53788174])
```

Note that you could instead use `SGDRegressor(penalty="l1")`.

Elastic Net

Elastic Net is a middle ground between Ridge Regression and Lasso Regression. The regularization term is a simple mix of both Ridge and Lasso's regularization terms, and you can control the mix ratio r. When $r = 0$, Elastic Net is equivalent to Ridge Regression, and when $r = 1$, it is equivalent to Lasso Regression (see Equation 4-12).

Equation 4-12. Elastic Net cost function

$$J(\boldsymbol{\theta}) = \text{MSE}(\boldsymbol{\theta}) + r\alpha \sum_{i=1}^{n} |\theta_i| + \frac{1-r}{2} \alpha \sum_{i=1}^{n} \theta_i^2$$

So when should you use plain Linear Regression (i.e., without any regularization), Ridge, Lasso, or Elastic Net? It is almost always preferable to have at least a little bit of regularization, so generally you should avoid plain Linear Regression. Ridge is a good default, but if you suspect that only a few features are useful, you should prefer Lasso or Elastic Net because they tend to reduce the useless features' weights down to zero, as we have discussed. In general, Elastic Net is preferred over Lasso because Lasso

---

13 You can think of a subgradient vector at a nondifferentiable point as an intermediate vector between the gradient vectors around that point.

may behave erratically when the number of features is greater than the number of training instances or when several features are strongly correlated.

Here is a short example that uses Scikit-Learn's `ElasticNet` (`l1_ratio` corresponds to the mix ratio *r*):

```
>>> from sklearn.linear_model import ElasticNet
>>> elastic_net = ElasticNet(alpha=0.1, l1_ratio=0.5)
>>> elastic_net.fit(X, y)
>>> elastic_net.predict([[1.5]])
array([1.54333232])
```

Early Stopping

A very different way to regularize iterative learning algorithms such as Gradient Descent is to stop training as soon as the validation error reaches a minimum. This is called *early stopping*. Figure 4-20 shows a complex model (in this case, a high-degree Polynomial Regression model) being trained with Batch Gradient Descent. As the epochs go by the algorithm learns, and its prediction error (RMSE) on the training set goes down, along with its prediction error on the validation set. After a while though, the validation error stops decreasing and starts to go back up. This indicates that the model has started to overfit the training data. With early stopping you just stop training as soon as the validation error reaches the minimum. It is such a simple and efficient regularization technique that Geoffrey Hinton called it a "beautiful free lunch."

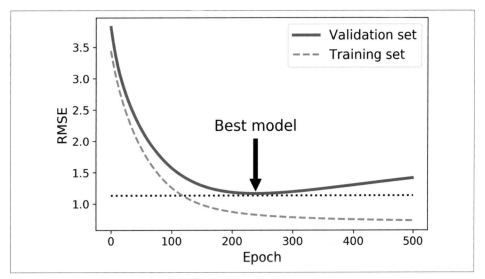

Figure 4-20. Early stopping regularization

With Stochastic and Mini-batch Gradient Descent, the curves are not so smooth, and it may be hard to know whether you have reached the minimum or not. One solution is to stop only after the validation error has been above the minimum for some time (when you are confident that the model will not do any better), then roll back the model parameters to the point where the validation error was at a minimum.

Here is a basic implementation of early stopping:

```python
from sklearn.base import clone

# prepare the data
poly_scaler = Pipeline([
        ("poly_features", PolynomialFeatures(degree=90, include_bias=False)),
        ("std_scaler", StandardScaler())
    ])
X_train_poly_scaled = poly_scaler.fit_transform(X_train)
X_val_poly_scaled = poly_scaler.transform(X_val)

sgd_reg = SGDRegressor(max_iter=1, tol=-np.infty, warm_start=True,
                       penalty=None, learning_rate="constant", eta0=0.0005)

minimum_val_error = float("inf")
best_epoch = None
best_model = None
for epoch in range(1000):
    sgd_reg.fit(X_train_poly_scaled, y_train)  # continues where it left off
    y_val_predict = sgd_reg.predict(X_val_poly_scaled)
    val_error = mean_squared_error(y_val, y_val_predict)
    if val_error < minimum_val_error:
        minimum_val_error = val_error
        best_epoch = epoch
        best_model = clone(sgd_reg)
```

Note that with `warm_start=True`, when the `fit()` method is called it continues training where it left off, instead of restarting from scratch.

Logistic Regression

As we discussed in Chapter 1, some regression algorithms can be used for classification (and vice versa). *Logistic Regression* (also called *Logit Regression*) is commonly used to estimate the probability that an instance belongs to a particular class (e.g., what is the probability that this email is spam?). If the estimated probability is greater than 50%, then the model predicts that the instance belongs to that class (called the *positive class*, labeled "1"), and otherwise it predicts that it does not (i.e., it belongs to the *negative class*, labeled "0"). This makes it a binary classifier.

Estimating Probabilities

So how does Logistic Regression work? Just like a Linear Regression model, a Logistic Regression model computes a weighted sum of the input features (plus a bias term), but instead of outputting the result directly like the Linear Regression model does, it outputs the *logistic* of this result (see Equation 4-13).

Equation 4-13. Logistic Regression model estimated probability (vectorized form)

$$\hat{p} = h_\theta(\mathbf{x}) = \sigma(\mathbf{x}^\mathsf{T}\theta)$$

The logistic—noted $\sigma(\cdot)$—is a *sigmoid function* (i.e., S-shaped) that outputs a number between 0 and 1. It is defined as shown in Equation 4-14 and Figure 4-21.

Equation 4-14. Logistic function

$$\sigma(t) = \frac{1}{1 + \exp(-t)}$$

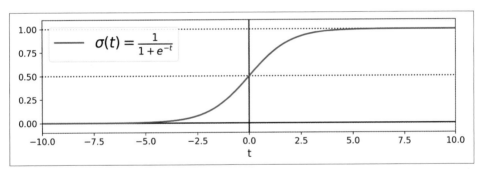

Figure 4-21. Logistic function

Once the Logistic Regression model has estimated the probability $\hat{p} = h_\theta(\mathbf{x})$ that an instance $\mathbf{x}$ belongs to the positive class, it can make its prediction $\hat{y}$ easily (see Equation 4-15).

Equation 4-15. Logistic Regression model prediction

$$\hat{y} = \begin{cases} 0 & \text{if } \hat{p} < 0.5 \\ 1 & \text{if } \hat{p} \geq 0.5 \end{cases}$$

Notice that $\sigma(t) < 0.5$ when $t < 0$, and $\sigma(t) \geq 0.5$ when $t \geq 0$, so a Logistic Regression model predicts 1 if $\mathbf{x}^\mathsf{T}\,\theta$ is positive and 0 if it is negative.

The score t is often called the *logit*. The name comes from the fact that the logit function, defined as $\text{logit}(p) = \log(p / (1 - p))$, is the inverse of the logistic function. Indeed, if you compute the logit of the estimated probability p, you will find that the result is t. The logit is also called the *log-odds*, since it is the log of the ratio between the estimated probability for the positive class and the estimated probability for the negative class.

Training and Cost Function

Now you know how a Logistic Regression model estimates probabilities and makes predictions. But how is it trained? The objective of training is to set the parameter vector $\boldsymbol{\theta}$ so that the model estimates high probabilities for positive instances ($y = 1$) and low probabilities for negative instances ($y = 0$). This idea is captured by the cost function shown in Equation 4-16 for a single training instance **x**.

Equation 4-16. Cost function of a single training instance

$$c(\boldsymbol{\theta}) = \begin{cases} -\log(\hat{p}) & \text{if } y = 1 \\ -\log(1 - \hat{p}) & \text{if } y = 0 \end{cases}$$

This cost function makes sense because $-\log(t)$ grows very large when t approaches 0, so the cost will be large if the model estimates a probability close to 0 for a positive instance, and it will also be very large if the model estimates a probability close to 1 for a negative instance. On the other hand, $-\log(t)$ is close to 0 when t is close to 1, so the cost will be close to 0 if the estimated probability is close to 0 for a negative instance or close to 1 for a positive instance, which is precisely what we want.

The cost function over the whole training set is the average cost over all training instances. It can be written in a single expression called the *log loss*, shown in Equation 4-17.

Equation 4-17. Logistic Regression cost function (log loss)

$$J(\boldsymbol{\theta}) = -\frac{1}{m}\sum_{i=1}^{m}\left[y^{(i)}\log\left(\hat{p}^{(i)}\right) + \left(1 - y^{(i)}\right)\log\left(1 - \hat{p}^{(i)}\right)\right]$$

The bad news is that there is no known closed-form equation to compute the value of $\boldsymbol{\theta}$ that minimizes this cost function (there is no equivalent of the Normal Equation). The good news is that this cost function is convex, so Gradient Descent (or any other optimization algorithm) is guaranteed to find the global minimum (if the learning rate is not too large and you wait long enough). The partial derivatives of the cost function with regard to the jth model parameter θ_j are given by Equation 4-18.

Equation 4-18. Logistic cost function partial derivatives

$$\frac{\partial}{\partial \theta_j} J(\mathbf{\theta}) = \frac{1}{m} \sum_{i=1}^{m} \left(\sigma\left(\mathbf{\theta}^{\mathsf{T}} \mathbf{x}^{(i)}\right) - y^{(i)} \right) x_j^{(i)}$$

This equation looks very much like Equation 4-5: for each instance it computes the prediction error and multiplies it by the jth feature value, and then it computes the average over all training instances. Once you have the gradient vector containing all the partial derivatives, you can use it in the Batch Gradient Descent algorithm. That's it: you now know how to train a Logistic Regression model. For Stochastic GD you would take one instance at a time, and for Mini-batch GD you would use a mini-batch at a time.

Decision Boundaries

Let's use the iris dataset to illustrate Logistic Regression. This is a famous dataset that contains the sepal and petal length and width of 150 iris flowers of three different species: *Iris setosa*, *Iris versicolor*, and *Iris virginica* (see Figure 4-22).

Figure 4-22. Flowers of three iris plant species[14]

---

14 Photos reproduced from the corresponding Wikipedia pages. *Iris virginica* photo by Frank Mayfield (Creative Commons BY-SA 2.0 (*https://creativecommons.org/licenses/by-sa/2.0/*)), *Iris versicolor* photo by D. Gordon E. Robertson (Creative Commons BY-SA 3.0 (*https://creativecommons.org/licenses/by-sa/3.0/*)), *Iris setosa* photo public domain.

Let's try to build a classifier to detect the *Iris virginica* type based only on the petal width feature. First let's load the data:

```
>>> from sklearn import datasets
>>> iris = datasets.load_iris()
>>> list(iris.keys())
['data', 'target', 'target_names', 'DESCR', 'feature_names', 'filename']
>>> X = iris["data"][:, 3:]  # petal width
>>> y = (iris["target"] == 2).astype(np.int)  # 1 if Iris virginica, else 0
```

Now let's train a Logistic Regression model:

```
from sklearn.linear_model import LogisticRegression

log_reg = LogisticRegression()
log_reg.fit(X, y)
```

Let's look at the model's estimated probabilities for flowers with petal widths varying from 0 cm to 3 cm (Figure 4-23):[15]

```
X_new = np.linspace(0, 3, 1000).reshape(-1, 1)
y_proba = log_reg.predict_proba(X_new)
plt.plot(X_new, y_proba[:, 1], "g-", label="Iris virginica")
plt.plot(X_new, y_proba[:, 0], "b--", label="Not Iris virginica")
# + more Matplotlib code to make the image look pretty
```

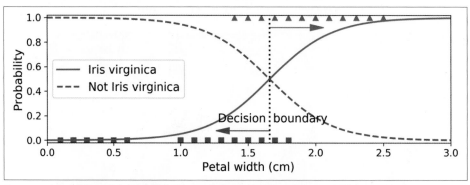

Figure 4-23. Estimated probabilities and decision boundary

The petal width of *Iris virginica* flowers (represented by triangles) ranges from 1.4 cm to 2.5 cm, while the other iris flowers (represented by squares) generally have a smaller petal width, ranging from 0.1 cm to 1.8 cm. Notice that there is a bit of overlap. Above about 2 cm the classifier is highly confident that the flower is an *Iris virginica* (it outputs a high probability for that class), while below 1 cm it is highly confident that it is not an *Iris virginica* (high probability for the "Not Iris virginica"

---

15 NumPy's reshape() function allows one dimension to be –1, which means "unspecified": the value is inferred from the length of the array and the remaining dimensions.

class). In between these extremes, the classifier is unsure. However, if you ask it to predict the class (using the `predict()` method rather than the `predict_proba()` method), it will return whichever class is the most likely. Therefore, there is a *decision boundary* at around 1.6 cm where both probabilities are equal to 50%: if the petal width is higher than 1.6 cm, the classifier will predict that the flower is an *Iris virginica*, and otherwise it will predict that it is not (even if it is not very confident):

```
>>> log_reg.predict([[1.7], [1.5]])
array([1, 0])
```

Figure 4-24 shows the same dataset, but this time displaying two features: petal width and length. Once trained, the Logistic Regression classifier can, based on these two features, estimate the probability that a new flower is an *Iris virginica*. The dashed line represents the points where the model estimates a 50% probability: this is the model's decision boundary. Note that it is a linear boundary.[16] Each parallel line represents the points where the model outputs a specific probability, from 15% (bottom left) to 90% (top right). All the flowers beyond the top-right line have an over 90% chance of being *Iris virginica*, according to the model.

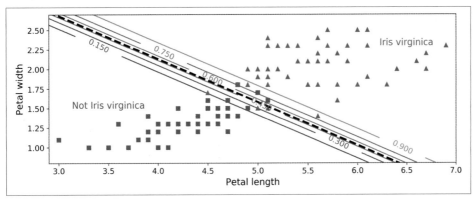

Figure 4-24. Linear decision boundary

Just like the other linear models, Logistic Regression models can be regularized using ℓ_1 or ℓ_2 penalties. Scikit-Learn actually adds an ℓ_2 penalty by default.

> The hyperparameter controlling the regularization strength of a Scikit-Learn `LogisticRegression` model is not `alpha` (as in other linear models), but its inverse: `C`. The higher the value of `C`, the *less* the model is regularized.

---

16 It is the the set of points **x** such that $\theta_0 + \theta_1 x_1 + \theta_2 x_2 = 0$, which defines a straight line.

Softmax Regression

The Logistic Regression model can be generalized to support multiple classes directly, without having to train and combine multiple binary classifiers (as discussed in Chapter 3). This is called *Softmax Regression*, or *Multinomial Logistic Regression*.

The idea is simple: when given an instance **x**, the Softmax Regression model first computes a score $s_k(\mathbf{x})$ for each class k, then estimates the probability of each class by applying the *softmax function* (also called the *normalized exponential*) to the scores. The equation to compute $s_k(\mathbf{x})$ should look familiar, as it is just like the equation for Linear Regression prediction (see Equation 4-19).

Equation 4-19. Softmax score for class k

$$s_k(\mathbf{x}) = \mathbf{x}^\mathsf{T}\boldsymbol{\theta}^{(k)}$$

Note that each class has its own dedicated parameter vector $\boldsymbol{\theta}^{(k)}$. All these vectors are typically stored as rows in a *parameter matrix* $\boldsymbol{\Theta}$.

Once you have computed the score of every class for the instance **x**, you can estimate the probability $\hat{p}_k$ that the instance belongs to class k by running the scores through the softmax function (Equation 4-20). The function computes the exponential of every score, then normalizes them (dividing by the sum of all the exponentials). The scores are generally called logits or log-odds (although they are actually unnormalized log-odds).

Equation 4-20. Softmax function

$$\hat{p}_k = \sigma(\mathbf{s}(\mathbf{x}))_k = \frac{\exp\left(s_k(\mathbf{x})\right)}{\sum_{j=1}^{K} \exp\left(s_j(\mathbf{x})\right)}$$

In this equation:

- K is the number of classes.
- $\mathbf{s}(\mathbf{x})$ is a vector containing the scores of each class for the instance **x**.
- $\sigma(\mathbf{s}(\mathbf{x}))_k$ is the estimated probability that the instance **x** belongs to class k, given the scores of each class for that instance.

Just like the Logistic Regression classifier, the Softmax Regression classifier predicts the class with the highest estimated probability (which is simply the class with the highest score), as shown in Equation 4-21.

Equation 4-21. Softmax Regression classifier prediction

$$\hat{y} = \underset{k}{\operatorname{argmax}} \ \sigma(\mathbf{s}(\mathbf{x}))_k = \underset{k}{\operatorname{argmax}} \ s_k(\mathbf{x}) = \underset{k}{\operatorname{argmax}} \left(\left(\boldsymbol{\theta}^{(k)}\right)^{\mathsf{T}} \mathbf{x} \right)$$

The *argmax* operator returns the value of a variable that maximizes a function. In this equation, it returns the value of k that maximizes the estimated probability $\sigma(\mathbf{s}(\mathbf{x}))_k$.

The Softmax Regression classifier predicts only one class at a time (i.e., it is multiclass, not multioutput), so it should be used only with mutually exclusive classes, such as different types of plants. You cannot use it to recognize multiple people in one picture.

Now that you know how the model estimates probabilities and makes predictions, let's take a look at training. The objective is to have a model that estimates a high probability for the target class (and consequently a low probability for the other classes). Minimizing the cost function shown in Equation 4-22, called the *cross entropy*, should lead to this objective because it penalizes the model when it estimates a low probability for a target class. Cross entropy is frequently used to measure how well a set of estimated class probabilities matches the target classes.

Equation 4-22. Cross entropy cost function

$$J(\boldsymbol{\Theta}) = -\frac{1}{m}\sum_{i=1}^{m}\sum_{k=1}^{K}y_k^{(i)}\log\left(\hat{p}_k^{(i)}\right)$$

In this equation:

- $y_k^{(i)}$ is the target probability that the ith instance belongs to class k. In general, it is either equal to 1 or 0, depending on whether the instance belongs to the class or not.

Notice that when there are just two classes ($K = 2$), this cost function is equivalent to the Logistic Regression's cost function (log loss; see Equation 4-17).

> # Cross Entropy
>
> Cross entropy originated from information theory. Suppose you want to efficiently transmit information about the weather every day. If there are eight options (sunny, rainy, etc.), you could encode each option using three bits because $2^3 = 8$. However, if you think it will be sunny almost every day, it would be much more efficient to code "sunny" on just one bit (0) and the other seven options on four bits (starting with a 1). Cross entropy measures the average number of bits you actually send per option. If your assumption about the weather is perfect, cross entropy will be equal to the entropy of the weather itself (i.e., its intrinsic unpredictability). But if your assumptions are wrong (e.g., if it rains often), cross entropy will be greater by an amount called the *Kullback–Leibler (KL) divergence*.
>
> The cross entropy between two probability distributions p and q is defined as $H(p,q) = -\Sigma_x p(x) \log q(x)$ (at least when the distributions are discrete). For more details, check out my video on the subject (*https://homl.info/xentropy*).

The gradient vector of this cost function with regard to $\theta^{(k)}$ is given by Equation 4-23.

Equation 4-23. Cross entropy gradient vector for class k

$$\nabla_{\theta^{(k)}} J(\Theta) = \frac{1}{m} \sum_{i=1}^{m} \left(\hat{p}_k^{(i)} - y_k^{(i)} \right) \mathbf{x}^{(i)}$$

Now you can compute the gradient vector for every class, then use Gradient Descent (or any other optimization algorithm) to find the parameter matrix Θ that minimizes the cost function.

Let's use Softmax Regression to classify the iris flowers into all three classes. Scikit-Learn's `LogisticRegression` uses one-versus-the-rest by default when you train it on more than two classes, but you can set the `multi_class` hyperparameter to `"multinomial"` to switch it to Softmax Regression. You must also specify a solver that supports Softmax Regression, such as the `"lbfgs"` solver (see Scikit-Learn's documentation for more details). It also applies ℓ_2 regularization by default, which you can control using the hyperparameter `C`:

```
X = iris["data"][:, (2, 3)]  # petal length, petal width
y = iris["target"]

softmax_reg = LogisticRegression(multi_class="multinomial",solver="lbfgs", C=10)
softmax_reg.fit(X, y)
```

So the next time you find an iris with petals that are 5 cm long and 2 cm wide, you can ask your model to tell you what type of iris it is, and it will answer *Iris virginica* (class 2) with 94.2% probability (or *Iris versicolor* with 5.8% probability):

```
>>> softmax_reg.predict([[5, 2]])
array([2])
>>> softmax_reg.predict_proba([[5, 2]])
array([[6.38014896e-07, 5.74929995e-02, 9.42506362e-01]])
```

Figure 4-25 shows the resulting decision boundaries, represented by the background colors. Notice that the decision boundaries between any two classes are linear. The figure also shows the probabilities for the *Iris versicolor* class, represented by the curved lines (e.g., the line labeled with 0.450 represents the 45% probability boundary). Notice that the model can predict a class that has an estimated probability below 50%. For example, at the point where all decision boundaries meet, all classes have an equal estimated probability of 33%.

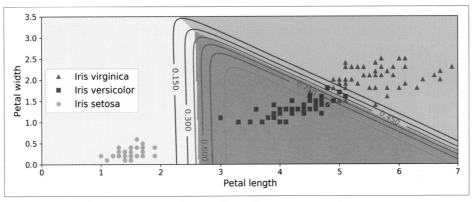

Figure 4-25. Softmax Regression decision boundaries

Exercises

1. Which Linear Regression training algorithm can you use if you have a training set with millions of features?

2. Suppose the features in your training set have very different scales. Which algorithms might suffer from this, and how? What can you do about it?

3. Can Gradient Descent get stuck in a local minimum when training a Logistic Regression model?

4. Do all Gradient Descent algorithms lead to the same model, provided you let them run long enough?

5. Suppose you use Batch Gradient Descent and you plot the validation error at every epoch. If you notice that the validation error consistently goes up, what is likely going on? How can you fix this?

6. Is it a good idea to stop Mini-batch Gradient Descent immediately when the validation error goes up?

7. Which Gradient Descent algorithm (among those we discussed) will reach the vicinity of the optimal solution the fastest? Which will actually converge? How can you make the others converge as well?

8. Suppose you are using Polynomial Regression. You plot the learning curves and you notice that there is a large gap between the training error and the validation error. What is happening? What are three ways to solve this?

9. Suppose you are using Ridge Regression and you notice that the training error and the validation error are almost equal and fairly high. Would you say that the model suffers from high bias or high variance? Should you increase the regularization hyperparameter α or reduce it?

10. Why would you want to use:

    a. Ridge Regression instead of plain Linear Regression (i.e., without any regularization)?

    b. Lasso instead of Ridge Regression?

    c. Elastic Net instead of Lasso?

11. Suppose you want to classify pictures as outdoor/indoor and daytime/nighttime. Should you implement two Logistic Regression classifiers or one Softmax Regression classifier?

12. Implement Batch Gradient Descent with early stopping for Softmax Regression (without using Scikit-Learn).

Solutions to these exercises are available in Appendix A.

## CHAPTER 5

Support Vector Machines

A *Support Vector Machine* (SVM) is a powerful and versatile Machine Learning model, capable of performing linear or nonlinear classification, regression, and even outlier detection. It is one of the most popular models in Machine Learning, and anyone interested in Machine Learning should have it in their toolbox. SVMs are particularly well suited for classification of complex small- or medium-sized datasets.

This chapter will explain the core concepts of SVMs, how to use them, and how they work.

Linear SVM Classification

The fundamental idea behind SVMs is best explained with some pictures. Figure 5-1 shows part of the iris dataset that was introduced at the end of Chapter 4. The two classes can clearly be separated easily with a straight line (they are *linearly separable*). The left plot shows the decision boundaries of three possible linear classifiers. The model whose decision boundary is represented by the dashed line is so bad that it does not even separate the classes properly. The other two models work perfectly on this training set, but their decision boundaries come so close to the instances that these models will probably not perform as well on new instances. In contrast, the solid line in the plot on the right represents the decision boundary of an SVM classifier; this line not only separates the two classes but also stays as far away from the closest training instances as possible. You can think of an SVM classifier as fitting the widest possible street (represented by the parallel dashed lines) between the classes. This is called *large margin classification*.

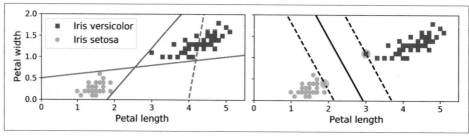

Figure 5-1. Large margin classification

Notice that adding more training instances "off the street" will not affect the decision boundary at all: it is fully determined (or "supported") by the instances located on the edge of the street. These instances are called the *support vectors* (they are circled in Figure 5-1).

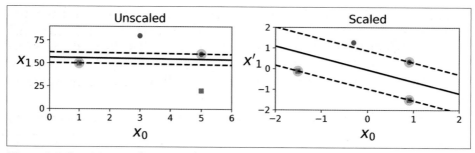

Figure 5-2. Sensitivity to feature scales

 SVMs are sensitive to the feature scales, as you can see in Figure 5-2: in the left plot, the vertical scale is much larger than the horizontal scale, so the widest possible street is close to horizontal. After feature scaling (e.g., using Scikit-Learn's `StandardScaler`), the decision boundary in the right plot looks much better.

Soft Margin Classification

If we strictly impose that all instances must be off the street and on the right side, this is called *hard margin classification*. There are two main issues with hard margin classification. First, it only works if the data is linearly separable. Second, it is sensitive to outliers. Figure 5-3 shows the iris dataset with just one additional outlier: on the left, it is impossible to find a hard margin; on the right, the decision boundary ends up very different from the one we saw in Figure 5-1 without the outlier, and it will probably not generalize as well.

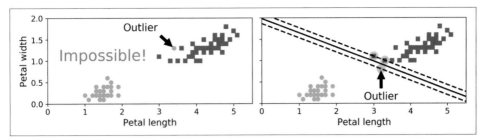

Figure 5-3. Hard margin sensitivity to outliers

To avoid these issues, use a more flexible model. The objective is to find a good balance between keeping the street as large as possible and limiting the *margin violations* (i.e., instances that end up in the middle of the street or even on the wrong side). This is called *soft margin classification*.

When creating an SVM model using Scikit-Learn, we can specify a number of hyperparameters. C is one of those hyperparameters. If we set it to a low value, then we end up with the model on the left of Figure 5-4. With a high value, we get the model on the right. Margin violations are bad. It's usually better to have few of them. However, in this case the model on the left has a lot of margin violations but will probably generalize better.

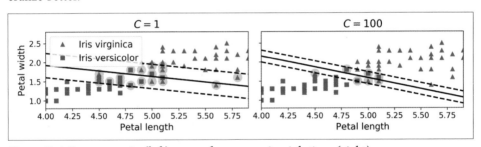

Figure 5-4. Large margin (left) versus fewer margin violations (right)

If your SVM model is overfitting, you can try regularizing it by reducing C.

The following Scikit-Learn code loads the iris dataset, scales the features, and then trains a linear SVM model (using the LinearSVC class with C=1 and the *hinge loss* function, described shortly) to detect *Iris virginica* flowers:

```
import numpy as np
from sklearn import datasets
from sklearn.pipeline import Pipeline
```

```
from sklearn.preprocessing import StandardScaler
from sklearn.svm import LinearSVC

iris = datasets.load_iris()
X = iris["data"][:, (2, 3)]  # petal length, petal width
y = (iris["target"] == 2).astype(np.float64)  # Iris virginica

svm_clf = Pipeline([
        ("scaler", StandardScaler()),
        ("linear_svc", LinearSVC(C=1, loss="hinge")),
    ])

svm_clf.fit(X, y)
```

The resulting model is represented on the left in Figure 5-4.

Then, as usual, you can use the model to make predictions:

```
>>> svm_clf.predict([[5.5, 1.7]])
array([1.])
```

> Unlike Logistic Regression classifiers, SVM classifiers do not out-
> put probabilities for each class.

Instead of using the LinearSVC class, we could use the SVC class with a linear kernel.
When creating the SVC model, we would write SVC(kernel="linear", C=1). Or we
could use the SGDClassifier class, with SGDClassifier(loss="hinge", alpha=1/
(m*C)). This applies regular Stochastic Gradient Descent (see Chapter 4) to train a
linear SVM classifier. It does not converge as fast as the LinearSVC class, but it can be
useful to handle online classification tasks or huge datasets that do not fit in memory
(out-of-core training).

> The LinearSVC class regularizes the bias term, so you should center
> the training set first by subtracting its mean. This is automatic if
> you scale the data using the StandardScaler. Also make sure you
> set the loss hyperparameter to "hinge", as it is not the default
> value. Finally, for better performance, you should set the dual
> hyperparameter to False, unless there are more features than
> training instances (we will discuss duality later in the chapter).

Nonlinear SVM Classification

Although linear SVM classifiers are efficient and work surprisingly well in many cases, many datasets are not even close to being linearly separable. One approach to handling nonlinear datasets is to add more features, such as polynomial features (as you did in Chapter 4); in some cases this can result in a linearly separable dataset. Consider the left plot in Figure 5-5: it represents a simple dataset with just one feature, x_1. This dataset is not linearly separable, as you can see. But if you add a second feature $x_2 = (x_1)^2$, the resulting 2D dataset is perfectly linearly separable.

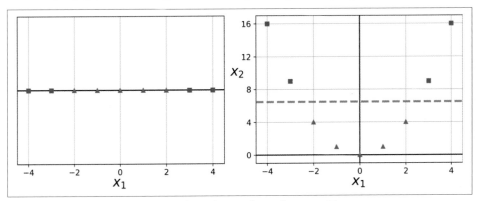

Figure 5-5. Adding features to make a dataset linearly separable

To implement this idea using Scikit-Learn, create a `Pipeline` containing a `Polyno mialFeatures` transformer (discussed in "Polynomial Regression" on page 128), followed by a `StandardScaler` and a `LinearSVC`. Let's test this on the moons dataset: this is a toy dataset for binary classification in which the data points are shaped as two interleaving half circles (see Figure 5-6). You can generate this dataset using the `make_moons()` function:

```
from sklearn.datasets import make_moons
from sklearn.pipeline import Pipeline
from sklearn.preprocessing import PolynomialFeatures

X, y = make_moons(n_samples=100, noise=0.15)
polynomial_svm_clf = Pipeline([
        ("poly_features", PolynomialFeatures(degree=3)),
        ("scaler", StandardScaler()),
        ("svm_clf", LinearSVC(C=10, loss="hinge"))
    ])

polynomial_svm_clf.fit(X, y)
```

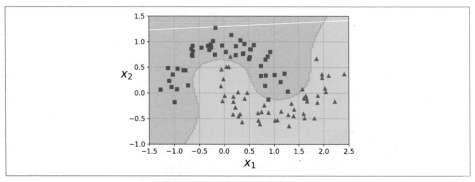

Figure 5-6. Linear SVM classifier using polynomial features

Polynomial Kernel

Adding polynomial features is simple to implement and can work great with all sorts of Machine Learning algorithms (not just SVMs). That said, at a low polynomial degree, this method cannot deal with very complex datasets, and with a high polynomial degree it creates a huge number of features, making the model too slow.

Fortunately, when using SVMs you can apply an almost miraculous mathematical technique called the *kernel trick* (explained in a moment). The kernel trick makes it possible to get the same result as if you had added many polynomial features, even with very high-degree polynomials, without actually having to add them. So there is no combinatorial explosion of the number of features because you don't actually add any features. This trick is implemented by the SVC class. Let's test it on the moons dataset:

```
from sklearn.svm import SVC
poly_kernel_svm_clf = Pipeline([
        ("scaler", StandardScaler()),
        ("svm_clf", SVC(kernel="poly", degree=3, coef0=1, C=5))
    ])
poly_kernel_svm_clf.fit(X, y)
```

This code trains an SVM classifier using a third-degree polynomial kernel. It is represented on the left in Figure 5-7. On the right is another SVM classifier using a 10th-degree polynomial kernel. Obviously, if your model is overfitting, you might want to reduce the polynomial degree. Conversely, if it is underfitting, you can try increasing it. The hyperparameter coef0 controls how much the model is influenced by high-degree polynomials versus low-degree polynomials.

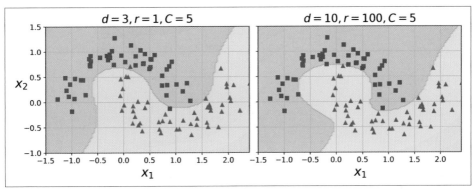

Figure 5-7. SVM classifiers with a polynomial kernel

A common approach to finding the right hyperparameter values is to use grid search (see Chapter 2). It is often faster to first do a very coarse grid search, then a finer grid search around the best values found. Having a good sense of what each hyperparameter actually does can also help you search in the right part of the hyperparameter space.

Similarity Features

Another technique to tackle nonlinear problems is to add features computed using a *similarity function*, which measures how much each instance resembles a particular *landmark*. For example, let's take the 1D dataset discussed earlier and add two landmarks to it at $x_1 = -2$ and $x_1 = 1$ (see the left plot in Figure 5-8). Next, let's define the similarity function to be the Gaussian *Radial Basis Function* (RBF) with $\gamma = 0.3$ (see Equation 5-1).

Equation 5-1. Gaussian RBF

$$\phi_\gamma(\mathbf{x}, \ell) = \exp\left(-\gamma \| \mathbf{x} - \ell \|^2\right)$$

This is a bell-shaped function varying from 0 (very far away from the landmark) to 1 (at the landmark). Now we are ready to compute the new features. For example, let's look at the instance $x_1 = -1$: it is located at a distance of 1 from the first landmark and 2 from the second landmark. Therefore its new features are $x_2 = \exp(-0.3 \times 1^2) \approx 0.74$ and $x_3 = \exp(-0.3 \times 2^2) \approx 0.30$. The plot on the right in Figure 5-8 shows the transformed dataset (dropping the original features). As you can see, it is now linearly separable.

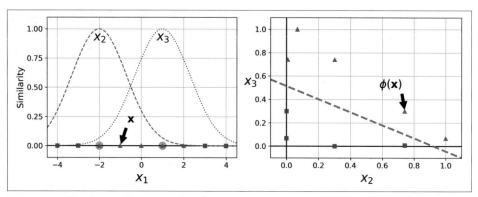

Figure 5-8. Similarity features using the Gaussian RBF

You may wonder how to select the landmarks. The simplest approach is to create a landmark at the location of each and every instance in the dataset. Doing that creates many dimensions and thus increases the chances that the transformed training set will be linearly separable. The downside is that a training set with *m* instances and *n* features gets transformed into a training set with *m* instances and *m* features (assuming you drop the original features). If your training set is very large, you end up with an equally large number of features.

Gaussian RBF Kernel

Just like the polynomial features method, the similarity features method can be useful with any Machine Learning algorithm, but it may be computationally expensive to compute all the additional features, especially on large training sets. Once again the kernel trick does its SVM magic, making it possible to obtain a similar result as if you had added many similarity features. Let's try the SVC class with the Gaussian RBF kernel:

```
rbf_kernel_svm_clf = Pipeline([
        ("scaler", StandardScaler()),
        ("svm_clf", SVC(kernel="rbf", gamma=5, C=0.001))
    ])
rbf_kernel_svm_clf.fit(X, y)
```

This model is represented at the bottom left in Figure 5-9. The other plots show models trained with different values of hyperparameters gamma (γ) and C. Increasing gamma makes the bell-shaped curve narrower (see the lefthand plots in Figure 5-8). As a result, each instance's range of influence is smaller: the decision boundary ends up being more irregular, wiggling around individual instances. Conversely, a small gamma value makes the bell-shaped curve wider: instances have a larger range of influence, and the decision boundary ends up smoother. So γ acts like a regularization hyperparameter: if your model is overfitting, you should reduce it; if it is underfitting, you should increase it (similar to the C hyperparameter).

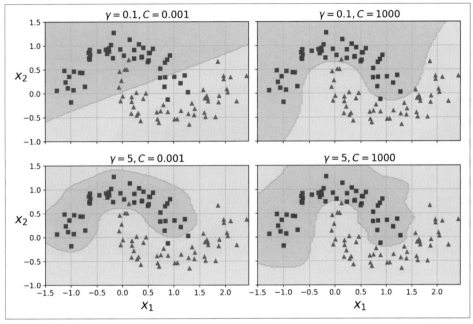

Figure 5-9. SVM classifiers using an RBF kernel

Other kernels exist but are used much more rarely. Some kernels are specialized for specific data structures. *String kernels* are sometimes used when classifying text documents or DNA sequences (e.g., using the *string subsequence kernel* or kernels based on the *Levenshtein distance*).

With so many kernels to choose from, how can you decide which one to use? As a rule of thumb, you should always try the linear kernel first (remember that LinearSVC is much faster than SVC(kernel="linear")), especially if the training set is very large or if it has plenty of features. If the training set is not too large, you should also try the Gaussian RBF kernel; it works well in most cases. Then if you have spare time and computing power, you can experiment with a few other kernels, using cross-validation and grid search. You'd want to experiment like that especially if there are kernels specialized for your training set's data structure.

Computational Complexity

The LinearSVC class is based on the liblinear library, which implements an optimized algorithm (*https://homl.info/13*) for linear SVMs.[1] It does not support the kernel trick, but it scales almost linearly with the number of training instances and the number of features. Its training time complexity is roughly $O(m \times n)$.

The algorithm takes longer if you require very high precision. This is controlled by the tolerance hyperparameter ϵ (called tol in Scikit-Learn). In most classification tasks, the default tolerance is fine.

The SVC class is based on the libsvm library, which implements an algorithm (*https://homl.info/14*) that supports the kernel trick.[2] The training time complexity is usually between $O(m^2 \times n)$ and $O(m^3 \times n)$. Unfortunately, this means that it gets dreadfully slow when the number of training instances gets large (e.g., hundreds of thousands of instances). This algorithm is perfect for complex small or medium-sized training sets. It scales well with the number of features, especially with *sparse features* (i.e., when each instance has few nonzero features). In this case, the algorithm scales roughly with the average number of nonzero features per instance. Table 5-1 compares Scikit-Learn's SVM classification classes.

Table 5-1. Comparison of Scikit-Learn classes for SVM classification

Class	Time complexity	Out-of-core support	Scaling required	Kernel trick
LinearSVC	$O(m \times n)$	No	Yes	No
SGDClassifier	$O(m \times n)$	Yes	Yes	No
SVC	$O(m^2 \times n)$ to $O(m^3 \times n)$	No	Yes	Yes

SVM Regression

As mentioned earlier, the SVM algorithm is versatile: not only does it support linear and nonlinear classification, but it also supports linear and nonlinear regression. To use SVMs for regression instead of classification, the trick is to reverse the objective: instead of trying to fit the largest possible street between two classes while limiting margin violations, SVM Regression tries to fit as many instances as possible *on* the street while limiting margin violations (i.e., instances *off* the street). The width of the street is controlled by a hyperparameter, ϵ. Figure 5-10 shows two linear SVM

---

1 Chih-Jen Lin et al., "A Dual Coordinate Descent Method for Large-Scale Linear SVM," *Proceedings of the 25th International Conference on Machine Learning* (2008): 408–415.

2 John Platt, "Sequential Minimal Optimization: A Fast Algorithm for Training Support Vector Machines" (Microsoft Research technical report, April 21, 1998), *https://www.microsoft.com/en-us/research/wp-content/uploads/2016/02/tr-98-14.pdf*.

Regression models trained on some random linear data, one with a large margin (ε = 1.5) and the other with a small margin (ε = 0.5).

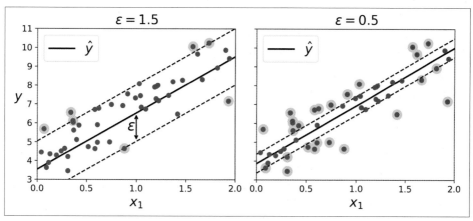

Figure 5-10. SVM Regression

Adding more training instances within the margin does not affect the model's predictions; thus, the model is said to be *ε-insensitive*.

You can use Scikit-Learn's `LinearSVR` class to perform linear SVM Regression. The following code produces the model represented on the left in Figure 5-10 (the training data should be scaled and centered first):

```
from sklearn.svm import LinearSVR

svm_reg = LinearSVR(epsilon=1.5)
svm_reg.fit(X, y)
```

To tackle nonlinear regression tasks, you can use a kernelized SVM model. Figure 5-11 shows SVM Regression on a random quadratic training set, using a second-degree polynomial kernel. There is little regularization in the left plot (i.e., a large C value), and much more regularization in the right plot (i.e., a small C value).

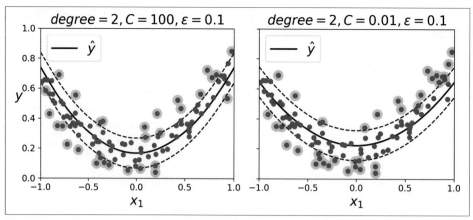

Figure 5-11. SVM Regression using a second-degree polynomial kernel

The following code uses Scikit-Learn's `SVR` class (which supports the kernel trick) to produce the model represented on the left in Figure 5-11:

```
from sklearn.svm import SVR

svm_poly_reg = SVR(kernel="poly", degree=2, C=100, epsilon=0.1)
svm_poly_reg.fit(X, y)
```

The `SVR` class is the regression equivalent of the `SVC` class, and the `LinearSVR` class is the regression equivalent of the `LinearSVC` class. The `LinearSVR` class scales linearly with the size of the training set (just like the `LinearSVC` class), while the `SVR` class gets much too slow when the training set grows large (just like the `SVC` class).

> SVMs can also be used for outlier detection; see Scikit-Learn's documentation for more details.

Under the Hood

This section explains how SVMs make predictions and how their training algorithms work, starting with linear SVM classifiers. If you are just getting started with Machine Learning, you can safely skip it and go straight to the exercises at the end of this chapter, and come back later when you want to get a deeper understanding of SVMs.

First, a word about notations. In Chapter 4 we used the convention of putting all the model parameters in one vector $\boldsymbol{\theta}$, including the bias term θ_0 and the input feature weights θ_1 to θ_n, and adding a bias input $x_0 = 1$ to all instances. In this chapter we will use a convention that is more convenient (and more common) when dealing with

SVMs: the bias term will be called *b*, and the feature weights vector will be called **w**. No bias feature will be added to the input feature vectors.

Decision Function and Predictions

The linear SVM classifier model predicts the class of a new instance **x** by simply computing the decision function $\mathbf{w}^\intercal \mathbf{x} + b = w_1 x_1 + \cdots + w_n x_n + b$. If the result is positive, the predicted class $\hat{y}$ is the positive class (1), and otherwise it is the negative class (0); see Equation 5-2.

Equation 5-2. Linear SVM classifier prediction

$$\hat{y} = \begin{cases} 0 & \text{if } \mathbf{w}^\intercal \mathbf{x} + b < 0, \\ 1 & \text{if } \mathbf{w}^\intercal \mathbf{x} + b \geq 0 \end{cases}$$

Figure 5-12 shows the decision function that corresponds to the model in the left in Figure 5-4: it is a 2D plane because this dataset has two features (petal width and petal length). The decision boundary is the set of points where the decision function is equal to 0: it is the intersection of two planes, which is a straight line (represented by the thick solid line).[3]

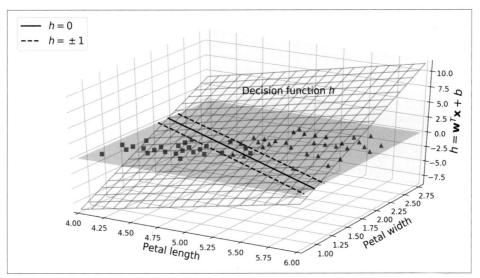

Figure 5-12. Decision function for the iris dataset

---

3 More generally, when there are *n* features, the decision function is an *n*-dimensional *hyperplane*, and the decision boundary is an (*n* − 1)-dimensional hyperplane.

The dashed lines represent the points where the decision function is equal to 1 or –1: they are parallel and at equal distance to the decision boundary, and they form a margin around it. Training a linear SVM classifier means finding the values of **w** and b that make this margin as wide as possible while avoiding margin violations (hard margin) or limiting them (soft margin).

Training Objective

Consider the slope of the decision function: it is equal to the norm of the weight vector, $\| \mathbf{w} \|$. If we divide this slope by 2, the points where the decision function is equal to ±1 are going to be twice as far away from the decision boundary. In other words, dividing the slope by 2 will multiply the margin by 2. This may be easier to visualize in 2D, as shown in Figure 5-13. The smaller the weight vector **w**, the larger the margin.

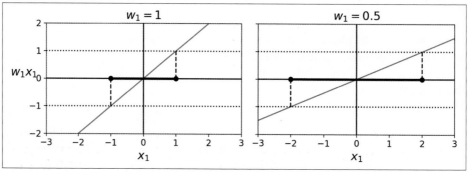

Figure 5-13. A smaller weight vector results in a larger margin

So we want to minimize $\| \mathbf{w} \|$ to get a large margin. If we also want to avoid any margin violations (hard margin), then we need the decision function to be greater than 1 for all positive training instances and lower than –1 for negative training instances. If we define $t^{(i)} = -1$ for negative instances (if $y^{(i)} = 0$) and $t^{(i)} = 1$ for positive instances (if $y^{(i)} = 1$), then we can express this constraint as $t^{(i)}(\mathbf{w}^\top \mathbf{x}^{(i)} + b) \geq 1$ for all instances.

We can therefore express the hard margin linear SVM classifier objective as the constrained optimization problem in Equation 5-3.

Equation 5-3. Hard margin linear SVM classifier objective

$$\underset{\mathbf{w}, b}{\text{minimize}} \quad \frac{1}{2}\mathbf{w}^\top\mathbf{w}$$

$$\text{subject to} \quad t^{(i)}\left(\mathbf{w}^\top\mathbf{x}^{(i)} + b\right) \geq 1 \quad \text{for } i = 1, 2, \cdots, m$$

We are minimizing ½ $\mathbf{w}^\mathsf{T}$ $\mathbf{w}$, which is equal to ½$\| \mathbf{w} \|^2$, rather than minimizing $\| \mathbf{w} \|$. Indeed, ½$\| \mathbf{w} \|^2$ has a nice, simple derivative (it is just $\mathbf{w}$), while $\| \mathbf{w} \|$ is not differentiable at $\mathbf{w} = 0$. Optimization algorithms work much better on differentiable functions.

To get the soft margin objective, we need to introduce a *slack variable* $\zeta^{(i)} \geq 0$ for each instance:[4] $\zeta^{(i)}$ measures how much the ith instance is allowed to violate the margin. We now have two conflicting objectives: make the slack variables as small as possible to reduce the margin violations, and make ½ $\mathbf{w}^\mathsf{T}$ $\mathbf{w}$ as small as possible to increase the margin. This is where the C hyperparameter comes in: it allows us to define the trade-off between these two objectives. This gives us the constrained optimization problem in Equation 5-4.

Equation 5-4. Soft margin linear SVM classifier objective

$$\underset{\mathbf{w}, b, \zeta}{\text{minimize}} \quad \frac{1}{2}\mathbf{w}^\mathsf{T}\mathbf{w} + C \sum_{i=1}^{m} \zeta^{(i)}$$

$$\text{subject to} \quad t^{(i)}\left(\mathbf{w}^\mathsf{T}\mathbf{x}^{(i)} + b\right) \geq 1 - \zeta^{(i)} \quad \text{and} \quad \zeta^{(i)} \geq 0 \quad \text{for } i = 1, 2, \cdots, m$$

Quadratic Programming

The hard margin and soft margin problems are both convex quadratic optimization problems with linear constraints. Such problems are known as *Quadratic Programming* (QP) problems. Many off-the-shelf solvers are available to solve QP problems by using a variety of techniques that are outside the scope of this book.[5]

---

4 Zeta (ζ) is the sixth letter of the Greek alphabet.

5 To learn more about Quadratic Programming, you can start by reading Stephen Boyd and Lieven Vandenberghe's book *Convex Optimization* (*https://homl.info/15*) (Cambridge University Press, 2004) or watch Richard Brown's series of video lectures (*https://homl.info/16*).

The general problem formulation is given by Equation 5-5.

Equation 5-5. Quadratic Programming problem

$$\text{Minimize}_{\mathbf{p}} \quad \frac{1}{2}\mathbf{p}^\top\mathbf{H}\mathbf{p} \quad + \quad \mathbf{f}^\top\mathbf{p}$$

subject to $\quad \mathbf{A}\mathbf{p} \leq \mathbf{b}$

where $\begin{cases} \mathbf{p} & \text{is an } n_p\text{-dimensional vector } (n_p = \text{number of parameters}), \\ \mathbf{H} & \text{is an } n_p \times n_p \text{ matrix}, \\ \mathbf{f} & \text{is an } n_p\text{-dimensional vector}, \\ \mathbf{A} & \text{is an } n_c \times n_p \text{ matrix } (n_c = \text{number of constraints}), \\ \mathbf{b} & \text{is an } n_c\text{-dimensional vector}. \end{cases}$

Note that the expression $\mathbf{A}\,\mathbf{p} \leq \mathbf{b}$ defines n_c constraints: $\mathbf{p}^\top\,\mathbf{a}^{(i)} \leq b^{(i)}$ for $i = 1, 2, \cdots, n_c$, where $\mathbf{a}^{(i)}$ is the vector containing the elements of the ith row of $\mathbf{A}$ and $b^{(i)}$ is the ith element of $\mathbf{b}$.

You can easily verify that if you set the QP parameters in the following way, you get the hard margin linear SVM classifier objective:

- $n_p = n + 1$, where n is the number of features (the +1 is for the bias term).
- $n_c = m$, where m is the number of training instances.
- $\mathbf{H}$ is the $n_p \times n_p$ identity matrix, except with a zero in the top-left cell (to ignore the bias term).
- $\mathbf{f} = 0$, an n_p-dimensional vector full of 0s.
- $\mathbf{b} = -1$, an n_c-dimensional vector full of –1s.
- $\mathbf{a}^{(i)} = -t^{(i)}\,\dot{\mathbf{x}}^{(i)}$, where $\dot{\mathbf{x}}^{(i)}$ is equal to $\mathbf{x}^{(i)}$ with an extra bias feature $\dot{x}_0 = 1$.

One way to train a hard margin linear SVM classifier is to use an off-the-shelf QP solver and pass it the preceding parameters. The resulting vector $\mathbf{p}$ will contain the bias term $b = p_0$ and the feature weights $w_i = p_i$ for $i = 1, 2, \cdots, n$. Similarly, you can use a QP solver to solve the soft margin problem (see the exercises at the end of the chapter).

To use the kernel trick, we are going to look at a different constrained optimization problem.

The Dual Problem

Given a constrained optimization problem, known as the *primal problem*, it is possible to express a different but closely related problem, called its *dual problem*. The solution to the dual problem typically gives a lower bound to the solution of the primal problem, but under some conditions it can have the same solution as the primal problem. Luckily, the SVM problem happens to meet these conditions,[6] so you can choose to solve the primal problem or the dual problem; both will have the same solution. Equation 5-6 shows the dual form of the linear SVM objective (if you are interested in knowing how to derive the dual problem from the primal problem, see Appendix C).

Equation 5-6. Dual form of the linear SVM objective

$$\underset{\alpha}{\text{minimize}} \frac{1}{2} \sum_{i=1}^{m} \sum_{j=1}^{m} \alpha^{(i)} \alpha^{(j)} t^{(i)} t^{(j)} \mathbf{x}^{(i)\top} \mathbf{x}^{(j)} \quad - \quad \sum_{i=1}^{m} \alpha^{(i)}$$

$$\text{subject to} \quad \alpha^{(i)} \geq 0 \quad \text{for } i = 1, 2, \cdots, m$$

Once you find the vector $\hat{\alpha}$ that minimizes this equation (using a QP solver), use Equation 5-7 to compute $\hat{\mathbf{w}}$ and $\hat{b}$ that minimize the primal problem.

Equation 5-7. From the dual solution to the primal solution

$$\hat{\mathbf{w}} = \sum_{i=1}^{m} \hat{\alpha}^{(i)} t^{(i)} \mathbf{x}^{(i)}$$

$$\hat{b} = \frac{1}{n_s} \sum_{\substack{i=1 \\ \hat{\alpha}^{(i)} > 0}}^{m} \left(t^{(i)} - \hat{\mathbf{w}}^{\top} \mathbf{x}^{(i)} \right)$$

The dual problem is faster to solve than the primal one when the number of training instances is smaller than the number of features. More importantly, the dual problem makes the kernel trick possible, while the primal does not. So what is this kernel trick, anyway?

Kernelized SVMs

Suppose you want to apply a second-degree polynomial transformation to a two-dimensional training set (such as the moons training set), then train a linear SVM classifier on the transformed training set. Equation 5-8 shows the second-degree polynomial mapping function ϕ that you want to apply.

---

6 The objective function is convex, and the inequality constraints are continuously differentiable and convex functions.

Equation 5-8. Second-degree polynomial mapping

$$\phi(\mathbf{x}) = \phi\left(\begin{pmatrix} x_1 \\ x_2 \end{pmatrix}\right) = \begin{pmatrix} x_1^2 \\ \sqrt{2}\,x_1 x_2 \\ x_2^2 \end{pmatrix}$$

Notice that the transformed vector is 3D instead of 2D. Now let's look at what happens to a couple of 2D vectors, **a** and **b**, if we apply this second-degree polynomial mapping and then compute the dot product[7] of the transformed vectors (See Equation 5-9).

Equation 5-9. Kernel trick for a second-degree polynomial mapping

$$\phi(\mathbf{a})^\mathsf{T}\phi(\mathbf{b}) = \begin{pmatrix} a_1^2 \\ \sqrt{2}\,a_1 a_2 \\ a_2^2 \end{pmatrix}^\mathsf{T} \begin{pmatrix} b_1^2 \\ \sqrt{2}\,b_1 b_2 \\ b_2^2 \end{pmatrix} = a_1^2 b_1^2 + 2a_1 b_1 a_2 b_2 + a_2^2 b_2^2$$

$$= \left(a_1 b_1 + a_2 b_2\right)^2 = \left(\begin{pmatrix} a_1 \\ a_2 \end{pmatrix}^\mathsf{T} \begin{pmatrix} b_1 \\ b_2 \end{pmatrix}\right)^2 = \left(\mathbf{a}^\mathsf{T}\mathbf{b}\right)^2$$

How about that? The dot product of the transformed vectors is equal to the square of the dot product of the original vectors: $\phi(\mathbf{a})^\mathsf{T}\,\phi(\mathbf{b}) = (\mathbf{a}^\mathsf{T}\,\mathbf{b})^2$.

Here is the key insight: if you apply the transformation ϕ to all training instances, then the dual problem (see Equation 5-6) will contain the dot product $\phi(\mathbf{x}^{(i)})^\mathsf{T}\,\phi(\mathbf{x}^{(j)})$. But if ϕ is the second-degree polynomial transformation defined in Equation 5-8, then you can replace this dot product of transformed vectors simply by $\left(\mathbf{x}^{(i)\mathsf{T}}\mathbf{x}^{(j)}\right)^2$. So, you don't need to transform the training instances at all; just replace the dot product by its square in Equation 5-6. The result will be strictly the same as if you had gone through the trouble of transforming the training set then fitting a linear SVM algorithm, but this trick makes the whole process much more computationally efficient.

The function $K(\mathbf{a}, \mathbf{b}) = (\mathbf{a}^\mathsf{T}\,\mathbf{b})^2$ is a second-degree polynomial kernel. In Machine Learning, a *kernel* is a function capable of computing the dot product $\phi(\mathbf{a})^\mathsf{T}\,\phi(\mathbf{b})$,

---

7 As explained in Chapter 4, the dot product of two vectors **a** and **b** is normally noted **a** · **b**. However, in Machine Learning, vectors are frequently represented as column vectors (i.e., single-column matrices), so the dot product is achieved by computing $\mathbf{a}^\mathsf{T}\mathbf{b}$. To remain consistent with the rest of the book, we will use this notation here, ignoring the fact that this technically results in a single-cell matrix rather than a scalar value.

based only on the original vectors **a** and **b**, without having to compute (or even to know about) the transformation ϕ. Equation 5-10 lists some of the most commonly used kernels.

Equation 5-10. Common kernels

Linear: $\quad K(\mathbf{a}, \mathbf{b}) = \mathbf{a}^\mathsf{T}\mathbf{b}$

Polynomial: $\quad K(\mathbf{a}, \mathbf{b}) = \left(\gamma \mathbf{a}^\mathsf{T}\mathbf{b} + r\right)^d$

Gaussian RBF: $\quad K(\mathbf{a}, \mathbf{b}) = \exp\left(-\gamma \|\mathbf{a} - \mathbf{b}\|^2\right)$

Sigmoid: $\quad K(\mathbf{a}, \mathbf{b}) = \tanh\left(\gamma \mathbf{a}^\mathsf{T}\mathbf{b} + r\right)$

Mercer's Theorem

According to *Mercer's theorem*, if a function $K(\mathbf{a}, \mathbf{b})$ respects a few mathematical conditions called *Mercer's conditions* (e.g., K must be continuous and symmetric in its arguments so that $K(\mathbf{a}, \mathbf{b}) = K(\mathbf{b}, \mathbf{a})$, etc.), then there exists a function ϕ that maps **a** and **b** into another space (possibly with much higher dimensions) such that $K(\mathbf{a}, \mathbf{b}) = \phi(\mathbf{a})^\mathsf{T} \phi(\mathbf{b})$. You can use K as a kernel because you know ϕ exists, even if you don't know what ϕ is. In the case of the Gaussian RBF kernel, it can be shown that ϕ maps each training instance to an infinite-dimensional space, so it's a good thing you don't need to actually perform the mapping!

Note that some frequently used kernels (such as the sigmoid kernel) don't respect all of Mercer's conditions, yet they generally work well in practice.

There is still one loose end we must tie up. Equation 5-7 shows how to go from the dual solution to the primal solution in the case of a linear SVM classifier. But if you apply the kernel trick, you end up with equations that include $\phi(x^{(i)})$. In fact, $\widehat{\mathbf{w}}$ must have the same number of dimensions as $\phi(x^{(i)})$, which may be huge or even infinite, so you can't compute it. But how can you make predictions without knowing $\widehat{\mathbf{w}}$? Well, the good news is that you can plug the formula for $\widehat{\mathbf{w}}$ from Equation 5-7 into the decision function for a new instance $\mathbf{x}^{(n)}$, and you get an equation with only dot products between input vectors. This makes it possible to use the kernel trick (Equation 5-11).

Equation 5-11. Making predictions with a kernelized SVM

$$h_{\widehat{\mathbf{w}}, \hat{b}}\left(\phi\left(\mathbf{x}^{(n)}\right)\right) = \widehat{\mathbf{w}}^{\mathsf{T}}\phi\left(\mathbf{x}^{(n)}\right) + \hat{b} = \left(\sum_{i=1}^{m} \hat{\alpha}^{(i)}t^{(i)}\phi\left(\mathbf{x}^{(i)}\right)\right)^{\mathsf{T}}\phi\left(\mathbf{x}^{(n)}\right) + \hat{b}$$

$$= \sum_{i=1}^{m} \hat{\alpha}^{(i)}t^{(i)}\left(\phi\left(\mathbf{x}^{(i)}\right)^{\mathsf{T}}\phi\left(\mathbf{x}^{(n)}\right)\right) + \hat{b}$$

$$= \sum_{\substack{i=1 \\ \hat{\alpha}^{(i)} > 0}}^{m} \hat{\alpha}^{(i)}t^{(i)}K\left(\mathbf{x}^{(i)}, \mathbf{x}^{(n)}\right) + \hat{b}$$

Note that since $\alpha^{(i)} \neq 0$ only for support vectors, making predictions involves computing the dot product of the new input vector $\mathbf{x}^{(n)}$ with only the support vectors, not all the training instances. Of course, you need to use the same trick to compute the bias term $\hat{b}$ (Equation 5-12).

Equation 5-12. Using the kernel trick to compute the bias term

$$\hat{b} = \frac{1}{n_s}\sum_{\substack{i=1 \\ \hat{\alpha}^{(i)} > 0}}^{m}\left(t^{(i)} - \widehat{\mathbf{w}}^{\mathsf{T}}\phi\left(\mathbf{x}^{(i)}\right)\right) = \frac{1}{n_s}\sum_{\substack{i=1 \\ \hat{\alpha}^{(i)} > 0}}^{m}\left(t^{(i)} - \left(\sum_{j=1}^{m}\hat{\alpha}^{(j)}t^{(j)}\phi\left(\mathbf{x}^{(j)}\right)\right)^{\mathsf{T}}\phi\left(\mathbf{x}^{(i)}\right)\right)$$

$$= \frac{1}{n_s}\sum_{\substack{i=1 \\ \hat{\alpha}^{(i)} > 0}}^{m}\left(t^{(i)} - \sum_{\substack{j=1 \\ \hat{\alpha}^{(j)} > 0}}^{m}\hat{\alpha}^{(j)}t^{(j)}K\left(\mathbf{x}^{(i)}, \mathbf{x}^{(j)}\right)\right)$$

If you are starting to get a headache, it's perfectly normal: it's an unfortunate side effect of the kernel trick.

Online SVMs

Before concluding this chapter, let's take a quick look at online SVM classifiers (recall that online learning means learning incrementally, typically as new instances arrive).

For linear SVM classifiers, one method for implementing an online SVM classifier is to use Gradient Descent (e.g., using SGDClassifier) to minimize the cost function in Equation 5-13, which is derived from the primal problem. Unfortunately, Gradient Descent converges much more slowly than the methods based on QP.

Equation 5-13. Linear SVM classifier cost function

$$J(\mathbf{w}, b) = \frac{1}{2}\mathbf{w}^{\mathsf{T}}\mathbf{w} \quad + \quad C\sum_{i=1}^{m} max\left(0, 1 - t^{(i)}\left(\mathbf{w}^{\mathsf{T}}\mathbf{x}^{(i)} + b\right)\right)$$

The first sum in the cost function will push the model to have a small weight vector **w**, leading to a larger margin. The second sum computes the total of all margin violations. An instance's margin violation is equal to 0 if it is located off the street and on the correct side, or else it is proportional to the distance to the correct side of the street. Minimizing this term ensures that the model makes the margin violations as small and as few as possible.

Hinge Loss

The function $max(0, 1 - t)$ is called the *hinge loss* function (see the following image). It is equal to 0 when $t \geq 1$. Its derivative (slope) is equal to –1 if $t < 1$ and 0 if $t > 1$. It is not differentiable at $t = 1$, but just like for Lasso Regression (see "Lasso Regression" on page 137), you can still use Gradient Descent using any *subderivative* at $t = 1$ (i.e., any value between –1 and 0).

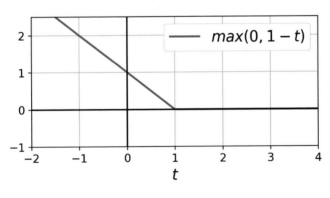

It is also possible to implement online kernelized SVMs, as described in the papers "Incremental and Decremental Support Vector Machine Learning" (*https://homl.info/17*)[8] and "Fast Kernel Classifiers with Online and Active Learning" (*https://homl.info/18*).[9] These kernelized SVMs are implemented in Matlab and C++. For

8 Gert Cauwenberghs and Tomaso Poggio, "Incremental and Decremental Support Vector Machine Learning," *Proceedings of the 13th International Conference on Neural Information Processing Systems* (2000): 388–394.

9 Antoine Bordes et al., "Fast Kernel Classifiers with Online and Active Learning," *Journal of Machine Learning Research* 6 (2005): 1579–1619.

large-scale nonlinear problems, you may want to consider using neural networks instead (see Part II).

Exercises

1. What is the fundamental idea behind Support Vector Machines?

2. What is a support vector?

3. Why is it important to scale the inputs when using SVMs?

4. Can an SVM classifier output a confidence score when it classifies an instance? What about a probability?

5. Should you use the primal or the dual form of the SVM problem to train a model on a training set with millions of instances and hundreds of features?

6. Say you've trained an SVM classifier with an RBF kernel, but it seems to underfit the training set. Should you increase or decrease γ (gamma)? What about C?

7. How should you set the QP parameters (**H**, **f**, **A**, and **b**) to solve the soft margin linear SVM classifier problem using an off-the-shelf QP solver?

8. Train a LinearSVC on a linearly separable dataset. Then train an SVC and a SGDClassifier on the same dataset. See if you can get them to produce roughly the same model.

9. Train an SVM classifier on the MNIST dataset. Since SVM classifiers are binary classifiers, you will need to use one-versus-the-rest to classify all 10 digits. You may want to tune the hyperparameters using small validation sets to speed up the process. What accuracy can you reach?

10. Train an SVM regressor on the California housing dataset.

Solutions to these exercises are available in Appendix A.

## CHAPTER 6

Decision Trees

Like SVMs, *Decision Trees* are versatile Machine Learning algorithms that can perform both classification and regression tasks, and even multioutput tasks. They are powerful algorithms, capable of fitting complex datasets. For example, in Chapter 2 you trained a `DecisionTreeRegressor` model on the California housing dataset, fitting it perfectly (actually, overfitting it).

Decision Trees are also the fundamental components of Random Forests (see Chapter 7), which are among the most powerful Machine Learning algorithms available today.

In this chapter we will start by discussing how to train, visualize, and make predictions with Decision Trees. Then we will go through the CART training algorithm used by Scikit-Learn, and we will discuss how to regularize trees and use them for regression tasks. Finally, we will discuss some of the limitations of Decision Trees.

Training and Visualizing a Decision Tree

To understand Decision Trees, let's build one and take a look at how it makes predictions. The following code trains a `DecisionTreeClassifier` on the iris dataset (see Chapter 4):

```python
from sklearn.datasets import load_iris
from sklearn.tree import DecisionTreeClassifier

iris = load_iris()
X = iris.data[:, 2:] # petal length and width
y = iris.target

tree_clf = DecisionTreeClassifier(max_depth=2)
tree_clf.fit(X, y)
```

You can visualize the trained Decision Tree by first using the `export_graphviz()` method to output a graph definition file called *iris_tree.dot*:

```
from sklearn.tree import export_graphviz

export_graphviz(
        tree_clf,
        out_file=image_path("iris_tree.dot"),
        feature_names=iris.feature_names[2:],
        class_names=iris.target_names,
        rounded=True,
        filled=True
    )
```

Then you can use the `dot` command-line tool from the Graphviz package to convert this *.dot* file to a variety of formats, such as PDF or PNG.[1] This command line converts the *.dot* file to a *.png* image file:

```
$ dot -Tpng iris_tree.dot -o iris_tree.png
```

Your first Decision Tree looks like Figure 6-1.

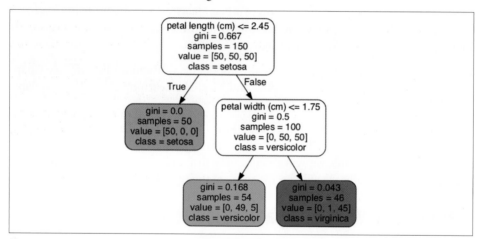

Figure 6-1. Iris Decision Tree

Making Predictions

Let's see how the tree represented in Figure 6-1 makes predictions. Suppose you find an iris flower and you want to classify it. You start at the *root node* (depth 0, at the top): this node asks whether the flower's petal length is smaller than 2.45 cm. If it is, then you move down to the root's left child node (depth 1, left). In this case, it is a *leaf*

---

1 Graphviz is an open source graph visualization software package, available at *http://www.graphviz.org/*.

node (i.e., it does not have any child nodes), so it does not ask any questions: simply look at the predicted class for that node, and the Decision Tree predicts that your flower is an *Iris setosa* (class=setosa).

Now suppose you find another flower, and this time the petal length is greater than 2.45 cm. You must move down to the root's right child node (depth 1, right), which is not a leaf node, so the node asks another question: is the petal width smaller than 1.75 cm? If it is, then your flower is most likely an *Iris versicolor* (depth 2, left). If not, it is likely an *Iris virginica* (depth 2, right). It's really that simple.

One of the many qualities of Decision Trees is that they require very little data preparation. In fact, they don't require feature scaling or centering at all.

A node's samples attribute counts how many training instances it applies to. For example, 100 training instances have a petal length greater than 2.45 cm (depth 1, right), and of those 100, 54 have a petal width smaller than 1.75 cm (depth 2, left). A node's value attribute tells you how many training instances of each class this node applies to: for example, the bottom-right node applies to 0 *Iris setosa*, 1 *Iris versicolor*, and 45 *Iris virginica*. Finally, a node's gini attribute measures its *impurity*: a node is "pure" (gini=0) if all training instances it applies to belong to the same class. For example, since the depth-1 left node applies only to *Iris setosa* training instances, it is pure and its gini score is 0. Equation 6-1 shows how the training algorithm computes the gini score G_i of the ith node. The depth-2 left node has a gini score equal to $1 - (0/54)^2 - (49/54)^2 - (5/54)^2 \approx 0.168$.

Equation 6-1. Gini impurity

$$G_i = 1 - \sum_{k=1}^{n} p_{i,k}^{\ 2}$$

In this equation:

- $p_{i,k}$ is the ratio of class k instances among the training instances in the ith node.

Scikit-Learn uses the CART algorithm, which produces only *binary trees*: nonleaf nodes always have two children (i.e., questions only have yes/no answers). However, other algorithms such as ID3 can produce Decision Trees with nodes that have more than two children.

Figure 6-2 shows this Decision Tree's decision boundaries. The thick vertical line represents the decision boundary of the root node (depth 0): petal length = 2.45 cm. Since the lefthand area is pure (only *Iris setosa*), it cannot be split any further. However, the righthand area is impure, so the depth-1 right node splits it at petal width = 1.75 cm (represented by the dashed line). Since max_depth was set to 2, the Decision Tree stops right there. If you set max_depth to 3, then the two depth-2 nodes would each add another decision boundary (represented by the dotted lines).

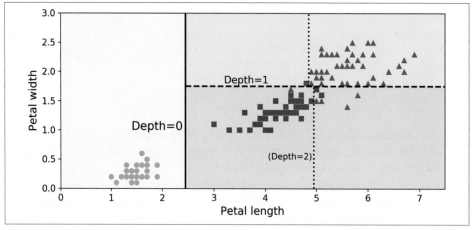

Figure 6-2. Decision Tree decision boundaries

Model Interpretation: White Box Versus Black Box

Decision Trees are intuitive, and their decisions are easy to interpret. Such models are often called *white box models*. In contrast, as we will see, Random Forests or neural networks are generally considered *black box models*. They make great predictions, and you can easily check the calculations that they performed to make these predictions; nevertheless, it is usually hard to explain in simple terms why the predictions were made. For example, if a neural network says that a particular person appears on a picture, it is hard to know what contributed to this prediction: did the model recognize that person's eyes? Their mouth? Their nose? Their shoes? Or even the couch that they were sitting on? Conversely, Decision Trees provide nice, simple classification rules that can even be applied manually if need be (e.g., for flower classification).

Estimating Class Probabilities

A Decision Tree can also estimate the probability that an instance belongs to a particular class k. First it traverses the tree to find the leaf node for this instance, and then it returns the ratio of training instances of class k in this node. For example, suppose you have found a flower whose petals are 5 cm long and 1.5 cm wide. The corresponding leaf node is the depth-2 left node, so the Decision Tree should output the following probabilities: 0% for *Iris setosa* (0/54), 90.7% for *Iris versicolor* (49/54), and 9.3% for *Iris virginica* (5/54). And if you ask it to predict the class, it should output *Iris versicolor* (class 1) because it has the highest probability. Let's check this:

```
>>> tree_clf.predict_proba([[5, 1.5]])
array([[0.       , 0.90740741, 0.09259259]])
>>> tree_clf.predict([[5, 1.5]])
array([1])
```

Perfect! Notice that the estimated probabilities would be identical anywhere else in the bottom-right rectangle of Figure 6-2—for example, if the petals were 6 cm long and 1.5 cm wide (even though it seems obvious that it would most likely be an *Iris virginica* in this case).

The CART Training Algorithm

Scikit-Learn uses the *Classification and Regression Tree* (CART) algorithm to train Decision Trees (also called "growing" trees). The algorithm works by first splitting the training set into two subsets using a single feature k and a threshold t_k (e.g., "petal length ≤ 2.45 cm"). How does it choose k and t_k? It searches for the pair (k, t_k) that produces the purest subsets (weighted by their size). Equation 6-2 gives the cost function that the algorithm tries to minimize.

Equation 6-2. CART cost function for classification

$$J(k, t_k) = \frac{m_{\text{left}}}{m} G_{\text{left}} + \frac{m_{\text{right}}}{m} G_{\text{right}}$$

$$\text{where} \begin{cases} G_{\text{left/right}} \text{ measures the impurity of the left/right subset,} \\ m_{\text{left/right}} \text{ is the number of instances in the left/right subset.} \end{cases}$$

Once the CART algorithm has successfully split the training set in two, it splits the subsets using the same logic, then the sub-subsets, and so on, recursively. It stops recursing once it reaches the maximum depth (defined by the max_depth hyperparameter), or if it cannot find a split that will reduce impurity. A few other hyperparameters (described in a moment) control additional stopping conditions (min_samples_split, min_samples_leaf, min_weight_fraction_leaf, and max_leaf_nodes).

As you can see, the CART algorithm is a *greedy algorithm*: it greedily searches for an optimum split at the top level, then repeats the process at each subsequent level. It does not check whether or not the split will lead to the lowest possible impurity several levels down. A greedy algorithm often produces a solution that's reasonably good but not guaranteed to be optimal.

Unfortunately, finding the optimal tree is known to be an *NP-Complete* problem:[2] it requires $O(\exp(m))$ time, making the problem intractable even for small training sets. This is why we must settle for a "reasonably good" solution.

Computational Complexity

Making predictions requires traversing the Decision Tree from the root to a leaf. Decision Trees generally are approximately balanced, so traversing the Decision Tree requires going through roughly $O(\log_2(m))$ nodes.[3] Since each node only requires checking the value of one feature, the overall prediction complexity is $O(\log_2(m))$, independent of the number of features. So predictions are very fast, even when dealing with large training sets.

The training algorithm compares all features (or less if `max_features` is set) on all samples at each node. Comparing all features on all samples at each node results in a training complexity of $O(n \times m \log_2(m))$. For small training sets (less than a few thousand instances), Scikit-Learn can speed up training by presorting the data (set `presort=True`), but doing that slows down training considerably for larger training sets.

Gini Impurity or Entropy?

By default, the Gini impurity measure is used, but you can select the *entropy* impurity measure instead by setting the `criterion` hyperparameter to `"entropy"`. The concept of entropy originated in thermodynamics as a measure of molecular disorder: entropy approaches zero when molecules are still and well ordered. Entropy later spread to a wide variety of domains, including Shannon's *information theory*, where it measures the average information content of a message:[4] entropy is zero when all messages are identical. In Machine Learning, entropy is frequently used as an

---

2 P is the set of problems that can be solved in polynomial time. NP is the set of problems whose solutions can be verified in polynomial time. An NP-Hard problem is a problem to which any NP problem can be reduced in polynomial time. An NP-Complete problem is both NP and NP-Hard. A major open mathematical question is whether or not P = NP. If P ≠ NP (which seems likely), then no polynomial algorithm will ever be found for any NP-Complete problem (except perhaps on a quantum computer).

3 $\log_2$ is the binary logarithm. It is equal to $\log_2(m) = \log(m) / \log(2)$.

4 A reduction of entropy is often called an *information gain*.

impurity measure: a set's entropy is zero when it contains instances of only one class. Equation 6-3 shows the definition of the entropy of the ith node. For example, the depth-2 left node in Figure 6-1 has an entropy equal to $-(49/54) \log_2 (49/54) - (5/54) \log_2 (5/54) \approx 0.445$.

Equation 6-3. Entropy

$$H_i = - \sum_{\substack{k = 1 \\ p_{i,k} \neq 0}}^{n} p_{i,k} \log_2 \left(p_{i,k}\right)$$

So, should you use Gini impurity or entropy? The truth is, most of the time it does not make a big difference: they lead to similar trees. Gini impurity is slightly faster to compute, so it is a good default. However, when they differ, Gini impurity tends to isolate the most frequent class in its own branch of the tree, while entropy tends to produce slightly more balanced trees.[5]

Regularization Hyperparameters

Decision Trees make very few assumptions about the training data (as opposed to linear models, which assume that the data is linear, for example). If left unconstrained, the tree structure will adapt itself to the training data, fitting it very closely—indeed, most likely overfitting it. Such a model is often called a *nonparametric model*, not because it does not have any parameters (it often has a lot) but because the number of parameters is not determined prior to training, so the model structure is free to stick closely to the data. In contrast, a *parametric model*, such as a linear model, has a predetermined number of parameters, so its degree of freedom is limited, reducing the risk of overfitting (but increasing the risk of underfitting).

To avoid overfitting the training data, you need to restrict the Decision Tree's freedom during training. As you know by now, this is called regularization. The regularization hyperparameters depend on the algorithm used, but generally you can at least restrict the maximum depth of the Decision Tree. In Scikit-Learn, this is controlled by the max_depth hyperparameter (the default value is None, which means unlimited). Reducing max_depth will regularize the model and thus reduce the risk of overfitting.

The DecisionTreeClassifier class has a few other parameters that similarly restrict the shape of the Decision Tree: min_samples_split (the minimum number of samples a node must have before it can be split), min_samples_leaf (the minimum number of samples a leaf node must have), min_weight_fraction_leaf (same as

---

5 See Sebastian Raschka's interesting analysis for more details (*https://homl.info/19*).

`min_samples_leaf` but expressed as a fraction of the total number of weighted instances), `max_leaf_nodes` (the maximum number of leaf nodes), and `max_features` (the maximum number of features that are evaluated for splitting at each node). Increasing `min_*` hyperparameters or reducing `max_*` hyperparameters will regularize the model.

 Other algorithms work by first training the Decision Tree without restrictions, then *pruning* (deleting) unnecessary nodes. A node whose children are all leaf nodes is considered unnecessary if the purity improvement it provides is not statistically significant. Standard statistical tests, such as the χ^2 *test* (chi-squared test), are used to estimate the probability that the improvement is purely the result of chance (which is called the *null hypothesis*). If this probability, called the *p-value*, is higher than a given threshold (typically 5%, controlled by a hyperparameter), then the node is considered unnecessary and its children are deleted. The pruning continues until all unnecessary nodes have been pruned.

Figure 6-3 shows two Decision Trees trained on the moons dataset (introduced in Chapter 5). On the left the Decision Tree is trained with the default hyperparameters (i.e., no restrictions), and on the right it's trained with `min_samples_leaf=4`. It is quite obvious that the model on the left is overfitting, and the model on the right will probably generalize better.

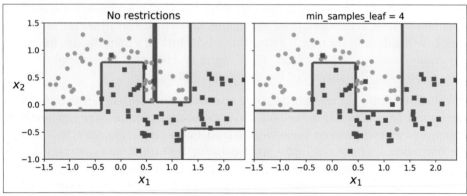

Figure 6-3. Regularization using `min_samples_leaf`

Regression

Decision Trees are also capable of performing regression tasks. Let's build a regression tree using Scikit-Learn's `DecisionTreeRegressor` class, training it on a noisy quadratic dataset with `max_depth=2`:

```
from sklearn.tree import DecisionTreeRegressor

tree_reg = DecisionTreeRegressor(max_depth=2)
tree_reg.fit(X, y)
```

The resulting tree is represented in Figure 6-4.

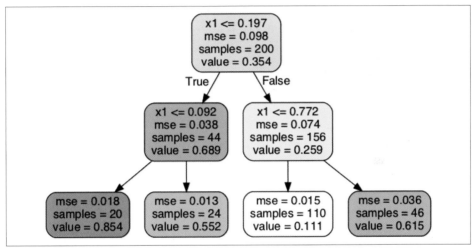

Figure 6-4. A Decision Tree for regression

This tree looks very similar to the classification tree you built earlier. The main difference is that instead of predicting a class in each node, it predicts a value. For example, suppose you want to make a prediction for a new instance with $x_1 = 0.6$. You traverse the tree starting at the root, and you eventually reach the leaf node that predicts `value=0.111`. This prediction is the average target value of the 110 training instances associated with this leaf node, and it results in a mean squared error equal to 0.015 over these 110 instances.

This model's predictions are represented on the left in Figure 6-5. If you set `max_depth=3`, you get the predictions represented on the right. Notice how the predicted value for each region is always the average target value of the instances in that region. The algorithm splits each region in a way that makes most training instances as close as possible to that predicted value.

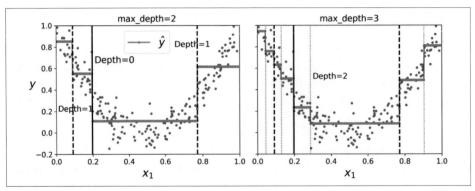

Figure 6-5. Predictions of two Decision Tree regression models

The CART algorithm works mostly the same way as earlier, except that instead of trying to split the training set in a way that minimizes impurity, it now tries to split the
training set in a way that minimizes the MSE. Equation 6-4 shows the cost function
that the algorithm tries to minimize.

Equation 6-4. CART cost function for regression

$$J(k, t_k) = \frac{m_{left}}{m} MSE_{left} + \frac{m_{right}}{m} MSE_{right} \quad \text{where} \begin{cases} MSE_{node} = \sum_{i \in node} \left(\hat{y}_{node} - y^{(i)} \right)^2 \\ \hat{y}_{node} = \frac{1}{m_{node}} \sum_{i \in node} y^{(i)} \end{cases}$$

Just like for classification tasks, Decision Trees are prone to overfitting when dealing
with regression tasks. Without any regularization (i.e., using the default hyperparameters), you get the predictions on the left in Figure 6-6. These predictions are obviously overfitting the training set very badly. Just setting `min_samples_leaf=10` results
in a much more reasonable model, represented on the right in Figure 6-6.

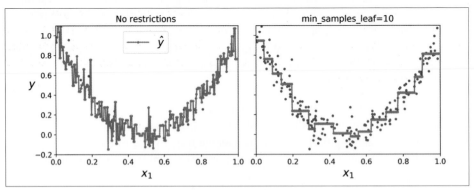

Figure 6-6. Regularizing a Decision Tree regressor

Instability

Hopefully by now you are convinced that Decision Trees have a lot going for them: they are simple to understand and interpret, easy to use, versatile, and powerful. However, they do have a few limitations. First, as you may have noticed, Decision Trees love orthogonal decision boundaries (all splits are perpendicular to an axis), which makes them sensitive to training set rotation. For example, Figure 6-7 shows a simple linearly separable dataset: on the left, a Decision Tree can split it easily, while on the right, after the dataset is rotated by 45°, the decision boundary looks unnecessarily convoluted. Although both Decision Trees fit the training set perfectly, it is very likely that the model on the right will not generalize well. One way to limit this problem is to use Principal Component Analysis (see Chapter 8), which often results in a better orientation of the training data.

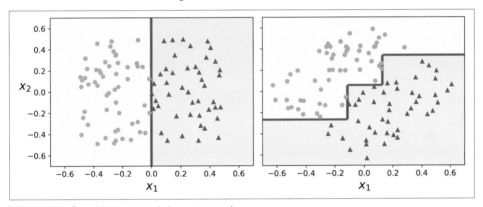

Figure 6-7. Sensitivity to training set rotation

More generally, the main issue with Decision Trees is that they are very sensitive to small variations in the training data. For example, if you just remove the widest *Iris versicolor* from the iris training set (the one with petals 4.8 cm long and 1.8 cm wide) and train a new Decision Tree, you may get the model represented in Figure 6-8. As you can see, it looks very different from the previous Decision Tree (Figure 6-2). Actually, since the training algorithm used by Scikit-Learn is stochastic,[6] you may get very different models even on the same training data (unless you set the random_state hyperparameter).

---

6 It randomly selects the set of features to evaluate at each node.

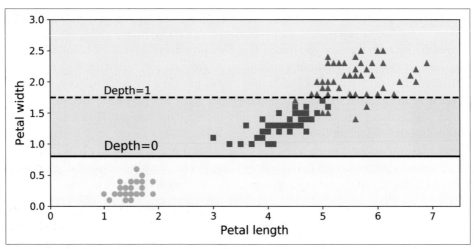

Figure 6-8. Sensitivity to training set details

Random Forests can limit this instability by averaging predictions over many trees, as we will see in the next chapter.

Exercises

1. What is the approximate depth of a Decision Tree trained (without restrictions) on a training set with one million instances?

2. Is a node's Gini impurity generally lower or greater than its parent's? Is it *generally* lower/greater, or *always* lower/greater?

3. If a Decision Tree is overfitting the training set, is it a good idea to try decreasing `max_depth`?

4. If a Decision Tree is underfitting the training set, is it a good idea to try scaling the input features?

5. If it takes one hour to train a Decision Tree on a training set containing 1 million instances, roughly how much time will it take to train another Decision Tree on a training set containing 10 million instances?

6. If your training set contains 100,000 instances, will setting `presort=True` speed up training?

7. Train and fine-tune a Decision Tree for the moons dataset by following these steps:

   a. Use `make_moons(n_samples=10000, noise=0.4)` to generate a moons dataset.

   b. Use `train_test_split()` to split the dataset into a training set and a test set.

c. Use grid search with cross-validation (with the help of the GridSearchCV class) to find good hyperparameter values for a DecisionTreeClassifier. Hint: try various values for max_leaf_nodes.

d. Train it on the full training set using these hyperparameters, and measure your model's performance on the test set. You should get roughly 85% to 87% accuracy.

8. Grow a forest by following these steps:

a. Continuing the previous exercise, generate 1,000 subsets of the training set, each containing 100 instances selected randomly. Hint: you can use Scikit-Learn's ShuffleSplit class for this.

b. Train one Decision Tree on each subset, using the best hyperparameter values found in the previous exercise. Evaluate these 1,000 Decision Trees on the test set. Since they were trained on smaller sets, these Decision Trees will likely perform worse than the first Decision Tree, achieving only about 80% accuracy.

c. Now comes the magic. For each test set instance, generate the predictions of the 1,000 Decision Trees, and keep only the most frequent prediction (you can use SciPy's mode() function for this). This approach gives you *majority-vote predictions* over the test set.

d. Evaluate these predictions on the test set: you should obtain a slightly higher accuracy than your first model (about 0.5 to 1.5% higher). Congratulations, you have trained a Random Forest classifier!

Solutions to these exercises are available in Appendix A.

## CHAPTER 7

Ensemble Learning and Random Forests

Suppose you pose a complex question to thousands of random people, then aggregate their answers. In many cases you will find that this aggregated answer is better than an expert's answer. This is called the *wisdom of the crowd*. Similarly, if you aggregate the predictions of a group of predictors (such as classifiers or regressors), you will often get better predictions than with the best individual predictor. A group of predictors is called an *ensemble*; thus, this technique is called *Ensemble Learning*, and an Ensemble Learning algorithm is called an *Ensemble method*.

As an example of an Ensemble method, you can train a group of Decision Tree classifiers, each on a different random subset of the training set. To make predictions, you obtain the predictions of all the individual trees, then predict the class that gets the most votes (see the last exercise in Chapter 6). Such an ensemble of Decision Trees is called a *Random Forest*, and despite its simplicity, this is one of the most powerful Machine Learning algorithms available today.

As discussed in Chapter 2, you will often use Ensemble methods near the end of a project, once you have already built a few good predictors, to combine them into an even better predictor. In fact, the winning solutions in Machine Learning competitions often involve several Ensemble methods (most famously in the Netflix Prize competition (*http://netflixprize.com/*)).

In this chapter we will discuss the most popular Ensemble methods, including *bagging*, *boosting*, and *stacking*. We will also explore Random Forests.

Voting Classifiers

Suppose you have trained a few classifiers, each one achieving about 80% accuracy. You may have a Logistic Regression classifier, an SVM classifier, a Random Forest classifier, a K-Nearest Neighbors classifier, and perhaps a few more (see Figure 7-1).

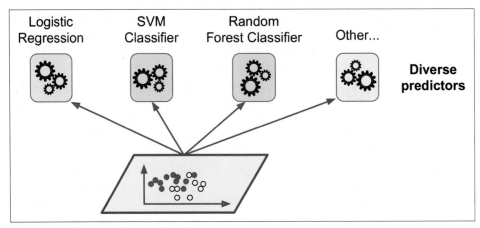

Figure 7-1. Training diverse classifiers

A very simple way to create an even better classifier is to aggregate the predictions of each classifier and predict the class that gets the most votes. This majority-vote classifier is called a *hard voting* classifier (see Figure 7-2).

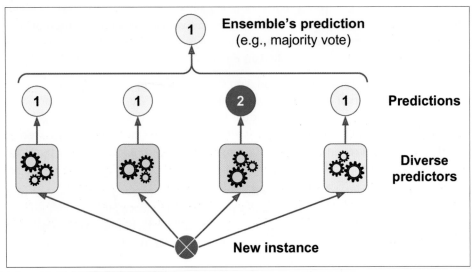

Figure 7-2. Hard voting classifier predictions

Somewhat surprisingly, this voting classifier often achieves a higher accuracy than the best classifier in the ensemble. In fact, even if each classifier is a *weak learner* (meaning it does only slightly better than random guessing), the ensemble can still be a *strong learner* (achieving high accuracy), provided there are a sufficient number of weak learners and they are sufficiently diverse.

How is this possible? The following analogy can help shed some light on this mystery. Suppose you have a slightly biased coin that has a 51% chance of coming up heads and 49% chance of coming up tails. If you toss it 1,000 times, you will generally get more or less 510 heads and 490 tails, and hence a majority of heads. If you do the math, you will find that the probability of obtaining a majority of heads after 1,000 tosses is close to 75%. The more you toss the coin, the higher the probability (e.g., with 10,000 tosses, the probability climbs over 97%). This is due to the *law of large numbers*: as you keep tossing the coin, the ratio of heads gets closer and closer to the probability of heads (51%). Figure 7-3 shows 10 series of biased coin tosses. You can see that as the number of tosses increases, the ratio of heads approaches 51%. Eventually all 10 series end up so close to 51% that they are consistently above 50%.

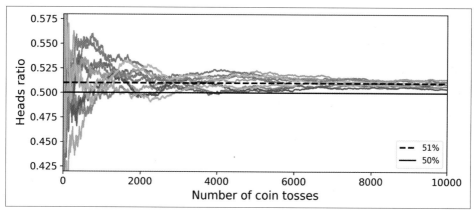

Figure 7-3. The law of large numbers

Similarly, suppose you build an ensemble containing 1,000 classifiers that are individually correct only 51% of the time (barely better than random guessing). If you predict the majority voted class, you can hope for up to 75% accuracy! However, this is only true if all classifiers are perfectly independent, making uncorrelated errors, which is clearly not the case because they are trained on the same data. They are likely to make the same types of errors, so there will be many majority votes for the wrong class, reducing the ensemble's accuracy.

> Ensemble methods work best when the predictors are as independent from one another as possible. One way to get diverse classifiers is to train them using very different algorithms. This increases the chance that they will make very different types of errors, improving the ensemble's accuracy.

The following code creates and trains a voting classifier in Scikit-Learn, composed of three diverse classifiers (the training set is the moons dataset, introduced in Chapter 5):

```
from sklearn.ensemble import RandomForestClassifier
from sklearn.ensemble import VotingClassifier
from sklearn.linear_model import LogisticRegression
from sklearn.svm import SVC

log_clf = LogisticRegression()
rnd_clf = RandomForestClassifier()
svm_clf = SVC()

voting_clf = VotingClassifier(
    estimators=[('lr', log_clf), ('rf', rnd_clf), ('svc', svm_clf)],
    voting='hard')
voting_clf.fit(X_train, y_train)
```

Let's look at each classifier's accuracy on the test set:

```
>>> from sklearn.metrics import accuracy_score
>>> for clf in (log_clf, rnd_clf, svm_clf, voting_clf):
...     clf.fit(X_train, y_train)
...     y_pred = clf.predict(X_test)
...     print(clf.__class__.__name__, accuracy_score(y_test, y_pred))
...
LogisticRegression 0.864
RandomForestClassifier 0.896
SVC 0.888
VotingClassifier 0.904
```

There you have it! The voting classifier slightly outperforms all the individual classifiers.

If all classifiers are able to estimate class probabilities (i.e., they all have a `pre dict_proba()` method), then you can tell Scikit-Learn to predict the class with the highest class probability, averaged over all the individual classifiers. This is called *soft voting*. It often achieves higher performance than hard voting because it gives more weight to highly confident votes. All you need to do is replace `voting="hard"` with `voting="soft"` and ensure that all classifiers can estimate class probabilities. This is not the case for the SVC class by default, so you need to set its `probability` hyper-parameter to `True` (this will make the SVC class use cross-validation to estimate class probabilities, slowing down training, and it will add a `predict_proba()` method). If you modify the preceding code to use soft voting, you will find that the voting classi-fier achieves over 91.2% accuracy!

Bagging and Pasting

One way to get a diverse set of classifiers is to use very different training algorithms, as just discussed. Another approach is to use the same training algorithm for every predictor and train them on different random subsets of the training set. When sam-pling is performed *with* replacement, this method is called *bagging* (*https://*

---

homl.info/20)[1] (short for *bootstrap aggregating*[2]). When sampling is performed *without* replacement, it is called *pasting* (*https://homl.info/21*).[3]

In other words, both bagging and pasting allow training instances to be sampled several times across multiple predictors, but only bagging allows training instances to be sampled several times for the same predictor. This sampling and training process is represented in Figure 7-4.

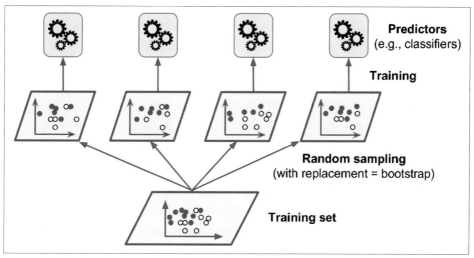

Figure 7-4. Bagging and pasting involves training several predictors on different random samples of the training set

Once all predictors are trained, the ensemble can make a prediction for a new instance by simply aggregating the predictions of all predictors. The aggregation function is typically the *statistical mode* (i.e., the most frequent prediction, just like a hard voting classifier) for classification, or the average for regression. Each individual predictor has a higher bias than if it were trained on the original training set, but aggregation reduces both bias and variance.[4] Generally, the net result is that the ensemble has a similar bias but a lower variance than a single predictor trained on the original training set.

As you can see in Figure 7-4, predictors can all be trained in parallel, via different CPU cores or even different servers. Similarly, predictions can be made in parallel.

---

1 Leo Breiman, "Bagging Predictors," *Machine Learning* 24, no. 2 (1996): 123–140.

2 In statistics, resampling with replacement is called *bootstrapping*.

3 Leo Breiman, "Pasting Small Votes for Classification in Large Databases and On-Line," *Machine Learning* 36, no. 1–2 (1999): 85–103.

4 Bias and variance were introduced in Chapter 4.

This is one of the reasons bagging and pasting are such popular methods: they scale very well.

Bagging and Pasting in Scikit-Learn

Scikit-Learn offers a simple API for both bagging and pasting with the `BaggingClas sifier` class (or `BaggingRegressor` for regression). The following code trains an ensemble of 500 Decision Tree classifiers:[5] each is trained on 100 training instances randomly sampled from the training set with replacement (this is an example of bagging, but if you want to use pasting instead, just set `bootstrap=False`). The `n_jobs` parameter tells Scikit-Learn the number of CPU cores to use for training and predictions (`-1` tells Scikit-Learn to use all available cores):

```
from sklearn.ensemble import BaggingClassifier
from sklearn.tree import DecisionTreeClassifier

bag_clf = BaggingClassifier(
    DecisionTreeClassifier(), n_estimators=500,
    max_samples=100, bootstrap=True, n_jobs=-1)
bag_clf.fit(X_train, y_train)
y_pred = bag_clf.predict(X_test)
```

The `BaggingClassifier` automatically performs soft voting instead of hard voting if the base classifier can estimate class probabilities (i.e., if it has a `predict_proba()` method), which is the case with Decision Tree classifiers.

Figure 7-5 compares the decision boundary of a single Decision Tree with the decision boundary of a bagging ensemble of 500 trees (from the preceding code), both trained on the moons dataset. As you can see, the ensemble's predictions will likely generalize much better than the single Decision Tree's predictions: the ensemble has a comparable bias but a smaller variance (it makes roughly the same number of errors on the training set, but the decision boundary is less irregular).

---

5 `max_samples` can alternatively be set to a float between 0.0 and 1.0, in which case the max number of instances to sample is equal to the size of the training set times `max_samples`.

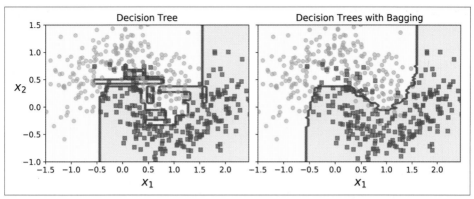

Figure 7-5. A single Decision Tree (left) versus a bagging ensemble of 500 trees (right)

Bootstrapping introduces a bit more diversity in the subsets that each predictor is trained on, so bagging ends up with a slightly higher bias than pasting; but the extra diversity also means that the predictors end up being less correlated, so the ensemble's variance is reduced. Overall, bagging often results in better models, which explains why it is generally preferred. However, if you have spare time and CPU power, you can use cross-validation to evaluate both bagging and pasting and select the one that works best.

Out-of-Bag Evaluation

With bagging, some instances may be sampled several times for any given predictor, while others may not be sampled at all. By default a `BaggingClassifier` samples m training instances with replacement (`bootstrap=True`), where m is the size of the training set. This means that only about 63% of the training instances are sampled on average for each predictor.[6] The remaining 37% of the training instances that are not sampled are called *out-of-bag* (oob) instances. Note that they are not the same 37% for all predictors.

Since a predictor never sees the oob instances during training, it can be evaluated on these instances, without the need for a separate validation set. You can evaluate the ensemble itself by averaging out the oob evaluations of each predictor.

In Scikit-Learn, you can set `oob_score=True` when creating a `BaggingClassifier` to request an automatic oob evaluation after training. The following code demonstrates this. The resulting evaluation score is available through the `oob_score_` variable:

---

6 As m grows, this ratio approaches $1 - \exp(-1) \approx 63.212\%$.

```
>>> bag_clf = BaggingClassifier(
...     DecisionTreeClassifier(), n_estimators=500,
...     bootstrap=True, n_jobs=-1, oob_score=True)
...
>>> bag_clf.fit(X_train, y_train)
>>> bag_clf.oob_score_
0.90133333333333332
```

According to this oob evaluation, this `BaggingClassifier` is likely to achieve about 90.1% accuracy on the test set. Let's verify this:

```
>>> from sklearn.metrics import accuracy_score
>>> y_pred = bag_clf.predict(X_test)
>>> accuracy_score(y_test, y_pred)
0.91200000000000003
```

We get 91.2% accuracy on the test set—close enough!

The oob decision function for each training instance is also available through the `oob_decision_function_` variable. In this case (since the base estimator has a `predict_proba()` method), the decision function returns the class probabilities for each training instance. For example, the oob evaluation estimates that the first training instance has a 68.25% probability of belonging to the positive class (and 31.75% of belonging to the negative class):

```
>>> bag_clf.oob_decision_function_
array([[0.31746032, 0.68253968],
       [0.34117647, 0.65882353],
       [1.        , 0.        ],
       ...
       [1.        , 0.        ],
       [0.03108808, 0.96891192],
       [0.57291667, 0.42708333]])
```

Random Patches and Random Subspaces

The `BaggingClassifier` class supports sampling the features as well. Sampling is controlled by two hyperparameters: `max_features` and `bootstrap_features`. They work the same way as `max_samples` and `bootstrap`, but for feature sampling instead of instance sampling. Thus, each predictor will be trained on a random subset of the input features.

This technique is particularly useful when you are dealing with high-dimensional inputs (such as images). Sampling both training instances and features is called the *Random Patches* method (*https://homl.info/22*).[7] Keeping all training instances (by

---

[7] Gilles Louppe and Pierre Geurts, "Ensembles on Random Patches," *Lecture Notes in Computer Science* 7523 (2012): 346–361.

setting `bootstrap=False` and `max_samples=1.0`) but sampling features (by setting `bootstrap_features` to `True` and/or `max_features` to a value smaller than `1.0`) is called the *Random Subspaces* method (*https://homl.info/23*).[8]

Sampling features results in even more predictor diversity, trading a bit more bias for a lower variance.

Random Forests

As we have discussed, a Random Forest (*https://homl.info/24*)[9] is an ensemble of Decision Trees, generally trained via the bagging method (or sometimes pasting), typically with `max_samples` set to the size of the training set. Instead of building a `BaggingClassifier` and passing it a `DecisionTreeClassifier`, you can instead use the `RandomForestClassifier` class, which is more convenient and optimized for Decision Trees[10] (similarly, there is a `RandomForestRegressor` class for regression tasks). The following code uses all available CPU cores to train a Random Forest classifier with 500 trees (each limited to maximum 16 nodes):

```
from sklearn.ensemble import RandomForestClassifier

rnd_clf = RandomForestClassifier(n_estimators=500, max_leaf_nodes=16, n_jobs=-1)
rnd_clf.fit(X_train, y_train)

y_pred_rf = rnd_clf.predict(X_test)
```

With a few exceptions, a `RandomForestClassifier` has all the hyperparameters of a `DecisionTreeClassifier` (to control how trees are grown), plus all the hyperparameters of a `BaggingClassifier` to control the ensemble itself.[11]

The Random Forest algorithm introduces extra randomness when growing trees; instead of searching for the very best feature when splitting a node (see Chapter 6), it searches for the best feature among a random subset of features. The algorithm results in greater tree diversity, which (again) trades a higher bias for a lower variance, generally yielding an overall better model. The following `BaggingClassifier` is roughly equivalent to the previous `RandomForestClassifier`:

---

8 Tin Kam Ho, "The Random Subspace Method for Constructing Decision Forests," *IEEE Transactions on Pattern Analysis and Machine Intelligence* 20, no. 8 (1998): 832–844.

9 Tin Kam Ho, "Random Decision Forests," *Proceedings of the Third International Conference on Document Analysis and Recognition* 1 (1995): 278.

10 The `BaggingClassifier` class remains useful if you want a bag of something other than Decision Trees.

11 There are a few notable exceptions: `splitter` is absent (forced to `"random"`), `presort` is absent (forced to `False`), `max_samples` is absent (forced to `1.0`), and `base_estimator` is absent (forced to `DecisionTreeClassifier` with the provided hyperparameters).

```
bag_clf = BaggingClassifier(
    DecisionTreeClassifier(splitter="random", max_leaf_nodes=16),
    n_estimators=500, max_samples=1.0, bootstrap=True, n_jobs=-1)
```

Extra-Trees

When you are growing a tree in a Random Forest, at each node only a random subset of the features is considered for splitting (as discussed earlier). It is possible to make trees even more random by also using random thresholds for each feature rather than searching for the best possible thresholds (like regular Decision Trees do).

A forest of such extremely random trees is called an *Extremely Randomized Trees* (*https://homl.info/25*) ensemble[12] (or *Extra-Trees* for short). Once again, this technique trades more bias for a lower variance. It also makes Extra-Trees much faster to train than regular Random Forests, because finding the best possible threshold for each feature at every node is one of the most time-consuming tasks of growing a tree.

You can create an Extra-Trees classifier using Scikit-Learn's `ExtraTreesClassifier` class. Its API is identical to the `RandomForestClassifier` class. Similarly, the `Extra TreesRegressor` class has the same API as the `RandomForestRegressor` class.

> It is hard to tell in advance whether a `RandomForestClassifier` will perform better or worse than an `ExtraTreesClassifier`. Generally, the only way to know is to try both and compare them using cross-validation (tuning the hyperparameters using grid search).

Feature Importance

Yet another great quality of Random Forests is that they make it easy to measure the relative importance of each feature. Scikit-Learn measures a feature's importance by looking at how much the tree nodes that use that feature reduce impurity on average (across all trees in the forest). More precisely, it is a weighted average, where each node's weight is equal to the number of training samples that are associated with it (see Chapter 6).

Scikit-Learn computes this score automatically for each feature after training, then it scales the results so that the sum of all importances is equal to 1. You can access the result using the `feature_importances_` variable. For example, the following code trains a `RandomForestClassifier` on the iris dataset (introduced in Chapter 4) and outputs each feature's importance. It seems that the most important features are the petal length (44%) and width (42%), while sepal length and width are rather unimportant in comparison (11% and 2%, respectively):

---

12 Pierre Geurts et al., "Extremely Randomized Trees," *Machine Learning* 63, no. 1 (2006): 3–42.

```
>>> from sklearn.datasets import load_iris
>>> iris = load_iris()
>>> rnd_clf = RandomForestClassifier(n_estimators=500, n_jobs=-1)
>>> rnd_clf.fit(iris["data"], iris["target"])
>>> for name, score in zip(iris["feature_names"], rnd_clf.feature_importances_):
...     print(name, score)
...
sepal length (cm) 0.112492250999
sepal width (cm) 0.0231192882825
petal length (cm) 0.441030464364
petal width (cm) 0.423357996355
```

Similarly, if you train a Random Forest classifier on the MNIST dataset (introduced in Chapter 3) and plot each pixel's importance, you get the image represented in Figure 7-6.

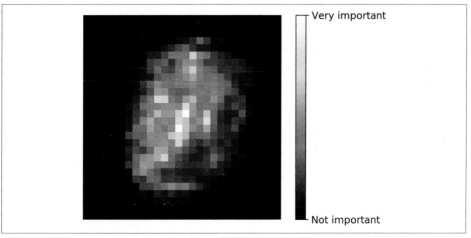

Figure 7-6. MNIST pixel importance (according to a Random Forest classifier)

Random Forests are very handy to get a quick understanding of what features actually matter, in particular if you need to perform feature selection.

Boosting

Boosting (originally called *hypothesis boosting*) refers to any Ensemble method that can combine several weak learners into a strong learner. The general idea of most boosting methods is to train predictors sequentially, each trying to correct its predecessor. There are many boosting methods available, but by far the most popular are

AdaBoost (https://homl.info/26)[13] (short for *Adaptive Boosting*) and *Gradient Boosting*. Let's start with AdaBoost.

AdaBoost

One way for a new predictor to correct its predecessor is to pay a bit more attention to the training instances that the predecessor underfitted. This results in new predictors focusing more and more on the hard cases. This is the technique used by AdaBoost.

For example, when training an AdaBoost classifier, the algorithm first trains a base classifier (such as a Decision Tree) and uses it to make predictions on the training set. The algorithm then increases the relative weight of misclassified training instances. Then it trains a second classifier, using the updated weights, and again makes predictions on the training set, updates the instance weights, and so on (see Figure 7-7).

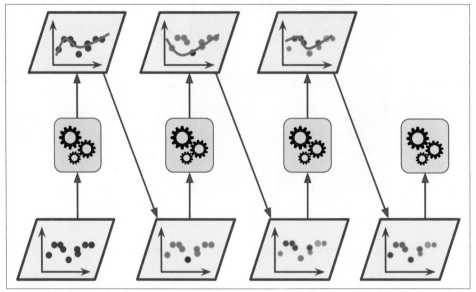

Figure 7-7. AdaBoost sequential training with instance weight updates

Figure 7-8 shows the decision boundaries of five consecutive predictors on the moons dataset (in this example, each predictor is a highly regularized SVM classifier with an RBF kernel[14]). The first classifier gets many instances wrong, so their weights

---

13 Yoav Freund and Robert E. Schapire, "A Decision-Theoretic Generalization of On-Line Learning and an Application to Boosting," *Journal of Computer and System Sciences* 55, no. 1 (1997): 119–139.

14 This is just for illustrative purposes. SVMs are generally not good base predictors for AdaBoost; they are slow and tend to be unstable with it.

get boosted. The second classifier therefore does a better job on these instances, and so on. The plot on the right represents the same sequence of predictors, except that the learning rate is halved (i.e., the misclassified instance weights are boosted half as much at every iteration). As you can see, this sequential learning technique has some similarities with Gradient Descent, except that instead of tweaking a single predictor's parameters to minimize a cost function, AdaBoost adds predictors to the ensemble, gradually making it better.

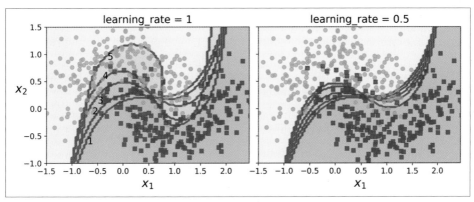

Figure 7-8. Decision boundaries of consecutive predictors

Once all predictors are trained, the ensemble makes predictions very much like bagging or pasting, except that predictors have different weights depending on their overall accuracy on the weighted training set.

 There is one important drawback to this sequential learning technique: it cannot be parallelized (or only partially), since each predictor can only be trained after the previous predictor has been trained and evaluated. As a result, it does not scale as well as bagging or pasting.

Let's take a closer look at the AdaBoost algorithm. Each instance weight $w^{(i)}$ is initially set to $1/m$. A first predictor is trained, and its weighted error rate r_1 is computed on the training set; see Equation 7-1.

Equation 7-1. Weighted error rate of the jth predictor

$$r_j = \frac{\displaystyle\sum_{\substack{i=1 \\ \hat{y}_j^{(i)} \neq y^{(i)}}}^{m} w^{(i)}}{\displaystyle\sum_{i=1}^{m} w^{(i)}} \quad \text{where } \hat{y}_j^{(i)} \text{ is the } j^{th} \text{ predictor's prediction for the } i^{th} \text{ instance.}$$

The predictor's weight α_j is then computed using Equation 7-2, where η is the learning rate hyperparameter (defaults to 1).[15] The more accurate the predictor is, the higher its weight will be. If it is just guessing randomly, then its weight will be close to zero. However, if it is most often wrong (i.e., less accurate than random guessing), then its weight will be negative.

Equation 7-2. Predictor weight

$$\alpha_j = \eta \log \frac{1 - r_j}{r_j}$$

Next, the AdaBoost algorithm updates the instance weights, using Equation 7-3, which boosts the weights of the misclassified instances.

Equation 7-3. Weight update rule

$$\text{for } i = 1, 2, \cdots, m$$

$$w^{(i)} \leftarrow \begin{cases} w^{(i)} & \text{if } \widehat{y}_j^{(i)} = y^{(i)} \\ w^{(i)} \exp\left(\alpha_j\right) & \text{if } \widehat{y}_j^{(i)} \neq y^{(i)} \end{cases}$$

Then all the instance weights are normalized (i.e., divided by $\sum_{i=1}^m w^{(i)}$).

Finally, a new predictor is trained using the updated weights, and the whole process is repeated (the new predictor's weight is computed, the instance weights are updated, then another predictor is trained, and so on). The algorithm stops when the desired number of predictors is reached, or when a perfect predictor is found.

To make predictions, AdaBoost simply computes the predictions of all the predictors and weighs them using the predictor weights α_j. The predicted class is the one that receives the majority of weighted votes (see Equation 7-4).

Equation 7-4. AdaBoost predictions

$$\widehat{y}(\mathbf{x}) = \underset{k}{\text{argmax}} \sum_{\substack{j=1 \\ \widehat{y}_j(\mathbf{x}) = k}}^{N} \alpha_j \quad \text{where } N \text{ is the number of predictors.}$$

---

15 The original AdaBoost algorithm does not use a learning rate hyperparameter.

Scikit-Learn uses a multiclass version of AdaBoost called *SAMME* (*https://homl.info/27*)[16] (which stands for *Stagewise Additive Modeling using a Multiclass Exponential loss function*). When there are just two classes, SAMME is equivalent to Ada-Boost. If the predictors can estimate class probabilities (i.e., if they have a `predict_proba()` method), Scikit-Learn can use a variant of SAMME called *SAMME.R* (the *R* stands for "Real"), which relies on class probabilities rather than predictions and generally performs better.

The following code trains an AdaBoost classifier based on 200 *Decision Stumps* using Scikit-Learn's `AdaBoostClassifier` class (as you might expect, there is also an `AdaBoostRegressor` class). A Decision Stump is a Decision Tree with `max_depth=1`—in other words, a tree composed of a single decision node plus two leaf nodes. This is the default base estimator for the `AdaBoostClassifier` class:

```
from sklearn.ensemble import AdaBoostClassifier

ada_clf = AdaBoostClassifier(
    DecisionTreeClassifier(max_depth=1), n_estimators=200,
    algorithm="SAMME.R", learning_rate=0.5)
ada_clf.fit(X_train, y_train)
```

 If your AdaBoost ensemble is overfitting the training set, you can try reducing the number of estimators or more strongly regularizing the base estimator.

Gradient Boosting

Another very popular boosting algorithm is *Gradient Boosting* (*https://homl.info/28*).[17] Just like AdaBoost, Gradient Boosting works by sequentially adding predictors to an ensemble, each one correcting its predecessor. However, instead of tweaking the instance weights at every iteration like AdaBoost does, this method tries to fit the new predictor to the *residual errors* made by the previous predictor.

Let's go through a simple regression example, using Decision Trees as the base predictors (of course, Gradient Boosting also works great with regression tasks). This is called *Gradient Tree Boosting*, or *Gradient Boosted Regression Trees* (GBRT). First, let's fit a `DecisionTreeRegressor` to the training set (for example, a noisy quadratic training set):

---

16 For more details, see Ji Zhu et al., "Multi-Class AdaBoost," *Statistics and Its Interface* 2, no. 3 (2009): 349–360.

17 Gradient Boosting was first introduced in Leo Breiman's 1997 paper "Arcing the Edge" (*https://homl.info/arcing*) and was further developed in the 1999 paper (*https://homl.info/gradboost*) "Greedy Function Approximation: A Gradient Boosting Machine" by Jerome H. Friedman.

```
from sklearn.tree import DecisionTreeRegressor

tree_reg1 = DecisionTreeRegressor(max_depth=2)
tree_reg1.fit(X, y)
```

Next, we'll train a second `DecisionTreeRegressor` on the residual errors made by the first predictor:

```
y2 = y - tree_reg1.predict(X)
tree_reg2 = DecisionTreeRegressor(max_depth=2)
tree_reg2.fit(X, y2)
```

Then we train a third regressor on the residual errors made by the second predictor:

```
y3 = y2 - tree_reg2.predict(X)
tree_reg3 = DecisionTreeRegressor(max_depth=2)
tree_reg3.fit(X, y3)
```

Now we have an ensemble containing three trees. It can make predictions on a new instance simply by adding up the predictions of all the trees:

```
y_pred = sum(tree.predict(X_new) for tree in (tree_reg1, tree_reg2, tree_reg3))
```

Figure 7-9 represents the predictions of these three trees in the left column, and the ensemble's predictions in the right column. In the first row, the ensemble has just one tree, so its predictions are exactly the same as the first tree's predictions. In the second row, a new tree is trained on the residual errors of the first tree. On the right you can see that the ensemble's predictions are equal to the sum of the predictions of the first two trees. Similarly, in the third row another tree is trained on the residual errors of the second tree. You can see that the ensemble's predictions gradually get better as trees are added to the ensemble.

A simpler way to train GBRT ensembles is to use Scikit-Learn's `GradientBoostingRe gressor` class. Much like the `RandomForestRegressor` class, it has hyperparameters to control the growth of Decision Trees (e.g., `max_depth`, `min_samples_leaf`), as well as hyperparameters to control the ensemble training, such as the number of trees (`n_estimators`). The following code creates the same ensemble as the previous one:

```
from sklearn.ensemble import GradientBoostingRegressor

gbrt = GradientBoostingRegressor(max_depth=2, n_estimators=3, learning_rate=1.0)
gbrt.fit(X, y)
```

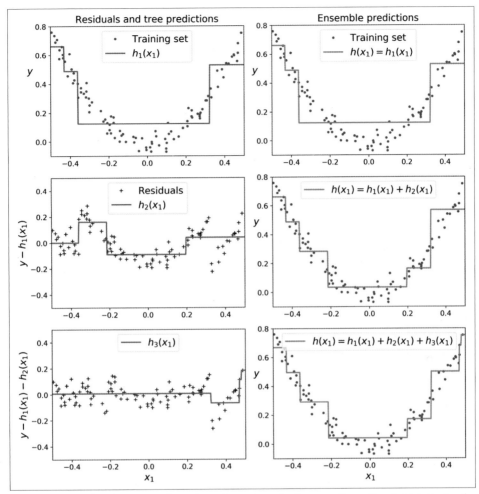

Figure 7-9. In this depiction of Gradient Boosting, the first predictor (top left) is trained normally, then each consecutive predictor (middle left and lower left) is trained on the previous predictor's residuals; the right column shows the resulting ensemble's predictions

The `learning_rate` hyperparameter scales the contribution of each tree. If you set it to a low value, such as `0.1`, you will need more trees in the ensemble to fit the training set, but the predictions will usually generalize better. This is a regularization technique called *shrinkage*. Figure 7-10 shows two GBRT ensembles trained with a low learning rate: the one on the left does not have enough trees to fit the training set, while the one on the right has too many trees and overfits the training set.

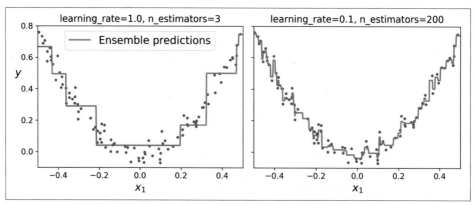

Figure 7-10. GBRT ensembles with not enough predictors (left) and too many (right)

In order to find the optimal number of trees, you can use early stopping (see Chapter 4). A simple way to implement this is to use the `staged_predict()` method: it returns an iterator over the predictions made by the ensemble at each stage of training (with one tree, two trees, etc.). The following code trains a GBRT ensemble with 120 trees, then measures the validation error at each stage of training to find the optimal number of trees, and finally trains another GBRT ensemble using the optimal number of trees:

```
import numpy as np
from sklearn.model_selection import train_test_split
from sklearn.metrics import mean_squared_error

X_train, X_val, y_train, y_val = train_test_split(X, y)

gbrt = GradientBoostingRegressor(max_depth=2, n_estimators=120)
gbrt.fit(X_train, y_train)

errors = [mean_squared_error(y_val, y_pred)
          for y_pred in gbrt.staged_predict(X_val)]
bst_n_estimators = np.argmin(errors) + 1

gbrt_best = GradientBoostingRegressor(max_depth=2,n_estimators=bst_n_estimators)
gbrt_best.fit(X_train, y_train)
```

The validation errors are represented on the left of Figure 7-11, and the best model's predictions are represented on the right.

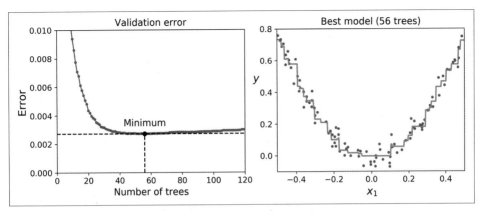

Figure 7-11. Tuning the number of trees using early stopping

It is also possible to implement early stopping by actually stopping training early (instead of training a large number of trees first and then looking back to find the optimal number). You can do so by setting warm_start=True, which makes Scikit-Learn keep existing trees when the fit() method is called, allowing incremental training. The following code stops training when the validation error does not improve for five iterations in a row:

```
gbrt = GradientBoostingRegressor(max_depth=2, warm_start=True)

min_val_error = float("inf")
error_going_up = 0
for n_estimators in range(1, 120):
    gbrt.n_estimators = n_estimators
    gbrt.fit(X_train, y_train)
    y_pred = gbrt.predict(X_val)
    val_error = mean_squared_error(y_val, y_pred)
    if val_error < min_val_error:
        min_val_error = val_error
        error_going_up = 0
    else:
        error_going_up += 1
        if error_going_up == 5:
            break  # early stopping
```

The GradientBoostingRegressor class also supports a subsample hyperparameter, which specifies the fraction of training instances to be used for training each tree. For example, if subsample=0.25, then each tree is trained on 25% of the training instances, selected randomly. As you can probably guess by now, this technique trades a higher bias for a lower variance. It also speeds up training considerably. This is called *Stochastic Gradient Boosting*.

It is possible to use Gradient Boosting with other cost functions. This is controlled by the `loss` hyperparameter (see Scikit-Learn's documentation for more details).

It is worth noting that an optimized implementation of Gradient Boosting is available in the popular Python library XGBoost (*https://github.com/dmlc/xgboost*), which stands for Extreme Gradient Boosting. This package was initially developed by Tianqi Chen as part of the Distributed (Deep) Machine Learning Community (DMLC), and it aims to be extremely fast, scalable, and portable. In fact, XGBoost is often an important component of the winning entries in ML competitions. XGBoost's API is quite similar to Scikit-Learn's:

```
import xgboost

xgb_reg = xgboost.XGBRegressor()
xgb_reg.fit(X_train, y_train)
y_pred = xgb_reg.predict(X_val)
```

XGBoost also offers several nice features, such as automatically taking care of early stopping:

```
xgb_reg.fit(X_train, y_train,
            eval_set=[(X_val, y_val)], early_stopping_rounds=2)
y_pred = xgb_reg.predict(X_val)
```

You should definitely check it out!

Stacking

The last Ensemble method we will discuss in this chapter is called *stacking* (short for *stacked generalization* (*https://homl.info/29*)).[18] It is based on a simple idea: instead of using trivial functions (such as hard voting) to aggregate the predictions of all predictors in an ensemble, why don't we train a model to perform this aggregation? Figure 7-12 shows such an ensemble performing a regression task on a new instance. Each of the bottom three predictors predicts a different value (3.1, 2.7, and 2.9), and then the final predictor (called a *blender*, or a *meta learner*) takes these predictions as inputs and makes the final prediction (3.0).

---

18 David H. Wolpert, "Stacked Generalization," *Neural Networks* 5, no. 2 (1992): 241–259.

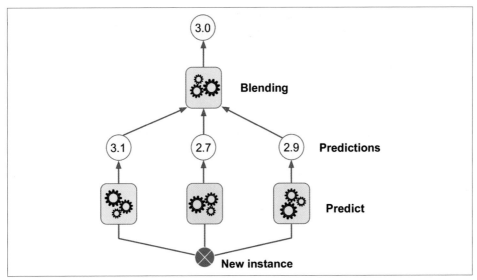

Figure 7-12. Aggregating predictions using a blending predictor

To train the blender, a common approach is to use a hold-out set.[19] Let's see how it works. First, the training set is split into two subsets. The first subset is used to train the predictors in the first layer (see Figure 7-13).

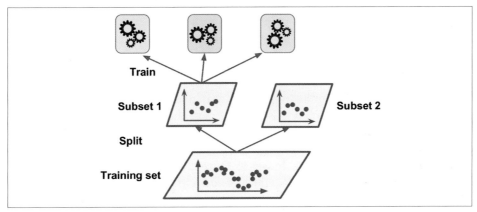

Figure 7-13. Training the first layer

Next, the first layer's predictors are used to make predictions on the second (held-out) set (see Figure 7-14). This ensures that the predictions are "clean," since the predictors never saw these instances during training. For each instance in the hold-out

---

19 Alternatively, it is possible to use out-of-fold predictions. In some contexts this is called *stacking*, while using a hold-out set is called *blending*. For many people these terms are synonymous.

set, there are three predicted values. We can create a new training set using these predicted values as input features (which makes this new training set 3D), and keeping the target values. The blender is trained on this new training set, so it learns to predict the target value, given the first layer's predictions.

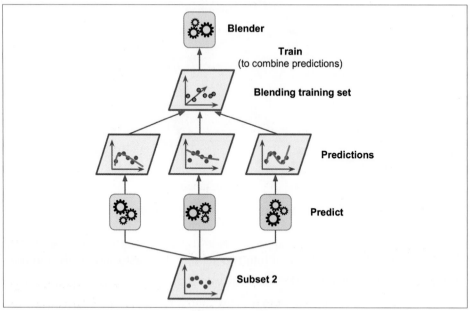

Figure 7-14. Training the blender

It is actually possible to train several different blenders this way (e.g., one using Linear Regression, another using Random Forest Regression), to get a whole layer of blenders. The trick is to split the training set into three subsets: the first one is used to train the first layer, the second one is used to create the training set used to train the second layer (using predictions made by the predictors of the first layer), and the third one is used to create the training set to train the third layer (using predictions made by the predictors of the second layer). Once this is done, we can make a prediction for a new instance by going through each layer sequentially, as shown in Figure 7-15.

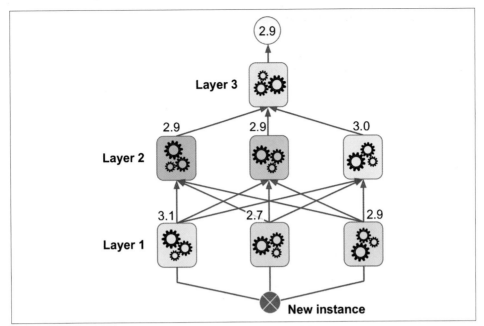

Figure 7-15. Predictions in a multilayer stacking ensemble

Unfortunately, Scikit-Learn does not support stacking directly, but it is not too hard to roll out your own implementation (see the following exercises). Alternatively, you can use an open source implementation such as `DESlib` (*https://github.com/Menelau/DESlib*).

Exercises

1. If you have trained five different models on the exact same training data, and they all achieve 95% precision, is there any chance that you can combine these models to get better results? If so, how? If not, why?

2. What is the difference between hard and soft voting classifiers?

3. Is it possible to speed up training of a bagging ensemble by distributing it across multiple servers? What about pasting ensembles, boosting ensembles, Random Forests, or stacking ensembles?

4. What is the benefit of out-of-bag evaluation?

5. What makes Extra-Trees more random than regular Random Forests? How can this extra randomness help? Are Extra-Trees slower or faster than regular Random Forests?

6. If your AdaBoost ensemble underfits the training data, which hyperparameters should you tweak and how?

7. If your Gradient Boosting ensemble overfits the training set, should you increase or decrease the learning rate?

8. Load the MNIST data (introduced in Chapter 3), and split it into a training set, a validation set, and a test set (e.g., use 50,000 instances for training, 10,000 for validation, and 10,000 for testing). Then train various classifiers, such as a Random Forest classifier, an Extra-Trees classifier, and an SVM classifier. Next, try to combine them into an ensemble that outperforms each individual classifier on the validation set, using soft or hard voting. Once you have found one, try it on the test set. How much better does it perform compared to the individual classifiers?

9. Run the individual classifiers from the previous exercise to make predictions on the validation set, and create a new training set with the resulting predictions: each training instance is a vector containing the set of predictions from all your classifiers for an image, and the target is the image's class. Train a classifier on this new training set. Congratulations, you have just trained a blender, and together with the classifiers it forms a stacking ensemble! Now evaluate the ensemble on the test set. For each image in the test set, make predictions with all your classifiers, then feed the predictions to the blender to get the ensemble's predictions. How does it compare to the voting classifier you trained earlier?

Solutions to these exercises are available in Appendix A.

CHAPTER 8

Dimensionality Reduction

Many Machine Learning problems involve thousands or even millions of features for each training instance. Not only do all these features make training extremely slow, but they can also make it much harder to find a good solution, as we will see. This problem is often referred to as the *curse of dimensionality*.

Fortunately, in real-world problems, it is often possible to reduce the number of features considerably, turning an intractable problem into a tractable one. For example, consider the MNIST images (introduced in Chapter 3): the pixels on the image borders are almost always white, so you could completely drop these pixels from the training set without losing much information. Figure 7-6 confirms that these pixels are utterly unimportant for the classification task. Additionally, two neighboring pixels are often highly correlated: if you merge them into a single pixel (e.g., by taking the mean of the two pixel intensities), you will not lose much information.

Reducing dimensionality does cause some information loss (just like compressing an image to JPEG can degrade its quality), so even though it will speed up training, it may make your system perform slightly worse. It also makes your pipelines a bit more complex and thus harder to maintain. So, if training is too slow, you should first try to train your system with the original data before considering using dimensionality reduction. In some cases, reducing the dimensionality of the training data may filter out some noise and unnecessary details and thus result in higher performance, but in general it won't; it will just speed up training.

Apart from speeding up training, dimensionality reduction is also extremely useful for data visualization (or *DataViz*). Reducing the number of dimensions down to two (or three) makes it possible to plot a condensed view of a high-dimensional training set on a graph and often gain some important insights by visually detecting patterns, such as clusters. Moreover, DataViz is essential to communicate your conclusions to people who are not data scientists—in particular, decision makers who will use your results.

In this chapter we will discuss the curse of dimensionality and get a sense of what goes on in high-dimensional space. Then, we will consider the two main approaches to dimensionality reduction (projection and Manifold Learning), and we will go through three of the most popular dimensionality reduction techniques: PCA, Kernel PCA, and LLE.

The Curse of Dimensionality

We are so used to living in three dimensions[1] that our intuition fails us when we try to imagine a high-dimensional space. Even a basic 4D hypercube is incredibly hard to picture in our minds (see Figure 8-1), let alone a 200-dimensional ellipsoid bent in a 1,000-dimensional space.

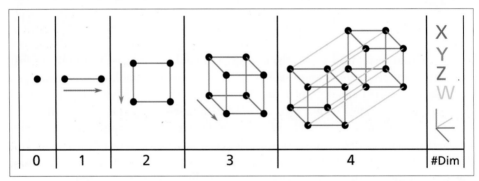

Figure 8-1. Point, segment, square, cube, and tesseract (0D to 4D hypercubes)[2]

It turns out that many things behave very differently in high-dimensional space. For example, if you pick a random point in a unit square (a 1 × 1 square), it will have only about a 0.4% chance of being located less than 0.001 from a border (in other words, it is very unlikely that a random point will be "extreme" along any dimension). But in a 10,000-dimensional unit hypercube, this probability is greater than 99.999999%. Most points in a high-dimensional hypercube are very close to the border.[3]

---

1 Well, four dimensions if you count time, and a few more if you are a string theorist.

2 Watch a rotating tesseract projected into 3D space at *https://homl.info/30*. Image by Wikipedia user Nerd-Boy1392 (Creative Commons BY-SA 3.0 (*https://creativecommons.org/licenses/by-sa/3.0/*)). Reproduced from *https://en.wikipedia.org/wiki/Tesseract*.

3 Fun fact: anyone you know is probably an extremist in at least one dimension (e.g., how much sugar they put in their coffee), if you consider enough dimensions.

Here is a more troublesome difference: if you pick two points randomly in a unit square, the distance between these two points will be, on average, roughly 0.52. If you pick two random points in a unit 3D cube, the average distance will be roughly 0.66. But what about two points picked randomly in a 1,000,000-dimensional hypercube? The average distance, believe it or not, will be about 408.25 (roughly $\sqrt{1,000,000/6}$)! This is counterintuitive: how can two points be so far apart when they both lie within the same unit hypercube? Well, there's just plenty of space in high dimensions. As a result, high-dimensional datasets are at risk of being very sparse: most training instances are likely to be far away from each other. This also means that a new instance will likely be far away from any training instance, making predictions much less reliable than in lower dimensions, since they will be based on much larger extrapolations. In short, the more dimensions the training set has, the greater the risk of overfitting it.

In theory, one solution to the curse of dimensionality could be to increase the size of the training set to reach a sufficient density of training instances. Unfortunately, in practice, the number of training instances required to reach a given density grows exponentially with the number of dimensions. With just 100 features (significantly fewer than in the MNIST problem), you would need more training instances than atoms in the observable universe in order for training instances to be within 0.1 of each other on average, assuming they were spread out uniformly across all dimensions.

Main Approaches for Dimensionality Reduction

Before we dive into specific dimensionality reduction algorithms, let's take a look at the two main approaches to reducing dimensionality: projection and Manifold Learning.

Projection

In most real-world problems, training instances are *not* spread out uniformly across all dimensions. Many features are almost constant, while others are highly correlated (as discussed earlier for MNIST). As a result, all training instances lie within (or close to) a much lower-dimensional *subspace* of the high-dimensional space. This sounds very abstract, so let's look at an example. In Figure 8-2 you can see a 3D dataset represented by circles.

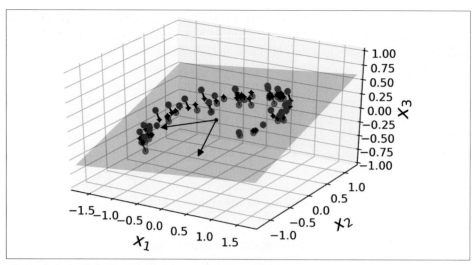

Figure 8-2. A 3D dataset lying close to a 2D subspace

Notice that all training instances lie close to a plane: this is a lower-dimensional (2D) subspace of the high-dimensional (3D) space. If we project every training instance perpendicularly onto this subspace (as represented by the short lines connecting the instances to the plane), we get the new 2D dataset shown in Figure 8-3. Ta-da! We have just reduced the dataset's dimensionality from 3D to 2D. Note that the axes correspond to new features z_1 and z_2 (the coordinates of the projections on the plane).

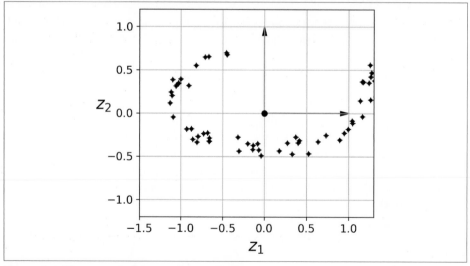

Figure 8-3. The new 2D dataset after projection

However, projection is not always the best approach to dimensionality reduction. In many cases the subspace may twist and turn, such as in the famous *Swiss roll* toy dataset represented in Figure 8-4.

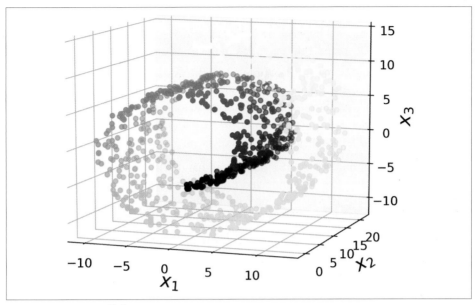

Figure 8-4. Swiss roll dataset

Simply projecting onto a plane (e.g., by dropping x_3) would squash different layers of the Swiss roll together, as shown on the left side of Figure 8-5. What you really want is to unroll the Swiss roll to obtain the 2D dataset on the right side of Figure 8-5.

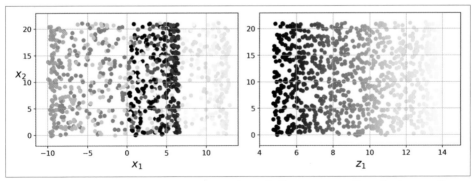

Figure 8-5. Squashing by projecting onto a plane (left) versus unrolling the Swiss roll (right)

Manifold Learning

The Swiss roll is an example of a 2D *manifold*. Put simply, a 2D manifold is a 2D shape that can be bent and twisted in a higher-dimensional space. More generally, a *d*-dimensional manifold is a part of an *n*-dimensional space (where $d < n$) that locally resembles a *d*-dimensional hyperplane. In the case of the Swiss roll, $d = 2$ and $n = 3$: it locally resembles a 2D plane, but it is rolled in the third dimension.

Many dimensionality reduction algorithms work by modeling the manifold on which the training instances lie; this is called *Manifold Learning*. It relies on the *manifold assumption*, also called the *manifold hypothesis*, which holds that most real-world high-dimensional datasets lie close to a much lower-dimensional manifold. This assumption is very often empirically observed.

Once again, think about the MNIST dataset: all handwritten digit images have some similarities. They are made of connected lines, the borders are white, and they are more or less centered. If you randomly generated images, only a ridiculously tiny fraction of them would look like handwritten digits. In other words, the degrees of freedom available to you if you try to create a digit image are dramatically lower than the degrees of freedom you would have if you were allowed to generate any image you wanted. These constraints tend to squeeze the dataset into a lower-dimensional manifold.

The manifold assumption is often accompanied by another implicit assumption: that the task at hand (e.g., classification or regression) will be simpler if expressed in the lower-dimensional space of the manifold. For example, in the top row of Figure 8-6 the Swiss roll is split into two classes: in the 3D space (on the left), the decision boundary would be fairly complex, but in the 2D unrolled manifold space (on the right), the decision boundary is a straight line.

However, this implicit assumption does not always hold. For example, in the bottom row of Figure 8-6, the decision boundary is located at $x_1 = 5$. This decision boundary looks very simple in the original 3D space (a vertical plane), but it looks more complex in the unrolled manifold (a collection of four independent line segments).

In short, reducing the dimensionality of your training set before training a model will usually speed up training, but it may not always lead to a better or simpler solution; it all depends on the dataset.

Hopefully you now have a good sense of what the curse of dimensionality is and how dimensionality reduction algorithms can fight it, especially when the manifold assumption holds. The rest of this chapter will go through some of the most popular algorithms.

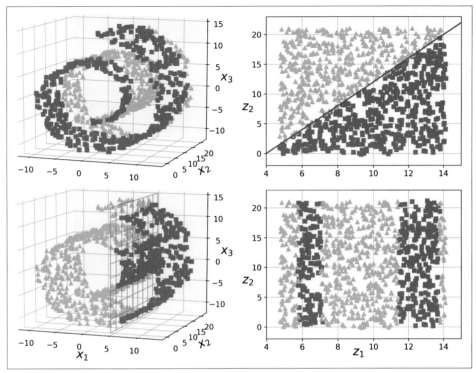

Figure 8-6. The decision boundary may not always be simpler with lower dimensions

PCA

Principal Component Analysis (PCA) is by far the most popular dimensionality reduction algorithm. First it identifies the hyperplane that lies closest to the data, and then it projects the data onto it, just like in Figure 8-2.

Preserving the Variance

Before you can project the training set onto a lower-dimensional hyperplane, you first need to choose the right hyperplane. For example, a simple 2D dataset is represented on the left in Figure 8-7, along with three different axes (i.e., 1D hyperplanes). On the right is the result of the projection of the dataset onto each of these axes. As you can see, the projection onto the solid line preserves the maximum variance, while the projection onto the dotted line preserves very little variance and the projection onto the dashed line preserves an intermediate amount of variance.

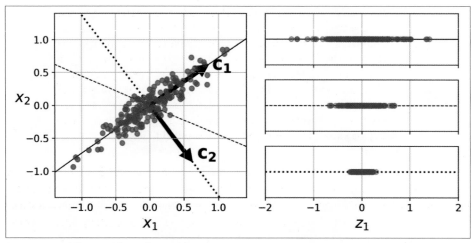

Figure 8-7. Selecting the subspace to project on

It seems reasonable to select the axis that preserves the maximum amount of variance, as it will most likely lose less information than the other projections. Another way to justify this choice is that it is the axis that minimizes the mean squared distance between the original dataset and its projection onto that axis. This is the rather simple idea behind PCA.[4]

Principal Components

PCA identifies the axis that accounts for the largest amount of variance in the training set. In Figure 8-7, it is the solid line. It also finds a second axis, orthogonal to the first one, that accounts for the largest amount of remaining variance. In this 2D example there is no choice: it is the dotted line. If it were a higher-dimensional dataset, PCA would also find a third axis, orthogonal to both previous axes, and a fourth, a fifth, and so on—as many axes as the number of dimensions in the dataset.

The ith axis is called the ith *principal component* (PC) of the data. In Figure 8-7, the first PC is the axis on which vector $\mathbf{c}_1$ lies, and the second PC is the axis on which vector $\mathbf{c}_2$ lies. In Figure 8-2 the first two PCs are the orthogonal axes on which the two arrows lie, on the plane, and the third PC is the axis orthogonal to that plane.

---

4 Karl Pearson, "On Lines and Planes of Closest Fit to Systems of Points in Space," *The London, Edinburgh, and Dublin Philosophical Magazine and Journal of Science* 2, no. 11 (1901): 559-572, *https://homl.info/pca*.

For each principal component, PCA finds a zero-centered unit vector pointing in the direction of the PC. Since two opposing unit vectors lie on the same axis, the direction of the unit vectors returned by PCA is not stable: if you perturb the training set slightly and run PCA again, the unit vectors may point in the opposite direction as the original vectors. However, they will generally still lie on the same axes. In some cases, a pair of unit vectors may even rotate or swap (if the variances along these two axes are close), but the plane they define will generally remain the same.

So how can you find the principal components of a training set? Luckily, there is a standard matrix factorization technique called *Singular Value Decomposition* (SVD) that can decompose the training set matrix $\mathbf{X}$ into the matrix multiplication of three matrices $\mathbf{U} \, \boldsymbol{\Sigma} \, \mathbf{V}^{\mathsf{T}}$, where $\mathbf{V}$ contains the unit vectors that define all the principal components that we are looking for, as shown in Equation 8-1.

Equation 8-1. Principal components matrix

$$\mathbf{V} = \begin{pmatrix} | & | & & | \\ \mathbf{c}_1 & \mathbf{c}_2 & \cdots & \mathbf{c}_n \\ | & | & & | \end{pmatrix}$$

The following Python code uses NumPy's svd() function to obtain all the principal components of the training set, then extracts the two unit vectors that define the first two PCs:

```
X_centered = X - X.mean(axis=0)
U, s, Vt = np.linalg.svd(X_centered)
c1 = Vt.T[:, 0]
c2 = Vt.T[:, 1]
```

PCA assumes that the dataset is centered around the origin. As we will see, Scikit-Learn's PCA classes take care of centering the data for you. If you implement PCA yourself (as in the preceding example), or if you use other libraries, don't forget to center the data first.

Projecting Down to d Dimensions

Once you have identified all the principal components, you can reduce the dimensionality of the dataset down to d dimensions by projecting it onto the hyperplane defined by the first d principal components. Selecting this hyperplane ensures that the projection will preserve as much variance as possible. For example, in Figure 8-2 the 3D dataset is projected down to the 2D plane defined by the first two principal components, preserving a large part of the dataset's variance. As a result, the 2D projection looks very much like the original 3D dataset.

To project the training set onto the hyperplane and obtain a reduced dataset $\mathbf{X}_{d\text{-proj}}$ of dimensionality d, compute the matrix multiplication of the training set matrix $\mathbf{X}$ by the matrix $\mathbf{W}_d$, defined as the matrix containing the first d columns of $\mathbf{V}$, as shown in Equation 8-2.

Equation 8-2. Projecting the training set down to d dimensions

$$\mathbf{X}_{d\text{-proj}} = \mathbf{X}\mathbf{W}_d$$

The following Python code projects the training set onto the plane defined by the first two principal components:

```
W2 = Vt.T[:, :2]
X2D = X_centered.dot(W2)
```

There you have it! You now know how to reduce the dimensionality of any dataset down to any number of dimensions, while preserving as much variance as possible.

Using Scikit-Learn

Scikit-Learn's PCA class uses SVD decomposition to implement PCA, just like we did earlier in this chapter. The following code applies PCA to reduce the dimensionality of the dataset down to two dimensions (note that it automatically takes care of centering the data):

```
from sklearn.decomposition import PCA

pca = PCA(n_components = 2)
X2D = pca.fit_transform(X)
```

After fitting the PCA transformer to the dataset, its components_ attribute holds the transpose of $\mathbf{W}_d$ (e.g., the unit vector that defines the first principal component is equal to pca.components_.T[:, 0]).

Explained Variance Ratio

Another useful piece of information is the *explained variance ratio* of each principal component, available via the explained_variance_ratio_ variable. The ratio indicates the proportion of the dataset's variance that lies along each principal component. For example, let's look at the explained variance ratios of the first two components of the 3D dataset represented in Figure 8-2:

```
>>> pca.explained_variance_ratio_
array([0.84248607, 0.14631839])
```

This output tells you that 84.2% of the dataset's variance lies along the first PC, and 14.6% lies along the second PC. This leaves less than 1.2% for the third PC, so it is reasonable to assume that the third PC probably carries little information.

Choosing the Right Number of Dimensions

Instead of arbitrarily choosing the number of dimensions to reduce down to, it is simpler to choose the number of dimensions that add up to a sufficiently large portion of the variance (e.g., 95%). Unless, of course, you are reducing dimensionality for data visualization—in that case you will want to reduce the dimensionality down to 2 or 3.

The following code performs PCA without reducing dimensionality, then computes the minimum number of dimensions required to preserve 95% of the training set's variance:

```
pca = PCA()
pca.fit(X_train)
cumsum = np.cumsum(pca.explained_variance_ratio_)
d = np.argmax(cumsum >= 0.95) + 1
```

You could then set n_components=d and run PCA again. But there is a much better option: instead of specifying the number of principal components you want to preserve, you can set n_components to be a float between 0.0 and 1.0, indicating the ratio of variance you wish to preserve:

```
pca = PCA(n_components=0.95)
X_reduced = pca.fit_transform(X_train)
```

Yet another option is to plot the explained variance as a function of the number of dimensions (simply plot cumsum; see Figure 8-8). There will usually be an elbow in the curve, where the explained variance stops growing fast. In this case, you can see that reducing the dimensionality down to about 100 dimensions wouldn't lose too much explained variance.

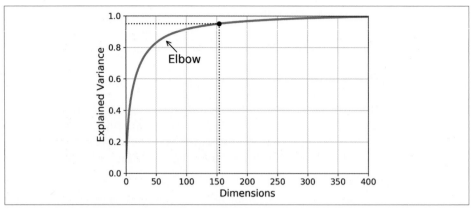

Figure 8-8. Explained variance as a function of the number of dimensions

PCA for Compression

After dimensionality reduction, the training set takes up much less space. As an example, try applying PCA to the MNIST dataset while preserving 95% of its variance. You should find that each instance will have just over 150 features, instead of the original 784 features. So, while most of the variance is preserved, the dataset is now less than 20% of its original size! This is a reasonable compression ratio, and you can see how this size reduction can speed up a classification algorithm (such as an SVM classifier) tremendously.

It is also possible to decompress the reduced dataset back to 784 dimensions by applying the inverse transformation of the PCA projection. This won't give you back the original data, since the projection lost a bit of information (within the 5% variance that was dropped), but it will likely be close to the original data. The mean squared distance between the original data and the reconstructed data (compressed and then decompressed) is called the *reconstruction error*.

The following code compresses the MNIST dataset down to 154 dimensions, then uses the `inverse_transform()` method to decompress it back to 784 dimensions:

```
pca = PCA(n_components = 154)
X_reduced = pca.fit_transform(X_train)
X_recovered = pca.inverse_transform(X_reduced)
```

Figure 8-9 shows a few digits from the original training set (on the left), and the corresponding digits after compression and decompression. You can see that there is a slight image quality loss, but the digits are still mostly intact.

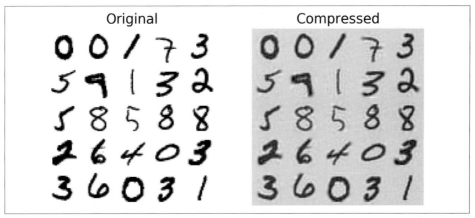

Figure 8-9. MNIST compression that preserves 95% of the variance

The equation of the inverse transformation is shown in Equation 8-3.

Equation 8-3. PCA inverse transformation, back to the original number of dimensions

$$X_{\text{recovered}} = X_{d\text{-proj}} W_d^T$$

Randomized PCA

If you set the svd_solver hyperparameter to "randomized", Scikit-Learn uses a stochastic algorithm called *Randomized PCA* that quickly finds an approximation of the first d principal components. Its computational complexity is $O(m \times d^2) + O(d^3)$, instead of $O(m \times n^2) + O(n^3)$ for the full SVD approach, so it is dramatically faster than full SVD when d is much smaller than n:

```
rnd_pca = PCA(n_components=154, svd_solver="randomized")
X_reduced = rnd_pca.fit_transform(X_train)
```

By default, svd_solver is actually set to "auto": Scikit-Learn automatically uses the randomized PCA algorithm if m or n is greater than 500 and d is less than 80% of m or n, or else it uses the full SVD approach. If you want to force Scikit-Learn to use full SVD, you can set the svd_solver hyperparameter to "full".

Incremental PCA

One problem with the preceding implementations of PCA is that they require the whole training set to fit in memory in order for the algorithm to run. Fortunately, *Incremental PCA* (IPCA) algorithms have been developed. They allow you to split the training set into mini-batches and feed an IPCA algorithm one mini-batch at a time.

This is useful for large training sets and for applying PCA online (i.e., on the fly, as new instances arrive).

The following code splits the MNIST dataset into 100 mini-batches (using NumPy's `array_split()` function) and feeds them to Scikit-Learn's `IncrementalPCA` class (*https://homl.info/32*)[5] to reduce the dimensionality of the MNIST dataset down to 154 dimensions (just like before). Note that you must call the `partial_fit()` method with each mini-batch, rather than the `fit()` method with the whole training set:

```
from sklearn.decomposition import IncrementalPCA

n_batches = 100
inc_pca = IncrementalPCA(n_components=154)
for X_batch in np.array_split(X_train, n_batches):
    inc_pca.partial_fit(X_batch)

X_reduced = inc_pca.transform(X_train)
```

Alternatively, you can use NumPy's `memmap` class, which allows you to manipulate a large array stored in a binary file on disk as if it were entirely in memory; the class loads only the data it needs in memory, when it needs it. Since the `IncrementalPCA` class uses only a small part of the array at any given time, the memory usage remains under control. This makes it possible to call the usual `fit()` method, as you can see in the following code:

```
X_mm = np.memmap(filename, dtype="float32", mode="readonly", shape=(m, n))

batch_size = m // n_batches
inc_pca = IncrementalPCA(n_components=154, batch_size=batch_size)
inc_pca.fit(X_mm)
```

Kernel PCA

In Chapter 5 we discussed the kernel trick, a mathematical technique that implicitly maps instances into a very high-dimensional space (called the *feature space*), enabling nonlinear classification and regression with Support Vector Machines. Recall that a linear decision boundary in the high-dimensional feature space corresponds to a complex nonlinear decision boundary in the *original space*.

It turns out that the same trick can be applied to PCA, making it possible to perform complex nonlinear projections for dimensionality reduction. This is called *Kernel PCA* (kPCA) (*https://homl.info/33*).[6] It is often good at preserving clusters of

---

5 Scikit-Learn uses the algorithm described in David A. Ross et al., "Incremental Learning for Robust Visual Tracking," *International Journal of Computer Vision* 77, no. 1–3 (2008): 125–141.

6 Bernhard Schölkopf et al., "Kernel Principal Component Analysis," in *Lecture Notes in Computer Science* 1327 (Berlin: Springer, 1997): 583–588.

instances after projection, or sometimes even unrolling datasets that lie close to a twisted manifold.

The following code uses Scikit-Learn's KernelPCA class to perform kPCA with an RBF kernel (see Chapter 5 for more details about the RBF kernel and other kernels):

```
from sklearn.decomposition import KernelPCA

rbf_pca = KernelPCA(n_components = 2, kernel="rbf", gamma=0.04)
X_reduced = rbf_pca.fit_transform(X)
```

Figure 8-10 shows the Swiss roll, reduced to two dimensions using a linear kernel (equivalent to simply using the PCA class), an RBF kernel, and a sigmoid kernel.

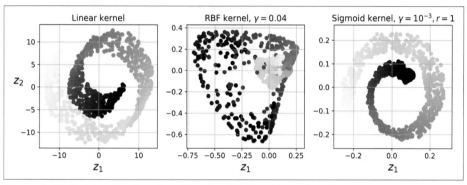

Figure 8-10. Swiss roll reduced to 2D using kPCA with various kernels

Selecting a Kernel and Tuning Hyperparameters

As kPCA is an unsupervised learning algorithm, there is no obvious performance measure to help you select the best kernel and hyperparameter values. That said, dimensionality reduction is often a preparation step for a supervised learning task (e.g., classification), so you can use grid search to select the kernel and hyperparameters that lead to the best performance on that task. The following code creates a two-step pipeline, first reducing dimensionality to two dimensions using kPCA, then applying Logistic Regression for classification. Then it uses GridSearchCV to find the best kernel and gamma value for kPCA in order to get the best classification accuracy at the end of the pipeline:

```
from sklearn.model_selection import GridSearchCV
from sklearn.linear_model import LogisticRegression
from sklearn.pipeline import Pipeline

clf = Pipeline([
        ("kpca", KernelPCA(n_components=2)),
        ("log_reg", LogisticRegression())
    ])
```

```
param_grid = [{
        "kpca__gamma": np.linspace(0.03, 0.05, 10),
        "kpca__kernel": ["rbf", "sigmoid"]
    }]

grid_search = GridSearchCV(clf, param_grid, cv=3)
grid_search.fit(X, y)
```

The best kernel and hyperparameters are then available through the `best_params_`
variable:

```
>>> print(grid_search.best_params_)
{'kpca__gamma': 0.043333333333333335, 'kpca__kernel': 'rbf'}
```

Another approach, this time entirely unsupervised, is to select the kernel and hyperparameters that yield the lowest reconstruction error. Note that reconstruction is not
as easy as with linear PCA. Here's why. Figure 8-11 shows the original Swiss roll 3D
dataset (top left) and the resulting 2D dataset after kPCA is applied using an RBF kernel (top right). Thanks to the kernel trick, this transformation is mathematically
equivalent to using the *feature map* φ to map the training set to an infinite-
dimensional feature space (bottom right), then projecting the transformed training
set down to 2D using linear PCA.

Notice that if we could invert the linear PCA step for a given instance in the reduced
space, the reconstructed point would lie in feature space, not in the original space
(e.g., like the one represented by an X in the diagram). Since the feature space is
infinite-dimensional, we cannot compute the reconstructed point, and therefore we
cannot compute the true reconstruction error. Fortunately, it is possible to find a
point in the original space that would map close to the reconstructed point. This
point is called the reconstruction *pre-image*. Once you have this pre-image, you can
measure its squared distance to the original instance. You can then select the kernel
and hyperparameters that minimize this reconstruction pre-image error.

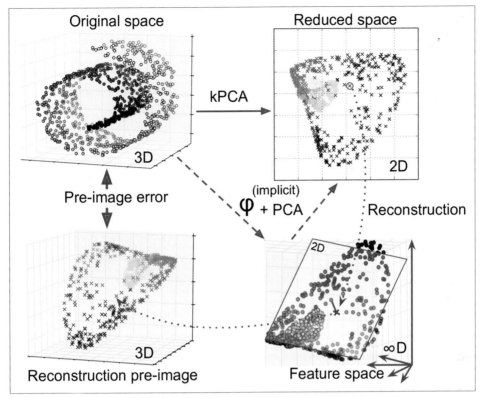

Figure 8-11. Kernel PCA and the reconstruction pre-image error

You may be wondering how to perform this reconstruction. One solution is to train a supervised regression model, with the projected instances as the training set and the original instances as the targets. Scikit-Learn will do this automatically if you set fit_inverse_transform=True, as shown in the following code:[7]

```
rbf_pca = KernelPCA(n_components = 2, kernel="rbf", gamma=0.0433,
                    fit_inverse_transform=True)
X_reduced = rbf_pca.fit_transform(X)
X_preimage = rbf_pca.inverse_transform(X_reduced)
```

By default, fit_inverse_transform=False and KernelPCA has no inverse_transform() method. This method only gets created when you set fit_inverse_transform=True.

7 If you set fit_inverse_transform=True, Scikit-Learn will use the algorithm (based on Kernel Ridge Regression) described in Gokhan H. Bakır et al., "Learning to Find Pre-Images" (*https://homl.info/34*), *Proceedings of the 16th International Conference on Neural Information Processing Systems* (2004): 449–456.

You can then compute the reconstruction pre-image error:

```
>>> from sklearn.metrics import mean_squared_error
>>> mean_squared_error(X, X_preimage)
32.786308795766132
```

Now you can use grid search with cross-validation to find the kernel and hyperparameters that minimize this error.

LLE

Locally Linear Embedding (LLE) (*https://homl.info/lle*)[8] is another powerful *nonlinear dimensionality reduction* (NLDR) technique. It is a Manifold Learning technique that does not rely on projections, like the previous algorithms do. In a nutshell, LLE works by first measuring how each training instance linearly relates to its closest neighbors (c.n.), and then looking for a low-dimensional representation of the training set where these local relationships are best preserved (more details shortly). This approach makes it particularly good at unrolling twisted manifolds, especially when there is not too much noise.

The following code uses Scikit-Learn's `LocallyLinearEmbedding` class to unroll the Swiss roll:

```
from sklearn.manifold import LocallyLinearEmbedding

lle = LocallyLinearEmbedding(n_components=2, n_neighbors=10)
X_reduced = lle.fit_transform(X)
```

The resulting 2D dataset is shown in Figure 8-12. As you can see, the Swiss roll is completely unrolled, and the distances between instances are locally well preserved. However, distances are not preserved on a larger scale: the left part of the unrolled Swiss roll is stretched, while the right part is squeezed. Nevertheless, LLE did a pretty good job at modeling the manifold.

---

8 Sam T. Roweis and Lawrence K. Saul, "Nonlinear Dimensionality Reduction by Locally Linear Embedding," *Science* 290, no. 5500 (2000): 2323–2326.

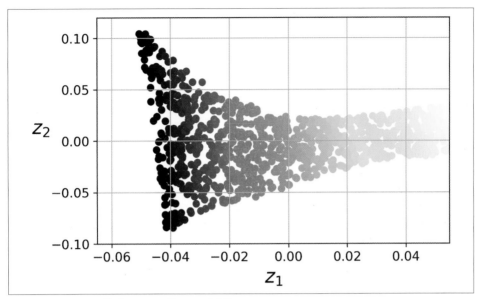

Figure 8-12. Unrolled Swiss roll using LLE

Here's how LLE works: for each training instance $\mathbf{x}^{(i)}$, the algorithm identifies its k closest neighbors (in the preceding code $k = 10$), then tries to reconstruct $\mathbf{x}^{(i)}$ as a linear function of these neighbors. More specifically, it finds the weights $w_{i,j}$ such that the squared distance between $\mathbf{x}^{(i)}$ and $\sum_{j=1}^{m} w_{i,j}\mathbf{x}^{(j)}$ is as small as possible, assuming $w_{i,j}$ = 0 if $\mathbf{x}^{(j)}$ is not one of the k closest neighbors of $\mathbf{x}^{(i)}$. Thus the first step of LLE is the constrained optimization problem described in Equation 8-4, where $\mathbf{W}$ is the weight matrix containing all the weights $w_{i,j}$. The second constraint simply normalizes the weights for each training instance $\mathbf{x}^{(i)}$.

Equation 8-4. LLE step one: linearly modeling local relationships

$$\widehat{\mathbf{W}} = \underset{\mathbf{W}}{\operatorname{argmin}} \sum_{i=1}^{m} \left(\mathbf{x}^{(i)} - \sum_{j=1}^{m} w_{i,j}\mathbf{x}^{(j)} \right)^2$$

$$\text{subject to } \begin{cases} w_{i,j} = 0 & \text{if } \mathbf{x}^{(j)} \text{ is not one of the } k \text{ c.n. of } \mathbf{x}^{(i)} \\ \sum_{j=1}^{m} w_{i,j} = 1 & \text{for } i = 1, 2, \cdots, m \end{cases}$$

After this step, the weight matrix $\widehat{\mathbf{W}}$ (containing the weights $\widehat{w}_{i,j}$) encodes the local linear relationships between the training instances. The second step is to map the training instances into a d-dimensional space (where $d < n$) while preserving these local relationships as much as possible. If $\mathbf{z}^{(i)}$ is the image of $\mathbf{x}^{(i)}$ in this d-dimensional space, then we want the squared distance between $\mathbf{z}^{(i)}$ and $\sum_{j=1}^{m} \hat{w}_{i,j} \mathbf{z}^{(j)}$ to be as small as possible. This idea leads to the unconstrained optimization problem described in Equation 8-5. It looks very similar to the first step, but instead of keeping the instances fixed and finding the optimal weights, we are doing the reverse: keeping the weights fixed and finding the optimal position of the instances' images in the low-dimensional space. Note that $\mathbf{Z}$ is the matrix containing all $\mathbf{z}^{(i)}$.

Equation 8-5. LLE step two: reducing dimensionality while preserving relationships

$$\hat{\mathbf{Z}} = \underset{\mathbf{Z}}{\mathrm{argmin}} \sum_{i=1}^{m} \left(\mathbf{z}^{(i)} - \sum_{j=1}^{m} \hat{w}_{i,j} \mathbf{z}^{(j)} \right)^2$$

Scikit-Learn's LLE implementation has the following computational complexity: $O(m \log(m)n \log(k))$ for finding the k nearest neighbors, $O(mnk^3)$ for optimizing the weights, and $O(dm^2)$ for constructing the low-dimensional representations. Unfortunately, the m^2 in the last term makes this algorithm scale poorly to very large datasets.

Other Dimensionality Reduction Techniques

There are many other dimensionality reduction techniques, several of which are available in Scikit-Learn. Here are some of the most popular ones:

Random Projections
> As its name suggests, projects the data to a lower-dimensional space using a random linear projection. This may sound crazy, but it turns out that such a random projection is actually very likely to preserve distances well, as was demonstrated mathematically by William B. Johnson and Joram Lindenstrauss in a famous lemma. The quality of the dimensionality reduction depends on the number of instances and the target dimensionality, but surprisingly not on the initial dimensionality. Check out the documentation for the `sklearn.random_projection` package for more details.

Multidimensional Scaling (MDS)
> Reduces dimensionality while trying to preserve the distances between the instances.

Isomap

Creates a graph by connecting each instance to its nearest neighbors, then reduces dimensionality while trying to preserve the *geodesic distances*[9] between the instances.

t-Distributed Stochastic Neighbor Embedding (t-SNE)

Reduces dimensionality while trying to keep similar instances close and dissimilar instances apart. It is mostly used for visualization, in particular to visualize clusters of instances in high-dimensional space (e.g., to visualize the MNIST images in 2D).

Linear Discriminant Analysis (LDA)

Is a classification algorithm, but during training it learns the most discriminative axes between the classes, and these axes can then be used to define a hyperplane onto which to project the data. The benefit of this approach is that the projection will keep classes as far apart as possible, so LDA is a good technique to reduce dimensionality before running another classification algorithm such as an SVM classifier.

Figure 8-13 shows the results of a few of these techniques.

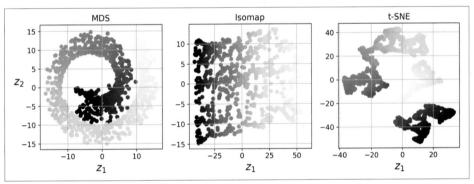

Figure 8-13. Using various techniques to reduce the Swill roll to 2D

Exercises

1. What are the main motivations for reducing a dataset's dimensionality? What are the main drawbacks?

2. What is the curse of dimensionality?

---

9 The geodesic distance between two nodes in a graph is the number of nodes on the shortest path between these nodes.

3. Once a dataset's dimensionality has been reduced, is it possible to reverse the operation? If so, how? If not, why?

4. Can PCA be used to reduce the dimensionality of a highly nonlinear dataset?

5. Suppose you perform PCA on a 1,000-dimensional dataset, setting the explained variance ratio to 95%. How many dimensions will the resulting dataset have?

6. In what cases would you use vanilla PCA, Incremental PCA, Randomized PCA, or Kernel PCA?

7. How can you evaluate the performance of a dimensionality reduction algorithm on your dataset?

8. Does it make any sense to chain two different dimensionality reduction algorithms?

9. Load the MNIST dataset (introduced in Chapter 3) and split it into a training set and a test set (take the first 60,000 instances for training, and the remaining 10,000 for testing). Train a Random Forest classifier on the dataset and time how long it takes, then evaluate the resulting model on the test set. Next, use PCA to reduce the dataset's dimensionality, with an explained variance ratio of 95%. Train a new Random Forest classifier on the reduced dataset and see how long it takes. Was training much faster? Next, evaluate the classifier on the test set. How does it compare to the previous classifier?

10. Use t-SNE to reduce the MNIST dataset down to two dimensions and plot the result using Matplotlib. You can use a scatterplot using 10 different colors to represent each image's target class. Alternatively, you can replace each dot in the scatterplot with the corresponding instance's class (a digit from 0 to 9), or even plot scaled-down versions of the digit images themselves (if you plot all digits, the visualization will be too cluttered, so you should either draw a random sample or plot an instance only if no other instance has already been plotted at a close distance). You should get a nice visualization with well-separated clusters of digits. Try using other dimensionality reduction algorithms such as PCA, LLE, or MDS and compare the resulting visualizations.

Solutions to these exercises are available in Appendix A.

## CHAPTER 9

Unsupervised Learning Techniques

Although most of the applications of Machine Learning today are based on supervised learning (and as a result, this is where most of the investments go to), the vast majority of the available data is unlabeled: we have the input features **X**, but we do not have the labels **y**. The computer scientist Yann LeCun famously said that "if intelligence was a cake, unsupervised learning would be the cake, supervised learning would be the icing on the cake, and reinforcement learning would be the cherry on the cake." In other words, there is a huge potential in unsupervised learning that we have only barely started to sink our teeth into.

Say you want to create a system that will take a few pictures of each item on a manufacturing production line and detect which items are defective. You can fairly easily create a system that will take pictures automatically, and this might give you thousands of pictures every day. You can then build a reasonably large dataset in just a few weeks. But wait, there are no labels! If you want to train a regular binary classifier that will predict whether an item is defective or not, you will need to label every single picture as "defective" or "normal." This will generally require human experts to sit down and manually go through all the pictures. This is a long, costly, and tedious task, so it will usually only be done on a small subset of the available pictures. As a result, the labeled dataset will be quite small, and the classifier's performance will be disappointing. Moreover, every time the company makes any change to its products, the whole process will need to be started over from scratch. Wouldn't it be great if the algorithm could just exploit the unlabeled data without needing humans to label every picture? Enter unsupervised learning.

In Chapter 8 we looked at the most common unsupervised learning task: dimensionality reduction. In this chapter we will look at a few more unsupervised learning tasks and algorithms:

Clustering

The goal is to group similar instances together into *clusters*. Clustering is a great tool for data analysis, customer segmentation, recommender systems, search engines, image segmentation, semi-supervised learning, dimensionality reduction, and more.

Anomaly detection

The objective is to learn what "normal" data looks like, and then use that to detect abnormal instances, such as defective items on a production line or a new trend in a time series.

Density estimation

This is the task of estimating the *probability density function* (PDF) of the random process that generated the dataset. Density estimation is commonly used for anomaly detection: instances located in very low-density regions are likely to be anomalies. It is also useful for data analysis and visualization.

Ready for some cake? We will start with clustering, using K-Means and DBSCAN, and then we will discuss Gaussian mixture models and see how they can be used for density estimation, clustering, and anomaly detection.

Clustering

As you enjoy a hike in the mountains, you stumble upon a plant you have never seen before. You look around and you notice a few more. They are not identical, yet they are sufficiently similar for you to know that they most likely belong to the same species (or at least the same genus). You may need a botanist to tell you what species that is, but you certainly don't need an expert to identify groups of similar-looking objects. This is called *clustering*: it is the task of identifying similar instances and assigning them to *clusters*, or groups of similar instances.

Just like in classification, each instance gets assigned to a group. However, unlike classification, clustering is an unsupervised task. Consider Figure 9-1: on the left is the iris dataset (introduced in Chapter 4), where each instance's species (i.e., its class) is represented with a different marker. It is a labeled dataset, for which classification algorithms such as Logistic Regression, SVMs, or Random Forest classifiers are well suited. On the right is the same dataset, but without the labels, so you cannot use a classification algorithm anymore. This is where clustering algorithms step in: many of them can easily detect the lower-left cluster. It is also quite easy to see with our own eyes, but it is not so obvious that the upper-right cluster is composed of two distinct sub-clusters. That said, the dataset has two additional features (sepal length and width), not represented here, and clustering algorithms can make good use of all features, so in fact they identify the three clusters fairly well (e.g., using a Gaussian mixture model, only 5 instances out of 150 are assigned to the wrong cluster).

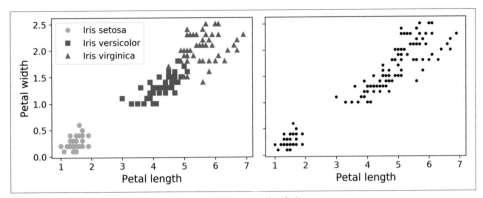

Figure 9-1. Classification (left) versus clustering (right)

Clustering is used in a wide variety of applications, including these:

For customer segmentation

You can cluster your customers based on their purchases and their activity on your website. This is useful to understand who your customers are and what they need, so you can adapt your products and marketing campaigns to each segment. For example, customer segmentation can be useful in *recommender systems* to suggest content that other users in the same cluster enjoyed.

For data analysis

When you analyze a new dataset, it can be helpful to run a clustering algorithm, and then analyze each cluster separately.

As a dimensionality reduction technique

Once a dataset has been clustered, it is usually possible to measure each instance's *affinity* with each cluster (affinity is any measure of how well an instance fits into a cluster). Each instance's feature vector **x** can then be replaced with the vector of its cluster affinities. If there are k clusters, then this vector is k-dimensional. This vector is typically much lower-dimensional than the original feature vector, but it can preserve enough information for further processing.

For anomaly detection (also called outlier detection)

Any instance that has a low affinity to all the clusters is likely to be an anomaly. For example, if you have clustered the users of your website based on their behavior, you can detect users with unusual behavior, such as an unusual number of requests per second. Anomaly detection is particularly useful in detecting defects in manufacturing, or for *fraud detection*.

For semi-supervised learning

If you only have a few labels, you could perform clustering and propagate the labels to all the instances in the same cluster. This technique can greatly increase the number of labels available for a subsequent supervised learning algorithm, and thus improve its performance.

For search engines

Some search engines let you search for images that are similar to a reference image. To build such a system, you would first apply a clustering algorithm to all the images in your database; similar images would end up in the same cluster. Then when a user provides a reference image, all you need to do is use the trained clustering model to find this image's cluster, and you can then simply return all the images from this cluster.

To segment an image

By clustering pixels according to their color, then replacing each pixel's color with the mean color of its cluster, it is possible to considerably reduce the number of different colors in the image. Image segmentation is used in many object detection and tracking systems, as it makes it easier to detect the contour of each object.

There is no universal definition of what a cluster is: it really depends on the context, and different algorithms will capture different kinds of clusters. Some algorithms look for instances centered around a particular point, called a *centroid*. Others look for continuous regions of densely packed instances: these clusters can take on any shape. Some algorithms are hierarchical, looking for clusters of clusters. And the list goes on.

In this section, we will look at two popular clustering algorithms, K-Means and DBSCAN, and explore some of their applications, such as nonlinear dimensionality reduction, semi-supervised learning, and anomaly detection.

K-Means

Consider the unlabeled dataset represented in Figure 9-2: you can clearly see five blobs of instances. The K-Means algorithm is a simple algorithm capable of clustering this kind of dataset very quickly and efficiently, often in just a few iterations. It was proposed by Stuart Lloyd at Bell Labs in 1957 as a technique for pulse-code modulation, but it was only published outside of the company in 1982 (*https://homl.info/36*).[1] In 1965, Edward W. Forgy had published virtually the same algorithm, so K-Means is sometimes referred to as Lloyd–Forgy.

---

[1] Stuart P. Lloyd, "Least Squares Quantization in PCM," *IEEE Transactions on Information Theory* 28, no. 2 (1982): 129–137.

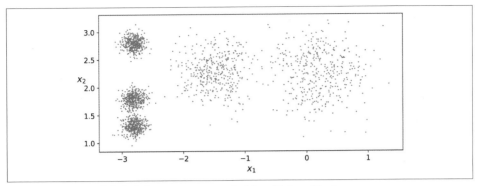

Figure 9-2. An unlabeled dataset composed of five blobs of instances

Let's train a K-Means clusterer on this dataset. It will try to find each blob's center and assign each instance to the closest blob:

```
from sklearn.cluster import KMeans
k = 5
kmeans = KMeans(n_clusters=k)
y_pred = kmeans.fit_predict(X)
```

Note that you have to specify the number of clusters k that the algorithm must find. In this example, it is pretty obvious from looking at the data that k should be set to 5, but in general it is not that easy. We will discuss this shortly.

Each instance was assigned to one of the five clusters. In the context of clustering, an instance's *label* is the index of the cluster that this instance gets assigned to by the algorithm: this is not to be confused with the class labels in classification (remember that clustering is an unsupervised learning task). The KMeans instance preserves a copy of the labels of the instances it was trained on, available via the labels_ instance variable:

```
>>> y_pred
array([4, 0, 1, ..., 2, 1, 0], dtype=int32)
>>> y_pred is kmeans.labels_
True
```

We can also take a look at the five centroids that the algorithm found:

```
>>> kmeans.cluster_centers_
array([[-2.80389616,  1.80117999],
       [ 0.20876306,  2.25551336],
       [-2.79290307,  2.79641063],
       [-1.46679593,  2.28585348],
       [-2.80037642,  1.30082566]])
```

You can easily assign new instances to the cluster whose centroid is closest:

```
>>> X_new = np.array([[0, 2], [3, 2], [-3, 3], [-3, 2.5]])
>>> kmeans.predict(X_new)
array([1, 1, 2, 2], dtype=int32)
```

If you plot the cluster's decision boundaries, you get a Voronoi tessellation (see Figure 9-3, where each centroid is represented with an X).

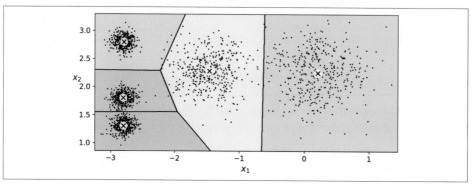

Figure 9-3. K-Means decision boundaries (Voronoi tessellation)

The vast majority of the instances were clearly assigned to the appropriate cluster, but a few instances were probably mislabeled (especially near the boundary between the top-left cluster and the central cluster). Indeed, the K-Means algorithm does not behave very well when the blobs have very different diameters because all it cares about when assigning an instance to a cluster is the distance to the centroid.

Instead of assigning each instance to a single cluster, which is called *hard clustering*, it can be useful to give each instance a score per cluster, which is called *soft clustering*. The score can be the distance between the instance and the centroid; conversely, it can be a similarity score (or affinity), such as the Gaussian Radial Basis Function (introduced in Chapter 5). In the KMeans class, the transform() method measures the distance from each instance to every centroid:

```
>>> kmeans.transform(X_new)
array([[2.81093633, 0.32995317, 2.9042344 , 1.49439034, 2.88633901],
       [5.80730058, 2.80290755, 5.84739223, 4.4759332 , 5.84236351],
       [1.21475352, 3.29399768, 0.29040966, 1.69136631, 1.71086031],
       [0.72581411, 3.21806371, 0.36159148, 1.54808703, 1.21567622]])
```

In this example, the first instance in X_new is located at a distance of 2.81 from the first centroid, 0.33 from the second centroid, 2.90 from the third centroid, 1.49 from the fourth centroid, and 2.89 from the fifth centroid. If you have a high-dimensional dataset and you transform it this way, you end up with a *k*-dimensional dataset: this transformation can be a very efficient nonlinear dimensionality reduction technique.

The K-Means algorithm

So, how does the algorithm work? Well, suppose you were given the centroids. You could easily label all the instances in the dataset by assigning each of them to the cluster whose centroid is closest. Conversely, if you were given all the instance labels, you could easily locate all the centroids by computing the mean of the instances for each cluster. But you are given neither the labels nor the centroids, so how can you proceed? Well, just start by placing the centroids randomly (e.g., by picking k instances at random and using their locations as centroids). Then label the instances, update the centroids, label the instances, update the centroids, and so on until the centroids stop moving. The algorithm is guaranteed to converge in a finite number of steps (usually quite small); it will not oscillate forever.[2]

You can see the algorithm in action in Figure 9-4: the centroids are initialized randomly (top left), then the instances are labeled (top right), then the centroids are updated (center left), the instances are relabeled (center right), and so on. As you can see, in just three iterations, the algorithm has reached a clustering that seems close to optimal.

> The computational complexity of the algorithm is generally linear with regard to the number of instances m, the number of clusters k, and the number of dimensions n. However, this is only true when the data has a clustering structure. If it does not, then in the worst-case scenario the complexity can increase exponentially with the number of instances. In practice, this rarely happens, and K-Means is generally one of the fastest clustering algorithms.

---

2 That's because the mean squared distance between the instances and their closest centroid can only go down at each step.

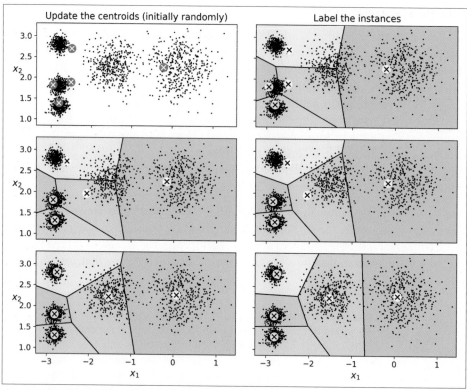

Figure 9-4. The K-Means algorithm

Although the algorithm is guaranteed to converge, it may not converge to the right solution (i.e., it may converge to a local optimum): whether it does or not depends on the centroid initialization. Figure 9-5 shows two suboptimal solutions that the algorithm can converge to if you are not lucky with the random initialization step.

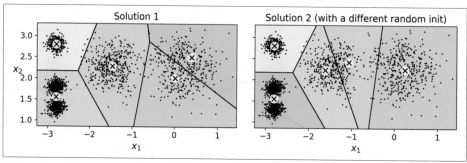

Figure 9-5. Suboptimal solutions due to unlucky centroid initializations

Let's look at a few ways you can mitigate this risk by improving the centroid initialization.

Centroid initialization methods

If you happen to know approximately where the centroids should be (e.g., if you ran another clustering algorithm earlier), then you can set the `init` hyperparameter to a NumPy array containing the list of centroids, and set `n_init` to 1:

```
good_init = np.array([[-3, 3], [-3, 2], [-3, 1], [-1, 2], [0, 2]])
kmeans = KMeans(n_clusters=5, init=good_init, n_init=1)
```

Another solution is to run the algorithm multiple times with different random initializations and keep the best solution. The number of random initializations is controlled by the `n_init` hyperparameter: by default, it is equal to `10`, which means that the whole algorithm described earlier runs 10 times when you call `fit()`, and Scikit-Learn keeps the best solution. But how exactly does it know which solution is the best? It uses a performance metric! That metric is called the model's *inertia*, which is the mean squared distance between each instance and its closest centroid. It is roughly equal to 223.3 for the model on the left in Figure 9-5, 237.5 for the model on the right in Figure 9-5, and 211.6 for the model in Figure 9-3. The `KMeans` class runs the algorithm `n_init` times and keeps the model with the lowest inertia. In this example, the model in Figure 9-3 will be selected (unless we are very unlucky with `n_init` consecutive random initializations). If you are curious, a model's inertia is accessible via the `inertia_` instance variable:

```
>>> kmeans.inertia_
211.59853725816856
```

The `score()` method returns the negative inertia. Why negative? Because a predictor's `score()` method must always respect Scikit-Learn's "greater is better" rule: if a predictor is better than another, its `score()` method should return a greater score.

```
>>> kmeans.score(X)
-211.59853725816856
```

An important improvement to the K-Means algorithm, *K-Means++*, was proposed in a 2006 paper (*https://homl.info/37*) by David Arthur and Sergei Vassilvitskii.[3] They introduced a smarter initialization step that tends to select centroids that are distant from one another, and this improvement makes the K-Means algorithm much less likely to converge to a suboptimal solution. They showed that the additional computation required for the smarter initialization step is well worth it because it makes it possible to drastically reduce the number of times the algorithm needs to be run to find the optimal solution. Here is the K-Means++ initialization algorithm:

1. Take one centroid $c^{(1)}$, chosen uniformly at random from the dataset.

---

3 David Arthur and Sergei Vassilvitskii, "k-Means++: The Advantages of Careful Seeding," *Proceedings of the 18th Annual ACM-SIAM Symposium on Discrete Algorithms* (2007): 1027–1035.

2. Take a new centroid $\mathbf{c}^{(i)}$, choosing an instance $\mathbf{x}^{(i)}$ with probability $D(\mathbf{x}^{(i)})^2$ / $\sum_{j=1}^{m} D(\mathbf{x}^{(j)})^2$, where $D(\mathbf{x}^{(i)})$ is the distance between the instance $\mathbf{x}^{(i)}$ and the closest centroid that was already chosen. This probability distribution ensures that instances farther away from already chosen centroids are much more likely be selected as centroids.

3. Repeat the previous step until all k centroids have been chosen.

The KMeans class uses this initialization method by default. If you want to force it to use the original method (i.e., picking k instances randomly to define the initial centroids), then you can set the init hyperparameter to "random". You will rarely need to do this.

Accelerated K-Means and mini-batch K-Means

Another important improvement to the K-Means algorithm was proposed in a 2003 paper (*https://homl.info/38*) by Charles Elkan.[4] It considerably accelerates the algorithm by avoiding many unnecessary distance calculations. Elkan achieved this by exploiting the triangle inequality (i.e., that a straight line is always the shortest distance between two points[5]) and by keeping track of lower and upper bounds for distances between instances and centroids. This is the algorithm the KMeans class uses by default (you can force it to use the original algorithm by setting the algorithm hyperparameter to "full", although you probably will never need to).

Yet another important variant of the K-Means algorithm was proposed in a 2010 paper (*https://homl.info/39*) by David Sculley.[6] Instead of using the full dataset at each iteration, the algorithm is capable of using mini-batches, moving the centroids just slightly at each iteration. This speeds up the algorithm typically by a factor of three or four and makes it possible to cluster huge datasets that do not fit in memory. Scikit-Learn implements this algorithm in the MiniBatchKMeans class. You can just use this class like the KMeans class:

```
from sklearn.cluster import MiniBatchKMeans

minibatch_kmeans = MiniBatchKMeans(n_clusters=5)
minibatch_kmeans.fit(X)
```

---

4 Charles Elkan, "Using the Triangle Inequality to Accelerate k-Means," *Proceedings of the 20th International Conference on Machine Learning* (2003): 147–153.

5 The triangle inequality is AC ≤ AB + BC where A, B and C are three points and AB, AC, and BC are the distances between these points.

6 David Sculley, "Web-Scale K-Means Clustering," *Proceedings of the 19th International Conference on World Wide Web* (2010): 1177–1178.

If the dataset does not fit in memory, the simplest option is to use the `memmap` class, as we did for incremental PCA in Chapter 8. Alternatively, you can pass one mini-batch at a time to the `partial_fit()` method, but this will require much more work, since you will need to perform multiple initializations and select the best one yourself (see the mini-batch K-Means section of the notebook for an example).

Although the Mini-batch K-Means algorithm is much faster than the regular K-Means algorithm, its inertia is generally slightly worse, especially as the number of clusters increases. You can see this in Figure 9-6: the plot on the left compares the inertias of Mini-batch K-Means and regular K-Means models trained on the previous dataset using various numbers of clusters k. The difference between the two curves remains fairly constant, but this difference becomes more and more significant as k increases, since the inertia becomes smaller and smaller. In the plot on the right, you can see that Mini-batch K-Means is much faster than regular K-Means, and this difference increases with k.

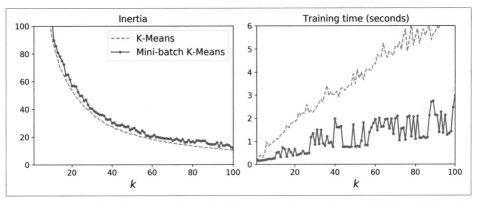

Figure 9-6. Mini-batch K-Means has a higher inertia than K-Means (left) but it is much faster (right), especially as k increases

Finding the optimal number of clusters

So far, we have set the number of clusters k to 5 because it was obvious by looking at the data that this was the correct number of clusters. But in general, it will not be so easy to know how to set k, and the result might be quite bad if you set it to the wrong value. As you can see in Figure 9-7, setting k to 3 or 8 results in fairly bad models.

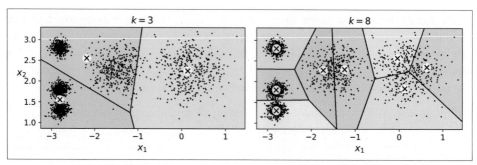

Figure 9-7. Bad choices for the number of clusters: when k is too small, separate clusters get merged (left), and when k is too large, some clusters get chopped into multiple pieces (right)

You might be thinking that we could just pick the model with the lowest inertia, right? Unfortunately, it is not that simple. The inertia for $k=3$ is 653.2, which is much higher than for $k=5$ (which was 211.6). But with $k=8$, the inertia is just 119.1. The inertia is not a good performance metric when trying to choose k because it keeps getting lower as we increase k. Indeed, the more clusters there are, the closer each instance will be to its closest centroid, and therefore the lower the inertia will be. Let's plot the inertia as a function of k (see Figure 9-8).

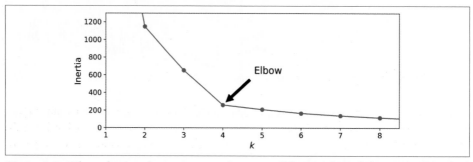

Figure 9-8. When plotting the inertia as a function of the number of clusters k, the curve often contains an inflexion point called the "elbow"

As you can see, the inertia drops very quickly as we increase k up to 4, but then it decreases much more slowly as we keep increasing k. This curve has roughly the shape of an arm, and there is an "elbow" at $k = 4$. So, if we did not know better, 4 would be a good choice: any lower value would be dramatic, while any higher value would not help much, and we might just be splitting perfectly good clusters in half for no good reason.

This technique for choosing the best value for the number of clusters is rather coarse. A more precise approach (but also more computationally expensive) is to use the *silhouette score*, which is the mean *silhouette coefficient* over all the instances. An instance's silhouette coefficient is equal to $(b - a) / \max(a, b)$, where a is the mean distance to the other instances in the same cluster (i.e., the mean intra-cluster distance) and b is the mean nearest-cluster distance (i.e., the mean distance to the instances of the next closest cluster, defined as the one that minimizes b, excluding the instance's own cluster). The silhouette coefficient can vary between –1 and +1. A coefficient close to +1 means that the instance is well inside its own cluster and far from other clusters, while a coefficient close to 0 means that it is close to a cluster boundary, and finally a coefficient close to –1 means that the instance may have been assigned to the wrong cluster.

To compute the silhouette score, you can use Scikit-Learn's `silhouette_score()` function, giving it all the instances in the dataset and the labels they were assigned:

```
>>> from sklearn.metrics import silhouette_score
>>> silhouette_score(X, kmeans.labels_)
0.655517642572828
```

Let's compare the silhouette scores for different numbers of clusters (see Figure 9-9).

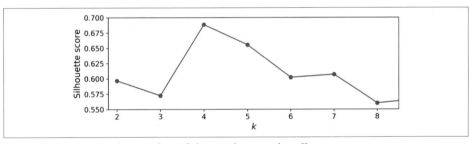

Figure 9-9. Selecting the number of clusters k using the silhouette score

As you can see, this visualization is much richer than the previous one: although it confirms that $k = 4$ is a very good choice, it also underlines the fact that $k = 5$ is quite good as well, and much better than $k = 6$ or 7. This was not visible when comparing inertias.

An even more informative visualization is obtained when you plot every instance's silhouette coefficient, sorted by the cluster they are assigned to and by the value of the coefficient. This is called a *silhouette diagram* (see Figure 9-10). Each diagram contains one knife shape per cluster. The shape's height indicates the number of instances the cluster contains, and its width represents the sorted silhouette coefficients of the instances in the cluster (wider is better). The dashed line indicates the mean silhouette coefficient.

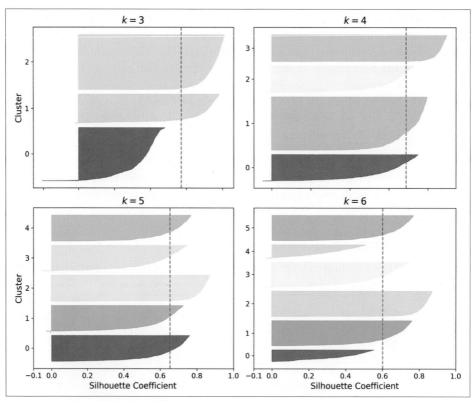

Figure 9-10. Analyzing the silhouette diagrams for various values of k

The vertical dashed lines represent the silhouette score for each number of clusters. When most of the instances in a cluster have a lower coefficient than this score (i.e., if many of the instances stop short of the dashed line, ending to the left of it), then the cluster is rather bad since this means its instances are much too close to other clusters. We can see that when $k = 3$ and when $k = 6$, we get bad clusters. But when $k = 4$ or $k = 5$, the clusters look pretty good: most instances extend beyond the dashed line, to the right and closer to 1.0. When $k = 4$, the cluster at index 1 (the third from the top) is rather big. When $k = 5$, all clusters have similar sizes. So, even though the overall silhouette score from $k = 4$ is slightly greater than for $k = 5$, it seems like a good idea to use $k = 5$ to get clusters of similar sizes.

Limits of K-Means

Despite its many merits, most notably being fast and scalable, K-Means is not perfect. As we saw, it is necessary to run the algorithm several times to avoid suboptimal solutions, plus you need to specify the number of clusters, which can be quite a hassle. Moreover, K-Means does not behave very well when the clusters have varying sizes, different densities, or nonspherical shapes. For example, Figure 9-11 shows how K-Means clusters a dataset containing three ellipsoidal clusters of different dimensions, densities, and orientations.

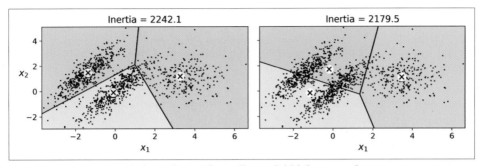

Figure 9-11. K-Means fails to cluster these ellipsoidal blobs properly

As you can see, neither of these solutions is any good. The solution on the left is better, but it still chops off 25% of the middle cluster and assigns it to the cluster on the right. The solution on the right is just terrible, even though its inertia is lower. So, depending on the data, different clustering algorithms may perform better. On these types of elliptical clusters, Gaussian mixture models work great.

 It is important to scale the input features before you run K-Means, or the clusters may be very stretched and K-Means will perform poorly. Scaling the features does not guarantee that all the clusters will be nice and spherical, but it generally improves things.

Now let's look at a few ways we can benefit from clustering. We will use K-Means, but feel free to experiment with other clustering algorithms.

Using Clustering for Image Segmentation

Image segmentation is the task of partitioning an image into multiple segments. In *semantic segmentation*, all pixels that are part of the same object type get assigned to the same segment. For example, in a self-driving car's vision system, all pixels that are part of a pedestrian's image might be assigned to the "pedestrian" segment (there would be one segment containing all the pedestrians). In *instance segmentation*, all pixels that are part of the same individual object are assigned to the same segment. In this case there would be a different segment for each pedestrian. The state of the art in semantic or instance segmentation today is achieved using complex architectures based on convolutional neural networks (see Chapter 14). Here, we are going to do something much simpler: *color segmentation*. We will simply assign pixels to the same segment if they have a similar color. In some applications, this may be sufficient. For example, if you want to analyze satellite images to measure how much total forest area there is in a region, color segmentation may be just fine.

First, use Matplotlib's `imread()` function to load the image (see the upper-left image in Figure 9-12):

```
>>> from matplotlib.image import imread  # or `from imageio import imread`
>>> image = imread(os.path.join("images","unsupervised_learning","ladybug.png"))
>>> image.shape
(533, 800, 3)
```

The image is represented as a 3D array. The first dimension's size is the height; the second is the width; and the third is the number of color channels, in this case red, green, and blue (RGB). In other words, for each pixel there is a 3D vector containing the intensities of red, green, and blue, each between 0.0 and 1.0 (or between 0 and 255, if you use `imageio.imread()`). Some images may have fewer channels, such as grayscale images (one channel). And some images may have more channels, such as images with an additional *alpha channel* for transparency or satellite images, which often contain channels for many light frequencies (e.g., infrared). The following code reshapes the array to get a long list of RGB colors, then it clusters these colors using K-Means:

```
X = image.reshape(-1, 3)
kmeans = KMeans(n_clusters=8).fit(X)
segmented_img = kmeans.cluster_centers_[kmeans.labels_]
segmented_img = segmented_img.reshape(image.shape)
```

For example, it may identify a color cluster for all shades of green. Next, for each color (e.g., dark green), it looks for the mean color of the pixel's color cluster. For example, all shades of green may be replaced with the same light green color (assuming the mean color of the green cluster is light green). Finally, it reshapes this long list of colors to get the same shape as the original image. And we're done!

This outputs the image shown in the upper right of Figure 9-12. You can experiment with various numbers of clusters, as shown in the figure. When you use fewer than eight clusters, notice that the ladybug's flashy red color fails to get a cluster of its own: it gets merged with colors from the environment. This is because K-Means prefers clusters of similar sizes. The ladybug is small—much smaller than the rest of the image—so even though its color is flashy, K-Means fails to dedicate a cluster to it.

Figure 9-12. Image segmentation using K-Means with various numbers of color clusters

That wasn't too hard, was it? Now let's look at another application of clustering: preprocessing.

Using Clustering for Preprocessing

Clustering can be an efficient approach to dimensionality reduction, in particular as a preprocessing step before a supervised learning algorithm. As an example of using clustering for dimensionality reduction, let's tackle the digits dataset, which is a simple MNIST-like dataset containing 1,797 grayscale 8 × 8 images representing the digits 0 to 9. First, load the dataset:

```
from sklearn.datasets import load_digits

X_digits, y_digits = load_digits(return_X_y=True)
```

Now, split it into a training set and a test set:

```
from sklearn.model_selection import train_test_split

X_train, X_test, y_train, y_test = train_test_split(X_digits, y_digits)
```

Next, fit a Logistic Regression model:

```
from sklearn.linear_model import LogisticRegression

log_reg = LogisticRegression()
log_reg.fit(X_train, y_train)
```

Let's evaluate its accuracy on the test set:

```
>>> log_reg.score(X_test, y_test)
0.9688888888888889
```

OK, that's our baseline: 96.9% accuracy. Let's see if we can do better by using K-Means as a preprocessing step. We will create a pipeline that will first cluster the training set into 50 clusters and replace the images with their distances to these 50 clusters, then apply a Logistic Regression model:

```
from sklearn.pipeline import Pipeline

pipeline = Pipeline([
    ("kmeans", KMeans(n_clusters=50)),
    ("log_reg", LogisticRegression()),
])
pipeline.fit(X_train, y_train)
```

 Since there are 10 different digits, it is tempting to set the number of clusters to 10. However, each digit can be written several different ways, so it is preferable to use a larger number of clusters, such as 50.

Now let's evaluate this classification pipeline:

```
>>> pipeline.score(X_test, y_test)
0.9777777777777777
```

How about that? We reduced the error rate by almost 30% (from about 3.1% to about 2.2%)!

But we chose the number of clusters k arbitrarily; we can surely do better. Since K-Means is just a preprocessing step in a classification pipeline, finding a good value for k is much simpler than earlier. There's no need to perform silhouette analysis or minimize the inertia; the best value of k is simply the one that results in the best classification performance during cross-validation. We can use GridSearchCV to find the optimal number of clusters:

```
from sklearn.model_selection import GridSearchCV

param_grid = dict(kmeans__n_clusters=range(2, 100))
grid_clf = GridSearchCV(pipeline, param_grid, cv=3, verbose=2)
grid_clf.fit(X_train, y_train)
```

Let's look at the best value for k and the performance of the resulting pipeline:

```
>>> grid_clf.best_params_
{'kmeans__n_clusters': 99}
>>> grid_clf.score(X_test, y_test)
0.9822222222222222
```

With $k = 99$ clusters, we get a significant accuracy boost, reaching 98.22% accuracy on the test set. Cool! You may want to keep exploring higher values for k, since 99 was the largest value in the range we explored.

Using Clustering for Semi-Supervised Learning

Another use case for clustering is in semi-supervised learning, when we have plenty of unlabeled instances and very few labeled instances. Let's train a Logistic Regression model on a sample of 50 labeled instances from the digits dataset:

```
n_labeled = 50
log_reg = LogisticRegression()
log_reg.fit(X_train[:n_labeled], y_train[:n_labeled])
```

What is the performance of this model on the test set?

```
>>> log_reg.score(X_test, y_test)
0.8333333333333334
```

The accuracy is just 83.3%. It should come as no surprise that this is much lower than earlier, when we trained the model on the full training set. Let's see how we can do better. First, let's cluster the training set into 50 clusters. Then for each cluster, let's find the image closest to the centroid. We will call these images the *representative images*:

```
k = 50
kmeans = KMeans(n_clusters=k)
X_digits_dist = kmeans.fit_transform(X_train)
representative_digit_idx = np.argmin(X_digits_dist, axis=0)
X_representative_digits = X_train[representative_digit_idx]
```

Figure 9-13 shows these 50 representative images.

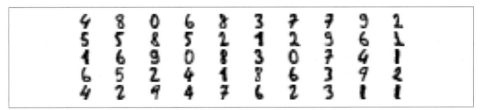

Figure 9-13. Fifty representative digit images (one per cluster)

Let's look at each image and manually label it:

```
y_representative_digits = np.array([4, 8, 0, 6, 8, 3, ..., 7, 6, 2, 3, 1, 1])
```

Now we have a dataset with just 50 labeled instances, but instead of being random instances, each of them is a representative image of its cluster. Let's see if the performance is any better:

```
>>> log_reg = LogisticRegression()
>>> log_reg.fit(X_representative_digits, y_representative_digits)
>>> log_reg.score(X_test, y_test)
0.9222222222222223
```

Wow! We jumped from 83.3% accuracy to 92.2%, although we are still only training the model on 50 instances. Since it is often costly and painful to label instances, especially when it has to be done manually by experts, it is a good idea to label representative instances rather than just random instances.

But perhaps we can go one step further: what if we propagated the labels to all the other instances in the same cluster? This is called *label propagation*:

```
y_train_propagated = np.empty(len(X_train), dtype=np.int32)
for i in range(k):
    y_train_propagated[kmeans.labels_==i] = y_representative_digits[i]
```

Now let's train the model again and look at its performance:

```
>>> log_reg = LogisticRegression()
>>> log_reg.fit(X_train, y_train_propagated)
>>> log_reg.score(X_test, y_test)
0.9333333333333333
```

We got a reasonable accuracy boost, but nothing absolutely astounding. The problem is that we propagated each representative instance's label to all the instances in the same cluster, including the instances located close to the cluster boundaries, which are more likely to be mislabeled. Let's see what happens if we only propagate the labels to the 20% of the instances that are closest to the centroids:

```
percentile_closest = 20

X_cluster_dist = X_digits_dist[np.arange(len(X_train)), kmeans.labels_]
for i in range(k):
    in_cluster = (kmeans.labels_ == i)
    cluster_dist = X_cluster_dist[in_cluster]
    cutoff_distance = np.percentile(cluster_dist, percentile_closest)
    above_cutoff = (X_cluster_dist > cutoff_distance)
    X_cluster_dist[in_cluster & above_cutoff] = -1

partially_propagated = (X_cluster_dist != -1)
X_train_partially_propagated = X_train[partially_propagated]
y_train_partially_propagated = y_train_propagated[partially_propagated]
```

Now let's train the model again on this partially propagated dataset:

```
>>> log_reg = LogisticRegression()
>>> log_reg.fit(X_train_partially_propagated, y_train_partially_propagated)
>>> log_reg.score(X_test, y_test)
0.94
```

Nice! With just 50 labeled instances (only 5 examples per class on average!), we got 94.0% accuracy, which is pretty close to the performance of Logistic Regression on the fully labeled digits dataset (which was 96.9%). This good performance is due to the fact that the propagated labels are actually pretty good—their accuracy is very close to 99%, as the following code shows:

```
>>> np.mean(y_train_partially_propagated == y_train[partially_propagated])
0.9896907216494846
```

Active Learning

To continue improving your model and your training set, the next step could be to do a few rounds of *active learning*, which is when a human expert interacts with the learning algorithm, providing labels for specific instances when the algorithm requests them. There are many different strategies for active learning, but one of the most common ones is called *uncertainty sampling*. Here is how it works:

1. The model is trained on the labeled instances gathered so far, and this model is used to make predictions on all the unlabeled instances.

2. The instances for which the model is most uncertain (i.e., when its estimated probability is lowest) are given to the expert to be labeled.

3. You iterate this process until the performance improvement stops being worth the labeling effort.

Other strategies include labeling the instances that would result in the largest model change, or the largest drop in the model's validation error, or the instances that different models disagree on (e.g., an SVM or a Random Forest).

Before we move on to Gaussian mixture models, let's take a look at DBSCAN, another popular clustering algorithm that illustrates a very different approach based on local density estimation. This approach allows the algorithm to identify clusters of arbitrary shapes.

DBSCAN

This algorithm defines clusters as continuous regions of high density. Here is how it works:

- For each instance, the algorithm counts how many instances are located within a small distance ε (epsilon) from it. This region is called the instance's *ε-neighborhood*.

- If an instance has at least min_samples instances in its ε-neighborhood (including itself), then it is considered a *core instance*. In other words, core instances are those that are located in dense regions.

- All instances in the neighborhood of a core instance belong to the same cluster. This neighborhood may include other core instances; therefore, a long sequence of neighboring core instances forms a single cluster.

- Any instance that is not a core instance and does not have one in its neighborhood is considered an anomaly.

This algorithm works well if all the clusters are dense enough and if they are well separated by low-density regions. The DBSCAN class in Scikit-Learn is as simple to use as you might expect. Let's test it on the moons dataset, introduced in Chapter 5:

```
from sklearn.cluster import DBSCAN
from sklearn.datasets import make_moons

X, y = make_moons(n_samples=1000, noise=0.05)
dbscan = DBSCAN(eps=0.05, min_samples=5)
dbscan.fit(X)
```

The labels of all the instances are now available in the labels_ instance variable:

```
>>> dbscan.labels_
array([ 0,  2, -1, -1,  1,  0,  0,  0, ...,  3,  2,  3,  3,  4,  2,  6,  3])
```

Notice that some instances have a cluster index equal to –1, which means that they are considered as anomalies by the algorithm. The indices of the core instances are available in the core_sample_indices_ instance variable, and the core instances themselves are available in the components_ instance variable:

```
>>> len(dbscan.core_sample_indices_)
808
>>> dbscan.core_sample_indices_
array([  0,   4,   5,   6,   7,   8,  10,  11, ..., 992, 993, 995, 997, 998, 999])
>>> dbscan.components_
array([[-0.02137124,  0.40618608],
       [-0.84192557,  0.53058695],
                   ...
       [-0.94355873,  0.3278936 ],
       [ 0.79419406,  0.60777171]])
```

This clustering is represented in the lefthand plot of Figure 9-14. As you can see, it identified quite a lot of anomalies, plus seven different clusters. How disappointing! Fortunately, if we widen each instance's neighborhood by increasing eps to 0.2, we get the clustering on the right, which looks perfect. Let's continue with this model.

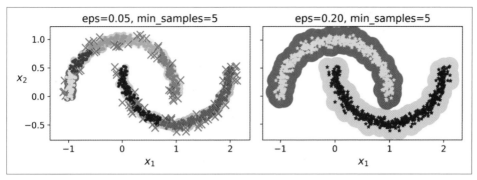

Figure 9-14. DBSCAN clustering using two different neighborhood radiuses

Somewhat surprisingly, the DBSCAN class does not have a predict() method, although it has a fit_predict() method. In other words, it cannot predict which cluster a new instance belongs to. This implementation decision was made because different classification algorithms can be better for different tasks, so the authors decided to let the user choose which one to use. Moreover, it's not hard to implement. For example, let's train a KNeighborsClassifier:

```
from sklearn.neighbors import KNeighborsClassifier

knn = KNeighborsClassifier(n_neighbors=50)
knn.fit(dbscan.components_, dbscan.labels_[dbscan.core_sample_indices_])
```

Now, given a few new instances, we can predict which cluster they most likely belong to and even estimate a probability for each cluster:

```
>>> X_new = np.array([[-0.5, 0], [0, 0.5], [1, -0.1], [2, 1]])
>>> knn.predict(X_new)
array([1, 0, 1, 0])
>>> knn.predict_proba(X_new)
array([[0.18, 0.82],
       [1.  , 0.  ],
       [0.12, 0.88],
       [1.  , 0.  ]])
```

Note that we only trained the classifier on the core instances, but we could also have chosen to train it on all the instances, or all but the anomalies: this choice depends on the final task.

The decision boundary is represented in Figure 9-15 (the crosses represent the four instances in X_new). Notice that since there is no anomaly in the training set, the classifier always chooses a cluster, even when that cluster is far away. It is fairly straightforward to introduce a maximum distance, in which case the two instances that are far away from both clusters are classified as anomalies. To do this, use the kneigh bors() method of the KNeighborsClassifier. Given a set of instances, it returns the distances and the indices of the *k* nearest neighbors in the training set (two matrices, each with *k* columns):

```
>>> y_dist, y_pred_idx = knn.kneighbors(X_new, n_neighbors=1)
>>> y_pred = dbscan.labels_[dbscan.core_sample_indices_][y_pred_idx]
>>> y_pred[y_dist > 0.2] = -1
>>> y_pred.ravel()
array([-1,  0,  1, -1])
```

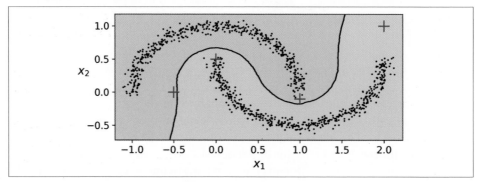

Figure 9-15. Decision boundary between two clusters

In short, DBSCAN is a very simple yet powerful algorithm capable of identifying any number of clusters of any shape. It is robust to outliers, and it has just two hyperparameters (`eps` and `min_samples`). If the density varies significantly across the clusters, however, it can be impossible for it to capture all the clusters properly. Its computational complexity is roughly $O(m \log m)$, making it pretty close to linear with regard to the number of instances, but Scikit-Learn's implementation can require up to $O(m^2)$ memory if `eps` is large.

 You may also want to try *Hierarchical DBSCAN* (HDBSCAN), which is implemented in the scikit-learn-contrib project (*https:// github.com/scikit-learn-contrib/hdbscan/*).

Other Clustering Algorithms

Scikit-Learn implements several more clustering algorithms that you should take a look at. We cannot cover them all in detail here, but here is a brief overview:

Agglomerative clustering
   A hierarchy of clusters is built from the bottom up. Think of many tiny bubbles floating on water and gradually attaching to each other until there's one big group of bubbles. Similarly, at each iteration, agglomerative clustering connects the nearest pair of clusters (starting with individual instances). If you drew a tree

with a branch for every pair of clusters that merged, you would get a binary tree of clusters, where the leaves are the individual instances. This approach scales very well to large numbers of instances or clusters. It can capture clusters of various shapes, it produces a flexible and informative cluster tree instead of forcing you to choose a particular cluster scale, and it can be used with any pairwise distance. It can scale nicely to large numbers of instances if you provide a connectivity matrix, which is a sparse $m \times m$ matrix that indicates which pairs of instances are neighbors (e.g., returned by `sklearn.neighbors.kneighbors_graph()`). Without a connectivity matrix, the algorithm does not scale well to large datasets.

BIRCH

The BIRCH (Balanced Iterative Reducing and Clustering using Hierarchies) algorithm was designed specifically for very large datasets, and it can be faster than batch K-Means, with similar results, as long as the number of features is not too large (<20). During training, it builds a tree structure containing just enough information to quickly assign each new instance to a cluster, without having to store all the instances in the tree: this approach allows it to use limited memory, while handling huge datasets.

Mean-Shift

This algorithm starts by placing a circle centered on each instance; then for each circle it computes the mean of all the instances located within it, and it shifts the circle so that it is centered on the mean. Next, it iterates this mean-shifting step until all the circles stop moving (i.e., until each of them is centered on the mean of the instances it contains). Mean-Shift shifts the circles in the direction of higher density, until each of them has found a local density maximum. Finally, all the instances whose circles have settled in the same place (or close enough) are assigned to the same cluster. Mean-Shift has some of the same features as DBSCAN, like how it can find any number of clusters of any shape, it has very few hyperparameters (just one—the radius of the circles, called the *bandwidth*), and it relies on local density estimation. But unlike DBSCAN, Mean-Shift tends to chop clusters into pieces when they have internal density variations. Unfortunately, its computational complexity is $O(m^2)$, so it is not suited for large datasets.

Affinity propagation

This algorithm uses a voting system, where instances vote for similar instances to be their representatives, and once the algorithm converges, each representative and its voters form a cluster. Affinity propagation can detect any number of clusters of different sizes. Unfortunately, this algorithm has a computational complexity of $O(m^2)$, so it too is not suited for large datasets.

Spectral clustering

This algorithm takes a similarity matrix between the instances and creates a low-dimensional embedding from it (i.e., it reduces its dimensionality), then it uses another clustering algorithm in this low-dimensional space (Scikit-Learn's implementation uses K-Means.) Spectral clustering can capture complex cluster structures, and it can also be used to cut graphs (e.g., to identify clusters of friends on a social network). It does not scale well to large numbers of instances, and it does not behave well when the clusters have very different sizes.

Now let's dive into Gaussian mixture models, which can be used for density estimation, clustering, and anomaly detection.

Gaussian Mixtures

A *Gaussian mixture model* (GMM) is a probabilistic model that assumes that the instances were generated from a mixture of several Gaussian distributions whose parameters are unknown. All the instances generated from a single Gaussian distribution form a cluster that typically looks like an ellipsoid. Each cluster can have a different ellipsoidal shape, size, density, and orientation, just like in Figure 9-11. When you observe an instance, you know it was generated from one of the Gaussian distributions, but you are not told which one, and you do not know what the parameters of these distributions are.

There are several GMM variants. In the simplest variant, implemented in the `Gaus sianMixture` class, you must know in advance the number k of Gaussian distributions. The dataset $\mathbf{X}$ is assumed to have been generated through the following probabilistic process:

- For each instance, a cluster is picked randomly from among k clusters. The probability of choosing the jth cluster is defined by the cluster's weight, $\phi^{(j)}$.[7] The index of the cluster chosen for the ith instance is noted $z^{(i)}$.

- If $z^{(i)}=j$, meaning the ith instance has been assigned to the jth cluster, the location $\mathbf{x}^{(i)}$ of this instance is sampled randomly from the Gaussian distribution with mean $\mathbf{\mu}^{(j)}$ and covariance matrix $\mathbf{\Sigma}^{(j)}$. This is noted $\mathbf{x}^{(i)} \sim \mathcal{N}\left(\mathbf{\mu}^{(j)}, \mathbf{\Sigma}^{(j)}\right)$.

This generative process can be represented as a graphical model. Figure 9-16 represents the structure of the conditional dependencies between random variables.

---

7 Phi (ϕ or φ) is the 21st letter of the Greek alphabet.

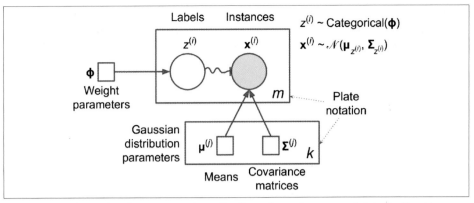

Figure 9-16. A graphical representation of a Gaussian mixture model, including its parameters (squares), random variables (circles), and their conditional dependencies (solid arrows)

Here is how to interpret the figure:[8]

- The circles represent random variables.

- The squares represent fixed values (i.e., parameters of the model).

- The large rectangles are called *plates*. They indicate that their content is repeated several times.

- The number at the bottom right of each plate indicates how many times its content is repeated. So, there are m random variables $z^{(i)}$ (from $z^{(1)}$ to $z^{(m)}$) and m random variables $x^{(i)}$. There are also k means $\mu^{(j)}$ and k covariance matrices $\Sigma^{(j)}$. Lastly, there is just one weight vector ϕ (containing all the weights $\phi^{(1)}$ to $\phi^{(k)}$).

- Each variable $z^{(i)}$ is drawn from the *categorical distribution* with weights ϕ. Each variable $x^{(i)}$ is drawn from the normal distribution, with the mean and covariance matrix defined by its cluster $z^{(i)}$.

- The solid arrows represent conditional dependencies. For example, the probability distribution for each random variable $z^{(i)}$ depends on the weight vector ϕ. Note that when an arrow crosses a plate boundary, it means that it applies to all the repetitions of that plate. For example, the weight vector ϕ conditions the probability distributions of all the random variables $x^{(1)}$ to $x^{(m)}$.

- The squiggly arrow from $z^{(i)}$ to $x^{(i)}$ represents a switch: depending on the value of $z^{(i)}$, the instance $x^{(i)}$ will be sampled from a different Gaussian distribution. For example, if $z^{(i)}=j$, then $x^{(i)} \sim \mathcal{N}\left(\mu^{(j)}, \Sigma^{(j)}\right)$.

---

8 Most of these notations are standard, but a few additional notations were taken from the Wikipedia article on plate notation (*https://en.wikipedia.org/wiki/Plate_notation*).

- Shaded nodes indicate that the value is known. So, in this case, only the random variables $\mathbf{x}^{(i)}$ have known values: they are called *observed variables*. The unknown random variables $z^{(i)}$ are called *latent variables*.

So, what can you do with such a model? Well, given the dataset $\mathbf{X}$, you typically want to start by estimating the weights $\boldsymbol{\phi}$ and all the distribution parameters $\boldsymbol{\mu}^{(1)}$ to $\boldsymbol{\mu}^{(k)}$ and $\boldsymbol{\Sigma}^{(1)}$ to $\boldsymbol{\Sigma}^{(k)}$. Scikit-Learn's `GaussianMixture` class makes this super easy:

```
from sklearn.mixture import GaussianMixture

gm = GaussianMixture(n_components=3, n_init=10)
gm.fit(X)
```

Let's look at the parameters that the algorithm estimated:

```
>>> gm.weights_
array([0.20965228, 0.4000662 , 0.39028152])
>>> gm.means_
array([[ 3.39909717,  1.05933727],
       [-1.40763984,  1.42710194],
       [ 0.05135313,  0.07524095]])
>>> gm.covariances_
array([[[ 1.14807234, -0.03270354],
        [-0.03270354,  0.95496237]],

       [[ 0.63478101,  0.72969804],
        [ 0.72969804,  1.1609872 ]],

       [[ 0.68809572,  0.79608475],
        [ 0.79608475,  1.21234145]]])
```

Great, it worked fine! Indeed, the weights that were used to generate the data were 0.2, 0.4, and 0.4; and similarly, the means and covariance matrices were very close to those found by the algorithm. But how? This class relies on the *Expectation-Maximization* (EM) algorithm, which has many similarities with the K-Means algorithm: it also initializes the cluster parameters randomly, then it repeats two steps until convergence, first assigning instances to clusters (this is called the *expectation step*) and then updating the clusters (this is called the *maximization step*). Sounds familiar, right? In the context of clustering, you can think of EM as a generalization of K-Means that not only finds the cluster centers ($\boldsymbol{\mu}^{(1)}$ to $\boldsymbol{\mu}^{(k)}$), but also their size, shape, and orientation ($\boldsymbol{\Sigma}^{(1)}$ to $\boldsymbol{\Sigma}^{(k)}$), as well as their relative weights ($\phi^{(1)}$ to $\phi^{(k)}$). Unlike K-Means, though, EM uses soft cluster assignments, not hard assignments. For each instance, during the expectation step, the algorithm estimates the probability that it belongs to each cluster (based on the current cluster parameters). Then, during the maximization step, each cluster is updated using *all* the instances in the dataset, with each instance weighted by the estimated probability that it belongs to that cluster. These probabilities are called the *responsibilities* of the clusters for the instances.

During the maximization step, each cluster's update will mostly be impacted by the instances it is most responsible for.

 Unfortunately, just like K-Means, EM can end up converging to poor solutions, so it needs to be run several times, keeping only the best solution. This is why we set n_init to 10. Be careful: by default n_init is set to 1.

You can check whether or not the algorithm converged and how many iterations it took:

```
>>> gm.converged_
True
>>> gm.n_iter_
3
```

Now that you have an estimate of the location, size, shape, orientation, and relative weight of each cluster, the model can easily assign each instance to the most likely cluster (hard clustering) or estimate the probability that it belongs to a particular cluster (soft clustering). Just use the predict() method for hard clustering, or the predict_proba() method for soft clustering:

```
>>> gm.predict(X)
array([2, 2, 1, ..., 0, 0, 0])
>>> gm.predict_proba(X)
array([[2.32389467e-02, 6.77397850e-07, 9.76760376e-01],
       [1.64685609e-02, 6.75361303e-04, 9.82856078e-01],
       [2.01535333e-06, 9.99923053e-01, 7.49319577e-05],
       ...,
       [9.99999571e-01, 2.13946075e-26, 4.28788333e-07],
       [1.00000000e+00, 1.46454409e-41, 5.12459171e-16],
       [1.00000000e+00, 8.02006365e-41, 2.27626238e-15]])
```

A Gaussian mixture model is a *generative model*, meaning you can sample new instances from it (note that they are ordered by cluster index):

```
>>> X_new, y_new = gm.sample(6)
>>> X_new
array([[ 2.95400315,  2.63680992],
       [-1.16654575,  1.62792705],
       [-1.39477712, -1.48511338],
       [ 0.27221525,  0.690366  ],
       [ 0.54095936,  0.48591934],
       [ 0.38064009, -0.56240465]])

>>> y_new
array([0, 1, 2, 2, 2, 2])
```

It is also possible to estimate the density of the model at any given location. This is achieved using the score_samples() method: for each instance it is given, this method estimates the log of the *probability density function* (PDF) at that location. The greater the score, the higher the density:

```
>>> gm.score_samples(X)
array([-2.60782346, -3.57106041, -3.33003479, ..., -3.51352783,
       -4.39802535, -3.80743859])
```

If you compute the exponential of these scores, you get the value of the PDF at the location of the given instances. These are not probabilities, but probability *densities*: they can take on any positive value, not just a value between 0 and 1. To estimate the probability that an instance will fall within a particular region, you would have to integrate the PDF over that region (if you do so over the entire space of possible instance locations, the result will be 1).

Figure 9-17 shows the cluster means, the decision boundaries (dashed lines), and the density contours of this model.

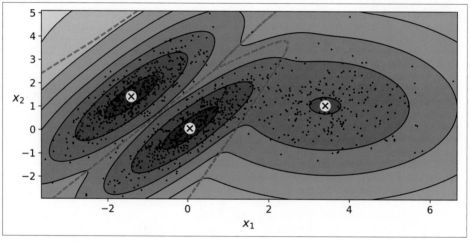

Figure 9-17. Cluster means, decision boundaries, and density contours of a trained Gaussian mixture model

Nice! The algorithm clearly found an excellent solution. Of course, we made its task easy by generating the data using a set of 2D Gaussian distributions (unfortunately, real-life data is not always so Gaussian and low-dimensional). We also gave the algorithm the correct number of clusters. When there are many dimensions, or many clusters, or few instances, EM can struggle to converge to the optimal solution. You might need to reduce the difficulty of the task by limiting the number of parameters that the algorithm has to learn. One way to do this is to limit the range of shapes and orientations that the clusters can have. This can be achieved by imposing constraints on the covariance matrices. To do this, set the `covariance_type` hyperparameter to one of the following values:

`"spherical"`
> All clusters must be spherical, but they can have different diameters (i.e., different variances).

`"diag"`
> Clusters can take on any ellipsoidal shape of any size, but the ellipsoid's axes must be parallel to the coordinate axes (i.e., the covariance matrices must be diagonal).

`"tied"`
> All clusters must have the same ellipsoidal shape, size, and orientation (i.e., all clusters share the same covariance matrix).

By default, `covariance_type` is equal to `"full"`, which means that each cluster can take on any shape, size, and orientation (it has its own unconstrained covariance matrix). Figure 9-18 plots the solutions found by the EM algorithm when `cova riance_type` is set to `"tied"` or `"spherical."`

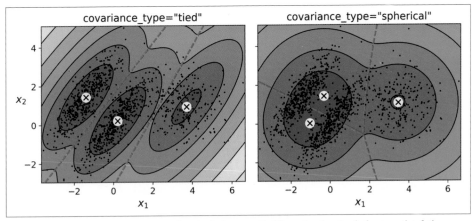

Figure 9-18. Gaussian mixtures for tied clusters (left) and spherical clusters (right)

 The computational complexity of training a `GaussianMixture` model depends on the number of instances m, the number of dimensions n, the number of clusters k, and the constraints on the covariance matrices. If `covariance_type` is `"spherical` or `"diag"`, it is $O(kmn)$, assuming the data has a clustering structure. If `cova riance_type` is `"tied"` or `"full"`, it is $O(kmn^2 + kn^3)$, so it will not scale to large numbers of features.

Gaussian mixture models can also be used for anomaly detection. Let's see how.

Anomaly Detection Using Gaussian Mixtures

Anomaly detection (also called *outlier detection*) is the task of detecting instances that deviate strongly from the norm. These instances are called *anomalies*, or *outliers*, while the normal instances are called *inliers*. Anomaly detection is useful in a wide variety of applications, such as fraud detection, detecting defective products in manufacturing, or removing outliers from a dataset before training another model (which can significantly improve the performance of the resulting model).

Using a Gaussian mixture model for anomaly detection is quite simple: any instance located in a low-density region can be considered an anomaly. You must define what density threshold you want to use. For example, in a manufacturing company that tries to detect defective products, the ratio of defective products is usually well known. Say it is equal to 4%. You then set the density threshold to be the value that results in having 4% of the instances located in areas below that threshold density. If you notice that you get too many false positives (i.e., perfectly good products that are flagged as defective), you can lower the threshold. Conversely, if you have too many false negatives (i.e., defective products that the system does not flag as defective), you can increase the threshold. This is the usual precision/recall trade-off (see Chapter 3). Here is how you would identify the outliers using the fourth percentile lowest density as the threshold (i.e., approximately 4% of the instances will be flagged as anomalies):

```
densities = gm.score_samples(X)
density_threshold = np.percentile(densities, 4)
anomalies = X[densities < density_threshold]
```

Figure 9-19 represents these anomalies as stars.

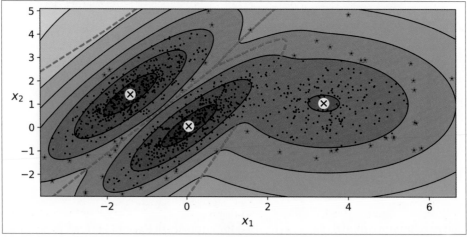

Figure 9-19. Anomaly detection using a Gaussian mixture model

A closely related task is *novelty detection*: it differs from anomaly detection in that the algorithm is assumed to be trained on a "clean" dataset, uncontaminated by outliers, whereas anomaly detection does not make this assumption. Indeed, outlier detection is often used to clean up a dataset.

 Gaussian mixture models try to fit all the data, including the outliers, so if you have too many of them, this will bias the model's view of "normality," and some outliers may wrongly be considered as normal. If this happens, you can try to fit the model once, use it to detect and remove the most extreme outliers, then fit the model again on the cleaned-up dataset. Another approach is to use robust covariance estimation methods (see the `EllipticEnvelope` class).

Just like K-Means, the `GaussianMixture` algorithm requires you to specify the number of clusters. So, how can you find it?

Selecting the Number of Clusters

With K-Means, you could use the inertia or the silhouette score to select the appropriate number of clusters. But with Gaussian mixtures, it is not possible to use these metrics because they are not reliable when the clusters are not spherical or have different sizes. Instead, you can try to find the model that minimizes a *theoretical information criterion*, such as the *Bayesian information criterion* (BIC) or the *Akaike information criterion* (AIC), defined in Equation 9-1.

Equation 9-1. Bayesian information criterion (BIC) and Akaike information criterion (AIC)

$$BIC = \log(m)p - 2\log(\hat{L})$$
$$AIC = 2p - 2\log(\hat{L})$$

In these equations:

- m is the number of instances, as always.
- p is the number of parameters learned by the model.
- $\hat{L}$ is the maximized value of the *likelihood function* of the model.

Both the BIC and the AIC penalize models that have more parameters to learn (e.g., more clusters) and reward models that fit the data well. They often end up selecting the same model. When they differ, the model selected by the BIC tends to be simpler (fewer parameters) than the one selected by the AIC, but tends to not fit the data quite as well (this is especially true for larger datasets).

Likelihood Function

The terms "probability" and "likelihood" are often used interchangeably in the English language, but they have very different meanings in statistics. Given a statistical model with some parameters θ, the word "probability" is used to describe how plausible a future outcome x is (knowing the parameter values θ), while the word "likelihood" is used to describe how plausible a particular set of parameter values θ are, after the outcome x is known.

Consider a 1D mixture model of two Gaussian distributions centered at −4 and +1. For simplicity, this toy model has a single parameter θ that controls the standard deviations of both distributions. The top-left contour plot in Figure 9-20 shows the entire model $f(x; \theta)$ as a function of both x and θ. To estimate the probability distribution of a future outcome x, you need to set the model parameter θ. For example, if you set θ to 1.3 (the horizontal line), you get the probability density function $f(x; \theta=1.3)$ shown in the lower-left plot. Say you want to estimate the probability that x will fall between −2 and +2. You must calculate the integral of the PDF on this range (i.e., the surface of the shaded region). But what if you don't know θ, and instead if you have observed a single instance $x=2.5$ (the vertical line in the upper-left plot)? In this case, you get the likelihood function $\mathscr{L}(\theta|x=2.5)=f(x=2.5; \theta)$, represented in the upper-right plot.

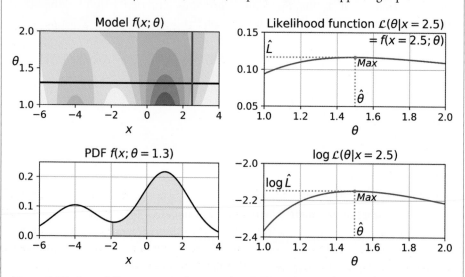

Figure 9-20. A model's parametric function (top left), and some derived functions: a PDF (lower left), a likelihood function (top right), and a log likelihood function (lower right)

In short, the PDF is a function of x (with θ fixed), while the likelihood function is a function of θ (with x fixed). It is important to understand that the likelihood function is *not* a probability distribution: if you integrate a probability distribution over all possible values of x, you always get 1; but if you integrate the likelihood function over all possible values of θ, the result can be any positive value.

Given a dataset $\mathbf{X}$, a common task is to try to estimate the most likely values for the model parameters. To do this, you must find the values that maximize the likelihood function, given $\mathbf{X}$. In this example, if you have observed a single instance $x=2.5$, the *maximum likelihood estimate* (MLE) of θ is $\hat{\theta}=1.5$. If a prior probability distribution g over θ exists, it is possible to take it into account by maximizing $\mathscr{L}(\theta|x)g(\theta)$ rather than just maximizing $\mathscr{L}(\theta|x)$. This is called *maximum a-posteriori* (MAP) estimation. Since MAP constrains the parameter values, you can think of it as a regularized version of MLE.

Notice that maximizing the likelihood function is equivalent to maximizing its logarithm (represented in the lower-righthand plot in Figure 9-20). Indeed the logarithm is a strictly increasing function, so if θ maximizes the log likelihood, it also maximizes the likelihood. It turns out that it is generally easier to maximize the log likelihood. For example, if you observed several independent instances $x^{(1)}$ to $x^{(m)}$, you would need to find the value of θ that maximizes the product of the individual likelihood functions. But it is equivalent, and much simpler, to maximize the sum (not the product) of the log likelihood functions, thanks to the magic of the logarithm which converts products into sums: $\log(ab)=\log(a)+\log(b)$.

Once you have estimated $\hat{\theta}$, the value of θ that maximizes the likelihood function, then you are ready to compute $\hat{L} = \mathscr{L}(\hat{\theta}, \mathbf{X})$, which is the value used to compute the AIC and BIC; you can think of it as a measure of how well the model fits the data.

To compute the BIC and AIC, call the `bic()` and `aic()` methods:

```
>>> gm.bic(X)
8189.74345832983
>>> gm.aic(X)
8102.518178214792
```

Figure 9-21 shows the BIC for different numbers of clusters k. As you can see, both the BIC and the AIC are lowest when $k=3$, so it is most likely the best choice. Note that we could also search for the best value for the `covariance_type` hyperparameter. For example, if it is `"spherical"` rather than `"full"`, then the model has significantly fewer parameters to learn, but it does not fit the data as well.

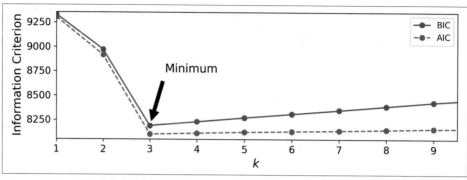

Figure 9-21. AIC and BIC for different numbers of clusters k

Bayesian Gaussian Mixture Models

Rather than manually searching for the optimal number of clusters, you can use the
BayesianGaussianMixture class, which is capable of giving weights equal (or close)
to zero to unnecessary clusters. Set the number of clusters n_components to a value
that you have good reason to believe is greater than the optimal number of clusters
(this assumes some minimal knowledge about the problem at hand), and the algorithm will eliminate the unnecessary clusters automatically. For example, let's set the
number of clusters to 10 and see what happens:

```
>>> from sklearn.mixture import BayesianGaussianMixture
>>> bgm = BayesianGaussianMixture(n_components=10, n_init=10)
>>> bgm.fit(X)
>>> np.round(bgm.weights_, 2)
array([0.4 , 0.21, 0.4 , 0.  , 0.  , 0.  , 0.  , 0.  , 0.  , 0.  ])
```

Perfect: the algorithm automatically detected that only three clusters are needed, and
the resulting clusters are almost identical to the ones in Figure 9-17.

In this model, the cluster parameters (including the weights, means, and covariance
matrices) are not treated as fixed model parameters anymore, but as latent random
variables, like the cluster assignments (see Figure 9-22). So z now includes both the
cluster parameters and the cluster assignments.

The Beta distribution is commonly used to model random variables whose values lie
within a fixed range. In this case, the range is from 0 to 1. The Stick-Breaking Process
(SBP) is best explained through an example: suppose Φ=[0.3, 0.6, 0.5,...], then 30% of
the instances will be assigned to cluster 0, then 60% of the remaining instances will be
assigned to cluster 1, then 50% of the remaining instances will be assigned to cluster
2, and so on. This process is a good model for datasets where new instances are more
likely to join large clusters than small clusters (e.g., people are more likely to move to
larger cities). If the concentration α is high, then Φ values will likely be close to 0, and
the SBP generate many clusters. Conversely, if the concentration is low, then Φ values

will likely be close to 1, and there will be few clusters. Finally, the Wishart distribution is used to sample covariance matrices: the parameters d and V control the distribution of cluster shapes.

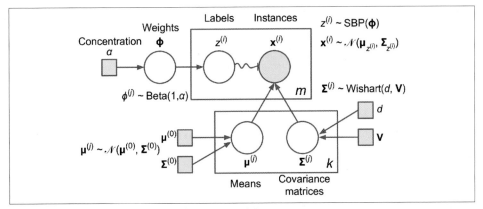

Figure 9-22. Bayesian Gaussian mixture model

Prior knowledge about the latent variables **z** can be encoded in a probability distribution $p(\mathbf{z})$ called the *prior*. For example, we may have a prior belief that the clusters are likely to be few (low concentration), or conversely, that they are likely to be plentiful (high concentration). This prior belief about the number of clusters can be adjusted using the `weight_concentration_prior` hyperparameter. Setting it to 0.01 or 10,000 gives very different clusterings (see Figure 9-23). The more data we have, however, the less the priors matter. In fact, to plot diagrams with such large differences, you must use very strong priors and little data.

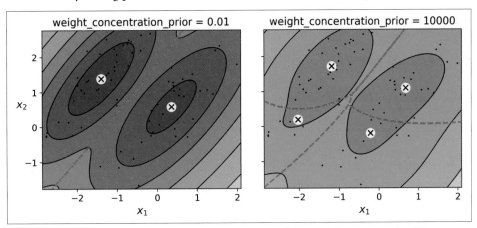

Figure 9-23. Using different concentration priors on the same data results in different numbers of clusters

Bayes' theorem (Equation 9-2) tells us how to update the probability distribution over the latent variables after we observe some data **X**. It computes the *posterior* distribution $p(z|X)$, which is the conditional probability of **z** given **X**.

Equation 9-2. Bayes' theorem

$$p(z|X) = \text{posterior} = \frac{\text{likelihood} \times \text{prior}}{\text{evidence}} = \frac{p(X|z)\,p(z)}{p(X)}$$

Unfortunately, in a Gaussian mixture model (and many other problems), the denominator $p(x)$ is intractable, as it requires integrating over all the possible values of **z** (Equation 9-3), which would require considering all possible combinations of cluster parameters and cluster assignments.

Equation 9-3. The evidence p(X) is often intractable

$$p\left(X\right) = \int p(X|z)p(z)dz$$

This intractability is one of the central problems in Bayesian statistics, and there are several approaches to solving it. One of them is *variational inference*, which picks a family of distributions $q(z; \lambda)$ with its own *variational parameters* λ (lambda), then optimizes these parameters to make $q(z)$ a good approximation of $p(z|X)$. This is achieved by finding the value of λ that minimizes the KL divergence from $q(z)$ to $p(z|X)$, noted $D_{KL}(q\|p)$. The KL divergence equation is shown in Equation 9-4, and it can be rewritten as the log of the evidence ($\log p(X)$) minus the *evidence lower bound* (ELBO). Since the log of the evidence does not depend on q, it is a constant term, so minimizing the KL divergence just requires maximizing the ELBO.

Equation 9-4. KL divergence from q(z) to p(z|X)

$$
\begin{aligned}
D_{KL}(q \| p) &= \mathbb{E}_q\left[\log \frac{q(z)}{p(z \mid X)} \right] \\
&= \mathbb{E}_q[\log q(z) - \log p(z \mid X)] \\
&= \mathbb{E}_q\left[\log q(z) - \log \frac{p(z, X)}{p(X)} \right] \\
&= \mathbb{E}_q[\log q(z) - \log p(z, X) + \log p(X)] \\
&= \mathbb{E}_q[\log q(z)] - \mathbb{E}_q[\log p(z, X)] + \mathbb{E}_q[\log p(X)] \\
&= \mathbb{E}_q[\log p(X)] - \left(\mathbb{E}_q[\log p(z, X)] - \mathbb{E}_q[\log q(z)] \right) \\
&= \log p(X) - \text{ELBO} \\
&\quad \text{where ELBO} = \mathbb{E}_q[\log p(z, X)] - \mathbb{E}_q[\log q(z)]
\end{aligned}
$$

In practice, there are different techniques to maximize the ELBO. In *mean field variational inference*, it is necessary to pick the family of distributions $q(\mathbf{z}; \boldsymbol{\lambda})$ and the prior $p(z)$ very carefully to ensure that the equation for the ELBO simplifies to a form that can be computed. Unfortunately, there is no general way to do this. Picking the right family of distributions and the right prior depends on the task and requires some mathematical skills. For example, the distributions and lower-bound equations used in Scikit-Learn's `BayesianGaussianMixture` class are presented in the documentation (*https://homl.info/40*). From these equations it is possible to derive update equations for the cluster parameters and assignment variables: these are then used very much like in the Expectation-Maximization algorithm. In fact, the computational complexity of the `BayesianGaussianMixture` class is similar to that of the `GaussianMixture` class (but generally significantly slower). A simpler approach to maximizing the ELBO is called *black box stochastic variational inference* (BBSVI): at each iteration, a few samples are drawn from q, and they are used to estimate the gradients of the ELBO with regard to the variational parameters $\boldsymbol{\lambda}$, which are then used in a gradient ascent step. This approach makes it possible to use Bayesian inference with any kind of model (provided it is differentiable), even deep neural networks; using Bayesian inference with deep neural networks is called Bayesian Deep Learning.

If you want to dive deeper into Bayesian statistics, check out the book *Bayesian Data Analysis* (*https://homl.info/bda*) by Andrew Gelman et al. (Chapman & Hall).

Gaussian mixture models work great on clusters with ellipsoidal shapes, but if you try to fit a dataset with different shapes, you may have bad surprises. For example, let's see what happens if we use a Bayesian Gaussian mixture model to cluster the moons dataset (see Figure 9-24).

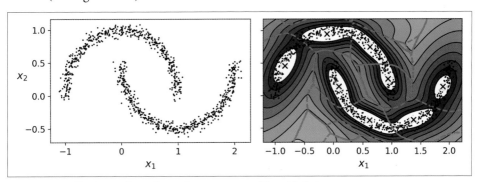

Figure 9-24. Fitting a Gaussian mixture to nonellipsoidal clusters

Oops! The algorithm desperately searched for ellipsoids, so it found eight different clusters instead of two. The density estimation is not too bad, so this model could perhaps be used for anomaly detection, but it failed to identify the two moons. Let's now look at a few clustering algorithms capable of dealing with arbitrarily shaped clusters.

Other Algorithms for Anomaly and Novelty Detection

Scikit-Learn implements other algorithms dedicated to anomaly detection or novelty detection:

PCA (and other dimensionality reduction techniques with an `inverse_transform()` method)

If you compare the reconstruction error of a normal instance with the reconstruction error of an anomaly, the latter will usually be much larger. This is a simple and often quite efficient anomaly detection approach (see this chapter's exercises for an application of this approach).

Fast-MCD (minimum covariance determinant)

Implemented by the `EllipticEnvelope` class, this algorithm is useful for outlier detection, in particular to clean up a dataset. It assumes that the normal instances (inliers) are generated from a single Gaussian distribution (not a mixture). It also assumes that the dataset is contaminated with outliers that were not generated from this Gaussian distribution. When the algorithm estimates the parameters of the Gaussian distribution (i.e., the shape of the elliptic envelope around the inliers), it is careful to ignore the instances that are most likely outliers. This technique gives a better estimation of the elliptic envelope and thus makes the algorithm better at identifying the outliers.

Isolation Forest

This is an efficient algorithm for outlier detection, especially in high-dimensional datasets. The algorithm builds a Random Forest in which each Decision Tree is grown randomly: at each node, it picks a feature randomly, then it picks a random threshold value (between the min and max values) to split the dataset in two. The dataset gradually gets chopped into pieces this way, until all instances end up isolated from the other instances. Anomalies are usually far from other instances, so on average (across all the Decision Trees) they tend to get isolated in fewer steps than normal instances.

Local Outlier Factor (LOF)

This algorithm is also good for outlier detection. It compares the density of instances around a given instance to the density around its neighbors. An anomaly is often more isolated than its *k* nearest neighbors.

One-class SVM

This algorithm is better suited for novelty detection. Recall that a kernelized SVM classifier separates two classes by first (implicitly) mapping all the instances to a high-dimensional space, then separating the two classes using a linear SVM classifier within this high-dimensional space (see Chapter 5). Since we just have one class of instances, the one-class SVM algorithm instead tries to separate the instances in high-dimensional space from the origin. In the original space, this will correspond to finding a small region that encompasses all the instances. If a new instance does not fall within this region, it is an anomaly. There are a few hyperparameters to tweak: the usual ones for a kernelized SVM, plus a margin hyperparameter that corresponds to the probability of a new instance being mistakenly considered as novel when it is in fact normal. It works great, especially with high-dimensional datasets, but like all SVMs it does not scale to large datasets.

Exercises

1. How would you define clustering? Can you name a few clustering algorithms?

2. What are some of the main applications of clustering algorithms?

3. Describe two techniques to select the right number of clusters when using K-Means.

4. What is label propagation? Why would you implement it, and how?

5. Can you name two clustering algorithms that can scale to large datasets? And two that look for regions of high density?

6. Can you think of a use case where active learning would be useful? How would you implement it?

7. What is the difference between anomaly detection and novelty detection?

8. What is a Gaussian mixture? What tasks can you use it for?

9. Can you name two techniques to find the right number of clusters when using a Gaussian mixture model?

10. The classic Olivetti faces dataset contains 400 grayscale 64 × 64–pixel images of faces. Each image is flattened to a 1D vector of size 4,096. 40 different people were photographed (10 times each), and the usual task is to train a model that can predict which person is represented in each picture. Load the dataset using the sklearn.datasets.fetch_olivetti_faces() function, then split it into a training set, a validation set, and a test set (note that the dataset is already scaled between 0 and 1). Since the dataset is quite small, you probably want to use stratified sampling to ensure that there are the same number of images per person in each set. Next, cluster the images using K-Means, and ensure that you have a

good number of clusters (using one of the techniques discussed in this chapter). Visualize the clusters: do you see similar faces in each cluster?

11. Continuing with the Olivetti faces dataset, train a classifier to predict which person is represented in each picture, and evaluate it on the validation set. Next, use K-Means as a dimensionality reduction tool, and train a classifier on the reduced set. Search for the number of clusters that allows the classifier to get the best performance: what performance can you reach? What if you append the features from the reduced set to the original features (again, searching for the best number of clusters)?

12. Train a Gaussian mixture model on the Olivetti faces dataset. To speed up the algorithm, you should probably reduce the dataset's dimensionality (e.g., use PCA, preserving 99% of the variance). Use the model to generate some new faces (using the `sample()` method), and visualize them (if you used PCA, you will need to use its `inverse_transform()` method). Try to modify some images (e.g., rotate, flip, darken) and see if the model can detect the anomalies (i.e., compare the output of the `score_samples()` method for normal images and for anomalies).

13. Some dimensionality reduction techniques can also be used for anomaly detection. For example, take the Olivetti faces dataset and reduce it with PCA, preserving 99% of the variance. Then compute the reconstruction error for each image. Next, take some of the modified images you built in the previous exercise, and look at their reconstruction error: notice how much larger the reconstruction error is. If you plot a reconstructed image, you will see why: it tries to reconstruct a normal face.

Solutions to these exercises are available in Appendix A.

Neural Networks and Deep Learning

Introduction to Artificial Neural Networks with Keras

Birds inspired us to fly, burdock plants inspired Velcro, and nature has inspired countless more inventions. It seems only logical, then, to look at the brain's architecture for inspiration on how to build an intelligent machine. This is the logic that sparked *artificial neural networks* (ANNs): an ANN is a Machine Learning model inspired by the networks of biological neurons found in our brains. However, although planes were inspired by birds, they don't have to flap their wings. Similarly, ANNs have gradually become quite different from their biological cousins. Some researchers even argue that we should drop the biological analogy altogether (e.g., by saying "units" rather than "neurons"), lest we restrict our creativity to biologically plausible systems.[1]

ANNs are at the very core of Deep Learning. They are versatile, powerful, and scalable, making them ideal to tackle large and highly complex Machine Learning tasks such as classifying billions of images (e.g., Google Images), powering speech recognition services (e.g., Apple's Siri), recommending the best videos to watch to hundreds of millions of users every day (e.g., YouTube), or learning to beat the world champion at the game of Go (DeepMind's AlphaGo).

The first part of this chapter introduces artificial neural networks, starting with a quick tour of the very first ANN architectures and leading up to *Multilayer Perceptrons* (MLPs), which are heavily used today (other architectures will be explored in the next chapters). In the second part, we will look at how to implement neural networks using the popular Keras API. This is a beautifully designed and simple high-

---

1 You can get the best of both worlds by being open to biological inspirations without being afraid to create biologically unrealistic models, as long as they work well.

level API for building, training, evaluating, and running neural networks. But don't be fooled by its simplicity: it is expressive and flexible enough to let you build a wide variety of neural network architectures. In fact, it will probably be sufficient for most of your use cases. And should you ever need extra flexibility, you can always write custom Keras components using its lower-level API, as we will see in Chapter 12.

But first, let's go back in time to see how artificial neural networks came to be!

From Biological to Artificial Neurons

Surprisingly, ANNs have been around for quite a while: they were first introduced back in 1943 by the neurophysiologist Warren McCulloch and the mathematician Walter Pitts. In their landmark paper (*https://homl.info/43*)[2] "A Logical Calculus of Ideas Immanent in Nervous Activity," McCulloch and Pitts presented a simplified computational model of how biological neurons might work together in animal brains to perform complex computations using *propositional logic*. This was the first artificial neural network architecture. Since then many other architectures have been invented, as we will see.

The early successes of ANNs led to the widespread belief that we would soon be conversing with truly intelligent machines. When it became clear in the 1960s that this promise would go unfulfilled (at least for quite a while), funding flew elsewhere, and ANNs entered a long winter. In the early 1980s, new architectures were invented and better training techniques were developed, sparking a revival of interest in *connectionism* (the study of neural networks). But progress was slow, and by the 1990s other powerful Machine Learning techniques were invented, such as Support Vector Machines (see Chapter 5). These techniques seemed to offer better results and stronger theoretical foundations than ANNs, so once again the study of neural networks was put on hold.

We are now witnessing yet another wave of interest in ANNs. Will this wave die out like the previous ones did? Well, here are a few good reasons to believe that this time is different and that the renewed interest in ANNs will have a much more profound impact on our lives:

- There is now a huge quantity of data available to train neural networks, and ANNs frequently outperform other ML techniques on very large and complex problems.

- The tremendous increase in computing power since the 1990s now makes it possible to train large neural networks in a reasonable amount of time. This is in

---

2 Warren S. McCulloch and Walter Pitts, "A Logical Calculus of the Ideas Immanent in Nervous Activity," *The Bulletin of Mathematical Biology* 5, no. 4 (1943): 115–113.

part due to Moore's law (the number of components in integrated circuits has doubled about every 2 years over the last 50 years), but also thanks to the gaming industry, which has stimulated the production of powerful GPU cards by the millions. Moreover, cloud platforms have made this power accessible to everyone.

- The training algorithms have been improved. To be fair they are only slightly different from the ones used in the 1990s, but these relatively small tweaks have had a huge positive impact.

- Some theoretical limitations of ANNs have turned out to be benign in practice. For example, many people thought that ANN training algorithms were doomed because they were likely to get stuck in local optima, but it turns out that this is rather rare in practice (and when it is the case, they are usually fairly close to the global optimum).

- ANNs seem to have entered a virtuous circle of funding and progress. Amazing products based on ANNs regularly make the headline news, which pulls more and more attention and funding toward them, resulting in more and more progress and even more amazing products.

Biological Neurons

Before we discuss artificial neurons, let's take a quick look at a biological neuron (represented in Figure 10-1). It is an unusual-looking cell mostly found in animal brains. It's composed of a *cell body* containing the nucleus and most of the cell's complex components, many branching extensions called *dendrites*, plus one very long extension called the *axon*. The axon's length may be just a few times longer than the cell body, or up to tens of thousands of times longer. Near its extremity the axon splits off into many branches called *telodendria*, and at the tip of these branches are minuscule structures called *synaptic terminals* (or simply *synapses*), which are connected to the dendrites or cell bodies of other neurons.[3] Biological neurons produce short electrical impulses called *action potentials* (APs, or just *signals*) which travel along the axons and make the synapses release chemical signals called *neurotransmitters*. When a neuron receives a sufficient amount of these neurotransmitters within a few milliseconds, it fires its own electrical impulses (actually, it depends on the neurotransmitters, as some of them inhibit the neuron from firing).

---

3 They are not actually attached, just so close that they can very quickly exchange chemical signals.

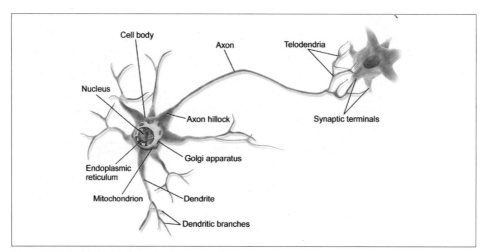

Figure 10-1. Biological neuron[4]

Thus, individual biological neurons seem to behave in a rather simple way, but they are organized in a vast network of billions, with each neuron typically connected to thousands of other neurons. Highly complex computations can be performed by a network of fairly simple neurons, much like a complex anthill can emerge from the combined efforts of simple ants. The architecture of biological neural networks (BNNs)[5] is still the subject of active research, but some parts of the brain have been mapped, and it seems that neurons are often organized in consecutive layers, especially in the cerebral cortex (i.e., the outer layer of your brain), as shown in Figure 10-2.

---

4 Image by Bruce Blaus (Creative Commons 3.0 (*https://creativecommons.org/licenses/by/3.0/*)). Reproduced from *https://en.wikipedia.org/wiki/Neuron*.

5 In the context of Machine Learning, the phrase "neural networks" generally refers to ANNs, not BNNs.

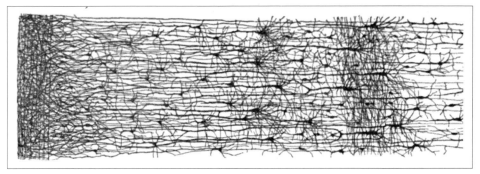

Figure 10-2. Multiple layers in a biological neural network (human cortex)[6]

Logical Computations with Neurons

McCulloch and Pitts proposed a very simple model of the biological neuron, which later became known as an *artificial neuron*: it has one or more binary (on/off) inputs and one binary output. The artificial neuron activates its output when more than a certain number of its inputs are active. In their paper, they showed that even with such a simplified model it is possible to build a network of artificial neurons that computes any logical proposition you want. To see how such a network works, let's build a few ANNs that perform various logical computations (see Figure 10-3), assuming that a neuron is activated when at least two of its inputs are active.

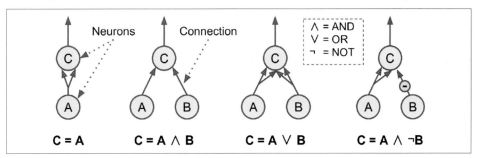

Figure 10-3. ANNs performing simple logical computations

---

6 Drawing of a cortical lamination by S. Ramon y Cajal (public domain). Reproduced from *https://en.wikipe dia.org/wiki/Cerebral_cortex*.

Let's see what these networks do:

- The first network on the left is the identity function: if neuron A is activated, then neuron C gets activated as well (since it receives two input signals from neuron A); but if neuron A is off, then neuron C is off as well.
- The second network performs a logical AND: neuron C is activated only when both neurons A and B are activated (a single input signal is not enough to activate neuron C).
- The third network performs a logical OR: neuron C gets activated if either neuron A or neuron B is activated (or both).
- Finally, if we suppose that an input connection can inhibit the neuron's activity (which is the case with biological neurons), then the fourth network computes a slightly more complex logical proposition: neuron C is activated only if neuron A is active and neuron B is off. If neuron A is active all the time, then you get a logical NOT: neuron C is active when neuron B is off, and vice versa.

You can imagine how these networks can be combined to compute complex logical expressions (see the exercises at the end of the chapter for an example).

The Perceptron

The *Perceptron* is one of the simplest ANN architectures, invented in 1957 by Frank Rosenblatt. It is based on a slightly different artificial neuron (see Figure 10-4) called a *threshold logic unit* (TLU), or sometimes a *linear threshold unit* (LTU). The inputs and output are numbers (instead of binary on/off values), and each input connection is associated with a weight. The TLU computes a weighted sum of its inputs ($z = w_1 x_1 + w_2 x_2 + \cdots + w_n x_n = \mathbf{x}^\mathsf{T} \mathbf{w}$), then applies a *step function* to that sum and outputs the result: $h_\mathbf{w}(\mathbf{x}) = \text{step}(z)$, where $z = \mathbf{x}^\mathsf{T} \mathbf{w}$.

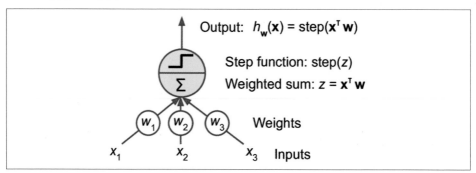

Figure 10-4. Threshold logic unit: an artificial neuron which computes a weighted sum of its inputs then applies a step function

The most common step function used in Perceptrons is the *Heaviside step function* (see Equation 10-1). Sometimes the sign function is used instead.

Equation 10-1. Common step functions used in Perceptrons (assuming threshold = 0)

$$\text{heaviside}\ (z) = \begin{cases} 0 & \text{if}\ z < 0 \\ 1 & \text{if}\ z \geq 0 \end{cases} \qquad \text{sgn}\ (z) = \begin{cases} -1 & \text{if}\ z < 0 \\ 0 & \text{if}\ z = 0 \\ +1 & \text{if}\ z > 0 \end{cases}$$

A single TLU can be used for simple linear binary classification. It computes a linear combination of the inputs, and if the result exceeds a threshold, it outputs the positive class. Otherwise it outputs the negative class (just like a Logistic Regression or linear SVM classifier). You could, for example, use a single TLU to classify iris flowers based on petal length and width (also adding an extra bias feature $x_0 = 1$, just like we did in previous chapters). Training a TLU in this case means finding the right values for w_0, w_1, and w_2 (the training algorithm is discussed shortly).

A Perceptron is simply composed of a single layer of TLUs,[7] with each TLU connected to all the inputs. When all the neurons in a layer are connected to every neuron in the previous layer (i.e., its input neurons), the layer is called a *fully connected layer*, or a *dense layer*. The inputs of the Perceptron are fed to special passthrough neurons called *input neurons*: they output whatever input they are fed. All the input neurons form the *input layer*. Moreover, an extra bias feature is generally added ($x_0 = 1$): it is typically represented using a special type of neuron called a *bias neuron*, which outputs 1 all the time. A Perceptron with two inputs and three outputs is represented in Figure 10-5. This Perceptron can classify instances simultaneously into three different binary classes, which makes it a multioutput classifier.

---

7 The name *Perceptron* is sometimes used to mean a tiny network with a single TLU.

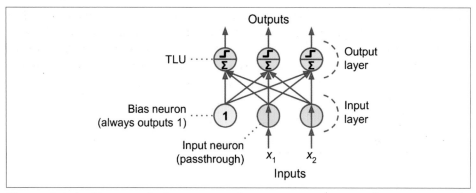

Figure 10-5. Architecture of a Perceptron with two input neurons, one bias neuron, and three output neurons

Thanks to the magic of linear algebra, Equation 10-2 makes it possible to efficiently compute the outputs of a layer of artificial neurons for several instances at once.

Equation 10-2. Computing the outputs of a fully connected layer

$$h_{W,b}(X) = \phi(XW + b)$$

In this equation:

- As always, **X** represents the matrix of input features. It has one row per instance and one column per feature.
- The weight matrix **W** contains all the connection weights except for the ones from the bias neuron. It has one row per input neuron and one column per artificial neuron in the layer.
- The bias vector **b** contains all the connection weights between the bias neuron and the artificial neurons. It has one bias term per artificial neuron.
- The function ϕ is called the *activation function*: when the artificial neurons are TLUs, it is a step function (but we will discuss other activation functions shortly).

So, how is a Perceptron trained? The Perceptron training algorithm proposed by Rosenblatt was largely inspired by *Hebb's rule*. In his 1949 book *The Organization of Behavior* (Wiley), Donald Hebb suggested that when a biological neuron triggers another neuron often, the connection between these two neurons grows stronger. Sie-grid Löwel later summarized Hebb's idea in the catchy phrase, "Cells that fire together, wire together"; that is, the connection weight between two neurons tends to increase when they fire simultaneously. This rule later became known as Hebb's rule (or *Hebbian learning*). Perceptrons are trained using a variant of this rule that takes into account the error made by the network when it makes a prediction; the

Perceptron learning rule reinforces connections that help reduce the error. More specifically, the Perceptron is fed one training instance at a time, and for each instance it makes its predictions. For every output neuron that produced a wrong prediction, it reinforces the connection weights from the inputs that would have contributed to the correct prediction. The rule is shown in Equation 10-3.

Equation 10-3. Perceptron learning rule (weight update)

$$w_{i,j}^{(\text{next step})} = w_{i,j} + \eta\left(y_j - \hat{y}_j\right)x_i$$

In this equation:

- $w_{i,j}$ is the connection weight between the ith input neuron and the jth output neuron.
- x_i is the ith input value of the current training instance.
- $\hat{y}_j$ is the output of the jth output neuron for the current training instance.
- y_j is the target output of the jth output neuron for the current training instance.
- η is the learning rate.

The decision boundary of each output neuron is linear, so Perceptrons are incapable of learning complex patterns (just like Logistic Regression classifiers). However, if the training instances are linearly separable, Rosenblatt demonstrated that this algorithm would converge to a solution.[8] This is called the *Perceptron convergence theorem*.

Scikit-Learn provides a `Perceptron` class that implements a single-TLU network. It can be used pretty much as you would expect—for example, on the iris dataset (introduced in Chapter 4):

```python
import numpy as np
from sklearn.datasets import load_iris
from sklearn.linear_model import Perceptron

iris = load_iris()
X = iris.data[:, (2, 3)]  # petal length, petal width
y = (iris.target == 0).astype(np.int)  # Iris setosa?

per_clf = Perceptron()
per_clf.fit(X, y)

y_pred = per_clf.predict([[2, 0.5]])
```

---

8 Note that this solution is not unique: when data points are linearly separable, there is an infinity of hyperplanes that can separate them.

You may have noticed that the Perceptron learning algorithm strongly resembles Stochastic Gradient Descent. In fact, Scikit-Learn's `Perceptron` class is equivalent to using an `SGDClassifier` with the following hyperparameters: `loss="perceptron"`, `learning_rate="constant"`, `eta0=1` (the learning rate), and `penalty=None` (no regularization).

Note that contrary to Logistic Regression classifiers, Perceptrons do not output a class probability; rather, they make predictions based on a hard threshold. This is one reason to prefer Logistic Regression over Perceptrons.

In their 1969 monograph *Perceptrons*, Marvin Minsky and Seymour Papert highlighted a number of serious weaknesses of Perceptrons—in particular, the fact that they are incapable of solving some trivial problems (e.g., the *Exclusive OR* (XOR) classification problem; see the left side of Figure 10-6). This is true of any other linear classification model (such as Logistic Regression classifiers), but researchers had expected much more from Perceptrons, and some were so disappointed that they dropped neural networks altogether in favor of higher-level problems such as logic, problem solving, and search.

It turns out that some of the limitations of Perceptrons can be eliminated by stacking multiple Perceptrons. The resulting ANN is called a *Multilayer Perceptron* (MLP). An MLP can solve the XOR problem, as you can verify by computing the output of the MLP represented on the right side of Figure 10-6: with inputs (0, 0) or (1, 1), the network outputs 0, and with inputs (0, 1) or (1, 0) it outputs 1. All connections have a weight equal to 1, except the four connections where the weight is shown. Try verifying that this network indeed solves the XOR problem!

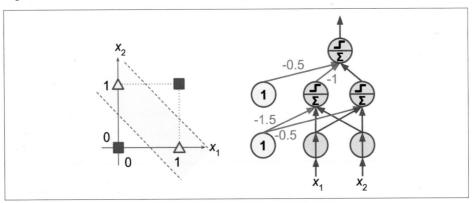

Figure 10-6. XOR classification problem and an MLP that solves it

The Multilayer Perceptron and Backpropagation

An MLP is composed of one (passthrough) *input layer*, one or more layers of TLUs, called *hidden layers*, and one final layer of TLUs called the *output layer* (see Figure 10-7). The layers close to the input layer are usually called the *lower layers*, and the ones close to the outputs are usually called the *upper layers*. Every layer except the output layer includes a bias neuron and is fully connected to the next layer.

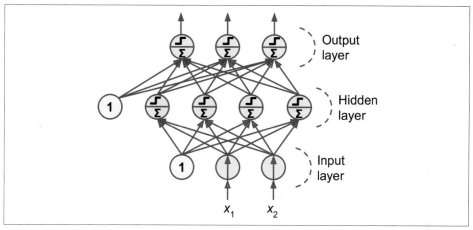

Figure 10-7. Architecture of a Multilayer Perceptron with two inputs, one hidden layer of four neurons, and three output neurons (the bias neurons are shown here, but usually they are implicit)

 The signal flows only in one direction (from the inputs to the outputs), so this architecture is an example of a *feedforward neural network* (FNN).

When an ANN contains a deep stack of hidden layers,[9] it is called a *deep neural network* (DNN). The field of Deep Learning studies DNNs, and more generally models containing deep stacks of computations. Even so, many people talk about Deep Learning whenever neural networks are involved (even shallow ones).

For many years researchers struggled to find a way to train MLPs, without success. But in 1986, David Rumelhart, Geoffrey Hinton, and Ronald Williams published a

---

9 In the 1990s, an ANN with more than two hidden layers was considered deep. Nowadays, it is common to see ANNs with dozens of layers, or even hundreds, so the definition of "deep" is quite fuzzy.

groundbreaking paper (*https://homl.info/44*)[10] that introduced the *backpropagation* training algorithm, which is still used today. In short, it is Gradient Descent (introduced in Chapter 4) using an efficient technique for computing the gradients automatically:[11] in just two passes through the network (one forward, one backward), the backpropagation algorithm is able to compute the gradient of the network's error with regard to every single model parameter. In other words, it can find out how each connection weight and each bias term should be tweaked in order to reduce the error. Once it has these gradients, it just performs a regular Gradient Descent step, and the whole process is repeated until the network converges to the solution.

Automatically computing gradients is called *automatic differentiation*, or *autodiff*. There are various autodiff techniques, with different pros and cons. The one used by backpropagation is called *reverse-mode autodiff*. It is fast and precise, and is well suited when the function to differentiate has many variables (e.g., connection weights) and few outputs (e.g., one loss). If you want to learn more about autodiff, check out Appendix D.

Let's run through this algorithm in a bit more detail:

- It handles one mini-batch at a time (for example, containing 32 instances each), and it goes through the full training set multiple times. Each pass is called an *epoch*.

- Each mini-batch is passed to the network's input layer, which sends it to the first hidden layer. The algorithm then computes the output of all the neurons in this layer (for every instance in the mini-batch). The result is passed on to the next layer, its output is computed and passed to the next layer, and so on until we get the output of the last layer, the output layer. This is the *forward pass*: it is exactly like making predictions, except all intermediate results are preserved since they are needed for the backward pass.

- Next, the algorithm measures the network's output error (i.e., it uses a loss function that compares the desired output and the actual output of the network, and returns some measure of the error).

- Then it computes how much each output connection contributed to the error. This is done analytically by applying the *chain rule* (perhaps the most fundamental rule in calculus), which makes this step fast and precise.

---

10 David Rumelhart et al. "Learning Internal Representations by Error Propagation," (Defense Technical Information Center technical report, September 1985).

11 This technique was actually independently invented several times by various researchers in different fields, starting with Paul Werbos in 1974.

- The algorithm then measures how much of these error contributions came from each connection in the layer below, again using the chain rule, working backward until the algorithm reaches the input layer. As explained earlier, this reverse pass efficiently measures the error gradient across all the connection weights in the network by propagating the error gradient backward through the network (hence the name of the algorithm).

- Finally, the algorithm performs a Gradient Descent step to tweak all the connection weights in the network, using the error gradients it just computed.

This algorithm is so important that it's worth summarizing it again: for each training instance, the backpropagation algorithm first makes a prediction (forward pass) and measures the error, then goes through each layer in reverse to measure the error contribution from each connection (reverse pass), and finally tweaks the connection weights to reduce the error (Gradient Descent step).

 It is important to initialize all the hidden layers' connection weights randomly, or else training will fail. For example, if you initialize all weights and biases to zero, then all neurons in a given layer will be perfectly identical, and thus backpropagation will affect them in exactly the same way, so they will remain identical. In other words, despite having hundreds of neurons per layer, your model will act as if it had only one neuron per layer: it won't be too smart. If instead you randomly initialize the weights, you *break the symmetry* and allow backpropagation to train a diverse team of neurons.

In order for this algorithm to work properly, its authors made a key change to the MLP's architecture: they replaced the step function with the logistic (sigmoid) function, $\sigma(z) = 1 / (1 + \exp(-z))$. This was essential because the step function contains only flat segments, so there is no gradient to work with (Gradient Descent cannot move on a flat surface), while the logistic function has a well-defined nonzero derivative everywhere, allowing Gradient Descent to make some progress at every step. In fact, the backpropagation algorithm works well with many other activation functions, not just the logistic function. Here are two other popular choices:

The hyperbolic tangent function: $tanh(z) = 2\sigma(2z) – 1$
Just like the logistic function, this activation function is S-shaped, continuous, and differentiable, but its output value ranges from –1 to 1 (instead of 0 to 1 in the case of the logistic function). That range tends to make each layer's output more or less centered around 0 at the beginning of training, which often helps speed up convergence.

The Rectified Linear Unit function: ReLU(z) = max(0, z)

The ReLU function is continuous but unfortunately not differentiable at $z = 0$ (the slope changes abruptly, which can make Gradient Descent bounce around), and its derivative is 0 for $z < 0$. In practice, however, it works very well and has the advantage of being fast to compute, so it has become the default.[12] Most importantly, the fact that it does not have a maximum output value helps reduce some issues during Gradient Descent (we will come back to this in Chapter 11).

These popular activation functions and their derivatives are represented in Figure 10-8. But wait! Why do we need activation functions in the first place? Well, if you chain several linear transformations, all you get is a linear transformation. For example, if $f(x) = 2x + 3$ and $g(x) = 5x - 1$, then chaining these two linear functions gives you another linear function: $f(g(x)) = 2(5x - 1) + 3 = 10x + 1$. So if you don't have some nonlinearity between layers, then even a deep stack of layers is equivalent to a single layer, and you can't solve very complex problems with that. Conversely, a large enough DNN with nonlinear activations can theoretically approximate any continuous function.

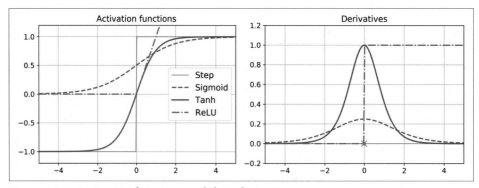

Figure 10-8. Activation functions and their derivatives

OK! You know where neural nets came from, what their architecture is, and how to compute their outputs. You've also learned about the backpropagation algorithm. But what exactly can you do with them?

Regression MLPs

First, MLPs can be used for regression tasks. If you want to predict a single value (e.g., the price of a house, given many of its features), then you just need a single output neuron: its output is the predicted value. For multivariate regression (i.e., to predict

---

12 Biological neurons seem to implement a roughly sigmoid (S-shaped) activation function, so researchers stuck to sigmoid functions for a very long time. But it turns out that ReLU generally works better in ANNs. This is one of the cases where the biological analogy was misleading.

multiple values at once), you need one output neuron per output dimension. For example, to locate the center of an object in an image, you need to predict 2D coordinates, so you need two output neurons. If you also want to place a bounding box around the object, then you need two more numbers: the width and the height of the object. So, you end up with four output neurons.

In general, when building an MLP for regression, you do not want to use any activation function for the output neurons, so they are free to output any range of values. If you want to guarantee that the output will always be positive, then you can use the ReLU activation function in the output layer. Alternatively, you can use the *softplus* activation function, which is a smooth variant of ReLU: softplus$(z) = \log(1 + \exp(z))$. It is close to 0 when z is negative, and close to z when z is positive. Finally, if you want to guarantee that the predictions will fall within a given range of values, then you can use the logistic function or the hyperbolic tangent, and then scale the labels to the appropriate range: 0 to 1 for the logistic function and –1 to 1 for the hyperbolic tangent.

The loss function to use during training is typically the mean squared error, but if you have a lot of outliers in the training set, you may prefer to use the mean absolute error instead. Alternatively, you can use the Huber loss, which is a combination of both.

The Huber loss is quadratic when the error is smaller than a threshold δ (typically 1) but linear when the error is larger than δ. The linear part makes it less sensitive to outliers than the mean squared error, and the quadratic part allows it to converge faster and be more precise than the mean absolute error.

Table 10-1 summarizes the typical architecture of a regression MLP.

Table 10-1. Typical regression MLP architecture

Hyperparameter	Typical value
# input neurons	One per input feature (e.g., 28 x 28 = 784 for MNIST)
# hidden layers	Depends on the problem, but typically 1 to 5
# neurons per hidden layer	Depends on the problem, but typically 10 to 100
# output neurons	1 per prediction dimension
Hidden activation	ReLU (or SELU, see Chapter 11)
Output activation	None, or ReLU/softplus (if positive outputs) or logistic/tanh (if bounded outputs)
Loss function	MSE or MAE/Huber (if outliers)

Classification MLPs

MLPs can also be used for classification tasks. For a binary classification problem, you just need a single output neuron using the logistic activation function: the output will be a number between 0 and 1, which you can interpret as the estimated probability of the positive class. The estimated probability of the negative class is equal to one minus that number.

MLPs can also easily handle multilabel binary classification tasks (see Chapter 3). For example, you could have an email classification system that predicts whether each incoming email is ham or spam, and simultaneously predicts whether it is an urgent or nonurgent email. In this case, you would need two output neurons, both using the logistic activation function: the first would output the probability that the email is spam, and the second would output the probability that it is urgent. More generally, you would dedicate one output neuron for each positive class. Note that the output probabilities do not necessarily add up to 1. This lets the model output any combination of labels: you can have nonurgent ham, urgent ham, nonurgent spam, and perhaps even urgent spam (although that would probably be an error).

If each instance can belong only to a single class, out of three or more possible classes (e.g., classes 0 through 9 for digit image classification), then you need to have one output neuron per class, and you should use the softmax activation function for the whole output layer (see Figure 10-9). The softmax function (introduced in Chapter 4) will ensure that all the estimated probabilities are between 0 and 1 and that they add up to 1 (which is required if the classes are exclusive). This is called multiclass classification.

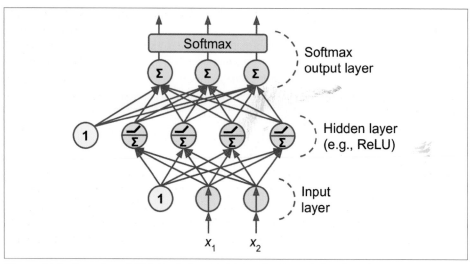

Figure 10-9. A modern MLP (including ReLU and softmax) for classification

---

Regarding the loss function, since we are predicting probability distributions, the cross-entropy loss (also called the log loss, see Chapter 4) is generally a good choice.

Table 10-2 summarizes the typical architecture of a classification MLP.

Table 10-2. Typical classification MLP architecture

Hyperparameter	Binary classification	Multilabel binary classification	Multiclass classification
Input and hidden layers	Same as regression	Same as regression	Same as regression
# output neurons	1	1 per label	1 per class
Output layer activation	Logistic	Logistic	Softmax
Loss function	Cross entropy	Cross entropy	Cross entropy

 Before we go on, I recommend you go through exercise 1 at the end of this chapter. You will play with various neural network architectures and visualize their outputs using the *TensorFlow Playground*. This will be very useful to better understand MLPs, including the effects of all the hyperparameters (number of layers and neurons, activation functions, and more).

Now you have all the concepts you need to start implementing MLPs with Keras!

Implementing MLPs with Keras

Keras is a high-level Deep Learning API that allows you to easily build, train, evaluate, and execute all sorts of neural networks. Its documentation (or specification) is available at *https://keras.io/*. The reference implementation (*https://github.com/keras-team/keras*), also called Keras, was developed by François Chollet as part of a research project[13] and was released as an open source project in March 2015. It quickly gained popularity, owing to its ease of use, flexibility, and beautiful design. To perform the heavy computations required by neural networks, this reference implementation relies on a computation backend. At present, you can choose from three popular open source Deep Learning libraries: TensorFlow, Microsoft Cognitive Toolkit (CNTK), and Theano. Therefore, to avoid any confusion, we will refer to this reference implementation as *multibackend Keras*.

Since late 2016, other implementations have been released. You can now run Keras on Apache MXNet, Apple's Core ML, JavaScript or TypeScript (to run Keras code in a web browser), and PlaidML (which can run on all sorts of GPU devices, not just Nvidia). Moreover, TensorFlow itself now comes bundled with its own Keras implementation, tf.keras. It only supports TensorFlow as the backend, but it has the advantage

---

13 Project ONEIROS (Open-ended Neuro-Electronic Intelligent Robot Operating System).

of offering some very useful extra features (see Figure 10-10): for example, it supports TensorFlow's Data API, which makes it easy to load and preprocess data efficiently. For this reason, we will use tf.keras in this book. However, in this chapter we will not use any of the TensorFlow-specific features, so the code should run fine on other Keras implementations as well (at least in Python), with only minor modifications, such as changing the imports.

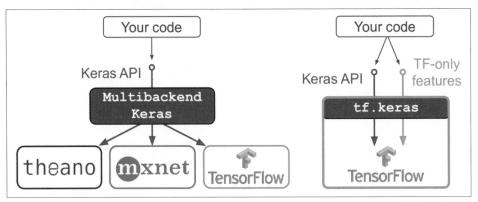

Figure 10-10. Two implementations of the Keras API: multibackend Keras (left) and tf.keras (right)

The most popular Deep Learning library, after Keras and TensorFlow, is Facebook's PyTorch (*https://pytorch.org/*) library. The good news is that its API is quite similar to Keras's (in part because both APIs were inspired by Scikit-Learn and Chainer (*https://chainer.org/*)), so once you know Keras, it is not difficult to switch to PyTorch, if you ever want to. PyTorch's popularity grew exponentially in 2018, largely thanks to its simplicity and excellent documentation, which were not TensorFlow 1.x's main strengths. However, TensorFlow 2 is arguably just as simple as PyTorch, as it has adopted Keras as its official high-level API and its developers have greatly simplified and cleaned up the rest of the API. The documentation has also been completely reorganized, and it is much easier to find what you need now. Similarly, PyTorch's main weaknesses (e.g., limited portability and no computation graph analysis) have been largely addressed in PyTorch 1.0. Healthy competition is beneficial to everyone.

All right, it's time to code! As tf.keras is bundled with TensorFlow, let's start by installing TensorFlow.

Installing TensorFlow 2

Assuming you installed Jupyter and Scikit-Learn by following the installation instructions in Chapter 2, use pip to install TensorFlow. If you created an isolated environment using virtualenv, you first need to activate it:

```
$ cd $ML_PATH              # Your ML working directory (e.g., $HOME/ml)
$ source my_env/bin/activate # on Linux or macOS
$ .\my_env\Scripts\activate  # on Windows
```

Next, install TensorFlow 2 (if you are not using a virtualenv, you will need administrator rights, or to add the --user option):

```
$ python3 -m pip install --upgrade tensorflow
```

 For GPU support, at the time of this writing you need to install tensorflow-gpu instead of tensorflow, but the TensorFlow team is working on having a single library that will support both CPU-only and GPU-equipped systems. You will still need to install extra libraries for GPU support (see *https://tensorflow.org/install* for more details). We will look at GPUs in more depth in Chapter 19.

To test your installation, open a Python shell or a Jupyter notebook, then import TensorFlow and tf.keras and print their versions:

```
>>> import tensorflow as tf
>>> from tensorflow import keras
>>> tf.__version__
'2.0.0'
>>> keras.__version__
'2.2.4-tf'
```

The second version is the version of the Keras API implemented by tf.keras. Note that it ends with -tf, highlighting the fact that tf.keras implements the Keras API, plus some extra TensorFlow-specific features.

Now let's use tf.keras! We'll start by building a simple image classifier.

Building an Image Classifier Using the Sequential API

First, we need to load a dataset. In this chapter we will tackle Fashion MNIST, which is a drop-in replacement of MNIST (introduced in Chapter 3). It has the exact same format as MNIST (70,000 grayscale images of 28 × 28 pixels each, with 10 classes), but the images represent fashion items rather than handwritten digits, so each class is more diverse, and the problem turns out to be significantly more challenging than MNIST. For example, a simple linear model reaches about 92% accuracy on MNIST, but only about 83% on Fashion MNIST.

Using Keras to load the dataset

Keras provides some utility functions to fetch and load common datasets, including MNIST, Fashion MNIST, and the California housing dataset we used in Chapter 2. Let's load Fashion MNIST:

```
fashion_mnist = keras.datasets.fashion_mnist
(X_train_full, y_train_full), (X_test, y_test) = fashion_mnist.load_data()
```

When loading MNIST or Fashion MNIST using Keras rather than Scikit-Learn, one important difference is that every image is represented as a 28 × 28 array rather than a 1D array of size 784. Moreover, the pixel intensities are represented as integers (from 0 to 255) rather than floats (from 0.0 to 255.0). Let's take a look at the shape and data type of the training set:

```
>>> X_train_full.shape
(60000, 28, 28)
>>> X_train_full.dtype
dtype('uint8')
```

Note that the dataset is already split into a training set and a test set, but there is no validation set, so we'll create one now. Additionally, since we are going to train the neural network using Gradient Descent, we must scale the input features. For simplicity, we'll scale the pixel intensities down to the 0–1 range by dividing them by 255.0 (this also converts them to floats):

```
X_valid, X_train = X_train_full[:5000] / 255.0, X_train_full[5000:] / 255.0
y_valid, y_train = y_train_full[:5000], y_train_full[5000:]
```

With MNIST, when the label is equal to 5, it means that the image represents the handwritten digit 5. Easy. For Fashion MNIST, however, we need the list of class names to know what we are dealing with:

```
class_names = ["T-shirt/top", "Trouser", "Pullover", "Dress", "Coat",
               "Sandal", "Shirt", "Sneaker", "Bag", "Ankle boot"]
```

For example, the first image in the training set represents a coat:

```
>>> class_names[y_train[0]]
'Coat'
```

Figure 10-11 shows some samples from the Fashion MNIST dataset.

Figure 10-11. Samples from Fashion MNIST

Creating the model using the Sequential API

Now let's build the neural network! Here is a classification MLP with two hidden layers:

```
model = keras.models.Sequential()
model.add(keras.layers.Flatten(input_shape=[28, 28]))
model.add(keras.layers.Dense(300, activation="relu"))
model.add(keras.layers.Dense(100, activation="relu"))
model.add(keras.layers.Dense(10, activation="softmax"))
```

Let's go through this code line by line:

- The first line creates a `Sequential` model. This is the simplest kind of Keras model for neural networks that are just composed of a single stack of layers connected sequentially. This is called the Sequential API.

- Next, we build the first layer and add it to the model. It is a `Flatten` layer whose role is to convert each input image into a 1D array: if it receives input data X, it computes `X.reshape(-1, 1)`. This layer does not have any parameters; it is just there to do some simple preprocessing. Since it is the first layer in the model, you should specify the `input_shape`, which doesn't include the batch size, only the shape of the instances. Alternatively, you could add a `keras.layers.InputLayer` as the first layer, setting `input_shape=[28,28]`.

- Next we add a `Dense` hidden layer with 300 neurons. It will use the ReLU activation function. Each `Dense` layer manages its own weight matrix, containing all the connection weights between the neurons and their inputs. It also manages a vector of bias terms (one per neuron). When it receives some input data, it computes Equation 10-2.

- Then we add a second `Dense` hidden layer with 100 neurons, also using the ReLU activation function.

- Finally, we add a `Dense` output layer with 10 neurons (one per class), using the softmax activation function (because the classes are exclusive).

> Specifying `activation="relu"` is equivalent to specifying `activation=keras.activations.relu`. Other activation functions are available in the `keras.activations` package, we will use many of them in this book. See *https://keras.io/activations/* for the full list.

Instead of adding the layers one by one as we just did, you can pass a list of layers when creating the `Sequential` model:

```
model = keras.models.Sequential([
    keras.layers.Flatten(input_shape=[28, 28]),
    keras.layers.Dense(300, activation="relu"),
    keras.layers.Dense(100, activation="relu"),
    keras.layers.Dense(10, activation="softmax")
])
```

Using Code Examples from keras.io

Code examples documented on keras.io will work fine with tf.keras, but you need to change the imports. For example, consider this keras.io code:

```
from keras.layers import Dense
output_layer = Dense(10)
```

You must change the imports like this:

```
from tensorflow.keras.layers import Dense
output_layer = Dense(10)
```

Or simply use full paths, if you prefer:

```
from tensorflow import keras
output_layer = keras.layers.Dense(10)
```

This approach is more verbose, but I use it in this book so you can easily see which packages to use, and to avoid confusion between standard classes and custom classes. In production code, I prefer the previous approach. Many people also use `from ten sorflow.keras import layers` followed by `layers.Dense(10)`.

The model's `summary()` method displays all the model's layers,[14] including each layer's name (which is automatically generated unless you set it when creating the layer), its output shape (`None` means the batch size can be anything), and its number of parameters. The summary ends with the total number of parameters, including trainable and non-trainable parameters. Here we only have trainable parameters (we will see examples of non-trainable parameters in Chapter 11):

```
>>> model.summary()
Model: "sequential"
```

Layer (type)	Output Shape	Param #
flatten (Flatten)	(None, 784)	0
dense (Dense)	(None, 300)	235500

---

14 You can use `keras.utils.plot_model()` to generate an image of your model.

```
dense_1 (Dense)              (None, 100)              30100
_____
dense_2 (Dense)              (None, 10)               1010
=================================================================
Total params: 266,610
Trainable params: 266,610
Non-trainable params: 0
_____
```

Note that Dense layers often have a *lot* of parameters. For example, the first hidden layer has 784 × 300 connection weights, plus 300 bias terms, which adds up to 235,500 parameters! This gives the model quite a lot of flexibility to fit the training data, but it also means that the model runs the risk of overfitting, especially when you do not have a lot of training data. We will come back to this later.

You can easily get a model's list of layers, to fetch a layer by its index, or you can fetch it by name:

```
>>> model.layers
[<tensorflow.python.keras.layers.core.Flatten at 0x132414e48>,
 <tensorflow.python.keras.layers.core.Dense at 0x1324149b0>,
 <tensorflow.python.keras.layers.core.Dense at 0x1356ba8d0>,
 <tensorflow.python.keras.layers.core.Dense at 0x13240d240>]
>>> hidden1 = model.layers[1]
>>> hidden1.name
'dense'
>>> model.get_layer('dense') is hidden1
True
```

All the parameters of a layer can be accessed using its get_weights() and set_weights() methods. For a Dense layer, this includes both the connection weights and the bias terms:

```
>>> weights, biases = hidden1.get_weights()
>>> weights
array([[ 0.02448617, -0.00877795, -0.02189048, ..., -0.02766046,
         0.03859074, -0.06889391],
       ...,
       [-0.06022581,  0.01577859, -0.02585464, ..., -0.00527829,
         0.00272203, -0.06793761]], dtype=float32)
>>> weights.shape
(784, 300)
>>> biases
array([0., 0., 0., 0., 0., 0., 0., 0., 0., ...,  0., 0., 0.], dtype=float32)
>>> biases.shape
(300,)
```

Notice that the Dense layer initialized the connection weights randomly (which is needed to break symmetry, as we discussed earlier), and the biases were initialized to zeros, which is fine. If you ever want to use a different initialization method, you can set kernel_initializer (*kernel* is another name for the matrix of connection weights) or `bias_initializer` when creating the layer. We will discuss initializers further in Chapter 11, but if you want the full list, see *https://keras.io/initializers/*.

> The shape of the weight matrix depends on the number of inputs. This is why it is recommended to specify the `input_shape` when creating the first layer in a `Sequential` model. However, if you do not specify the input shape, it's OK: Keras will simply wait until it knows the input shape before it actually builds the model. This will happen either when you feed it actual data (e.g., during training), or when you call its `build()` method. Until the model is really built, the layers will not have any weights, and you will not be able to do certain things (such as print the model summary or save the model). So, if you know the input shape when creating the model, it is best to specify it.

Compiling the model

After a model is created, you must call its `compile()` method to specify the loss function and the optimizer to use. Optionally, you can specify a list of extra metrics to compute during training and evaluation:

```
model.compile(loss="sparse_categorical_crossentropy",
              optimizer="sgd",
              metrics=["accuracy"])
```

> Using `loss="sparse_categorical_crossentropy"` is equivalent to using `loss=keras.losses.sparse_categorical_crossentropy`. Similarly, specifying `optimizer="sgd"` is equivalent to specifying `optimizer=keras.optimizers.SGD()`, and `metrics=["accuracy"]` is equivalent to `metrics=[keras.metrics.sparse_categorical_accuracy]` (when using this loss). We will use many other losses, optimizers, and metrics in this book; for the full lists, see *https://keras.io/losses*, *https://keras.io/optimizers*, and *https://keras.io/metrics*.

This code requires some explanation. First, we use the `"sparse_categorical_crossentropy"` loss because we have sparse labels (i.e., for each instance, there is just a target class index, from 0 to 9 in this case), and the classes are exclusive. If instead we had one target probability per class for each instance (such as one-hot vectors, e.g. `[0., 0., 0., 1., 0., 0., 0., 0., 0., 0.]` to represent class 3), then we would need to use the `"categorical_crossentropy"` loss instead. If we were doing binary classification (with one or more binary labels), then we would use the `"sigmoid"` (i.e., logistic) activation function in the output layer instead of the `"softmax"` activation function, and we would use the `"binary_crossentropy"` loss.

If you want to convert sparse labels (i.e., class indices) to one-hot vector labels, use the keras.utils.to_categorical() function. To go the other way round, use the np.argmax() function with axis=1.

Regarding the optimizer, "sgd" means that we will train the model using simple Stochastic Gradient Descent. In other words, Keras will perform the backpropagation algorithm described earlier (i.e., reverse-mode autodiff plus Gradient Descent). We will discuss more efficient optimizers in Chapter 11 (they improve the Gradient Descent part, not the autodiff).

 When using the SGD optimizer, it is important to tune the learning rate. So, you will generally want to use optimizer=keras.optimiz ers.SGD(lr=???) to set the learning rate, rather than opti mizer="sgd", which defaults to lr=0.01.

Finally, since this is a classifier, it's useful to measure its "accuracy" during training and evaluation.

Training and evaluating the model

Now the model is ready to be trained. For this we simply need to call its fit() method:

```
>>> history = model.fit(X_train, y_train, epochs=30,
...                     validation_data=(X_valid, y_valid))
...
Train on 55000 samples, validate on 5000 samples
Epoch 1/30
55000/55000 [======] - 3s 49us/sample - loss: 0.7218     - accuracy: 0.7660
                                       - val_loss: 0.4973 - val_accuracy: 0.8366
Epoch 2/30
55000/55000 [======] - 2s 45us/sample - loss: 0.4840     - accuracy: 0.8327
                                       - val_loss: 0.4456 - val_accuracy: 0.8480
[...]
Epoch 30/30
55000/55000 [======] - 3s 53us/sample - loss: 0.2252     - accuracy: 0.9192
                                       - val_loss: 0.2999 - val_accuracy: 0.8926
```

We pass it the input features (X_train) and the target classes (y_train), as well as the number of epochs to train (or else it would default to just 1, which would definitely not be enough to converge to a good solution). We also pass a validation set (this is optional). Keras will measure the loss and the extra metrics on this set at the end of each epoch, which is very useful to see how well the model really performs. If the performance on the training set is much better than on the validation set, your model is probably overfitting the training set (or there is a bug, such as a data mismatch between the training set and the validation set).

And that's it! The neural network is trained.[15] At each epoch during training, Keras displays the number of instances processed so far (along with a progress bar), the mean training time per sample, and the loss and accuracy (or any other extra metrics you asked for) on both the training set and the validation set. You can see that the training loss went down, which is a good sign, and the validation accuracy reached 89.26% after 30 epochs. That's not too far from the training accuracy, so there does not seem to be much overfitting going on.

> Instead of passing a validation set using the `validation_data` argument, you could set `validation_split` to the ratio of the training set that you want Keras to use for validation. For example, `validation_split=0.1` tells Keras to use the last 10% of the data (before shuffling) for validation.

If the training set was very skewed, with some classes being overrepresented and others underrepresented, it would be useful to set the `class_weight` argument when calling the `fit()` method, which would give a larger weight to underrepresented classes and a lower weight to overrepresented classes. These weights would be used by Keras when computing the loss. If you need per-instance weights, set the `sam ple_weight` argument (it supersedes `class_weight`). Per-instance weights could be useful if some instances were labeled by experts while others were labeled using a crowdsourcing platform: you might want to give more weight to the former. You can also provide sample weights (but not class weights) for the validation set by adding them as a third item in the `validation_data` tuple.

The `fit()` method returns a `History` object containing the training parameters (`history.params`), the list of epochs it went through (`history.epoch`), and most importantly a dictionary (`history.history`) containing the loss and extra metrics it measured at the end of each epoch on the training set and on the validation set (if any). If you use this dictionary to create a pandas DataFrame and call its `plot()` method, you get the learning curves shown in Figure 10-12:

---

15 If your training or validation data does not match the expected shape, you will get an exception. This is perhaps the most common error, so you should get familiar with the error message. The message is actually quite clear: for example, if you try to train this model with an array containing flattened images (`X_train.reshape(-1, 784)`), then you will get the following exception: "ValueError: Error when checking input: expected flatten_input to have 3 dimensions, but got array with shape (60000, 784)."

```
import pandas as pd
import matplotlib.pyplot as plt

pd.DataFrame(history.history).plot(figsize=(8, 5))
plt.grid(True)
plt.gca().set_ylim(0, 1) # set the vertical range to [0-1]
plt.show()
```

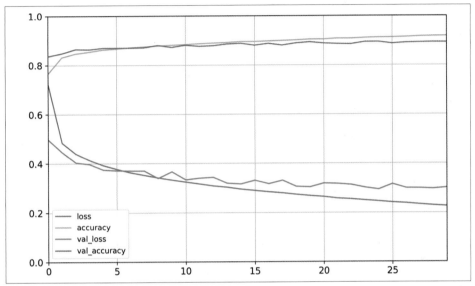

Figure 10-12. Learning curves: the mean training loss and accuracy measured over each epoch, and the mean validation loss and accuracy measured at the end of each epoch

You can see that both the training accuracy and the validation accuracy steadily increase during training, while the training loss and the validation loss decrease. Good! Moreover, the validation curves are close to the training curves, which means that there is not too much overfitting. In this particular case, the model looks like it performed better on the validation set than on the training set at the beginning of training. But that's not the case: indeed, the validation error is computed at the *end* of each epoch, while the training error is computed using a running mean *during* each epoch. So the training curve should be shifted by half an epoch to the left. If you do that, you will see that the training and validation curves overlap almost perfectly at the beginning of training.

> When plotting the training curve, it should be shifted by half an epoch to the left.

The training set performance ends up beating the validation performance, as is generally the case when you train for long enough. You can tell that the model has not quite converged yet, as the validation loss is still going down, so you should probably continue training. It's as simple as calling the `fit()` method again, since Keras just continues training where it left off (you should be able to reach close to 89% validation accuracy).

If you are not satisfied with the performance of your model, you should go back and tune the hyperparameters. The first one to check is the learning rate. If that doesn't help, try another optimizer (and always retune the learning rate after changing any hyperparameter). If the performance is still not great, then try tuning model hyperparameters such as the number of layers, the number of neurons per layer, and the types of activation functions to use for each hidden layer. You can also try tuning other hyperparameters, such as the batch size (it can be set in the `fit()` method using the `batch_size` argument, which defaults to 32). We will get back to hyperparameter tuning at the end of this chapter. Once you are satisfied with your model's validation accuracy, you should evaluate it on the test set to estimate the generalization error before you deploy the model to production. You can easily do this using the `evaluate()` method (it also supports several other arguments, such as `batch_size` and `sample_weight`; please check the documentation for more details):

```
>>> model.evaluate(X_test, y_test)
10000/10000 [==========] - 0s 29us/sample - loss: 0.3340 - accuracy: 0.8851
[0.3339798209667206, 0.8851]
```

As we saw in Chapter 2, it is common to get slightly lower performance on the test set than on the validation set, because the hyperparameters are tuned on the validation set, not the test set (however, in this example, we did not do any hyperparameter tuning, so the lower accuracy is just bad luck). Remember to resist the temptation to tweak the hyperparameters on the test set, or else your estimate of the generalization error will be too optimistic.

Using the model to make predictions

Next, we can use the model's `predict()` method to make predictions on new instances. Since we don't have actual new instances, we will just use the first three instances of the test set:

```
>>> X_new = X_test[:3]
>>> y_proba = model.predict(X_new)
>>> y_proba.round(2)
array([[0.  , 0.  , 0.  , 0.  , 0.  , 0.03, 0.  , 0.01, 0.  , 0.96],
       [0.  , 0.  , 0.98, 0.  , 0.02, 0.  , 0.  , 0.  , 0.  , 0.  ],
       [0.  , 1.  , 0.  , 0.  , 0.  , 0.  , 0.  , 0.  , 0.  , 0.  ]],
      dtype=float32)
```

As you can see, for each instance the model estimates one probability per class, from class 0 to class 9. For example, for the first image it estimates that the probability of class 9 (ankle boot) is 96%, the probability of class 5 (sandal) is 3%, the probability of class 7 (sneaker) is 1%, and the probabilities of the other classes are negligible. In other words, it "believes" the first image is footwear, most likely ankle boots but possibly sandals or sneakers. If you only care about the class with the highest estimated probability (even if that probability is quite low), then you can use the `pre dict_classes()` method instead:

```
>>> y_pred = model.predict_classes(X_new)
>>> y_pred
array([9, 2, 1])
>>> np.array(class_names)[y_pred]
array(['Ankle boot', 'Pullover', 'Trouser'], dtype='<U11')
```

Here, the classifier actually classified all three images correctly (these images are shown in Figure 10-13):

```
>>> y_new = y_test[:3]
>>> y_new
array([9, 2, 1])
```

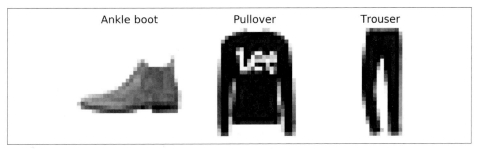

Figure 10-13. Correctly classified Fashion MNIST images

Now you know how to use the Sequential API to build, train, evaluate, and use a classification MLP. But what about regression?

Building a Regression MLP Using the Sequential API

Let's switch to the California housing problem and tackle it using a regression neural network. For simplicity, we will use Scikit-Learn's `fetch_california_housing()` function to load the data. This dataset is simpler than the one we used in Chapter 2, since it contains only numerical features (there is no `ocean_proximity` feature), and there is no missing value. After loading the data, we split it into a training set, a validation set, and a test set, and we scale all the features:

```
from sklearn.datasets import fetch_california_housing
from sklearn.model_selection import train_test_split
from sklearn.preprocessing import StandardScaler
```

```
housing = fetch_california_housing()

X_train_full, X_test, y_train_full, y_test = train_test_split(
    housing.data, housing.target)
X_train, X_valid, y_train, y_valid = train_test_split(
    X_train_full, y_train_full)

scaler = StandardScaler()
X_train = scaler.fit_transform(X_train)
X_valid = scaler.transform(X_valid)
X_test = scaler.transform(X_test)
```

Using the Sequential API to build, train, evaluate, and use a regression MLP to make predictions is quite similar to what we did for classification. The main differences are the fact that the output layer has a single neuron (since we only want to predict a single value) and uses no activation function, and the loss function is the mean squared error. Since the dataset is quite noisy, we just use a single hidden layer with fewer neurons than before, to avoid overfitting:

```
model = keras.models.Sequential([
    keras.layers.Dense(30, activation="relu", input_shape=X_train.shape[1:]),
    keras.layers.Dense(1)
])
model.compile(loss="mean_squared_error", optimizer="sgd")
history = model.fit(X_train, y_train, epochs=20,
                    validation_data=(X_valid, y_valid))
mse_test = model.evaluate(X_test, y_test)
X_new = X_test[:3] # pretend these are new instances
y_pred = model.predict(X_new)
```

As you can see, the Sequential API is quite easy to use. However, although Sequen tial models are extremely common, it is sometimes useful to build neural networks with more complex topologies, or with multiple inputs or outputs. For this purpose, Keras offers the Functional API.

Building Complex Models Using the Functional API

One example of a nonsequential neural network is a *Wide & Deep* neural network. This neural network architecture was introduced in a 2016 paper (*https://homl.info/widedeep*) by Heng-Tze Cheng et al.[16] It connects all or part of the inputs directly to the output layer, as shown in Figure 10-14. This architecture makes it possible for the neural network to learn both deep patterns (using the deep path) and simple rules (through the short path).[17] In contrast, a regular MLP forces all the data to flow

---

16 Heng-Tze Cheng et al., "Wide & Deep Learning for Recommender Systems," *Proceedings of the First Workshop on Deep Learning for Recommender Systems* (2016): 7–10.

17 The short path can also be used to provide manually engineered features to the neural network.

through the full stack of layers; thus, simple patterns in the data may end up being distorted by this sequence of transformations.

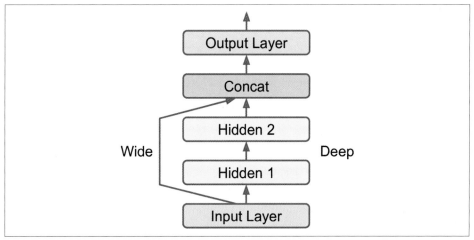

Figure 10-14. Wide & Deep neural network

Let's build such a neural network to tackle the California housing problem:

```python
input_ = keras.layers.Input(shape=X_train.shape[1:])
hidden1 = keras.layers.Dense(30, activation="relu")(input_)
hidden2 = keras.layers.Dense(30, activation="relu")(hidden1)
concat = keras.layers.Concatenate()([input_, hidden2])
output = keras.layers.Dense(1)(concat)
model = keras.Model(inputs=[input_], outputs=[output])
```

Let's go through each line of this code:

- First, we need to create an Input object.[18] This is a specification of the kind of input the model will get, including its shape and dtype. A model may actually have multiple inputs, as we will see shortly.

- Next, we create a Dense layer with 30 neurons, using the ReLU activation function. As soon as it is created, notice that we call it like a function, passing it the input. This is why this is called the Functional API. Note that we are just telling Keras how it should connect the layers together; no actual data is being processed yet.

- We then create a second hidden layer, and again we use it as a function. Note that we pass it the output of the first hidden layer.

---

18 The name input_ is used to avoid overshadowing Python's built-in input() function.

- Next, we create a `Concatenate` layer, and once again we immediately use it like a function, to concatenate the input and the output of the second hidden layer. You may prefer the `keras.layers.concatenate()` function, which creates a `Concatenate` layer and immediately calls it with the given inputs.

- Then we create the output layer, with a single neuron and no activation function, and we call it like a function, passing it the result of the concatenation.

- Lastly, we create a Keras `Model`, specifying which inputs and outputs to use.

Once you have built the Keras model, everything is exactly like earlier, so there's no need to repeat it here: you must compile the model, train it, evaluate it, and use it to make predictions.

But what if you want to send a subset of the features through the wide path and a different subset (possibly overlapping) through the deep path (see Figure 10-15)? In this case, one solution is to use multiple inputs. For example, suppose we want to send five features through the wide path (features 0 to 4), and six features through the deep path (features 2 to 7):

```
input_A = keras.layers.Input(shape=[5], name="wide_input")
input_B = keras.layers.Input(shape=[6], name="deep_input")
hidden1 = keras.layers.Dense(30, activation="relu")(input_B)
hidden2 = keras.layers.Dense(30, activation="relu")(hidden1)
concat = keras.layers.concatenate([input_A, hidden2])
output = keras.layers.Dense(1, name="output")(concat)
model = keras.Model(inputs=[input_A, input_B], outputs=[output])
```

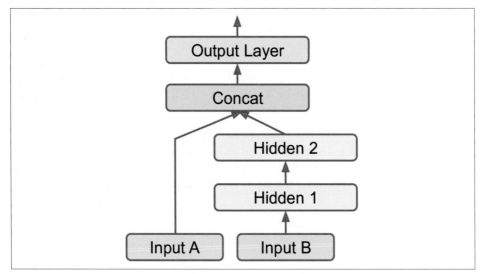

Figure 10-15. Handling multiple inputs

The code is self-explanatory. You should name at least the most important layers, especially when the model gets a bit complex like this. Note that we specified `inputs=[input_A, input_B]` when creating the model. Now we can compile the model as usual, but when we call the `fit()` method, instead of passing a single input matrix `X_train`, we must pass a pair of matrices (`X_train_A`, `X_train_B`): one per input.[19] The same is true for `X_valid`, and also for `X_test` and `X_new` when you call `evaluate()` or `predict()`:

```
model.compile(loss="mse", optimizer=keras.optimizers.SGD(lr=1e-3))

X_train_A, X_train_B = X_train[:, :5], X_train[:, 2:]
X_valid_A, X_valid_B = X_valid[:, :5], X_valid[:, 2:]
X_test_A, X_test_B = X_test[:, :5], X_test[:, 2:]
X_new_A, X_new_B = X_test_A[:3], X_test_B[:3]

history = model.fit((X_train_A, X_train_B), y_train, epochs=20,
                    validation_data=((X_valid_A, X_valid_B), y_valid))
mse_test = model.evaluate((X_test_A, X_test_B), y_test)
y_pred = model.predict((X_new_A, X_new_B))
```

There are many use cases in which you may want to have multiple outputs:

- The task may demand it. For instance, you may want to locate and classify the main object in a picture. This is both a regression task (finding the coordinates of the object's center, as well as its width and height) and a classification task.

- Similarly, you may have multiple independent tasks based on the same data. Sure, you could train one neural network per task, but in many cases you will get better results on all tasks by training a single neural network with one output per task. This is because the neural network can learn features in the data that are useful across tasks. For example, you could perform *multitask classification* on pictures of faces, using one output to classify the person's facial expression (smiling, surprised, etc.) and another output to identify whether they are wearing glasses or not.

- Another use case is as a regularization technique (i.e., a training constraint whose objective is to reduce overfitting and thus improve the model's ability to generalize). For example, you may want to add some auxiliary outputs in a neural network architecture (see Figure 10-16) to ensure that the underlying part of the network learns something useful on its own, without relying on the rest of the network.

---

19 Alternatively, you can pass a dictionary mapping the input names to the input values, like {"wide_input": X_train_A, "deep_input": X_train_B}. This is especially useful when there are many inputs, to avoid getting the order wrong.

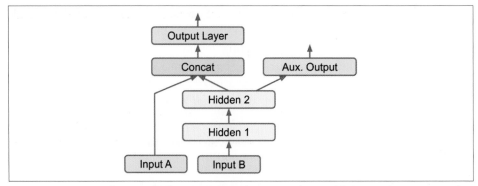

Figure 10-16. Handling multiple outputs, in this example to add an auxiliary output for regularization

Adding extra outputs is quite easy: just connect them to the appropriate layers and add them to your model's list of outputs. For example, the following code builds the network represented in Figure 10-16:

```
[...] # Same as above, up to the main output layer
output = keras.layers.Dense(1, name="main_output")(concat)
aux_output = keras.layers.Dense(1, name="aux_output")(hidden2)
model = keras.Model(inputs=[input_A, input_B], outputs=[output, aux_output])
```

Each output will need its own loss function. Therefore, when we compile the model, we should pass a list of losses[20] (if we pass a single loss, Keras will assume that the same loss must be used for all outputs). By default, Keras will compute all these losses and simply add them up to get the final loss used for training. We care much more about the main output than about the auxiliary output (as it is just used for regularization), so we want to give the main output's loss a much greater weight. Fortunately, it is possible to set all the loss weights when compiling the model:

```
model.compile(loss=["mse", "mse"], loss_weights=[0.9, 0.1], optimizer="sgd")
```

Now when we train the model, we need to provide labels for each output. In this example, the main output and the auxiliary output should try to predict the same thing, so they should use the same labels. So instead of passing y_train, we need to pass (y_train, y_train) (and the same goes for y_valid and y_test):

```
history = model.fit(
    [X_train_A, X_train_B], [y_train, y_train], epochs=20,
    validation_data=([X_valid_A, X_valid_B], [y_valid, y_valid]))
```

---

20 Alternatively, you can pass a dictionary that maps each output name to the corresponding loss. Just like for the inputs, this is useful when there are multiple outputs, to avoid getting the order wrong. The loss weights and metrics (discussed shortly) can also be set using dictionaries.

When we evaluate the model, Keras will return the total loss, as well as all the individual losses:

```
total_loss, main_loss, aux_loss = model.evaluate(
    [X_test_A, X_test_B], [y_test, y_test])
```

Similarly, the predict() method will return predictions for each output:

```
y_pred_main, y_pred_aux = model.predict([X_new_A, X_new_B])
```

As you can see, you can build any sort of architecture you want quite easily with the Functional API. Let's look at one last way you can build Keras models.

Using the Subclassing API to Build Dynamic Models

Both the Sequential API and the Functional API are declarative: you start by declaring which layers you want to use and how they should be connected, and only then can you start feeding the model some data for training or inference. This has many advantages: the model can easily be saved, cloned, and shared; its structure can be displayed and analyzed; the framework can infer shapes and check types, so errors can be caught early (i.e., before any data ever goes through the model). It's also fairly easy to debug, since the whole model is a static graph of layers. But the flip side is just that: it's static. Some models involve loops, varying shapes, conditional branching, and other dynamic behaviors. For such cases, or simply if you prefer a more imperative programming style, the Subclassing API is for you.

Simply subclass the Model class, create the layers you need in the constructor, and use them to perform the computations you want in the call() method. For example, creating an instance of the following WideAndDeepModel class gives us an equivalent model to the one we just built with the Functional API. You can then compile it, evaluate it, and use it to make predictions, exactly like we just did:

```
class WideAndDeepModel(keras.Model):
    def __init__(self, units=30, activation="relu", **kwargs):
        super().__init__(**kwargs) # handles standard args (e.g., name)
        self.hidden1 = keras.layers.Dense(units, activation=activation)
        self.hidden2 = keras.layers.Dense(units, activation=activation)
        self.main_output = keras.layers.Dense(1)
        self.aux_output = keras.layers.Dense(1)

    def call(self, inputs):
        input_A, input_B = inputs
        hidden1 = self.hidden1(input_B)
        hidden2 = self.hidden2(hidden1)
        concat = keras.layers.concatenate([input_A, hidden2])
        main_output = self.main_output(concat)
        aux_output = self.aux_output(hidden2)
        return main_output, aux_output

model = WideAndDeepModel()
```

This example looks very much like the Functional API, except we do not need to create the inputs; we just use the `input` argument to the `call()` method, and we separate the creation of the layers[21] in the constructor from their usage in the `call()` method. The big difference is that you can do pretty much anything you want in the `call()` method: `for` loops, `if` statements, low-level TensorFlow operations—your imagination is the limit (see Chapter 12)! This makes it a great API for researchers experimenting with new ideas.

This extra flexibility does come at a cost: your model's architecture is hidden within the `call()` method, so Keras cannot easily inspect it; it cannot save or clone it; and when you call the `summary()` method, you only get a list of layers, without any information on how they are connected to each other. Moreover, Keras cannot check types and shapes ahead of time, and it is easier to make mistakes. So unless you really need that extra flexibility, you should probably stick to the Sequential API or the Functional API.

Keras models can be used just like regular layers, so you can easily combine them to build complex architectures.

Now that you know how to build and train neural nets using Keras, you will want to save them!

Saving and Restoring a Model

When using the Sequential API or the Functional API, saving a trained Keras model is as simple as it gets:

```
model = keras.layers.Sequential([...]) # or keras.Model([...])
model.compile([...])
model.fit([...])
model.save("my_keras_model.h5")
```

Keras will use the HDF5 format to save both the model's architecture (including every layer's hyperparameters) and the values of all the model parameters for every layer (e.g., connection weights and biases). It also saves the optimizer (including its hyperparameters and any state it may have).

---

21 Keras models have an `output` attribute, so we cannot use that name for the main output layer, which is why we renamed it to `main_output`.

You will typically have a script that trains a model and saves it, and one or more scripts (or web services) that load the model and use it to make predictions. Loading the model is just as easy:

```
model = keras.models.load_model("my_keras_model.h5")
```

 This will work when using the Sequential API or the Functional API, but unfortunately not when using model subclassing. You can use `save_weights()` and `load_weights()` to at least save and restore the model parameters, but you will need to save and restore everything else yourself.

But what if training lasts several hours? This is quite common, especially when training on large datasets. In this case, you should not only save your model at the end of training, but also save checkpoints at regular intervals during training, to avoid losing everything if your computer crashes. But how can you tell the `fit()` method to save checkpoints? Use callbacks.

Using Callbacks

The `fit()` method accepts a `callbacks` argument that lets you specify a list of objects that Keras will call at the start and end of training, at the start and end of each epoch, and even before and after processing each batch. For example, the `ModelCheckpoint` callback saves checkpoints of your model at regular intervals during training, by default at the end of each epoch:

```
[...] # build and compile the model
checkpoint_cb = keras.callbacks.ModelCheckpoint("my_keras_model.h5")
history = model.fit(X_train, y_train, epochs=10, callbacks=[checkpoint_cb])
```

Moreover, if you use a validation set during training, you can set `save_best_only=True` when creating the `ModelCheckpoint`. In this case, it will only save your model when its performance on the validation set is the best so far. This way, you do not need to worry about training for too long and overfitting the training set: simply restore the last model saved after training, and this will be the best model on the validation set. The following code is a simple way to implement early stopping (introduced in Chapter 4):

```
checkpoint_cb = keras.callbacks.ModelCheckpoint("my_keras_model.h5",
                                                save_best_only=True)
history = model.fit(X_train, y_train, epochs=10,
                    validation_data=(X_valid, y_valid),
                    callbacks=[checkpoint_cb])
model = keras.models.load_model("my_keras_model.h5") # roll back to best model
```

Another way to implement early stopping is to simply use the `EarlyStopping` callback. It will interrupt training when it measures no progress on the validation set for a number of epochs (defined by the `patience` argument), and it will optionally roll back to the best model. You can combine both callbacks to save checkpoints of your model (in case your computer crashes) and interrupt training early when there is no more progress (to avoid wasting time and resources):

```
early_stopping_cb = keras.callbacks.EarlyStopping(patience=10,
                                                  restore_best_weights=True)
history = model.fit(X_train, y_train, epochs=100,
                    validation_data=(X_valid, y_valid),
                    callbacks=[checkpoint_cb, early_stopping_cb])
```

The number of epochs can be set to a large value since training will stop automatically when there is no more progress. In this case, there is no need to restore the best model saved because the `EarlyStopping` callback will keep track of the best weights and restore them for you at the end of training.

 There are many other callbacks available in the `keras.callbacks` package (*https://keras.io/callbacks/*).

If you need extra control, you can easily write your own custom callbacks. As an example of how to do that, the following custom callback will display the ratio between the validation loss and the training loss during training (e.g., to detect over-fitting):

```
class PrintValTrainRatioCallback(keras.callbacks.Callback):
    def on_epoch_end(self, epoch, logs):
        print("\nval/train: {:.2f}".format(logs["val_loss"] / logs["loss"]))
```

As you might expect, you can implement `on_train_begin()`, `on_train_end()`, `on_epoch_begin()`, `on_epoch_end()`, `on_batch_begin()`, and `on_batch_end()`. Callbacks can also be used during evaluation and predictions, should you ever need them (e.g., for debugging). For evaluation, you should implement `on_test_begin()`, `on_test_end()`, `on_test_batch_begin()`, or `on_test_batch_end()` (called by `evaluate()`), and for prediction you should implement `on_predict_begin()`, `on_predict_end()`, `on_predict_batch_begin()`, or `on_predict_batch_end()` (called by `predict()`).

Now let's take a look at one more tool you should definitely have in your toolbox when using tf.keras: TensorBoard.

Using TensorBoard for Visualization

TensorBoard is a great interactive visualization tool that you can use to view the learning curves during training, compare learning curves between multiple runs, visualize the computation graph, analyze training statistics, view images generated by your model, visualize complex multidimensional data projected down to 3D and automatically clustered for you, and more! This tool is installed automatically when you install TensorFlow, so you already have it.

To use it, you must modify your program so that it outputs the data you want to visualize to special binary log files called *event files*. Each binary data record is called a *summary*. The TensorBoard server will monitor the log directory, and it will automatically pick up the changes and update the visualizations: this allows you to visualize live data (with a short delay), such as the learning curves during training. In general, you want to point the TensorBoard server to a root log directory and configure your program so that it writes to a different subdirectory every time it runs. This way, the same TensorBoard server instance will allow you to visualize and compare data from multiple runs of your program, without getting everything mixed up.

Let's start by defining the root log directory we will use for our TensorBoard logs, plus a small function that will generate a subdirectory path based on the current date and time so that it's different at every run. You may want to include extra information in the log directory name, such as hyperparameter values that you are testing, to make it easier to know what you are looking at in TensorBoard:

```
import os
root_logdir = os.path.join(os.curdir, "my_logs")

def get_run_logdir():
    import time
    run_id = time.strftime("run_%Y_%m_%d-%H_%M_%S")
    return os.path.join(root_logdir, run_id)

run_logdir = get_run_logdir() # e.g., './my_logs/run_2019_06_07-15_15_22'
```

The good news is that Keras provides a nice `TensorBoard()` callback:

```
[...] # Build and compile your model
tensorboard_cb = keras.callbacks.TensorBoard(run_logdir)
history = model.fit(X_train, y_train, epochs=30,
                    validation_data=(X_valid, y_valid),
                    callbacks=[tensorboard_cb])
```

And that's all there is to it! It could hardly be easier to use. If you run this code, the `TensorBoard()` callback will take care of creating the log directory for you (along with its parent directories if needed), and during training it will create event files and write summaries to them. After running the program a second time (perhaps changing some hyperparameter value), you will end up with a directory structure similar to this one:

```
my_logs/
├── run_2019_06_07-15_15_22
│   ├── train
│   │   ├── events.out.tfevents.1559891732.mycomputer.local.38511.694049.v2
│   │   ├── events.out.tfevents.1559891732.mycomputer.local.profile-empty
│   │   └── plugins/profile/2019-06-07_15-15-32
│   │       └── local.trace
│   └── validation
│       └── events.out.tfevents.1559891733.mycomputer.local.38511.696430.v2
└── run_2019_06_07-15_15_49
    └── [...]
```

There's one directory per run, each containing one subdirectory for training logs and one for validation logs. Both contain event files, but the training logs also include profiling traces: this allows TensorBoard to show you exactly how much time the model spent on each part of your model, across all your devices, which is great for locating performance bottlenecks.

Next you need to start the TensorBoard server. One way to do this is by running a command in a terminal. If you installed TensorFlow within a virtualenv, you should activate it. Next, run the following command at the root of the project (or from any-where else, as long as you point to the appropriate log directory):

```
$ tensorboard --logdir=./my_logs --port=6006
TensorBoard 2.0.0 at http://mycomputer.local:6006/ (Press CTRL+C to quit)
```

If your shell cannot find the *tensorboard* script, then you must update your PATH environment variable so that it contains the directory in which the script was installed (alternatively, you can just replace `tensorboard` in the command line with `python3 -m tensorboard.main`). Once the server is up, you can open a web browser and go to *http://localhost:6006*.

Alternatively, you can use TensorBoard directly within Jupyter, by running the following commands. The first line loads the TensorBoard extension, and the second line starts a TensorBoard server on port 6006 (unless it is already started) and connects to it:

```
%load_ext tensorboard
%tensorboard --logdir=./my_logs --port=6006
```

Either way, you should see TensorBoard's web interface. Click the SCALARS tab to view the learning curves (see Figure 10-17). At the bottom left, select the logs you want to visualize (e.g., the training logs from the first and second run), and click the `epoch_loss` scalar. Notice that the training loss went down nicely during both runs, but the second run went down much faster. Indeed, we used a learning rate of 0.05 (`optimizer=keras.optimizers.SGD(lr=0.05)`) instead of 0.001.

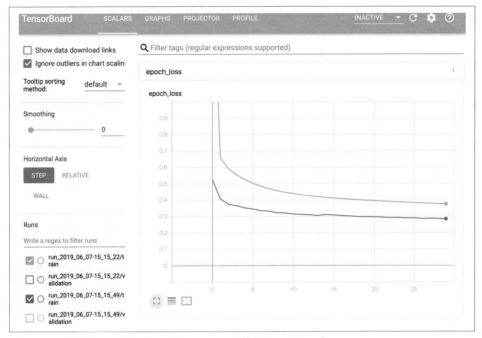

Figure 10-17. Visualizing learning curves with TensorBoard

You can also visualize the whole graph, the learned weights (projected to 3D), or the profiling traces. The TensorBoard() callback has options to log extra data too, such as embeddings (see Chapter 13).

Additionally, TensorFlow offers a lower-level API in the tf.summary package. The following code creates a SummaryWriter using the create_file_writer() function, and it uses this writer as a context to log scalars, histograms, images, audio, and text, all of which can then be visualized using TensorBoard (give it a try!):

```
test_logdir = get_run_logdir()
writer = tf.summary.create_file_writer(test_logdir)
with writer.as_default():
    for step in range(1, 1000 + 1):
        tf.summary.scalar("my_scalar", np.sin(step / 10), step=step)
        data = (np.random.randn(100) + 2) * step / 100 # some random data
        tf.summary.histogram("my_hist", data, buckets=50, step=step)
        images = np.random.rand(2, 32, 32, 3) # random 32×32 RGB images
        tf.summary.image("my_images", images * step / 1000, step=step)
        texts = ["The step is " + str(step), "Its square is " + str(step**2)]
        tf.summary.text("my_text", texts, step=step)
        sine_wave = tf.math.sin(tf.range(12000) / 48000 * 2 * np.pi * step)
        audio = tf.reshape(tf.cast(sine_wave, tf.float32), [1, -1, 1])
        tf.summary.audio("my_audio", audio, sample_rate=48000, step=step)
```

This is actually a useful visualization tool to have, even beyond TensorFlow or Deep Learning.

Let's summarize what you've learned so far in this chapter: we saw where neural nets came from, what an MLP is and how you can use it for classification and regression, how to use tf.keras's Sequential API to build MLPs, and how to use the Functional API or the Subclassing API to build more complex model architectures. You learned how to save and restore a model and how to use callbacks for checkpointing, early stopping, and more. Finally, you learned how to use TensorBoard for visualization. You can already go ahead and use neural networks to tackle many problems! However, you may wonder how to choose the number of hidden layers, the number of neurons in the network, and all the other hyperparameters. Let's look at this now.

Fine-Tuning Neural Network Hyperparameters

The flexibility of neural networks is also one of their main drawbacks: there are many hyperparameters to tweak. Not only can you use any imaginable network architecture, but even in a simple MLP you can change the number of layers, the number of neurons per layer, the type of activation function to use in each layer, the weight initialization logic, and much more. How do you know what combination of hyperparameters is the best for your task?

One option is to simply try many combinations of hyperparameters and see which one works best on the validation set (or use K-fold cross-validation). For example, we can use GridSearchCV or RandomizedSearchCV to explore the hyperparameter space, as we did in Chapter 2. To do this, we need to wrap our Keras models in objects that mimic regular Scikit-Learn regressors. The first step is to create a function that will build and compile a Keras model, given a set of hyperparameters:

```
def build_model(n_hidden=1, n_neurons=30, learning_rate=3e-3, input_shape=[8]):
    model = keras.models.Sequential()
    model.add(keras.layers.InputLayer(input_shape=input_shape))
    for layer in range(n_hidden):
        model.add(keras.layers.Dense(n_neurons, activation="relu"))
    model.add(keras.layers.Dense(1))
    optimizer = keras.optimizers.SGD(lr=learning_rate)
    model.compile(loss="mse", optimizer=optimizer)
    return model
```

This function creates a simple Sequential model for univariate regression (only one output neuron), with the given input shape and the given number of hidden layers and neurons, and it compiles it using an SGD optimizer configured with the specified learning rate. It is good practice to provide reasonable defaults to as many hyperparameters as you can, as Scikit-Learn does.

Next, let's create a KerasRegressor based on this build_model() function:

```
keras_reg = keras.wrappers.scikit_learn.KerasRegressor(build_model)
```

The KerasRegressor object is a thin wrapper around the Keras model built using build_model(). Since we did not specify any hyperparameters when creating it, it will use the default hyperparameters we defined in build_model(). Now we can use this object like a regular Scikit-Learn regressor: we can train it using its fit() method, then evaluate it using its score() method, and use it to make predictions using its predict() method, as you can see in the following code:

```
keras_reg.fit(X_train, y_train, epochs=100,
              validation_data=(X_valid, y_valid),
              callbacks=[keras.callbacks.EarlyStopping(patience=10)])
mse_test = keras_reg.score(X_test, y_test)
y_pred = keras_reg.predict(X_new)
```

Note that any extra parameter you pass to the fit() method will get passed to the underlying Keras model. Also note that the score will be the opposite of the MSE because Scikit-Learn wants scores, not losses (i.e., higher should be better).

We don't want to train and evaluate a single model like this, though we want to train hundreds of variants and see which one performs best on the validation set. Since there are many hyperparameters, it is preferable to use a randomized search rather than grid search (as we discussed in Chapter 2). Let's try to explore the number of hidden layers, the number of neurons, and the learning rate:

```
from scipy.stats import reciprocal
from sklearn.model_selection import RandomizedSearchCV

param_distribs = {
    "n_hidden": [0, 1, 2, 3],
    "n_neurons": np.arange(1, 100),
    "learning_rate": reciprocal(3e-4, 3e-2),
}

rnd_search_cv = RandomizedSearchCV(keras_reg, param_distribs, n_iter=10, cv=3)
rnd_search_cv.fit(X_train, y_train, epochs=100,
                  validation_data=(X_valid, y_valid),
                  callbacks=[keras.callbacks.EarlyStopping(patience=10)])
```

This is identical to what we did in Chapter 2, except here we pass extra parameters to the fit() method, and they get relayed to the underlying Keras models. Note that RandomizedSearchCV uses K-fold cross-validation, so it does not use X_valid and y_valid, which are only used for early stopping.

The exploration may last many hours, depending on the hardware, the size of the dataset, the complexity of the model, and the values of n_iter and cv. When it's over, you can access the best parameters found, the best score, and the trained Keras model like this:

```
>>> rnd_search_cv.best_params_
{'learning_rate': 0.0033625641252688094, 'n_hidden': 2, 'n_neurons': 42}
>>> rnd_search_cv.best_score_
-0.3189529188278931
>>> model = rnd_search_cv.best_estimator_.model
```

You can now save this model, evaluate it on the test set, and, if you are satisfied with its performance, deploy it to production. Using randomized search is not too hard, and it works well for many fairly simple problems. When training is slow, however (e.g., for more complex problems with larger datasets), this approach will only explore a tiny portion of the hyperparameter space. You can partially alleviate this problem by assisting the search process manually: first run a quick random search using wide ranges of hyperparameter values, then run another search using smaller ranges of values centered on the best ones found during the first run, and so on. This approach will hopefully zoom in on a good set of hyperparameters. However, it's very time consuming, and probably not the best use of your time.

Fortunately, there are many techniques to explore a search space much more efficiently than randomly. Their core idea is simple: when a region of the space turns out to be good, it should be explored more. Such techniques take care of the "zooming" process for you and lead to much better solutions in much less time. Here are some Python libraries you can use to optimize hyperparameters:

Hyperopt (https://github.com/hyperopt/hyperopt)
    A popular library for optimizing over all sorts of complex search spaces (including real values, such as the learning rate, and discrete values, such as the number of layers).

Hyperas (https://github.com/maxpumperla/hyperas), kopt (https://github.com/Avsecz/kopt), or Talos (https://github.com/autonomio/talos)
    Useful libraries for optimizing hyperparameters for Keras models (the first two are based on Hyperopt).

Keras Tuner (https://homl.info/kerastuner)
    An easy-to-use hyperparameter optimization library by Google for Keras models, with a hosted service for visualization and analysis.

Scikit-Optimize (skopt) (https://scikit-optimize.github.io/)
    A general-purpose optimization library. The BayesSearchCV class performs Bayesian optimization using an interface similar to GridSearchCV.

Spearmint (https://github.com/JasperSnoek/spearmint)
    A Bayesian optimization library.

Hyperband (https://github.com/zygmuntz/hyperband)
> A fast hyperparameter tuning library based on the recent Hyperband paper (*https://homl.info/hyperband*)[22] by Lisha Li et al.

Sklearn-Deap (https://github.com/rsteca/sklearn-deap)
> A hyperparameter optimization library based on evolutionary algorithms, with a `GridSearchCV`-like interface.

Moreover, many companies offer services for hyperparameter optimization. We'll discuss Google Cloud AI Platform's hyperparameter tuning service (*https://homl.info/googletuning*) in Chapter 19. Other options include services by Arimo (*https://arimo.com/*) and SigOpt (*https://sigopt.com/*), and CallDesk's Oscar (*http://oscar.call desk.ai/*).

Hyperparameter tuning is still an active area of research, and evolutionary algorithms are making a comeback. For example, check out DeepMind's excellent 2017 paper (*https://homl.info/pbt*),[23] where the authors jointly optimize a population of models and their hyperparameters. Google has also used an evolutionary approach, not just to search for hyperparameters but also to look for the best neural network architecture for the problem; their AutoML suite is already available as a cloud service (*https://cloud.google.com/automl/*). Perhaps the days of building neural networks manually will soon be over? Check out Google's post (*https://homl.info/automlpost*) on this topic. In fact, evolutionary algorithms have been used successfully to train individual neural networks, replacing the ubiquitous Gradient Descent! For an example, see the 2017 post (*https://homl.info/neuroevol*) by Uber where the authors introduce their *Deep Neuroevolution* technique.

But despite all this exciting progress and all these tools and services, it still helps to have an idea of what values are reasonable for each hyperparameter so that you can build a quick prototype and restrict the search space. The following sections provide guidelines for choosing the number of hidden layers and neurons in an MLP and for selecting good values for some of the main hyperparameters.

Number of Hidden Layers

For many problems, you can begin with a single hidden layer and get reasonable results. An MLP with just one hidden layer can theoretically model even the most complex functions, provided it has enough neurons. But for complex problems, deep networks have a much higher *parameter efficiency* than shallow ones: they can model

---

22 Lisha Li et al., "Hyperband: A Novel Bandit-Based Approach to Hyperparameter Optimization," *Journal of Machine Learning Research* 18 (April 2018): 1–52.

23 Max Jaderberg et al., "Population Based Training of Neural Networks," arXiv preprint arXiv:1711.09846 (2017).

complex functions using exponentially fewer neurons than shallow nets, allowing them to reach much better performance with the same amount of training data.

To understand why, suppose you are asked to draw a forest using some drawing software, but you are forbidden to copy and paste anything. It would take an enormous amount of time: you would have to draw each tree individually, branch by branch, leaf by leaf. If you could instead draw one leaf, copy and paste it to draw a branch, then copy and paste that branch to create a tree, and finally copy and paste this tree to make a forest, you would be finished in no time. Real-world data is often structured in such a hierarchical way, and deep neural networks automatically take advantage of this fact: lower hidden layers model low-level structures (e.g., line segments of various shapes and orientations), intermediate hidden layers combine these low-level structures to model intermediate-level structures (e.g., squares, circles), and the highest hidden layers and the output layer combine these intermediate structures to model high-level structures (e.g., faces).

Not only does this hierarchical architecture help DNNs converge faster to a good solution, but it also improves their ability to generalize to new datasets. For example, if you have already trained a model to recognize faces in pictures and you now want to train a new neural network to recognize hairstyles, you can kickstart the training by reusing the lower layers of the first network. Instead of randomly initializing the weights and biases of the first few layers of the new neural network, you can initialize them to the values of the weights and biases of the lower layers of the first network. This way the network will not have to learn from scratch all the low-level structures that occur in most pictures; it will only have to learn the higher-level structures (e.g., hairstyles). This is called *transfer learning*.

In summary, for many problems you can start with just one or two hidden layers and the neural network will work just fine. For instance, you can easily reach above 97% accuracy on the MNIST dataset using just one hidden layer with a few hundred neurons, and above 98% accuracy using two hidden layers with the same total number of neurons, in roughly the same amount of training time. For more complex problems, you can ramp up the number of hidden layers until you start overfitting the training set. Very complex tasks, such as large image classification or speech recognition, typically require networks with dozens of layers (or even hundreds, but not fully connected ones, as we will see in Chapter 14), and they need a huge amount of training data. You will rarely have to train such networks from scratch: it is much more common to reuse parts of a pretrained state-of-the-art network that performs a similar task. Training will then be a lot faster and require much less data (we will discuss this in Chapter 11).

Number of Neurons per Hidden Layer

The number of neurons in the input and output layers is determined by the type of input and output your task requires. For example, the MNIST task requires $28 \times 28 = 784$ input neurons and 10 output neurons.

As for the hidden layers, it used to be common to size them to form a pyramid, with fewer and fewer neurons at each layer—the rationale being that many low-level features can coalesce into far fewer high-level features. A typical neural network for MNIST might have 3 hidden layers, the first with 300 neurons, the second with 200, and the third with 100. However, this practice has been largely abandoned because it seems that using the same number of neurons in all hidden layers performs just as well in most cases, or even better; plus, there is only one hyperparameter to tune, instead of one per layer. That said, depending on the dataset, it can sometimes help to make the first hidden layer bigger than the others.

Just like the number of layers, you can try increasing the number of neurons gradually until the network starts overfitting. But in practice, it's often simpler and more efficient to pick a model with more layers and neurons than you actually need, then use early stopping and other regularization techniques to prevent it from overfitting. Vincent Vanhoucke, a scientist at Google, has dubbed this the "stretch pants" approach: instead of wasting time looking for pants that perfectly match your size, just use large stretch pants that will shrink down to the right size. With this approach, you avoid bottleneck layers that could ruin your model. On the flip side, if a layer has too few neurons, it will not have enough representational power to preserve all the useful information from the inputs (e.g., a layer with two neurons can only output 2D data, so if it processes 3D data, some information will be lost). No matter how big and powerful the rest of the network is, that information will never be recovered.

> In general you will get more bang for your buck by increasing the number of layers instead of the number of neurons per layer.

Learning Rate, Batch Size, and Other Hyperparameters

The numbers of hidden layers and neurons are not the only hyperparameters you can tweak in an MLP. Here are some of the most important ones, as well as tips on how to set them:

Learning rate
> The learning rate is arguably the most important hyperparameter. In general, the optimal learning rate is about half of the maximum learning rate (i.e., the learning rate above which the training algorithm diverges, as we saw in Chapter 4).

One way to find a good learning rate is to train the model for a few hundred iterations, starting with a very low learning rate (e.g., 10^{-5}) and gradually increasing it up to a very large value (e.g., 10). This is done by multiplying the learning rate by a constant factor at each iteration (e.g., by $\exp(\log(10^6)/500)$ to go from 10^{-5} to 10 in 500 iterations). If you plot the loss as a function of the learning rate (using a log scale for the learning rate), you should see it dropping at first. But after a while, the learning rate will be too large, so the loss will shoot back up: the optimal learning rate will be a bit lower than the point at which the loss starts to climb (typically about 10 times lower than the turning point). You can then reinitialize your model and train it normally using this good learning rate. We will look at more learning rate techniques in Chapter 11.

Optimizer

Choosing a better optimizer than plain old Mini-batch Gradient Descent (and tuning its hyperparameters) is also quite important. We will see several advanced optimizers in Chapter 11.

Batch size

The batch size can have a significant impact on your model's performance and training time. The main benefit of using large batch sizes is that hardware accelerators like GPUs can process them efficiently (see Chapter 19), so the training algorithm will see more instances per second. Therefore, many researchers and practitioners recommend using the largest batch size that can fit in GPU RAM. There's a catch, though: in practice, large batch sizes often lead to training instabilities, especially at the beginning of training, and the resulting model may not generalize as well as a model trained with a small batch size. In April 2018, Yann LeCun even tweeted "Friends don't let friends use mini-batches larger than 32," citing a 2018 paper (*https://homl.info/smallbatch*)[24] by Dominic Masters and Carlo Luschi which concluded that using small batches (from 2 to 32) was preferable because small batches led to better models in less training time. Other papers point in the opposite direction, however; in 2017, papers by Elad Hoffer et al. (*https://homl.info/largebatch*)[25] and Priya Goyal et al. (*https://homl.info/largebatch2*)[26] showed that it was possible to use very large batch sizes (up to 8,192) using various techniques such as warming up the learning rate (i.e., starting training with a small learning rate, then ramping it up, as we will see in Chap-

---

24 Dominic Masters and Carlo Luschi, "Revisiting Small Batch Training for Deep Neural Networks," arXiv preprint arXiv:1804.07612 (2018).

25 Elad Hoffer et al., "Train Longer, Generalize Better: Closing the Generalization Gap in Large Batch Training of Neural Networks," *Proceedings of the 31st International Conference on Neural Information Processing Systems* (2017): 1729–1739.

26 Priya Goyal et al., "Accurate, Large Minibatch SGD: Training ImageNet in 1 Hour," arXiv preprint arXiv:1706.02677 (2017).

ter 11). This led to a very short training time, without any generalization gap. So, one strategy is to try to use a large batch size, using learning rate warmup, and if training is unstable or the final performance is disappointing, then try using a small batch size instead.

Activation function

We discussed how to choose the activation function earlier in this chapter: in general, the ReLU activation function will be a good default for all hidden layers. For the output layer, it really depends on your task.

Number of iterations

In most cases, the number of training iterations does not actually need to be tweaked: just use early stopping instead.

> The optimal learning rate depends on the other hyperparameters— especially the batch size—so if you modify any hyperparameter, make sure to update the learning rate as well.

For more best practices regarding tuning neural network hyperparameters, check out the excellent 2018 paper (*https://homl.info/1cycle*)[27] by Leslie Smith.

This concludes our introduction to artificial neural networks and their implementation with Keras. In the next few chapters, we will discuss techniques to train very deep nets. We will also explore how to customize models using TensorFlow's lower-level API and how to load and preprocess data efficiently using the Data API. And we will dive into other popular neural network architectures: convolutional neural networks for image processing, recurrent neural networks for sequential data, autoencoders for representation learning, and generative adversarial networks to model and generate data.[28]

Exercises

1. The TensorFlow Playground (*https://playground.tensorflow.org/*) is a handy neural network simulator built by the TensorFlow team. In this exercise, you will train several binary classifiers in just a few clicks, and tweak the model's architecture and its hyperparameters to gain some intuition on how neural networks

---

27 Leslie N. Smith, "A Disciplined Approach to Neural Network Hyper-Parameters: Part 1—Learning Rate, Batch Size, Momentum, and Weight Decay," arXiv preprint arXiv:1803.09820 (2018).

28 A few extra ANN architectures are presented in Appendix E.

work and what their hyperparameters do. Take some time to explore the following:

a. The patterns learned by a neural net. Try training the default neural network by clicking the Run button (top left). Notice how it quickly finds a good solution for the classification task. The neurons in the first hidden layer have learned simple patterns, while the neurons in the second hidden layer have learned to combine the simple patterns of the first hidden layer into more complex patterns. In general, the more layers there are, the more complex the patterns can be.

b. Activation functions. Try replacing the tanh activation function with a ReLU activation function, and train the network again. Notice that it finds a solution even faster, but this time the boundaries are linear. This is due to the shape of the ReLU function.

c. The risk of local minima. Modify the network architecture to have just one hidden layer with three neurons. Train it multiple times (to reset the network weights, click the Reset button next to the Play button). Notice that the training time varies a lot, and sometimes it even gets stuck in a local minimum.

d. What happens when neural nets are too small. Remove one neuron to keep just two. Notice that the neural network is now incapable of finding a good solution, even if you try multiple times. The model has too few parameters and systematically underfits the training set.

e. What happens when neural nets are large enough. Set the number of neurons to eight, and train the network several times. Notice that it is now consistently fast and never gets stuck. This highlights an important finding in neural network theory: large neural networks almost never get stuck in local minima, and even when they do these local optima are almost as good as the global optimum. However, they can still get stuck on long plateaus for a long time.

f. The risk of vanishing gradients in deep networks. Select the spiral dataset (the bottom-right dataset under "DATA"), and change the network architecture to have four hidden layers with eight neurons each. Notice that training takes much longer and often gets stuck on plateaus for long periods of time. Also notice that the neurons in the highest layers (on the right) tend to evolve faster than the neurons in the lowest layers (on the left). This problem, called the "vanishing gradients" problem, can be alleviated with better weight initialization and other techniques, better optimizers (such as AdaGrad or Adam), or Batch Normalization (discussed in Chapter 11).

g. Go further. Take an hour or so to play around with other parameters and get a feel for what they do, to build an intuitive understanding about neural networks.

2. Draw an ANN using the original artificial neurons (like the ones in Figure 10-3) that computes $A \oplus B$ (where $\oplus$ represents the XOR operation). Hint: $A \oplus B = (A \wedge \neg B) \vee (\neg A \wedge B)$.

3. Why is it generally preferable to use a Logistic Regression classifier rather than a classical Perceptron (i.e., a single layer of threshold logic units trained using the Perceptron training algorithm)? How can you tweak a Perceptron to make it equivalent to a Logistic Regression classifier?

4. Why was the logistic activation function a key ingredient in training the first MLPs?

5. Name three popular activation functions. Can you draw them?

6. Suppose you have an MLP composed of one input layer with 10 passthrough neurons, followed by one hidden layer with 50 artificial neurons, and finally one output layer with 3 artificial neurons. All artificial neurons use the ReLU activation function.

   - What is the shape of the input matrix $\mathbf{X}$?
   - What are the shapes of the hidden layer's weight vector $\mathbf{W}_h$ and its bias vector $\mathbf{b}_h$?
   - What are the shapes of the output layer's weight vector $\mathbf{W}_o$ and its bias vector $\mathbf{b}_o$?
   - What is the shape of the network's output matrix $\mathbf{Y}$?
   - Write the equation that computes the network's output matrix $\mathbf{Y}$ as a function of $\mathbf{X}$, $\mathbf{W}_h$, $\mathbf{b}_h$, $\mathbf{W}_o$, and $\mathbf{b}_o$.

7. How many neurons do you need in the output layer if you want to classify email into spam or ham? What activation function should you use in the output layer? If instead you want to tackle MNIST, how many neurons do you need in the output layer, and which activation function should you use? What about for getting your network to predict housing prices, as in Chapter 2?

8. What is backpropagation and how does it work? What is the difference between backpropagation and reverse-mode autodiff?

9. Can you list all the hyperparameters you can tweak in a basic MLP? If the MLP overfits the training data, how could you tweak these hyperparameters to try to solve the problem?

10. Train a deep MLP on the MNIST dataset (you can load it using `keras.data sets.mnist.load_data()`. See if you can get over 98% precision. Try searching for the optimal learning rate by using the approach presented in this chapter (i.e., by growing the learning rate exponentially, plotting the error, and finding the

point where the error shoots up). Try adding all the bells and whistles—save checkpoints, use early stopping, and plot learning curves using TensorBoard.

Solutions to these exercises are available in Appendix A.

## CHAPTER 11

Training Deep Neural Networks

In Chapter 10 we introduced artificial neural networks and trained our first deep neural networks. But they were shallow nets, with just a few hidden layers. What if you need to tackle a complex problem, such as detecting hundreds of types of objects in high-resolution images? You may need to train a much deeper DNN, perhaps with 10 layers or many more, each containing hundreds of neurons, linked by hundreds of thousands of connections. Training a deep DNN isn't a walk in the park. Here are some of the problems you could run into:

- You may be faced with the tricky *vanishing gradients* problem or the related *exploding gradients* problem. This is when the gradients grow smaller and smaller, or larger and larger, when flowing backward through the DNN during training. Both of these problems make lower layers very hard to train.

- You might not have enough training data for such a large network, or it might be too costly to label.

- Training may be extremely slow.

- A model with millions of parameters would severely risk overfitting the training set, especially if there are not enough training instances or if they are too noisy.

In this chapter we will go through each of these problems and present techniques to solve them. We will start by exploring the vanishing and exploding gradients problems and some of their most popular solutions. Next, we will look at transfer learning and unsupervised pretraining, which can help you tackle complex tasks even when you have little labeled data. Then we will discuss various optimizers that can speed up training large models tremendously. Finally, we will go through a few popular regularization techniques for large neural networks.

With these tools, you will be able to train very deep nets. Welcome to Deep Learning!

The Vanishing/Exploding Gradients Problems

As we discussed in Chapter 10, the backpropagation algorithm works by going from the output layer to the input layer, propagating the error gradient along the way. Once the algorithm has computed the gradient of the cost function with regard to each parameter in the network, it uses these gradients to update each parameter with a Gradient Descent step.

Unfortunately, gradients often get smaller and smaller as the algorithm progresses down to the lower layers. As a result, the Gradient Descent update leaves the lower layers' connection weights virtually unchanged, and training never converges to a good solution. We call this the *vanishing gradients* problem. In some cases, the opposite can happen: the gradients can grow bigger and bigger until layers get insanely large weight updates and the algorithm diverges. This is the *exploding gradients* problem, which surfaces in recurrent neural networks (see Chapter 15). More generally, deep neural networks suffer from unstable gradients; different layers may learn at widely different speeds.

This unfortunate behavior was empirically observed long ago, and it was one of the reasons deep neural networks were mostly abandoned in the early 2000s. It wasn't clear what caused the gradients to be so unstable when training a DNN, but some light was shed in a 2010 paper (*https://homl.info/47*) by Xavier Glorot and Yoshua Bengio.[1] The authors found a few suspects, including the combination of the popular logistic sigmoid activation function and the weight initialization technique that was most popular at the time (i.e., a normal distribution with a mean of 0 and a standard deviation of 1). In short, they showed that with this activation function and this initialization scheme, the variance of the outputs of each layer is much greater than the variance of its inputs. Going forward in the network, the variance keeps increasing after each layer until the activation function saturates at the top layers. This saturation is actually made worse by the fact that the logistic function has a mean of 0.5, not 0 (the hyperbolic tangent function has a mean of 0 and behaves slightly better than the logistic function in deep networks).

Looking at the logistic activation function (see Figure 11-1), you can see that when inputs become large (negative or positive), the function saturates at 0 or 1, with a derivative extremely close to 0. Thus, when backpropagation kicks in it has virtually no gradient to propagate back through the network; and what little gradient exists keeps getting diluted as backpropagation progresses down through the top layers, so there is really nothing left for the lower layers.

---

1 Xavier Glorot and Yoshua Bengio, "Understanding the Difficulty of Training Deep Feedforward Neural Networks," *Proceedings of the 13th International Conference on Artificial Intelligence and Statistics* (2010): 249–256.

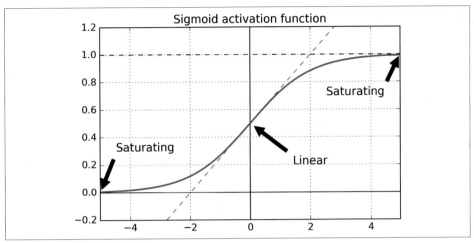

Figure 11-1. Logistic activation function saturation

Glorot and He Initialization

In their paper, Glorot and Bengio propose a way to significantly alleviate the unstable gradients problem. They point out that we need the signal to flow properly in both directions: in the forward direction when making predictions, and in the reverse direction when backpropagating gradients. We don't want the signal to die out, nor do we want it to explode and saturate. For the signal to flow properly, the authors argue that we need the variance of the outputs of each layer to be equal to the variance of its inputs,[2] and we need the gradients to have equal variance before and after flowing through a layer in the reverse direction (please check out the paper if you are interested in the mathematical details). It is actually not possible to guarantee both unless the layer has an equal number of inputs and neurons (these numbers are called the *fan-in* and *fan-out* of the layer), but Glorot and Bengio proposed a good compromise that has proven to work very well in practice: the connection weights of each layer must be initialized randomly as described in Equation 11-1, where $fan_{avg} = (fan_{in} + fan_{out})/2$. This initialization strategy is called *Xavier initialization* or *Glorot initialization*, after the paper's first author.

---

2 Here's an analogy: if you set a microphone amplifier's knob too close to zero, people won't hear your voice, but if you set it too close to the max, your voice will be saturated and people won't understand what you are saying. Now imagine a chain of such amplifiers: they all need to be set properly in order for your voice to come out loud and clear at the end of the chain. Your voice has to come out of each amplifier at the same amplitude as it came in.

Equation 11-1. Glorot initialization (when using the logistic activation function)

Normal distribution with mean 0 and variance $\sigma^2 = \dfrac{1}{fan_{avg}}$

Or a uniform distribution between $-r$ and $+r$, with $r = \sqrt{\dfrac{3}{fan_{avg}}}$

If you replace fan_{avg} with fan_{in} in Equation 11-1, you get an initialization strategy that Yann LeCun proposed in the 1990s. He called it *LeCun initialization*. Genevieve Orr and Klaus-Robert Müller even recommended it in their 1998 book *Neural Networks: Tricks of the Trade* (Springer). LeCun initialization is equivalent to Glorot initialization when $fan_{in} = fan_{out}$. It took over a decade for researchers to realize how important this trick is. Using Glorot initialization can speed up training considerably, and it is one of the tricks that led to the success of Deep Learning.

Some papers[3] have provided similar strategies for different activation functions. These strategies differ only by the scale of the variance and whether they use fan_{avg} or fan_{in}, as shown in Table 11-1 (for the uniform distribution, just compute $r = \sqrt{3\sigma^2}$). The initialization strategy (*https://homl.info/48*) for the ReLU activation function (and its variants, including the ELU activation described shortly) is sometimes called *He initialization*, after the paper's first author. The SELU activation function will be explained later in this chapter. It should be used with LeCun initialization (preferably with a normal distribution, as we will see).

Table 11-1. Initialization parameters for each type of activation function

Initialization	Activation functions	σ^2 (Normal)
Glorot	None, tanh, logistic, softmax	$1 / fan_{avg}$
He	ReLU and variants	$2 / fan_{in}$
LeCun	SELU	$1 / fan_{in}$

By default, Keras uses Glorot initialization with a uniform distribution. When creating a layer, you can change this to He initialization by setting `kernel_initial izer="he_uniform"` or `kernel_initializer="he_normal"` like this:

```
keras.layers.Dense(10, activation="relu", kernel_initializer="he_normal")
```

If you want He initialization with a uniform distribution but based on fan_{avg} rather than fan_{in}, you can use the `VarianceScaling` initializer like this:

---

3 E.g., Kaiming He et al., "Delving Deep into Rectifiers: Surpassing Human-Level Performance on ImageNet Classification," *Proceedings of the 2015 IEEE International Conference on Computer Vision* (2015): 1026–1034.

```
he_avg_init = keras.initializers.VarianceScaling(scale=2., mode='fan_avg',
                                                 distribution='uniform')
keras.layers.Dense(10, activation="sigmoid", kernel_initializer=he_avg_init)
```

Nonsaturating Activation Functions

One of the insights in the 2010 paper by Glorot and Bengio was that the problems
with unstable gradients were in part due to a poor choice of activation function. Until
then most people had assumed that if Mother Nature had chosen to use roughly sigmoid activation functions in biological neurons, they must be an excellent choice. But
it turns out that other activation functions behave much better in deep neural networks—in particular, the ReLU activation function, mostly because it does not saturate for positive values (and because it is fast to compute).

Unfortunately, the ReLU activation function is not perfect. It suffers from a problem
known as the *dying ReLUs*: during training, some neurons effectively "die," meaning
they stop outputting anything other than 0. In some cases, you may find that half of
your network's neurons are dead, especially if you used a large learning rate. A neuron dies when its weights get tweaked in such a way that the weighted sum of its
inputs are negative for all instances in the training set. When this happens, it just
keeps outputting zeros, and Gradient Descent does not affect it anymore because the
gradient of the ReLU function is zero when its input is negative.[4]

To solve this problem, you may want to use a variant of the ReLU function, such as
the *leaky ReLU*. This function is defined as LeakyReLU$_\alpha(z)$ = max(αz, z) (see
Figure 11-2). The hyperparameter α defines how much the function "leaks": it is the
slope of the function for $z < 0$ and is typically set to 0.01. This small slope ensures that
leaky ReLUs never die; they can go into a long coma, but they have a chance to eventually wake up. A 2015 paper (*https://homl.info/49*)[5] compared several variants of the
ReLU activation function, and one of its conclusions was that the leaky variants
always outperformed the strict ReLU activation function. In fact, setting α = 0.2 (a
huge leak) seemed to result in better performance than α = 0.01 (a small leak). The
paper also evaluated the *randomized leaky ReLU* (RReLU), where α is picked randomly in a given range during training and is fixed to an average value during testing.
RReLU also performed fairly well and seemed to act as a regularizer (reducing the
risk of overfitting the training set). Finally, the paper evaluated the *parametric leaky
ReLU* (PReLU), where α is authorized to be learned during training (instead of being
a hyperparameter, it becomes a parameter that can be modified by backpropagation

---

4 Unless it is part of the first hidden layer, a dead neuron may sometimes come back to life: Gradient Descent
  may indeed tweak neurons in the layers below in such a way that the weighted sum of the dead neuron's
  inputs is positive again.

5 Bing Xu et al., "Empirical Evaluation of Rectified Activations in Convolutional Network," arXiv preprint
  arXiv:1505.00853 (2015).

like any other parameter). PReLU was reported to strongly outperform ReLU on large image datasets, but on smaller datasets it runs the risk of overfitting the training set.

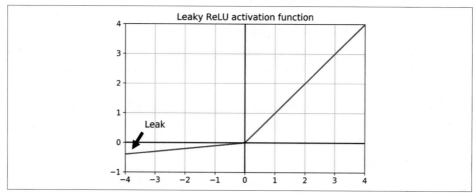

Figure 11-2. Leaky ReLU: like ReLU, but with a small slope for negative values

Last but not least, a 2015 paper (*https://homl.info/50*) by Djork-Arné Clevert et al.[6] proposed a new activation function called the *exponential linear unit* (ELU) that out-performed all the ReLU variants in the authors' experiments: training time was reduced, and the neural network performed better on the test set. Figure 11-3 graphs the function, and Equation 11-2 shows its definition.

Equation 11-2. ELU activation function

$$\text{ELU}_\alpha (z) = \begin{cases} \alpha(\exp(z) - 1) & \text{if } z < 0 \\ z & \text{if } z \geq 0 \end{cases}$$

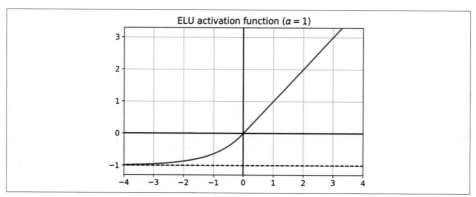

Figure 11-3. ELU activation function

---

6 Djork-Arné Clevert et al., "Fast and Accurate Deep Network Learning by Exponential Linear Units (ELUs)," *Proceedings of the International Conference on Learning Representations* (2016).

The ELU activation function looks a lot like the ReLU function, with a few major differences:

- It takes on negative values when $z < 0$, which allows the unit to have an average output closer to 0 and helps alleviate the vanishing gradients problem. The hyperparameter α defines the value that the ELU function approaches when z is a large negative number. It is usually set to 1, but you can tweak it like any other hyperparameter.

- It has a nonzero gradient for $z < 0$, which avoids the dead neurons problem.

- If α is equal to 1 then the function is smooth everywhere, including around $z = 0$, which helps speed up Gradient Descent since it does not bounce as much to the left and right of $z = 0$.

The main drawback of the ELU activation function is that it is slower to compute than the ReLU function and its variants (due to the use of the exponential function). Its faster convergence rate during training compensates for that slow computation, but still, at test time an ELU network will be slower than a ReLU network.

Then, a 2017 paper (*https://homl.info/selu*)[7] by Günter Klambauer et al. introduced the Scaled ELU (SELU) activation function: as its name suggests, it is a scaled variant of the ELU activation function. The authors showed that if you build a neural network composed exclusively of a stack of dense layers, and if all hidden layers use the SELU activation function, then the network will *self-normalize*: the output of each layer will tend to preserve a mean of 0 and standard deviation of 1 during training, which solves the vanishing/exploding gradients problem. As a result, the SELU activation function often significantly outperforms other activation functions for such neural nets (especially deep ones). There are, however, a few conditions for self-normalization to happen (see the paper for the mathematical justification):

- The input features must be standardized (mean 0 and standard deviation 1).

- Every hidden layer's weights must be initialized with LeCun normal initialization. In Keras, this means setting `kernel_initializer="lecun_normal"`.

- The network's architecture must be sequential. Unfortunately, if you try to use SELU in nonsequential architectures, such as recurrent networks (see Chapter 15) or networks with *skip connections* (i.e., connections that skip layers, such as in Wide & Deep nets), self-normalization will not be guaranteed, so SELU will not necessarily outperform other activation functions.

---

[7] Günter Klambauer et al., "Self-Normalizing Neural Networks," *Proceedings of the 31st International Conference on Neural Information Processing Systems* (2017): 972–981.

- The paper only guarantees self-normalization if all layers are dense, but some researchers have noted that the SELU activation function can improve performance in convolutional neural nets as well (see Chapter 14).

So, which activation function should you use for the hidden layers of your deep neural networks? Although your mileage will vary, in general SELU > ELU > leaky ReLU (and its variants) > ReLU > tanh > logistic. If the network's architecture prevents it from self-normalizing, then ELU may perform better than SELU (since SELU is not smooth at $z = 0$). If you care a lot about runtime latency, then you may prefer leaky ReLU. If you don't want to tweak yet another hyperparameter, you may use the default α values used by Keras (e.g., 0.3 for leaky ReLU). If you have spare time and computing power, you can use cross-validation to evaluate other activation functions, such as RReLU if your network is overfitting or PReLU if you have a huge training set. That said, because ReLU is the most used activation function (by far), many libraries and hardware accelerators provide ReLU-specific optimizations; therefore, if speed is your priority, ReLU might still be the best choice.

To use the leaky ReLU activation function, create a `LeakyReLU` layer and add it to your model just after the layer you want to apply it to:

```
model = keras.models.Sequential([
    [...]
    keras.layers.Dense(10, kernel_initializer="he_normal"),
    keras.layers.LeakyReLU(alpha=0.2),
    [...]
])
```

For PReLU, replace `LeakyRelu(alpha=0.2)` with `PReLU()`. There is currently no official implementation of RReLU in Keras, but you can fairly easily implement your own (to learn how to do that, see the exercises at the end of Chapter 12).

For SELU activation, set `activation="selu"` and `kernel_initializer="lecun_nor mal"` when creating a layer:

```
layer = keras.layers.Dense(10, activation="selu",
                           kernel_initializer="lecun_normal")
```

Batch Normalization

Although using He initialization along with ELU (or any variant of ReLU) can significantly reduce the danger of the vanishing/exploding gradients problems at the beginning of training, it doesn't guarantee that they won't come back during training.

In a 2015 paper (*https://homl.info/51*),[8] Sergey Ioffe and Christian Szegedy proposed a technique called *Batch Normalization* (BN) that addresses these problems. The technique consists of adding an operation in the model just before or after the activation function of each hidden layer. This operation simply zero-centers and normalizes each input, then scales and shifts the result using two new parameter vectors per layer: one for scaling, the other for shifting. In other words, the operation lets the model learn the optimal scale and mean of each of the layer's inputs. In many cases, if you add a BN layer as the very first layer of your neural network, you do not need to standardize your training set (e.g., using a `StandardScaler`); the BN layer will do it for you (well, approximately, since it only looks at one batch at a time, and it can also rescale and shift each input feature).

In order to zero-center and normalize the inputs, the algorithm needs to estimate each input's mean and standard deviation. It does so by evaluating the mean and standard deviation of the input over the current mini-batch (hence the name "Batch Normalization"). The whole operation is summarized step by step in Equation 11-3.

Equation 11-3. Batch Normalization algorithm

1. $$\mu_B = \frac{1}{m_B} \sum_{i=1}^{m_B} \mathbf{x}^{(i)}$$

2. $${\sigma_B}^2 = \frac{1}{m_B} \sum_{i=1}^{m_B} \left(\mathbf{x}^{(i)} - \mu_B \right)^2$$

3. $$\widehat{\mathbf{x}}^{(i)} = \frac{\mathbf{x}^{(i)} - \mu_B}{\sqrt{{\sigma_B}^2 + \varepsilon}}$$

4. $$\mathbf{z}^{(i)} = \gamma \otimes \widehat{\mathbf{x}}^{(i)} + \beta$$

In this algorithm:

- μ_B is the vector of input means, evaluated over the whole mini-batch B (it contains one mean per input).

- σ_B is the vector of input standard deviations, also evaluated over the whole mini-batch (it contains one standard deviation per input).

- m_B is the number of instances in the mini-batch.

- $\widehat{\mathbf{x}}^{(i)}$ is the vector of zero-centered and normalized inputs for instance i.

---

8 Sergey Ioffe and Christian Szegedy, "Batch Normalization: Accelerating Deep Network Training by Reducing Internal Covariate Shift," *Proceedings of the 32nd International Conference on Machine Learning* (2015): 448–456.

- **γ** is the output scale parameter vector for the layer (it contains one scale parameter per input).

- ⊗ represents element-wise multiplication (each input is multiplied by its corresponding output scale parameter).

- **β** is the output shift (offset) parameter vector for the layer (it contains one offset parameter per input). Each input is offset by its corresponding shift parameter.

- ε is a tiny number that avoids division by zero (typically 10^{-5}). This is called a *smoothing term*.

- $\mathbf{z}^{(i)}$ is the output of the BN operation. It is a rescaled and shifted version of the inputs.

So during training, BN standardizes its inputs, then rescales and offsets them. Good! What about at test time? Well, it's not that simple. Indeed, we may need to make predictions for individual instances rather than for batches of instances: in this case, we will have no way to compute each input's mean and standard deviation. Moreover, even if we do have a batch of instances, it may be too small, or the instances may not be independent and identically distributed, so computing statistics over the batch instances would be unreliable. One solution could be to wait until the end of training, then run the whole training set through the neural network and compute the mean and standard deviation of each input of the BN layer. These "final" input means and standard deviations could then be used instead of the batch input means and standard deviations when making predictions. However, most implementations of Batch Normalization estimate these final statistics during training by using a moving average of the layer's input means and standard deviations. This is what Keras does automatically when you use the `BatchNormalization` layer. To sum up, four parameter vectors are learned in each batch-normalized layer: **γ** (the output scale vector) and **β** (the output offset vector) are learned through regular backpropagation, and **μ** (the final input mean vector) and **σ** (the final input standard deviation vector) are estimated using an exponential moving average. Note that **μ** and **σ** are estimated during training, but they are used only after training (to replace the batch input means and standard deviations in Equation 11-3).

Ioffe and Szegedy demonstrated that Batch Normalization considerably improved all the deep neural networks they experimented with, leading to a huge improvement in the ImageNet classification task (ImageNet is a large database of images classified into many classes, commonly used to evaluate computer vision systems). The vanishing gradients problem was strongly reduced, to the point that they could use saturating activation functions such as the tanh and even the logistic activation function. The networks were also much less sensitive to the weight initialization. The authors were able to use much larger learning rates, significantly speeding up the learning process. Specifically, they note that:

Applied to a state-of-the-art image classification model, Batch Normalization achieves the same accuracy with 14 times fewer training steps, and beats the original model by a significant margin. [...] Using an ensemble of batch-normalized networks, we improve upon the best published result on ImageNet classification: reaching 4.9% top-5 validation error (and 4.8% test error), exceeding the accuracy of human raters.

Finally, like a gift that keeps on giving, Batch Normalization acts like a regularizer, reducing the need for other regularization techniques (such as dropout, described later in this chapter).

Batch Normalization does, however, add some complexity to the model (although it can remove the need for normalizing the input data, as we discussed earlier). Moreover, there is a runtime penalty: the neural network makes slower predictions due to the extra computations required at each layer. Fortunately, it's often possible to fuse the BN layer with the previous layer, after training, thereby avoiding the runtime penalty. This is done by updating the previous layer's weights and biases so that it directly produces outputs of the appropriate scale and offset. For example, if the previous layer computes $\mathbf{XW} + \mathbf{b}$, then the BN layer will compute $\gamma \otimes (\mathbf{XW} + \mathbf{b} - \mu)/\sigma + \beta$ (ignoring the smoothing term ε in the denominator). If we define $\mathbf{W}' = \gamma \otimes \mathbf{W}/\sigma$ and $\mathbf{b}' = \gamma \otimes (\mathbf{b} - \mu)/\sigma + \beta$, the equation simplifies to $\mathbf{XW}' + \mathbf{b}'$. So if we replace the previous layer's weights and biases ($\mathbf{W}$ and $\mathbf{b}$) with the updated weights and biases ($\mathbf{W}'$ and $\mathbf{b}'$), we can get rid of the BN layer (TFLite's optimizer does this automatically; see Chapter 19).

You may find that training is rather slow, because each epoch takes much more time when you use Batch Normalization. This is usually counterbalanced by the fact that convergence is much faster with BN, so it will take fewer epochs to reach the same performance. All in all, *wall time* will usually be shorter (this is the time measured by the clock on your wall).

Implementing Batch Normalization with Keras

As with most things with Keras, implementing Batch Normalization is simple and intuitive. Just add a `BatchNormalization` layer before or after each hidden layer's activation function, and optionally add a BN layer as well as the first layer in your model. For example, this model applies BN after every hidden layer and as the first layer in the model (after flattening the input images):

```
model = keras.models.Sequential([
    keras.layers.Flatten(input_shape=[28, 28]),
    keras.layers.BatchNormalization(),
    keras.layers.Dense(300, activation="elu", kernel_initializer="he_normal"),
    keras.layers.BatchNormalization(),
    keras.layers.Dense(100, activation="elu", kernel_initializer="he_normal"),
    keras.layers.BatchNormalization(),
    keras.layers.Dense(10, activation="softmax")
])
```

That's all! In this tiny example with just two hidden layers, it's unlikely that Batch Normalization will have a very positive impact; but for deeper networks it can make a tremendous difference.

Let's display the model summary:

```
>>> model.summary()
Model: "sequential_3"
```

Layer (type)	Output Shape	Param #
flatten_3 (Flatten)	(None, 784)	0
batch_normalization_v2 (Batc	(None, 784)	3136
dense_50 (Dense)	(None, 300)	235500
batch_normalization_v2_1 (Ba	(None, 300)	1200
dense_51 (Dense)	(None, 100)	30100
batch_normalization_v2_2 (Ba	(None, 100)	400
dense_52 (Dense)	(None, 10)	1010

```
Total params: 271,346
Trainable params: 268,978
Non-trainable params: 2,368
```

As you can see, each BN layer adds four parameters per input: γ, β, μ, and σ (for example, the first BN layer adds 3,136 parameters, which is 4 × 784). The last two parameters, μ and σ, are the moving averages; they are not affected by backpropagation, so Keras calls them "non-trainable"[9] (if you count the total number of BN parameters, 3,136 + 1,200 + 400, and divide by 2, you get 2,368, which is the total number of non-trainable parameters in this model).

---

9 However, they are estimated during training, based on the training data, so arguably they *are* trainable. In Keras, "non-trainable" really means "untouched by backpropagation."

Let's look at the parameters of the first BN layer. Two are trainable (by backpropagation), and two are not:

```
>>> [(var.name, var.trainable) for var in model.layers[1].variables]
[('batch_normalization_v2/gamma:0', True),
 ('batch_normalization_v2/beta:0', True),
 ('batch_normalization_v2/moving_mean:0', False),
 ('batch_normalization_v2/moving_variance:0', False)]
```

Now when you create a BN layer in Keras, it also creates two operations that will be called by Keras at each iteration during training. These operations will update the moving averages. Since we are using the TensorFlow backend, these operations are TensorFlow operations (we will discuss TF operations in Chapter 12):

```
>>> model.layers[1].updates
[<tf.Operation 'cond_2/Identity' type=Identity>,
 <tf.Operation 'cond_3/Identity' type=Identity>]
```

The authors of the BN paper argued in favor of adding the BN layers before the activation functions, rather than after (as we just did). There is some debate about this, as which is preferable seems to depend on the task—you can experiment with this too to see which option works best on your dataset. To add the BN layers before the activation functions, you must remove the activation function from the hidden layers and add them as separate layers after the BN layers. Moreover, since a Batch Normalization layer includes one offset parameter per input, you can remove the bias term from the previous layer (just pass use_bias=False when creating it):

```
model = keras.models.Sequential([
    keras.layers.Flatten(input_shape=[28, 28]),
    keras.layers.BatchNormalization(),
    keras.layers.Dense(300, kernel_initializer="he_normal", use_bias=False),
    keras.layers.BatchNormalization(),
    keras.layers.Activation("elu"),
    keras.layers.Dense(100, kernel_initializer="he_normal", use_bias=False),
    keras.layers.BatchNormalization(),
    keras.layers.Activation("elu"),
    keras.layers.Dense(10, activation="softmax")
])
```

The BatchNormalization class has quite a few hyperparameters you can tweak. The defaults will usually be fine, but you may occasionally need to tweak the momentum. This hyperparameter is used by the BatchNormalization layer when it updates the exponential moving averages; given a new value $\mathbf{v}$ (i.e., a new vector of input means or standard deviations computed over the current batch), the layer updates the running average $\widehat{\mathbf{v}}$ using the following equation:

$$\widehat{\mathbf{v}} \leftarrow \widehat{\mathbf{v}} \times \text{momentum} + \mathbf{v} \times (1 - \text{momentum})$$

A good momentum value is typically close to 1; for example, 0.9, 0.99, or 0.999 (you want more 9s for larger datasets and smaller mini-batches).

Another important hyperparameter is `axis`: it determines which axis should be normalized. It defaults to –1, meaning that by default it will normalize the last axis (using the means and standard deviations computed across the *other* axes). When the input batch is 2D (i.e., the batch shape is [*batch size, features*]), this means that each input feature will be normalized based on the mean and standard deviation computed across all the instances in the batch. For example, the first BN layer in the previous code example will independently normalize (and rescale and shift) each of the 784 input features. If we move the first BN layer before the `Flatten` layer, then the input batches will be 3D, with shape [*batch size, height, width*]; therefore, the BN layer will compute 28 means and 28 standard deviations (1 per column of pixels, computed across all instances in the batch and across all rows in the column), and it will normalize all pixels in a given column using the same mean and standard deviation. There will also be just 28 scale parameters and 28 shift parameters. If instead you still want to treat each of the 784 pixels independently, then you should set `axis=[1, 2]`.

Notice that the BN layer does not perform the same computation during training and after training: it uses batch statistics during training and the "final" statistics after training (i.e., the final values of the moving averages). Let's take a peek at the source code of this class to see how this is handled:

```
class BatchNormalization(keras.layers.Layer):
    [...]
    def call(self, inputs, training=None):
        [...]
```

The `call()` method is the one that performs the computations; as you can see, it has an extra `training` argument, which is set to `None` by default, but the `fit()` method sets to it to `1` during training. If you ever need to write a custom layer, and it must behave differently during training and testing, add a `training` argument to the `call()` method and use this argument in the method to decide what to compute[10] (we will discuss custom layers in Chapter 12).

`BatchNormalization` has become one of the most-used layers in deep neural networks, to the point that it is often omitted in the diagrams, as it is assumed that BN is added after every layer. But a recent paper (*https://homl.info/fixup*)[11] by Hongyi Zhang et al. may change this assumption: by using a novel *fixed-update* (fixup) weight initialization technique, the authors managed to train a very deep neural

---

10 The Keras API also specifies a `keras.backend.learning_phase()` function that should return 1 during training and 0 otherwise.

11 Hongyi Zhang et al., "Fixup Initialization: Residual Learning Without Normalization," arXiv preprint arXiv: 1901.09321 (2019).

network (10,000 layers!) without BN, achieving state-of-the-art performance on complex image classification tasks. As this is bleeding-edge research, however, you may want to wait for additional research to confirm this finding before you drop Batch Normalization.

Gradient Clipping

Another popular technique to mitigate the exploding gradients problem is to clip the gradients during backpropagation so that they never exceed some threshold. This is called *Gradient Clipping* (*https://homl.info/52*).[12] This technique is most often used in recurrent neural networks, as Batch Normalization is tricky to use in RNNs, as we will see in Chapter 15. For other types of networks, BN is usually sufficient.

In Keras, implementing Gradient Clipping is just a matter of setting the `clipvalue` or `clipnorm` argument when creating an optimizer, like this:

```
optimizer = keras.optimizers.SGD(clipvalue=1.0)
model.compile(loss="mse", optimizer=optimizer)
```

This optimizer will clip every component of the gradient vector to a value between −1.0 and 1.0. This means that all the partial derivatives of the loss (with regard to each and every trainable parameter) will be clipped between −1.0 and 1.0. The threshold is a hyperparameter you can tune. Note that it may change the orientation of the gradient vector. For instance, if the original gradient vector is [0.9, 100.0], it points mostly in the direction of the second axis; but once you clip it by value, you get [0.9, 1.0], which points roughly in the diagonal between the two axes. In practice, this approach works well. If you want to ensure that Gradient Clipping does not change the direction of the gradient vector, you should clip by norm by setting `clipnorm` instead of `clipvalue`. This will clip the whole gradient if its ℓ_2 norm is greater than the threshold you picked. For example, if you set `clipnorm=1.0`, then the vector [0.9, 100.0] will be clipped to [0.00899964, 0.9999595], preserving its orientation but almost eliminating the first component. If you observe that the gradients explode during training (you can track the size of the gradients using TensorBoard), you may want to try both clipping by value and clipping by norm, with different thresholds, and see which option performs best on the validation set.

Reusing Pretrained Layers

It is generally not a good idea to train a very large DNN from scratch: instead, you should always try to find an existing neural network that accomplishes a similar task to the one you are trying to tackle (we will discuss how to find them in Chapter 14),

---

12 Razvan Pascanu et al., "On the Difficulty of Training Recurrent Neural Networks," *Proceedings of the 30th International Conference on Machine Learning* (2013): 1310–1318.

then reuse the lower layers of this network. This technique is called *transfer learning*. It will not only speed up training considerably, but also require significantly less training data.

Suppose you have access to a DNN that was trained to classify pictures into 100 different categories, including animals, plants, vehicles, and everyday objects. You now want to train a DNN to classify specific types of vehicles. These tasks are very similar, even partly overlapping, so you should try to reuse parts of the first network (see Figure 11-4).

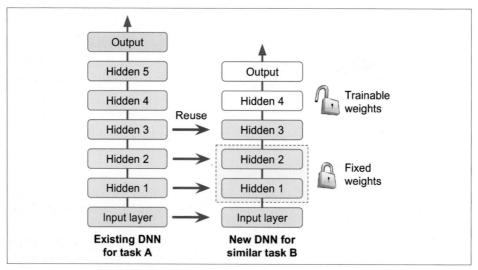

Figure 11-4. Reusing pretrained layers

 If the input pictures of your new task don't have the same size as the ones used in the original task, you will usually have to add a preprocessing step to resize them to the size expected by the original model. More generally, transfer learning will work best when the inputs have similar low-level features.

The output layer of the original model should usually be replaced because it is most likely not useful at all for the new task, and it may not even have the right number of outputs for the new task.

Similarly, the upper hidden layers of the original model are less likely to be as useful as the lower layers, since the high-level features that are most useful for the new task may differ significantly from the ones that were most useful for the original task. You want to find the right number of layers to reuse.

The more similar the tasks are, the more layers you want to reuse (starting with the lower layers). For very similar tasks, try keeping all the hidden layers and just replacing the output layer.

Try freezing all the reused layers first (i.e., make their weights non-trainable so that Gradient Descent won't modify them), then train your model and see how it performs. Then try unfreezing one or two of the top hidden layers to let backpropagation tweak them and see if performance improves. The more training data you have, the more layers you can unfreeze. It is also useful to reduce the learning rate when you unfreeze reused layers: this will avoid wrecking their fine-tuned weights.

If you still cannot get good performance, and you have little training data, try dropping the top hidden layer(s) and freezing all the remaining hidden layers again. You can iterate until you find the right number of layers to reuse. If you have plenty of training data, you may try replacing the top hidden layers instead of dropping them, and even adding more hidden layers.

Transfer Learning with Keras

Let's look at an example. Suppose the Fashion MNIST dataset only contained eight classes—for example, all the classes except for sandal and shirt. Someone built and trained a Keras model on that set and got reasonably good performance (>90% accuracy). Let's call this model A. You now want to tackle a different task: you have images of sandals and shirts, and you want to train a binary classifier (positive=shirt, negative=sandal). Your dataset is quite small; you only have 200 labeled images. When you train a new model for this task (let's call it model B) with the same architecture as model A, it performs reasonably well (97.2% accuracy). But since it's a much easier task (there are just two classes), you were hoping for more. While drinking your morning coffee, you realize that your task is quite similar to task A, so perhaps transfer learning can help? Let's find out!

First, you need to load model A and create a new model based on that model's layers. Let's reuse all the layers except for the output layer:

```
model_A = keras.models.load_model("my_model_A.h5")
model_B_on_A = keras.models.Sequential(model_A.layers[:-1])
model_B_on_A.add(keras.layers.Dense(1, activation="sigmoid"))
```

Note that model_A and model_B_on_A now share some layers. When you train model_B_on_A, it will also affect model_A. If you want to avoid that, you need to *clone* model_A before you reuse its layers. To do this, you clone model A's architecture with clone.model(), then copy its weights (since clone_model() does not clone the weights):

```
model_A_clone = keras.models.clone_model(model_A)
model_A_clone.set_weights(model_A.get_weights())
```

Now you could train `model_B_on_A` for task B, but since the new output layer was initialized randomly it will make large errors (at least during the first few epochs), so there will be large error gradients that may wreck the reused weights. To avoid this, one approach is to freeze the reused layers during the first few epochs, giving the new layer some time to learn reasonable weights. To do this, set every layer's `trainable` attribute to `False` and compile the model:

```
for layer in model_B_on_A.layers[:-1]:
    layer.trainable = False

model_B_on_A.compile(loss="binary_crossentropy", optimizer="sgd",
                     metrics=["accuracy"])
```

 You must always compile your model after you freeze or unfreeze layers.

Now you can train the model for a few epochs, then unfreeze the reused layers (which requires compiling the model again) and continue training to fine-tune the reused layers for task B. After unfreezing the reused layers, it is usually a good idea to reduce the learning rate, once again to avoid damaging the reused weights:

```
history = model_B_on_A.fit(X_train_B, y_train_B, epochs=4,
                           validation_data=(X_valid_B, y_valid_B))

for layer in model_B_on_A.layers[:-1]:
    layer.trainable = True

optimizer = keras.optimizers.SGD(lr=1e-4) # the default lr is 1e-2
model_B_on_A.compile(loss="binary_crossentropy", optimizer=optimizer,
                     metrics=["accuracy"])
history = model_B_on_A.fit(X_train_B, y_train_B, epochs=16,
                           validation_data=(X_valid_B, y_valid_B))
```

So, what's the final verdict? Well, this model's test accuracy is 99.25%, which means that transfer learning reduced the error rate from 2.8% down to almost 0.7%! That's a factor of four!

```
>>> model_B_on_A.evaluate(X_test_B, y_test_B)
[0.06887910133600235, 0.9925]
```

Are you convinced? You shouldn't be: I cheated! I tried many configurations until I found one that demonstrated a strong improvement. If you try to change the classes or the random seed, you will see that the improvement generally drops, or even vanishes or reverses. What I did is called "torturing the data until it confesses." When a paper just looks too positive, you should be suspicious: perhaps the flashy new technique does not actually help much (in fact, it may even degrade performance), but the authors tried many variants and reported only the best results (which may be due to sheer luck), without mentioning how many failures they encountered on the way. Most of the time, this is not malicious at all, but it is part of the reason so many results in science can never be reproduced.

Why did I cheat? It turns out that transfer learning does not work very well with small dense networks, presumably because small networks learn few patterns, and dense networks learn very specific patterns, which are unlikely to be useful in other tasks. Transfer learning works best with deep convolutional neural networks, which tend to learn feature detectors that are much more general (especially in the lower layers). We will revisit transfer learning in Chapter 14, using the techniques we just discussed (and this time there will be no cheating, I promise!).

Unsupervised Pretraining

Suppose you want to tackle a complex task for which you don't have much labeled training data, but unfortunately you cannot find a model trained on a similar task. Don't lose hope! First, you should try to gather more labeled training data, but if you can't, you may still be able to perform *unsupervised pretraining* (see Figure 11-5). Indeed, it is often cheap to gather unlabeled training examples, but expensive to label them. If you can gather plenty of unlabeled training data, you can try to use it to train an unsupervised model, such as an autoencoder or a generative adversarial network (see Chapter 17). Then you can reuse the lower layers of the autoencoder or the lower layers of the GAN's discriminator, add the output layer for your task on top, and fine-tune the final network using supervised learning (i.e., with the labeled training examples).

It is this technique that Geoffrey Hinton and his team used in 2006 and which led to the revival of neural networks and the success of Deep Learning. Until 2010, unsupervised pretraining—typically with restricted Boltzmann machines (RBMs; see Appendix E)—was the norm for deep nets, and only after the vanishing gradients problem was alleviated did it become much more common to train DNNs purely using supervised learning. Unsupervised pretraining (today typically using autoencoders or GANs rather than RBMs) is still a good option when you have a complex task to solve, no similar model you can reuse, and little labeled training data but plenty of unlabeled training data.

Note that in the early days of Deep Learning it was difficult to train deep models, so people would use a technique called *greedy layer-wise pretraining* (depicted in Figure 11-5). They would first train an unsupervised model with a single layer, typically an RBM, then they would freeze that layer and add another one on top of it, then train the model again (effectively just training the new layer), then freeze the new layer and add another layer on top of it, train the model again, and so on. Nowadays, things are much simpler: people generally train the full unsupervised model in one shot (i.e., in Figure 11-5, just start directly at step three) and use autoencoders or GANs rather than RBMs.

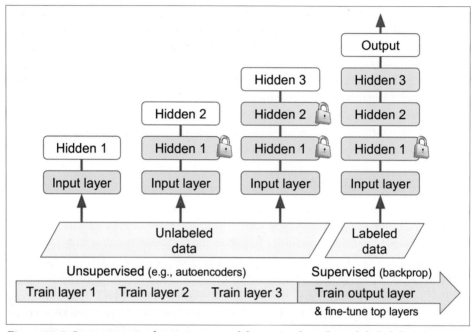

Figure 11-5. In unsupervised training, a model is trained on the unlabeled data (or on all the data) using an unsupervised learning technique, then it is fine-tuned for the final task on the labeled data using a supervised learning technique; the unsupervised part may train one layer at a time as shown here, or it may train the full model directly

Pretraining on an Auxiliary Task

If you do not have much labeled training data, one last option is to train a first neural network on an auxiliary task for which you can easily obtain or generate labeled training data, then reuse the lower layers of that network for your actual task. The first neural network's lower layers will learn feature detectors that will likely be reusable by the second neural network.

For example, if you want to build a system to recognize faces, you may only have a few pictures of each individual—clearly not enough to train a good classifier. Gathering hundreds of pictures of each person would not be practical. You could, however, gather a lot of pictures of random people on the web and train a first neural network to detect whether or not two different pictures feature the same person. Such a network would learn good feature detectors for faces, so reusing its lower layers would allow you to train a good face classifier that uses little training data.

For *natural language processing* (NLP) applications, you can download a corpus of millions of text documents and automatically generate labeled data from it. For example, you could randomly mask out some words and train a model to predict what the missing words are (e.g., it should predict that the missing word in the sentence "What ___ you saying?" is probably "are" or "were"). If you can train a model to reach good performance on this task, then it will already know quite a lot about language, and you can certainly reuse it for your actual task and fine-tune it on your labeled data (we will discuss more pretraining tasks in Chapter 15).

Self-supervised learning is when you automatically generate the labels from the data itself, then you train a model on the resulting "labeled" dataset using supervised learning techniques. Since this approach requires no human labeling whatsoever, it is best classified as a form of unsupervised learning.

Faster Optimizers

Training a very large deep neural network can be painfully slow. So far we have seen four ways to speed up training (and reach a better solution): applying a good initialization strategy for the connection weights, using a good activation function, using Batch Normalization, and reusing parts of a pretrained network (possibly built on an auxiliary task or using unsupervised learning). Another huge speed boost comes from using a faster optimizer than the regular Gradient Descent optimizer. In this section we will present the most popular algorithms: momentum optimization, Nesterov Accelerated Gradient, AdaGrad, RMSProp, and finally Adam and Nadam optimization.

Momentum Optimization

Imagine a bowling ball rolling down a gentle slope on a smooth surface: it will start out slowly, but it will quickly pick up momentum until it eventually reaches terminal velocity (if there is some friction or air resistance). This is the very simple idea behind *momentum optimization*, proposed by Boris Polyak in 1964 (*https://homl.info/54*).[13] In contrast, regular Gradient Descent will simply take small, regular steps down the slope, so the algorithm will take much more time to reach the bottom.

---

13 Boris T. Polyak, "Some Methods of Speeding Up the Convergence of Iteration Methods," *USSR Computational Mathematics and Mathematical Physics* 4, no. 5 (1964): 1–17.

Recall that Gradient Descent updates the weights θ by directly subtracting the gradient of the cost function $J(\theta)$ with regard to the weights ($\nabla_\theta J(\theta)$) multiplied by the learning rate η. The equation is: $\theta \leftarrow \theta - \eta \nabla_\theta J(\theta)$. It does not care about what the earlier gradients were. If the local gradient is tiny, it goes very slowly.

Momentum optimization cares a great deal about what previous gradients were: at each iteration, it subtracts the local gradient from the *momentum vector* **m** (multiplied by the learning rate η), and it updates the weights by adding this momentum vector (see Equation 11-4). In other words, the gradient is used for acceleration, not for speed. To simulate some sort of friction mechanism and prevent the momentum from growing too large, the algorithm introduces a new hyperparameter β, called the *momentum*, which must be set between 0 (high friction) and 1 (no friction). A typical momentum value is 0.9.

Equation 11-4. Momentum algorithm

1. $\mathbf{m} \leftarrow \beta \mathbf{m} - \eta \nabla_\theta J(\theta)$

2. $\theta \leftarrow \theta + \mathbf{m}$

You can easily verify that if the gradient remains constant, the terminal velocity (i.e., the maximum size of the weight updates) is equal to that gradient multiplied by the learning rate η multiplied by $1/(1-\beta)$ (ignoring the sign). For example, if $\beta = 0.9$, then the terminal velocity is equal to 10 times the gradient times the learning rate, so momentum optimization ends up going 10 times faster than Gradient Descent! This allows momentum optimization to escape from plateaus much faster than Gradient Descent. We saw in Chapter 4 that when the inputs have very different scales, the cost function will look like an elongated bowl (see Figure 4-7). Gradient Descent goes down the steep slope quite fast, but then it takes a very long time to go down the valley. In contrast, momentum optimization will roll down the valley faster and faster until it reaches the bottom (the optimum). In deep neural networks that don't use Batch Normalization, the upper layers will often end up having inputs with very different scales, so using momentum optimization helps a lot. It can also help roll past local optima.

Due to the momentum, the optimizer may overshoot a bit, then come back, overshoot again, and oscillate like this many times before stabilizing at the minimum. This is one of the reasons it's good to have a bit of friction in the system: it gets rid of these oscillations and thus speeds up convergence.

Implementing momentum optimization in Keras is a no-brainer: just use the SGD optimizer and set its momentum hyperparameter, then lie back and profit!

```
optimizer = keras.optimizers.SGD(lr=0.001, momentum=0.9)
```

The one drawback of momentum optimization is that it adds yet another hyperparameter to tune. However, the momentum value of 0.9 usually works well in practice and almost always goes faster than regular Gradient Descent.

Nesterov Accelerated Gradient

One small variant to momentum optimization, proposed by Yurii Nesterov in 1983 (*https://homl.info/55*),[14] is almost always faster than vanilla momentum optimization. The *Nesterov Accelerated Gradient* (NAG) method, also known as *Nesterov momentum optimization*, measures the gradient of the cost function not at the local position θ but slightly ahead in the direction of the momentum, at $\theta + \beta\mathbf{m}$ (see Equation 11-5).

Equation 11-5. Nesterov Accelerated Gradient algorithm

1. $\mathbf{m} \leftarrow \beta\mathbf{m} - \eta\nabla_{\theta}J(\theta + \beta\mathbf{m})$
2. $\theta \leftarrow \theta + \mathbf{m}$

This small tweak works because in general the momentum vector will be pointing in the right direction (i.e., toward the optimum), so it will be slightly more accurate to use the gradient measured a bit farther in that direction rather than the gradient at the original position, as you can see in Figure 11-6 (where ∇_1 represents the gradient of the cost function measured at the starting point θ, and ∇_2 represents the gradient at the point located at $\theta + \beta\mathbf{m}$).

As you can see, the Nesterov update ends up slightly closer to the optimum. After a while, these small improvements add up and NAG ends up being significantly faster than regular momentum optimization. Moreover, note that when the momentum pushes the weights across a valley, ∇_1 continues to push farther across the valley, while ∇_2 pushes back toward the bottom of the valley. This helps reduce oscillations and thus NAG converges faster.

NAG is generally faster than regular momentum optimization. To use it, simply set nesterov=True when creating the SGD optimizer:

```
optimizer = keras.optimizers.SGD(lr=0.001, momentum=0.9, nesterov=True)
```

---

14 Yurii Nesterov, "A Method for Unconstrained Convex Minimization Problem with the Rate of Convergence $O(1/k^2)$," *Doklady AN USSR* 269 (1983): 543–547.

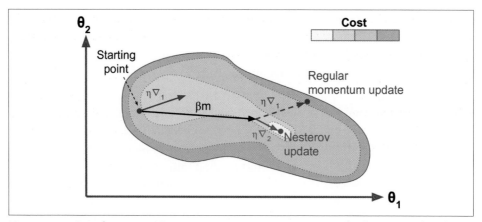

Figure 11-6. Regular versus Nesterov momentum optimization: the former applies the gradients computed before the momentum step, while the latter applies the gradients computed after

AdaGrad

Consider the elongated bowl problem again: Gradient Descent starts by quickly going down the steepest slope, which does not point straight toward the global optimum, then it very slowly goes down to the bottom of the valley. It would be nice if the algorithm could correct its direction earlier to point a bit more toward the global optimum. The *AdaGrad* algorithm (*https://homl.info/56*)[15] achieves this correction by scaling down the gradient vector along the steepest dimensions (see Equation 11-6).

Equation 11-6. AdaGrad algorithm

1. $\mathbf{s} \leftarrow \mathbf{s} + \nabla_\theta J(\theta) \otimes \nabla_\theta J(\theta)$

2. $\theta \leftarrow \theta - \eta \, \nabla_\theta J(\theta) \oslash \sqrt{\mathbf{s} + \varepsilon}$

The first step accumulates the square of the gradients into the vector $\mathbf{s}$ (recall that the $\otimes$ symbol represents the element-wise multiplication). This vectorized form is equivalent to computing $s_i \leftarrow s_i + (\partial J(\theta) / \partial \theta_i)^2$ for each element s_i of the vector $\mathbf{s}$; in other words, each s_i accumulates the squares of the partial derivative of the cost function with regard to parameter θ_i. If the cost function is steep along the ith dimension, then s_i will get larger and larger at each iteration.

The second step is almost identical to Gradient Descent, but with one big difference: the gradient vector is scaled down by a factor of $\sqrt{\mathbf{s} + \varepsilon}$ (the $\oslash$ symbol represents the

---

15 John Duchi et al., "Adaptive Subgradient Methods for Online Learning and Stochastic Optimization," *Journal of Machine Learning Research* 12 (2011): 2121–2159.

element-wise division, and ε is a smoothing term to avoid division by zero, typically set to 10^{-10}). This vectorized form is equivalent to simultaneously computing $\theta_i \leftarrow \theta_i - \eta\, \partial J(\boldsymbol{\theta})/\partial\theta_i/\sqrt{s_i + \varepsilon}$ for all parameters θ_i.

In short, this algorithm decays the learning rate, but it does so faster for steep dimensions than for dimensions with gentler slopes. This is called an *adaptive learning rate*. It helps point the resulting updates more directly toward the global optimum (see Figure 11-7). One additional benefit is that it requires much less tuning of the learning rate hyperparameter η.

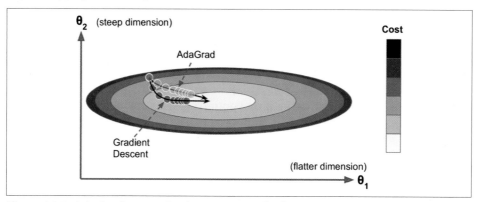

Figure 11-7. AdaGrad versus Gradient Descent: the former can correct its direction earlier to point to the optimum

AdaGrad frequently performs well for simple quadratic problems, but it often stops too early when training neural networks. The learning rate gets scaled down so much that the algorithm ends up stopping entirely before reaching the global optimum. So even though Keras has an `Adagrad` optimizer, you should not use it to train deep neural networks (it may be efficient for simpler tasks such as Linear Regression, though). Still, understanding AdaGrad is helpful to grasp the other adaptive learning rate optimizers.

RMSProp

As we've seen, AdaGrad runs the risk of slowing down a bit too fast and never converging to the global optimum. The *RMSProp* algorithm[16] fixes this by accumulating only the gradients from the most recent iterations (as opposed to all the gradients

---

16 This algorithm was created by Geoffrey Hinton and Tijmen Tieleman in 2012 and presented by Geoffrey Hinton in his Coursera class on neural networks (slides: *https://homl.info/57*; video: *https://homl.info/58*). Amusingly, since the authors did not write a paper to describe the algorithm, researchers often cite "slide 29 in lecture 6" in their papers.

since the beginning of training). It does so by using exponential decay in the first step (see Equation 11-7).

Equation 11-7. RMSProp algorithm

1. $\mathbf{s} \leftarrow \beta \mathbf{s} + (1 - \beta)\nabla_\theta J(\theta) \otimes \nabla_\theta J(\theta)$
2. $\theta \leftarrow \theta - \eta \nabla_\theta J(\theta) \oslash \sqrt{\mathbf{s} + \varepsilon}$

The decay rate β is typically set to 0.9. Yes, it is once again a new hyperparameter, but this default value often works well, so you may not need to tune it at all.

As you might expect, Keras has an `RMSprop` optimizer:

```
optimizer = keras.optimizers.RMSprop(lr=0.001, rho=0.9)
```

Except on very simple problems, this optimizer almost always performs much better than AdaGrad. In fact, it was the preferred optimization algorithm of many researchers until Adam optimization came around.

Adam and Nadam Optimization

Adam (*https://homl.info/59*),[17] which stands for *adaptive moment estimation*, combines the ideas of momentum optimization and RMSProp: just like momentum optimization, it keeps track of an exponentially decaying average of past gradients; and just like RMSProp, it keeps track of an exponentially decaying average of past squared gradients (see Equation 11-8).[18]

Equation 11-8. Adam algorithm

1. $\mathbf{m} \leftarrow \beta_1 \mathbf{m} - (1 - \beta_1)\nabla_\theta J(\theta)$
2. $\mathbf{s} \leftarrow \beta_2 \mathbf{s} + (1 - \beta_2)\nabla_\theta J(\theta) \otimes \nabla_\theta J(\theta)$
3. $\widehat{\mathbf{m}} \leftarrow \dfrac{\mathbf{m}}{1 - \beta_1^{\mathrm{T}}}$
4. $\widehat{\mathbf{s}} \leftarrow \dfrac{\mathbf{s}}{1 - \beta_2^{\mathrm{T}}}$
5. $\theta \leftarrow \theta + \eta \widehat{\mathbf{m}} \oslash \sqrt{\widehat{\mathbf{s}} + \varepsilon}$

---

[17] Diederik P. Kingma and Jimmy Ba, "Adam: A Method for Stochastic Optimization," arXiv preprint arXiv: 1412.6980 (2014).

[18] These are estimations of the mean and (uncentered) variance of the gradients. The mean is often called the *first moment* while the variance is often called the *second moment*, hence the name of the algorithm.

In this equation, t represents the iteration number (starting at 1).

If you just look at steps 1, 2, and 5, you will notice Adam's close similarity to both momentum optimization and RMSProp. The only difference is that step 1 computes an exponentially decaying average rather than an exponentially decaying sum, but these are actually equivalent except for a constant factor (the decaying average is just $1 - \beta_1$ times the decaying sum). Steps 3 and 4 are somewhat of a technical detail: since **m** and **s** are initialized at 0, they will be biased toward 0 at the beginning of training, so these two steps will help boost **m** and **s** at the beginning of training.

The momentum decay hyperparameter β_1 is typically initialized to 0.9, while the scaling decay hyperparameter β_2 is often initialized to 0.999. As earlier, the smoothing term ε is usually initialized to a tiny number such as 10^{-7}. These are the default values for the Adam class (to be precise, epsilon defaults to None, which tells Keras to use `keras.backend.epsilon()`, which defaults to 10^{-7}; you can change it using `keras.backend.set_epsilon()`). Here is how to create an Adam optimizer using Keras:

```
optimizer = keras.optimizers.Adam(lr=0.001, beta_1=0.9, beta_2=0.999)
```

Since Adam is an adaptive learning rate algorithm (like AdaGrad and RMSProp), it requires less tuning of the learning rate hyperparameter η. You can often use the default value $\eta = 0.001$, making Adam even easier to use than Gradient Descent.

 If you are starting to feel overwhelmed by all these different techniques and are wondering how to choose the right ones for your task, don't worry: some practical guidelines are provided at the end of this chapter.

Finally, two variants of Adam are worth mentioning:

AdaMax

Notice that in step 2 of Equation 11-8, Adam accumulates the squares of the gradients in **s** (with a greater weight for more recent weights). In step 5, if we ignore ε and steps 3 and 4 (which are technical details anyway), Adam scales down the parameter updates by the square root of **s**. In short, Adam scales down the parameter updates by the ℓ_2 norm of the time-decayed gradients (recall that the ℓ_2 norm is the square root of the sum of squares). AdaMax, introduced in the same paper as Adam, replaces the ℓ_2 norm with the ℓ_∞ norm (a fancy way of saying the max). Specifically, it replaces step 2 in Equation 11-8 with $\mathbf{s} \leftarrow \max(\beta_2\mathbf{s}, \nabla_\theta J(\boldsymbol{\theta}))$, it drops step 4, and in step 5 it scales down the gradient updates by a factor of **s**, which is just the max of the time-decayed gradients. In practice, this can make AdaMax more stable than Adam, but it really depends on the dataset, and in general Adam performs better. So, this is just one more optimizer you can try if you experience problems with Adam on some task.

Nadam

Nadam optimization is Adam optimization plus the Nesterov trick, so it will often converge slightly faster than Adam. In his report introducing this technique (*https://homl.info/nadam*),[19] the researcher Timothy Dozat compares many different optimizers on various tasks and finds that Nadam generally outperforms Adam but is sometimes outperformed by RMSProp.

 Adaptive optimization methods (including RMSProp, Adam, and Nadam optimization) are often great, converging fast to a good solution. However, a 2017 paper (*https://homl.info/60*)[20] by Ashia C. Wilson et al. showed that they can lead to solutions that generalize poorly on some datasets. So when you are disappointed by your model's performance, try using plain Nesterov Accelerated Gradient instead: your dataset may just be allergic to adaptive gradients. Also check out the latest research, because it's moving fast.

All the optimization techniques discussed so far only rely on the *first-order partial derivatives* (*Jacobians*). The optimization literature also contains amazing algorithms based on the *second-order partial derivatives* (the *Hessians*, which are the partial derivatives of the Jacobians). Unfortunately, these algorithms are very hard to apply to deep neural networks because there are n^2 Hessians per output (where n is the number of parameters), as opposed to just n Jacobians per output. Since DNNs typically have tens of thousands of parameters, the second-order optimization algorithms often don't even fit in memory, and even when they do, computing the Hessians is just too slow.

---

19 Timothy Dozat, "Incorporating Nesterov Momentum into Adam" (2016).

20 Ashia C. Wilson et al., "The Marginal Value of Adaptive Gradient Methods in Machine Learning," *Advances in Neural Information Processing Systems* 30 (2017): 4148–4158.

Training Sparse Models

All the optimization algorithms just presented produce dense models, meaning that most parameters will be nonzero. If you need a blazingly fast model at runtime, or if you need it to take up less memory, you may prefer to end up with a sparse model instead.

One easy way to achieve this is to train the model as usual, then get rid of the tiny weights (set them to zero). Note that this will typically not lead to a very sparse model, and it may degrade the model's performance.

A better option is to apply strong ℓ_1 regularization during training (we will see how later in this chapter), as it pushes the optimizer to zero out as many weights as it can (as discussed in "Lasso Regression" on page 137 in Chapter 4).

If these techniques remain insufficient, check out the TensorFlow Model Optimization Toolkit (TF-MOT) (*https://homl.info/tfmot*), which provides a pruning API capable of iteratively removing connections during training based on their magnitude.

Table 11-2 compares all the optimizers we've discussed so far (* is bad, ** is average, and *** is good).

Table 11-2. Optimizer comparison

Class	Convergence speed	Convergence quality
SGD	*	***
SGD(momentum=...)	**	***
SGD(momentum=..., nesterov=True)	**	***
Adagrad	***	* (stops too early)
RMSprop	***	** or ***
Adam	***	** or ***
Nadam	***	** or ***
AdaMax	***	** or ***

Learning Rate Scheduling

Finding a good learning rate is very important. If you set it much too high, training may diverge (as we discussed in "Gradient Descent" on page 118). If you set it too low, training will eventually converge to the optimum, but it will take a very long time. If you set it slightly too high, it will make progress very quickly at first, but it will end up dancing around the optimum, never really settling down. If you have a limited computing budget, you may have to interrupt training before it has converged properly, yielding a suboptimal solution (see Figure 11-8).

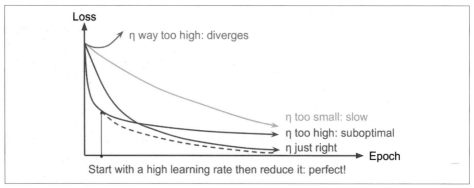

Figure 11-8. Learning curves for various learning rates η

As we discussed in Chapter 10, you can find a good learning rate by training the model for a few hundred iterations, exponentially increasing the learning rate from a very small value to a very large value, and then looking at the learning curve and picking a learning rate slightly lower than the one at which the learning curve starts shooting back up. You can then reinitialize your model and train it with that learning rate.

But you can do better than a constant learning rate: if you start with a large learning rate and then reduce it once training stops making fast progress, you can reach a good solution faster than with the optimal constant learning rate. There are many different strategies to reduce the learning rate during training. It can also be beneficial to start with a low learning rate, increase it, then drop it again. These strategies are called *learning schedules* (we briefly introduced this concept in Chapter 4). These are the most commonly used learning schedules:

Power scheduling

Set the learning rate to a function of the iteration number t: $\eta(t) = \eta_0 / (1 + t/s)^c$. The initial learning rate η_0, the power c (typically set to 1), and the steps s are hyperparameters. The learning rate drops at each step. After s steps, it is down to $\eta_0 / 2$. After s more steps, it is down to $\eta_0 / 3$, then it goes down to $\eta_0 / 4$, then $\eta_0 / 5$, and so on. As you can see, this schedule first drops quickly, then more and more slowly. Of course, power scheduling requires tuning η_0 and s (and possibly c).

Exponential scheduling

Set the learning rate to $\eta(t) = \eta_0 \, 0.1^{t/s}$. The learning rate will gradually drop by a factor of 10 every s steps. While power scheduling reduces the learning rate more and more slowly, exponential scheduling keeps slashing it by a factor of 10 every s steps.

Piecewise constant scheduling

Use a constant learning rate for a number of epochs (e.g., $\eta_0 = 0.1$ for 5 epochs), then a smaller learning rate for another number of epochs (e.g., $\eta_1 = 0.001$ for 50 epochs), and so on. Although this solution can work very well, it requires fiddling around to figure out the right sequence of learning rates and how long to use each of them.

Performance scheduling

Measure the validation error every N steps (just like for early stopping), and reduce the learning rate by a factor of λ when the error stops dropping.

1cycle scheduling

Contrary to the other approaches, *1cycle* (introduced in a 2018 paper (*https://homl.info/1cycle*)[21] by Leslie Smith) starts by increasing the initial learning rate η_0, growing linearly up to η_1 halfway through training. Then it decreases the learning rate linearly down to η_0 again during the second half of training, finishing the last few epochs by dropping the rate down by several orders of magnitude (still linearly). The maximum learning rate η_1 is chosen using the same approach we used to find the optimal learning rate, and the initial learning rate η_0 is chosen to be roughly 10 times lower. When using a momentum, we start with a high momentum first (e.g., 0.95), then drop it down to a lower momentum during the first half of training (e.g., down to 0.85, linearly), and then bring it back up to the maximum value (e.g., 0.95) during the second half of training, finishing the last few epochs with that maximum value. Smith did many experiments showing that this approach was often able to speed up training considerably and reach better performance. For example, on the popular CIFAR10 image dataset, this approach reached 91.9% validation accuracy in just 100 epochs, instead of 90.3% accuracy in 800 epochs through a standard approach (with the same neural network architecture).

A 2013 paper (*https://homl.info/63*)[22] by Andrew Senior et al. compared the performance of some of the most popular learning schedules when using momentum optimization to train deep neural networks for speech recognition. The authors concluded that, in this setting, both performance scheduling and exponential scheduling performed well. They favored exponential scheduling because it was easy to tune and it converged slightly faster to the optimal solution (they also mentioned that

---

21 Leslie N. Smith, "A Disciplined Approach to Neural Network Hyper-Parameters: Part 1—Learning Rate, Batch Size, Momentum, and Weight Decay," arXiv preprint arXiv:1803.09820 (2018).

22 Andrew Senior et al., "An Empirical Study of Learning Rates in Deep Neural Networks for Speech Recognition," *Proceedings of the IEEE International Conference on Acoustics, Speech, and Signal Processing* (2013): 6724–6728.

it was easier to implement than performance scheduling, but in Keras both options are easy). That said, the 1cycle approach seems to perform even better.

Implementing power scheduling in Keras is the easiest option: just set the `decay` hyperparameter when creating an optimizer:

```
optimizer = keras.optimizers.SGD(lr=0.01, decay=1e-4)
```

The `decay` is the inverse of s (the number of steps it takes to divide the learning rate by one more unit), and Keras assumes that c is equal to 1.

Exponential scheduling and piecewise scheduling are quite simple too. You first need to define a function that takes the current epoch and returns the learning rate. For example, let's implement exponential scheduling:

```
def exponential_decay_fn(epoch):
    return 0.01 * 0.1**(epoch / 20)
```

If you do not want to hardcode η_0 and s, you can create a function that returns a configured function:

```
def exponential_decay(lr0, s):
    def exponential_decay_fn(epoch):
        return lr0 * 0.1**(epoch / s)
    return exponential_decay_fn

exponential_decay_fn = exponential_decay(lr0=0.01, s=20)
```

Next, create a `LearningRateScheduler` callback, giving it the schedule function, and pass this callback to the `fit()` method:

```
lr_scheduler = keras.callbacks.LearningRateScheduler(exponential_decay_fn)
history = model.fit(X_train_scaled, y_train, [...], callbacks=[lr_scheduler])
```

The `LearningRateScheduler` will update the optimizer's `learning_rate` attribute at the beginning of each epoch. Updating the learning rate once per epoch is usually enough, but if you want it to be updated more often, for example at every step, you can always write your own callback (see the "Exponential Scheduling" section of the notebook for an example). Updating the learning rate at every step makes sense if there are many steps per epoch. Alternatively, you can use the `keras.optimiz ers.schedules` approach, described shortly.

The schedule function can optionally take the current learning rate as a second argument. For example, the following schedule function multiplies the previous learning rate by $0.1^{1/20}$, which results in the same exponential decay (except the decay now starts at the beginning of epoch 0 instead of 1):

```
def exponential_decay_fn(epoch, lr):
    return lr * 0.1**(1 / 20)
```

This implementation relies on the optimizer's initial learning rate (contrary to the previous implementation), so make sure to set it appropriately.

When you save a model, the optimizer and its learning rate get saved along with it. This means that with this new schedule function, you could just load a trained model and continue training where it left off, no problem. Things are not so simple if your schedule function uses the epoch argument, however: the epoch does not get saved, and it gets reset to 0 every time you call the fit() method. If you were to continue training a model where it left off, this could lead to a very large learning rate, which would likely damage your model's weights. One solution is to manually set the fit() method's initial_epoch argument so the epoch starts at the right value.

For piecewise constant scheduling, you can use a schedule function like the following one (as earlier, you can define a more general function if you want; see the "Piecewise Constant Scheduling" section of the notebook for an example), then create a Lear ningRateScheduler callback with this function and pass it to the fit() method, just like we did for exponential scheduling:

```
def piecewise_constant_fn(epoch):
    if epoch < 5:
        return 0.01
    elif epoch < 15:
        return 0.005
    else:
        return 0.001
```

For performance scheduling, use the ReduceLROnPlateau callback. For example, if you pass the following callback to the fit() method, it will multiply the learning rate by 0.5 whenever the best validation loss does not improve for five consecutive epochs (other options are available; please check the documentation for more details):

```
lr_scheduler = keras.callbacks.ReduceLROnPlateau(factor=0.5, patience=5)
```

Lastly, tf.keras offers an alternative way to implement learning rate scheduling: define the learning rate using one of the schedules available in keras.optimizers.sched ules, then pass this learning rate to any optimizer. This approach updates the learning rate at each step rather than at each epoch. For example, here is how to implement the same exponential schedule as the exponential_decay_fn() function we defined earlier:

```
s = 20 * len(X_train) // 32 # number of steps in 20 epochs (batch size = 32)
learning_rate = keras.optimizers.schedules.ExponentialDecay(0.01, s, 0.1)
optimizer = keras.optimizers.SGD(learning_rate)
```

This is nice and simple, plus when you save the model, the learning rate and its schedule (including its state) get saved as well. This approach, however, is not part of the Keras API; it is specific to tf.keras.

As for the 1cycle approach, the implementation poses no particular difficulty: just create a custom callback that modifies the learning rate at each iteration (you can update the optimizer's learning rate by changing `self.model.optimizer.lr`). See the "1Cycle scheduling" section of the notebook for an example.

To sum up, exponential decay, performance scheduling, and 1cycle can considerably speed up convergence, so give them a try!

Avoiding Overfitting Through Regularization

> With four parameters I can fit an elephant and with five I can make him wiggle his trunk.
>
> —John von Neumann, cited by Enrico Fermi in *Nature* 427

With thousands of parameters, you can fit the whole zoo. Deep neural networks typically have tens of thousands of parameters, sometimes even millions. This gives them an incredible amount of freedom and means they can fit a huge variety of complex datasets. But this great flexibility also makes the network prone to overfitting the training set. We need regularization.

We already implemented one of the best regularization techniques in Chapter 10: early stopping. Moreover, even though Batch Normalization was designed to solve the unstable gradients problems, it also acts like a pretty good regularizer. In this section we will examine other popular regularization techniques for neural networks: ℓ_1 and ℓ_2 regularization, dropout, and max-norm regularization.

ℓ_1 and ℓ_2 Regularization

Just like you did in Chapter 4 for simple linear models, you can use ℓ_2 regularization to constrain a neural network's connection weights, and/or ℓ_1 regularization if you want a sparse model (with many weights equal to 0). Here is how to apply ℓ_2 regularization to a Keras layer's connection weights, using a regularization factor of 0.01:

```
layer = keras.layers.Dense(100, activation="elu",
                           kernel_initializer="he_normal",
                           kernel_regularizer=keras.regularizers.l2(0.01))
```

The l2() function returns a regularizer that will be called at each step during training to compute the regularization loss. This is then added to the final loss. As you might expect, you can just use `keras.regularizers.l1()` if you want ℓ_1 regularization; if you want both ℓ_1 and ℓ_2 regularization, use `keras.regularizers.l1_l2()` (specifying both regularization factors).

Since you will typically want to apply the same regularizer to all layers in your network, as well as using the same activation function and the same initialization strategy in all hidden layers, you may find yourself repeating the same arguments. This makes the code ugly and error-prone. To avoid this, you can try refactoring your code to use loops. Another option is to use Python's `functools.partial()` function, which lets you create a thin wrapper for any callable, with some default argument values:

```
from functools import partial

RegularizedDense = partial(keras.layers.Dense,
                           activation="elu",
                           kernel_initializer="he_normal",
                           kernel_regularizer=keras.regularizers.l2(0.01))

model = keras.models.Sequential([
    keras.layers.Flatten(input_shape=[28, 28]),
    RegularizedDense(300),
    RegularizedDense(100),
    RegularizedDense(10, activation="softmax",
                     kernel_initializer="glorot_uniform")
])
```

Dropout

Dropout is one of the most popular regularization techniques for deep neural networks. It was proposed in a paper (*https://homl.info/64*)[23] by Geoffrey Hinton in 2012 and further detailed in a 2014 paper (*https://homl.info/65*)[24] by Nitish Srivastava et al., and it has proven to be highly successful: even the state-of-the-art neural networks get a 1–2% accuracy boost simply by adding dropout. This may not sound like a lot, but when a model already has 95% accuracy, getting a 2% accuracy boost means dropping the error rate by almost 40% (going from 5% error to roughly 3%).

It is a fairly simple algorithm: at every training step, every neuron (including the input neurons, but always excluding the output neurons) has a probability p of being temporarily "dropped out," meaning it will be entirely ignored during this training step, but it may be active during the next step (see Figure 11-9). The hyperparameter p is called the *dropout rate*, and it is typically set between 10% and 50%: closer to 20–30% in recurrent neural nets (see Chapter 15), and closer to 40–50% in convolutional neural networks (see Chapter 14). After training, neurons don't get dropped anymore. And that's all (except for a technical detail we will discuss momentarily).

---

23 Geoffrey E. Hinton et al., "Improving Neural Networks by Preventing Co-Adaptation of Feature Detectors," arXiv preprint arXiv:1207.0580 (2012).

24 Nitish Srivastava et al., "Dropout: A Simple Way to Prevent Neural Networks from Overfitting," *Journal of Machine Learning Research* 15 (2014): 1929–1958.

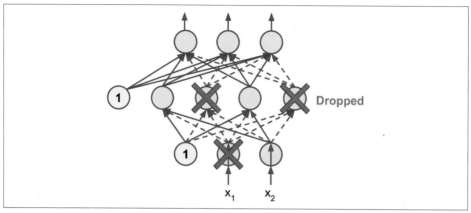

Figure 11-9. With dropout regularization, at each training iteration a random subset of all neurons in one or more layers—except the output layer—are "dropped out"; these neurons output 0 at this iteration (represented by the dashed arrows)

It's surprising at first that this destructive technique works at all. Would a company perform better if its employees were told to toss a coin every morning to decide whether or not to go to work? Well, who knows; perhaps it would! The company would be forced to adapt its organization; it could not rely on any single person to work the coffee machine or perform any other critical tasks, so this expertise would have to be spread across several people. Employees would have to learn to cooperate with many of their coworkers, not just a handful of them. The company would become much more resilient. If one person quit, it wouldn't make much of a difference. It's unclear whether this idea would actually work for companies, but it certainly does for neural networks. Neurons trained with dropout cannot co-adapt with their neighboring neurons; they have to be as useful as possible on their own. They also cannot rely excessively on just a few input neurons; they must pay attention to each of their input neurons. They end up being less sensitive to slight changes in the inputs. In the end, you get a more robust network that generalizes better.

Another way to understand the power of dropout is to realize that a unique neural network is generated at each training step. Since each neuron can be either present or absent, there are a total of 2^N possible networks (where N is the total number of droppable neurons). This is such a huge number that it is virtually impossible for the same neural network to be sampled twice. Once you have run 10,000 training steps, you have essentially trained 10,000 different neural networks (each with just one training instance). These neural networks are obviously not independent because they share many of their weights, but they are nevertheless all different. The resulting neural network can be seen as an averaging ensemble of all these smaller neural networks.

In practice, you can usually apply dropout only to the neurons in the top one to three layers (excluding the output layer).

There is one small but important technical detail. Suppose p = 50%, in which case during testing a neuron would be connected to twice as many input neurons as it would be (on average) during training. To compensate for this fact, we need to multiply each neuron's input connection weights by 0.5 after training. If we don't, each neuron will get a total input signal roughly twice as large as what the network was trained on and will be unlikely to perform well. More generally, we need to multiply each input connection weight by the *keep probability* $(1 - p)$ after training. Alternatively, we can divide each neuron's output by the keep probability during training (these alternatives are not perfectly equivalent, but they work equally well).

To implement dropout using Keras, you can use the `keras.layers.Dropout` layer. During training, it randomly drops some inputs (setting them to 0) and divides the remaining inputs by the keep probability. After training, it does nothing at all; it just passes the inputs to the next layer. The following code applies dropout regularization before every `Dense` layer, using a dropout rate of 0.2:

```python
model = keras.models.Sequential([
    keras.layers.Flatten(input_shape=[28, 28]),
    keras.layers.Dropout(rate=0.2),
    keras.layers.Dense(300, activation="elu", kernel_initializer="he_normal"),
    keras.layers.Dropout(rate=0.2),
    keras.layers.Dense(100, activation="elu", kernel_initializer="he_normal"),
    keras.layers.Dropout(rate=0.2),
    keras.layers.Dense(10, activation="softmax")
])
```

Since dropout is only active during training, comparing the training loss and the validation loss can be misleading. In particular, a model may be overfitting the training set and yet have similar training and validation losses. So make sure to evaluate the training loss without dropout (e.g., after training).

If you observe that the model is overfitting, you can increase the dropout rate. Conversely, you should try decreasing the dropout rate if the model underfits the training set. It can also help to increase the dropout rate for large layers, and reduce it for small ones. Moreover, many state-of-the-art architectures only use dropout after the last hidden layer, so you may want to try this if full dropout is too strong.

Dropout does tend to significantly slow down convergence, but it usually results in a much better model when tuned properly. So, it is generally well worth the extra time and effort.

> If you want to regularize a self-normalizing network based on the SELU activation function (as discussed earlier), you should use *alpha dropout*: this is a variant of dropout that preserves the mean and standard deviation of its inputs (it was introduced in the same paper as SELU, as regular dropout would break self-normalization).

Monte Carlo (MC) Dropout

In 2016, a paper (*https://homl.info/mcdropout*)[25] by Yarin Gal and Zoubin Ghahramani added a few more good reasons to use dropout:

- First, the paper established a profound connection between dropout networks (i.e., neural networks containing a `Dropout` layer before every weight layer) and approximate Bayesian inference,[26] giving dropout a solid mathematical justification.

- Second, the authors introduced a powerful technique called *MC Dropout*, which can boost the performance of any trained dropout model without having to retrain it or even modify it at all, provides a much better measure of the model's uncertainty, and is also amazingly simple to implement.

If this all sounds like a "one weird trick" advertisement, then take a look at the following code. It is the full implementation of *MC Dropout*, boosting the dropout model we trained earlier without retraining it:

```
y_probas = np.stack([model(X_test_scaled, training=True)
                     for sample in range(100)])
y_proba = y_probas.mean(axis=0)
```

We just make 100 predictions over the test set, setting `training=True` to ensure that the `Dropout` layer is active, and stack the predictions. Since dropout is active, all the predictions will be different. Recall that `predict()` returns a matrix with one row per instance and one column per class. Because there are 10,000 instances in the test set and 10 classes, this is a matrix of shape [10000, 10]. We stack 100 such matrices, so `y_probas` is an array of shape [100, 10000, 10]. Once we average over the first

---

25 Yarin Gal and Zoubin Ghahramani, "Dropout as a Bayesian Approximation: Representing Model Uncertainty in Deep Learning," *Proceedings of the 33rd International Conference on Machine Learning* (2016): 1050–1059.

26 Specifically, they show that training a dropout network is mathematically equivalent to approximate Bayesian inference in a specific type of probabilistic model called a *Deep Gaussian Process*.

dimension (`axis=0`), we get `y_proba`, an array of shape [10000, 10], like we would get with a single prediction. That's all! Averaging over multiple predictions with dropout on gives us a Monte Carlo estimate that is generally more reliable than the result of a single prediction with dropout off. For example, let's look at the model's prediction for the first instance in the test set, with dropout off:

```
>>> np.round(model.predict(X_test_scaled[:1]), 2)
array([[0.  , 0.  , 0.  , 0.  , 0.  , 0.  , 0.  , 0.01, 0.  , 0.99]],
      dtype=float32)
```

The model seems almost certain that this image belongs to class 9 (ankle boot). Should you trust it? Is there really so little room for doubt? Compare this with the predictions made when dropout is activated:

```
>>> np.round(y_probas[:, :1], 2)
array([[[0.  , 0.  , 0.  , 0.  , 0.  , 0.14, 0.  , 0.17, 0.  , 0.68]],
       [[0.  , 0.  , 0.  , 0.  , 0.  , 0.16, 0.  , 0.2 , 0.  , 0.64]],
       [[0.  , 0.  , 0.  , 0.  , 0.  , 0.02, 0.  , 0.01, 0.  , 0.97]],
       [...]
```

This tells a very different story: apparently, when we activate dropout, the model is not sure anymore. It still seems to prefer class 9, but sometimes it hesitates with classes 5 (sandal) and 7 (sneaker), which makes sense given they're all footwear. Once we average over the first dimension, we get the following MC Dropout predictions:

```
>>> np.round(y_proba[:1], 2)
array([[0.  , 0.  , 0.  , 0.  , 0.  , 0.22, 0.  , 0.16, 0.  , 0.62]],
      dtype=float32)
```

The model still thinks this image belongs to class 9, but only with a 62% confidence, which seems much more reasonable than 99%. Plus it's useful to know exactly which other classes it thinks are likely. And you can also take a look at the standard deviation of the probability estimates (*https://xkcd.com/2110*):

```
>>> y_std = y_probas.std(axis=0)
>>> np.round(y_std[:1], 2)
array([[0.  , 0.  , 0.  , 0.  , 0.  , 0.28, 0.  , 0.21, 0.02, 0.32]],
      dtype=float32)
```

Apparently there's quite a lot of variance in the probability estimates: if you were building a risk-sensitive system (e.g., a medical or financial system), you should probably treat such an uncertain prediction with extreme caution. You definitely would not treat it like a 99% confident prediction. Moreover, the model's accuracy got a small boost from 86.8 to 86.9:

```
>>> accuracy = np.sum(y_pred == y_test) / len(y_test)
>>> accuracy
0.8694
```

The number of Monte Carlo samples you use (100 in this example) is a hyperparameter you can tweak. The higher it is, the more accurate the predictions and their uncertainty estimates will be. However, if you double it, inference time will also be doubled. Moreover, above a certain number of samples, you will notice little improvement. So your job is to find the right trade-off between latency and accuracy, depending on your application.

If your model contains other layers that behave in a special way during training (such as `BatchNormalization` layers), then you should not force training mode like we just did. Instead, you should replace the `Dropout` layers with the following `MCDropout` class:[27]

```
class MCDropout(keras.layers.Dropout):
    def call(self, inputs):
        return super().call(inputs, training=True)
```

Here, we just subclass the `Dropout` layer and override the `call()` method to force its `training` argument to `True` (see Chapter 12). Similarly, you could define an `MCAlpha Dropout` class by subclassing `AlphaDropout` instead. If you are creating a model from scratch, it's just a matter of using `MCDropout` rather than `Dropout`. But if you have a model that was already trained using `Dropout`, you need to create a new model that's identical to the existing model except that it replaces the `Dropout` layers with `MCDrop out`, then copy the existing model's weights to your new model.

In short, MC Dropout is a fantastic technique that boosts dropout models and provides better uncertainty estimates. And of course, since it is just regular dropout during training, it also acts like a regularizer.

Max-Norm Regularization

Another regularization technique that is popular for neural networks is called *max-norm regularization*: for each neuron, it constrains the weights $\mathbf{w}$ of the incoming connections such that $\| \mathbf{w} \|_2 \le r$, where r is the max-norm hyperparameter and $\| \cdot \|_2$ is the ℓ_2 norm.

Max-norm regularization does not add a regularization loss term to the overall loss function. Instead, it is typically implemented by computing $\|\mathbf{w}\|_2$ after each training step and rescaling $\mathbf{w}$ if needed ($\mathbf{w} \leftarrow \mathbf{w} \, r/\| \mathbf{w} \|_2$).

---

27 This `MCDropout` class will work with all Keras APIs, including the Sequential API. If you only care about the Functional API or the Subclassing API, you do not have to create an `MCDropout` class; you can create a regular `Dropout` layer and call it with `training=True`.

Reducing *r* increases the amount of regularization and helps reduce overfitting. Max-norm regularization can also help alleviate the unstable gradients problems (if you are not using Batch Normalization).

To implement max-norm regularization in Keras, set the `kernel_constraint` argument of each hidden layer to a `max_norm()` constraint with the appropriate max value, like this:

```
keras.layers.Dense(100, activation="elu", kernel_initializer="he_normal",
                   kernel_constraint=keras.constraints.max_norm(1.))
```

After each training iteration, the model's `fit()` method will call the object returned by `max_norm()`, passing it the layer's weights and getting rescaled weights in return, which then replace the layer's weights. As you'll see in Chapter 12, you can define your own custom constraint function if necessary and use it as the `kernel_con straint`. You can also constrain the bias terms by setting the `bias_constraint` argument.

The `max_norm()` function has an `axis` argument that defaults to `0`. A `Dense` layer usually has weights of shape [*number of inputs, number of neurons*], so using `axis=0` means that the max-norm constraint will apply independently to each neuron's weight vector. If you want to use max-norm with convolutional layers (see Chapter 14), make sure to set the `max_norm()` constraint's `axis` argument appropriately (usually `axis=[0, 1, 2]`).

Summary and Practical Guidelines

In this chapter we have covered a wide range of techniques, and you may be wondering which ones you should use. This depends on the task, and there is no clear consensus yet, but I have found the configuration in Table 11-3 to work fine in most cases, without requiring much hyperparameter tuning. That said, please do not consider these defaults as hard rules!

Table 11-3. Default DNN configuration

Hyperparameter	Default value
Kernel initializer	He initialization
Activation function	ELU
Normalization	None if shallow; Batch Norm if deep
Regularization	Early stopping ($+\ell_2$ reg. if needed)
Optimizer	Momentum optimization (or RMSProp or Nadam)
Learning rate schedule	1cycle

If the network is a simple stack of dense layers, then it can self-normalize, and you should use the configuration in Table 11-4 instead.

Table 11-4. DNN configuration for a self-normalizing net

Hyperparameter	Default value
Kernel initializer	LeCun initialization
Activation function	SELU
Normalization	None (self-normalization)
Regularization	Alpha dropout if needed
Optimizer	Momentum optimization (or RMSProp or Nadam)
Learning rate schedule	1cycle

Don't forget to normalize the input features! You should also try to reuse parts of a pretrained neural network if you can find one that solves a similar problem, or use unsupervised pretraining if you have a lot of unlabeled data, or use pretraining on an auxiliary task if you have a lot of labeled data for a similar task.

While the previous guidelines should cover most cases, here are some exceptions:

- If you need a sparse model, you can use ℓ_1 regularization (and optionally zero out the tiny weights after training). If you need an even sparser model, you can use the TensorFlow Model Optimization Toolkit. This will break self-normalization, so you should use the default configuration in this case.

- If you need a low-latency model (one that performs lightning-fast predictions), you may need to use fewer layers, fold the Batch Normalization layers into the previous layers, and possibly use a faster activation function such as leaky ReLU or just ReLU. Having a sparse model will also help. Finally, you may want to reduce the float precision from 32 bits to 16 or even 8 bits (see "Deploying a Model to a Mobile or Embedded Device" on page 685). Again, check out TF-MOT.

- If you are building a risk-sensitive application, or inference latency is not very important in your application, you can use MC Dropout to boost performance and get more reliable probability estimates, along with uncertainty estimates.

With these guidelines, you are now ready to train very deep nets! I hope you are now convinced that you can go quite a long way using just Keras. There may come a time, however, when you need to have even more control; for example, to write a custom loss function or to tweak the training algorithm. For such cases you will need to use TensorFlow's lower-level API, as you will see in the next chapter.

Exercises

1. Is it OK to initialize all the weights to the same value as long as that value is selected randomly using He initialization?

2. Is it OK to initialize the bias terms to 0?

3. Name three advantages of the SELU activation function over ReLU.

4. In which cases would you want to use each of the following activation functions: SELU, leaky ReLU (and its variants), ReLU, tanh, logistic, and softmax?

5. What may happen if you set the `momentum` hyperparameter too close to 1 (e.g., 0.99999) when using an `SGD` optimizer?

6. Name three ways you can produce a sparse model.

7. Does dropout slow down training? Does it slow down inference (i.e., making predictions on new instances)? What about MC Dropout?

8. Practice training a deep neural network on the CIFAR10 image dataset:

   a. Build a DNN with 20 hidden layers of 100 neurons each (that's too many, but it's the point of this exercise). Use He initialization and the ELU activation function.

   b. Using Nadam optimization and early stopping, train the network on the CIFAR10 dataset. You can load it with `keras.datasets.cifar10.load_data()`. The dataset is composed of 60,000 32 × 32–pixel color images (50,000 for training, 10,000 for testing) with 10 classes, so you'll need a softmax output layer with 10 neurons. Remember to search for the right learning rate each time you change the model's architecture or hyperparameters.

   c. Now try adding Batch Normalization and compare the learning curves: Is it converging faster than before? Does it produce a better model? How does it affect training speed?

   d. Try replacing Batch Normalization with SELU, and make the necessary adjustements to ensure the network self-normalizes (i.e., standardize the input features, use LeCun normal initialization, make sure the DNN contains only a sequence of dense layers, etc.).

   e. Try regularizing the model with alpha dropout. Then, without retraining your model, see if you can achieve better accuracy using MC Dropout.

   f. Retrain your model using 1cycle scheduling and see if it improves training speed and model accuracy.

Solutions to these exercises are available in Appendix A.

## CHAPTER 12

Custom Models and Training with TensorFlow

Up until now, we've used only TensorFlow's high-level API, tf.keras, but it already got us pretty far: we built various neural network architectures, including regression and classification nets, Wide & Deep nets, and self-normalizing nets, using all sorts of techniques, such as Batch Normalization, dropout, and learning rate schedules. In fact, 95% of the use cases you will encounter will not require anything other than tf.keras (and tf.data; see Chapter 13). But now it's time to dive deeper into TensorFlow and take a look at its lower-level Python API (*https://homl.info/tf2api*). This will be useful when you need extra control to write custom loss functions, custom metrics, layers, models, initializers, regularizers, weight constraints, and more. You may even need to fully control the training loop itself, for example to apply special transformations or constraints to the gradients (beyond just clipping them) or to use multiple optimizers for different parts of the network. We will cover all these cases in this chapter, and we will also look at how you can boost your custom models and training algorithms using TensorFlow's automatic graph generation feature. But first, let's take a quick tour of TensorFlow.

TensorFlow 2.0 (beta) was released in June 2019, making Tensor-Flow much easier to use. The first edition of this book used TF 1, while this edition uses TF 2.

A Quick Tour of TensorFlow

As you know, TensorFlow is a powerful library for numerical computation, particularly well suited and fine-tuned for large-scale Machine Learning (but you could use it for anything else that requires heavy computations). It was developed by the Google Brain team and it powers many of Google's large-scale services, such as Google Cloud Speech, Google Photos, and Google Search. It was open sourced in November 2015, and it is now the most popular Deep Learning library (in terms of citations in papers, adoption in companies, stars on GitHub, etc.). Countless projects use TensorFlow for all sorts of Machine Learning tasks, such as image classification, natural language processing, recommender systems, and time series forecasting.

So what does TensorFlow offer? Here's a summary:

- Its core is very similar to NumPy, but with GPU support.
- It supports distributed computing (across multiple devices and servers).
- It includes a kind of just-in-time (JIT) compiler that allows it to optimize computations for speed and memory usage. It works by extracting the *computation graph* from a Python function, then optimizing it (e.g., by pruning unused nodes), and finally running it efficiently (e.g., by automatically running independent operations in parallel).
- Computation graphs can be exported to a portable format, so you can train a TensorFlow model in one environment (e.g., using Python on Linux) and run it in another (e.g., using Java on an Android device).
- It implements autodiff (see Chapter 10 and Appendix D) and provides some excellent optimizers, such as RMSProp and Nadam (see Chapter 11), so you can easily minimize all sorts of loss functions.

TensorFlow offers many more features built on top of these core features: the most important is of course `tf.keras`,[1] but it also has data loading and preprocessing ops (`tf.data`, `tf.io`, etc.), image processing ops (`tf.image`), signal processing ops (`tf.signal`), and more (see Figure 12-1 for an overview of TensorFlow's Python API).

---

[1] TensorFlow includes another Deep Learning API called the *Estimators API*, but the TensorFlow team recommends using `tf.keras` instead.

We will cover many of the packages and functions of the Tensor-Flow API, but it's impossible to cover them all, so you should really take some time to browse through the API; you will find that it is quite rich and well documented.

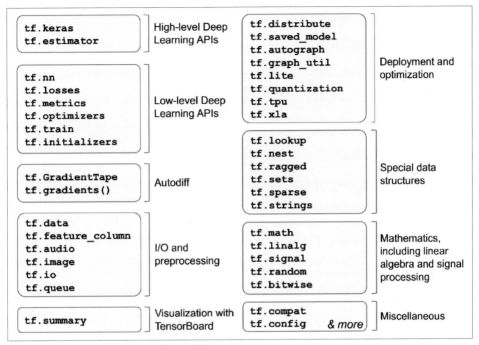

Figure 12-1. TensorFlow's Python API

At the lowest level, each TensorFlow operation (*op* for short) is implemented using highly efficient C++ code.[2] Many operations have multiple implementations called *kernels*: each kernel is dedicated to a specific device type, such as CPUs, GPUs, or even TPUs (*tensor processing units*). As you may know, GPUs can dramatically speed up computations by splitting them into many smaller chunks and running them in parallel across many GPU threads. TPUs are even faster: they are custom ASIC chips built specifically for Deep Learning operations[3] (we will discuss how to use Tensor-Flow with GPUs or TPUs in Chapter 19).

TensorFlow's architecture is shown in Figure 12-2. Most of the time your code will use the high-level APIs (especially tf.keras and tf.data); but when you need more flexibility, you will use the lower-level Python API, handling tensors directly. Note that

---

2 If you ever need to (but you probably won't), you can write your own operations using the C++ API.

3 To learn more about TPUs and how they work, check out *https://homl.info/tpus*.

APIs for other languages are also available. In any case, TensorFlow's execution engine will take care of running the operations efficiently, even across multiple devices and machines if you tell it to.

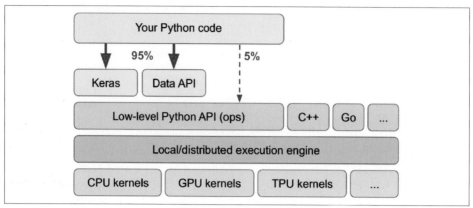

Figure 12-2. TensorFlow's architecture

TensorFlow runs not only on Windows, Linux, and macOS, but also on mobile devices (using *TensorFlow Lite*), including both iOS and Android (see Chapter 19). If you do not want to use the Python API, there are C++, Java, Go, and Swift APIs. There is even a JavaScript implementation called *TensorFlow.js* that makes it possible to run your models directly in your browser.

There's more to TensorFlow than the library. TensorFlow is at the center of an extensive ecosystem of libraries. First, there's TensorBoard for visualization (see Chapter 10). Next, there's TensorFlow Extended (TFX) (*https://tensorflow.org/tfx*), which is a set of libraries built by Google to productionize TensorFlow projects: it includes tools for data validation, preprocessing, model analysis, and serving (with TF Serving; see Chapter 19). Google's *TensorFlow Hub* provides a way to easily download and reuse pretrained neural networks. You can also get many neural network architectures, some of them pretrained, in TensorFlow's model garden (*https://github.com/ tensorflow/models/*). Check out the TensorFlow Resources (*https://www.tensor flow.org/resources*) and *https://github.com/jtoy/awesome-tensorflow* for more TensorFlow-based projects. You will find hundreds of TensorFlow projects on GitHub, so it is often easy to find existing code for whatever you are trying to do.

> More and more ML papers are released along with their implementations, and sometimes even with pretrained models. Check out *https://paperswithcode.com/* to easily find them.

Last but not least, TensorFlow has a dedicated team of passionate and helpful developers, as well as a large community contributing to improving it. To ask technical questions, you should use *http://stackoverflow.com/* and tag your question with *tensorflow* and *python*. You can file bugs and feature requests through GitHub (*https://github.com/tensorflow/tensorflow*). For general discussions, join the Google group (*https://homl.info/41*).

OK, it's time to start coding!

Using TensorFlow like NumPy

TensorFlow's API revolves around *tensors*, which flow from operation to operation—hence the name Tensor*Flow*. A tensor is usually a multidimensional array (exactly like a NumPy ndarray), but it can also hold a scalar (a simple value, such as 42). These tensors will be important when we create custom cost functions, custom metrics, custom layers, and more, so let's see how to create and manipulate them.

Tensors and Operations

You can create a tensor with tf.constant(). For example, here is a tensor representing a matrix with two rows and three columns of floats:

```
>>> tf.constant([[1., 2., 3.], [4., 5., 6.]]) # matrix
<tf.Tensor: id=0, shape=(2, 3), dtype=float32, numpy=
array([[1., 2., 3.],
       [4., 5., 6.]], dtype=float32)>
>>> tf.constant(42) # scalar
<tf.Tensor: id=1, shape=(), dtype=int32, numpy=42>
```

Just like an ndarray, a tf.Tensor has a shape and a data type (dtype):

```
>>> t = tf.constant([[1., 2., 3.], [4., 5., 6.]])
>>> t.shape
TensorShape([2, 3])
>>> t.dtype
tf.float32
```

Indexing works much like in NumPy:

```
>>> t[:, 1:]
<tf.Tensor: id=5, shape=(2, 2), dtype=float32, numpy=
array([[2., 3.],
       [5., 6.]], dtype=float32)>
>>> t[..., 1, tf.newaxis]
<tf.Tensor: id=15, shape=(2, 1), dtype=float32, numpy=
array([[2.],
       [5.]], dtype=float32)>
```

Most importantly, all sorts of tensor operations are available:

```
>>> t + 10
<tf.Tensor: id=18, shape=(2, 3), dtype=float32, numpy=
array([[11., 12., 13.],
       [14., 15., 16.]], dtype=float32)>
>>> tf.square(t)
<tf.Tensor: id=20, shape=(2, 3), dtype=float32, numpy=
array([[ 1.,  4.,  9.],
       [16., 25., 36.]], dtype=float32)>
>>> t @ tf.transpose(t)
<tf.Tensor: id=24, shape=(2, 2), dtype=float32, numpy=
array([[14., 32.],
       [32., 77.]], dtype=float32)>
```

Note that writing t + 10 is equivalent to calling tf.add(t, 10) (indeed, Python calls the magic method t.__add__(10), which just calls tf.add(t, 10)). Other operators like - and * are also supported. The @ operator was added in Python 3.5, for matrix multiplication: it is equivalent to calling the tf.matmul() function.

You will find all the basic math operations you need (tf.add(), tf.multiply(), tf.square(), tf.exp(), tf.sqrt(), etc.) and most operations that you can find in NumPy (e.g., tf.reshape(), tf.squeeze(), tf.tile()). Some functions have a different name than in NumPy; for instance, tf.reduce_mean(), tf.reduce_sum(), tf.reduce_max(), and tf.math.log() are the equivalent of np.mean(), np.sum(), np.max() and np.log(). When the name differs, there is often a good reason for it. For example, in TensorFlow you must write tf.transpose(t); you cannot just write t.T like in NumPy. The reason is that the tf.transpose() function does not do exactly the same thing as NumPy's T attribute: in TensorFlow, a new tensor is created with its own copy of the transposed data, while in NumPy, t.T is just a transposed view on the same data. Similarly, the tf.reduce_sum() operation is named this way because its GPU kernel (i.e., GPU implementation) uses a reduce algorithm that does not guarantee the order in which the elements are added: because 32-bit floats have limited precision, the result may change ever so slightly every time you call this operation. The same is true of tf.reduce_mean() (but of course tf.reduce_max() is deterministic).

Many functions and classes have aliases. For example, tf.add() and tf.math.add() are the same function. This allows TensorFlow to have concise names for the most common operations[4] while preserving well-organized packages.

---

4 A notable exception is tf.math.log(), which is commonly used but doesn't have a tf.log() alias (as it might be confused with logging).

Tensors and NumPy

Tensors play nice with NumPy: you can create a tensor from a NumPy array, and vice versa. You can even apply TensorFlow operations to NumPy arrays and NumPy operations to tensors:

```
>>> a = np.array([2., 4., 5.])
>>> tf.constant(a)
<tf.Tensor: id=111, shape=(3,), dtype=float64, numpy=array([2., 4., 5.])>
>>> t.numpy() # or np.array(t)
array([[1., 2., 3.],
       [4., 5., 6.]], dtype=float32)
>>> tf.square(a)
<tf.Tensor: id=116, shape=(3,), dtype=float64, numpy=array([4., 16., 25.])>
>>> np.square(t)
array([[ 1.,  4.,  9.],
       [16., 25., 36.]], dtype=float32)
```

Notice that NumPy uses 64-bit precision by default, while TensorFlow uses 32-bit. This is because 32-bit precision is generally more than enough for neural networks, plus it runs faster and uses less RAM. So when you create a tensor from a NumPy array, make sure to set `dtype=tf.float32`.

Type Conversions

Type conversions can significantly hurt performance, and they can easily go unnoticed when they are done automatically. To avoid this, TensorFlow does not perform any type conversions automatically: it just raises an exception if you try to execute an operation on tensors with incompatible types. For example, you cannot add a float tensor and an integer tensor, and you cannot even add a 32-bit float and a 64-bit float:

```
>>> tf.constant(2.) + tf.constant(40)
Traceback[...]InvalidArgumentError[...]expected to be a float[...]
>>> tf.constant(2.) + tf.constant(40., dtype=tf.float64)
Traceback[...]InvalidArgumentError[...]expected to be a double[...]
```

This may be a bit annoying at first, but remember that it's for a good cause! And of course you can use tf.cast() when you really need to convert types:

```
>>> t2 = tf.constant(40., dtype=tf.float64)
>>> tf.constant(2.0) + tf.cast(t2, tf.float32)
<tf.Tensor: id=136, shape=(), dtype=float32, numpy=42.0>
```

Variables

The tf.Tensor values we've seen so far are immutable: you cannot modify them. This means that we cannot use regular tensors to implement weights in a neural network, since they need to be tweaked by backpropagation. Plus, other parameters may also need to change over time (e.g., a momentum optimizer keeps track of past gradients). What we need is a tf.Variable:

```
>>> v = tf.Variable([[1., 2., 3.], [4., 5., 6.]])
>>> v
<tf.Variable 'Variable:0' shape=(2, 3) dtype=float32, numpy=
array([[1., 2., 3.],
       [4., 5., 6.]], dtype=float32)>
```

A tf.Variable acts much like a tf.Tensor: you can perform the same operations with it, it plays nicely with NumPy as well, and it is just as picky with types. But it can also be modified in place using the assign() method (or assign_add() or assign_sub(), which increment or decrement the variable by the given value). You can also modify individual cells (or slices), by using the cell's (or slice's) assign() method (direct item assignment will not work) or by using the scatter_update() or scatter_nd_update() methods:

```
v.assign(2 * v)           # => [[2., 4., 6.], [8., 10., 12.]]
v[0, 1].assign(42)        # => [[2., 42., 6.], [8., 10., 12.]]
v[:, 2].assign([0., 1.])  # => [[2., 42., 0.], [8., 10., 1.]]
v.scatter_nd_update(indices=[[0, 0], [1, 2]], updates=[100., 200.])
                          # => [[100., 42., 0.], [8., 10., 200.]]
```

In practice you will rarely have to create variables manually, since Keras provides an add_weight() method that will take care of it for you, as we will see. Moreover, model parameters will generally be updated directly by the optimizers, so you will rarely need to update variables manually.

Other Data Structures

TensorFlow supports several other data structures, including the following (please see the "Tensors and Operations" section in the notebook or Appendix F for more details):

Sparse tensors (tf.SparseTensor)
Efficiently represent tensors containing mostly zeros. The tf.sparse package contains operations for sparse tensors.

Tensor arrays (tf.TensorArray)
Are lists of tensors. They have a fixed size by default but can optionally be made dynamic. All tensors they contain must have the same shape and data type.

Ragged tensors (tf.RaggedTensor)
Represent static lists of lists of tensors, where every tensor has the same shape and data type. The tf.ragged package contains operations for ragged tensors.

String tensors
Are regular tensors of type tf.string. These represent byte strings, not Unicode strings, so if you create a string tensor using a Unicode string (e.g., a regular Python 3 string like "café"), then it will get encoded to UTF-8 automatically (e.g., b"caf\xc3\xa9"). Alternatively, you can represent Unicode strings using tensors of type tf.int32, where each item represents a Unicode code point (e.g., [99, 97, 102, 233]). The tf.strings package (with an s) contains ops for byte strings and Unicode strings (and to convert one into the other). It's important to note that a tf.string is atomic, meaning that its length does not appear in the tensor's shape. Once you convert it to a Unicode tensor (i.e., a tensor of type tf.int32 holding Unicode code points), the length appears in the shape.

Sets
Are represented as regular tensors (or sparse tensors). For example, tf.con stant([[1, 2], [3, 4]]) represents the two sets {1, 2} and {3, 4}. More gener- ally, each set is represented by a vector in the tensor's last axis. You can manipulate sets using operations from the tf.sets package.

Queues
Store tensors across multiple steps. TensorFlow offers various kinds of queues: simple First In, First Out (FIFO) queues (FIFOQueue), queues that can prioritize

some items (`PriorityQueue`), shuffle their items (`RandomShuffleQueue`), and batch items of different shapes by padding (`PaddingFIFOQueue`). These classes are all in the `tf.queue` package.

With tensors, operations, variables, and various data structures at your disposal, you are now ready to customize your models and training algorithms!

Customizing Models and Training Algorithms

Let's start by creating a custom loss function, which is a simple and common use case.

Custom Loss Functions

Suppose you want to train a regression model, but your training set is a bit noisy. Of course, you start by trying to clean up your dataset by removing or fixing the outliers, but that turns out to be insufficient; the dataset is still noisy. Which loss function should you use? The mean squared error might penalize large errors too much and cause your model to be imprecise. The mean absolute error would not penalize outliers as much, but training might take a while to converge, and the trained model might not be very precise. This is probably a good time to use the Huber loss (introduced in Chapter 10) instead of the good old MSE. The Huber loss is not currently part of the official Keras API, but it is available in tf.keras (just use an instance of the `keras.losses.Huber` class). But let's pretend it's not there: implementing it is easy as pie! Just create a function that takes the labels and predictions as arguments, and use TensorFlow operations to compute every instance's loss:

```
def huber_fn(y_true, y_pred):
    error = y_true - y_pred
    is_small_error = tf.abs(error) < 1
    squared_loss = tf.square(error) / 2
    linear_loss  = tf.abs(error) - 0.5
    return tf.where(is_small_error, squared_loss, linear_loss)
```

> For better performance, you should use a vectorized implementation, as in this example. Moreover, if you want to benefit from TensorFlow's graph features, you should use only TensorFlow operations.

It is also preferable to return a tensor containing one loss per instance, rather than returning the mean loss. This way, Keras can apply class weights or sample weights when requested (see Chapter 10).

Now you can use this loss when you compile the Keras model, then train your model:

```
model.compile(loss=huber_fn, optimizer="nadam")
model.fit(X_train, y_train, [...])
```

And that's it! For each batch during training, Keras will call the huber_fn() function to compute the loss and use it to perform a Gradient Descent step. Moreover, it will keep track of the total loss since the beginning of the epoch, and it will display the mean loss.

But what happens to this custom loss when you save the model?

Saving and Loading Models That Contain Custom Components

Saving a model containing a custom loss function works fine, as Keras saves the name of the function. Whenever you load it, you'll need to provide a dictionary that maps the function name to the actual function. More generally, when you load a model containing custom objects, you need to map the names to the objects:

```
model = keras.models.load_model("my_model_with_a_custom_loss.h5",
                                custom_objects={"huber_fn": huber_fn})
```

With the current implementation, any error between –1 and 1 is considered "small." But what if you want a different threshold? One solution is to create a function that creates a configured loss function:

```
def create_huber(threshold=1.0):
    def huber_fn(y_true, y_pred):
        error = y_true - y_pred
        is_small_error = tf.abs(error) < threshold
        squared_loss = tf.square(error) / 2
        linear_loss  = threshold * tf.abs(error) - threshold**2 / 2
        return tf.where(is_small_error, squared_loss, linear_loss)
    return huber_fn

model.compile(loss=create_huber(2.0), optimizer="nadam")
```

Unfortunately, when you save the model, the threshold will not be saved. This means that you will have to specify the threshold value when loading the model (note that the name to use is "huber_fn", which is the name of the function you gave Keras, not the name of the function that created it):

```
model = keras.models.load_model("my_model_with_a_custom_loss_threshold_2.h5",
                                custom_objects={"huber_fn": create_huber(2.0)})
```

You can solve this by creating a subclass of the keras.losses.Loss class, and then implementing its get_config() method:

```
class HuberLoss(keras.losses.Loss):
    def __init__(self, threshold=1.0, **kwargs):
        self.threshold = threshold
        super().__init__(**kwargs)
    def call(self, y_true, y_pred):
        error = y_true - y_pred
        is_small_error = tf.abs(error) < self.threshold
        squared_loss = tf.square(error) / 2
        linear_loss  = self.threshold * tf.abs(error) - self.threshold**2 / 2
        return tf.where(is_small_error, squared_loss, linear_loss)
    def get_config(self):
        base_config = super().get_config()
        return {**base_config, "threshold": self.threshold}
```

The Keras API currently only specifies how to use subclassing to define layers, models, callbacks, and regularizers. If you build other components (such as losses, metrics, initializers, or constraints) using subclassing, they may not be portable to other Keras implementations. It's likely that the Keras API will be updated to specify subclassing for all these components as well.

Let's walk through this code:

- The constructor accepts **kwargs and passes them to the parent constructor, which handles standard hyperparameters: the name of the loss and the reduction algorithm to use to aggregate the individual instance losses. By default, it is "sum_over_batch_size", which means that the loss will be the sum of the instance losses, weighted by the sample weights, if any, and divided by the batch size (not by the sum of weights, so this is *not* the weighted mean).[5] Other possible values are "sum" and None.

- The call() method takes the labels and predictions, computes all the instance losses, and returns them.

- The get_config() method returns a dictionary mapping each hyperparameter name to its value. It first calls the parent class's get_config() method, then adds the new hyperparameters to this dictionary (note that the convenient {**x} syntax was added in Python 3.5).

You can then use any instance of this class when you compile the model:

```
model.compile(loss=HuberLoss(2.), optimizer="nadam")
```

---

5 It would not be a good idea to use a weighted mean: if you did, then two instances with the same weight but in different batches would have a different impact on training, depending on the total weight of each batch.

When you save the model, the threshold will be saved along with it; and when you load the model, you just need to map the class name to the class itself:

```
model = keras.models.load_model("my_model_with_a_custom_loss_class.h5",
                                custom_objects={"HuberLoss": HuberLoss})
```

When you save a model, Keras calls the loss instance's `get_config()` method and saves the config as JSON in the HDF5 file. When you load the model, it calls the `from_config()` class method on the `HuberLoss` class: this method is implemented by the base class (`Loss`) and creates an instance of the class, passing `**config` to the constructor.

That's it for losses! That wasn't too hard, was it? Just as simple are custom activation functions, initializers, regularizers, and constraints. Let's look at these now.

Custom Activation Functions, Initializers, Regularizers, and Constraints

Most Keras functionalities, such as losses, regularizers, constraints, initializers, metrics, activation functions, layers, and even full models, can be customized in very much the same way. Most of the time, you will just need to write a simple function with the appropriate inputs and outputs. Here are examples of a custom activation function (equivalent to `keras.activations.softplus()` or `tf.nn.softplus()`), a custom Glorot initializer (equivalent to `keras.initializers.glorot_normal()`), a custom ℓ_1 regularizer (equivalent to `keras.regularizers.l1(0.01)`), and a custom constraint that ensures weights are all positive (equivalent to `keras.con straints.nonneg()` or `tf.nn.relu()`):

```
def my_softplus(z): # return value is just tf.nn.softplus(z)
    return tf.math.log(tf.exp(z) + 1.0)

def my_glorot_initializer(shape, dtype=tf.float32):
    stddev = tf.sqrt(2. / (shape[0] + shape[1]))
    return tf.random.normal(shape, stddev=stddev, dtype=dtype)

def my_l1_regularizer(weights):
    return tf.reduce_sum(tf.abs(0.01 * weights))

def my_positive_weights(weights): # return value is just tf.nn.relu(weights)
    return tf.where(weights < 0., tf.zeros_like(weights), weights)
```

As you can see, the arguments depend on the type of custom function. These custom functions can then be used normally; for example:

```
layer = keras.layers.Dense(30, activation=my_softplus,
                           kernel_initializer=my_glorot_initializer,
                           kernel_regularizer=my_l1_regularizer,
                           kernel_constraint=my_positive_weights)
```

The activation function will be applied to the output of this `Dense` layer, and its result will be passed on to the next layer. The layer's weights will be initialized using the value returned by the initializer. At each training step the weights will be passed to the regularization function to compute the regularization loss, which will be added to the main loss to get the final loss used for training. Finally, the constraint function will be called after each training step, and the layer's weights will be replaced by the constrained weights.

If a function has hyperparameters that need to be saved along with the model, then you will want to subclass the appropriate class, such as `keras.regularizers.Regular` `izer`, `keras.constraints.Constraint`, `keras.initializers.Initializer`, or `keras.layers.Layer` (for any layer, including activation functions). Much like we did for the custom loss, here is a simple class for ℓ_1 regularization that saves its `factor` hyperparameter (this time we do not need to call the parent constructor or the `get_config()` method, as they are not defined by the parent class):

```
class MyL1Regularizer(keras.regularizers.Regularizer):
    def __init__(self, factor):
        self.factor = factor
    def __call__(self, weights):
        return tf.reduce_sum(tf.abs(self.factor * weights))
    def get_config(self):
        return {"factor": self.factor}
```

Note that you must implement the `call()` method for losses, layers (including activation functions), and models, or the `__call__()` method for regularizers, initializers, and constraints. For metrics, things are a bit different, as we will see now.

Custom Metrics

Losses and metrics are conceptually not the same thing: losses (e.g., cross entropy) are used by Gradient Descent to *train* a model, so they must be differentiable (at least where they are evaluated), and their gradients should not be 0 everywhere. Plus, it's OK if they are not easily interpretable by humans. In contrast, metrics (e.g., accuracy) are used to *evaluate* a model: they must be more easily interpretable, and they can be non-differentiable or have 0 gradients everywhere.

That said, in most cases, defining a custom metric function is exactly the same as defining a custom loss function. In fact, we could even use the Huber loss function we created earlier as a metric;[6] it would work just fine (and persistence would also work the same way, in this case only saving the name of the function, `"huber_fn"`):

---

6 However, the Huber loss is seldom used as a metric (the MAE or MSE is preferred).

```
model.compile(loss="mse", optimizer="nadam", metrics=[create_huber(2.0)])
```

For each batch during training, Keras will compute this metric and keep track of its mean since the beginning of the epoch. Most of the time, this is exactly what you want. But not always! Consider a binary classifier's precision, for example. As we saw in Chapter 3, precision is the number of true positives divided by the number of positive predictions (including both true positives and false positives). Suppose the model made five positive predictions in the first batch, four of which were correct: that's 80% precision. Then suppose the model made three positive predictions in the second batch, but they were all incorrect: that's 0% precision for the second batch. If you just compute the mean of these two precisions, you get 40%. But wait a second—that's *not* the model's precision over these two batches! Indeed, there were a total of four true positives (4 + 0) out of eight positive predictions (5 + 3), so the overall precision is 50%, not 40%. What we need is an object that can keep track of the number of true positives and the number of false positives and that can compute their ratio when requested. This is precisely what the `keras.metrics.Precision` class does:

```
>>> precision = keras.metrics.Precision()
>>> precision([0, 1, 1, 1, 0, 1, 0, 1], [1, 1, 0, 1, 0, 1, 0, 1])
<tf.Tensor: id=581729, shape=(), dtype=float32, numpy=0.8>
>>> precision([0, 1, 0, 0, 1, 0, 1, 1], [1, 0, 1, 1, 0, 0, 0, 0])
<tf.Tensor: id=581780, shape=(), dtype=float32, numpy=0.5>
```

In this example, we created a `Precision` object, then we used it like a function, passing it the labels and predictions for the first batch, then for the second batch (note that we could also have passed sample weights). We used the same number of true and false positives as in the example we just discussed. After the first batch, it returns a precision of 80%; then after the second batch, it returns 50% (which is the overall precision so far, not the second batch's precision). This is called a *streaming metric* (or *stateful metric*), as it is gradually updated, batch after batch.

At any point, we can call the `result()` method to get the current value of the metric. We can also look at its variables (tracking the number of true and false positives) by using the `variables` attribute, and we can reset these variables using the `reset_states()` method:

```
>>> p.result()
<tf.Tensor: id=581794, shape=(), dtype=float32, numpy=0.5>
>>> p.variables
[<tf.Variable 'true_positives:0' [...] numpy=array([4.], dtype=float32)>,
 <tf.Variable 'false_positives:0' [...] numpy=array([4.], dtype=float32)>]
>>> p.reset_states() # both variables get reset to 0.0
```

If you need to create such a streaming metric, create a subclass of the `keras.metrics.Metric` class. Here is a simple example that keeps track of the total Huber loss and the number of instances seen so far. When asked for the result, it returns the ratio, which is simply the mean Huber loss:

```python
class HuberMetric(keras.metrics.Metric):
    def __init__(self, threshold=1.0, **kwargs):
        super().__init__(**kwargs) # handles base args (e.g., dtype)
        self.threshold = threshold
        self.huber_fn = create_huber(threshold)
        self.total = self.add_weight("total", initializer="zeros")
        self.count = self.add_weight("count", initializer="zeros")
    def update_state(self, y_true, y_pred, sample_weight=None):
        metric = self.huber_fn(y_true, y_pred)
        self.total.assign_add(tf.reduce_sum(metric))
        self.count.assign_add(tf.cast(tf.size(y_true), tf.float32))
    def result(self):
        return self.total / self.count
    def get_config(self):
        base_config = super().get_config()
        return {**base_config, "threshold": self.threshold}
```

Let's walk through this code:[7]

- The constructor uses the `add_weight()` method to create the variables needed to keep track of the metric's state over multiple batches—in this case, the sum of all Huber losses (`total`) and the number of instances seen so far (`count`). You could just create variables manually if you preferred. Keras tracks any `tf.Variable` that is set as an attribute (and more generally, any "trackable" object, such as layers or models).

- The `update_state()` method is called when you use an instance of this class as a function (as we did with the `Precision` object). It updates the variables, given the labels and predictions for one batch (and sample weights, but in this case we ignore them).

- The `result()` method computes and returns the final result, in this case the mean Huber metric over all instances. When you use the metric as a function, the `update_state()` method gets called first, then the `result()` method is called, and its output is returned.

- We also implement the `get_config()` method to ensure the `threshold` gets saved along with the model.

- The default implementation of the `reset_states()` method resets all variables to 0.0 (but you can override it if needed).

---

7 This class is for illustration purposes only. A simpler and better implementation would just subclass the `keras.metrics.Mean` class; see the "Streaming metrics" section of the notebook for an example.

Keras will take care of variable persistence seamlessly; no action is required.

When you define a metric using a simple function, Keras automatically calls it for each batch, and it keeps track of the mean during each epoch, just like we did manually. So the only benefit of our `HuberMetric` class is that the `threshold` will be saved. But of course, some metrics, like precision, cannot simply be averaged over batches: in those cases, there's no other option than to implement a streaming metric.

Now that we have built a streaming metric, building a custom layer will seem like a walk in the park!

Custom Layers

You may occasionally want to build an architecture that contains an exotic layer for which TensorFlow does not provide a default implementation. In this case, you will need to create a custom layer. Or you may simply want to build a very repetitive architecture, containing identical blocks of layers repeated many times, and it would be convenient to treat each block of layers as a single layer. For example, if the model is a sequence of layers A, B, C, A, B, C, A, B, C, then you might want to define a custom layer D containing layers A, B, C, so your model would then simply be D, D, D. Let's see how to build custom layers.

First, some layers have no weights, such as `keras.layers.Flatten` or `keras.lay ers.ReLU`. If you want to create a custom layer without any weights, the simplest option is to write a function and wrap it in a `keras.layers.Lambda` layer. For example, the following layer will apply the exponential function to its inputs:

```
exponential_layer = keras.layers.Lambda(lambda x: tf.exp(x))
```

This custom layer can then be used like any other layer, using the Sequential API, the Functional API, or the Subclassing API. You can also use it as an activation function (or you could use `activation=tf.exp`, `activation=keras.activations.exponen tial`, or simply `activation="exponential"`). The exponential layer is sometimes used in the output layer of a regression model when the values to predict have very different scales (e.g., 0.001, 10., 1,000.).

As you've probably guessed by now, to build a custom stateful layer (i.e., a layer with weights), you need to create a subclass of the `keras.layers.Layer` class. For example, the following class implements a simplified version of the `Dense` layer:

```
class MyDense(keras.layers.Layer):
    def __init__(self, units, activation=None, **kwargs):
        super().__init__(**kwargs)
        self.units = units
        self.activation = keras.activations.get(activation)

    def build(self, batch_input_shape):
        self.kernel = self.add_weight(
            name="kernel", shape=[batch_input_shape[-1], self.units],
            initializer="glorot_normal")
        self.bias = self.add_weight(
            name="bias", shape=[self.units], initializer="zeros")
        super().build(batch_input_shape) # must be at the end

    def call(self, X):
        return self.activation(X @ self.kernel + self.bias)

    def compute_output_shape(self, batch_input_shape):
        return tf.TensorShape(batch_input_shape.as_list()[:-1] + [self.units])

    def get_config(self):
        base_config = super().get_config()
        return {**base_config, "units": self.units,
                "activation": keras.activations.serialize(self.activation)}
```

Let's walk through this code:

- The constructor takes all the hyperparameters as arguments (in this example, units and activation), and importantly it also takes a **kwargs argument. It calls the parent constructor, passing it the kwargs: this takes care of standard arguments such as input_shape, trainable, and name. Then it saves the hyperparameters as attributes, converting the activation argument to the appropriate activation function using the keras.activations.get() function (it accepts functions, standard strings like "relu" or "selu", or simply None).[8]

- The build() method's role is to create the layer's variables by calling the add_weight() method for each weight. The build() method is called the first time the layer is used. At that point, Keras will know the shape of this layer's inputs, and it will pass it to the build() method,[9] which is often necessary to create some of the weights. For example, we need to know the number of neurons in the previous layer in order to create the connection weights matrix (i.e., the "kernel"): this corresponds to the size of the last dimension of the inputs. At the end of the build() method (and only at the end), you must call the parent's

---

8 This function is specific to tf.keras. You could use keras.activations.Activation instead.

9 The Keras API calls this argument input_shape, but since it also includes the batch dimension, I prefer to call it batch_input_shape. Same for compute_output_shape().

build() method: this tells Keras that the layer is built (it just sets self.built=True).

- The call() method performs the desired operations. In this case, we compute the matrix multiplication of the inputs X and the layer's kernel, we add the bias vector, and we apply the activation function to the result, and this gives us the output of the layer.

- The compute_output_shape() method simply returns the shape of this layer's outputs. In this case, it is the same shape as the inputs, except the last dimension is replaced with the number of neurons in the layer. Note that in tf.keras, shapes are instances of the tf.TensorShape class, which you can convert to Python lists using as_list().

- The get_config() method is just like in the previous custom classes. Note that we save the activation function's full configuration by calling keras.activa tions.serialize().

You can now use a MyDense layer just like any other layer!

You can generally omit the compute_output_shape() method, as tf.keras automatically infers the output shape, except when the layer is dynamic (as we will see shortly). In other Keras implementations, this method is either required or its default implementation assumes the output shape is the same as the input shape.

To create a layer with multiple inputs (e.g., Concatenate), the argument to the call() method should be a tuple containing all the inputs, and similarly the argument to the compute_output_shape() method should be a tuple containing each input's batch shape. To create a layer with multiple outputs, the call() method should return the list of outputs, and compute_output_shape() should return the list of batch output shapes (one per output). For example, the following toy layer takes two inputs and returns three outputs:

```
class MyMultiLayer(keras.layers.Layer):
    def call(self, X):
        X1, X2 = X
        return [X1 + X2, X1 * X2, X1 / X2]

    def compute_output_shape(self, batch_input_shape):
        b1, b2 = batch_input_shape
        return [b1, b1, b1] # should probably handle broadcasting rules
```

This layer may now be used like any other layer, but of course only using the Functional and Subclassing APIs, not the Sequential API (which only accepts layers with one input and one output).

If your layer needs to have a different behavior during training and during testing (e.g., if it uses Dropout or BatchNormalization layers), then you must add a `train ing` argument to the `call()` method and use this argument to decide what to do. For example, let's create a layer that adds Gaussian noise during training (for regularization) but does nothing during testing (Keras has a layer that does the same thing, `keras.layers.GaussianNoise`):

```
class MyGaussianNoise(keras.layers.Layer):
    def __init__(self, stddev, **kwargs):
        super().__init__(**kwargs)
        self.stddev = stddev

    def call(self, X, training=None):
        if training:
            noise = tf.random.normal(tf.shape(X), stddev=self.stddev)
            return X + noise
        else:
            return X

    def compute_output_shape(self, batch_input_shape):
        return batch_input_shape
```

With that, you can now build any custom layer you need! Now let's create custom models.

Custom Models

We already looked at creating custom model classes in Chapter 10, when we discussed the Subclassing API.[10] It's straightforward: subclass the `keras.Model` class, create layers and variables in the constructor, and implement the `call()` method to do whatever you want the model to do. Suppose you want to build the model represented in Figure 12-3.

---

10 The name "Subclassing API" usually refers only to the creation of custom models by subclassing, although many other things can be created by subclassing, as we saw in this chapter.

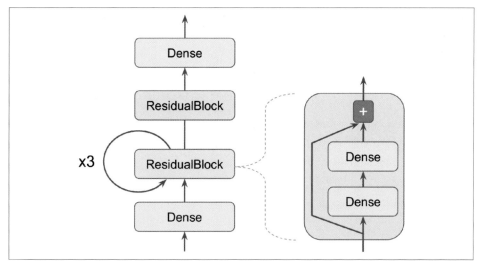

Figure 12-3. Custom model example: an arbitrary model with a custom ResidualBlock layer containing a skip connection

The inputs go through a first dense layer, then through a *residual block* composed of two dense layers and an addition operation (as we will see in Chapter 14, a residual block adds its inputs to its outputs), then through this same residual block three more times, then through a second residual block, and the final result goes through a dense output layer. Note that this model does not make much sense; it's just an example to illustrate the fact that you can easily build any kind of model you want, even one that contains loops and skip connections. To implement this model, it is best to first create a `ResidualBlock` layer, since we are going to create a couple of identical blocks (and we might want to reuse it in another model):

```python
class ResidualBlock(keras.layers.Layer):
    def __init__(self, n_layers, n_neurons, **kwargs):
        super().__init__(**kwargs)
        self.hidden = [keras.layers.Dense(n_neurons, activation="elu",
                                          kernel_initializer="he_normal")
                       for _ in range(n_layers)]

    def call(self, inputs):
        Z = inputs
        for layer in self.hidden:
            Z = layer(Z)
        return inputs + Z
```

This layer is a bit special since it contains other layers. This is handled transparently by Keras: it automatically detects that the `hidden` attribute contains trackable objects (layers in this case), so their variables are automatically added to this layer's list of variables. The rest of this class is self-explanatory. Next, let's use the Subclassing API to define the model itself:

```python
class ResidualRegressor(keras.Model):
    def __init__(self, output_dim, **kwargs):
        super().__init__(**kwargs)
        self.hidden1 = keras.layers.Dense(30, activation="elu",
                                          kernel_initializer="he_normal")
        self.block1 = ResidualBlock(2, 30)
        self.block2 = ResidualBlock(2, 30)
        self.out = keras.layers.Dense(output_dim)

    def call(self, inputs):
        Z = self.hidden1(inputs)
        for _ in range(1 + 3):
            Z = self.block1(Z)
        Z = self.block2(Z)
        return self.out(Z)
```

We create the layers in the constructor and use them in the `call()` method. This model can then be used like any other model (compile it, fit it, evaluate it, and use it to make predictions). If you also want to be able to save the model using the `save()` method and load it using the `keras.models.load_model()` function, you must implement the `get_config()` method (as we did earlier) in both the `ResidualBlock` class and the `ResidualRegressor` class. Alternatively, you can save and load the weights using the `save_weights()` and `load_weights()` methods.

The `Model` class is a subclass of the `Layer` class, so models can be defined and used exactly like layers. But a model has some extra functionalities, including of course its `compile()`, `fit()`, `evaluate()`, and `predict()` methods (and a few variants), plus the `get_layers()` method (which can return any of the model's layers by name or by index) and the `save()` method (and support for `keras.models.load_model()` and `keras.models.clone_model()`).

 If models provide more functionality than layers, why not just define every layer as a model? Well, technically you could, but it is usually cleaner to distinguish the internal components of your model (i.e., layers or reusable blocks of layers) from the model itself (i.e., the object you will train). The former should subclass the `Layer` class, while the latter should subclass the `Model` class.

With that, you can naturally and concisely build almost any model that you find in a paper, using the Sequential API, the Functional API, the Subclassing API, or even a mix of these. "Almost" any model? Yes, there are still a few things that we need to look at: first, how to define losses or metrics based on model internals, and second, how to build a custom training loop.

Losses and Metrics Based on Model Internals

The custom losses and metrics we defined earlier were all based on the labels and the predictions (and optionally sample weights). There will be times when you want to define losses based on other parts of your model, such as the weights or activations of its hidden layers. This may be useful for regularization purposes or to monitor some internal aspect of your model.

To define a custom loss based on model internals, compute it based on any part of the model you want, then pass the result to the add_loss() method.For example, let's build a custom regression MLP model composed of a stack of five hidden layers plus an output layer. This custom model will also have an auxiliary output on top of the upper hidden layer. The loss associated to this auxiliary output will be called the *reconstruction loss* (see Chapter 17): it is the mean squared difference between the reconstruction and the inputs. By adding this reconstruction loss to the main loss, we will encourage the model to preserve as much information as possible through the hidden layers—even information that is not directly useful for the regression task itself. In practice, this loss sometimes improves generalization (it is a regularization loss). Here is the code for this custom model with a custom reconstruction loss:

```
class ReconstructingRegressor(keras.Model):
    def __init__(self, output_dim, **kwargs):
        super().__init__(**kwargs)
        self.hidden = [keras.layers.Dense(30, activation="selu",
                                          kernel_initializer="lecun_normal")
                       for _ in range(5)]
        self.out = keras.layers.Dense(output_dim)

    def build(self, batch_input_shape):
        n_inputs = batch_input_shape[-1]
        self.reconstruct = keras.layers.Dense(n_inputs)
        super().build(batch_input_shape)

    def call(self, inputs):
        Z = inputs
        for layer in self.hidden:
            Z = layer(Z)
        reconstruction = self.reconstruct(Z)
        recon_loss = tf.reduce_mean(tf.square(reconstruction - inputs))
        self.add_loss(0.05 * recon_loss)
        return self.out(Z)
```

Let's go through this code:

- The constructor creates the DNN with five dense hidden layers and one dense output layer.

- The build() method creates an extra dense layer which will be used to reconstruct the inputs of the model. It must be created here because its number of units must be equal to the number of inputs, and this number is unknown before the build() method is called.

- The call() method processes the inputs through all five hidden layers, then passes the result through the reconstruction layer, which produces the reconstruction.

- Then the call() method computes the reconstruction loss (the mean squared difference between the reconstruction and the inputs), and adds it to the model's list of losses using the add_loss() method.[11] Notice that we scale down the reconstruction loss by multiplying it by 0.05 (this is a hyperparameter you can tune). This ensures that the reconstruction loss does not dominate the main loss.

- Finally, the call() method passes the output of the hidden layers to the output layer and returns its output.

Similarly, you can add a custom metric based on model internals by computing it in any way you want, as long as the result is the output of a metric object. For example, you can create a keras.metrics.Mean object in the constructor, then call it in the call() method, passing it the recon_loss, and finally add it to the model by calling the model's add_metric() method. This way, when you train the model, Keras will display both the mean loss over each epoch (the loss is the sum of the main loss plus 0.05 times the reconstruction loss) and the mean reconstruction error over each epoch. Both will go down during training:

```
Epoch 1/5
11610/11610 [==============] [...] loss: 4.3092 - reconstruction_error: 1.7360
Epoch 2/5
11610/11610 [==============] [...] loss: 1.1232 - reconstruction_error: 0.8964
[...]
```

In over 99% of cases, everything we have discussed so far will be sufficient to implement whatever model you want to build, even with complex architectures, losses, and metrics. However, in some rare cases you may need to customize the training loop

---

11 You can also call add_loss() on any layer inside the model, as the model recursively gathers losses from all of its layers.

itself. Before we get there, we need to look at how to compute gradients automatically in TensorFlow.

Computing Gradients Using Autodiff

To understand how to use autodiff (see Chapter 10 and Appendix D) to compute gradients automatically, let's consider a simple toy function:

```
def f(w1, w2):
    return 3 * w1 ** 2 + 2 * w1 * w2
```

If you know calculus, you can analytically find that the partial derivative of this function with regard to `w1` is `6 * w1 + 2 * w2`. You can also find that its partial derivative with regard to `w2` is `2 * w1`. For example, at the point `(w1, w2)` = (5, 3), these partial derivatives are equal to 36 and 10, respectively, so the gradient vector at this point is (36, 10). But if this were a neural network, the function would be much more complex, typically with tens of thousands of parameters, and finding the partial derivatives analytically by hand would be an almost impossible task. One solution could be to compute an approximation of each partial derivative by measuring how much the function's output changes when you tweak the corresponding parameter:

```
>>> w1, w2 = 5, 3
>>> eps = 1e-6
>>> (f(w1 + eps, w2) - f(w1, w2)) / eps
36.000003007075065
>>> (f(w1, w2 + eps) - f(w1, w2)) / eps
10.000000003174137
```

Looks about right! This works rather well and is easy to implement, but it is just an approximation, and importantly you need to call `f()` at least once per parameter (not twice, since we could compute `f(w1, w2)` just once). Needing to call `f()` at least once per parameter makes this approach intractable for large neural networks. So instead, we should use autodiff. TensorFlow makes this pretty simple:

```
w1, w2 = tf.Variable(5.), tf.Variable(3.)
with tf.GradientTape() as tape:
    z = f(w1, w2)

gradients = tape.gradient(z, [w1, w2])
```

We first define two variables `w1` and `w2`, then we create a `tf.GradientTape` context that will automatically record every operation that involves a variable, and finally we ask this tape to compute the gradients of the result `z` with regard to both variables `[w1, w2]`. Let's take a look at the gradients that TensorFlow computed:

```
>>> gradients
[<tf.Tensor: id=828234, shape=(), dtype=float32, numpy=36.0>,
 <tf.Tensor: id=828229, shape=(), dtype=float32, numpy=10.0>]
```

Perfect! Not only is the result accurate (the precision is only limited by the floating-point errors), but the gradient() method only goes through the recorded computations once (in reverse order), no matter how many variables there are, so it is incredibly efficient. It's like magic!

 To save memory, only put the strict minimum inside the tf.Gra dientTape() block. Alternatively, pause recording by creating a with tape.stop_recording() block inside the tf.Gradient Tape() block.

The tape is automatically erased immediately after you call its gradient() method, so you will get an exception if you try to call gradient() twice:

```
with tf.GradientTape() as tape:
    z = f(w1, w2)

dz_dw1 = tape.gradient(z, w1) # => tensor 36.0
dz_dw2 = tape.gradient(z, w2) # RuntimeError!
```

If you need to call gradient() more than once, you must make the tape persistent and delete it each time you are done with it to free resources:[12]

```
with tf.GradientTape(persistent=True) as tape:
    z = f(w1, w2)

dz_dw1 = tape.gradient(z, w1) # => tensor 36.0
dz_dw2 = tape.gradient(z, w2) # => tensor 10.0, works fine now!
del tape
```

By default, the tape will only track operations involving variables, so if you try to compute the gradient of z with regard to anything other than a variable, the result will be None:

```
c1, c2 = tf.constant(5.), tf.constant(3.)
with tf.GradientTape() as tape:
    z = f(c1, c2)

gradients = tape.gradient(z, [c1, c2]) # returns [None, None]
```

However, you can force the tape to watch any tensors you like, to record every operation that involves them. You can then compute gradients with regard to these tensors, as if they were variables:

---

12 If the tape goes out of scope, for example when the function that used it returns, Python's garbage collector will delete it for you.

```
with tf.GradientTape() as tape:
    tape.watch(c1)
    tape.watch(c2)
    z = f(c1, c2)

gradients = tape.gradient(z, [c1, c2]) # returns [tensor 36., tensor 10.]
```

This can be useful in some cases, like if you want to implement a regularization loss that penalizes activations that vary a lot when the inputs vary little: the loss will be based on the gradient of the activations with regard to the inputs. Since the inputs are not variables, you would need to tell the tape to watch them.

Most of the time a gradient tape is used to compute the gradients of a single value (usually the loss) with regard to a set of values (usually the model parameters). This is where reverse-mode autodiff shines, as it just needs to do one forward pass and one reverse pass to get all the gradients at once. If you try to compute the gradients of a vector, for example a vector containing multiple losses, then TensorFlow will compute the gradients of the vector's sum. So if you ever need to get the individual gradients (e.g., the gradients of each loss with regard to the model parameters), you must call the tape's jacobian() method: it will perform reverse-mode autodiff once for each loss in the vector (all in parallel by default). It is even possible to compute second-order partial derivatives (the Hessians, i.e., the partial derivatives of the partial derivatives), but this is rarely needed in practice (see the "Computing Gradients with Autodiff" section of the notebook for an example).

In some cases you may want to stop gradients from backpropagating through some part of your neural network. To do this, you must use the tf.stop_gradient() function. The function returns its inputs during the forward pass (like tf.identity()), but it does not let gradients through during backpropagation (it acts like a constant):

```
def f(w1, w2):
    return 3 * w1 ** 2 + tf.stop_gradient(2 * w1 * w2)

with tf.GradientTape() as tape:
    z = f(w1, w2) # same result as without stop_gradient()

gradients = tape.gradient(z, [w1, w2]) # => returns [tensor 30., None]
```

Finally, you may occasionally run into some numerical issues when computing gradients. For example, if you compute the gradients of the my_softplus() function for large inputs, the result will be NaN:

```
>>> x = tf.Variable([100.])
>>> with tf.GradientTape() as tape:
...     z = my_softplus(x)
...
>>> tape.gradient(z, [x])
<tf.Tensor: [...] numpy=array([nan], dtype=float32)>
```

This is because computing the gradients of this function using autodiff leads to some numerical difficulties: due to floating-point precision errors, autodiff ends up computing infinity divided by infinity (which returns NaN). Fortunately, we can analytically find that the derivative of the softplus function is just $1 / (1 + 1 / \exp(x))$, which is numerically stable. Next, we can tell TensorFlow to use this stable function when computing the gradients of the `my_softplus()` function by decorating it with `@tf.custom_gradient` and making it return both its normal output and the function that computes the derivatives (note that it will receive as input the gradients that were backpropagated so far, down to the softplus function; and according to the chain rule, we should multiply them with this function's gradients):

```
@tf.custom_gradient
def my_better_softplus(z):
    exp = tf.exp(z)
    def my_softplus_gradients(grad):
        return grad / (1 + 1 / exp)
    return tf.math.log(exp + 1), my_softplus_gradients
```

Now when we compute the gradients of the `my_better_softplus()` function, we get the proper result, even for large input values (however, the main output still explodes because of the exponential; one workaround is to use `tf.where()` to return the inputs when they are large).

Congratulations! You can now compute the gradients of any function (provided it is differentiable at the point where you compute it), even blocking backpropagation when needed, and write your own gradient functions! This is probably more flexibility than you will ever need, even if you build your own custom training loops, as we will see now.

Custom Training Loops

In some rare cases, the `fit()` method may not be flexible enough for what you need to do. For example, the Wide & Deep paper (*https://homl.info/widedeep*) we discussed in Chapter 10 uses two different optimizers: one for the wide path and the other for the deep path. Since the `fit()` method only uses one optimizer (the one that we specify when compiling the model), implementing this paper requires writing your own custom loop.

You may also like to write custom training loops simply to feel more confident that they do precisely what you intend them to do (perhaps you are unsure about some details of the `fit()` method). It can sometimes feel safer to make everything explicit. However, remember that writing a custom training loop will make your code longer, more error-prone, and harder to maintain.

Unless you really need the extra flexibility, you should prefer using the `fit()` method rather than implementing your own training loop, especially if you work in a team.

First, let's build a simple model. No need to compile it, since we will handle the training loop manually:

```
l2_reg = keras.regularizers.l2(0.05)
model = keras.models.Sequential([
    keras.layers.Dense(30, activation="elu", kernel_initializer="he_normal",
                       kernel_regularizer=l2_reg),
    keras.layers.Dense(1, kernel_regularizer=l2_reg)
])
```

Next, let's create a tiny function that will randomly sample a batch of instances from the training set (in Chapter 13 we will discuss the Data API, which offers a much better alternative):

```
def random_batch(X, y, batch_size=32):
    idx = np.random.randint(len(X), size=batch_size)
    return X[idx], y[idx]
```

Let's also define a function that will display the training status, including the number of steps, the total number of steps, the mean loss since the start of the epoch (i.e., we will use the Mean metric to compute it), and other metrics:

```
def print_status_bar(iteration, total, loss, metrics=None):
    metrics = " - ".join(["{}: {:.4f}".format(m.name, m.result())
                          for m in [loss] + (metrics or [])])
    end = "" if iteration < total else "\n"
    print("\r{}/{} - ".format(iteration, total) + metrics,
          end=end)
```

This code is self-explanatory, unless you are unfamiliar with Python string formatting: `{:.4f}` will format a float with four digits after the decimal point, and using `\r` (carriage return) along with `end=""` ensures that the status bar always gets printed on the same line. In the notebook, the `print_status_bar()` function includes a progress bar, but you could use the handy tqdm library instead.

With that, let's get down to business! First, we need to define some hyperparameters and choose the optimizer, the loss function, and the metrics (just the MAE in this example):

```
n_epochs = 5
batch_size = 32
n_steps = len(X_train) // batch_size
optimizer = keras.optimizers.Nadam(lr=0.01)
loss_fn = keras.losses.mean_squared_error
```

```
mean_loss = keras.metrics.Mean()
metrics = [keras.metrics.MeanAbsoluteError()]
```

And now we are ready to build the custom loop!

```
for epoch in range(1, n_epochs + 1):
    print("Epoch {}/{}".format(epoch, n_epochs))
    for step in range(1, n_steps + 1):
        X_batch, y_batch = random_batch(X_train_scaled, y_train)
        with tf.GradientTape() as tape:
            y_pred = model(X_batch, training=True)
            main_loss = tf.reduce_mean(loss_fn(y_batch, y_pred))
            loss = tf.add_n([main_loss] + model.losses)
        gradients = tape.gradient(loss, model.trainable_variables)
        optimizer.apply_gradients(zip(gradients, model.trainable_variables))
        mean_loss(loss)
        for metric in metrics:
            metric(y_batch, y_pred)
        print_status_bar(step * batch_size, len(y_train), mean_loss, metrics)
    print_status_bar(len(y_train), len(y_train), mean_loss, metrics)
    for metric in [mean_loss] + metrics:
        metric.reset_states()
```

There's a lot going on in this code, so let's walk through it:

- We create two nested loops: one for the epochs, the other for the batches within an epoch.

- Then we sample a random batch from the training set.

- Inside the tf.GradientTape() block, we make a prediction for one batch (using the model as a function), and we compute the loss: it is equal to the main loss plus the other losses (in this model, there is one regularization loss per layer). Since the mean_squared_error() function returns one loss per instance, we compute the mean over the batch using tf.reduce_mean() (if you wanted to apply different weights to each instance, this is where you would do it). The regularization losses are already reduced to a single scalar each, so we just need to sum them (using tf.add_n(), which sums multiple tensors of the same shape and data type).

- Next, we ask the tape to compute the gradient of the loss with regard to each trainable variable (*not* all variables!), and we apply them to the optimizer to perform a Gradient Descent step.

- Then we update the mean loss and the metrics (over the current epoch), and we display the status bar.

- At the end of each epoch, we display the status bar again to make it look complete[13] and to print a line feed, and we reset the states of the mean loss and the metrics.

If you set the optimizer's `clipnorm` or `clipvalue` hyperparameter, it will take care of this for you. If you want to apply any other transformation to the gradients, simply do so before calling the `apply_gradients()` method.

If you add weight constraints to your model (e.g., by setting `kernel_constraint` or `bias_constraint` when creating a layer), you should update the training loop to apply these constraints just after `apply_gradients()`:

```
for variable in model.variables:
    if variable.constraint is not None:
        variable.assign(variable.constraint(variable))
```

Most importantly, this training loop does not handle layers that behave differently during training and testing (e.g., `BatchNormalization` or `Dropout`). To handle these, you need to call the model with `training=True` and make sure it propagates this to every layer that needs it.

As you can see, there are quite a lot of things you need to get right, and it's easy to make a mistake. But on the bright side, you get full control, so it's your call.

Now that you know how to customize any part of your models[14] and training algorithms, let's see how you can use TensorFlow's automatic graph generation feature: it can speed up your custom code considerably, and it will also make it portable to any platform supported by TensorFlow (see Chapter 19).

TensorFlow Functions and Graphs

In TensorFlow 1, graphs were unavoidable (as were the complexities that came with them) because they were a central part of TensorFlow's API. In TensorFlow 2, they are still there, but not as central, and they're much (much!) simpler to use. To show just how simple, let's start with a trivial function that computes the cube of its input:

```
def cube(x):
    return x ** 3
```

---

13 The truth is we did not process every single instance in the training set, because we sampled instances randomly: some were processed more than once, while others were not processed at all. Likewise, if the training set size is not a multiple of the batch size, we will miss a few instances. In practice that's fine.

14 With the exception of optimizers, as very few people ever customize these; see the "Custom Optimizers" section in the notebook for an example.

We can obviously call this function with a Python value, such as an int or a float, or we can call it with a tensor:

```
>>> cube(2)
8
>>> cube(tf.constant(2.0))
<tf.Tensor: id=18634148, shape=(), dtype=float32, numpy=8.0>
```

Now, let's use `tf.function()` to convert this Python function to a *TensorFlow Function*:

```
>>> tf_cube = tf.function(cube)
>>> tf_cube
<tensorflow.python.eager.def_function.Function at 0x1546fc080>
```

This TF Function can then be used exactly like the original Python function, and it will return the same result (but as tensors):

```
>>> tf_cube(2)
<tf.Tensor: id=18634201, shape=(), dtype=int32, numpy=8>
>>> tf_cube(tf.constant(2.0))
<tf.Tensor: id=18634211, shape=(), dtype=float32, numpy=8.0>
```

Under the hood, `tf.function()` analyzed the computations performed by the `cube()` function and generated an equivalent computation graph! As you can see, it was rather painless (we will see how this works shortly). Alternatively, we could have used `tf.function` as a decorator; this is actually more common:

```
@tf.function
def tf_cube(x):
    return x ** 3
```

The original Python function is still available via the TF Function's `python_function` attribute, in case you ever need it:

```
>>> tf_cube.python_function(2)
8
```

TensorFlow optimizes the computation graph, pruning unused nodes, simplifying expressions (e.g., 1 + 2 would get replaced with 3), and more. Once the optimized graph is ready, the TF Function efficiently executes the operations in the graph, in the appropriate order (and in parallel when it can). As a result, a TF Function will usually run much faster than the original Python function, especially if it performs complex computations.[15] Most of the time you will not really need to know more than that: when you want to boost a Python function, just transform it into a TF Function. That's all!

---

15 However, in this trivial example, the computation graph is so small that there is nothing at all to optimize, so `tf_cube()` actually runs much slower than `cube()`.

Moreover, when you write a custom loss function, a custom metric, a custom layer, or any other custom function and you use it in a Keras model (as we did throughout this chapter), Keras automatically converts your function into a TF Function—no need to use tf.function(). So most of the time, all this magic is 100% transparent.

> You can tell Keras *not* to convert your Python functions to TF Functions by setting dynamic=True when creating a custom layer or a custom model. Alternatively, you can set run_eagerly=True when calling the model's compile() method.

By default, a TF Function generates a new graph for every unique set of input shapes and data types and caches it for subsequent calls. For example, if you call tf_cube(tf.constant(10)), a graph will be generated for int32 tensors of shape []. Then if you call tf_cube(tf.constant(20)), the same graph will be reused. But if you then call tf_cube(tf.constant([10, 20])), a new graph will be generated for int32 tensors of shape [2]. This is how TF Functions handle polymorphism (i.e., varying argument types and shapes). However, this is only true for tensor arguments: if you pass numerical Python values to a TF Function, a new graph will be generated for every distinct value: for example, calling tf_cube(10) and tf_cube(20) will generate two graphs.

> If you call a TF Function many times with different numerical Python values, then many graphs will be generated, slowing down your program and using up a lot of RAM (you must delete the TF Function to release it). Python values should be reserved for arguments that will have few unique values, such as hyperparameters like the number of neurons per layer. This allows TensorFlow to better optimize each variant of your model.

AutoGraph and Tracing

So how does TensorFlow generate graphs? It starts by analyzing the Python function's source code to capture all the control flow statements, such as for loops, while loops, and if statements, as well as break, continue, and return statements. This first step is called *AutoGraph*. The reason TensorFlow has to analyze the source code is that Python does not provide any other way to capture control flow statements: it offers magic methods like __add__() and __mul__() to capture operators like + and *, but there are no __while__() or __if__() magic methods. After analyzing the function's code, AutoGraph outputs an upgraded version of that function in which all the control flow statements are replaced by the appropriate TensorFlow operations, such as tf.while_loop() for loops and tf.cond() for if statements. For example, in Figure 12-4, AutoGraph analyzes the source code of the sum_squares() Python function, and it generates the tf__sum_squares() function. In this function, the for loop is replaced by the definition of the loop_body() function (containing the body of the original for loop), followed by a call to the for_stmt() function. This call will build the appropriate tf.while_loop() operation in the computation graph.

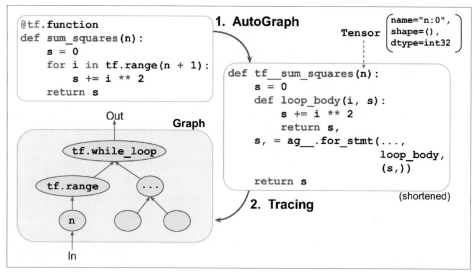

Figure 12-4. How TensorFlow generates graphs using AutoGraph and tracing

Next, TensorFlow calls this "upgraded" function, but instead of passing the argument, it passes a *symbolic tensor*—a tensor without any actual value, only a name, a data type, and a shape. For example, if you call sum_squares(tf.constant(10)), then the tf__sum_squares() function will be called with a symbolic tensor of type int32 and shape []. The function will run in *graph mode*, meaning that each TensorFlow operation will add a node in the graph to represent itself and its output tensor(s) (as opposed to the regular mode, called *eager execution*, or *eager mode*). In graph mode, TF operations do not perform any computations. This should feel familiar if you know TensorFlow 1, as graph mode was the default mode. In Figure 12-4, you can see the tf__sum_squares() function being called with a symbolic tensor as its argument (in this case, an int32 tensor of shape []) and the final graph being generated during tracing. The nodes represent operations, and the arrows represent tensors (both the generated function and the graph are simplified).

To view the generated function's source code, you can call `tf.auto graph.to_code(sum_squares.python_function)`. The code is not meant to be pretty, but it can sometimes help for debugging.

TF Function Rules

Most of the time, converting a Python function that performs TensorFlow operations into a TF Function is trivial: decorate it with `@tf.function` or let Keras take care of it for you. However, there are a few rules to respect:

- If you call any external library, including NumPy or even the standard library, this call will run only during tracing; it will not be part of the graph. Indeed, a TensorFlow graph can only include TensorFlow constructs (tensors, operations, variables, datasets, and so on). So, make sure you use `tf.reduce_sum()` instead of `np.sum()`, `tf.sort()` instead of the built-in `sorted()` function, and so on (unless you really want the code to run only during tracing). This has a few additional implications:

  — If you define a TF Function `f(x)` that just returns `np.random.rand()`, a random number will only be generated when the function is traced, so `f(tf.con stant(2.))` and `f(tf.constant(3.))` will return the same random number, but `f(tf.constant([2., 3.]))` will return a different one. If you replace `np.random.rand()` with `tf.random.uniform([])`, then a new random number will be generated upon every call, since the operation will be part of the graph.

  — If your non-TensorFlow code has side effects (such as logging something or updating a Python counter), then you should not expect those side effects to occur every time you call the TF Function, as they will only occur when the function is traced.

  — You can wrap arbitrary Python code in a `tf.py_function()` operation, but doing so will hinder performance, as TensorFlow will not be able to do any graph optimization on this code. It will also reduce portability, as the graph will only run on platforms where Python is available (and where the right libraries are installed).

- You can call other Python functions or TF Functions, but they should follow the same rules, as TensorFlow will capture their operations in the computation graph. Note that these other functions do not need to be decorated with `@tf.function`.

- If the function creates a TensorFlow variable (or any other stateful TensorFlow object, such as a dataset or a queue), it must do so upon the very first call, and only then, or else you will get an exception. It is usually preferable to create variables outside of the TF Function (e.g., in the `build()` method of a custom layer). If you want to assign a new value to the variable, make sure you call its `assign()` method, instead of using the = operator.

- The source code of your Python function should be available to TensorFlow. If the source code is unavailable (for example, if you define your function in the Python shell, which does not give access to the source code, or if you deploy only the compiled *.pyc Python files to production), then the graph generation process will fail or have limited functionality.

- TensorFlow will only capture `for` loops that iterate over a tensor or a dataset. So make sure you use `for i in tf.range(x)` rather than `for i in range(x)`, or else the loop will not be captured in the graph. Instead, it will run during tracing. (This may be what you want if the `for` loop is meant to build the graph, for example to create each layer in a neural network.)

- As always, for performance reasons, you should prefer a vectorized implementation whenever you can, rather than using loops.

It's time to sum up! In this chapter we started with a brief overview of TensorFlow, then we looked at TensorFlow's low-level API, including tensors, operations, variables, and special data structures. We then used these tools to customize almost every component in tf.keras. Finally, we looked at how TF Functions can boost performance, how graphs are generated using AutoGraph and tracing, and what rules to follow when you write TF Functions (if you would like to open the black box a bit further, for example to explore the generated graphs, you will find technical details in Appendix G).

In the next chapter, we will look at how to efficiently load and preprocess data with TensorFlow.

Exercises

1. How would you describe TensorFlow in a short sentence? What are its main features? Can you name other popular Deep Learning libraries?

2. Is TensorFlow a drop-in replacement for NumPy? What are the main differences between the two?

3. Do you get the same result with `tf.range(10)` and `tf.constant(np.ara nge(10))`?

4. Can you name six other data structures available in TensorFlow, beyond regular tensors?

5. A custom loss function can be defined by writing a function or by subclassing the keras.losses.Loss class. When would you use each option?

6. Similarly, a custom metric can be defined in a function or a subclass of keras.metrics.Metric. When would you use each option?

7. When should you create a custom layer versus a custom model?

8. What are some use cases that require writing your own custom training loop?

9. Can custom Keras components contain arbitrary Python code, or must they be convertible to TF Functions?

10. What are the main rules to respect if you want a function to be convertible to a TF Function?

11. When would you need to create a dynamic Keras model? How do you do that? Why not make all your models dynamic?

12. Implement a custom layer that performs *Layer Normalization* (we will use this type of layer in Chapter 15):

    a. The build() method should define two trainable weights α and β, both of shape input_shape[-1:] and data type tf.float32. α should be initialized with 1s, and β with 0s.

    b. The call() method should compute the mean μ and standard deviation σ of each instance's features. For this, you can use tf.nn.moments(inputs, axes=-1, keepdims=True), which returns the mean μ and the variance σ^2 of all instances (compute the square root of the variance to get the standard deviation). Then the function should compute and return $\alpha \otimes (X - \mu)/(\sigma + \varepsilon) + \beta$, where $\otimes$ represents itemwise multiplication (*) and ε is a smoothing term (small constant to avoid division by zero, e.g., 0.001).

    c. Ensure that your custom layer produces the same (or very nearly the same) output as the keras.layers.LayerNormalization layer.

13. Train a model using a custom training loop to tackle the Fashion MNIST dataset (see Chapter 10).

    a. Display the epoch, iteration, mean training loss, and mean accuracy over each epoch (updated at each iteration), as well as the validation loss and accuracy at the end of each epoch.

    b. Try using a different optimizer with a different learning rate for the upper layers and the lower layers.

Solutions to these exercises are available in Appendix A.

## CHAPTER 13

Loading and Preprocessing Data with TensorFlow

So far we have used only datasets that fit in memory, but Deep Learning systems are often trained on very large datasets that will not fit in RAM. Ingesting a large dataset and preprocessing it efficiently can be tricky to implement with other Deep Learning libraries, but TensorFlow makes it easy thanks to the *Data API*: you just create a dataset object, and tell it where to get the data and how to transform it. TensorFlow takes care of all the implementation details, such as multithreading, queuing, batching, and prefetching. Moreover, the Data API works seamlessly with tf.keras!

Off the shelf, the Data API can read from text files (such as CSV files), binary files with fixed-size records, and binary files that use TensorFlow's TFRecord format, which supports records of varying sizes. TFRecord is a flexible and efficient binary format based on Protocol Buffers (an open source binary format). The Data API also has support for reading from SQL databases. Moreover, many open source extensions are available to read from all sorts of data sources, such as Google's BigQuery service.

Reading huge datasets efficiently is not the only difficulty: the data also needs to be preprocessed, usually normalized. Moreover, it is not always composed strictly of convenient numerical fields: there may be text features, categorical features, and so on. These need to be encoded, for example using one-hot encoding, bag-of-words encoding, or *embeddings* (as we will see, an embedding is a trainable dense vector that represents a category or token). One option to handle all this preprocessing is to write your own custom preprocessing layers. Another is to use the standard preprocessing layers provided by Keras.

In this chapter, we will cover the Data API, the TFRecord format, and how to create custom preprocessing layers and use the standard Keras ones. We will also take a quick look at a few related projects from TensorFlow's ecosystem:

TF Transform (tf.Transform)
> Makes it possible to write a single preprocessing function that can be run in batch mode on your full training set, before training (to speed it up), and then exported to a TF Function and incorporated into your trained model so that once it is deployed in production it can take care of preprocessing new instances on the fly.

TF Datasets (TFDS)
> Provides a convenient function to download many common datasets of all kinds, including large ones like ImageNet, as well as convenient dataset objects to manipulate them using the Data API.

So let's get started!

The Data API

The whole Data API revolves around the concept of a *dataset*: as you might suspect, this represents a sequence of data items. Usually you will use datasets that gradually read data from disk, but for simplicity let's create a dataset entirely in RAM using `tf.data.Dataset.from_tensor_slices()`:

```
>>> X = tf.range(10)  # any data tensor
>>> dataset = tf.data.Dataset.from_tensor_slices(X)
>>> dataset
<TensorSliceDataset shapes: (), types: tf.int32>
```

The `from_tensor_slices()` function takes a tensor and creates a `tf.data.Dataset` whose elements are all the slices of X (along the first dimension), so this dataset contains 10 items: tensors 0, 1, 2, …, 9. In this case we would have obtained the same dataset if we had used `tf.data.Dataset.range(10)`.

You can simply iterate over a dataset's items like this:

```
>>> for item in dataset:
...     print(item)
...
tf.Tensor(0, shape=(), dtype=int32)
tf.Tensor(1, shape=(), dtype=int32)
tf.Tensor(2, shape=(), dtype=int32)
[...]
tf.Tensor(9, shape=(), dtype=int32)
```

Chaining Transformations

Once you have a dataset, you can apply all sorts of transformations to it by calling its transformation methods. Each method returns a new dataset, so you can chain transformations like this (this chain is illustrated in Figure 13-1):

```
>>> dataset = dataset.repeat(3).batch(7)
>>> for item in dataset:
...     print(item)
...
tf.Tensor([0 1 2 3 4 5 6], shape=(7,), dtype=int32)
tf.Tensor([7 8 9 0 1 2 3], shape=(7,), dtype=int32)
tf.Tensor([4 5 6 7 8 9 0], shape=(7,), dtype=int32)
tf.Tensor([1 2 3 4 5 6 7], shape=(7,), dtype=int32)
tf.Tensor([8 9], shape=(2,), dtype=int32)
```

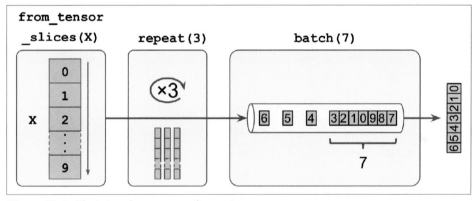

Figure 13-1. Chaining dataset transformations

In this example, we first call the repeat() method on the original dataset, and it returns a new dataset that will repeat the items of the original dataset three times. Of course, this will not copy all the data in memory three times! (If you call this method with no arguments, the new dataset will repeat the source dataset forever, so the code that iterates over the dataset will have to decide when to stop.) Then we call the batch() method on this new dataset, and again this creates a new dataset. This one will group the items of the previous dataset in batches of seven items. Finally, we iterate over the items of this final dataset. As you can see, the batch() method had to output a final batch of size two instead of seven, but you can call it with drop_remainder=True if you want it to drop this final batch so that all batches have the exact same size.

The dataset methods do *not* modify datasets, they create new ones, so make sure to keep a reference to these new datasets (e.g., with `dataset = ...`), or else nothing will happen.

You can also transform the items by calling the `map()` method. For example, this creates a new dataset with all items doubled:

```
>>> dataset = dataset.map(lambda x: x * 2) # Items: [0,2,4,6,8,10,12]
```

This function is the one you will call to apply any preprocessing you want to your data. Sometimes this will include computations that can be quite intensive, such as reshaping or rotating an image, so you will usually want to spawn multiple threads to speed things up: it's as simple as setting the `num_parallel_calls` argument. Note that the function you pass to the `map()` method must be convertible to a TF Function (see Chapter 12).

While the `map()` method applies a transformation to each item, the `apply()` method applies a transformation to the dataset as a whole. For example, the following code applies the `unbatch()` function to the dataset (this function is currently experimental, but it will most likely move to the core API in a future release). Each item in the new dataset will be a single-integer tensor instead of a batch of seven integers:

```
>>> dataset = dataset.apply(tf.data.experimental.unbatch()) # Items: 0,2,4,...
```

It is also possible to simply filter the dataset using the `filter()` method:

```
>>> dataset = dataset.filter(lambda x: x < 10) # Items: 0 2 4 6 8 0 2 4 6...
```

You will often want to look at just a few items from a dataset. You can use the `take()` method for that:

```
>>> for item in dataset.take(3):
...     print(item)
...
tf.Tensor(0, shape=(), dtype=int64)
tf.Tensor(2, shape=(), dtype=int64)
tf.Tensor(4, shape=(), dtype=int64)
```

Shuffling the Data

As you know, Gradient Descent works best when the instances in the training set are independent and identically distributed (see Chapter 4). A simple way to ensure this is to shuffle the instances, using the `shuffle()` method. It will create a new dataset that will start by filling up a buffer with the first items of the source dataset. Then, whenever it is asked for an item, it will pull one out randomly from the buffer and replace it with a fresh one from the source dataset, until it has iterated entirely through the source dataset. At this point it continues to pull out items randomly from the buffer until it is empty. You must specify the buffer size, and it is important to make it large enough, or else shuffling will not be very effective.[1] Just don't exceed the amount of RAM you have, and even if you have plenty of it, there's no need to go beyond the dataset's size. You can provide a random seed if you want the same random order every time you run your program. For example, the following code creates and displays a dataset containing the integers 0 to 9, repeated 3 times, shuffled using a buffer of size 5 and a random seed of 42, and batched with a batch size of 7:

```
>>> dataset = tf.data.Dataset.range(10).repeat(3)  # 0 to 9, three times
>>> dataset = dataset.shuffle(buffer_size=5, seed=42).batch(7)
>>> for item in dataset:
...     print(item)
...
tf.Tensor([0 2 3 6 7 9 4], shape=(7,), dtype=int64)
tf.Tensor([5 0 1 1 8 6 5], shape=(7,), dtype=int64)
tf.Tensor([4 8 7 1 2 3 0], shape=(7,), dtype=int64)
tf.Tensor([5 4 2 7 8 9 9], shape=(7,), dtype=int64)
tf.Tensor([3 6], shape=(2,), dtype=int64)
```

> If you call repeat() on a shuffled dataset, by default it will generate a new order at every iteration. This is generally a good idea, but if you prefer to reuse the same order at each iteration (e.g., for tests or debugging), you can set reshuffle_each_iteration=False.

For a large dataset that does not fit in memory, this simple shuffling-buffer approach may not be sufficient, since the buffer will be small compared to the dataset. One solution is to shuffle the source data itself (for example, on Linux you can shuffle text files using the shuf command). This will definitely improve shuffling a lot! Even if the source data is shuffled, you will usually want to shuffle it some more, or else the same order will be repeated at each epoch, and the model may end up being biased (e.g., due to some spurious patterns present by chance in the source data's order). To shuffle the instances some more, a common approach is to split the source data into multiple files, then read them in a random order during training. However, instances located in the same file will still end up close to each other. To avoid this you can pick multiple files randomly and read them simultaneously, interleaving their records. Then on top of that you can add a shuffling buffer using the shuffle() method. If all

---

1 Imagine a sorted deck of cards on your left: suppose you just take the top three cards and shuffle them, then pick one randomly and put it to your right, keeping the other two in your hands. Take another card on your left, shuffle the three cards in your hands and pick one of them randomly, and put it on your right. When you are done going through all the cards like this, you will have a deck of cards on your right: do you think it will be perfectly shuffled?

this sounds like a lot of work, don't worry: the Data API makes all this possible in just a few lines of code. Let's see how to do this.

Interleaving lines from multiple files

First, let's suppose that you've loaded the California housing dataset, shuffled it (unless it was already shuffled), and split it into a training set, a validation set, and a test set. Then you split each set into many CSV files that each look like this (each row contains eight input features plus the target median house value):

```
MedInc,HouseAge,AveRooms,AveBedrms,Popul,AveOccup,Lat,Long,MedianHouseValue
3.5214,15.0,3.0499,1.1065,1447.0,1.6059,37.63,-122.43,1.442
5.3275,5.0,6.4900,0.9910,3464.0,3.4433,33.69,-117.39,1.687
3.1,29.0,7.5423,1.5915,1328.0,2.2508,38.44,-122.98,1.621
[...]
```

Let's also suppose `train_filepaths` contains the list of training file paths (and you also have `valid_filepaths` and `test_filepaths`):

```
>>> train_filepaths
['datasets/housing/my_train_00.csv', 'datasets/housing/my_train_01.csv',...]
```

Alternatively, you could use file patterns; for example, `train_filepaths = "data sets/housing/my_train_*.csv"`. Now let's create a dataset containing only these file paths:

```
filepath_dataset = tf.data.Dataset.list_files(train_filepaths, seed=42)
```

By default, the `list_files()` function returns a dataset that shuffles the file paths. In general this is a good thing, but you can set `shuffle=False` if you do not want that for some reason.

Next, you can call the `interleave()` method to read from five files at a time and interleave their lines (skipping the first line of each file, which is the header row, using the `skip()` method):

```
n_readers = 5
dataset = filepath_dataset.interleave(
    lambda filepath: tf.data.TextLineDataset(filepath).skip(1),
    cycle_length=n_readers)
```

The `interleave()` method will create a dataset that will pull five file paths from the `filepath_dataset`, and for each one it will call the function you gave it (a lambda in this example) to create a new dataset (in this case a `TextLineDataset`). To be clear, at this stage there will be seven datasets in all: the filepath dataset, the interleave dataset, and the five `TextLineDatasets` created internally by the interleave dataset. When we iterate over the interleave dataset, it will cycle through these five `TextLineDatasets`, reading one line at a time from each until all datasets are out of items. Then it will get the next five file paths from the `filepath_dataset` and interleave them the same way, and so on until it runs out of file paths.

 For interleaving to work best, it is preferable to have files of identical length; otherwise the ends of the longest files will not be interleaved.

By default, `interleave()` does not use parallelism; it just reads one line at a time from each file, sequentially. If you want it to actually read files in parallel, you can set the `num_parallel_calls` argument to the number of threads you want (note that the `map()` method also has this argument). You can even set it to `tf.data.experimen tal.AUTOTUNE` to make TensorFlow choose the right number of threads dynamically based on the available CPU (however, this is an experimental feature for now). Let's look at what the dataset contains now:

```
>>> for line in dataset.take(5):
...     print(line.numpy())
...
b'4.2083,44.0,5.3232,0.9171,846.0,2.3370,37.47,-122.2,2.782'
b'4.1812,52.0,5.7013,0.9965,692.0,2.4027,33.73,-118.31,3.215'
b'3.6875,44.0,4.5244,0.9930,457.0,3.1958,34.04,-118.15,1.625'
b'3.3456,37.0,4.5140,0.9084,458.0,3.2253,36.67,-121.7,2.526'
b'3.5214,15.0,3.0499,1.1065,1447.0,1.6059,37.63,-122.43,1.442'
```

These are the first rows (ignoring the header row) of five CSV files, chosen randomly. Looks good! But as you can see, these are just byte strings; we need to parse them and scale the data.

Preprocessing the Data

Let's implement a small function that will perform this preprocessing:

```
X_mean, X_std = [...] # mean and scale of each feature in the training set
n_inputs = 8

def preprocess(line):
    defs = [0.] * n_inputs + [tf.constant([], dtype=tf.float32)]
    fields = tf.io.decode_csv(line, record_defaults=defs)
    x = tf.stack(fields[:-1])
    y = tf.stack(fields[-1:])
    return (x - X_mean) / X_std, y
```

Let's walk through this code:

- First, the code assumes that we have precomputed the mean and standard deviation of each feature in the training set. X_mean and X_std are just 1D tensors (or NumPy arrays) containing eight floats, one per input feature.

- The preprocess() function takes one CSV line and starts by parsing it. For this it uses the tf.io.decode_csv() function, which takes two arguments: the first is the line to parse, and the second is an array containing the default value for each column in the CSV file. This array tells TensorFlow not only the default value for each column, but also the number of columns and their types. In this example, we tell it that all feature columns are floats and that missing values should default to 0, but we provide an empty array of type tf.float32 as the default value for the last column (the target): the array tells TensorFlow that this column contains floats, but that there is no default value, so it will raise an exception if it encounters a missing value.

- The decode_csv() function returns a list of scalar tensors (one per column), but we need to return 1D tensor arrays. So we call tf.stack() on all tensors except for the last one (the target): this will stack these tensors into a 1D array. We then do the same for the target value (this makes it a 1D tensor array with a single value, rather than a scalar tensor).

- Finally, we scale the input features by subtracting the feature means and then dividing by the feature standard deviations, and we return a tuple containing the scaled features and the target.

Let's test this preprocessing function:

```
>>> preprocess(b'4.2083,44.0,5.3232,0.9171,846.0,2.3370,37.47,-122.2,2.782')
(<tf.Tensor: id=6227, shape=(8,), dtype=float32, numpy=
array([ 0.16579159,  1.216324  , -0.05204564, -0.39215982, -0.5277444 ,
       -0.2633488 ,  0.8543046 , -1.3072058 ], dtype=float32)>,
 <tf.Tensor: [...], numpy=array([2.782], dtype=float32)>)
```

Looks good! We can now apply the function to the dataset.

Putting Everything Together

To make the code reusable, let's put together everything we have discussed so far into a small helper function: it will create and return a dataset that will efficiently load California housing data from multiple CSV files, preprocess it, shuffle it, optionally repeat it, and batch it (see Figure 13-2):

```
def csv_reader_dataset(filepaths, repeat=1, n_readers=5,
                       n_read_threads=None, shuffle_buffer_size=10000,
                       n_parse_threads=5, batch_size=32):
    dataset = tf.data.Dataset.list_files(filepaths)
    dataset = dataset.interleave(
        lambda filepath: tf.data.TextLineDataset(filepath).skip(1),
```

```
            cycle_length=n_readers, num_parallel_calls=n_read_threads)
    dataset = dataset.map(preprocess, num_parallel_calls=n_parse_threads)
    dataset = dataset.shuffle(shuffle_buffer_size).repeat(repeat)
    return dataset.batch(batch_size).prefetch(1)
```

Everything should make sense in this code, except the very last line (prefetch(1)), which is important for performance.

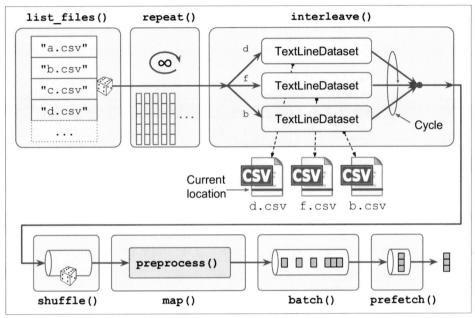

Figure 13-2. Loading and preprocessing data from multiple CSV files

Prefetching

By calling prefetch(1) at the end, we are creating a dataset that will do its best to always be one batch ahead.[2] In other words, while our training algorithm is working on one batch, the dataset will already be working in parallel on getting the next batch ready (e.g., reading the data from disk and preprocessing it). This can improve performance dramatically, as is illustrated in Figure 13-3. If we also ensure that loading and preprocessing are multithreaded (by setting num_parallel_calls when calling interleave() and map()), we can exploit multiple cores on the CPU and hopefully make preparing one batch of data shorter than running a training step on the GPU:

---

2 In general, just prefetching one batch is fine, but in some cases you may need to prefetch a few more. Alternatively, you can let TensorFlow decide automatically by passing tf.data.experimental.AUTOTUNE (this is an experimental feature for now).

this way the GPU will be almost 100% utilized (except for the data transfer time from the CPU to the GPU[3]), and training will run much faster.

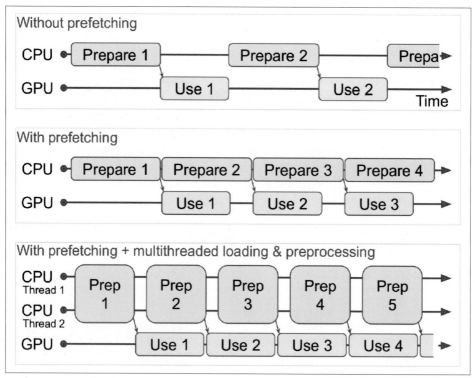

Figure 13-3. With prefetching, the CPU and the GPU work in parallel: as the GPU works on one batch, the CPU works on the next

If you plan to purchase a GPU card, its processing power and its memory size are of course very important (in particular, a large amount of RAM is crucial for computer vision). Just as important to get good performance is its *memory bandwidth*; this is the number of gigabytes of data it can get into or out of its RAM per second.

If the dataset is small enough to fit in memory, you can significantly speed up training by using the dataset's cache() method to cache its content to RAM. You should generally do this after loading and preprocessing the data, but before shuffling, repeating, batching, and prefetching. This way, each instance will only be read and

---

3 But check out the tf.data.experimental.prefetch_to_device() function, which can prefetch data directly to the GPU.

preprocessed once (instead of once per epoch), but the data will still be shuffled differently at each epoch, and the next batch will still be prepared in advance.

You now know how to build efficient input pipelines to load and preprocess data from multiple text files. We have discussed the most common dataset methods, but there are a few more you may want to look at: `concatenate()`, `zip()`, `window()`, `reduce()`, `shard()`, `flat_map()`, and `padded_batch()`. There are also a couple more class methods: `from_generator()` and `from_tensors()`, which create a new dataset from a Python generator or a list of tensors, respectively. Please check the API documentation for more details. Also note that there are experimental features available in `tf.data.experimental`, many of which will likely make it to the core API in future releases (e.g., check out the `CsvDataset` class, as well as the `make_csv_dataset()` method, which takes care of inferring the type of each column).

Using the Dataset with tf.keras

Now we can use the `csv_reader_dataset()` function to create a dataset for the training set. Note that we do not need to repeat it, as this will be taken care of by tf.keras. We also create datasets for the validation set and the test set:

```
train_set = csv_reader_dataset(train_filepaths)
valid_set = csv_reader_dataset(valid_filepaths)
test_set = csv_reader_dataset(test_filepaths)
```

And now we can simply build and train a Keras model using these datasets.[4] All we need to do is pass the training and validation datasets to the `fit()` method, instead of X_train, y_train, X_valid, and y_valid:[5]

```
model = keras.models.Sequential([...])
model.compile([...])
model.fit(train_set, epochs=10, validation_data=valid_set)
```

Similarly, we can pass a dataset to the `evaluate()` and `predict()` methods:

```
model.evaluate(test_set)
new_set = test_set.take(3).map(lambda X, y: X) # pretend we have 3 new instances
model.predict(new_set) # a dataset containing new instances
```

Unlike the other sets, the `new_set` will usually not contain labels (if it does, Keras will ignore them). Note that in all these cases, you can still use NumPy arrays instead of

---

[4] Support for datasets is specific to tf.keras; this will not work in other implementations of the Keras API.

[5] The `fit()` method will take care of repeating the training dataset. Alternatively, you could call `repeat()` on the training dataset so that it repeats forever and specify the `steps_per_epoch` argument when calling the `fit()` method. This may be useful in some rare cases, for example if you want to use a shuffle buffer that crosses over epochs.

datasets if you want (but of course they need to have been loaded and preprocessed first).

If you want to build your own custom training loop (as in Chapter 12), you can just iterate over the training set, very naturally:

```
for X_batch, y_batch in train_set:
    [...] # perform one Gradient Descent step
```

In fact, it is even possible to create a TF Function (see Chapter 12) that performs the whole training loop:

```
@tf.function
def train(model, optimizer, loss_fn, n_epochs, [...]):
    train_set = csv_reader_dataset(train_filepaths, repeat=n_epochs, [...])
    for X_batch, y_batch in train_set:
        with tf.GradientTape() as tape:
            y_pred = model(X_batch)
            main_loss = tf.reduce_mean(loss_fn(y_batch, y_pred))
            loss = tf.add_n([main_loss] + model.losses)
        grads = tape.gradient(loss, model.trainable_variables)
        optimizer.apply_gradients(zip(grads, model.trainable_variables))
```

Congratulations, you now know how to build powerful input pipelines using the Data API! However, so far we have used CSV files, which are common, simple, and convenient but not really efficient, and do not support large or complex data structures (such as images or audio) very well. So let's see how to use TFRecords instead.

 If you are happy with CSV files (or whatever other format you are using), you do not *have* to use TFRecords. As the saying goes, if it ain't broke, don't fix it! TFRecords are useful when the bottleneck during training is loading and parsing the data.

The TFRecord Format

The TFRecord format is TensorFlow's preferred format for storing large amounts of data and reading it efficiently. It is a very simple binary format that just contains a sequence of binary records of varying sizes (each record is comprised of a length, a CRC checksum to check that the length was not corrupted, then the actual data, and finally a CRC checksum for the data). You can easily create a TFRecord file using the `tf.io.TFRecordWriter` class:

```
with tf.io.TFRecordWriter("my_data.tfrecord") as f:
    f.write(b"This is the first record")
    f.write(b"And this is the second record")
```

And you can then use a `tf.data.TFRecordDataset` to read one or more TFRecord files:

```
filepaths = ["my_data.tfrecord"]
dataset = tf.data.TFRecordDataset(filepaths)
for item in dataset:
    print(item)
```

This will output:

```
tf.Tensor(b'This is the first record', shape=(), dtype=string)
tf.Tensor(b'And this is the second record', shape=(), dtype=string)
```

 By default, a TFRecordDataset will read files one by one, but you can make it read multiple files in parallel and interleave their records by setting num_parallel_reads. Alternatively, you could obtain the same result by using list_files() and interleave() as we did earlier to read multiple CSV files.

Compressed TFRecord Files

It can sometimes be useful to compress your TFRecord files, especially if they need to be loaded via a network connection. You can create a compressed TFRecord file by setting the options argument:

```
options = tf.io.TFRecordOptions(compression_type="GZIP")
with tf.io.TFRecordWriter("my_compressed.tfrecord", options) as f:
    [...]
```

When reading a compressed TFRecord file, you need to specify the compression type:

```
dataset = tf.data.TFRecordDataset(["my_compressed.tfrecord"],
                                  compression_type="GZIP")
```

A Brief Introduction to Protocol Buffers

Even though each record can use any binary format you want, TFRecord files usually contain serialized protocol buffers (also called *protobufs*). This is a portable, extensible, and efficient binary format developed at Google back in 2001 and made open source in 2008; protobufs are now widely used, in particular in gRPC (*https://grpc.io*), Google's remote procedure call system. They are defined using a simple language that looks like this:

```
syntax = "proto3";
message Person {
  string name = 1;
  int32 id = 2;
  repeated string email = 3;
}
```

This definition says we are using version 3 of the protobuf format, and it specifies that each `Person` object[6] may (optionally) have a `name` of type `string`, an id of type `int32`, and zero or more `email` fields, each of type `string`. The numbers 1, 2, and 3 are the field identifiers: they will be used in each record's binary representation. Once you have a definition in a *.proto* file, you can compile it. This requires `protoc`, the protobuf compiler, to generate access classes in Python (or some other language). Note that the protobuf definitions we will use have already been compiled for you, and their Python classes are part of TensorFlow, so you will not need to use `protoc`. All you need to know is how to use protobuf access classes in Python. To illustrate the basics, let's look at a simple example that uses the access classes generated for the `Person` protobuf (the code is explained in the comments):

```
>>> from person_pb2 import Person  # import the generated access class
>>> person = Person(name="Al", id=123, email=["a@b.com"])  # create a Person
>>> print(person)  # display the Person
name: "Al"
id: 123
email: "a@b.com"
>>> person.name  # read a field
"Al"
>>> person.name = "Alice"  # modify a field
>>> person.email[0]  # repeated fields can be accessed like arrays
"a@b.com"
>>> person.email.append("c@d.com")  # add an email address
>>> s = person.SerializeToString()  # serialize the object to a byte string
>>> s
b'\n\x05Alice\x10{\x1a\x07a@b.com\x1a\x07c@d.com'
>>> person2 = Person()  # create a new Person
>>> person2.ParseFromString(s)  # parse the byte string (27 bytes long)
27
>>> person == person2  # now they are equal
True
```

In short, we import the `Person` class generated by `protoc`, we create an instance and play with it, visualizing it and reading and writing some fields, then we serialize it using the `SerializeToString()` method. This is the binary data that is ready to be saved or transmitted over the network. When reading or receiving this binary data, we can parse it using the `ParseFromString()` method, and we get a copy of the object that was serialized.[7]

We could save the serialized `Person` object to a TFRecord file, then we could load and parse it: everything would work fine. However, `SerializeToString()` and `ParseFrom`

---

6 Since protobuf objects are meant to be serialized and transmitted, they are called *messages*.

7 This chapter contains the bare minimum you need to know about protobufs to use TFRecords. To learn more about protobufs, please visit *https://homl.info/protobuf*.

String() are not TensorFlow operations (and neither are the other operations in this code), so they cannot be included in a TensorFlow Function (except by wrapping them in a tf.py_function() operation, which would make the code slower and less portable, as we saw in Chapter 12). Fortunately, TensorFlow does include special protobuf definitions for which it provides parsing operations.

TensorFlow Protobufs

The main protobuf typically used in a TFRecord file is the Example protobuf, which represents one instance in a dataset. It contains a list of named features, where each feature can either be a list of byte strings, a list of floats, or a list of integers. Here is the protobuf definition:

```
syntax = "proto3";
message BytesList { repeated bytes value = 1; }
message FloatList { repeated float value = 1 [packed = true]; }
message Int64List { repeated int64 value = 1 [packed = true]; }
message Feature {
    oneof kind {
        BytesList bytes_list = 1;
        FloatList float_list = 2;
        Int64List int64_list = 3;
    }
};
message Features { map<string, Feature> feature = 1; };
message Example { Features features = 1; };
```

The definitions of BytesList, FloatList, and Int64List are straightforward enough. Note that [packed = true] is used for repeated numerical fields, for a more efficient encoding. A Feature contains either a BytesList, a FloatList, or an Int64List. A Features (with an s) contains a dictionary that maps a feature name to the corresponding feature value. And finally, an Example contains only a Features object.[8] Here is how you could create a tf.train.Example representing the same person as earlier and write it to a TFRecord file:

```
from tensorflow.train import BytesList, FloatList, Int64List
from tensorflow.train import Feature, Features, Example

person_example = Example(
    features=Features(
        feature={
            "name": Feature(bytes_list=BytesList(value=[b"Alice"])),
```

---

[8] Why was Example even defined, since it contains no more than a Features object? Well, TensorFlow's developers may one day decide to add more fields to it. As long as the new Example definition still contains the features field, with the same ID, it will be backward compatible. This extensibility is one of the great features of protobufs.

```
        "id": Feature(int64_list=Int64List(value=[123])),
        "emails": Feature(bytes_list=BytesList(value=[b"a@b.com",
                                                       b"c@d.com"]))
    }))
```

The code is a bit verbose and repetitive, but it's rather straightforward (and you could easily wrap it inside a small helper function). Now that we have an Example protobuf, we can serialize it by calling its SerializeToString() method, then write the resulting data to a TFRecord file:

```
with tf.io.TFRecordWriter("my_contacts.tfrecord") as f:
    f.write(person_example.SerializeToString())
```

Normally you would write much more than one Example! Typically, you would create a conversion script that reads from your current format (say, CSV files), creates an Example protobuf for each instance, serializes them, and saves them to several TFRecord files, ideally shuffling them in the process. This requires a bit of work, so once again make sure it is really necessary (perhaps your pipeline works fine with CSV files).

Now that we have a nice TFRecord file containing a serialized Example, let's try to load it.

Loading and Parsing Examples

To load the serialized Example protobufs, we will use a tf.data.TFRecordDataset once again, and we will parse each Example using tf.io.parse_single_example(). This is a TensorFlow operation, so it can be included in a TF Function. It requires at least two arguments: a string scalar tensor containing the serialized data, and a description of each feature. The description is a dictionary that maps each feature name to either a tf.io.FixedLenFeature descriptor indicating the feature's shape, type, and default value, or a tf.io.VarLenFeature descriptor indicating only the type (if the length of the feature's list may vary, such as for the "emails" feature).

The following code defines a description dictionary, then it iterates over the TFRecord Dataset and parses the serialized Example protobuf this dataset contains:

```
feature_description = {
    "name": tf.io.FixedLenFeature([], tf.string, default_value=""),
    "id": tf.io.FixedLenFeature([], tf.int64, default_value=0),
    "emails": tf.io.VarLenFeature(tf.string),
}

for serialized_example in tf.data.TFRecordDataset(["my_contacts.tfrecord"]):
    parsed_example = tf.io.parse_single_example(serialized_example,
                                                feature_description)
```

The fixed-length features are parsed as regular tensors, but the variable-length features are parsed as sparse tensors. You can convert a sparse tensor to a dense tensor using `tf.sparse.to_dense()`, but in this case it is simpler to just access its values:

```
>>> tf.sparse.to_dense(parsed_example["emails"], default_value=b"")
<tf.Tensor: [...] dtype=string, numpy=array([b'a@b.com', b'c@d.com'], [...])>
>>> parsed_example["emails"].values
<tf.Tensor: [...] dtype=string, numpy=array([b'a@b.com', b'c@d.com'], [...])>
```

A `BytesList` can contain any binary data you want, including any serialized object. For example, you can use `tf.io.encode_jpeg()` to encode an image using the JPEG format and put this binary data in a `BytesList`. Later, when your code reads the TFRecord, it will start by parsing the `Example`, then it will need to call `tf.io.decode_jpeg()` to parse the data and get the original image (or you can use `tf.io.decode_image()`, which can decode any BMP, GIF, JPEG, or PNG image). You can also store any tensor you want in a `BytesList` by serializing the tensor using `tf.io.serialize_tensor()` then putting the resulting byte string in a `BytesList` feature. Later, when you parse the TFRecord, you can parse this data using `tf.io.parse_tensor()`.

Instead of parsing examples one by one using `tf.io.parse_single_example()`, you may want to parse them batch by batch using `tf.io.parse_example()`:

```
dataset = tf.data.TFRecordDataset(["my_contacts.tfrecord"]).batch(10)
for serialized_examples in dataset:
    parsed_examples = tf.io.parse_example(serialized_examples,
                                          feature_description)
```

As you can see, the `Example` protobuf will probably be sufficient for most use cases. However, it may be a bit cumbersome to use when you are dealing with lists of lists. For example, suppose you want to classify text documents. Each document may be represented as a list of sentences, where each sentence is represented as a list of words. And perhaps each document also has a list of comments, where each comment is represented as a list of words. There may be some contextual data too, such as the document's author, title, and publication date. TensorFlow's `SequenceExample` protobuf is designed for such use cases.

Handling Lists of Lists Using the SequenceExample Protobuf

Here is the definition of the `SequenceExample` protobuf:

```
message FeatureList { repeated Feature feature = 1; };
message FeatureLists { map<string, FeatureList> feature_list = 1; };
message SequenceExample {
    Features context = 1;
    FeatureLists feature_lists = 2;
};
```

A `SequenceExample` contains a `Features` object for the contextual data and a `Fea tureLists` object that contains one or more named `FeatureList` objects (e.g., a `Fea tureList` named `"content"` and another named `"comments"`). Each `FeatureList` contains a list of `Feature` objects, each of which may be a list of byte strings, a list of 64-bit integers, or a list of floats (in this example, each `Feature` would represent a sentence or a comment, perhaps in the form of a list of word identifiers). Building a `SequenceExample`, serializing it, and parsing it is similar to building, serializing, and parsing an `Example`, but you must use `tf.io.parse_single_sequence_example()` to parse a single `SequenceExample` or `tf.io.parse_sequence_example()` to parse a batch. Both functions return a tuple containing the context features (as a dictionary) and the feature lists (also as a dictionary). If the feature lists contain sequences of varying sizes (as in the preceding example), you may want to convert them to ragged tensors, using `tf.RaggedTensor.from_sparse()` (see the notebook for the full code):

```
parsed_context, parsed_feature_lists = tf.io.parse_single_sequence_example(
    serialized_sequence_example, context_feature_descriptions,
    sequence_feature_descriptions)
parsed_content = tf.RaggedTensor.from_sparse(parsed_feature_lists["content"])
```

Now that you know how to efficiently store, load, and parse data, the next step is to prepare it so that it can be fed to a neural network.

Preprocessing the Input Features

Preparing your data for a neural network requires converting all features into numerical features, generally normalizing them, and more. In particular, if your data contains categorical features or text features, they need to be converted to numbers. This can be done ahead of time when preparing your data files, using any tool you like (e.g., NumPy, pandas, or Scikit-Learn). Alternatively, you can preprocess your data on the fly when loading it with the Data API (e.g., using the dataset's `map()` method, as we saw earlier), or you can include a preprocessing layer directly in your model. Let's look at this last option now.

For example, here is how you can implement a standardization layer using a `Lambda` layer. For each feature, it subtracts the mean and divides by its standard deviation (plus a tiny smoothing term to avoid division by zero):

```
means = np.mean(X_train, axis=0, keepdims=True)
stds = np.std(X_train, axis=0, keepdims=True)
eps = keras.backend.epsilon()
model = keras.models.Sequential([
    keras.layers.Lambda(lambda inputs: (inputs - means) / (stds + eps)),
    [...] # other layers
])
```

That's not too hard! However, you may prefer to use a nice self-contained custom layer (much like Scikit-Learn's `StandardScaler`), rather than having global variables like `means` and `stds` dangling around:

```python
class Standardization(keras.layers.Layer):
    def adapt(self, data_sample):
        self.means_ = np.mean(data_sample, axis=0, keepdims=True)
        self.stds_ = np.std(data_sample, axis=0, keepdims=True)
    def call(self, inputs):
        return (inputs - self.means_) / (self.stds_ + keras.backend.epsilon())
```

Before you can use this standardization layer, you will need to adapt it to your dataset by calling the `adapt()` method and passing it a data sample. This will allow it to use the appropriate mean and standard deviation for each feature:

```python
std_layer = Standardization()
std_layer.adapt(data_sample)
```

This sample must be large enough to be representative of your dataset, but it does not have to be the full training set: in general, a few hundred randomly selected instances will suffice (however, this depends on your task). Next, you can use this preprocessing layer like a normal layer:

```python
model = keras.Sequential()
model.add(std_layer)
[...] # create the rest of the model
model.compile([...])
model.fit([...])
```

If you are thinking that Keras should contain a standardization layer like this one, here's some good news for you: by the time you read this, the `keras.layers.Normalization` layer will probably be available. It will work very much like our custom `Standardization` layer: first, create the layer, then adapt it to your dataset by passing a data sample to the `adapt()` method, and finally use the layer normally.

Now let's look at categorical features. We will start by encoding them as one-hot vectors.

Encoding Categorical Features Using One-Hot Vectors

Consider the `ocean_proximity` feature in the California housing dataset we explored in Chapter 2: it is a categorical feature with five possible values: `"<1H OCEAN"`, `"INLAND"`, `"NEAR OCEAN"`, `"NEAR BAY"`, and `"ISLAND"`. We need to encode this feature before we feed it to a neural network. Since there are very few categories, we can use one-hot encoding. For this, we first need to map each category to its index (0 to 4), which can be done using a lookup table:

```python
vocab = ["<1H OCEAN", "INLAND", "NEAR OCEAN", "NEAR BAY", "ISLAND"]
indices = tf.range(len(vocab), dtype=tf.int64)
```

```
table_init = tf.lookup.KeyValueTensorInitializer(vocab, indices)
num_oov_buckets = 2
table = tf.lookup.StaticVocabularyTable(table_init, num_oov_buckets)
```

Let's go through this code:

- We first define the *vocabulary*: this is the list of all possible categories.
- Then we create a tensor with the corresponding indices (0 to 4).
- Next, we create an initializer for the lookup table, passing it the list of categories and their corresponding indices. In this example, we already have this data, so we use a `KeyValueTensorInitializer`; but if the categories were listed in a text file (with one category per line), we would use a `TextFileInitializer` instead.
- In the last two lines we create the lookup table, giving it the initializer and specifying the number of *out-of-vocabulary* (oov) buckets. If we look up a category that does not exist in the vocabulary, the lookup table will compute a hash of this category and use it to assign the unknown category to one of the oov buckets. Their indices start after the known categories, so in this example the indices of the two oov buckets are 5 and 6.

Why use oov buckets? Well, if the number of categories is large (e.g., zip codes, cities, words, products, or users) and the dataset is large as well, or it keeps changing, then getting the full list of categories may not be convenient. One solution is to define the vocabulary based on a data sample (rather than the whole training set) and add some oov buckets for the other categories that were not in the data sample. The more unknown categories you expect to find during training, the more oov buckets you should use. Indeed, if there are not enough oov buckets, there will be collisions: different categories will end up in the same bucket, so the neural network will not be able to distinguish them (at least not based on this feature).

Now let's use the lookup table to encode a small batch of categorical features to one-hot vectors:

```
>>> categories = tf.constant(["NEAR BAY", "DESERT", "INLAND", "INLAND"])
>>> cat_indices = table.lookup(categories)
>>> cat_indices
<tf.Tensor: id=514, shape=(4,), dtype=int64, numpy=array([3, 5, 1, 1])>
>>> cat_one_hot = tf.one_hot(cat_indices, depth=len(vocab) + num_oov_buckets)
>>> cat_one_hot
<tf.Tensor: id=524, shape=(4, 7), dtype=float32, numpy=
array([[0., 0., 0., 1., 0., 0., 0.],
       [0., 0., 0., 0., 0., 1., 0.],
       [0., 1., 0., 0., 0., 0., 0.],
       [0., 1., 0., 0., 0., 0., 0.]], dtype=float32)>
```

As you can see, "NEAR BAY" was mapped to index 3, the unknown category "DESERT" was mapped to one of the two oov buckets (at index 5), and "INLAND" was mapped to index 1, twice. Then we used `tf.one_hot()` to one-hot encode these indices. Notice that we have to tell this function the total number of indices, which is equal to the vocabulary size plus the number of oov buckets. Now you know how to encode categorical features to one-hot vectors using TensorFlow!

Just like earlier, it wouldn't be too difficult to bundle all of this logic into a nice self-contained class. Its `adapt()` method would take a data sample and extract all the distinct categories it contains. It would create a lookup table to map each category to its index (including unknown categories using oov buckets). Then its `call()` method would use the lookup table to map the input categories to their indices. Well, here's more good news: by the time you read this, Keras will probably include a layer called `keras.layers.TextVectorization`, which will be capable of doing exactly that: its `adapt()` method will extract the vocabulary from a data sample, and its `call()` method will convert each category to its index in the vocabulary. You could add this layer at the beginning of your model, followed by a `Lambda` layer that would apply the `tf.one_hot()` function, if you want to convert these indices to one-hot vectors.

This may not be the best solution, though. The size of each one-hot vector is the vocabulary length plus the number of oov buckets. This is fine when there are just a few possible categories, but if the vocabulary is large, it is much more efficient to encode them using *embeddings* instead.

 As a rule of thumb, if the number of categories is lower than 10, then one-hot encoding is generally the way to go (but your mileage may vary!). If the number of categories is greater than 50 (which is often the case when you use hash buckets), then embeddings are usually preferable. In between 10 and 50 categories, you may want to experiment with both options and see which one works best for your use case.

Encoding Categorical Features Using Embeddings

An embedding is a trainable dense vector that represents a category. By default, embeddings are initialized randomly, so for example the `"NEAR BAY"` category could be represented initially by a random vector such as `[0.131, 0.890]`, while the `"NEAR OCEAN"` category might be represented by another random vector such as `[0.631, 0.791]`. In this example, we use 2D embeddings, but the number of dimensions is a hyperparameter you can tweak. Since these embeddings are trainable, they will gradually improve during training; and as they represent fairly similar categories, Gradient Descent will certainly end up pushing them closer together, while it will tend to move them away from the `"INLAND"` category's embedding (see Figure 13-4). Indeed, the better the representation, the easier it will be for the neural network to make accurate predictions, so training tends to make embeddings useful representations of the categories. This is called *representation learning* (we will see other types of representation learning in Chapter 17).

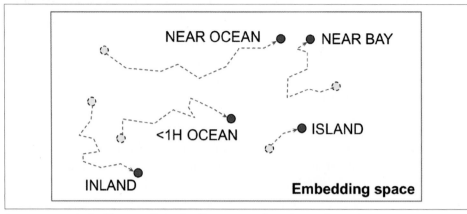

Figure 13-4. Embeddings will gradually improve during training

Word Embeddings

Not only will embeddings generally be useful representations for the task at hand, but quite often these same embeddings can be reused successfully for other tasks. The most common example of this is *word embeddings* (i.e., embeddings of individual words): when you are working on a natural language processing task, you are often better off reusing pretrained word embeddings than training your own.

The idea of using vectors to represent words dates back to the 1960s, and many sophisticated techniques have been used to generate useful vectors, including using neural networks. But things really took off in 2013, when Tomáš Mikolov and other Google researchers published a paper (*https://homl.info/word2vec*)[9] describing an efficient technique to learn word embeddings using neural networks, significantly outperforming previous attempts. This allowed them to learn embeddings on a very large corpus of text: they trained a neural network to predict the words near any given word, and obtained astounding word embeddings. For example, synonyms had very close embeddings, and semantically related words such as France, Spain, and Italy ended up clustered together.

It's not just about proximity, though: word embeddings were also organized along meaningful axes in the embedding space. Here is a famous example: if you compute King – Man + Woman (adding and subtracting the embedding vectors of these words), then the result will be very close to the embedding of the word Queen (see

---

9 Tomas Mikolov et al., "Distributed Representations of Words and Phrases and Their Compositionality," *Proceedings of the 26th International Conference on Neural Information Processing Systems* 2 (2013): 3111–3119.

Figure 13-5). In other words, the word embeddings encode the concept of gender! Similarly, you can compute Madrid – Spain + France, and the result is close to Paris, which seems to show that the notion of capital city was also encoded in the embeddings.

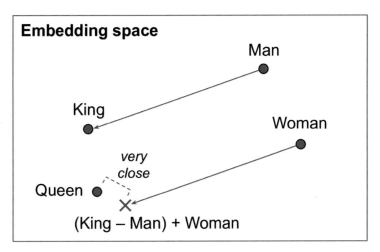

Figure 13-5. Word embeddings of similar words tend to be close, and some axes seem to encode meaningful concepts

Unfortunately, word embeddings sometimes capture our worst biases. For example, although they correctly learn that Man is to King as Woman is to Queen, they also seem to learn that Man is to Doctor as Woman is to Nurse: quite a sexist bias! To be fair, this particular example is probably exaggerated, as was pointed out in a 2019 paper (*https://homl.info/fairembeds*)[10] by Malvina Nissim et al. Nevertheless, ensuring fairness in Deep Learning algorithms is an important and active research topic.

Let's look at how we could implement embeddings manually, to understand how they work (then we will use a simple Keras layer instead). First, we need to create an *embedding matrix* containing each category's embedding, initialized randomly; it will have one row per category and per oov bucket, and one column per embedding dimension:

```
embedding_dim = 2
embed_init = tf.random.uniform([len(vocab) + num_oov_buckets, embedding_dim])
embedding_matrix = tf.Variable(embed_init)
```

---

10 Malvina Nissim et al., "Fair Is Better Than Sensational: Man Is to Doctor as Woman Is to Doctor," arXiv pre-print arXiv:1905.09866 (2019).

In this example we are using 2D embeddings, but as a rule of thumb embeddings typically have 10 to 300 dimensions, depending on the task and the vocabulary size (you will have to tune this hyperparameter).

This embedding matrix is a random 6 × 2 matrix, stored in a variable (so it can be tweaked by Gradient Descent during training):

```
>>> embedding_matrix
<tf.Variable 'Variable:0' shape=(6, 2) dtype=float32, numpy=
array([[0.6645621 , 0.44100678],
       [0.3528825 , 0.46448255],
       [0.03366041, 0.68467236],
       [0.74011743, 0.8724445 ],
       [0.22632635, 0.22319686],
       [0.3103881 , 0.7223358 ]], dtype=float32)>
```

Now let's encode the same batch of categorical features as earlier, but this time using these embeddings:

```
>>> categories = tf.constant(["NEAR BAY", "DESERT", "INLAND", "INLAND"])
>>> cat_indices = table.lookup(categories)
>>> cat_indices
<tf.Tensor: id=741, shape=(4,), dtype=int64, numpy=array([3, 5, 1, 1])>
>>> tf.nn.embedding_lookup(embedding_matrix, cat_indices)
<tf.Tensor: id=864, shape=(4, 2), dtype=float32, numpy=
array([[0.74011743, 0.8724445 ],
       [0.3103881 , 0.7223358 ],
       [0.3528825 , 0.46448255],
       [0.3528825 , 0.46448255]], dtype=float32)>
```

The `tf.nn.embedding_lookup()` function looks up the rows in the embedding matrix, at the given indices—that's all it does. For example, the lookup table says that the `"INLAND"` category is at index 1, so the `tf.nn.embedding_lookup()` function returns the embedding at row 1 in the embedding matrix (twice): [0.3528825, 0.46448255].

Keras provides a `keras.layers.Embedding` layer that handles the embedding matrix (trainable, by default); when the layer is created it initializes the embedding matrix randomly, and then when it is called with some category indices it returns the rows at those indices in the embedding matrix:

```
>>> embedding = keras.layers.Embedding(input_dim=len(vocab) + num_oov_buckets,
...                                    output_dim=embedding_dim)
...
>>> embedding(cat_indices)
<tf.Tensor: id=814, shape=(4, 2), dtype=float32, numpy=
array([[ 0.02401174,  0.03724445],
       [-0.01896119,  0.02223358],
       [-0.01471175, -0.00355174],
       [-0.01471175, -0.00355174]], dtype=float32)>
```

Putting everything together, we can now create a Keras model that can process categorical features (along with regular numerical features) and learn an embedding for each category (as well as for each oov bucket):

```
regular_inputs = keras.layers.Input(shape=[8])
categories = keras.layers.Input(shape=[], dtype=tf.string)
cat_indices = keras.layers.Lambda(lambda cats: table.lookup(cats))(categories)
cat_embed = keras.layers.Embedding(input_dim=6, output_dim=2)(cat_indices)
encoded_inputs = keras.layers.concatenate([regular_inputs, cat_embed])
outputs = keras.layers.Dense(1)(encoded_inputs)
model = keras.models.Model(inputs=[regular_inputs, categories],
                           outputs=[outputs])
```

This model takes two inputs: a regular input containing eight numerical features per instance, plus a categorical input (containing one categorical feature per instance). It uses a Lambda layer to look up each category's index, then it looks up the embeddings for these indices. Next, it concatenates the embeddings and the regular inputs in order to give the encoded inputs, which are ready to be fed to a neural network. We could add any kind of neural network at this point, but we just add a dense output layer, and we create the Keras model.

When the keras.layers.TextVectorization layer is available, you can call its adapt() method to make it extract the vocabulary from a data sample (it will take care of creating the lookup table for you). Then you can add it to your model, and it will perform the index lookup (replacing the Lambda layer in the previous code example).

> One-hot encoding followed by a Dense layer (with no activation function and no biases) is equivalent to an Embedding layer. However, the Embedding layer uses way fewer computations (the performance difference becomes clear when the size of the embedding matrix grows). The Dense layer's weight matrix plays the role of the embedding matrix. For example, using one-hot vectors of size 20 and a Dense layer with 10 units is equivalent to using an Embedding layer with input_dim=20 and output_dim=10. As a result, it would be wasteful to use more embedding dimensions than the number of units in the layer that follows the Embedding layer.

Now let's look a bit more closely at the Keras preprocessing layers.

Keras Preprocessing Layers

The TensorFlow team is working on providing a set of standard Keras preprocessing layers (*https://homl.info/preproc*). They will probably be available by the time you read this; however, the API may change slightly by then, so please refer to the notebook for this chapter if anything behaves unexpectedly. This new API will likely supersede the existing Feature Columns API, which is harder to use and less intuitive (if you want to learn more about the Feature Columns API anyway, please check out the notebook for this chapter).

We already discussed two of these layers: the keras.layers.Normalization layer that will perform feature standardization (it will be equivalent to the Standardization layer we defined earlier), and the TextVectorization layer that will be capable of encoding each word in the inputs into its index in the vocabulary. In both cases, you create the layer, you call its adapt() method with a data sample, and then you use the layer normally in your model. The other preprocessing layers will follow the same pattern.

The API will also include a keras.layers.Discretization layer that will chop continuous data into different bins and encode each bin as a one-hot vector. For example, you could use it to discretize prices into three categories, (low, medium, high), which would be encoded as [1, 0, 0], [0, 1, 0], and [0, 0, 1], respectively. Of course this loses a lot of information, but in some cases it can help the model detect patterns that would otherwise not be obvious when just looking at the continuous values.

The Discretization layer will not be differentiable, and it should only be used at the start of your model. Indeed, the model's preprocessing layers will be frozen during training, so their parameters will not be affected by Gradient Descent, and thus they do not need to be differentiable. This also means that you should not use an Embedding layer directly in a custom preprocessing layer, if you want it to be trainable: instead, it should be added separately to your model, as in the previous code example.

It will also be possible to chain multiple preprocessing layers using the PreprocessingStage class. For example, the following code will create a preprocessing pipeline that will first normalize the inputs, then discretize them (this may remind you of Scikit-Learn pipelines). After you adapt this pipeline to a data sample, you can use it like a regular layer in your models (but again, only at the start of the model, since it contains a nondifferentiable preprocessing layer):

```
normalization = keras.layers.Normalization()
discretization = keras.layers.Discretization([...])
pipeline = keras.layers.PreprocessingStage([normalization, discretization])
pipeline.adapt(data_sample)
```

The TextVectorization layer will also have an option to output word-count vectors instead of word indices. For example, if the vocabulary contains three words, say ["and", "basketball", "more"], then the text "more and more" will be mapped to the vector [1, 0, 2]: the word "and" appears once, the word "basketball" does not appear at all, and the word "more" appears twice. This text representation is called a *bag of words*, since it completely loses the order of the words. Common words like "and" will have a large value in most texts, even though they are usually the least interesting (e.g., in the text "more and more basketball" the word "basketball" is clearly the most important, precisely because it is not a very frequent word). So, the word counts should be normalized in a way that reduces the importance of frequent words. A common way to do this is to divide each word count by the log of the total number of training instances in which the word appears. This technique is called *Term-Frequency × Inverse-Document-Frequency* (TF-IDF). For example, let's imagine that the words "and", "basketball", and "more" appear respectively in 200, 10, and 100 text instances in the training set: in this case, the final vector will be [1/ log(200), 0/log(10), 2/log(100)], which is approximately equal to [0.19, 0., 0.43]. The TextVectorization layer will (likely) have an option to perform TF-IDF.

If the standard preprocessing layers are insufficient for your task, you will still have the option to create your own custom preprocessing layer, much like we did earlier with the Standardization class. Create a subclass of the keras.layers.PreprocessingLayer class with an adapt() method, which should take a data_sample argument and optionally an extra reset_state argument: if True, then the adapt() method should reset any existing state before computing the new state; if False, it should try to update the existing state.

As you can see, these Keras preprocessing layers will make preprocessing much easier! Now, whether you choose to write your own preprocessing layers or use Keras's (or even use the Feature Columns API), all the preprocessing will be done on the fly. During training, however, it may be preferable to perform preprocessing ahead of time. Let's see why we'd want to do that and how we'd go about it.

TF Transform

If preprocessing is computationally expensive, then handling it before training rather than on the fly may give you a significant speedup: the data will be preprocessed just once per instance *before* training, rather than once per instance and per epoch *during* training. As mentioned earlier, if the dataset is small enough to fit in RAM, you can use its cache() method. But if it is too large, then tools like Apache Beam or Spark will help. They let you run efficient data processing pipelines over large amounts of data, even distributed across multiple servers, so you can use them to preprocess all the training data before training.

This works great and indeed can speed up training, but there is one problem: once your model is trained, suppose you want to deploy it to a mobile app. In that case you will need to write some code in your app to take care of preprocessing the data before it is fed to the model. And suppose you also want to deploy the model to Tensor-Flow.js so that it runs in a web browser? Once again, you will need to write some pre-processing code. This can become a maintenance nightmare: whenever you want to change the preprocessing logic, you will need to update your Apache Beam code, your mobile app code, and your JavaScript code. This is not only time-consuming, but also error-prone: you may end up with subtle differences between the preprocessing operations performed before training and the ones performed in your app or in the browser. This *training/serving skew* will lead to bugs or degraded performance.

One improvement would be to take the trained model (trained on data that was preprocessed by your Apache Beam or Spark code) and, before deploying it to your app or the browser, add extra preprocessing layers to take care of preprocessing on the fly. That's definitely better, since now you just have two versions of your preprocessing code: the Apache Beam or Spark code, and the preprocessing layers' code.

But what if you could define your preprocessing operations just once? This is what TF Transform was designed for. It is part of TensorFlow Extended (TFX) (*https://tensorflow.org/tfx*), an end-to-end platform for productionizing TensorFlow models. First, to use a TFX component such as TF Transform, you must install it; it does not come bundled with TensorFlow. You then define your preprocessing function just once (in Python), by using TF Transform functions for scaling, bucketizing, and more. You can also use any TensorFlow operation you need. Here is what this preprocessing function might look like if we just had two features:

```python
import tensorflow_transform as tft

def preprocess(inputs):  # inputs = a batch of input features
    median_age = inputs["housing_median_age"]
    ocean_proximity = inputs["ocean_proximity"]
    standardized_age = tft.scale_to_z_score(median_age)
    ocean_proximity_id = tft.compute_and_apply_vocabulary(ocean_proximity)
    return {
        "standardized_median_age": standardized_age,
        "ocean_proximity_id": ocean_proximity_id
    }
```

Next, TF Transform lets you apply this `preprocess()` function to the whole training set using Apache Beam (it provides an `AnalyzeAndTransformDataset` class that you can use for this purpose in your Apache Beam pipeline). In the process, it will also compute all the necessary statistics over the whole training set: in this example, the mean and standard deviation of the `housing_median_age` feature, and the vocabulary for the `ocean_proximity` feature. The components that compute these statistics are called *analyzers*.

Importantly, TF Transform will also generate an equivalent TensorFlow Function that you can plug into the model you deploy. This TF Function includes some constants that correspond to all the all the necessary statistics computed by Apache Beam (the mean, standard deviation, and vocabulary).

With the Data API, TFRecords, the Keras preprocessing layers, and TF Transform, you can build highly scalable input pipelines for training and benefit from fast and portable data preprocessing in production.

But what if you just wanted to use a standard dataset? Well in that case, things are much simpler: just use TFDS!

The TensorFlow Datasets (TFDS) Project

The TensorFlow Datasets (*https://tensorflow.org/datasets*) project makes it very easy to download common datasets, from small ones like MNIST or Fashion MNIST to huge datasets like ImageNet (you will need quite a bit of disk space!). The list includes image datasets, text datasets (including translation datasets), and audio and video datasets. You can visit *https://homl.info/tfds* to view the full list, along with a description of each dataset.

TFDS is not bundled with TensorFlow, so you need to install the `tensorflow-datasets` library (e.g., using pip). Then call the `tfds.load()` function, and it will download the data you want (unless it was already downloaded earlier) and return the data as a dictionary of datasets (typically one for training and one for testing, but this depends on the dataset you choose). For example, let's download MNIST:

```
import tensorflow_datasets as tfds

dataset = tfds.load(name="mnist")
mnist_train, mnist_test = dataset["train"], dataset["test"]
```

You can then apply any transformation you want (typically shuffling, batching, and prefetching), and you're ready to train your model. Here is a simple example:

```
mnist_train = mnist_train.shuffle(10000).batch(32).prefetch(1)
for item in mnist_train:
    images = item["image"]
    labels = item["label"]
    [...]
```

 The `load()` function shuffles each data shard it downloads (only for the training set). This may not be sufficient, so it's best to shuffle the training data some more.

Note that each item in the dataset is a dictionary containing both the features and the labels. But Keras expects each item to be a tuple containing two elements (again, the features and the labels). You could transform the dataset using the `map()` method, like this:

```
mnist_train = mnist_train.shuffle(10000).batch(32)
mnist_train = mnist_train.map(lambda items: (items["image"], items["label"]))
mnist_train = mnist_train.prefetch(1)
```

But it's simpler to ask the `load()` function to do this for you by setting `as_super vised=True` (obviously this works only for labeled datasets). You can also specify the batch size if you want. Then you can pass the dataset directly to your tf.keras model:

```
dataset = tfds.load(name="mnist", batch_size=32, as_supervised=True)
mnist_train = dataset["train"].prefetch(1)
model = keras.models.Sequential([...])
model.compile(loss="sparse_categorical_crossentropy", optimizer="sgd")
model.fit(mnist_train, epochs=5)
```

This was quite a technical chapter, and you may feel that it is a bit far from the abstract beauty of neural networks, but the fact is Deep Learning often involves large amounts of data, and knowing how to load, parse, and preprocess it efficiently is a crucial skill to have. In the next chapter, we will look at convolutional neural networks, which are among the most successful neural net architectures for image processing and many other applications.

Exercises

1. Why would you want to use the Data API?
2. What are the benefits of splitting a large dataset into multiple files?
3. During training, how can you tell that your input pipeline is the bottleneck? What can you do to fix it?
4. Can you save any binary data to a TFRecord file, or only serialized protocol buffers?
5. Why would you go through the hassle of converting all your data to the `Example` protobuf format? Why not use your own protobuf definition?
6. When using TFRecords, when would you want to activate compression? Why not do it systematically?
7. Data can be preprocessed directly when writing the data files, or within the tf.data pipeline, or in preprocessing layers within your model, or using TF Transform. Can you list a few pros and cons of each option?
8. Name a few common techniques you can use to encode categorical features. What about text?
9. Load the Fashion MNIST dataset (introduced in Chapter 10); split it into a training set, a validation set, and a test set; shuffle the training set; and save each

dataset to multiple TFRecord files. Each record should be a serialized `Example` protobuf with two features: the serialized image (use `tf.io.serialize_tensor()` to serialize each image), and the label.[11] Then use tf.data to create an efficient dataset for each set. Finally, use a Keras model to train these datasets, including a preprocessing layer to standardize each input feature. Try to make the input pipeline as efficient as possible, using TensorBoard to visualize profiling data.

10. In this exercise you will download a dataset, split it, create a `tf.data.Dataset` to load it and preprocess it efficiently, then build and train a binary classification model containing an `Embedding` layer:

a. Download the Large Movie Review Dataset (*https://homl.info/imdb*), which contains 50,000 movies reviews from the Internet Movie Database (*https://imdb.com/*). The data is organized in two directories, *train* and *test*, each containing a *pos* subdirectory with 12,500 positive reviews and a *neg* subdirectory with 12,500 negative reviews. Each review is stored in a separate text file. There are other files and folders (including preprocessed bag-of-words), but we will ignore them in this exercise.

b. Split the test set into a validation set (15,000) and a test set (10,000).

c. Use tf.data to create an efficient dataset for each set.

d. Create a binary classification model, using a `TextVectorization` layer to preprocess each review. If the `TextVectorization` layer is not yet available (or if you like a challenge), try to create your own custom preprocessing layer: you can use the functions in the `tf.strings` package, for example `lower()` to make everything lowercase, `regex_replace()` to replace punctuation with spaces, and `split()` to split words on spaces. You should use a lookup table to output word indices, which must be prepared in the `adapt()` method.

e. Add an `Embedding` layer and compute the mean embedding for each review, multiplied by the square root of the number of words (see Chapter 16). This rescaled mean embedding can then be passed to the rest of your model.

f. Train the model and see what accuracy you get. Try to optimize your pipelines to make training as fast as possible.

g. Use TFDS to load the same dataset more easily: `tfds.load("imdb_reviews")`.

Solutions to these exercises are available in Appendix A.

---

11 For large images, you could use `tf.io.encode_jpeg()` instead. This would save a lot of space, but it would lose a bit of image quality.

## CHAPTER 14

Deep Computer Vision Using Convolutional Neural Networks

Although IBM's Deep Blue supercomputer beat the chess world champion Garry Kasparov back in 1996, it wasn't until fairly recently that computers were able to reliably perform seemingly trivial tasks such as detecting a puppy in a picture or recognizing spoken words. Why are these tasks so effortless to us humans? The answer lies in the fact that perception largely takes place outside the realm of our consciousness, within specialized visual, auditory, and other sensory modules in our brains. By the time sensory information reaches our consciousness, it is already adorned with high-level features; for example, when you look at a picture of a cute puppy, you cannot choose *not* to see the puppy, *not* to notice its cuteness. Nor can you explain *how* you recognize a cute puppy; it's just obvious to you. Thus, we cannot trust our subjective experience: perception is not trivial at all, and to understand it we must look at how the sensory modules work.

Convolutional neural networks (CNNs) emerged from the study of the brain's visual cortex, and they have been used in image recognition since the 1980s. In the last few years, thanks to the increase in computational power, the amount of available training data, and the tricks presented in Chapter 11 for training deep nets, CNNs have managed to achieve superhuman performance on some complex visual tasks. They power image search services, self-driving cars, automatic video classification systems, and more. Moreover, CNNs are not restricted to visual perception: they are also successful at many other tasks, such as voice recognition and natural language processing. However, we will focus on visual applications for now.

In this chapter we will explore where CNNs came from, what their building blocks look like, and how to implement them using TensorFlow and Keras. Then we will discuss some of the best CNN architectures, as well as other visual tasks, including object detection (classifying multiple objects in an image and placing bounding boxes around them) and semantic segmentation (classifying each pixel according to the class of the object it belongs to).

The Architecture of the Visual Cortex

David H. Hubel and Torsten Wiesel performed a series of experiments on cats in 1958 (*https://homl.info/71*)[1] and 1959 (*https://homl.info/72*)[2] (and a few years later on monkeys (*https://homl.info/73*)[3]), giving crucial insights into the structure of the visual cortex (the authors received the Nobel Prize in Physiology or Medicine in 1981 for their work). In particular, they showed that many neurons in the visual cortex have a small *local receptive field*, meaning they react only to visual stimuli located in a limited region of the visual field (see Figure 14-1, in which the local receptive fields of five neurons are represented by dashed circles). The receptive fields of different neurons may overlap, and together they tile the whole visual field.

Moreover, the authors showed that some neurons react only to images of horizontal lines, while others react only to lines with different orientations (two neurons may have the same receptive field but react to different line orientations). They also noticed that some neurons have larger receptive fields, and they react to more complex patterns that are combinations of the lower-level patterns. These observations led to the idea that the higher-level neurons are based on the outputs of neighboring lower-level neurons (in Figure 14-1, notice that each neuron is connected only to a few neurons from the previous layer). This powerful architecture is able to detect all sorts of complex patterns in any area of the visual field.

1 David H. Hubel, "Single Unit Activity in Striate Cortex of Unrestrained Cats," *The Journal of Physiology* 147 (1959): 226–238.

2 David H. Hubel and Torsten N. Wiesel, "Receptive Fields of Single Neurons in the Cat's Striate Cortex," *The Journal of Physiology* 148 (1959): 574–591.

3 David H. Hubel and Torsten N. Wiesel, "Receptive Fields and Functional Architecture of Monkey Striate Cortex," *The Journal of Physiology* 195 (1968): 215–243.

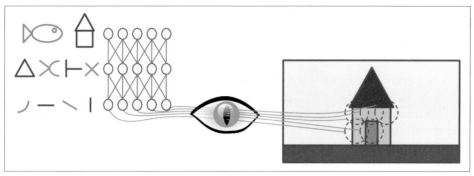

Figure 14-1. Biological neurons in the visual cortex respond to specific patterns in small regions of the visual field called receptive fields; as the visual signal makes its way through consecutive brain modules, neurons respond to more complex patterns in larger receptive fields.

These studies of the visual cortex inspired the neocognitron (*https://homl.info/74*),[4] introduced in 1980, which gradually evolved into what we now call *convolutional neural networks*. An important milestone was a 1998 paper (*https://homl.info/75*)[5] by Yann LeCun et al. that introduced the famous *LeNet-5* architecture, widely used by banks to recognize handwritten check numbers. This architecture has some building blocks that you already know, such as fully connected layers and sigmoid activation functions, but it also introduces two new building blocks: *convolutional layers* and *pooling layers*. Let's look at them now.

Why not simply use a deep neural network with fully connected layers for image recognition tasks? Unfortunately, although this works fine for small images (e.g., MNIST), it breaks down for larger images because of the huge number of parameters it requires. For example, a 100 × 100–pixel image has 10,000 pixels, and if the first layer has just 1,000 neurons (which already severely restricts the amount of information transmitted to the next layer), this means a total of 10 million connections. And that's just the first layer. CNNs solve this problem using partially connected layers and weight sharing.

---

4 Kunihiko Fukushima, "Neocognitron: A Self-Organizing Neural Network Model for a Mechanism of Pattern Recognition Unaffected by Shift in Position," *Biological Cybernetics* 36 (1980): 193–202.

5 Yann LeCun et al., "Gradient-Based Learning Applied to Document Recognition," *Proceedings of the IEEE* 86, no. 11 (1998): 2278–2324.

Convolutional Layers

The most important building block of a CNN is the *convolutional layer:*[6] neurons in the first convolutional layer are not connected to every single pixel in the input image (like they were in the layers discussed in previous chapters), but only to pixels in their receptive fields (see Figure 14-2). In turn, each neuron in the second convolutional layer is connected only to neurons located within a small rectangle in the first layer. This architecture allows the network to concentrate on small low-level features in the first hidden layer, then assemble them into larger higher-level features in the next hidden layer, and so on. This hierarchical structure is common in real-world images, which is one of the reasons why CNNs work so well for image recognition.

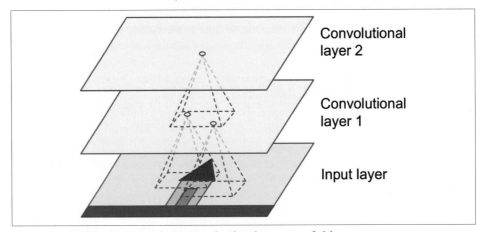

Figure 14-2. CNN layers with rectangular local receptive fields

 All the multilayer neural networks we've looked at so far had layers composed of a long line of neurons, and we had to flatten input images to 1D before feeding them to the neural network. In a CNN each layer is represented in 2D, which makes it easier to match neurons with their corresponding inputs.

---

6 A convolution is a mathematical operation that slides one function over another and measures the integral of their pointwise multiplication. It has deep connections with the Fourier transform and the Laplace transform and is heavily used in signal processing. Convolutional layers actually use cross-correlations, which are very similar to convolutions (see *https://homl.info/76* for more details).

A neuron located in row i, column j of a given layer is connected to the outputs of the neurons in the previous layer located in rows i to $i + f_h - 1$, columns j to $j + f_w - 1$, where f_h and f_w are the height and width of the receptive field (see Figure 14-3). In order for a layer to have the same height and width as the previous layer, it is common to add zeros around the inputs, as shown in the diagram. This is called *zero padding*.

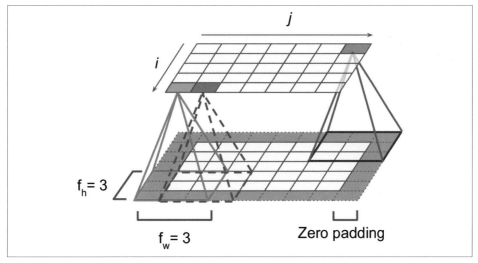

Figure 14-3. Connections between layers and zero padding

It is also possible to connect a large input layer to a much smaller layer by spacing out the receptive fields, as shown in Figure 14-4. This dramatically reduces the model's computational complexity. The shift from one receptive field to the next is called the *stride*. In the diagram, a 5 × 7 input layer (plus zero padding) is connected to a 3 × 4 layer, using 3 × 3 receptive fields and a stride of 2 (in this example the stride is the same in both directions, but it does not have to be so). A neuron located in row i, column j in the upper layer is connected to the outputs of the neurons in the previous layer located in rows $i \times s_h$ to $i \times s_h + f_h - 1$, columns $j \times s_w$ to $j \times s_w + f_w - 1$, where s_h and s_w are the vertical and horizontal strides.

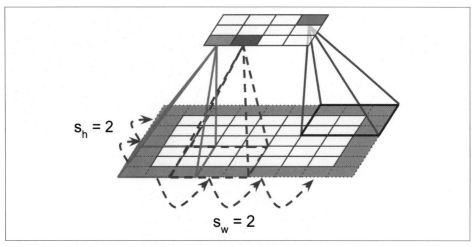

Figure 14-4. Reducing dimensionality using a stride of 2

Filters

A neuron's weights can be represented as a small image the size of the receptive field. For example, Figure 14-5 shows two possible sets of weights, called *filters* (or *convolution kernels*). The first one is represented as a black square with a vertical white line in the middle (it is a 7 × 7 matrix full of 0s except for the central column, which is full of 1s); neurons using these weights will ignore everything in their receptive field except for the central vertical line (since all inputs will get multiplied by 0, except for the ones located in the central vertical line). The second filter is a black square with a horizontal white line in the middle. Once again, neurons using these weights will ignore everything in their receptive field except for the central horizontal line.

Now if all neurons in a layer use the same vertical line filter (and the same bias term), and you feed the network the input image shown in Figure 14-5 (the bottom image), the layer will output the top-left image. Notice that the vertical white lines get enhanced while the rest gets blurred. Similarly, the upper-right image is what you get if all neurons use the same horizontal line filter; notice that the horizontal white lines get enhanced while the rest is blurred out. Thus, a layer full of neurons using the same filter outputs a *feature map*, which highlights the areas in an image that activate the filter the most. Of course, you do not have to define the filters manually: instead, during training the convolutional layer will automatically learn the most useful filters for its task, and the layers above will learn to combine them into more complex patterns.

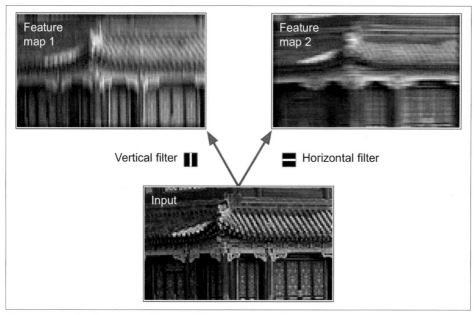

Figure 14-5. Applying two different filters to get two feature maps

Stacking Multiple Feature Maps

Up to now, for simplicity, I have represented the output of each convolutional layer as a 2D layer, but in reality a convolutional layer has multiple filters (you decide how many) and outputs one feature map per filter, so it is more accurately represented in 3D (see Figure 14-6). It has one neuron per pixel in each feature map, and all neurons within a given feature map share the same parameters (i.e., the same weights and bias term). Neurons in different feature maps use different parameters. A neuron's receptive field is the same as described earlier, but it extends across all the previous layers' feature maps. In short, a convolutional layer simultaneously applies multiple trainable filters to its inputs, making it capable of detecting multiple features anywhere in its inputs.

> The fact that all neurons in a feature map share the same parameters dramatically reduces the number of parameters in the model. Once the CNN has learned to recognize a pattern in one location, it can recognize it in any other location. In contrast, once a regular DNN has learned to recognize a pattern in one location, it can recognize it only in that particular location.

Input images are also composed of multiple sublayers: one per *color channel*. There are typically three: red, green, and blue (RGB). Grayscale images have just one channel, but some images may have much more—for example, satellite images that capture extra light frequencies (such as infrared).

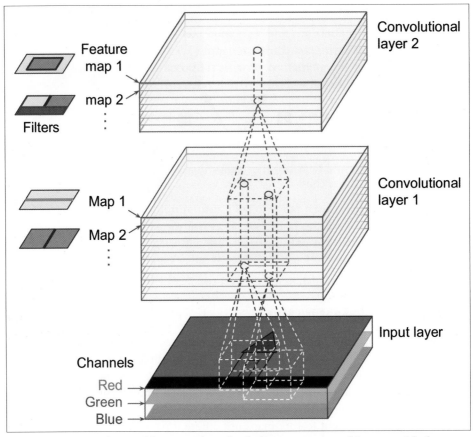

Figure 14-6. Convolutional layers with multiple feature maps, and images with three color channels

Specifically, a neuron located in row i, column j of the feature map k in a given convolutional layer l is connected to the outputs of the neurons in the previous layer $l - 1$, located in rows $i \times s_h$ to $i \times s_h + f_h - 1$ and columns $j \times s_w$ to $j \times s_w + f_w - 1$, across all feature maps (in layer $l - 1$). Note that all neurons located in the same row i and column j but in different feature maps are connected to the outputs of the exact same neurons in the previous layer.

Equation 14-1 summarizes the preceding explanations in one big mathematical equation: it shows how to compute the output of a given neuron in a convolutional layer.

It is a bit ugly due to all the different indices, but all it does is calculate the weighted sum of all the inputs, plus the bias term.

Equation 14-1. Computing the output of a neuron in a convolutional layer

$$z_{i,j,k} = b_k + \sum_{u=0}^{f_h-1} \sum_{v=0}^{f_w-1} \sum_{k'=0}^{f_{n'}-1} x_{i',j',k'} \cdot w_{u,v,k',k} \quad \text{with} \quad \begin{cases} i' = i \times s_h + u \\ j' = j \times s_w + v \end{cases}$$

In this equation:

- $z_{i,j,k}$ is the output of the neuron located in row i, column j in feature map k of the convolutional layer (layer l).

- As explained earlier, s_h and s_w are the vertical and horizontal strides, f_h and f_w are the height and width of the receptive field, and $f_{n'}$ is the number of feature maps in the previous layer (layer $l - 1$).

- $x_{i',j',k'}$ is the output of the neuron located in layer $l - 1$, row i', column j', feature map k' (or channel k' if the previous layer is the input layer).

- b_k is the bias term for feature map k (in layer l). You can think of it as a knob that tweaks the overall brightness of the feature map k.

- $w_{u,v,k',k}$ is the connection weight between any neuron in feature map k of the layer l and its input located at row u, column v (relative to the neuron's receptive field), and feature map k'.

TensorFlow Implementation

In TensorFlow, each input image is typically represented as a 3D tensor of shape [*height, width, channels*]. A mini-batch is represented as a 4D tensor of shape [*mini-batch size, height, width, channels*]. The weights of a convolutional layer are represented as a 4D tensor of shape [$f_h, f_w, f_{n'}, f_n$]. The bias terms of a convolutional layer are simply represented as a 1D tensor of shape [f_n].

Let's look at a simple example. The following code loads two sample images, using Scikit-Learn's load_sample_image() (which loads two color images, one of a Chinese temple, and the other of a flower), then it creates two filters and applies them to both images, and finally it displays one of the resulting feature maps:

```
from sklearn.datasets import load_sample_image

# Load sample images
china = load_sample_image("china.jpg") / 255
flower = load_sample_image("flower.jpg") / 255
images = np.array([china, flower])
batch_size, height, width, channels = images.shape
```

```
# Create 2 filters
filters = np.zeros(shape=(7, 7, channels, 2), dtype=np.float32)
filters[:, 3, :, 0] = 1  # vertical line
filters[3, :, :, 1] = 1  # horizontal line

outputs = tf.nn.conv2d(images, filters, strides=1, padding="same")

plt.imshow(outputs[0, :, :, 1], cmap="gray") # plot 1st image's 2nd feature map
plt.show()
```

Let's go through this code:

- The pixel intensity for each color channel is represented as a byte from 0 to 255, so we scale these features simply by dividing by 255, to get floats ranging from 0 to 1.

- Then we create two 7×7 filters (one with a vertical white line in the middle, and the other with a horizontal white line in the middle).

- We apply them to both images using the `tf.nn.conv2d()` function, which is part of TensorFlow's low-level Deep Learning API. In this example, we use zero padding (`padding="same"`) and a stride of 2.

- Finally, we plot one of the resulting feature maps (similar to the top-right image in Figure 14-5).

The `tf.nn.conv2d()` line deserves a bit more explanation:

- `images` is the input mini-batch (a 4D tensor, as explained earlier).

- `filters` is the set of filters to apply (also a 4D tensor, as explained earlier).

- `strides` is equal to 1, but it could also be a 1D array with four elements, where the two central elements are the vertical and horizontal strides (s_h and s_w). The first and last elements must currently be equal to 1. They may one day be used to specify a batch stride (to skip some instances) and a channel stride (to skip some of the previous layer's feature maps or channels).

- `padding` must be either `"same"` or `"valid"`:

  — If set to `"same"`, the convolutional layer uses zero padding if necessary. The output size is set to the number of input neurons divided by the stride, rounded up. For example, if the input size is 13 and the stride is 5 (see Figure 14-7), then the output size is 3 (i.e., 13 / 5 = 2.6, rounded up to 3). Then zeros are added as evenly as possible around the inputs, as needed. When `strides=1`, the layer's outputs will have the same spatial dimensions (width and height) as its inputs, hence the name *same*.

— If set to `"valid"`, the convolutional layer does *not* use zero padding and may ignore some rows and columns at the bottom and right of the input image, depending on the stride, as shown in Figure 14-7 (for simplicity, only the horizontal dimension is shown here, but of course the same logic applies to the vertical dimension). This means that every neuron's receptive field lies strictly within valid positions inside the input (it does not go out of bounds), hence the name *valid*.

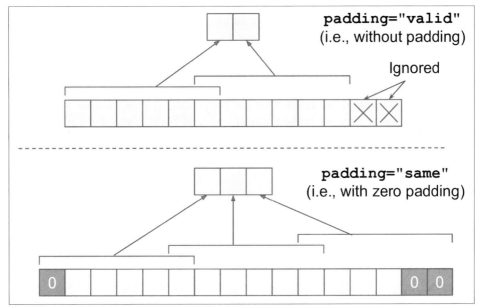

Figure 14-7. Padding="same" or "valid" (with input width 13, filter width 6, stride 5)

In this example we manually defined the filters, but in a real CNN you would normally define filters as trainable variables so the neural net can learn which filters work best, as explained earlier. Instead of manually creating the variables, use the `keras.layers.Conv2D` layer:

```
conv = keras.layers.Conv2D(filters=32, kernel_size=3, strides=1,
                           padding="same", activation="relu")
```

This code creates a `Conv2D` layer with 32 filters, each 3 × 3, using a stride of 1 (both horizontally and vertically) and `"same"` padding, and applying the ReLU activation function to its outputs. As you can see, convolutional layers have quite a few hyperparameters: you must choose the number of filters, their height and width, the strides, and the padding type. As always, you can use cross-validation to find the right hyperparameter values, but this is very time-consuming. We will discuss common CNN architectures later, to give you some idea of which hyperparameter values work best in practice.

Memory Requirements

Another problem with CNNs is that the convolutional layers require a huge amount of RAM. This is especially true during training, because the reverse pass of backpropagation requires all the intermediate values computed during the forward pass.

For example, consider a convolutional layer with 5 × 5 filters, outputting 200 feature maps of size 150 × 100, with stride 1 and "same" padding. If the input is a 150 × 100 RGB image (three channels), then the number of parameters is (5 × 5 × 3 + 1) × 200 = 15,200 (the + 1 corresponds to the bias terms), which is fairly small compared to a fully connected layer.[7] However, each of the 200 feature maps contains 150 × 100 neurons, and each of these neurons needs to compute a weighted sum of its 5 × 5 × 3 = 75 inputs: that's a total of 225 million float multiplications. Not as bad as a fully connected layer, but still quite computationally intensive. Moreover, if the feature maps are represented using 32-bit floats, then the convolutional layer's output will occupy 200 × 150 × 100 × 32 = 96 million bits (12 MB) of RAM.[8] And that's just for one instance—if a training batch contains 100 instances, then this layer will use up 1.2 GB of RAM!

During inference (i.e., when making a prediction for a new instance) the RAM occupied by one layer can be released as soon as the next layer has been computed, so you only need as much RAM as required by two consecutive layers. But during training everything computed during the forward pass needs to be preserved for the reverse pass, so the amount of RAM needed is (at least) the total amount of RAM required by all layers.

 If training crashes because of an out-of-memory error, you can try reducing the mini-batch size. Alternatively, you can try reducing dimensionality using a stride, or removing a few layers. Or you can try using 16-bit floats instead of 32-bit floats. Or you could distribute the CNN across multiple devices.

Now let's look at the second common building block of CNNs: the *pooling layer*.

Pooling Layers

Once you understand how convolutional layers work, the pooling layers are quite easy to grasp. Their goal is to *subsample* (i.e., shrink) the input image in order to

---

7 A fully connected layer with 150 × 100 neurons, each connected to all 150 × 100 × 3 inputs, would have 150^2 × 100^2 × 3 = 675 million parameters!

8 In the international system of units (SI), 1 MB = 1,000 KB = 1,000 × 1,000 bytes = 1,000 × 1,000 × 8 bits.

reduce the computational load, the memory usage, and the number of parameters (thereby limiting the risk of overfitting).

Just like in convolutional layers, each neuron in a pooling layer is connected to the outputs of a limited number of neurons in the previous layer, located within a small rectangular receptive field. You must define its size, the stride, and the padding type, just like before. However, a pooling neuron has no weights; all it does is aggregate the inputs using an aggregation function such as the max or mean. Figure 14-8 shows a *max pooling layer*, which is the most common type of pooling layer. In this example, we use a 2 × 2 *pooling kernel*,[9] with a stride of 2 and no padding. Only the max input value in each receptive field makes it to the next layer, while the other inputs are dropped. For example, in the lower-left receptive field in Figure 14-8, the input values are 1, 5, 3, 2, so only the max value, 5, is propagated to the next layer. Because of the stride of 2, the output image has half the height and half the width of the input image (rounded down since we use no padding).

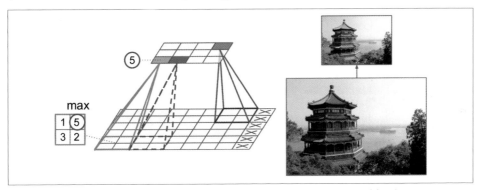

Figure 14-8. Max pooling layer (2 × 2 pooling kernel, stride 2, no padding)

A pooling layer typically works on every input channel independently, so the output depth is the same as the input depth.

Other than reducing computations, memory usage, and the number of parameters, a max pooling layer also introduces some level of *invariance* to small translations, as shown in Figure 14-9. Here we assume that the bright pixels have a lower value than dark pixels, and we consider three images (A, B, C) going through a max pooling layer with a 2 × 2 kernel and stride 2. Images B and C are the same as image A, but

---

9 Other kernels we've discussed so far had weights, but pooling kernels do not: they are just stateless sliding windows.

shifted by one and two pixels to the right. As you can see, the outputs of the max pooling layer for images A and B are identical. This is what translation invariance means. For image C, the output is different: it is shifted one pixel to the right (but there is still 75% invariance). By inserting a max pooling layer every few layers in a CNN, it is possible to get some level of translation invariance at a larger scale. Moreover, max pooling offers a small amount of rotational invariance and a slight scale invariance. Such invariance (even if it is limited) can be useful in cases where the prediction should not depend on these details, such as in classification tasks.

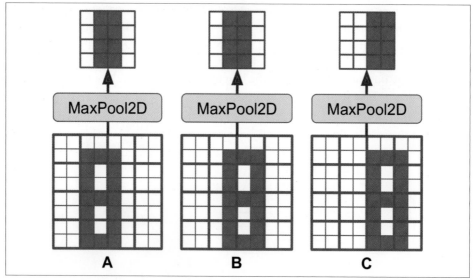

Figure 14-9. Invariance to small translations

However, max pooling has some downsides too. Firstly, it is obviously very destructive: even with a tiny 2 × 2 kernel and a stride of 2, the output will be two times smaller in both directions (so its area will be four times smaller), simply dropping 75% of the input values. And in some applications, invariance is not desirable. Take semantic segmentation (the task of classifying each pixel in an image according to the object that pixel belongs to, which we'll explore later in this chapter): obviously, if the input image is translated by one pixel to the right, the output should also be translated by one pixel to the right. The goal in this case is *equivariance*, not invariance: a small change to the inputs should lead to a corresponding small change in the output.

TensorFlow Implementation

Implementing a max pooling layer in TensorFlow is quite easy. The following code creates a max pooling layer using a 2 × 2 kernel. The strides default to the kernel size, so this layer will use a stride of 2 (both horizontally and vertically). By default, it uses "valid" padding (i.e., no padding at all):

```
max_pool = keras.layers.MaxPool2D(pool_size=2)
```

To create an *average pooling layer*, just use `AvgPool2D` instead of `MaxPool2D`. As you might expect, it works exactly like a max pooling layer, except it computes the mean rather than the max. Average pooling layers used to be very popular, but people mostly use max pooling layers now, as they generally perform better. This may seem surprising, since computing the mean generally loses less information than computing the max. But on the other hand, max pooling preserves only the strongest features, getting rid of all the meaningless ones, so the next layers get a cleaner signal to work with. Moreover, max pooling offers stronger translation invariance than average pooling, and it requires slightly less compute.

Note that max pooling and average pooling can be performed along the depth dimension rather than the spatial dimensions, although this is not as common. This can allow the CNN to learn to be invariant to various features. For example, it could learn multiple filters, each detecting a different rotation of the same pattern (such as handwritten digits; see Figure 14-10), and the depthwise max pooling layer would ensure that the output is the same regardless of the rotation. The CNN could similarly learn to be invariant to anything else: thickness, brightness, skew, color, and so on.

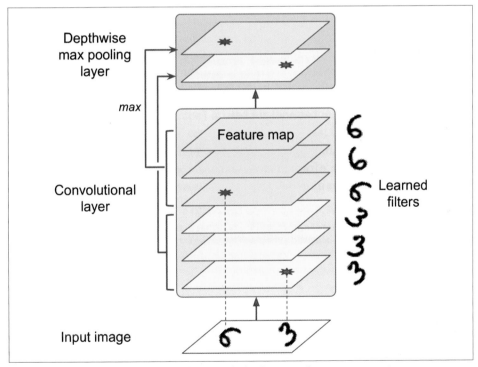

Figure 14-10. Depthwise max pooling can help the CNN learn any invariance

Keras does not include a depthwise max pooling layer, but TensorFlow's low-level Deep Learning API does: just use the tf.nn.max_pool() function, and specify the kernel size and strides as 4-tuples (i.e., tuples of size 4). The first three values of each should be 1: this indicates that the kernel size and stride along the batch, height, and width dimensions should be 1. The last value should be whatever kernel size and stride you want along the depth dimension—for example, 3 (this must be a divisor of the input depth; it will not work if the previous layer outputs 20 feature maps, since 20 is not a multiple of 3):

```
output = tf.nn.max_pool(images,
                        ksize=(1, 1, 1, 3),
                        strides=(1, 1, 1, 3),
                        padding="valid")
```

If you want to include this as a layer in your Keras models, wrap it in a Lambda layer (or create a custom Keras layer):

```
depth_pool = keras.layers.Lambda(
    lambda X: tf.nn.max_pool(X, ksize=(1, 1, 1, 3), strides=(1, 1, 1, 3),
                             padding="valid"))
```

One last type of pooling layer that you will often see in modern architectures is the *global average pooling layer*. It works very differently: all it does is compute the mean of each entire feature map (it's like an average pooling layer using a pooling kernel with the same spatial dimensions as the inputs). This means that it just outputs a single number per feature map and per instance. Although this is of course extremely destructive (most of the information in the feature map is lost), it can be useful as the output layer, as we will see later in this chapter. To create such a layer, simply use the keras.layers.GlobalAvgPool2D class:

```
global_avg_pool = keras.layers.GlobalAvgPool2D()
```

It's equivalent to this simple Lambda layer, which computes the mean over the spatial dimensions (height and width):

```
global_avg_pool = keras.layers.Lambda(lambda X: tf.reduce_mean(X, axis=[1, 2]))
```

Now you know all the building blocks to create convolutional neural networks. Let's see how to assemble them.

CNN Architectures

Typical CNN architectures stack a few convolutional layers (each one generally followed by a ReLU layer), then a pooling layer, then another few convolutional layers (+ReLU), then another pooling layer, and so on. The image gets smaller and smaller as it progresses through the network, but it also typically gets deeper and deeper (i.e., with more feature maps), thanks to the convolutional layers (see Figure 14-11). At the top of the stack, a regular feedforward neural network is added, composed of a few fully connected layers (+ReLUs), and the final layer outputs the prediction (e.g., a softmax layer that outputs estimated class probabilities).

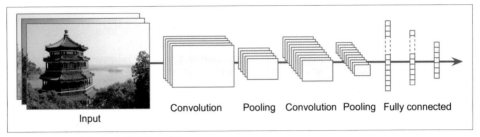

Figure 14-11. Typical CNN architecture

A common mistake is to use convolution kernels that are too large. For example, instead of using a convolutional layer with a 5 × 5 kernel, stack two layers with 3 × 3 kernels: it will use fewer parameters and require fewer computations, and it will usually perform better. One exception is for the first convolutional layer: it can typically have a large kernel (e.g., 5 × 5), usually with a stride of 2 or more: this will reduce the spatial dimension of the image without losing too much information, and since the input image only has three channels in general, it will not be too costly.

Here is how you can implement a simple CNN to tackle the Fashion MNIST dataset (introduced in Chapter 10):

```python
model = keras.models.Sequential([
    keras.layers.Conv2D(64, 7, activation="relu", padding="same",
                        input_shape=[28, 28, 1]),
    keras.layers.MaxPooling2D(2),
    keras.layers.Conv2D(128, 3, activation="relu", padding="same"),
    keras.layers.Conv2D(128, 3, activation="relu", padding="same"),
    keras.layers.MaxPooling2D(2),
    keras.layers.Conv2D(256, 3, activation="relu", padding="same"),
    keras.layers.Conv2D(256, 3, activation="relu", padding="same"),
    keras.layers.MaxPooling2D(2),
    keras.layers.Flatten(),
    keras.layers.Dense(128, activation="relu"),
    keras.layers.Dropout(0.5),
    keras.layers.Dense(64, activation="relu"),
    keras.layers.Dropout(0.5),
    keras.layers.Dense(10, activation="softmax")
])
```

Let's go through this model:

- The first layer uses 64 fairly large filters (7×7) but no stride because the input images are not very large. It also sets input_shape=[28, 28, 1], because the images are 28×28 pixels, with a single color channel (i.e., grayscale).

- Next we have a max pooling layer which uses a pool size of 2, so it divides each spatial dimension by a factor of 2.

- Then we repeat the same structure twice: two convolutional layers followed by a max pooling layer. For larger images, we could repeat this structure several more times (the number of repetitions is a hyperparameter you can tune).

- Note that the number of filters grows as we climb up the CNN toward the output layer (it is initially 64, then 128, then 256): it makes sense for it to grow, since the number of low-level features is often fairly low (e.g., small circles, horizontal lines), but there are many different ways to combine them into higher-level features. It is a common practice to double the number of filters after each pooling layer: since a pooling layer divides each spatial dimension by a factor of 2, we can afford to double the number of feature maps in the next layer without fear of exploding the number of parameters, memory usage, or computational load.

- Next is the fully connected network, composed of two hidden dense layers and a dense output layer. Note that we must flatten its inputs, since a dense network expects a 1D array of features for each instance. We also add two dropout layers, with a dropout rate of 50% each, to reduce overfitting.

This CNN reaches over 92% accuracy on the test set. It's not state of the art, but it is pretty good, and clearly much better than what we achieved with dense networks in Chapter 10.

Over the years, variants of this fundamental architecture have been developed, leading to amazing advances in the field. A good measure of this progress is the error rate in competitions such as the ILSVRC ImageNet challenge (*http://image-net.org/*). In this competition the top-five error rate for image classification fell from over 26% to less than 2.3% in just six years. The top-five error rate is the number of test images for which the system's top five predictions did not include the correct answer. The images are large (256 pixels high) and there are 1,000 classes, some of which are really subtle (try distinguishing 120 dog breeds). Looking at the evolution of the winning entries is a good way to understand how CNNs work.

We will first look at the classical LeNet-5 architecture (1998), then three of the winners of the ILSVRC challenge: AlexNet (2012), GoogLeNet (2014), and ResNet (2015).

LeNet-5

The LeNet-5 architecture (*https://homl.info/lenet5*)[10] is perhaps the most widely known CNN architecture. As mentioned earlier, it was created by Yann LeCun in 1998 and has been widely used for handwritten digit recognition (MNIST). It is composed of the layers shown in Table 14-1.

Table 14-1. LeNet-5 architecture

Layer	Type	Maps	Size	Kernel size	Stride	Activation
Out	Fully connected	–	10	–	–	RBF
F6	Fully connected	–	84	–	–	tanh
C5	Convolution	120	1×1	5×5	1	tanh
S4	Avg pooling	16	5×5	2×2	2	tanh
C3	Convolution	16	10×10	5×5	1	tanh
S2	Avg pooling	6	14×14	2×2	2	tanh
C1	Convolution	6	28×28	5×5	1	tanh
In	Input	1	32×32	–	–	–

There are a few extra details to be noted:

- MNIST images are 28 × 28 pixels, but they are zero-padded to 32 × 32 pixels and normalized before being fed to the network. The rest of the network does not use any padding, which is why the size keeps shrinking as the image progresses through the network.

- The average pooling layers are slightly more complex than usual: each neuron computes the mean of its inputs, then multiplies the result by a learnable coefficient (one per map) and adds a learnable bias term (again, one per map), then finally applies the activation function.

- Most neurons in C3 maps are connected to neurons in only three or four S2 maps (instead of all six S2 maps). See table 1 (page 8) in the original paper[10] for details.

- The output layer is a bit special: instead of computing the matrix multiplication of the inputs and the weight vector, each neuron outputs the square of the Euclidian distance between its input vector and its weight vector. Each output measures how much the image belongs to a particular digit class. The cross-entropy

---

10 Yann LeCun et al., "Gradient-Based Learning Applied to Document Recognition," *Proceedings of the IEEE* 86, no. 11 (1998): 2278–2324.

cost function is now preferred, as it penalizes bad predictions much more, producing larger gradients and converging faster.

Yann LeCun's website (*http://yann.lecun.com/exdb/lenet/index.html*) features great demos of LeNet-5 classifying digits.

AlexNet

The AlexNet CNN architecture (*https://homl.info/80*)[11] won the 2012 ImageNet ILSVRC challenge by a large margin: it achieved a top-five error rate of 17%, while the second best achieved only 26%! It was developed by Alex Krizhevsky (hence the name), Ilya Sutskever, and Geoffrey Hinton. It is similar to LeNet-5, only much larger and deeper, and it was the first to stack convolutional layers directly on top of one another, instead of stacking a pooling layer on top of each convolutional layer. Table 14-2 presents this architecture.

Table 14-2. AlexNet architecture

Layer	Type	Maps	Size	Kernel size	Stride	Padding	Activation
Out	Fully connected	–	1,000	–	–	–	Softmax
F9	Fully connected	–	4,096	–	–	–	ReLU
F8	Fully connected	–	4,096	–	–	–	ReLU
C7	Convolution	256	13×13	3×3	1	same	ReLU
C6	Convolution	384	13×13	3×3	1	same	ReLU
C5	Convolution	384	13×13	3×3	1	same	ReLU
S4	Max pooling	256	13×13	3×3	2	valid	–
C3	Convolution	256	27×27	5×5	1	same	ReLU
S2	Max pooling	96	27×27	3×3	2	valid	–
C1	Convolution	96	55×55	11×11	4	valid	ReLU
In	Input	3 (RGB)	227×227	–	–	–	–

To reduce overfitting, the authors used two regularization techniques. First, they applied dropout (introduced in Chapter 11) with a 50% dropout rate during training to the outputs of layers F8 and F9. Second, they performed *data augmentation* by randomly shifting the training images by various offsets, flipping them horizontally, and changing the lighting conditions.

---

11 Alex Krizhevsky et al., "ImageNet Classification with Deep Convolutional Neural Networks," _Proceedings of the 25th International Conference on Neural Information Processing Systems_ 1 (2012): 1097–1105.

Data Augmentation

Data augmentation artificially increases the size of the training set by generating many realistic variants of each training instance. This reduces overfitting, making this a regularization technique. The generated instances should be as realistic as possible: ideally, given an image from the augmented training set, a human should not be able to tell whether it was augmented or not. Simply adding white noise will not help; the modifications should be learnable (white noise is not).

For example, you can slightly shift, rotate, and resize every picture in the training set by various amounts and add the resulting pictures to the training set (see Figure 14-12). This forces the model to be more tolerant to variations in the position, orientation, and size of the objects in the pictures. For a model that's more tolerant of different lighting conditions, you can similarly generate many images with various contrasts. In general, you can also flip the pictures horizontally (except for text, and other asymmetrical objects). By combining these transformations, you can greatly increase the size of your training set.

Figure 14-12. Generating new training instances from existing ones

AlexNet also uses a competitive normalization step immediately after the ReLU step of layers C1 and C3, called *local response normalization* (LRN): the most strongly activated neurons inhibit other neurons located at the same position in neighboring feature maps (such competitive activation has been observed in biological neurons). This encourages different feature maps to specialize, pushing them apart and forcing them to explore a wider range of features, ultimately improving generalization. Equation 14-2 shows how to apply LRN.

Equation 14-2. Local response normalization (LRN)

$$b_i = a_i \left(k + \alpha \sum_{j=j_{\text{low}}}^{j_{\text{high}}} a_j^2 \right)^{-\beta} \quad \text{with} \quad \begin{cases} j_{\text{high}} = \min\left(i + \dfrac{r}{2}, f_n - 1\right) \\ j_{\text{low}} = \max\left(0, i - \dfrac{r}{2}\right) \end{cases}$$

In this equation:

- b_i is the normalized output of the neuron located in feature map i, at some row u and column v (note that in this equation we consider only neurons located at this row and column, so u and v are not shown).

- a_i is the activation of that neuron after the ReLU step, but before normalization.

- k, α, β, and r are hyperparameters. k is called the *bias*, and r is called the *depth radius*.

- f_n is the number of feature maps.

For example, if $r = 2$ and a neuron has a strong activation, it will inhibit the activation of the neurons located in the feature maps immediately above and below its own.

In AlexNet, the hyperparameters are set as follows: $r = 2$, $\alpha = 0.00002$, $\beta = 0.75$, and $k = 1$. This step can be implemented using the `tf.nn.local_response_normalization()` function (which you can wrap in a `Lambda` layer if you want to use it in a Keras model).

A variant of AlexNet called *ZF Net* (*https://homl.info/zfnet*)[12] was developed by Matthew Zeiler and Rob Fergus and won the 2013 ILSVRC challenge. It is essentially AlexNet with a few tweaked hyperparameters (number of feature maps, kernel size, stride, etc.).

GoogLeNet

The GoogLeNet architecture (*https://homl.info/81*) was developed by Christian Szegedy et al. from Google Research,[13] and it won the ILSVRC 2014 challenge by pushing the top-five error rate below 7%. This great performance came in large part from the

---

12 Matthew D. Zeiler and Rob Fergus, "Visualizing and Understanding Convolutional Networks," *Proceedings of the European Conference on Computer Vision* (2014): 818-833.

13 Christian Szegedy et al., "Going Deeper with Convolutions," *Proceedings of the IEEE Conference on Computer Vision and Pattern Recognition* (2015): 1–9.

fact that the network was much deeper than previous CNNs (as you'll see in Figure 14-14). This was made possible by subnetworks called *inception modules*,[14] which allow GoogLeNet to use parameters much more efficiently than previous architectures: GoogLeNet actually has 10 times fewer parameters than AlexNet (roughly 6 million instead of 60 million).

Figure 14-13 shows the architecture of an inception module. The notation "3 × 3 + 1(S)" means that the layer uses a 3 × 3 kernel, stride 1, and "same" padding. The input signal is first copied and fed to four different layers. All convolutional layers use the ReLU activation function. Note that the second set of convolutional layers uses different kernel sizes (1 × 1, 3 × 3, and 5 × 5), allowing them to capture patterns at different scales. Also note that every single layer uses a stride of 1 and "same" padding (even the max pooling layer), so their outputs all have the same height and width as their inputs. This makes it possible to concatenate all the outputs along the depth dimension in the final *depth concatenation layer* (i.e., stack the feature maps from all four top convolutional layers). This concatenation layer can be implemented in Tensor-Flow using the tf.concat() operation, with axis=3 (the axis is the depth).

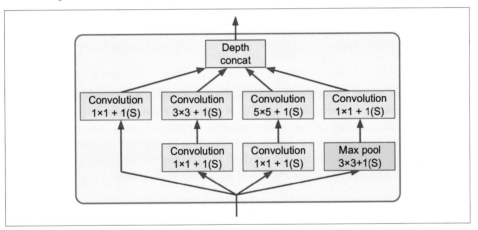

Figure 14-13. Inception module

You may wonder why inception modules have convolutional layers with 1 × 1 kernels. Surely these layers cannot capture any features because they look at only one pixel at a time? In fact, the layers serve three purposes:

- Although they cannot capture spatial patterns, they can capture patterns along the depth dimension.

---

14 In the 2010 movie *Inception*, the characters keep going deeper and deeper into multiple layers of dreams; hence the name of these modules.

- They are configured to output fewer feature maps than their inputs, so they serve as *bottleneck layers*, meaning they reduce dimensionality. This cuts the computational cost and the number of parameters, speeding up training and improving generalization.

- Each pair of convolutional layers ($[1 \times 1, 3 \times 3]$ and $[1 \times 1, 5 \times 5]$) acts like a single powerful convolutional layer, capable of capturing more complex patterns. Indeed, instead of sweeping a simple linear classifier across the image (as a single convolutional layer does), this pair of convolutional layers sweeps a two-layer neural network across the image.

In short, you can think of the whole inception module as a convolutional layer on steroids, able to output feature maps that capture complex patterns at various scales.

 The number of convolutional kernels for each convolutional layer is a hyperparameter. Unfortunately, this means that you have six more hyperparameters to tweak for every inception layer you add.

Now let's look at the architecture of the GoogLeNet CNN (see Figure 14-14). The number of feature maps output by each convolutional layer and each pooling layer is shown before the kernel size. The architecture is so deep that it has to be represented in three columns, but GoogLeNet is actually one tall stack, including nine inception modules (the boxes with the spinning tops). The six numbers in the inception modules represent the number of feature maps output by each convolutional layer in the module (in the same order as in Figure 14-13). Note that all the convolutional layers use the ReLU activation function.

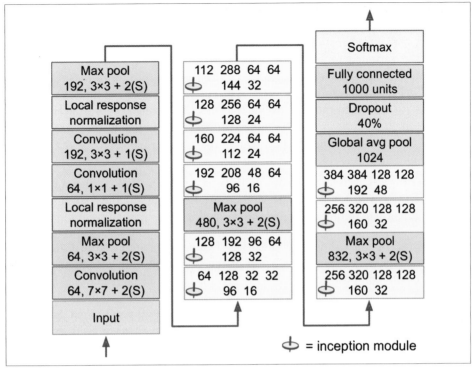

Figure 14-14. GoogLeNet architecture

Let's go through this network:

- The first two layers divide the image's height and width by 4 (so its area is divided by 16), to reduce the computational load. The first layer uses a large kernel size so that much of the information is preserved.

- Then the local response normalization layer ensures that the previous layers learn a wide variety of features (as discussed earlier).

- Two convolutional layers follow, where the first acts like a bottleneck layer. As explained earlier, you can think of this pair as a single smarter convolutional layer.

- Again, a local response normalization layer ensures that the previous layers capture a wide variety of patterns.

- Next, a max pooling layer reduces the image height and width by 2, again to speed up computations.

- Then comes the tall stack of nine inception modules, interleaved with a couple max pooling layers to reduce dimensionality and speed up the net.

- Next, the global average pooling layer outputs the mean of each feature map: this drops any remaining spatial information, which is fine because there was not much spatial information left at that point. Indeed, GoogLeNet input images are typically expected to be 224 × 224 pixels, so after 5 max pooling layers, each dividing the height and width by 2, the feature maps are down to 7 × 7. Moreover, it is a classification task, not localization, so it does not matter where the object is. Thanks to the dimensionality reduction brought by this layer, there is no need to have several fully connected layers at the top of the CNN (like in AlexNet), and this considerably reduces the number of parameters in the network and limits the risk of overfitting.

- The last layers are self-explanatory: dropout for regularization, then a fully connected layer with 1,000 units (since there are 1,000 classes) and a softmax activation function to output estimated class probabilities.

This diagram is slightly simplified: the original GoogLeNet architecture also included two auxiliary classifiers plugged on top of the third and sixth inception modules. They were both composed of one average pooling layer, one convolutional layer, two fully connected layers, and a softmax activation layer. During training, their loss (scaled down by 70%) was added to the overall loss. The goal was to fight the vanishing gradients problem and regularize the network. However, it was later shown that their effect was relatively minor.

Several variants of the GoogLeNet architecture were later proposed by Google researchers, including Inception-v3 and Inception-v4, using slightly different inception modules and reaching even better performance.

VGGNet

The runner-up in the ILSVRC 2014 challenge was VGGNet (*https://homl.info/83*),[15] developed by Karen Simonyan and Andrew Zisserman from the Visual Geometry Group (VGG) research lab at Oxford University. It had a very simple and classical architecture, with 2 or 3 convolutional layers and a pooling layer, then again 2 or 3 convolutional layers and a pooling layer, and so on (reaching a total of just 16 or 19 convolutional layers, depending on the VGG variant), plus a final dense network with 2 hidden layers and the output layer. It used only 3 × 3 filters, but many filters.

---

15 Karen Simonyan and Andrew Zisserman, "Very Deep Convolutional Networks for Large-Scale Image Recognition," arXiv preprint arXiv:1409.1556 (2014).

ResNet

Kaiming He et al. won the ILSVRC 2015 challenge using a *Residual Network* (or *ResNet*) (*https://homl.info/82*),[16] that delivered an astounding top-five error rate under 3.6%. The winning variant used an extremely deep CNN composed of 152 layers (other variants had 34, 50, and 101 layers). It confirmed the general trend: models are getting deeper and deeper, with fewer and fewer parameters. The key to being able to train such a deep network is to use *skip connections* (also called *shortcut connections*): the signal feeding into a layer is also added to the output of a layer located a bit higher up the stack. Let's see why this is useful.

When training a neural network, the goal is to make it model a target function $h(\mathbf{x})$. If you add the input $\mathbf{x}$ to the output of the network (i.e., you add a skip connection), then the network will be forced to model $f(\mathbf{x}) = h(\mathbf{x}) - \mathbf{x}$ rather than $h(\mathbf{x})$. This is called *residual learning* (see Figure 14-15).

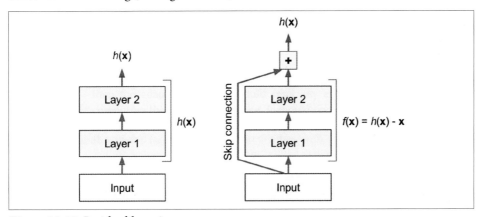

Figure 14-15. Residual learning

When you initialize a regular neural network, its weights are close to zero, so the network just outputs values close to zero. If you add a skip connection, the resulting network just outputs a copy of its inputs; in other words, it initially models the identity function. If the target function is fairly close to the identity function (which is often the case), this will speed up training considerably.

Moreover, if you add many skip connections, the network can start making progress even if several layers have not started learning yet (see Figure 14-16). Thanks to skip connections, the signal can easily make its way across the whole network. The deep residual network can be seen as a stack of *residual units* (RUs), where each residual unit is a small neural network with a skip connection.

---

16 Kaiming He et al., "Deep Residual Learning for Image Recognition," arXiv preprint arXiv:1512:03385 (2015).

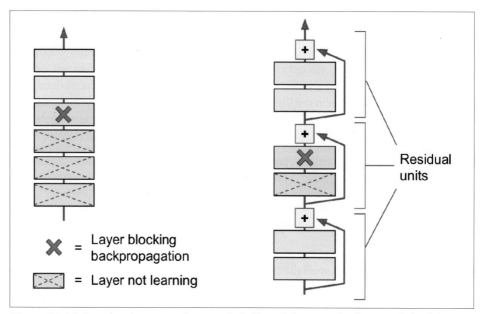

Figure 14-16. Regular deep neural network (left) and deep residual network (right)

Now let's look at ResNet's architecture (see Figure 14-17). It is surprisingly simple. It starts and ends exactly like GoogLeNet (except without a dropout layer), and in between is just a very deep stack of simple residual units. Each residual unit is composed of two convolutional layers (and no pooling layer!), with Batch Normalization (BN) and ReLU activation, using 3 × 3 kernels and preserving spatial dimensions (stride 1, "same" padding).

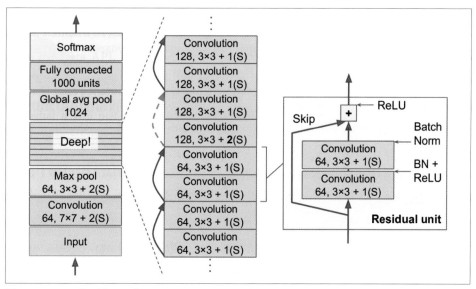

Figure 14-17. ResNet architecture

Note that the number of feature maps is doubled every few residual units, at the same time as their height and width are halved (using a convolutional layer with stride 2). When this happens, the inputs cannot be added directly to the outputs of the residual unit because they don't have the same shape (for example, this problem affects the skip connection represented by the dashed arrow in Figure 14-17). To solve this problem, the inputs are passed through a 1 × 1 convolutional layer with stride 2 and the right number of output feature maps (see Figure 14-18).

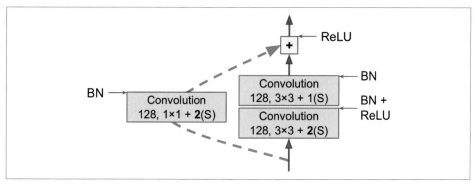

Figure 14-18. Skip connection when changing feature map size and depth

ResNet-34 is the ResNet with 34 layers (only counting the convolutional layers and the fully connected layer)[17] containing 3 residual units that output 64 feature maps, 4 RUs with 128 maps, 6 RUs with 256 maps, and 3 RUs with 512 maps. We will implement this architecture later in this chapter.

ResNets deeper than that, such as ResNet-152, use slightly different residual units. Instead of two 3 × 3 convolutional layers with, say, 256 feature maps, they use three convolutional layers: first a 1 × 1 convolutional layer with just 64 feature maps (4 times less), which acts as a bottleneck layer (as discussed already), then a 3 × 3 layer with 64 feature maps, and finally another 1 × 1 convolutional layer with 256 feature maps (4 times 64) that restores the original depth. ResNet-152 contains 3 such RUs that output 256 maps, then 8 RUs with 512 maps, a whopping 36 RUs with 1,024 maps, and finally 3 RUs with 2,048 maps.

Google's Inception-v4 (*https://homl.info/84*)[18] architecture merged the ideas of GoogLeNet and ResNet and achieved a top-five error rate of close to 3% on ImageNet classification.

Xception

Another variant of the GoogLeNet architecture is worth noting: Xception (*https://homl.info/xception*)[19] (which stands for *Extreme Inception*) was proposed in 2016 by François Chollet (the author of Keras), and it significantly outperformed Inception-v3 on a huge vision task (350 million images and 17,000 classes). Just like Inception-v4, it merges the ideas of GoogLeNet and ResNet, but it replaces the inception modules with a special type of layer called a *depthwise separable convolution layer* (or *separable convolution layer* for short[20]). These layers had been used before in some CNN architectures, but they were not as central as in the Xception architecture. While a regular convolutional layer uses filters that try to simultaneously capture spatial patterns (e.g., an oval) and cross-channel patterns (e.g., mouth + nose + eyes = face), a separable convolutional layer makes the strong assumption that spatial patterns and cross-channel patterns can be modeled separately (see Figure 14-19). Thus, it is composed of two parts: the first part applies a single spatial filter for each input

---

17 It is a common practice when describing a neural network to count only layers with parameters.

18 Christian Szegedy et al., "Inception–v4, Inception-ResNet and the Impact of Residual Connections on Learning," arXiv preprint arXiv:1602.07261 (2016).

19 François Chollet, "Xception: Deep Learning with Depthwise Separable Convolutions," arXiv preprint arXiv: 1610.02357 (2016).

20 This name can sometimes be ambiguous, since spatially separable convolutions are often called "separable convolutions" as well.

feature map, then the second part looks exclusively for cross-channel patterns—it is just a regular convolutional layer with 1 × 1 filters.

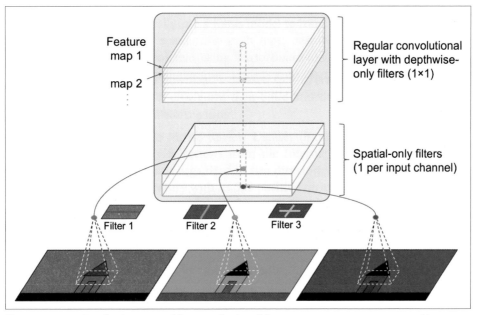

Figure 14-19. Depthwise separable convolutional layer

Since separable convolutional layers only have one spatial filter per input channel, you should avoid using them after layers that have too few channels, such as the input layer (granted, that's what Figure 14-19 represents, but it is just for illustration purposes). For this reason, the Xception architecture starts with 2 regular convolutional layers, but then the rest of the architecture uses only separable convolutions (34 in all), plus a few max pooling layers and the usual final layers (a global average pooling layer and a dense output layer).

You might wonder why Xception is considered a variant of GoogLeNet, since it contains no inception module at all. Well, as we discussed earlier, an inception module contains convolutional layers with 1 × 1 filters: these look exclusively for cross-channel patterns. However, the convolutional layers that sit on top of them are regular convolutional layers that look both for spatial and cross-channel patterns. So you can think of an inception module as an intermediate between a regular convolutional layer (which considers spatial patterns and cross-channel patterns jointly) and a separable convolutional layer (which considers them separately). In practice, it seems that separable convolutional layers generally perform better.

Separable convolutional layers use fewer parameters, less memory, and fewer computations than regular convolutional layers, and in general they even perform better, so you should consider using them by default (except after layers with few channels).

The ILSVRC 2016 challenge was won by the CUImage team from the Chinese University of Hong Kong. They used an ensemble of many different techniques, including a sophisticated object-detection system called GBD-Net (*https://homl.info/gbdnet*),[21] to achieve a top-five error rate below 3%. Although this result is unquestionably impressive, the complexity of the solution contrasted with the simplicity of ResNets. Moreover, one year later another fairly simple architecture performed even better, as we will see now.

SENet

The winning architecture in the ILSVRC 2017 challenge was the Squeeze-and-Excitation Network (SENet) (*https://homl.info/senet*).[22] This architecture extends existing architectures such as inception networks and ResNets, and boosts their performance. This allowed SENet to win the competition with an astonishing 2.25% top-five error rate! The extended versions of inception networks and ResNets are called *SE-Inception* and *SE-ResNet*, respectively. The boost comes from the fact that a SENet adds a small neural network, called an *SE block*, to every unit in the original architecture (i.e., every inception module or every residual unit), as shown in Figure 14-20.

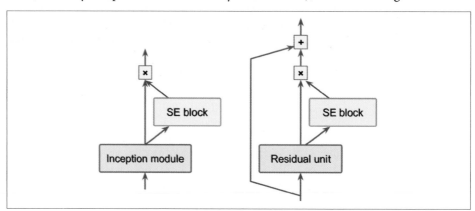

Figure 14-20. SE-Inception module (left) and SE-ResNet unit (right)

21 Xingyu Zeng et al., "Crafting GBD-Net for Object Detection," *IEEE Transactions on Pattern Analysis and Machine Intelligence* 40, no. 9 (2018): 2109–2123.

22 Jie Hu et al., "Squeeze-and-Excitation Networks," *Proceedings of the IEEE Conference on Computer Vision and Pattern Recognition* (2018): 7132–7141.

An SE block analyzes the output of the unit it is attached to, focusing exclusively on the depth dimension (it does not look for any spatial pattern), and it learns which features are usually most active together. It then uses this information to recalibrate the feature maps, as shown in Figure 14-21. For example, an SE block may learn that mouths, noses, and eyes usually appear together in pictures: if you see a mouth and a nose, you should expect to see eyes as well. So if the block sees a strong activation in the mouth and nose feature maps, but only mild activation in the eye feature map, it will boost the eye feature map (more accurately, it will reduce irrelevant feature maps). If the eyes were somewhat confused with something else, this feature map recalibration will help resolve the ambiguity.

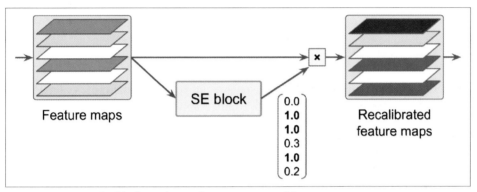

Figure 14-21. An SE block performs feature map recalibration

An SE block is composed of just three layers: a global average pooling layer, a hidden dense layer using the ReLU activation function, and a dense output layer using the sigmoid activation function (see Figure 14-22).

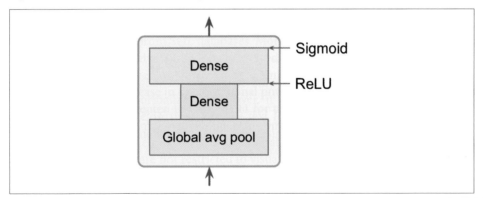

Figure 14-22. SE block architecture

As earlier, the global average pooling layer computes the mean activation for each feature map: for example, if its input contains 256 feature maps, it will output 256 numbers representing the overall level of response for each filter. The next layer is where the "squeeze" happens: this layer has significantly fewer than 256 neurons—typically 16 times fewer than the number of feature maps (e.g., 16 neurons)—so the 256 numbers get compressed into a small vector (e.g., 16 dimensions). This is a low-dimensional vector representation (i.e., an embedding) of the distribution of feature responses. This bottleneck step forces the SE block to learn a general representation of the feature combinations (we will see this principle in action again when we discuss autoencoders in Chapter 17). Finally, the output layer takes the embedding and outputs a recalibration vector containing one number per feature map (e.g., 256), each between 0 and 1. The feature maps are then multiplied by this recalibration vector, so irrelevant features (with a low recalibration score) get scaled down while relevant features (with a recalibration score close to 1) are left alone.

Implementing a ResNet-34 CNN Using Keras

Most CNN architectures described so far are fairly straightforward to implement (although generally you would load a pretrained network instead, as we will see). To illustrate the process, let's implement a ResNet-34 from scratch using Keras. First, let's create a ResidualUnit layer:

```python
class ResidualUnit(keras.layers.Layer):
    def __init__(self, filters, strides=1, activation="relu", **kwargs):
        super().__init__(**kwargs)
        self.activation = keras.activations.get(activation)
        self.main_layers = [
            keras.layers.Conv2D(filters, 3, strides=strides,
                                padding="same", use_bias=False),
            keras.layers.BatchNormalization(),
            self.activation,
            keras.layers.Conv2D(filters, 3, strides=1,
                                padding="same", use_bias=False),
            keras.layers.BatchNormalization()]
        self.skip_layers = []
        if strides > 1:
            self.skip_layers = [
                keras.layers.Conv2D(filters, 1, strides=strides,
                                    padding="same", use_bias=False),
                keras.layers.BatchNormalization()]

    def call(self, inputs):
        Z = inputs
        for layer in self.main_layers:
            Z = layer(Z)
        skip_Z = inputs
        for layer in self.skip_layers:
            skip_Z = layer(skip_Z)
        return self.activation(Z + skip_Z)
```

As you can see, this code matches Figure 14-18 pretty closely. In the constructor, we create all the layers we will need: the main layers are the ones on the right side of the diagram, and the skip layers are the ones on the left (only needed if the stride is greater than 1). Then in the `call()` method, we make the inputs go through the main layers and the skip layers (if any), then we add both outputs and apply the activation function.

Next, we can build the ResNet-34 using a `Sequential` model, since it's really just a long sequence of layers (we can treat each residual unit as a single layer now that we have the `ResidualUnit` class):

```
model = keras.models.Sequential()
model.add(keras.layers.Conv2D(64, 7, strides=2, input_shape=[224, 224, 3],
                              padding="same", use_bias=False))
model.add(keras.layers.BatchNormalization())
model.add(keras.layers.Activation("relu"))
model.add(keras.layers.MaxPool2D(pool_size=3, strides=2, padding="same"))
prev_filters = 64
for filters in [64] * 3 + [128] * 4 + [256] * 6 + [512] * 3:
    strides = 1 if filters == prev_filters else 2
    model.add(ResidualUnit(filters, strides=strides))
    prev_filters = filters
model.add(keras.layers.GlobalAvgPool2D())
model.add(keras.layers.Flatten())
model.add(keras.layers.Dense(10, activation="softmax"))
```

The only slightly tricky part in this code is the loop that adds the `ResidualUnit` layers to the model: as explained earlier, the first 3 RUs have 64 filters, then the next 4 RUs have 128 filters, and so on. We then set the stride to 1 when the number of filters is the same as in the previous RU, or else we set it to 2. Then we add the `ResidualUnit`, and finally we update `prev_filters`.

It is amazing that in fewer than 40 lines of code, we can build the model that won the ILSVRC 2015 challenge! This demonstrates both the elegance of the ResNet model and the expressiveness of the Keras API. Implementing the other CNN architectures is not much harder. However, Keras comes with several of these architectures built in, so why not use them instead?

Using Pretrained Models from Keras

In general, you won't have to implement standard models like GoogLeNet or ResNet manually, since pretrained networks are readily available with a single line of code in the `keras.applications` package. For example, you can load the ResNet-50 model, pretrained on ImageNet, with the following line of code:

```
model = keras.applications.resnet50.ResNet50(weights="imagenet")
```

That's all! This will create a ResNet-50 model and download weights pretrained on the ImageNet dataset. To use it, you first need to ensure that the images have the right size. A ResNet-50 model expects 224 × 224-pixel images (other models may expect other sizes, such as 299 × 299), so let's use TensorFlow's `tf.image.resize()` function to resize the images we loaded earlier:

```
images_resized = tf.image.resize(images, [224, 224])
```

 The `tf.image.resize()` will not preserve the aspect ratio. If this is a problem, try cropping the images to the appropriate aspect ratio before resizing. Both operations can be done in one shot with `tf.image.crop_and_resize()`.

The pretrained models assume that the images are preprocessed in a specific way. In some cases they may expect the inputs to be scaled from 0 to 1, or –1 to 1, and so on. Each model provides a `preprocess_input()` function that you can use to preprocess your images. These functions assume that the pixel values range from 0 to 255, so we must multiply them by 255 (since earlier we scaled them to the 0–1 range):

```
inputs = keras.applications.resnet50.preprocess_input(images_resized * 255)
```

Now we can use the pretrained model to make predictions:

```
Y_proba = model.predict(inputs)
```

As usual, the output `Y_proba` is a matrix with one row per image and one column per class (in this case, there are 1,000 classes). If you want to display the top *K* predictions, including the class name and the estimated probability of each predicted class, use the `decode_predictions()` function. For each image, it returns an array containing the top *K* predictions, where each prediction is represented as an array containing the class identifier,[23] its name, and the corresponding confidence score:

```
top_K = keras.applications.resnet50.decode_predictions(Y_proba, top=3)
for image_index in range(len(images)):
    print("Image #{}".format(image_index))
    for class_id, name, y_proba in top_K[image_index]:
        print("  {} - {:12s} {:.2f}%".format(class_id, name, y_proba * 100))
    print()
```

The output looks like this:

```
Image #0
  n03877845 - palace       42.87%
  n02825657 - bell_cote    40.57%
  n03781244 - monastery    14.56%
```

---

23 In the ImageNet dataset, each image is associated to a word in the WordNet dataset (*https://wordnet.prince ton.edu/*): the class ID is just a WordNet ID.

```
Image #1
  n04522168 - vase          46.83%
  n07930864 - cup            7.78%
  n11939491 - daisy          4.87%
```

The correct classes (monastery and daisy) appear in the top three results for both images. That's pretty good, considering that the model had to choose from among 1,000 classes.

As you can see, it is very easy to create a pretty good image classifier using a pre-trained model. Other vision models are available in keras.applications, including several ResNet variants, GoogLeNet variants like Inception-v3 and Xception, VGGNet variants, and MobileNet and MobileNetV2 (lightweight models for use in mobile applications).

But what if you want to use an image classifier for classes of images that are not part of ImageNet? In that case, you may still benefit from the pretrained models to perform transfer learning.

Pretrained Models for Transfer Learning

If you want to build an image classifier but you do not have enough training data, then it is often a good idea to reuse the lower layers of a pretrained model, as we discussed in Chapter 11. For example, let's train a model to classify pictures of flowers, reusing a pretrained Xception model. First, let's load the dataset using TensorFlow Datasets (see Chapter 13):

```python
import tensorflow_datasets as tfds

dataset, info = tfds.load("tf_flowers", as_supervised=True, with_info=True)
dataset_size = info.splits["train"].num_examples # 3670
class_names = info.features["label"].names # ["dandelion", "daisy", ...]
n_classes = info.features["label"].num_classes # 5
```

Note that you can get information about the dataset by setting with_info=True. Here, we get the dataset size and the names of the classes. Unfortunately, there is only a "train" dataset, no test set or validation set, so we need to split the training set. The TF Datasets project provides an API for this. For example, let's take the first 10% of the dataset for testing, the next 15% for validation, and the remaining 75% for training:

```python
test_split, valid_split, train_split = tfds.Split.TRAIN.subsplit([10, 15, 75])

test_set = tfds.load("tf_flowers", split=test_split, as_supervised=True)
valid_set = tfds.load("tf_flowers", split=valid_split, as_supervised=True)
train_set = tfds.load("tf_flowers", split=train_split, as_supervised=True)
```

Next we must preprocess the images. The CNN expects 224 × 224 images, so we need to resize them. We also need to run the images through Xception's `preprocess_input()` function:

```
def preprocess(image, label):
    resized_image = tf.image.resize(image, [224, 224])
    final_image = keras.applications.xception.preprocess_input(resized_image)
    return final_image, label
```

Let's apply this preprocessing function to all three datasets, shuffle the training set, and add batching and prefetching to all the datasets:

```
batch_size = 32
train_set = train_set.shuffle(1000)
train_set = train_set.map(preprocess).batch(batch_size).prefetch(1)
valid_set = valid_set.map(preprocess).batch(batch_size).prefetch(1)
test_set = test_set.map(preprocess).batch(batch_size).prefetch(1)
```

If you want to perform some data augmentation, change the preprocessing function for the training set, adding some random transformations to the training images. For example, use `tf.image.random_crop()` to randomly crop the images, use `tf.image.random_flip_left_right()` to randomly flip the images horizontally, and so on (see the "Pretrained Models for Transfer Learning" section of the notebook for an example).

 The `keras.preprocessing.image.ImageDataGenerator` class makes it easy to load images from disk and augment them in various ways: you can shift each image, rotate it, rescale it, flip it horizontally or vertically, shear it, or apply any transformation function you want to it. This is very convenient for simple projects. However, building a tf.data pipeline has many advantages: it can read the images efficiently (e.g., in parallel) from any source, not just the local disk; you can manipulate the `Dataset` as you wish; and if you write a preprocessing function based on `tf.image` operations, this function can be used both in the tf.data pipeline and in the model you will deploy to production (see Chapter 19).

Next let's load an Xception model, pretrained on ImageNet. We exclude the top of the network by setting `include_top=False`: this excludes the global average pooling layer and the dense output layer. We then add our own global average pooling layer, based on the output of the base model, followed by a dense output layer with one unit per class, using the softmax activation function. Finally, we create the Keras `Model`:

```
base_model = keras.applications.xception.Xception(weights="imagenet",
                                                   include_top=False)
avg = keras.layers.GlobalAveragePooling2D()(base_model.output)
output = keras.layers.Dense(n_classes, activation="softmax")(avg)
model = keras.Model(inputs=base_model.input, outputs=output)
```

As explained in Chapter 11, it's usually a good idea to freeze the weights of the pre-trained layers, at least at the beginning of training:

```
for layer in base_model.layers:
    layer.trainable = False
```

 Since our model uses the base model's layers directly, rather than the base_model object itself, setting base_model.trainable=False would have no effect.

Finally, we can compile the model and start training:

```
optimizer = keras.optimizers.SGD(lr=0.2, momentum=0.9, decay=0.01)
model.compile(loss="sparse_categorical_crossentropy", optimizer=optimizer,
              metrics=["accuracy"])
history = model.fit(train_set, epochs=5, validation_data=valid_set)
```

 This will be very slow, unless you have a GPU. If you do not, then you should run this chapter's notebook in Colab, using a GPU run-time (it's free!). See the instructions at *https://github.com/ageron/handson-ml2*.

After training the model for a few epochs, its validation accuracy should reach about 75–80% and stop making much progress. This means that the top layers are now pretty well trained, so we are ready to unfreeze all the layers (or you could try unfreezing just the top ones) and continue training (don't forget to compile the model when you freeze or unfreeze layers). This time we use a much lower learning rate to avoid damaging the pretrained weights:

```
for layer in base_model.layers:
    layer.trainable = True

optimizer = keras.optimizers.SGD(lr=0.01, momentum=0.9, decay=0.001)
model.compile(...)
history = model.fit(...)
```

It will take a while, but this model should reach around 95% accuracy on the test set. With that, you can start training amazing image classifiers! But there's more to computer vision than just classification. For example, what if you also want to know *where* the flower is in the picture? Let's look at this now.

Classification and Localization

Localizing an object in a picture can be expressed as a regression task, as discussed in Chapter 10: to predict a bounding box around the object, a common approach is to predict the horizontal and vertical coordinates of the object's center, as well as its height and width. This means we have four numbers to predict. It does not require much change to the model; we just need to add a second dense output layer with four units (typically on top of the global average pooling layer), and it can be trained using the MSE loss:

```
base_model = keras.applications.xception.Xception(weights="imagenet",
                                                  include_top=False)
avg = keras.layers.GlobalAveragePooling2D()(base_model.output)
class_output = keras.layers.Dense(n_classes, activation="softmax")(avg)
loc_output = keras.layers.Dense(4)(avg)
model = keras.Model(inputs=base_model.input,
                    outputs=[class_output, loc_output])
model.compile(loss=["sparse_categorical_crossentropy", "mse"],
              loss_weights=[0.8, 0.2], # depends on what you care most about
              optimizer=optimizer, metrics=["accuracy"])
```

But now we have a problem: the flowers dataset does not have bounding boxes around the flowers. So, we need to add them ourselves. This is often one of the hardest and most costly parts of a Machine Learning project: getting the labels. It's a good idea to spend time looking for the right tools. To annotate images with bounding boxes, you may want to use an open source image labeling tool like VGG Image Annotator, LabelImg, OpenLabeler, or ImgLab, or perhaps a commercial tool like LabelBox or Supervisely. You may also want to consider crowdsourcing platforms such as Amazon Mechanical Turk if you have a very large number of images to annotate. However, it is quite a lot of work to set up a crowdsourcing platform, prepare the form to be sent to the workers, supervise them, and ensure that the quality of the bounding boxes they produce is good, so make sure it is worth the effort. If there are just a few thousand images to label, and you don't plan to do this frequently, it may be preferable to do it yourself. Adriana Kovashka et al. wrote a very practical paper (*https://homl.info/crowd*)[24] about crowdsourcing in computer vision. I recommend you check it out, even if you do not plan to use crowdsourcing.

Let's suppose you've obtained the bounding boxes for every image in the flowers dataset (for now we will assume there is a single bounding box per image). You then need to create a dataset whose items will be batches of preprocessed images along with their class labels and their bounding boxes. Each item should be a tuple of the form (images, (class_labels, bounding_boxes)). Then you are ready to train your model!

---

24 Adriana Kovashka et al., "Crowdsourcing in Computer Vision," *Foundations and Trends in Computer Graphics and Vision* 10, no. 3 (2014): 177–243.

The bounding boxes should be normalized so that the horizontal and vertical coordinates, as well as the height and width, all range from 0 to 1. Also, it is common to predict the square root of the height and width rather than the height and width directly: this way, a 10-pixel error for a large bounding box will not be penalized as much as a 10-pixel error for a small bounding box.

The MSE often works fairly well as a cost function to train the model, but it is not a great metric to evaluate how well the model can predict bounding boxes. The most common metric for this is the *Intersection over Union* (IoU): the area of overlap between the predicted bounding box and the target bounding box, divided by the area of their union (see Figure 14-23). In tf.keras, it is implemented by the `tf.keras.metrics.MeanIoU` class.

Figure 14-23. Intersection over Union (IoU) metric for bounding boxes

Classifying and localizing a single object is nice, but what if the images contain multiple objects (as is often the case in the flowers dataset)?

Object Detection

The task of classifying and localizing multiple objects in an image is called *object detection*. Until a few years ago, a common approach was to take a CNN that was trained to classify and locate a single object, then slide it across the image, as shown in Figure 14-24. In this example, the image was chopped into a 6 × 8 grid, and we show a CNN (the thick black rectangle) sliding across all 3 × 3 regions. When the CNN was looking at the top left of the image, it detected part of the leftmost rose, and then it detected that same rose again when it was first shifted one step to the right. At the next step, it started detecting part of the topmost rose, and then it detected it again once it was shifted one more step to the right. You would then continue to slide the CNN through the whole image, looking at all 3 × 3 regions. Moreover, since objects can have varying sizes, you would also slide the CNN across regions of different sizes. For example, once you are done with the 3 × 3 regions, you might want to slide the CNN across all 4 × 4 regions as well.

Figure 14-24. Detecting multiple objects by sliding a CNN across the image

This technique is fairly straightforward, but as you can see it will detect the same object multiple times, at slightly different positions. Some post-processing will then be needed to get rid of all the unnecessary bounding boxes. A common approach for this is called *non-max suppression*. Here's how you do it:

1. First, you need to add an extra *objectness* output to your CNN, to estimate the probability that a flower is indeed present in the image (alternatively, you could add a "no-flower" class, but this usually does not work as well). It must use the sigmoid activation function, and you can train it using binary cross-entropy loss. Then get rid of all the bounding boxes for which the objectness score is below some threshold: this will drop all the bounding boxes that don't actually contain a flower.

2. Find the bounding box with the highest objectness score, and get rid of all the other bounding boxes that overlap a lot with it (e.g., with an IoU greater than 60%). For example, in Figure 14-24, the bounding box with the max objectness score is the thick bounding box over the topmost rose (the objectness score is represented by the thickness of the bounding boxes). The other bounding box

over that same rose overlaps a lot with the max bounding box, so we will get rid of it.

3. Repeat step two until there are no more bounding boxes to get rid of.

This simple approach to object detection works pretty well, but it requires running the CNN many times, so it is quite slow. Fortunately, there is a much faster way to slide a CNN across an image: using a *fully convolutional network* (FCN).

Fully Convolutional Networks

The idea of FCNs was first introduced in a 2015 paper (*https://homl.info/fcn*)[25] by Jonathan Long et al., for semantic segmentation (the task of classifying every pixel in an image according to the class of the object it belongs to). The authors pointed out that you could replace the dense layers at the top of a CNN by convolutional layers. To understand this, let's look at an example: suppose a dense layer with 200 neurons sits on top of a convolutional layer that outputs 100 feature maps, each of size 7 × 7 (this is the feature map size, not the kernel size). Each neuron will compute a weighted sum of all 100 × 7 × 7 activations from the convolutional layer (plus a bias term). Now let's see what happens if we replace the dense layer with a convolutional layer using 200 filters, each of size 7 × 7, and with "valid" padding. This layer will output 200 feature maps, each 1 × 1 (since the kernel is exactly the size of the input feature maps and we are using "valid" padding). In other words, it will output 200 numbers, just like the dense layer did; and if you look closely at the computations performed by a convolutional layer, you will notice that these numbers will be precisely the same as those the dense layer produced. The only difference is that the dense layer's output was a tensor of shape [*batch size*, 200], while the convolutional layer will output a tensor of shape [*batch size*, 1, 1, 200].

To convert a dense layer to a convolutional layer, the number of filters in the convolutional layer must be equal to the number of units in the dense layer, the filter size must be equal to the size of the input feature maps, and you must use "valid" padding. The stride may be set to 1 or more, as we will see shortly.

Why is this important? Well, while a dense layer expects a specific input size (since it has one weight per input feature), a convolutional layer will happily process images of any size[26] (however, it does expect its inputs to have a specific number of channels,

---

25 Jonathan Long et al., "Fully Convolutional Networks for Semantic Segmentation," *Proceedings of the IEEE Conference on Computer Vision and Pattern Recognition* (2015): 3431–3440.

26 There is one small exception: a convolutional layer using "valid" padding will complain if the input size is smaller than the kernel size.

since each kernel contains a different set of weights for each input channel). Since an FCN contains only convolutional layers (and pooling layers, which have the same property), it can be trained and executed on images of any size!

For example, suppose we'd already trained a CNN for flower classification and localization. It was trained on 224 × 224 images, and it outputs 10 numbers: outputs 0 to 4 are sent through the softmax activation function, and this gives the class probabilities (one per class); output 5 is sent through the logistic activation function, and this gives the objectness score; outputs 6 to 9 do not use any activation function, and they represent the bounding box's center coordinates, as well as its height and width. We can now convert its dense layers to convolutional layers. In fact, we don't even need to retrain it; we can just copy the weights from the dense layers to the convolutional layers! Alternatively, we could have converted the CNN into an FCN before training.

Now suppose the last convolutional layer before the output layer (also called the bottleneck layer) outputs 7 × 7 feature maps when the network is fed a 224 × 224 image (see the left side of Figure 14-25). If we feed the FCN a 448 × 448 image (see the right side of Figure 14-25), the bottleneck layer will now output 14 × 14 feature maps.[27] Since the dense output layer was replaced by a convolutional layer using 10 filters of size 7 × 7, with `"valid"` padding and stride 1, the output will be composed of 10 features maps, each of size 8 × 8 (since 14 − 7 + 1 = 8). In other words, the FCN will process the whole image only once, and it will output an 8 × 8 grid where each cell contains 10 numbers (5 class probabilities, 1 objectness score, and 4 bounding box coordinates). It's exactly like taking the original CNN and sliding it across the image using 8 steps per row and 8 steps per column. To visualize this, imagine chopping the original image into a 14 × 14 grid, then sliding a 7 × 7 window across this grid; there will be 8 × 8 = 64 possible locations for the window, hence 8 × 8 predictions. However, the FCN approach is *much* more efficient, since the network only looks at the image once. In fact, *You Only Look Once* (YOLO) is the name of a very popular object detection architecture, which we'll look at next.

---

27 This assumes we used only `"same"` padding in the network: indeed, `"valid"` padding would reduce the size of the feature maps. Moreover, 448 can be neatly divided by 2 several times until we reach 7, without any rounding error. If any layer uses a different stride than 1 or 2, then there may be some rounding error, so again the feature maps may end up being smaller.

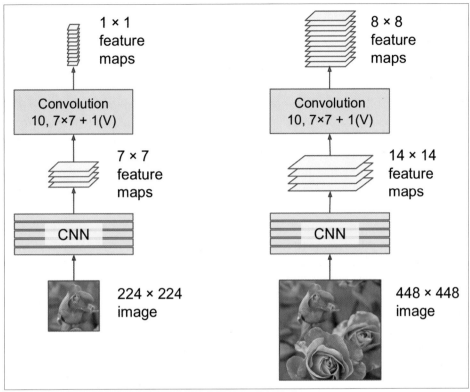

Figure 14-25. The same fully convolutional network processing a small image (left) and a large one (right)

You Only Look Once (YOLO)

YOLO is an extremely fast and accurate object detection architecture proposed by Joseph Redmon et al. in a 2015 paper (*https://homl.info/yolo*),[28] and subsequently improved in 2016 (*https://homl.info/yolo2*)[29] (YOLOv2) and in 2018 (*https://homl.info/yolo3*)[30] (YOLOv3). It is so fast that it can run in real time on a video, as seen in Redmon's demo (*https://homl.info/yolodemo*).

YOLOv3's architecture is quite similar to the one we just discussed, but with a few important differences:

---

28 Joseph Redmon et al., "You Only Look Once: Unified, Real-Time Object Detection," *Proceedings of the IEEE Conference on Computer Vision and Pattern Recognition* (2016): 779–788.

29 Joseph Redmon and Ali Farhadi, "YOLO9000: Better, Faster, Stronger," *Proceedings of the IEEE Conference on Computer Vision and Pattern Recognition* (2017): 6517–6525.

30 Joseph Redmon and Ali Farhadi, "YOLOv3: An Incremental Improvement," arXiv preprint arXiv:1804.02767 (2018).

- It outputs five bounding boxes for each grid cell (instead of just one), and each bounding box comes with an objectness score. It also outputs 20 class probabilities per grid cell, as it was trained on the PASCAL VOC dataset, which contains 20 classes. That's a total of 45 numbers per grid cell: 5 bounding boxes, each with 4 coordinates, plus 5 objectness scores, plus 20 class probabilities.

- Instead of predicting the absolute coordinates of the bounding box centers, YOLOv3 predicts an offset relative to the coordinates of the grid cell, where (0, 0) means the top left of that cell and (1, 1) means the bottom right. For each grid cell, YOLOv3 is trained to predict only bounding boxes whose center lies in that cell (but the bounding box itself generally extends well beyond the grid cell). YOLOv3 applies the logistic activation function to the bounding box coordinates to ensure they remain in the 0 to 1 range.

- Before training the neural net, YOLOv3 finds five representative bounding box dimensions, called *anchor boxes* (or *bounding box priors*). It does this by applying the K-Means algorithm (see Chapter 9) to the height and width of the training set bounding boxes. For example, if the training images contain many pedestrians, then one of the anchor boxes will likely have the dimensions of a typical pedestrian. Then when the neural net predicts five bounding boxes per grid cell, it actually predicts how much to rescale each of the anchor boxes. For example, suppose one anchor box is 100 pixels tall and 50 pixels wide, and the network predicts, say, a vertical rescaling factor of 1.5 and a horizontal rescaling of 0.9 (for one of the grid cells). This will result in a predicted bounding box of size 150 × 45 pixels. To be more precise, for each grid cell and each anchor box, the network predicts the log of the vertical and horizontal rescaling factors. Having these priors makes the network more likely to predict bounding boxes of the appropriate dimensions, and it also speeds up training because it will more quickly learn what reasonable bounding boxes look like.

- The network is trained using images of different scales: every few batches during training, the network randomly chooses a new image dimension (from 330 × 330 to 608 × 608 pixels). This allows the network to learn to detect objects at different scales. Moreover, it makes it possible to use YOLOv3 at different scales: the smaller scale will be less accurate but faster than the larger scale, so you can choose the right trade-off for your use case.

There are a few more innovations you might be interested in, such as the use of skip connections to recover some of the spatial resolution that is lost in the CNN (we will discuss this shortly, when we look at semantic segmentation). In the 2016 paper, the authors introduce the YOLO9000 model that uses hierarchical classification: the model predicts a probability for each node in a visual hierarchy called *WordTree*. This makes it possible for the network to predict with high confidence that an image represents, say, a dog, even though it is unsure what specific type of dog. I encourage you to go ahead and read all three papers: they are quite pleasant to read, and they provide excellent examples of how Deep Learning systems can be incrementally improved.

Mean Average Precision (mAP)

A very common metric used in object detection tasks is the *mean Average Precision* (mAP). "Mean Average" sounds a bit redundant, doesn't it? To understand this metric, let's go back to two classification metrics we discussed in Chapter 3: precision and recall. Remember the trade-off: the higher the recall, the lower the precision. You can visualize this in a precision/recall curve (see Figure 3-5). To summarize this curve into a single number, we could compute its area under the curve (AUC). But note that the precision/recall curve may contain a few sections where precision actually goes up when recall increases, especially at low recall values (you can see this at the top left of Figure 3-5). This is one of the motivations for the mAP metric.

Suppose the classifier has 90% precision at 10% recall, but 96% precision at 20% recall. There's really no trade-off here: it simply makes more sense to use the classifier at 20% recall rather than at 10% recall, as you will get both higher recall and higher precision. So instead of looking at the precision *at* 10% recall, we should really be looking at the *maximum* precision that the classifier can offer with *at least* 10% recall. It would be 96%, not 90%. Therefore, one way to get a fair idea of the model's performance is to compute the maximum precision you can get with at least 0% recall, then 10% recall, 20%, and so on up to 100%, and then calculate the mean of these maximum precisions. This is called the *Average Precision* (AP) metric. Now when there are more than two classes, we can compute the AP for each class, and then compute the mean AP (mAP). That's it!

In an object detection system, there is an additional level of complexity: what if the system detected the correct class, but at the wrong location (i.e., the bounding box is completely off)? Surely we should not count this as a positive prediction. One approach is to define an IOU threshold: for example, we may consider that a prediction is correct only if the IOU is greater than, say, 0.5, and the predicted class is correct. The corresponding mAP is generally noted mAP@0.5 (or mAP@50%, or sometimes just AP_{50}). In some competitions (such as the PASCAL VOC challenge), this is what is done. In others (such as the COCO competition), the mAP is computed for different IOU thresholds (0.50, 0.55, 0.60, ..., 0.95), and the final metric is the mean of all these mAPs (noted AP@[.50:.95] or AP@[.50:0.05:.95]). Yes, that's a mean mean average.

Several YOLO implementations built using TensorFlow are available on GitHub. In particular, check out Zihao Zang's TensorFlow 2 implementation (*https://homl.info/yolotf2*). Other object detection models are available in the TensorFlow Models project, many with pretrained weights; and some have even been ported to TF Hub, such as SSD (*https://homl.info/ssd*)[31] and Faster-RCNN (*https://homl.info/fasterrcnn*),[32] which are both quite popular. SSD is also a "single shot" detection model, similar to YOLO. Faster R-CNN is more complex: the image first goes through a CNN, then the output is passed to a *Region Proposal Network* (RPN) that proposes bounding boxes that are most likely to contain an object, and a classifier is run for each bounding box, based on the cropped output of the CNN.

The choice of detection system depends on many factors: speed, accuracy, available pretrained models, training time, complexity, etc. The papers contain tables of metrics, but there is quite a lot of variability in the testing environments, and the technologies evolve so fast that it is difficult to make a fair comparison that will be useful for most people and remain valid for more than a few months.

So, we can locate objects by drawing bounding boxes around them. Great! But perhaps you want to be a bit more precise. Let's see how to go down to the pixel level.

Semantic Segmentation

In *semantic segmentation*, each pixel is classified according to the class of the object it belongs to (e.g., road, car, pedestrian, building, etc.), as shown in Figure 14-26. Note that different objects of the same class are *not* distinguished. For example, all the bicycles on the right side of the segmented image end up as one big lump of pixels. The main difficulty in this task is that when images go through a regular CNN, they gradually lose their spatial resolution (due to the layers with strides greater than 1); so, a regular CNN may end up knowing that there's a person somewhere in the bottom left of the image, but it will not be much more precise than that.

Just like for object detection, there are many different approaches to tackle this problem, some quite complex. However, a fairly simple solution was proposed in the 2015 paper by Jonathan Long et al. we discussed earlier. The authors start by taking a pretrained CNN and turning it into an FCN. The CNN applies an overall stride of 32 to the input image (i.e., if you add up all the strides greater than 1), meaning the last layer outputs feature maps that are 32 times smaller than the input image. This is clearly too coarse, so they add a single *upsampling layer* that multiplies the resolution by 32.

---

31 Wei Liu et al., "SSD: Single Shot Multibox Detector," *Proceedings of the 14th European Conference on Computer Vision* 1 (2016): 21–37.

32 Shaoqing Ren et al., "Faster R-CNN: Towards Real-Time Object Detection with Region Proposal Networks," *Proceedings of the 28th International Conference on Neural Information Processing Systems* 1 (2015): 91–99.

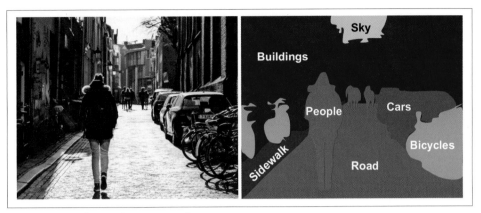

Figure 14-26. Semantic segmentation

There are several solutions available for upsampling (increasing the size of an image), such as bilinear interpolation, but that only works reasonably well up to ×4 or ×8. Instead, they use a *transposed convolutional layer:*[33] it is equivalent to first stretching the image by inserting empty rows and columns (full of zeros), then performing a regular convolution (see Figure 14-27). Alternatively, some people prefer to think of it as a regular convolutional layer that uses fractional strides (e.g., 1/2 in Figure 14-27). The transposed convolutional layer can be initialized to perform something close to linear interpolation, but since it is a trainable layer, it will learn to do better during training. In tf.keras, you can use the Conv2DTranspose layer.

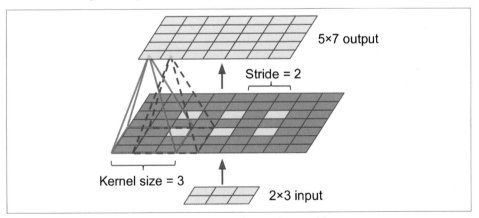

Figure 14-27. Upsampling using a transposed convolutional layer

---

33 This type of layer is sometimes referred to as a *deconvolution layer*, but it does *not* perform what mathematicians call a deconvolution, so this name should be avoided.

In a transposed convolutional layer, the stride defines how much the input will be stretched, not the size of the filter steps, so the larger the stride, the larger the output (unlike for convolutional layers or pooling layers).

TensorFlow Convolution Operations

TensorFlow also offers a few other kinds of convolutional layers:

`keras.layers.Conv1D`
> Creates a convolutional layer for 1D inputs, such as time series or text (sequences of letters or words), as we will see in Chapter 15.

`keras.layers.Conv3D`
> Creates a convolutional layer for 3D inputs, such as 3D PET scans.

`dilation_rate`
> Setting the `dilation_rate` hyperparameter of any convolutional layer to a value of 2 or more creates an *à-trous convolutional layer* ("à trous" is French for "with holes"). This is equivalent to using a regular convolutional layer with a filter dilated by inserting rows and columns of zeros (i.e., holes). For example, a 1 × 3 filter equal to `[[1,2,3]]` may be dilated with a *dilation rate* of 4, resulting in a *dilated filter* of `[[1, 0, 0, 0, 2, 0, 0, 0, 3]]`. This lets the convolutional layer have a larger receptive field at no computational price and using no extra parameters.

`tf.nn.depthwise_conv2d()`
> Can be used to create a *depthwise convolutional layer* (but you need to create the variables yourself). It applies every filter to every individual input channel independently. Thus, if there are f_n filters and $f_{n'}$ input channels, then this will output $f_n \times f_{n'}$ feature maps.

This solution is OK, but still too imprecise. To do better, the authors added skip connections from lower layers: for example, they upsampled the output image by a factor of 2 (instead of 32), and they added the output of a lower layer that had this double resolution. Then they upsampled the result by a factor of 16, leading to a total upsampling factor of 32 (see Figure 14-28). This recovered some of the spatial resolution that was lost in earlier pooling layers. In their best architecture, they used a second similar skip connection to recover even finer details from an even lower layer. In short, the output of the original CNN goes through the following extra steps: upscale ×2, add the output of a lower layer (of the appropriate scale), upscale ×2, add the output of an even lower layer, and finally upscale ×8. It is even possible to scale up beyond the size of the original image: this can be used to increase the resolution of an image, which is a technique called *super-resolution*.

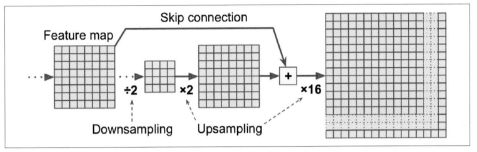

Figure 14-28. Skip layers recover some spatial resolution from lower layers

Once again, many GitHub repositories provide TensorFlow implementations of semantic segmentation (TensorFlow 1 for now), and you will even find pretrained *instance segmentation* models in the TensorFlow Models project. Instance segmentation is similar to semantic segmentation, but instead of merging all objects of the same class into one big lump, each object is distinguished from the others (e.g., it identifies each individual bicycle). At present, the instance segmentation models available in the TensorFlow Models project are based on the *Mask R-CNN* architecture, which was proposed in a 2017 paper (*https://homl.info/maskrcnn*):[34] it extends the Faster R-CNN model by additionally producing a pixel mask for each bounding box. So not only do you get a bounding box around each object, with a set of estimated class probabilities, but you also get a pixel mask that locates pixels in the bounding box that belong to the object.

As you can see, the field of Deep Computer Vision is vast and moving fast, with all sorts of architectures popping out every year, all based on convolutional neural networks. The progress made in just a few years has been astounding, and researchers are now focusing on harder and harder problems, such as *adversarial learning* (which attempts to make the network more resistant to images designed to fool it), explainability (understanding why the network makes a specific classification), realistic *image generation* (which we will come back to in Chapter 17), and *single-shot learning* (a system that can recognize an object after it has seen it just once). Some even explore completely novel architectures, such as Geoffrey Hinton's *capsule networks* (*https://homl.info/capsnet*)[35] (I presented them in a couple of videos (*https://homl.info/capsnet videos*), with the corresponding code in a notebook). Now on to the next chapter, where we will look at how to process sequential data such as time series using recurrent neural networks and convolutional neural networks.

34 Kaiming He et al., "Mask R-CNN," arXiv preprint arXiv:1703.06870 (2017).

35 Geoffrey Hinton et al., "Matrix Capsules with EM Routing," *Proceedings of the International Conference on Learning Representations* (2018).

Exercises

1. What are the advantages of a CNN over a fully connected DNN for image classification?

2. Consider a CNN composed of three convolutional layers, each with 3 × 3 kernels, a stride of 2, and `"same"` padding. The lowest layer outputs 100 feature maps, the middle one outputs 200, and the top one outputs 400. The input images are RGB images of 200 × 300 pixels.

   What is the total number of parameters in the CNN? If we are using 32-bit floats, at least how much RAM will this network require when making a prediction for a single instance? What about when training on a mini-batch of 50 images?

3. If your GPU runs out of memory while training a CNN, what are five things you could try to solve the problem?

4. Why would you want to add a max pooling layer rather than a convolutional layer with the same stride?

5. When would you want to add a local response normalization layer?

6. Can you name the main innovations in AlexNet, compared to LeNet-5? What about the main innovations in GoogLeNet, ResNet, SENet, and Xception?

7. What is a fully convolutional network? How can you convert a dense layer into a convolutional layer?

8. What is the main technical difficulty of semantic segmentation?

9. Build your own CNN from scratch and try to achieve the highest possible accuracy on MNIST.

10. Use transfer learning for large image classification, going through these steps:

    a. Create a training set containing at least 100 images per class. For example, you could classify your own pictures based on the location (beach, mountain, city, etc.), or alternatively you can use an existing dataset (e.g., from TensorFlow Datasets).

    b. Split it into a training set, a validation set, and a test set.

    c. Build the input pipeline, including the appropriate preprocessing operations, and optionally add data augmentation.

    d. Fine-tune a pretrained model on this dataset.

11. Go through TensorFlow's Style Transfer tutorial (*https://homl.info/styletuto*). It is a fun way to generate art using Deep Learning.

Solutions to these exercises are available in Appendix A.

## CHAPTER 15

Processing Sequences Using RNNs and CNNs

The batter hits the ball. The outfielder immediately starts running, anticipating the ball's trajectory. He tracks it, adapts his movements, and finally catches it (under a thunder of applause). Predicting the future is something you do all the time, whether you are finishing a friend's sentence or anticipating the smell of coffee at breakfast. In this chapter we will discuss recurrent neural networks (RNNs), a class of nets that can predict the future (well, up to a point, of course). They can analyze time series data such as stock prices, and tell you when to buy or sell. In autonomous driving systems, they can anticipate car trajectories and help avoid accidents. More generally, they can work on sequences of arbitrary lengths, rather than on fixed-sized inputs like all the nets we have considered so far. For example, they can take sentences, documents, or audio samples as input, making them extremely useful for natural language processing applications such as automatic translation or speech-to-text.

In this chapter we will first look at the fundamental concepts underlying RNNs and how to train them using backpropagation through time, then we will use them to forecast a time series. After that we'll explore the two main difficulties that RNNs face:

- Unstable gradients (discussed in Chapter 11), which can be alleviated using various techniques, including recurrent dropout and recurrent layer normalization

- A (very) limited short-term memory, which can be extended using LSTM and GRU cells

RNNs are not the only types of neural networks capable of handling sequential data: for small sequences, a regular dense network can do the trick; and for very long sequences, such as audio samples or text, convolutional neural networks can actually work quite well too. We will discuss both of these possibilities, and we will finish this chapter by implementing a *WaveNet*: this is a CNN architecture capable of handling sequences of tens of thousands of time steps. In Chapter 16, we will continue to explore RNNs and see how to use them for natural language processing, along with more recent architectures based on attention mechanisms. Let's get started!

Recurrent Neurons and Layers

Up to now we have focused on feedforward neural networks, where the activations flow only in one direction, from the input layer to the output layer (a few exceptions are discussed in Appendix E). A recurrent neural network looks very much like a feedforward neural network, except it also has connections pointing backward. Let's look at the simplest possible RNN, composed of one neuron receiving inputs, producing an output, and sending that output back to itself, as shown in Figure 15-1 (left). At each *time step t* (also called a *frame*), this *recurrent neuron* receives the inputs $\mathbf{x}_{(t)}$ as well as its own output from the previous time step, $y_{(t-1)}$. Since there is no previous output at the first time step, it is generally set to 0. We can represent this tiny network against the time axis, as shown in Figure 15-1 (right). This is called *unrolling the network through time* (it's the same recurrent neuron represented once per time step).

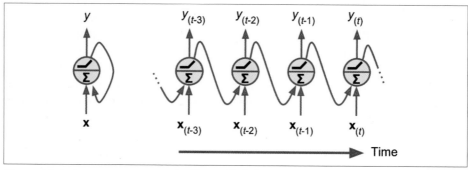

Figure 15-1. A recurrent neuron (left) unrolled through time (right)

You can easily create a layer of recurrent neurons. At each time step t, every neuron receives both the input vector $\mathbf{x}_{(t)}$ and the output vector from the previous time step $\mathbf{y}_{(t-1)}$, as shown in Figure 15-2. Note that both the inputs and outputs are vectors now (when there was just a single neuron, the output was a scalar).

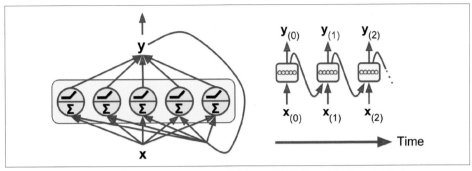

Figure 15-2. A layer of recurrent neurons (left) unrolled through time (right)

Each recurrent neuron has two sets of weights: one for the inputs $\mathbf{x}_{(t)}$ and the other for the outputs of the previous time step, $\mathbf{y}_{(t-1)}$. Let's call these weight vectors $\mathbf{w}_x$ and $\mathbf{w}_y$. If we consider the whole recurrent layer instead of just one recurrent neuron, we can place all the weight vectors in two weight matrices, $\mathbf{W}_x$ and $\mathbf{W}_y$. The output vector of the whole recurrent layer can then be computed pretty much as you might expect, as shown in Equation 15-1 ($\mathbf{b}$ is the bias vector and $\phi(\cdot)$ is the activation function (e.g., ReLU[1]).

Equation 15-1. Output of a recurrent layer for a single instance

$$\mathbf{y}_{(t)} = \phi\left(\mathbf{W}_x{}^\mathsf{T}\mathbf{x}_{(t)} + \mathbf{W}_y{}^\mathsf{T}\mathbf{y}_{(t-1)} + \mathbf{b}\right)$$

Just as with feedforward neural networks, we can compute a recurrent layer's output in one shot for a whole mini-batch by placing all the inputs at time step t in an input matrix $\mathbf{X}_{(t)}$ (see Equation 15-2).

Equation 15-2. Outputs of a layer of recurrent neurons for all instances in a mini-batch

$$\begin{aligned} \mathbf{Y}_{(t)} &= \phi\left(\mathbf{X}_{(t)}\mathbf{W}_x + \mathbf{Y}_{(t-1)}\mathbf{W}_y + \mathbf{b}\right) \\ &= \phi\left(\left[\mathbf{X}_{(t)} \quad \mathbf{Y}_{(t-1)}\right]\mathbf{W} + \mathbf{b}\right) \text{ with } \mathbf{W} = \begin{bmatrix} \mathbf{W}_x \\ \mathbf{W}_y \end{bmatrix} \end{aligned}$$

---

1 Note that many researchers prefer to use the hyperbolic tangent (tanh) activation function in RNNs rather than the ReLU activation function. For example, take a look at Vu Pham et al.'s 2013 paper "Dropout Improves Recurrent Neural Networks for Handwriting Recognition" (*https://homl.info/91*). ReLU-based RNNs are also possible, as shown in Quoc V. Le et al.'s 2015 paper "A Simple Way to Initialize Recurrent Networks of Rectified Linear Units" (*https://homl.info/92*).

In this equation:

- $\mathbf{Y}_{(t)}$ is an $m \times n_{\text{neurons}}$ matrix containing the layer's outputs at time step t for each instance in the mini-batch (m is the number of instances in the mini-batch and n_{neurons} is the number of neurons).

- $\mathbf{X}_{(t)}$ is an $m \times n_{\text{inputs}}$ matrix containing the inputs for all instances (n_{inputs} is the number of input features).

- $\mathbf{W}_x$ is an $n_{\text{inputs}} \times n_{\text{neurons}}$ matrix containing the connection weights for the inputs of the current time step.

- $\mathbf{W}_y$ is an $n_{\text{neurons}} \times n_{\text{neurons}}$ matrix containing the connection weights for the outputs of the previous time step.

- $\mathbf{b}$ is a vector of size n_{neurons} containing each neuron's bias term.

- The weight matrices $\mathbf{W}_x$ and $\mathbf{W}_y$ are often concatenated vertically into a single weight matrix $\mathbf{W}$ of shape $(n_{\text{inputs}} + n_{\text{neurons}}) \times n_{\text{neurons}}$ (see the second line of Equation 15-2).

- The notation $[\mathbf{X}_{(t)} \ \mathbf{Y}_{(t-1)}]$ represents the horizontal concatenation of the matrices $\mathbf{X}_{(t)}$ and $\mathbf{Y}_{(t-1)}$.

Notice that $\mathbf{Y}_{(t)}$ is a function of $\mathbf{X}_{(t)}$ and $\mathbf{Y}_{(t-1)}$, which is a function of $\mathbf{X}_{(t-1)}$ and $\mathbf{Y}_{(t-2)}$, which is a function of $\mathbf{X}_{(t-2)}$ and $\mathbf{Y}_{(t-3)}$, and so on. This makes $\mathbf{Y}_{(t)}$ a function of all the inputs since time $t = 0$ (that is, $\mathbf{X}_{(0)}, \mathbf{X}_{(1)}, \ldots, \mathbf{X}_{(t)}$). At the first time step, $t = 0$, there are no previous outputs, so they are typically assumed to be all zeros.

Memory Cells

Since the output of a recurrent neuron at time step t is a function of all the inputs from previous time steps, you could say it has a form of *memory*. A part of a neural network that preserves some state across time steps is called a *memory cell* (or simply a *cell*). A single recurrent neuron, or a layer of recurrent neurons, is a very basic cell, capable of learning only short patterns (typically about 10 steps long, but this varies depending on the task). Later in this chapter, we will look at some more complex and powerful types of cells capable of learning longer patterns (roughly 10 times longer, but again, this depends on the task).

In general a cell's state at time step t, denoted $\mathbf{h}_{(t)}$ (the "h" stands for "hidden"), is a function of some inputs at that time step and its state at the previous time step: $\mathbf{h}_{(t)} = f(\mathbf{h}_{(t-1)}, \mathbf{x}_{(t)})$. Its output at time step t, denoted $\mathbf{y}_{(t)}$, is also a function of the previous state and the current inputs. In the case of the basic cells we have discussed so far, the output is simply equal to the state, but in more complex cells this is not always the case, as shown in Figure 15-3.

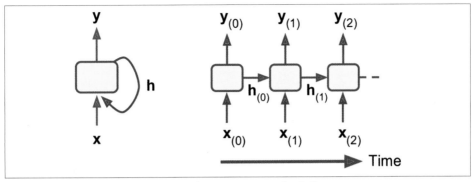

Figure 15-3. A cell's hidden state and its output may be different

Input and Output Sequences

An RNN can simultaneously take a sequence of inputs and produce a sequence of outputs (see the top-left network in Figure 15-4). This type of *sequence-to-sequence network* is useful for predicting time series such as stock prices: you feed it the prices over the last N days, and it must output the prices shifted by one day into the future (i.e., from N − 1 days ago to tomorrow).

Alternatively, you could feed the network a sequence of inputs and ignore all outputs except for the last one (see the top-right network in Figure 15-4). In other words, this is a *sequence-to-vector network*. For example, you could feed the network a sequence of words corresponding to a movie review, and the network would output a senti-ment score (e.g., from −1 [hate] to +1 [love]).

Conversely, you could feed the network the same input vector over and over again at each time step and let it output a sequence (see the bottom-left network of Figure 15-4). This is a *vector-to-sequence network*. For example, the input could be an image (or the output of a CNN), and the output could be a caption for that image.

Lastly, you could have a sequence-to-vector network, called an *encoder*, followed by a vector-to-sequence network, called a *decoder* (see the bottom-right network of Figure 15-4). For example, this could be used for translating a sentence from one lan-guage to another. You would feed the network a sentence in one language, the encoder would convert this sentence into a single vector representation, and then the decoder would decode this vector into a sentence in another language. This two-step model, called an *Encoder–Decoder*, works much better than trying to translate on the fly with a single sequence-to-sequence RNN (like the one represented at the top left): the last words of a sentence can affect the first words of the translation, so you need to wait until you have seen the whole sentence before translating it. We will see how to implement an Encoder–Decoder in Chapter 16 (as we will see, it is a bit more com-plex than in Figure 15-4 suggests).

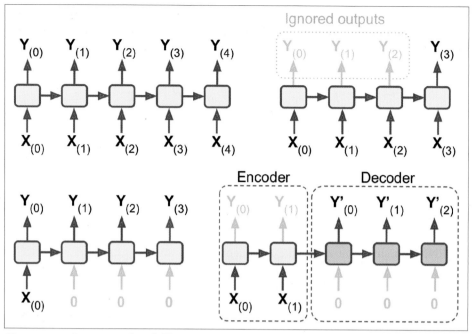

Figure 15-4. Seq-to-seq (top left), seq-to-vector (top right), vector-to-seq (bottom left), and Encoder–Decoder (bottom right) networks

Sounds promising, but how do you train a recurrent neural network?

Training RNNs

To train an RNN, the trick is to unroll it through time (like we just did) and then simply use regular backpropagation (see Figure 15-5). This strategy is called *backpropagation through time* (BPTT).

Just like in regular backpropagation, there is a first forward pass through the unrolled network (represented by the dashed arrows). Then the output sequence is evaluated using a cost function $C(\mathbf{Y}_{(0)}, \mathbf{Y}_{(1)}, \ldots \mathbf{Y}_{(T)})$ (where T is the max time step). Note that this cost function may ignore some outputs, as shown in Figure 15-5 (for example, in a sequence-to-vector RNN, all outputs are ignored except for the very last one). The gradients of that cost function are then propagated backward through the unrolled network (represented by the solid arrows). Finally the model parameters are updated using the gradients computed during BPTT. Note that the gradients flow backward through all the outputs used by the cost function, not just through the final output (for example, in Figure 15-5 the cost function is computed using the last three outputs of the network, $\mathbf{Y}_{(2)}$, $\mathbf{Y}_{(3)}$, and $\mathbf{Y}_{(4)}$, so gradients flow through these three outputs, but not through $\mathbf{Y}_{(0)}$ and $\mathbf{Y}_{(1)}$). Moreover, since the same parameters $\mathbf{W}$ and $\mathbf{b}$ are used at each time step, backpropagation will do the right thing and sum over all time steps.

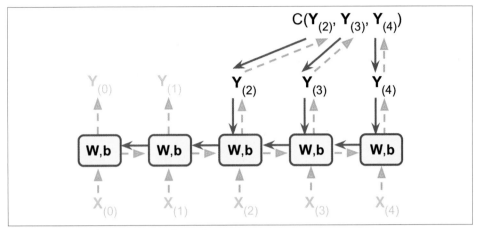

Figure 15-5. Backpropagation through time

Fortunately, tf.keras takes care of all of this complexity for you—so let's start coding!

Forecasting a Time Series

Suppose you are studying the number of active users per hour on your website, or the daily temperature in your city, or your company's financial health, measured quarterly using multiple metrics. In all these cases, the data will be a sequence of one or more values per time step. This is called a *time series*. In the first two examples there is a single value per time step, so these are *univariate time series*, while in the financial example there are multiple values per time step (e.g., the company's revenue, debt, and so on), so it is a *multivariate time series*. A typical task is to predict future values, which is called *forecasting*. Another common task is to fill in the blanks: to predict (or rather "postdict") missing values from the past. This is called *imputation*. For example, Figure 15-6 shows 3 univariate time series, each of them 50 time steps long, and the goal here is to forecast the value at the next time step (represented by the X) for each of them.

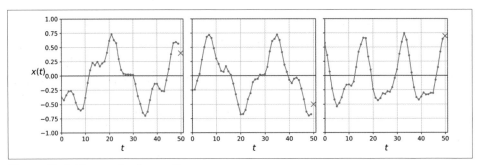

Figure 15-6. Time series forecasting

For simplicity, we are using a time series generated by the `generate_time_series()` function, shown here:

```
def generate_time_series(batch_size, n_steps):
    freq1, freq2, offsets1, offsets2 = np.random.rand(4, batch_size, 1)
    time = np.linspace(0, 1, n_steps)
    series = 0.5 * np.sin((time - offsets1) * (freq1 * 10 + 10))  #    wave 1
    series += 0.2 * np.sin((time - offsets2) * (freq2 * 20 + 20)) # + wave 2
    series += 0.1 * (np.random.rand(batch_size, n_steps) - 0.5)   # + noise
    return series[..., np.newaxis].astype(np.float32)
```

This function creates as many time series as requested (via the `batch_size` argument), each of length `n_steps`, and there is just one value per time step in each series (i.e., all series are univariate). The function returns a NumPy array of shape [*batch size*, *time steps*, 1], where each series is the sum of two sine waves of fixed amplitudes but random frequencies and phases, plus a bit of noise.

 When dealing with time series (and other types of sequences such as sentences), the input features are generally represented as 3D arrays of shape [*batch size*, *time steps*, *dimensionality*], where *dimensionality* is 1 for univariate time series and more for multivariate time series.

Now let's create a training set, a validation set, and a test set using this function:

```
n_steps = 50
series = generate_time_series(10000, n_steps + 1)
X_train, y_train = series[:7000, :n_steps], series[:7000, -1]
X_valid, y_valid = series[7000:9000, :n_steps], series[7000:9000, -1]
X_test, y_test = series[9000:, :n_steps], series[9000:, -1]
```

`X_train` contains 7,000 time series (i.e., its shape is [7000, 50, 1]), while `X_valid` contains 2,000 (from the 7,000th time series to the 8,999th) and `X_test` contains 1,000 (from the 9,000[th] to the 9,999[th]). Since we want to forecast a single value for each series, the targets are column vectors (e.g., `y_train` has a shape of [7000, 1]).

Baseline Metrics

Before we start using RNNs, it is often a good idea to have a few baseline metrics, or else we may end up thinking our model works great when in fact it is doing worse than basic models. For example, the simplest approach is to predict the last value in each series. This is called *naive forecasting*, and it is sometimes surprisingly difficult to outperform. In this case, it gives us a mean squared error of about 0.020:

```
>>> y_pred = X_valid[:, -1]
>>> np.mean(keras.losses.mean_squared_error(y_valid, y_pred))
0.020211367
```

Another simple approach is to use a fully connected network. Since it expects a flat list of features for each input, we need to add a `Flatten` layer. Let's just use a simple Linear Regression model so that each prediction will be a linear combination of the values in the time series:

```
model = keras.models.Sequential([
    keras.layers.Flatten(input_shape=[50, 1]),
    keras.layers.Dense(1)
])
```

If we compile this model using the MSE loss and the default Adam optimizer, then fit it on the training set for 20 epochs and evaluate it on the validation set, we get an MSE of about 0.004. That's much better than the naive approach!

Implementing a Simple RNN

Let's see if we can beat that with a simple RNN:

```
model = keras.models.Sequential([
    keras.layers.SimpleRNN(1, input_shape=[None, 1])
])
```

That's really the simplest RNN you can build. It just contains a single layer, with a single neuron, as we saw in Figure 15-1. We do not need to specify the length of the input sequences (unlike in the previous model), since a recurrent neural network can process any number of time steps (this is why we set the first input dimension to `None`). By default, the `SimpleRNN` layer uses the hyperbolic tangent activation function. It works exactly as we saw earlier: the initial state $h_{(init)}$ is set to 0, and it is passed to a single recurrent neuron, along with the value of the first time step, $x_{(0)}$. The neuron computes a weighted sum of these values and applies the hyperbolic tangent activation function to the result, and this gives the first output, y_0. In a simple RNN, this output is also the new state h_0. This new state is passed to the same recurrent neuron along with the next input value, $x_{(1)}$, and the process is repeated until the last time step. Then the layer just outputs the last value, y_{49}. All of this is performed simultaneously for every time series.

By default, recurrent layers in Keras only return the final output. To make them return one output per time step, you must set `return_sequences=True`, as we will see.

If you compile, fit, and evaluate this model (just like earlier, we train for 20 epochs using Adam), you will find that its MSE reaches only 0.014, so it is better than the naive approach but it does not beat a simple linear model. Note that for each neuron, a linear model has one parameter per input and per time step, plus a bias term (in the simple linear model we used, that's a total of 51 parameters). In contrast, for each recurrent neuron in a simple RNN, there is just one parameter per input and per hidden state dimension (in a simple RNN, that's just the number of recurrent neurons in the layer), plus a bias term. In this simple RNN, that's a total of just three parameters.

Trend and Seasonality

There are many other models to forecast time series, such as *weighted moving average* models or *autoregressive integrated moving average* (ARIMA) models. Some of them require you to first remove the trend and seasonality. For example, if you are studying the number of active users on your website, and it is growing by 10% every month, you would have to remove this trend from the time series. Once the model is trained and starts making predictions, you would have to add the trend back to get the final predictions. Similarly, if you are trying to predict the amount of sunscreen lotion sold every month, you will probably observe strong seasonality: since it sells well every summer, a similar pattern will be repeated every year. You would have to remove this seasonality from the time series, for example by computing the difference between the value at each time step and the value one year earlier (this technique is called *differencing*). Again, after the model is trained and makes predictions, you would have to add the seasonal pattern back to get the final predictions.

When using RNNs, it is generally not necessary to do all this, but it may improve performance in some cases, since the model will not have to learn the trend or the seasonality.

Apparently our simple RNN was too simple to get good performance. So let's try to add more recurrent layers!

Deep RNNs

It is quite common to stack multiple layers of cells, as shown in Figure 15-7. This gives you a *deep RNN*.

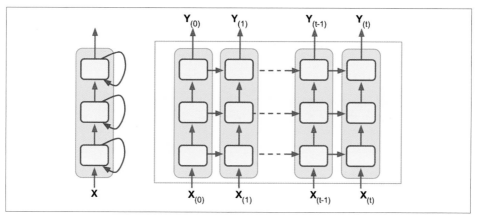

Figure 15-7. Deep RNN (left) unrolled through time (right)

Implementing a deep RNN with tf.keras is quite simple: just stack recurrent layers. In this example, we use three SimpleRNN layers (but we could add any other type of recurrent layer, such as an LSTM layer or a GRU layer, which we will discuss shortly):

```
model = keras.models.Sequential([
    keras.layers.SimpleRNN(20, return_sequences=True, input_shape=[None, 1]),
    keras.layers.SimpleRNN(20, return_sequences=True),
    keras.layers.SimpleRNN(1)
])
```

 Make sure to set return_sequences=True for all recurrent layers (except the last one, if you only care about the last output). If you don't, they will output a 2D array (containing only the output of the last time step) instead of a 3D array (containing outputs for all time steps), and the next recurrent layer will complain that you are not feeding it sequences in the expected 3D format.

If you compile, fit, and evaluate this model, you will find that it reaches an MSE of 0.003. We finally managed to beat the linear model!

Note that the last layer is not ideal: it must have a single unit because we want to forecast a univariate time series, and this means we must have a single output value per time step. However, having a single unit means that the hidden state is just a single number. That's really not much, and it's probably not that useful; presumably, the RNN will mostly use the hidden states of the other recurrent layers to carry over all the information it needs from time step to time step, and it will not use the final layer's hidden state very much. Moreover, since a SimpleRNN layer uses the tanh activation function by default, the predicted values must lie within the range –1 to 1. But what if you want to use another activation function? For both these reasons, it might be preferable to replace the output layer with a Dense layer: it would run slightly faster, the accuracy would be roughly the same, and it would allow us to choose any output activation function we want. If you make this change, also make sure to remove `return_sequences=True` from the second (now last) recurrent layer:

```
model = keras.models.Sequential([
    keras.layers.SimpleRNN(20, return_sequences=True, input_shape=[None, 1]),
    keras.layers.SimpleRNN(20),
    keras.layers.Dense(1)
])
```

If you train this model, you will see that it converges faster and performs just as well. Plus, you could change the output activation function if you wanted.

Forecasting Several Time Steps Ahead

So far we have only predicted the value at the next time step, but we could just as easily have predicted the value several steps ahead by changing the targets appropriately (e.g., to predict 10 steps ahead, just change the targets to be the value 10 steps ahead instead of 1 step ahead). But what if we want to predict the next 10 values?

The first option is to use the model we already trained, make it predict the next value, then add that value to the inputs (acting as if this predicted value had actually occurred), and use the model again to predict the following value, and so on, as in the following code:

```
series = generate_time_series(1, n_steps + 10)
X_new, Y_new = series[:, :n_steps], series[:, n_steps:]
X = X_new
for step_ahead in range(10):
    y_pred_one = model.predict(X[:, step_ahead:])[:, np.newaxis, :]
    X = np.concatenate([X, y_pred_one], axis=1)

Y_pred = X[:, n_steps:]
```

As you might expect, the prediction for the next step will usually be more accurate than the predictions for later time steps, since the errors might accumulate (as you can see in Figure 15-8). If you evaluate this approach on the validation set, you will find an MSE of about 0.029. This is much higher than the previous models, but it's also a much harder task, so the comparison doesn't mean much. It's much more meaningful to compare this performance with naive predictions (just forecasting that the time series will remain constant for 10 time steps) or with a simple linear model. The naive approach is terrible (it gives an MSE of about 0.223), but the linear model gives an MSE of about 0.0188: it's much better than using our RNN to forecast the future one step at a time, and also much faster to train and run. Still, if you only want to forecast a few time steps ahead, on more complex tasks, this approach may work well.

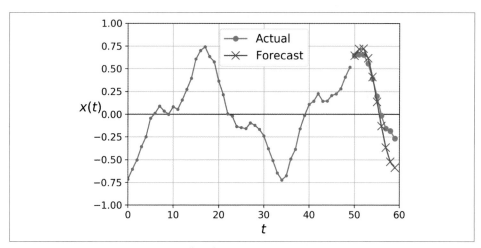

Figure 15-8. Forecasting 10 steps ahead, 1 step at a time

The second option is to train an RNN to predict all 10 next values at once. We can still use a sequence-to-vector model, but it will output 10 values instead of 1. However, we first need to change the targets to be vectors containing the next 10 values:

```
series = generate_time_series(10000, n_steps + 10)
X_train, Y_train = series[:7000, :n_steps], series[:7000, -10:, 0]
X_valid, Y_valid = series[7000:9000, :n_steps], series[7000:9000, -10:, 0]
X_test, Y_test = series[9000:, :n_steps], series[9000:, -10:, 0]
```

Now we just need the output layer to have 10 units instead of 1:

```
model = keras.models.Sequential([
    keras.layers.SimpleRNN(20, return_sequences=True, input_shape=[None, 1]),
    keras.layers.SimpleRNN(20),
    keras.layers.Dense(10)
])
```

After training this model, you can predict the next 10 values at once very easily:

```
Y_pred = model.predict(X_new)
```

This model works nicely: the MSE for the next 10 time steps is about 0.008. That's much better than the linear model. But we can still do better: indeed, instead of training the model to forecast the next 10 values only at the very last time step, we can train it to forecast the next 10 values at each and every time step. In other words, we can turn this sequence-to-vector RNN into a sequence-to-sequence RNN. The advantage of this technique is that the loss will contain a term for the output of the RNN at each and every time step, not just the output at the last time step. This means there will be many more error gradients flowing through the model, and they won't have to flow only through time; they will also flow from the output of each time step. This will both stabilize and speed up training.

To be clear, at time step 0 the model will output a vector containing the forecasts for time steps 1 to 10, then at time step 1 the model will forecast time steps 2 to 11, and so on. So each target must be a sequence of the same length as the input sequence, containing a 10-dimensional vector at each step. Let's prepare these target sequences:

```
Y = np.empty((10000, n_steps, 10)) # each target is a sequence of 10D vectors
for step_ahead in range(1, 10 + 1):
    Y[:, :, step_ahead - 1] = series[:, step_ahead:step_ahead + n_steps, 0]
Y_train = Y[:7000]
Y_valid = Y[7000:9000]
Y_test = Y[9000:]
```

 It may be surprising that the targets will contain values that appear in the inputs (there is a lot of overlap between X_train and Y_train). Isn't that cheating? Fortunately, not at all: at each time step, the model only knows about past time steps, so it cannot look ahead. It is said to be a *causal* model.

To turn the model into a sequence-to-sequence model, we must set return_sequen ces=True in all recurrent layers (even the last one), and we must apply the output Dense layer at every time step. Keras offers a TimeDistributed layer for this very purpose: it wraps any layer (e.g., a Dense layer) and applies it at every time step of its input sequence. It does this efficiently, by reshaping the inputs so that each time step is treated as a separate instance (i.e., it reshapes the inputs from [*batch size, time steps, input dimensions*] to [*batch size × time steps, input dimensions*]; in this example, the number of input dimensions is 20 because the previous SimpleRNN layer has 20 units), then it runs the Dense layer, and finally it reshapes the outputs back to sequences (i.e., it reshapes the outputs from [*batch size × time steps, output dimensions*] to [*batch size, time steps, output dimensions*]; in this example the number of output dimensions is 10, since the Dense layer has 10 units).[2] Here is the updated model:

```
model = keras.models.Sequential([
    keras.layers.SimpleRNN(20, return_sequences=True, input_shape=[None, 1]),
    keras.layers.SimpleRNN(20, return_sequences=True),
    keras.layers.TimeDistributed(keras.layers.Dense(10))
])
```

The Dense layer actually supports sequences as inputs (and even higher-dimensional inputs): it handles them just like TimeDistributed(Dense(...)), meaning it is applied to the last input dimension only (independently across all time steps). Thus, we could replace the last layer with just Dense(10). For the sake of clarity, however, we will keep using TimeDistributed(Dense(10)) because it makes it clear that the Dense

---

2 Note that a TimeDistributed(Dense(*n*)) layer is equivalent to a Conv1D(*n*, filter_size=1) layer.

layer is applied independently at each time step and that the model will output a sequence, not just a single vector.

All outputs are needed during training, but only the output at the last time step is useful for predictions and for evaluation. So although we will rely on the MSE over all the outputs for training, we will use a custom metric for evaluation, to only compute the MSE over the output at the last time step:

```
def last_time_step_mse(Y_true, Y_pred):
    return keras.metrics.mean_squared_error(Y_true[:, -1], Y_pred[:, -1])

optimizer = keras.optimizers.Adam(lr=0.01)
model.compile(loss="mse", optimizer=optimizer, metrics=[last_time_step_mse])
```

We get a validation MSE of about 0.006, which is 25% better than the previous model. You can combine this approach with the first one: just predict the next 10 values using this RNN, then concatenate these values to the input time series and use the model again to predict the next 10 values, and repeat the process as many times as needed. With this approach, you can generate arbitrarily long sequences. It may not be very accurate for long-term predictions, but it may be just fine if your goal is to generate original music or text, as we will see in Chapter 16.

> When forecasting time series, it is often useful to have some error bars along with your predictions. For this, an efficient technique is MC Dropout, introduced in Chapter 11: add an MC Dropout layer within each memory cell, dropping part of the inputs and hidden states. After training, to forecast a new time series, use the model many times and compute the mean and standard deviation of the predictions at each time step.

Simple RNNs can be quite good at forecasting time series or handling other kinds of sequences, but they do not perform as well on long time series or sequences. Let's discuss why and see what we can do about it.

Handling Long Sequences

To train an RNN on long sequences, we must run it over many time steps, making the unrolled RNN a very deep network. Just like any deep neural network it may suffer from the unstable gradients problem, discussed in Chapter 11: it may take forever to train, or training may be unstable. Moreover, when an RNN processes a long sequence, it will gradually forget the first inputs in the sequence. Let's look at both these problems, starting with the unstable gradients problem.

Fighting the Unstable Gradients Problem

Many of the tricks we used in deep nets to alleviate the unstable gradients problem can also be used for RNNs: good parameter initialization, faster optimizers, dropout, and so on. However, nonsaturating activation functions (e.g., ReLU) may not help as much here; in fact, they may actually lead the RNN to be even more unstable during training. Why? Well, suppose Gradient Descent updates the weights in a way that increases the outputs slightly at the first time step. Because the same weights are used at every time step, the outputs at the second time step may also be slightly increased, and those at the third, and so on until the outputs explode—and a nonsaturating activation function does not prevent that. You can reduce this risk by using a smaller learning rate, but you can also simply use a saturating activation function like the hyperbolic tangent (this explains why it is the default). In much the same way, the gradients themselves can explode. If you notice that training is unstable, you may want to monitor the size of the gradients (e.g., using TensorBoard) and perhaps use Gradient Clipping.

Moreover, Batch Normalization cannot be used as efficiently with RNNs as with deep feedforward nets. In fact, you cannot use it between time steps, only between recurrent layers. To be more precise, it is technically possible to add a BN layer to a memory cell (as we will see shortly) so that it will be applied at each time step (both on the inputs for that time step and on the hidden state from the previous step). However, the same BN layer will be used at each time step, with the same parameters, regardless of the actual scale and offset of the inputs and hidden state. In practice, this does not yield good results, as was demonstrated by César Laurent et al. in a 2015 paper (*https://homl.info/rnnbn*):[3] the authors found that BN was slightly beneficial only when it was applied to the inputs, not to the hidden states. In other words, it was slightly better than nothing when applied between recurrent layers (i.e., vertically in Figure 15-7), but not within recurrent layers (i.e., horizontally). In Keras this can be done simply by adding a `BatchNormalization` layer before each recurrent layer, but don't expect too much from it.

Another form of normalization often works better with RNNs: *Layer Normalization*. This idea was introduced by Jimmy Lei Ba et al. in a 2016 paper (*https://homl.info/layernorm*):[4] it is very similar to Batch Normalization, but instead of normalizing across the batch dimension, it normalizes across the features dimension. One advantage is that it can compute the required statistics on the fly, at each time step, independently for each instance. This also means that it behaves the same way during training and testing (as opposed to BN), and it does not need to use exponential

---

[3] César Laurent et al., "Batch Normalized Recurrent Neural Networks," *Proceedings of the IEEE International Conference on Acoustics, Speech, and Signal Processing* (2016): 2657–2661.

[4] Jimmy Lei Ba et al., "Layer Normalization," arXiv preprint arXiv:1607.06450 (2016).

moving averages to estimate the feature statistics across all instances in the training set. Like BN, Layer Normalization learns a scale and an offset parameter for each input. In an RNN, it is typically used right after the linear combination of the inputs and the hidden states.

Let's use tf.keras to implement Layer Normalization within a simple memory cell. For this, we need to define a custom memory cell. It is just like a regular layer, except its call() method takes two arguments: the inputs at the current time step and the hidden states from the previous time step. Note that the states argument is a list containing one or more tensors. In the case of a simple RNN cell it contains a single tensor equal to the outputs of the previous time step, but other cells may have multiple state tensors (e.g., an LSTMCell has a long-term state and a short-term state, as we will see shortly). A cell must also have a state_size attribute and an output_size attribute. In a simple RNN, both are simply equal to the number of units. The following code implements a custom memory cell which will behave like a SimpleRNNCell, except it will also apply Layer Normalization at each time step:

```
class LNSimpleRNNCell(keras.layers.Layer):
    def __init__(self, units, activation="tanh", **kwargs):
        super().__init__(**kwargs)
        self.state_size = units
        self.output_size = units
        self.simple_rnn_cell = keras.layers.SimpleRNNCell(units,
                                                          activation=None)
        self.layer_norm = keras.layers.LayerNormalization()
        self.activation = keras.activations.get(activation)
    def call(self, inputs, states):
        outputs, new_states = self.simple_rnn_cell(inputs, states)
        norm_outputs = self.activation(self.layer_norm(outputs))
        return norm_outputs, [norm_outputs]
```

The code is quite straightforward.[5] Our LNSimpleRNNCell class inherits from the keras.layers.Layer class, just like any custom layer. The constructor takes the number of units and the desired activation function, and it sets the state_size and output_size attributes, then creates a SimpleRNNCell with no activation function (because we want to perform Layer Normalization after the linear operation but before the activation function). Then the constructor creates the LayerNormaliza tion layer, and finally it fetches the desired activation function. The call() method starts by applying the simple RNN cell, which computes a linear combination of the current inputs and the previous hidden states, and it returns the result twice (indeed, in a SimpleRNNCell, the outputs are just equal to the hidden states: in other words,

---

5 It would have been simpler to inherit from SimpleRNNCell instead so that we wouldn't have to create an internal SimpleRNNCell or handle the state_size and output_size attributes, but the goal here was to show how to create a custom cell from scratch.

new_states[0] is equal to outputs, so we can safely ignore new_states in the rest of the call() method). Next, the call() method applies Layer Normalization, followed by the activation function. Finally, it returns the outputs twice (once as the outputs, and once as the new hidden states). To use this custom cell, all we need to do is create a keras.layers.RNN layer, passing it a cell instance:

```
model = keras.models.Sequential([
    keras.layers.RNN(LNSimpleRNNCell(20), return_sequences=True,
                     input_shape=[None, 1]),
    keras.layers.RNN(LNSimpleRNNCell(20), return_sequences=True),
    keras.layers.TimeDistributed(keras.layers.Dense(10))
])
```

Similarly, you could create a custom cell to apply dropout between each time step. But there's a simpler way: all recurrent layers (except for keras.layers.RNN) and all cells provided by Keras have a dropout hyperparameter and a recurrent_dropout hyperparameter: the former defines the dropout rate to apply to the inputs (at each time step), and the latter defines the dropout rate for the hidden states (also at each time step). No need to create a custom cell to apply dropout at each time step in an RNN.

With these techniques, you can alleviate the unstable gradients problem and train an RNN much more efficiently. Now let's look at how to deal with the short-term memory problem.

Tackling the Short-Term Memory Problem

Due to the transformations that the data goes through when traversing an RNN, some information is lost at each time step. After a while, the RNN's state contains virtually no trace of the first inputs. This can be a showstopper. Imagine Dory the fish[6] trying to translate a long sentence; by the time she's finished reading it, she has no clue how it started. To tackle this problem, various types of cells with long-term memory have been introduced. They have proven so successful that the basic cells are not used much anymore. Let's first look at the most popular of these long-term memory cells: the LSTM cell.

LSTM cells

The *Long Short-Term Memory* (LSTM) cell was proposed in 1997 (*https://homl.info/93*)[7] by Sepp Hochreiter and Jürgen Schmidhuber and gradually improved over the years by several researchers, such as Alex Graves (*https://homl.info/graves*),

---

6 A character from the animated movies *Finding Nemo* and *Finding Dory* who has short-term memory loss.

7 Sepp Hochreiter and Jürgen Schmidhuber, "Long Short-Term Memory," *Neural Computation* 9, no. 8 (1997): 1735–1780.

Haşim Sak (*https://homl.info/94*),[8] and Wojciech Zaremba (*https://homl.info/95*).[9] If you consider the LSTM cell as a black box, it can be used very much like a basic cell, except it will perform much better; training will converge faster, and it will detect long-term dependencies in the data. In Keras, you can simply use the LSTM layer instead of the SimpleRNN layer:

```
model = keras.models.Sequential([
    keras.layers.LSTM(20, return_sequences=True, input_shape=[None, 1]),
    keras.layers.LSTM(20, return_sequences=True),
    keras.layers.TimeDistributed(keras.layers.Dense(10))
])
```

Alternatively, you could use the general-purpose keras.layers.RNN layer, giving it an LSTMCell as an argument:

```
model = keras.models.Sequential([
    keras.layers.RNN(keras.layers.LSTMCell(20), return_sequences=True,
                     input_shape=[None, 1]),
    keras.layers.RNN(keras.layers.LSTMCell(20), return_sequences=True),
    keras.layers.TimeDistributed(keras.layers.Dense(10))
])
```

However, the LSTM layer uses an optimized implementation when running on a GPU (see Chapter 19), so in general it is preferable to use it (the RNN layer is mostly useful when you define custom cells, as we did earlier).

So how does an LSTM cell work? Its architecture is shown in Figure 15-9.

If you don't look at what's inside the box, the LSTM cell looks exactly like a regular cell, except that its state is split into two vectors: $\mathbf{h}_{(t)}$ and $\mathbf{c}_{(t)}$ ("c" stands for "cell"). You can think of $\mathbf{h}_{(t)}$ as the short-term state and $\mathbf{c}_{(t)}$ as the long-term state.

---

8 Haşim Sak et al., "Long Short-Term Memory Based Recurrent Neural Network Architectures for Large Vocabulary Speech Recognition," arXiv preprint arXiv:1402.1128 (2014).

9 Wojciech Zaremba et al., "Recurrent Neural Network Regularization," arXiv preprint arXiv:1409.2329 (2014).

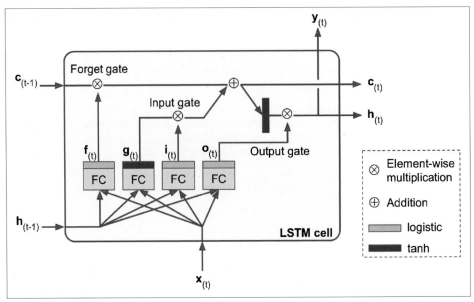

Figure 15-9. LSTM cell

Now let's open the box! The key idea is that the network can learn what to store in the long-term state, what to throw away, and what to read from it. As the long-term state $c_{(t-1)}$ traverses the network from left to right, you can see that it first goes through a *forget gate*, dropping some memories, and then it adds some new memories via the addition operation (which adds the memories that were selected by an *input gate*). The result $c_{(t)}$ is sent straight out, without any further transformation. So, at each time step, some memories are dropped and some memories are added. Moreover, after the addition operation, the long-term state is copied and passed through the tanh function, and then the result is filtered by the *output gate*. This produces the short-term state $h_{(t)}$ (which is equal to the cell's output for this time step, $y_{(t)}$). Now let's look at where new memories come from and how the gates work.

First, the current input vector $x_{(t)}$ and the previous short-term state $h_{(t-1)}$ are fed to four different fully connected layers. They all serve a different purpose:

- The main layer is the one that outputs $g_{(t)}$. It has the usual role of analyzing the current inputs $x_{(t)}$ and the previous (short-term) state $h_{(t-1)}$. In a basic cell, there is nothing other than this layer, and its output goes straight out to $y_{(t)}$ and $h_{(t)}$. In contrast, in an LSTM cell this layer's output does not go straight out, but instead its most important parts are stored in the long-term state (and the rest is dropped).

- The three other layers are *gate controllers*. Since they use the logistic activation function, their outputs range from 0 to 1. As you can see, their outputs are fed to element-wise multiplication operations, so if they output 0s they close the gate, and if they output 1s they open it. Specifically:

— The *forget gate* (controlled by $\mathbf{f}_{(t)}$) controls which parts of the long-term state should be erased.

— The *input gate* (controlled by $\mathbf{i}_{(t)}$) controls which parts of $\mathbf{g}_{(t)}$ should be added to the long-term state.

— Finally, the *output gate* (controlled by $\mathbf{o}_{(t)}$) controls which parts of the long-term state should be read and output at this time step, both to $\mathbf{h}_{(t)}$ and to $\mathbf{y}_{(t)}$.

In short, an LSTM cell can learn to recognize an important input (that's the role of the input gate), store it in the long-term state, preserve it for as long as it is needed (that's the role of the forget gate), and extract it whenever it is needed. This explains why these cells have been amazingly successful at capturing long-term patterns in time series, long texts, audio recordings, and more.

Equation 15-3 summarizes how to compute the cell's long-term state, its short-term state, and its output at each time step for a single instance (the equations for a whole mini-batch are very similar).

Equation 15-3. LSTM computations

$$\mathbf{i}_{(t)} = \sigma\left(\mathbf{W}_{xi}{}^{\mathsf{T}}\mathbf{x}_{(t)} + \mathbf{W}_{hi}{}^{\mathsf{T}}\mathbf{h}_{(t-1)} + \mathbf{b}_i\right)$$

$$\mathbf{f}_{(t)} = \sigma\left(\mathbf{W}_{xf}{}^{\mathsf{T}}\mathbf{x}_{(t)} + \mathbf{W}_{hf}{}^{\mathsf{T}}\mathbf{h}_{(t-1)} + \mathbf{b}_f\right)$$

$$\mathbf{o}_{(t)} = \sigma\left(\mathbf{W}_{xo}{}^{\mathsf{T}}\mathbf{x}_{(t)} + \mathbf{W}_{ho}{}^{\mathsf{T}}\mathbf{h}_{(t-1)} + \mathbf{b}_o\right)$$

$$\mathbf{g}_{(t)} = \tanh\left(\mathbf{W}_{xg}{}^{\mathsf{T}}\mathbf{x}_{(t)} + \mathbf{W}_{hg}{}^{\mathsf{T}}\mathbf{h}_{(t-1)} + \mathbf{b}_g\right)$$

$$\mathbf{c}_{(t)} = \mathbf{f}_{(t)} \otimes \mathbf{c}_{(t-1)} + \mathbf{i}_{(t)} \otimes \mathbf{g}_{(t)}$$

$$\mathbf{y}_{(t)} = \mathbf{h}_{(t)} = \mathbf{o}_{(t)} \otimes \tanh\left(\mathbf{c}_{(t)}\right)$$

In this equation:

- $\mathbf{W}_{xi}$, $\mathbf{W}_{xf}$, $\mathbf{W}_{xo}$, $\mathbf{W}_{xg}$ are the weight matrices of each of the four layers for their connection to the input vector $\mathbf{x}_{(t)}$.

- $\mathbf{W}_{hi}$, $\mathbf{W}_{hf}$, $\mathbf{W}_{ho}$, and $\mathbf{W}_{hg}$ are the weight matrices of each of the four layers for their connection to the previous short-term state $\mathbf{h}_{(t-1)}$.

- $\mathbf{b}_i$, $\mathbf{b}_f$, $\mathbf{b}_o$, and $\mathbf{b}_g$ are the bias terms for each of the four layers. Note that TensorFlow initializes $\mathbf{b}_f$ to a vector full of 1s instead of 0s. This prevents forgetting everything at the beginning of training.

Peephole connections

In a regular LSTM cell, the gate controllers can look only at the input $\mathbf{x}_{(t)}$ and the previous short-term state $\mathbf{h}_{(t-1)}$. It may be a good idea to give them a bit more context by letting them peek at the long-term state as well. This idea was proposed by Felix Gers and Jürgen Schmidhuber in 2000 (*https://homl.info/96*).[10] They proposed an LSTM variant with extra connections called *peephole connections*: the previous long-term state $\mathbf{c}_{(t-1)}$ is added as an input to the controllers of the forget gate and the input gate, and the current long-term state $\mathbf{c}_{(t)}$ is added as input to the controller of the output gate. This often improves performance, but not always, and there is no clear pattern for which tasks are better off with or without them: you will have to try it on your task and see if it helps.

In Keras, the `LSTM` layer is based on the `keras.layers.LSTMCell` cell, which does not support peepholes. The experimental `tf.keras.experimental.PeepholeLSTMCell` does, however, so you can create a `keras.layers.RNN` layer and pass a `PeepholeLSTMCell` to its constructor.

There are many other variants of the LSTM cell. One particularly popular variant is the GRU cell, which we will look at now.

GRU cells

The *Gated Recurrent Unit* (GRU) cell (see Figure 15-10) was proposed by Kyunghyun Cho et al. in a 2014 paper (*https://homl.info/97*)[11] that also introduced the Encoder–Decoder network we discussed earlier.

---

10 F. A. Gers and J. Schmidhuber, "Recurrent Nets That Time and Count," *Proceedings of the IEEE-INNS-ENNS International Joint Conference on Neural Networks* (2000): 189–194.

11 Kyunghyun Cho et al., "Learning Phrase Representations Using RNN Encoder-Decoder for Statistical Machine Translation," *Proceedings of the 2014 Conference on Empirical Methods in Natural Language Processing* (2014): 1724–1734.

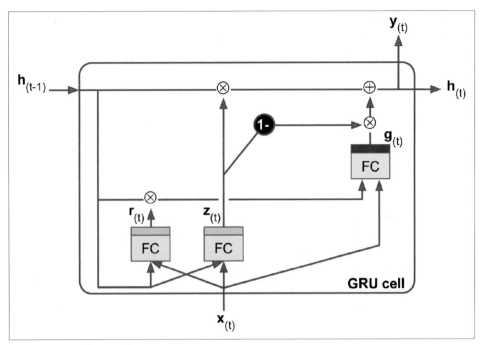

Figure 15-10. GRU cell

The GRU cell is a simplified version of the LSTM cell, and it seems to perform just as well[12] (which explains its growing popularity). These are the main simplifications:

- Both state vectors are merged into a single vector $\mathbf{h}_{(t)}$.

- A single gate controller $\mathbf{z}_{(t)}$ controls both the forget gate and the input gate. If the gate controller outputs a 1, the forget gate is open (= 1) and the input gate is closed (1 – 1 = 0). If it outputs a 0, the opposite happens. In other words, whenever a memory must be stored, the location where it will be stored is erased first. This is actually a frequent variant to the LSTM cell in and of itself.

- There is no output gate; the full state vector is output at every time step. However, there is a new gate controller $\mathbf{r}_{(t)}$ that controls which part of the previous state will be shown to the main layer ($\mathbf{g}_{(t)}$).

---

12 A 2015 paper by Klaus Greff et al., "LSTM: A Search Space Odyssey" (*https://homl.info/98*), seems to show that all LSTM variants perform roughly the same.

Equation 15-4 summarizes how to compute the cell's state at each time step for a single instance.

Equation 15-4. GRU computations

$$\mathbf{z}_{(t)} = \sigma\left(\mathbf{W}_{xz}{}^{\mathsf{T}}\mathbf{x}_{(t)} + \mathbf{W}_{hz}{}^{\mathsf{T}}\mathbf{h}_{(t-1)} + \mathbf{b}_z\right)$$

$$\mathbf{r}_{(t)} = \sigma\left(\mathbf{W}_{xr}{}^{\mathsf{T}}\mathbf{x}_{(t)} + \mathbf{W}_{hr}{}^{\mathsf{T}}\mathbf{h}_{(t-1)} + \mathbf{b}_r\right)$$

$$\mathbf{g}_{(t)} = \tanh\left(\mathbf{W}_{xg}{}^{\mathsf{T}}\mathbf{x}_{(t)} + \mathbf{W}_{hg}{}^{\mathsf{T}}\left(\mathbf{r}_{(t)} \otimes \mathbf{h}_{(t-1)}\right) + \mathbf{b}_g\right)$$

$$\mathbf{h}_{(t)} = \mathbf{z}_{(t)} \otimes \mathbf{h}_{(t-1)} + \left(1 - \mathbf{z}_{(t)}\right) \otimes \mathbf{g}_{(t)}$$

Keras provides a `keras.layers.GRU` layer (based on the `keras.layers.GRUCell` memory cell); using it is just a matter of replacing `SimpleRNN` or `LSTM` with `GRU`.

LSTM and GRU cells are one of the main reasons behind the success of RNNs. Yet while they can tackle much longer sequences than simple RNNs, they still have a fairly limited short-term memory, and they have a hard time learning long-term patterns in sequences of 100 time steps or more, such as audio samples, long time series, or long sentences. One way to solve this is to shorten the input sequences, for example using 1D convolutional layers.

Using 1D convolutional layers to process sequences

In Chapter 14, we saw that a 2D convolutional layer works by sliding several fairly small kernels (or filters) across an image, producing multiple 2D feature maps (one per kernel). Similarly, a 1D convolutional layer slides several kernels across a sequence, producing a 1D feature map per kernel. Each kernel will learn to detect a single very short sequential pattern (no longer than the kernel size). If you use 10 kernels, then the layer's output will be composed of 10 1-dimensional sequences (all of the same length), or equivalently you can view this output as a single 10-dimensional sequence. This means that you can build a neural network composed of a mix of recurrent layers and 1D convolutional layers (or even 1D pooling layers). If you use a 1D convolutional layer with a stride of 1 and `"same"` padding, then the output sequence will have the same length as the input sequence. But if you use `"valid"` padding or a stride greater than 1, then the output sequence will be shorter than the input sequence, so make sure you adjust the targets accordingly. For example, the following model is the same as earlier, except it starts with a 1D convolutional layer that downsamples the input sequence by a factor of 2, using a stride of 2. The kernel size is larger than the stride, so all inputs will be used to compute the layer's output, and therefore the model can learn to preserve the useful information, dropping only the unimportant details. By shortening the sequences, the convolutional layer may help the GRU layers detect longer patterns. Note that we must also crop off the first three time steps in the targets (since the kernel's size is 4, the first output of the convolutional layer will be based on the input time steps 0 to 3), and downsample the targets by a factor of 2:

```
model = keras.models.Sequential([
    keras.layers.Conv1D(filters=20, kernel_size=4, strides=2, padding="valid",
                        input_shape=[None, 1]),
    keras.layers.GRU(20, return_sequences=True),
    keras.layers.GRU(20, return_sequences=True),
    keras.layers.TimeDistributed(keras.layers.Dense(10))
])

model.compile(loss="mse", optimizer="adam", metrics=[last_time_step_mse])
history = model.fit(X_train, Y_train[:, 3::2], epochs=20,
                    validation_data=(X_valid, Y_valid[:, 3::2]))
```

If you train and evaluate this model, you will find that it is the best model so far. The convolutional layer really helps. In fact, it is actually possible to use only 1D convolutional layers and drop the recurrent layers entirely!

WaveNet

In a 2016 paper (*https://homl.info/wavenet*),[13] Aaron van den Oord and other Deep-Mind researchers introduced an architecture called *WaveNet*. They stacked 1D convolutional layers, doubling the dilation rate (how spread apart each neuron's inputs are) at every layer: the first convolutional layer gets a glimpse of just two time steps at a time, while the next one sees four time steps (its receptive field is four time steps long), the next one sees eight time steps, and so on (see Figure 15-11). This way, the lower layers learn short-term patterns, while the higher layers learn long-term patterns. Thanks to the doubling dilation rate, the network can process extremely large sequences very efficiently.

---

13 Aaron van den Oord et al., "WaveNet: A Generative Model for Raw Audio," arXiv preprint arXiv:1609.03499 (2016).

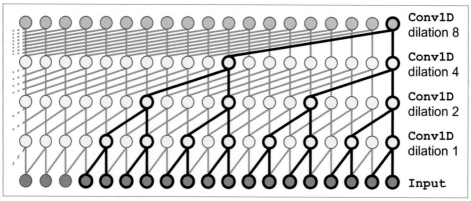

Figure 15-11. WaveNet architecture

In the WaveNet paper, the authors actually stacked 10 convolutional layers with dilation rates of 1, 2, 4, 8, …, 256, 512, then they stacked another group of 10 identical
layers (also with dilation rates 1, 2, 4, 8, …, 256, 512), then again another identical
group of 10 layers. They justified this architecture by pointing out that a single stack
of 10 convolutional layers with these dilation rates will act like a super-efficient convolutional layer with a kernel of size 1,024 (except way faster, more powerful, and
using significantly fewer parameters), which is why they stacked 3 such blocks. They
also left-padded the input sequences with a number of zeros equal to the dilation rate
before every layer, to preserve the same sequence length throughout the network.
Here is how to implement a simplified WaveNet to tackle the same sequences as
earlier:[14]

```
model = keras.models.Sequential()
model.add(keras.layers.InputLayer(input_shape=[None, 1]))
for rate in (1, 2, 4, 8) * 2:
    model.add(keras.layers.Conv1D(filters=20, kernel_size=2, padding="causal",
                                  activation="relu", dilation_rate=rate))
model.add(keras.layers.Conv1D(filters=10, kernel_size=1))
model.compile(loss="mse", optimizer="adam", metrics=[last_time_step_mse])
history = model.fit(X_train, Y_train, epochs=20,
                    validation_data=(X_valid, Y_valid))
```

This `Sequential` model starts with an explicit input layer (this is simpler than trying
to set `input_shape` only on the first layer), then continues with a 1D convolutional
layer using `"causal"` padding: this ensures that the convolutional layer does not peek
into the future when making predictions (it is equivalent to padding the inputs with
the right amount of zeros on the left and using `"valid"` padding). We then add

---

14 The complete WaveNet uses a few more tricks, such as skip connections like in a ResNet, and *Gated Activation
Units* similar to those found in a GRU cell. Please see the notebook for more details.

similar pairs of layers using growing dilation rates: 1, 2, 4, 8, and again 1, 2, 4, 8. Finally, we add the output layer: a convolutional layer with 10 filters of size 1 and without any activation function. Thanks to the padding layers, every convolutional layer outputs a sequence of the same length as the input sequences, so the targets we use during training can be the full sequences: no need to crop them or downsample them.

The last two models offer the best performance so far in forecasting our time series! In the WaveNet paper, the authors achieved state-of-the-art performance on various audio tasks (hence the name of the architecture), including text-to-speech tasks, producing incredibly realistic voices across several languages. They also used the model to generate music, one audio sample at a time. This feat is all the more impressive when you realize that a single second of audio can contain tens of thousands of time steps—even LSTMs and GRUs cannot handle such long sequences.

In Chapter 16, we will continue to explore RNNs, and we will see how they can tackle various NLP tasks.

Exercises

1. Can you think of a few applications for a sequence-to-sequence RNN? What about a sequence-to-vector RNN, and a vector-to-sequence RNN?

2. How many dimensions must the inputs of an RNN layer have? What does each dimension represent? What about its outputs?

3. If you want to build a deep sequence-to-sequence RNN, which RNN layers should have `return_sequences=True`? What about a sequence-to-vector RNN?

4. Suppose you have a daily univariate time series, and you want to forecast the next seven days. Which RNN architecture should you use?

5. What are the main difficulties when training RNNs? How can you handle them?

6. Can you sketch the LSTM cell's architecture?

7. Why would you want to use 1D convolutional layers in an RNN?

8. Which neural network architecture could you use to classify videos?

9. Train a classification model for the SketchRNN dataset, available in TensorFlow Datasets.

10. Download the Bach chorales (*https://homl.info/bach*) dataset and unzip it. It is composed of 382 chorales composed by Johann Sebastian Bach. Each chorale is 100 to 640 time steps long, and each time step contains 4 integers, where each integer corresponds to a note's index on a piano (except for the value 0, which means that no note is played). Train a model—recurrent, convolutional, or both—that can predict the next time step (four notes), given a sequence of time steps

from a chorale. Then use this model to generate Bach-like music, one note at a time: you can do this by giving the model the start of a chorale and asking it to predict the next time step, then appending these time steps to the input sequence and asking the model for the next note, and so on. Also make sure to check out Google's Coconet model (*https://homl.info/coconet*), which was used for a nice Google doodle about Bach.

Solutions to these exercises are available in Appendix A.

## CHAPTER 16

Natural Language Processing with RNNs and Attention

When Alan Turing imagined his famous Turing test (*https://homl.info/turingtest*)[1] in 1950, his objective was to evaluate a machine's ability to match human intelligence. He could have tested for many things, such as the ability to recognize cats in pictures, play chess, compose music, or escape a maze, but, interestingly, he chose a linguistic task. More specifically, he devised a *chatbot* capable of fooling its interlocutor into thinking it was human.[2] This test does have its weaknesses: a set of hardcoded rules can fool unsuspecting or naive humans (e.g., the machine could give vague predefined answers in response to some keywords; it could pretend that it is joking or drunk, to get a pass on its weirdest answers; or it could escape difficult questions by answering them with its own questions), and many aspects of human intelligence are utterly ignored (e.g., the ability to interpret nonverbal communication such as facial expressions, or to learn a manual task). But the test does highlight the fact that mastering language is arguably *Homo sapiens*'s greatest cognitive ability. Can we build a machine that can read and write natural language?

A common approach for natural language tasks is to use recurrent neural networks. We will therefore continue to explore RNNs (introduced in Chapter 15), starting with a *character RNN*, trained to predict the next character in a sentence. This will allow us to generate some original text, and in the process we will see how to build a Tensor-Flow Dataset on a very long sequence. We will first use a *stateless RNN* (which learns

---

1 Alan Turing, "Computing Machinery and Intelligence," *Mind* 49 (1950): 433–460.

2 Of course, the word *chatbot* came much later. Turing called his test the *imitation game*: machine A and human B chat with human interrogator C via text messages; the interrogator asks questions to figure out which one is the machine (A or B). The machine passes the test if it can fool the interrogator, while the human B must try to help the interrogator.

on random portions of text at each iteration, without any information on the rest of the text), then we will build a *stateful RNN* (which preserves the hidden state between training iterations and continues reading where it left off, allowing it to learn longer patterns). Next, we will build an RNN to perform sentiment analysis (e.g., reading movie reviews and extracting the rater's feeling about the movie), this time treating sentences as sequences of words, rather than characters. Then we will show how RNNs can be used to build an Encoder–Decoder architecture capable of performing neural machine translation (NMT). For this, we will use the seq2seq API provided by the TensorFlow Addons project.

In the second part of this chapter, we will look at *attention mechanisms*. As their name suggests, these are neural network components that learn to select the part of the inputs that the rest of the model should focus on at each time step. First we will see how to boost the performance of an RNN-based Encoder–Decoder architecture using attention, then we will drop RNNs altogether and look at a very successful attention-only architecture called the *Transformer*. Finally, we will take a look at some of the most important advances in NLP in 2018 and 2019, including incredibly powerful language models such as GPT-2 and BERT, both based on Transformers.

Let's start with a simple and fun model that can write like Shakespeare (well, sort of).

Generating Shakespearean Text Using a Character RNN

In a famous 2015 blog post (*https://homl.info/charrnn*) titled "The Unreasonable Effectiveness of Recurrent Neural Networks," Andrej Karpathy showed how to train an RNN to predict the next character in a sentence. This *Char-RNN* can then be used to generate novel text, one character at a time. Here is a small sample of the text generated by a Char-RNN model after it was trained on all of Shakespeare's work:

> PANDARUS:
>
> Alas, I think he shall be come approached and the day
>
> When little srain would be attain'd into being never fed,
>
> And who is but a chain and subjects of his death,
>
> I should not sleep.

Not exactly a masterpiece, but it is still impressive that the model was able to learn words, grammar, proper punctuation, and more, just by learning to predict the next character in a sentence. Let's look at how to build a Char-RNN, step by step, starting with the creation of the dataset.

Creating the Training Dataset

First, let's download all of Shakespeare's work, using Keras's handy `get_file()` function and downloading the data from Andrej Karpathy's Char-RNN project (*https://github.com/karpathy/char-rnn*):

```
shakespeare_url = "https://homl.info/shakespeare" # shortcut URL
filepath = keras.utils.get_file("shakespeare.txt", shakespeare_url)
with open(filepath) as f:
    shakespeare_text = f.read()
```

Next, we must encode every character as an integer. One option is to create a custom preprocessing layer, as we did in Chapter 13. But in this case, it will be simpler to use Keras's `Tokenizer` class. First we need to fit a tokenizer to the text: it will find all the characters used in the text and map each of them to a different character ID, from 1 to the number of distinct characters (it does not start at 0, so we can use that value for masking, as we will see later in this chapter):

```
tokenizer = keras.preprocessing.text.Tokenizer(char_level=True)
tokenizer.fit_on_texts([shakespeare_text])
```

We set `char_level=True` to get character-level encoding rather than the default word-level encoding. Note that this tokenizer converts the text to lowercase by default (but you can set `lower=False` if you do not want that). Now the tokenizer can encode a sentence (or a list of sentences) to a list of character IDs and back, and it tells us how many distinct characters there are and the total number of characters in the text:

```
>>> tokenizer.texts_to_sequences(["First"])
[[20, 6, 9, 8, 3]]
>>> tokenizer.sequences_to_texts([[20, 6, 9, 8, 3]])
['f i r s t']
>>> max_id = len(tokenizer.word_index) # number of distinct characters
>>> dataset_size = tokenizer.document_count # total number of characters
```

Let's encode the full text so each character is represented by its ID (we subtract 1 to get IDs from 0 to 38, rather than from 1 to 39):

```
[encoded] = np.array(tokenizer.texts_to_sequences([shakespeare_text])) - 1
```

Before we continue, we need to split the dataset into a training set, a validation set, and a test set. We can't just shuffle all the characters in the text, so how do you split a sequential dataset?

How to Split a Sequential Dataset

It is very important to avoid any overlap between the training set, the validation set, and the test set. For example, we can take the first 90% of the text for the training set, then the next 5% for the validation set, and the final 5% for the test set. It would also be a good idea to leave a gap between these sets to avoid the risk of a paragraph overlapping over two sets.

When dealing with time series, you would in general split across time,: for example, you might take the years 2000 to 2012 for the training set, the years 2013 to 2015 for the validation set, and the years 2016 to 2018 for the test set. However, in some cases you may be able to split along other dimensions, which will give you a longer time period to train on. For example, if you have data about the financial health of 10,000 companies from 2000 to 2018, you might be able to split this data across the different companies. It's very likely that many of these companies will be strongly correlated, though (e.g., whole economic sectors may go up or down jointly), and if you have correlated companies across the training set and the test set your test set will not be as useful, as its measure of the generalization error will be optimistically biased.

So, it is often safer to split across time—but this implicitly assumes that the patterns the RNN can learn in the past (in the training set) will still exist in the future. In other words, we assume that the time series is *stationary* (at least in a wide sense).[3] For many time series this assumption is reasonable (e.g., chemical reactions should be fine, since the laws of chemistry don't change every day), but for many others it is not (e.g., financial markets are notoriously not stationary since patterns disappear as soon as traders spot them and start exploiting them). To make sure the time series is indeed sufficiently stationary, you can plot the model's errors on the validation set across time: if the model performs much better on the first part of the validation set than on the last part, then the time series may not be stationary enough, and you might be better off training the model on a shorter time span.

In short, splitting a time series into a training set, a validation set, and a test set is not a trivial task, and how it's done will depend strongly on the task at hand.

Now back to Shakespeare! Let's take the first 90% of the text for the training set (keeping the rest for the validation set and the test set), and create a tf.data.Dataset that will return each character one by one from this set:

```
train_size = dataset_size * 90 // 100
dataset = tf.data.Dataset.from_tensor_slices(encoded[:train_size])
```

Chopping the Sequential Dataset into Multiple Windows

The training set now consists of a single sequence of over a million characters, so we can't just train the neural network directly on it: the RNN would be equivalent to a

---

3 By definition, a stationary time series's mean, variance, and *autocorrelations* (i.e., correlations between values in the time series separated by a given interval) do not change over time. This is quite restrictive; for example, it excludes time series with trends or cyclical patterns. RNNs are more tolerant in that they can learn trends and cyclical patterns.

deep net with over a million layers, and we would have a single (very long) instance to train it. Instead, we will use the dataset's `window()` method to convert this long sequence of characters into many smaller windows of text. Every instance in the dataset will be a fairly short substring of the whole text, and the RNN will be unrolled only over the length of these substrings. This is called *truncated backpropagation through time*. Let's call the `window()` method to create a dataset of short text windows:

```
n_steps = 100
window_length = n_steps + 1 # target = input shifted 1 character ahead
dataset = dataset.window(window_length, shift=1, drop_remainder=True)
```

 You can try tuning n_steps: it is easier to train RNNs on shorter input sequences, but of course the RNN will not be able to learn any pattern longer than n_steps, so don't make it too small.

By default, the `window()` method creates nonoverlapping windows, but to get the largest possible training set we use `shift=1` so that the first window contains characters 0 to 100, the second contains characters 1 to 101, and so on. To ensure that all windows are exactly 101 characters long (which will allow us to create batches without having to do any padding), we set `drop_remainder=True` (otherwise the last 100 windows will contain 100 characters, 99 characters, and so on down to 1 character).

The `window()` method creates a dataset that contains windows, each of which is also represented as a dataset. It's a *nested dataset*, analogous to a list of lists. This is useful when you want to transform each window by calling its dataset methods (e.g., to shuffle them or batch them). However, we cannot use a nested dataset directly for training, as our model will expect tensors as input, not datasets. So, we must call the `flat_map()` method: it converts a nested dataset into a *flat dataset* (one that does not contain datasets). For example, suppose {1, 2, 3} represents a dataset containing the sequence of tensors 1, 2, and 3. If you flatten the nested dataset {{1, 2}, {3, 4, 5, 6}}, you get back the flat dataset {1, 2, 3, 4, 5, 6}. Moreover, the `flat_map()` method takes a function as an argument, which allows you to transform each dataset in the nested dataset before flattening. For example, if you pass the function `lambda ds: ds.batch(2)` to `flat_map()`, then it will transform the nested dataset {{1, 2}, {3, 4, 5, 6}} into the flat dataset {[1, 2], [3, 4], [5, 6]}: it's a dataset of tensors of size 2. With that in mind, we are ready to flatten our dataset:

```
dataset = dataset.flat_map(lambda window: window.batch(window_length))
```

Notice that we call `batch(window_length)` on each window: since all windows have exactly that length, we will get a single tensor for each of them. Now the dataset contains consecutive windows of 101 characters each. Since Gradient Descent works best when the instances in the training set are independent and identically distributed (see Chapter 4), we need to shuffle these windows. Then we can batch the windows and separate the inputs (the first 100 characters) from the target (the last character):

```
batch_size = 32
dataset = dataset.shuffle(10000).batch(batch_size)
dataset = dataset.map(lambda windows: (windows[:, :-1], windows[:, 1:]))
```

Figure 16-1 summarizes the dataset preparation steps discussed so far (showing windows of length 11 rather than 101, and a batch size of 3 instead of 32).

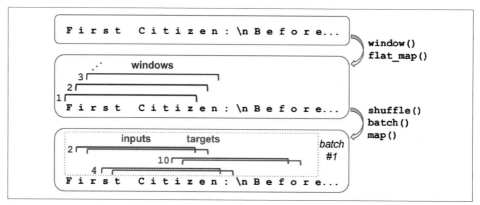

Figure 16-1. Preparing a dataset of shuffled windows

As discussed in Chapter 13, categorical input features should generally be encoded, usually as one-hot vectors or as embeddings. Here, we will encode each character using a one-hot vector because there are fairly few distinct characters (only 39):

```
dataset = dataset.map(
    lambda X_batch, Y_batch: (tf.one_hot(X_batch, depth=max_id), Y_batch))
```

Finally, we just need to add prefetching:

```
dataset = dataset.prefetch(1)
```

That's it! Preparing the dataset was the hardest part. Now let's create the model.

Building and Training the Char-RNN Model

To predict the next character based on the previous 100 characters, we can use an RNN with 2 GRU layers of 128 units each and 20% dropout on both the inputs (drop out) and the hidden states (recurrent_dropout). We can tweak these hyperparameters later, if needed. The output layer is a time-distributed Dense layer like we saw in Chapter 15. This time this layer must have 39 units (max_id) because there are 39 distinct characters in the text, and we want to output a probability for each possible character (at each time step). The output probabilities should sum up to 1 at each time step, so we apply the softmax activation function to the outputs of the Dense layer. We can then compile this model, using the `"sparse_categorical_crossen` `tropy"` loss and an Adam optimizer. Finally, we are ready to train the model for several epochs (this may take many hours, depending on your hardware):

```
model = keras.models.Sequential([
    keras.layers.GRU(128, return_sequences=True, input_shape=[None, max_id],
                     dropout=0.2, recurrent_dropout=0.2),
    keras.layers.GRU(128, return_sequences=True,
                     dropout=0.2, recurrent_dropout=0.2),
    keras.layers.TimeDistributed(keras.layers.Dense(max_id,
                                                    activation="softmax"))
])
model.compile(loss="sparse_categorical_crossentropy", optimizer="adam")
history = model.fit(dataset, epochs=20)
```

Using the Char-RNN Model

Now we have a model that can predict the next character in text written by Shakespeare. To feed it some text, we first need to preprocess it like we did earlier, so let's create a little function for this:

```
def preprocess(texts):
    X = np.array(tokenizer.texts_to_sequences(texts)) - 1
    return tf.one_hot(X, max_id)
```

Now let's use the model to predict the next letter in some text:

```
>>> X_new = preprocess(["How are yo"])
>>> Y_pred = model.predict_classes(X_new)
>>> tokenizer.sequences_to_texts(Y_pred + 1)[0][-1] # 1st sentence, last char
'u'
```

Success! The model guessed right. Now let's use this model to generate new text.

Generating Fake Shakespearean Text

To generate new text using the Char-RNN model, we could feed it some text, make the model predict the most likely next letter, add it at the end of the text, then give the extended text to the model to guess the next letter, and so on. But in practice this often leads to the same words being repeated over and over again. Instead, we can pick the next character randomly, with a probability equal to the estimated probability, using TensorFlow's `tf.random.categorical()` function. This will generate more diverse and interesting text. The `categorical()` function samples random class indices, given the class log probabilities (logits). To have more control over the diversity of the generated text, we can divide the logits by a number called the *temperature*, which we can tweak as we wish: a temperature close to 0 will favor the high-probability characters, while a very high temperature will give all characters an equal probability. The following `next_char()` function uses this approach to pick the next character to add to the input text:

```
def next_char(text, temperature=1):
    X_new = preprocess([text])
    y_proba = model.predict(X_new)[0, -1:, :]
    rescaled_logits = tf.math.log(y_proba) / temperature
    char_id = tf.random.categorical(rescaled_logits, num_samples=1) + 1
    return tokenizer.sequences_to_texts(char_id.numpy())[0]
```

Next, we can write a small function that will repeatedly call `next_char()` to get the next character and append it to the given text:

```
def complete_text(text, n_chars=50, temperature=1):
    for _ in range(n_chars):
        text += next_char(text, temperature)
    return text
```

We are now ready to generate some text! Let's try with different temperatures:

```
>>> print(complete_text("t", temperature=0.2))
the belly the great and who shall be the belly the
>>> print(complete_text("w", temperature=1))
thing? or why you gremio.
who make which the first
>>> print(complete_text("w", temperature=2))
th no cce:
yeolg-hormer firi. a play asks.
fol rusb
```

Apparently our Shakespeare model works best at a temperature close to 1. To gener‐ate more convincing text, you could try using more GRU layers and more neurons per layer, train for longer, and add some regularization (for example, you could set `recur rent_dropout=0.3` in the GRU layers). Moreover, the model is currently incapable of learning patterns longer than `n_steps`, which is just 100 characters. You could try making this window larger, but it will also make training harder, and even LSTM and GRU cells cannot handle very long sequences. Alternatively, you could use a stateful RNN.

Stateful RNN

Until now, we have used only *stateless RNNs*: at each training iteration the model starts with a hidden state full of zeros, then it updates this state at each time step, and after the last time step, it throws it away, as it is not needed anymore. What if we told the RNN to preserve this final state after processing one training batch and use it as the initial state for the next training batch? This way the model can learn long-term patterns despite only backpropagating through short sequences. This is called a *state‐ful RNN*. Let's see how to build one.

First, note that a stateful RNN only makes sense if each input sequence in a batch starts exactly where the corresponding sequence in the previous batch left off. So the first thing we need to do to build a stateful RNN is to use sequential and nonoverlapping input sequences (rather than the shuffled and overlapping sequences we used to train stateless RNNs). When creating the `Dataset`, we must therefore use `shift=n_steps` (instead of `shift=1`) when calling the `window()` method. Moreover, we must obviously *not* call the `shuffle()` method. Unfortunately, batching is much harder when preparing a dataset for a stateful RNN than it is for a stateless RNN. Indeed, if we were to call `batch(32)`, then 32 consecutive windows would be put in the same batch, and the following batch would not continue each of these window where it left off. The first batch would contain windows 1 to 32 and the second batch would contain windows 33 to 64, so if you consider, say, the first window of each batch (i.e., windows 1 and 33), you can see that they are not consecutive. The simplest solution to this problem is to just use "batches" containing a single window:

```
dataset = tf.data.Dataset.from_tensor_slices(encoded[:train_size])
dataset = dataset.window(window_length, shift=n_steps, drop_remainder=True)
dataset = dataset.flat_map(lambda window: window.batch(window_length))
dataset = dataset.batch(1)
dataset = dataset.map(lambda windows: (windows[:, :-1], windows[:, 1:]))
dataset = dataset.map(
    lambda X_batch, Y_batch: (tf.one_hot(X_batch, depth=max_id), Y_batch))
dataset = dataset.prefetch(1)
```

Figure 16-2 summarizes the first steps.

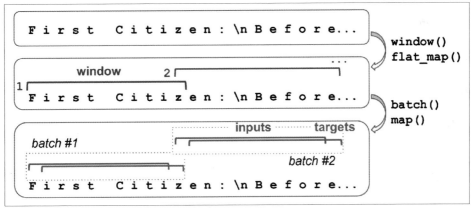

Figure 16-2. Preparing a dataset of consecutive sequence fragments for a stateful RNN

Batching is harder, but it is not impossible. For example, we could chop Shakespeare's text into 32 texts of equal length, create one dataset of consecutive input sequences for each of them, and finally use `tf.train.Dataset.zip(datasets).map(lambda *windows: tf.stack(windows))` to create proper consecutive batches, where the nth input sequence in a batch starts off exactly where the nth input sequence ended in the previous batch (see the notebook for the full code).

Now let's create the stateful RNN. First, we need to set `stateful=True` when creating every recurrent layer. Second, the stateful RNN needs to know the batch size (since it will preserve a state for each input sequence in the batch), so we must set the `batch_input_shape` argument in the first layer. Note that we can leave the second dimension unspecified, since the inputs could have any length:

```python
model = keras.models.Sequential([
    keras.layers.GRU(128, return_sequences=True, stateful=True,
                     dropout=0.2, recurrent_dropout=0.2,
                     batch_input_shape=[batch_size, None, max_id]),
    keras.layers.GRU(128, return_sequences=True, stateful=True,
                     dropout=0.2, recurrent_dropout=0.2),
    keras.layers.TimeDistributed(keras.layers.Dense(max_id,
                                                    activation="softmax"))
])
```

At the end of each epoch, we need to reset the states before we go back to the beginning of the text. For this, we can use a small callback:

```python
class ResetStatesCallback(keras.callbacks.Callback):
    def on_epoch_begin(self, epoch, logs):
        self.model.reset_states()
```

And now we can compile and fit the model (for more epochs, because each epoch is much shorter than earlier, and there is only one instance per batch):

```python
model.compile(loss="sparse_categorical_crossentropy", optimizer="adam")
model.fit(dataset, epochs=50, callbacks=[ResetStatesCallback()])
```

> After this model is trained, it will only be possible to use it to make predictions for batches of the same size as were used during training. To avoid this restriction, create an identical *stateless* model, and copy the stateful model's weights to this model.

Now that we have built a character-level model, it's time to look at word-level models and tackle a common natural language processing task: *sentiment analysis*. In the process we will learn how to handle sequences of variable lengths using masking.

Sentiment Analysis

If MNIST is the "hello world" of computer vision, then the IMDb reviews dataset is the "hello world" of natural language processing: it consists of 50,000 movie reviews in English (25,000 for training, 25,000 for testing) extracted from the famous Internet Movie Database (*https://imdb.com/*), along with a simple binary target for each review indicating whether it is negative (0) or positive (1). Just like MNIST, the IMDb reviews dataset is popular for good reasons: it is simple enough to be tackled on a laptop in a reasonable amount of time, but challenging enough to be fun and rewarding. Keras provides a simple function to load it:

```
>>> (X_train, y_train), (X_test, y_test) = keras.datasets.imdb.load_data()
>>> X_train[0][:10]
[1, 14, 22, 16, 43, 530, 973, 1622, 1385, 65]
```

Where are the movie reviews? Well, as you can see, the dataset is already preprocessed for you: X_train consists of a list of reviews, each of which is represented as a NumPy array of integers, where each integer represents a word. All punctuation was removed, and then words were converted to lowercase, split by spaces, and finally indexed by frequency (so low integers correspond to frequent words). The integers 0, 1, and 2 are special: they represent the padding token, the *start-of-sequence* (SSS) token, and unknown words, respectively. If you want to visualize a review, you can decode it like this:

```
>>> word_index = keras.datasets.imdb.get_word_index()
>>> id_to_word = {id_ + 3: word for word, id_ in word_index.items()}
>>> for id_, token in enumerate(("<pad>", "<sos>", "<unk>")):
...     id_to_word[id_] = token
...
>>> " ".join([id_to_word[id_] for id_ in X_train[0][:10]])
'<sos> this film was just brilliant casting location scenery story'
```

In a real project, you will have to preprocess the text yourself. You can do that using the same Tokenizer class we used earlier, but this time setting char_level=False (which is the default). When encoding words, it filters out a lot of characters, including most punctuation, line breaks, and tabs (but you can change this by setting the filters argument). Most importantly, it uses spaces to identify word boundaries. This is OK for English and many other scripts (written languages) that use spaces between words, but not all scripts use spaces this way. Chinese does not use spaces between words, Vietnamese uses spaces even within words, and languages such as German often attach multiple words together, without spaces. Even in English, spaces are not always the best way to tokenize text: think of "San Francisco" or "#ILoveDeepLearning."

Fortunately, there are better options! The 2018 paper (*https://homl.info/subword*)[4] by Taku Kudo introduced an unsupervised learning technique to tokenize and detokenize text at the subword level in a language-independent way, treating spaces like other characters. With this approach, even if your model encounters a word it has never seen before, it can still reasonably guess what it means. For example, it may never have seen the word "smartest" during training, but perhaps it learned the word "smart" and it also learned that the suffix "est" means "the most," so it can infer the

---

4 Taku Kudo, "Subword Regularization: Improving Neural Network Translation Models with Multiple Subword Candidates," arXiv preprint arXiv:1804.10959 (2018).

meaning of "smartest." Google's *SentencePiece* (*https://github.com/google/sentence piece*) project provides an open source implementation, described in a paper (*https://homl.info/sentencepiece*)[5] by Taku Kudo and John Richardson.

Another option was proposed in an earlier paper (*https://homl.info/rarewords*)[6] by Rico Sennrich et al. that explored other ways of creating subword encodings (e.g., using *byte pair encoding*). Last but not least, the TensorFlow team released the TF.Text (*https://homl.info/tftext*) library in June 2019, which implements various tokenization strategies, including WordPiece (*https://homl.info/wordpiece*)[7] (a variant of byte pair encoding).

If you want to deploy your model to a mobile device or a web browser, and you don't want to have to write a different preprocessing function every time, then you will want to handle preprocessing using only TensorFlow operations, so it can be included in the model itself. Let's see how. First, let's load the original IMDb reviews, as text (byte strings), using TensorFlow Datasets (introduced in Chapter 13):

```
import tensorflow_datasets as tfds

datasets, info = tfds.load("imdb_reviews", as_supervised=True, with_info=True)
train_size = info.splits["train"].num_examples
```

Next, let's write the preprocessing function:

```
def preprocess(X_batch, y_batch):
    X_batch = tf.strings.substr(X_batch, 0, 300)
    X_batch = tf.strings.regex_replace(X_batch, b"<br\\s*/?>", b" ")
    X_batch = tf.strings.regex_replace(X_batch, b"[^a-zA-Z']", b" ")
    X_batch = tf.strings.split(X_batch)
    return X_batch.to_tensor(default_value=b"<pad>"), y_batch
```

It starts by truncating the reviews, keeping only the first 300 characters of each: this will speed up training, and it won't impact performance too much because you can generally tell whether a review is positive or not in the first sentence or two. Then it uses *regular expressions* to replace `<br />` tags with spaces, and to replace any characters other than letters and quotes with spaces. For example, the text `"Well, I can't<br />"` will become `"Well I can't"`. Finally, the `preprocess()` function splits the reviews by the spaces, which returns a ragged tensor, and it converts this ragged tensor to a dense tensor, padding all reviews with the padding token `"<pad>"` so that they all have the same length.

---

5 Taku Kudo and John Richardson, "SentencePiece: A Simple and Language Independent Subword Tokenizer and Detokenizer for Neural Text Processing," arXiv preprint arXiv:1808.06226 (2018).

6 Rico Sennrich et al., "Neural Machine Translation of Rare Words with Subword Units," *Proceedings of the 54th Annual Meeting of the Association for Computational Linguistics* 1 (2016): 1715–1725.

7 Yonghui Wu et al., "Google's Neural Machine Translation System: Bridging the Gap Between Human and Machine Translation," arXiv preprint arXiv:1609.08144 (2016).

Next, we need to construct the vocabulary. This requires going through the whole training set once, applying our preprocess() function, and using a Counter to count the number of occurrences of each word:

```
from collections import Counter
vocabulary = Counter()
for X_batch, y_batch in datasets["train"].batch(32).map(preprocess):
    for review in X_batch:
        vocabulary.update(list(review.numpy()))
```

Let's look at the three most common words:

```
>>> vocabulary.most_common()[:3]
[(b'<pad>', 215797), (b'the', 61137), (b'a', 38564)]
```

Great! We probably don't need our model to know all the words in the dictionary to get good performance, though, so let's truncate the vocabulary, keeping only the 10,000 most common words:

```
vocab_size = 10000
truncated_vocabulary = [
    word for word, count in vocabulary.most_common()[:vocab_size]]
```

Now we need to add a preprocessing step to replace each word with its ID (i.e., its index in the vocabulary). Just like we did in Chapter 13, we will create a lookup table for this, using 1,000 out-of-vocabulary (oov) buckets:

```
words = tf.constant(truncated_vocabulary)
word_ids = tf.range(len(truncated_vocabulary), dtype=tf.int64)
vocab_init = tf.lookup.KeyValueTensorInitializer(words, word_ids)
num_oov_buckets = 1000
table = tf.lookup.StaticVocabularyTable(vocab_init, num_oov_buckets)
```

We can then use this table to look up the IDs of a few words:

```
>>> table.lookup(tf.constant([b"This movie was faaaaaantastic".split()]))
<tf.Tensor: [...], dtype=int64, numpy=array([[   22,    12,    11, 10054]])>
```

Note that the words "this," "movie," and "was" were found in the table, so their IDs are lower than 10,000, while the word "faaaaaantastic" was not found, so it was mapped to one of the oov buckets, with an ID greater than or equal to 10,000.

> TF Transform (introduced in Chapter 13) provides some useful functions to handle such vocabularies. For example, check out the tft.compute_and_apply_vocabulary() function: it will go through the dataset to find all distinct words and build the vocabulary, and it will generate the TF operations required to encode each word using this vocabulary.

Now we are ready to create the final training set. We batch the reviews, then convert them to short sequences of words using the preprocess() function, then encode these words using a simple encode_words() function that uses the table we just built, and finally prefetch the next batch:

```
def encode_words(X_batch, y_batch):
    return table.lookup(X_batch), y_batch

train_set = datasets["train"].batch(32).map(preprocess)
train_set = train_set.map(encode_words).prefetch(1)
```

At last we can create the model and train it:

```
embed_size = 128
model = keras.models.Sequential([
    keras.layers.Embedding(vocab_size + num_oov_buckets, embed_size,
                           input_shape=[None]),
    keras.layers.GRU(128, return_sequences=True),
    keras.layers.GRU(128),
    keras.layers.Dense(1, activation="sigmoid")
])
model.compile(loss="binary_crossentropy", optimizer="adam",
              metrics=["accuracy"])
history = model.fit(train_set, epochs=5)
```

The first layer is an Embedding layer, which will convert word IDs into embeddings (introduced in Chapter 13). The embedding matrix needs to have one row per word ID (vocab_size + num_oov_buckets) and one column per embedding dimension (this example uses 128 dimensions, but this is a hyperparameter you could tune). Whereas the inputs of the model will be 2D tensors of shape [*batch size, time steps*], the output of the Embedding layer will be a 3D tensor of shape [*batch size, time steps, embedding size*].

The rest of the model is fairly straightforward: it is composed of two GRU layers, with the second one returning only the output of the last time step. The output layer is just a single neuron using the sigmoid activation function to output the estimated probability that the review expresses a positive sentiment regarding the movie. We then compile the model quite simply, and we fit it on the dataset we prepared earlier, for a few epochs.

Masking

As it stands, the model will need to learn that the padding tokens should be ignored. But we already know that! Why don't we tell the model to ignore the padding tokens, so that it can focus on the data that actually matters? It's actually trivial: simply add `mask_zero=True` when creating the `Embedding` layer. This means that padding tokens (whose ID is 0)[8] will be ignored by all downstream layers. That's all!

The way this works is that the `Embedding` layer creates a *mask tensor* equal to `K.not_equal(inputs, 0)` (where `K = keras.backend`): it is a Boolean tensor with the same shape as the inputs, and it is equal to `False` anywhere the word IDs are 0, or `True` otherwise. This mask tensor is then automatically propagated by the model to all subsequent layers, as long as the time dimension is preserved. So in this example, both `GRU` layers will receive this mask automatically, but since the second `GRU` layer does not return sequences (it only returns the output of the last time step), the mask will not be transmitted to the `Dense` layer. Each layer may handle the mask differently, but in general they simply ignore masked time steps (i.e., time steps for which the mask is `False`). For example, when a recurrent layer encounters a masked time step, it simply copies the output from the previous time step. If the mask propagates all the way to the output (in models that output sequences, which is not the case in this example), then it will be applied to the losses as well, so the masked time steps will not contribute to the loss (their loss will be 0).

The `LSTM` and `GRU` layers have an optimized implementation for GPUs, based on Nvidia's cuDNN library. However, this implementation does not support masking. If your model uses a mask, then these layers will fall back to the (much slower) default implementation. Note that the optimized implementation also requires you to use the default values for several hyperparameters: `activation`, `recurrent_activation`, `recurrent_dropout`, `unroll`, `use_bias`, and `reset_after`.

All layers that receive the mask must support masking (or else an exception will be raised). This includes all recurrent layers, as well as the `TimeDistributed` layer and a few other layers. Any layer that supports masking must have a `supports_masking` attribute equal to `True`. If you want to implement your own custom layer with masking support, you should add a `mask` argument to the `call()` method (and obviously make the method use the mask somehow). Additionally, you should set `self.supports_masking = True` in the constructor. If your layer does not start with an `Embedding` layer, you may use the `keras.layers.Masking` layer instead: it sets the mask to `K.any(K.not_equal(inputs, 0), axis=-1)`, meaning that time steps where the last dimension is full of zeros will be masked out in subsequent layers (again, as long as the time dimension exists).

---

8 Their ID is 0 only because they are the most frequent "words" in the dataset. It would probably be a good idea to ensure that the padding tokens are always encoded as 0, even if they are not the most frequent.

Using masking layers and automatic mask propagation works best for simple `Sequential` models. It will not always work for more complex models, such as when you need to mix `Conv1D` layers with recurrent layers. In such cases, you will need to explicitly compute the mask and pass it to the appropriate layers, using either the Functional API or the Subclassing API. For example, the following model is identical to the previous model, except it is built using the Functional API and handles masking manually:

```
K = keras.backend
inputs = keras.layers.Input(shape=[None])
mask = keras.layers.Lambda(lambda inputs: K.not_equal(inputs, 0))(inputs)
z = keras.layers.Embedding(vocab_size + num_oov_buckets, embed_size)(inputs)
z = keras.layers.GRU(128, return_sequences=True)(z, mask=mask)
z = keras.layers.GRU(128)(z, mask=mask)
outputs = keras.layers.Dense(1, activation="sigmoid")(z)
model = keras.Model(inputs=[inputs], outputs=[outputs])
```

After training for a few epochs, this model will become quite good at judging whether a review is positive or not. If you use the `TensorBoard()` callback, you can visualize the embeddings in TensorBoard as they are being learned: it is fascinating to see words like "awesome" and "amazing" gradually cluster on one side of the embedding space, while words like "awful" and "terrible" cluster on the other side. Some words are not as positive as you might expect (at least with this model), such as the word "good," presumably because many negative reviews contain the phrase "not good." It's impressive that the model is able to learn useful word embeddings based on just 25,000 movie reviews. Imagine how good the embeddings would be if we had billions of reviews to train on! Unfortunately we don't, but perhaps we can reuse word embeddings trained on some other large text corpus (e.g., Wikipedia articles), even if it is not composed of movie reviews? After all, the word "amazing" generally has the same meaning whether you use it to talk about movies or anything else. Moreover, perhaps embeddings would be useful for sentiment analysis even if they were trained on another task: since words like "awesome" and "amazing" have a similar meaning, they will likely cluster in the embedding space even for other tasks (e.g., predicting the next word in a sentence). If all positive words and all negative words form clusters, then this will be helpful for sentiment analysis. So instead of using so many parameters to learn word embeddings, let's see if we can't just reuse pretrained embeddings.

Reusing Pretrained Embeddings

The TensorFlow Hub project makes it easy to reuse pretrained model components in your own models. These model components are called *modules*. Simply browse the TF Hub repository (*https://tfhub.dev*), find the one you need, and copy the code example into your project, and the module will be automatically downloaded, along with its pretrained weights, and included in your model. Easy!

For example, let's use the `nnlm-en-dim50` sentence embedding module, version 1, in our sentiment analysis model:

```
import tensorflow_hub as hub

model = keras.Sequential([
    hub.KerasLayer("https://tfhub.dev/google/tf2-preview/nnlm-en-dim50/1",
                   dtype=tf.string, input_shape=[], output_shape=[50]),
    keras.layers.Dense(128, activation="relu"),
    keras.layers.Dense(1, activation="sigmoid")
])
model.compile(loss="binary_crossentropy", optimizer="adam",
              metrics=["accuracy"])
```

The `hub.KerasLayer` layer downloads the module from the given URL. This particular module is a *sentence encoder*: it takes strings as input and encodes each one as a single vector (in this case, a 50-dimensional vector). Internally, it parses the string (splitting words on spaces) and embeds each word using an embedding matrix that was pretrained on a huge corpus: the Google News 7B corpus (seven billion words long!). Then it computes the mean of all the word embeddings, and the result is the sentence embedding.[9] We can then add two simple `Dense` layers to create a good sentiment analysis model. By default, a `hub.KerasLayer` is not trainable, but you can set `trainable=True` when creating it to change that so that you can fine-tune it for your task.

 Not all TF Hub modules support TensorFlow 2, so make sure you choose a module that does.

Next, we can just load the IMDb reviews dataset—no need to preprocess it (except for batching and prefetching)—and directly train the model:

```
datasets, info = tfds.load("imdb_reviews", as_supervised=True, with_info=True)
train_size = info.splits["train"].num_examples
batch_size = 32
train_set = datasets["train"].batch(batch_size).prefetch(1)
history = model.fit(train_set, epochs=5)
```

Note that the last part of the TF Hub module URL specified that we wanted version 1 of the model. This versioning ensures that if a new module version is released, it will not break our model. Conveniently, if you just enter this URL in a web browser, you

---

9 To be precise, the sentence embedding is equal to the mean word embedding multiplied by the square root of the number of words in the sentence. This compensates for the fact that the mean of *n* vectors gets shorter as *n* grows.

will get the documentation for this module. By default, TF Hub will cache the downloaded files into the local system's temporary directory. You may prefer to download them into a more permanent directory to avoid having to download them again after every system cleanup. To do that, set the `TFHUB_CACHE_DIR` environment variable to the directory of your choice (e.g., `os.environ["TFHUB_CACHE_DIR"]` = `"./my_tfhub_cache"`).

So far, we have looked at time series, text generation using Char-RNN, and sentiment analysis using word-level RNN models, training our own word embeddings or reusing pretrained embeddings. Let's now look at another important NLP task: *neural machine translation* (NMT), first using a pure Encoder–Decoder model, then improving it with attention mechanisms, and finally looking the extraordinary Transformer architecture.

An Encoder–Decoder Network for Neural Machine Translation

Let's take a look at a simple neural machine translation model (*https://homl.info/103*)[10] that will translate English sentences to French (see Figure 16-3).

In short, the English sentences are fed to the encoder, and the decoder outputs the French translations. Note that the French translations are also used as inputs to the decoder, but shifted back by one step. In other words, the decoder is given as input the word that it *should* have output at the previous step (regardless of what it actually output). For the very first word, it is given the start-of-sequence (SOS) token. The decoder is expected to end the sentence with an end-of-sequence (EOS) token.

Note that the English sentences are reversed before they are fed to the encoder. For example, "I drink milk" is reversed to "milk drink I." This ensures that the beginning of the English sentence will be fed last to the encoder, which is useful because that's generally the first thing that the decoder needs to translate.

Each word is initially represented by its ID (e.g., 288 for the word "milk"). Next, an `embedding` layer returns the word embedding. These word embeddings are what is actually fed to the encoder and the decoder.

---

10 Ilya Sutskever et al., "Sequence to Sequence Learning with Neural Networks," arXiv preprint arXiv:1409.3215 (2014).

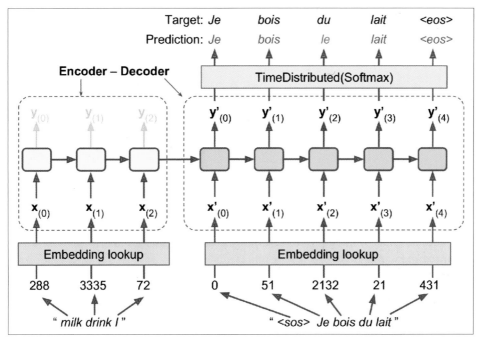

Figure 16-3. A simple machine translation model

At each step, the decoder outputs a score for each word in the output vocabulary (i.e., French), and then the softmax layer turns these scores into probabilities. For example, at the first step the word "Je" may have a probability of 20%, "Tu" may have a probability of 1%, and so on. The word with the highest probability is output. This is very much like a regular classification task, so you can train the model using the `"sparse_categorical_crossentropy"` loss, much like we did in the Char-RNN model.

Note that at inference time (after training), you will not have the target sentence to feed to the decoder. Instead, simply feed the decoder the word that it output at the previous step, as shown in Figure 16-4 (this will require an embedding lookup that is not shown in the diagram).

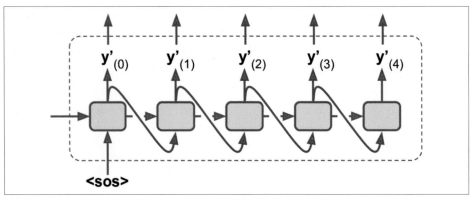

Figure 16-4. Feeding the previous output word as input at inference time

OK, now you have the big picture. Still, there are a few more details to handle if you implement this model:

- So far we have assumed that all input sequences (to the encoder and to the decoder) have a constant length. But obviously sentence lengths vary. Since regular tensors have fixed shapes, they can only contain sentences of the same length. You can use masking to handle this, as discussed earlier. However, if the sentences have very different lengths, you can't just crop them like we did for sentiment analysis (because we want full translations, not cropped translations). Instead, group sentences into buckets of similar lengths (e.g., a bucket for the 1- to 6-word sentences, another for the 7- to 12-word sentences, and so on), using padding for the shorter sequences to ensure all sentences in a bucket have the same length (check out the `tf.data.experimental.bucket_by_sequence_length()` function for this). For example, "I drink milk" becomes "<pad> <pad> <pad> milk drink I."

- We want to ignore any output past the EOS token, so these tokens should not contribute to the loss (they must be masked out). For example, if the model outputs "Je bois du lait <eos> oui," the loss for the last word should be ignored.

- When the output vocabulary is large (which is the case here), outputting a probability for each and every possible word would be terribly slow. If the target vocabulary contains, say, 50,000 French words, then the decoder would output 50,000-dimensional vectors, and then computing the softmax function over such a large vector would be very computationally intensive. To avoid this, one solution is to look only at the logits output by the model for the correct word and for a random sample of incorrect words, then compute an approximation of the loss based only on these logits. This *sampled softmax* technique was introduced in

2015 by Sébastien Jean et al. (*https://homl.info/104*).[11] In TensorFlow you can use the `tf.nn.sampled_softmax_loss()` function for this during training and use the normal softmax function at inference time (sampled softmax cannot be used at inference time because it requires knowing the target).

The TensorFlow Addons project includes many sequence-to-sequence tools to let you easily build production-ready Encoder–Decoders. For example, the following code creates a basic Encoder–Decoder model, similar to the one represented in Figure 16-3:

```python
import tensorflow_addons as tfa

encoder_inputs = keras.layers.Input(shape=[None], dtype=np.int32)
decoder_inputs = keras.layers.Input(shape=[None], dtype=np.int32)
sequence_lengths = keras.layers.Input(shape=[], dtype=np.int32)

embeddings = keras.layers.Embedding(vocab_size, embed_size)
encoder_embeddings = embeddings(encoder_inputs)
decoder_embeddings = embeddings(decoder_inputs)

encoder = keras.layers.LSTM(512, return_state=True)
encoder_outputs, state_h, state_c = encoder(encoder_embeddings)
encoder_state = [state_h, state_c]

sampler = tfa.seq2seq.sampler.TrainingSampler()

decoder_cell = keras.layers.LSTMCell(512)
output_layer = keras.layers.Dense(vocab_size)
decoder = tfa.seq2seq.basic_decoder.BasicDecoder(decoder_cell, sampler,
                                                 output_layer=output_layer)
final_outputs, final_state, final_sequence_lengths = decoder(
    decoder_embeddings, initial_state=encoder_state,
    sequence_length=sequence_lengths)
Y_proba = tf.nn.softmax(final_outputs.rnn_output)

model = keras.Model(inputs=[encoder_inputs, decoder_inputs, sequence_lengths],
                    outputs=[Y_proba])
```

The code is mostly self-explanatory, but there are a few points to note. First, we set `return_state=True` when creating the `LSTM` layer so that we can get its final hidden state and pass it to the decoder. Since we are using an LSTM cell, it actually returns two hidden states (short term and long term). The `TrainingSampler` is one of several samplers available in TensorFlow Addons: their role is to tell the decoder at each step what it should pretend the previous output was. During inference, this should be the

---

11 Sébastien Jean et al., "On Using Very Large Target Vocabulary for Neural Machine Translation," *Proceedings of the 53rd Annual Meeting of the Association for Computational Linguistics and the 7th International Joint Conference on Natural Language Processing of the Asian Federation of Natural Language Processing* 1 (2015): 1–10.

embedding of the token that was actually output. During training, it should be the embedding of the previous target token: this is why we used the `TrainingSampler`. In practice, it is often a good idea to start training with the embedding of the target of the previous time step and gradually transition to using the embedding of the actual token that was output at the previous step. This idea was introduced in a 2015 paper (*https://homl.info/scheduledsampling*)[12] by Samy Bengio et al. The `ScheduledEmbed dingTrainingSampler` will randomly choose between the target or the actual output, with a probability that you can gradually change during training.

Bidirectional RNNs

A each time step, a regular recurrent layer only looks at past and present inputs before generating its output. In other words, it is "causal," meaning it cannot look into the future. This type of RNN makes sense when forecasting time series, but for many NLP tasks, such as Neural Machine Translation, it is often preferable to look ahead at the next words before encoding a given word. For example, consider the phrases "the Queen of the United Kingdom," "the queen of hearts," and "the queen bee": to properly encode the word "queen," you need to look ahead. To implement this, run two recurrent layers on the same inputs, one reading the words from left to right and the other reading them from right to left. Then simply combine their outputs at each time step, typically by concatenating them. This is called a *bidirectional recurrent layer* (see Figure 16-5).

To implement a bidirectional recurrent layer in Keras, wrap a recurrent layer in a `keras.layers.Bidirectional` layer. For example, the following code creates a bidirectional GRU layer:

```
keras.layers.Bidirectional(keras.layers.GRU(10, return_sequences=True))
```

The `Bidirectional` layer will create a clone of the GRU layer (but in the reverse direction), and it will run both and concatenate their outputs. So although the GRU layer has 10 units, the `Bidirectional` layer will output 20 values per time step.

---

12 Samy Bengio et al., "Scheduled Sampling for Sequence Prediction with Recurrent Neural Networks," arXiv preprint arXiv:1506.03099 (2015).

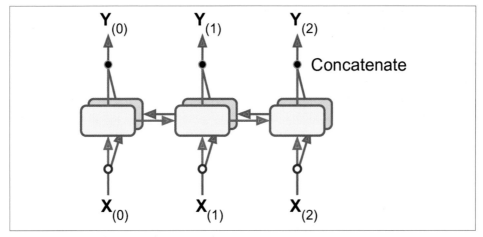

Figure 16-5. A bidirectional recurrent layer

Beam Search

Suppose you train an Encoder–Decoder model, and use it to translate the French sentence "Comment vas-tu?" to English. You are hoping that it will output the proper translation ("How are you?"), but unfortunately it outputs "How will you?" Looking at the training set, you notice many sentences such as "Comment vas-tu jouer?" which translates to "How will you play?" So it wasn't absurd for the model to output "How will" after seeing "Comment vas." Unfortunately, in this case it was a mistake, and the model could not go back and fix it, so it tried to complete the sentence as best it could. By greedily outputting the most likely word at every step, it ended up with a suboptimal translation. How can we give the model a chance to go back and fix mistakes it made earlier? One of the most common solutions is *beam search*: it keeps track of a short list of the k most promising sentences (say, the top three), and at each decoder step it tries to extend them by one word, keeping only the k most likely sentences. The parameter k is called the *beam width*.

For example, suppose you use the model to translate the sentence "Comment vas-tu?" using beam search with a beam width of 3. At the first decoder step, the model will output an estimated probability for each possible word. Suppose the top three words are "How" (75% estimated probability), "What" (3%), and "You" (1%). That's our short list so far. Next, we create three copies of our model and use them to find the next word for each sentence. Each model will output one estimated probability per word in the vocabulary. The first model will try to find the next word in the sentence "How," and perhaps it will output a probability of 36% for the word "will," 32% for the word "are," 16% for the word "do," and so on. Note that these are actually *conditional* probabilities, given that the sentence starts with "How." The second model will try to complete the sentence "What"; it might output a conditional probability of 50% for the word "are," and so on. Assuming the vocabulary has 10,000 words, each model will output 10,000 probabilities.

Next, we compute the probabilities of each of the 30,000 two-word sentences that these models considered ($3 \times 10,000$). We do this by multiplying the estimated conditional probability of each word by the estimated probability of the sentence it completes. For example, the estimated probability of the sentence "How" was 75%, while the estimated conditional probability of the word "will" (given that the first word is "How") was 36%, so the estimated probability of the sentence "How will" is 75% × 36% = 27%. After computing the probabilities of all 30,000 two-word sentences, we keep only the top 3. Perhaps they all start with the word "How": "How will" (27%), "How are" (24%), and "How do" (12%). Right now, the sentence "How will" is winning, but "How are" has not been eliminated.

Then we repeat the same process: we use three models to predict the next word in each of these three sentences, and we compute the probabilities of all 30,000 three-word sentences we considered. Perhaps the top three are now "How are you" (10%), "How do you" (8%), and "How will you" (2%). At the next step we may get "How do you do" (7%), "How are you <eos>" (6%), and "How are you doing" (3%). Notice that "How will" was eliminated, and we now have three perfectly reasonable translations. We boosted our Encoder–Decoder model's performance without any extra training, simply by using it more wisely.

You can implement beam search fairly easily using TensorFlow Addons:

```
beam_width = 10
decoder = tfa.seq2seq.beam_search_decoder.BeamSearchDecoder(
    cell=decoder_cell, beam_width=beam_width, output_layer=output_layer)
decoder_initial_state = tfa.seq2seq.beam_search_decoder.tile_batch(
    encoder_state, multiplier=beam_width)
outputs, _, _ = decoder(
    embedding_decoder, start_tokens=start_tokens, end_token=end_token,
    initial_state=decoder_initial_state)
```

We first create a `BeamSearchDecoder`, which wraps all the decoder clones (in this case 10 clones). Then we create one copy of the encoder's final state for each decoder clone, and we pass these states to the decoder, along with the start and end tokens.

With all this, you can get good translations for fairly short sentences (especially if you use pretrained word embeddings). Unfortunately, this model will be really bad at translating long sentences. Once again, the problem comes from the limited short-term memory of RNNs. *Attention mechanisms* are the game-changing innovation that addressed this problem.

Attention Mechanisms

Consider the path from the word "milk" to its translation "lait" in Figure 16-3: it is quite long! This means that a representation of this word (along with all the other words) needs to be carried over many steps before it is actually used. Can't we make this path shorter?

This was the core idea in a groundbreaking 2014 paper (*https://homl.info/attention*)[13] by Dzmitry Bahdanau et al. They introduced a technique that allowed the decoder to focus on the appropriate words (as encoded by the encoder) at each time step. For example, at the time step where the decoder needs to output the word "lait," it will focus its attention on the word "milk." This means that the path from an input word to its translation is now much shorter, so the short-term memory limitations of RNNs have much less impact. Attention mechanisms revolutionized neural machine translation (and NLP in general), allowing a significant improvement in the state of the art, especially for long sentences (over 30 words).[14]

Figure 16-6 shows this model's architecture (slightly simplified, as we will see). On the left, you have the encoder and the decoder. Instead of just sending the encoder's final hidden state to the decoder (which is still done, although it is not shown in the figure), we now send all of its outputs to the decoder. At each time step, the decoder's memory cell computes a weighted sum of all these encoder outputs: this determines which words it will focus on at this step. The weight $\alpha_{(t,i)}$ is the weight of the ith encoder output at the tth decoder time step. For example, if the weight $\alpha_{(3,2)}$ is much larger than the weights $\alpha_{(3,0)}$ and $\alpha_{(3,1)}$, then the decoder will pay much more attention to word number 2 ("milk") than to the other two words, at least at this time step. The rest of the decoder works just like earlier: at each time step the memory cell receives the inputs we just discussed, plus the hidden state from the previous time step, and finally (although it is not represented in the diagram) it receives the target word from the previous time step (or at inference time, the output from the previous time step).

---

13 Dzmitry Bahdanau et al., "Neural Machine Translation by Jointly Learning to Align and Translate," arXiv preprint arXiv:1409.0473 (2014).

14 The most common metric used in NMT is the BiLingual Evaluation Understudy (BLEU) score, which compares each translation produced by the model with several good translations produced by humans: it counts the number of *n*-grams (sequences of *n* words) that appear in any of the target translations and adjusts the score to take into account the frequency of the produced *n*-grams in the target translations.

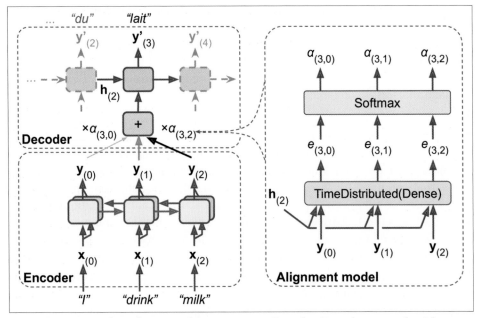

Figure 16-6. Neural machine translation using an Encoder–Decoder network with an attention model

But where do these $\alpha_{(t,i)}$ weights come from? It's actually pretty simple: they are generated by a type of small neural network called an *alignment model* (or an *attention layer*), which is trained jointly with the rest of the Encoder–Decoder model. This alignment model is illustrated on the righthand side of Figure 16-6. It starts with a time-distributed Dense layer[15] with a single neuron, which receives as input all the encoder outputs, concatenated with the decoder's previous hidden state (e.g., $h_{(2)}$). This layer outputs a score (or energy) for each encoder output (e.g., $e_{(3, 2)}$): this score measures how well each output is aligned with the decoder's previous hidden state. Finally, all the scores go through a softmax layer to get a final weight for each encoder output (e.g., $\alpha_{(3,2)}$). All the weights for a given decoder time step add up to 1 (since the softmax layer is not time-distributed). This particular attention mechanism is called *Bahdanau attention* (named after the paper's first author). Since it concatenates the encoder output with the decoder's previous hidden state, it is sometimes called *concatenative attention* (or *additive attention*).

---

15 Recall that a time-distributed Dense layer is equivalent to a regular Dense layer that you apply independently at each time step (only much faster).

If the input sentence is n words long, and assuming the output sentence is about as long, then this model will need to compute about n^2 weights. Fortunately, this quadratic computational complexity is still tractable because even long sentences don't have thousands of words.

Another common attention mechanism was proposed shortly after, in a 2015 paper (*https://homl.info/luongattention*)[16] by Minh-Thang Luong et al. Because the goal of the attention mechanism is to measure the similarity between one of the encoder's outputs and the decoder's previous hidden state, the authors proposed to simply compute the *dot product* (see Chapter 4) of these two vectors, as this is often a fairly good similarity measure, and modern hardware can compute it much faster. For this to be possible, both vectors must have the same dimensionality. This is called *Luong attention* (again, after the paper's first author), or sometimes *multiplicative attention*. The dot product gives a score, and all the scores (at a given decoder time step) go through a softmax layer to give the final weights, just like in Bahdanau attention. Another simplification they proposed was to use the decoder's hidden state at the current time step rather than at the previous time step (i.e., $\mathbf{h}_{(t)}$ rather than $\mathbf{h}_{(t-1)}$), then to use the output of the attention mechanism (noted $\tilde{\mathbf{h}}_{(t)}$) directly to compute the decoder's predictions (rather than using it to compute the decoder's current hidden state). They also proposed a variant of the dot product mechanism where the encoder outputs first go through a linear transformation (i.e., a time-distributed `Dense` layer without a bias term) before the dot products are computed. This is called the "general" dot product approach. They compared both dot product approaches to the concatenative attention mechanism (adding a rescaling parameter vector **v**), and they observed that the dot product variants performed better than concatenative attention. For this reason, concatenative attention is much less used now. The equations for these three attention mechanisms are summarized in Equation 16-1.

16 Minh-Thang Luong et al., "Effective Approaches to Attention-Based Neural Machine Translation," *Proceedings of the 2015 Conference on Empirical Methods in Natural Language Processing* (2015): 1412–1421.

Equation 16-1. Attention mechanisms

$$\widetilde{\mathbf{h}}_{(t)} = \sum_i \alpha_{(t,i)} \mathbf{y}_{(i)}$$

$$\text{with } \alpha_{(t,i)} = \frac{\exp\left(e_{(t,i)}\right)}{\sum_{i'} \exp\left(e_{(t,i')}\right)}$$

$$\text{and } e_{(t,i)} = \begin{cases} \mathbf{h}_{(t)}^{\mathsf{T}} \mathbf{y}_{(i)} & dot \\ \mathbf{h}_{(t)}^{\mathsf{T}} \mathbf{W} \mathbf{y}_{(i)} & general \\ \mathbf{v}^{\mathsf{T}} \tanh\left(\mathbf{W}\left[\mathbf{h}_{(t)}; \mathbf{y}_{(i)}\right]\right) & concat \end{cases}$$

Here is how you can add Luong attention to an Encoder–Decoder model using TensorFlow Addons:

```
attention_mechanism = tfa.seq2seq.attention_wrapper.LuongAttention(
    units, encoder_state, memory_sequence_length=encoder_sequence_length)
attention_decoder_cell = tfa.seq2seq.attention_wrapper.AttentionWrapper(
    decoder_cell, attention_mechanism, attention_layer_size=n_units)
```

We simply wrap the decoder cell in an `AttentionWrapper`, and we provide the desired attention mechanism (Luong attention in this example).

Visual Attention

Attention mechanisms are now used for a variety of purposes. One of their first applications beyond NMT was in generating image captions using visual attention (*https://homl.info/visualattention*):[17] a convolutional neural network first processes the image and outputs some feature maps, then a decoder RNN equipped with an attention mechanism generates the caption, one word at a time. At each decoder time step (each word), the decoder uses the attention model to focus on just the right part of the image. For example, in Figure 16-7, the model generated the caption "A woman is throwing a frisbee in a park," and you can see what part of the input image the decoder focused its attention on when it was about to output the word "frisbee": clearly, most of its attention was focused on the frisbee.

---

17 Kelvin Xu et al., "Show, Attend and Tell: Neural Image Caption Generation with Visual Attention," *Proceedings of the 32nd International Conference on Machine Learning* (2015): 2048–2057.

Figure 16-7. Visual attention: an input image (left) and the model's focus before producing the word "frisbee" (right)[18]

Explainability

One extra benefit of attention mechanisms is that they make it easier to understand what led the model to produce its output. This is called *explainability*. It can be especially useful when the model makes a mistake: for example, if an image of a dog walking in the snow is labeled as "a wolf walking in the snow," then you can go back and check what the model focused on when it output the word "wolf." You may find that it was paying attention not only to the dog, but also to the snow, hinting at a possible explanation: perhaps the way the model learned to distinguish dogs from wolves is by checking whether or not there's a lot of snow around. You can then fix this by training the model with more images of wolves without snow, and dogs with snow. This example comes from a great 2016 paper (*https://homl.info/explainclass*)[19] by Marco Tulio Ribeiro et al. that uses a different approach to explainability: learning an interpretable model locally around a classifier's prediction.

In some applications, explainability is not just a tool to debug a model; it can be a legal requirement (think of a system deciding whether or not it should grant you a loan).

---

18 This is a part of figure 3 from the paper. It is reproduced with the kind authorization of the authors.

19 Marco Tulio Ribeiro et al., "'Why Should I Trust You?': Explaining the Predictions of Any Classifier," *Proceedings of the 22nd ACM SIGKDD International Conference on Knowledge Discovery and Data Mining* (2016): 1135–1144.

Attention mechanisms are so powerful that you can actually build state-of-the-art models using only attention mechanisms.

Attention Is All You Need: The Transformer Architecture

In a groundbreaking 2017 paper (*https://homl.info/transformer*),[20] a team of Google researchers suggested that "Attention Is All You Need." They managed to create an architecture called the *Transformer*, which significantly improved the state of the art in NMT without using any recurrent or convolutional layers,[21] just attention mechanisms (plus embedding layers, dense layers, normalization layers, and a few other bits and pieces). As an extra bonus, this architecture was also much faster to train and easier to parallelize, so they managed to train it at a fraction of the time and cost of the previous state-of-the-art models.

The Transformer architecture is represented in Figure 16-8.

---

[20] Ashish Vaswani et al., "Attention Is All You Need," *Proceedings of the 31st International Conference on Neural Information Processing Systems* (2017): 6000–6010.

[21] Since the Transformer uses time-distributed `Dense` layers, you could argue that it uses 1D convolutional layers with a kernel size of 1.

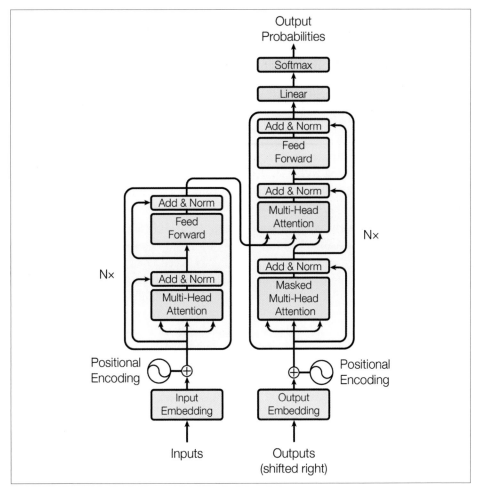

Figure 16-8. The Transformer architecture[22]

Let's walk through this figure:

- The lefthand part is the encoder. Just like earlier, it takes as input a batch of sentences represented as sequences of word IDs (the input shape is [*batch size, max input sentence length*]), and it encodes each word into a 512-dimensional representation (so the encoder's output shape is [*batch size, max input sentence length, 512*]). Note that the top part of the encoder is stacked N times (in the paper, $N = 6$).

---

22 This is figure 1 from the paper, reproduced with the kind authorization of the authors.

- The righthand part is the decoder. During training, it takes the target sentence as input (also represented as a sequence of word IDs), shifted one time step to the right (i.e., a start-of-sequence token is inserted at the beginning). It also receives the outputs of the encoder (i.e., the arrows coming from the left side). Note that the top part of the decoder is also stacked N times, and the encoder stack's final outputs are fed to the decoder at each of these N levels. Just like earlier, the decoder outputs a probability for each possible next word, at each time step (its output shape is [*batch size, max output sentence length, vocabulary length*]).

- During inference, the decoder cannot be fed targets, so we feed it the previously output words (starting with a start-of-sequence token). So the model needs to be called repeatedly, predicting one more word at every round (which is fed to the decoder at the next round, until the end-of-sequence token is output).

- Looking more closely, you can see that you are already familiar with most components: there are two embedding layers, $5 \times N$ skip connections, each of them followed by a layer normalization layer, $2 \times N$ "Feed Forward" modules that are composed of two dense layers each (the first one using the ReLU activation function, the second with no activation function), and finally the output layer is a dense layer using the softmax activation function. All of these layers are time-distributed, so each word is treated independently of all the others. But how can we translate a sentence by only looking at one word at a time? Well, that's where the new components come in:

  — The encoder's *Multi-Head Attention* layer encodes each word's relationship with every other word in the same sentence, paying more attention to the most relevant ones. For example, the output of this layer for the word "Queen" in the sentence "They welcomed the Queen of the United Kingdom" will depend on all the words in the sentence, but it will probably pay more attention to the words "United" and "Kingdom" than to the words "They" or "welcomed." This attention mechanism is called *self-attention* (the sentence is paying attention to itself). We will discuss exactly how it works shortly. The decoder's *Masked Multi-Head Attention* layer does the same thing, but each word is only allowed to attend to words located before it. Finally, the decoder's upper Multi-Head Attention layer is where the decoder pays attention to the words in the input sentence. For example, the decoder will probably pay close attention to the word "Queen" in the input sentence when it is about to output this word's translation.

  — The *positional embeddings* are simply dense vectors (much like word embeddings) that represent the position of a word in the sentence. The nth positional embedding is added to the word embedding of the nth word in each sentence. This gives the model access to each word's position, which is needed because the Multi-Head Attention layers do not consider the order or the position of the words; they only look at their relationships. Since all the other layers are time-distributed, they have no way of knowing the position of each word (either relative or absolute). Obviously, the relative and absolute word positions are important, so we need to give this information to the Transformer somehow, and positional embeddings are a good way to do this.

Let's look a bit closer at both these novel components of the Transformer architecture, starting with the positional embeddings.

Positional embeddings

A positional embedding is a dense vector that encodes the position of a word within a sentence: the ith positional embedding is simply added to the word embedding of the ith word in the sentence. These positional embeddings can be learned by the model, but in the paper the authors preferred to use fixed positional embeddings, defined using the sine and cosine functions of different frequencies. The positional embedding matrix **P** is defined in Equation 16-2 and represented at the bottom of Figure 16-9 (transposed), where $P_{p,i}$ is the ith component of the embedding for the word located at the pth position in the sentence.

Equation 16-2. Sine/cosine positional embeddings

$$P_{p,2i} = \sin\left(p/10000^{2i/d}\right)$$
$$P_{p,2i+1} = \cos\left(p/10000^{2i/d}\right)$$

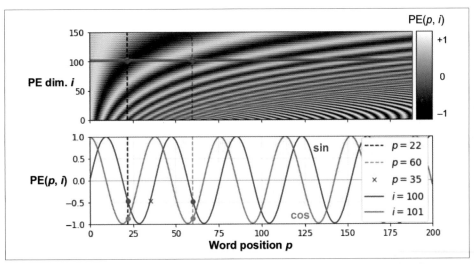

Figure 16-9. Sine/cosine positional embedding matrix (transposed, top) with a focus on two values of i (bottom)

This solution gives the same performance as learned positional embeddings do, but it can extend to arbitrarily long sentences, which is why it's favored. After the positional embeddings are added to the word embeddings, the rest of the model has access to the absolute position of each word in the sentence because there is a unique positional embedding for each position (e.g., the positional embedding for the word located at the 22nd position in a sentence is represented by the vertical dashed line at the bottom left of Figure 16-9, and you can see that it is unique to that position). Moreover, the choice of oscillating functions (sine and cosine) makes it possible for the model to learn relative positions as well. For example, words located 38 words apart (e.g., at positions $p = 22$ and $p = 60$) always have the same positional embedding values in the embedding dimensions $i = 100$ and $i = 101$, as you can see in Figure 16-9. This explains why we need both the sine and the cosine for each frequency: if we only used the sine (the blue wave at $i = 100$), the model would not be able to distinguish positions $p = 25$ and $p = 35$ (marked by a cross).

There is no `PositionalEmbedding` layer in TensorFlow, but it is easy to create one. For efficiency reasons, we precompute the positional embedding matrix in the constructor (so we need to know the maximum sentence length, `max_steps`, and the number of dimensions for each word representation, `max_dims`). Then the `call()` method crops this embedding matrix to the size of the inputs, and it adds it to the inputs. Since we added an extra first dimension of size 1 when creating the positional embedding matrix, the rules of broadcasting will ensure that the matrix gets added to every sentence in the inputs:

```python
class PositionalEncoding(keras.layers.Layer):
    def __init__(self, max_steps, max_dims, dtype=tf.float32, **kwargs):
        super().__init__(dtype=dtype, **kwargs)
        if max_dims % 2 == 1: max_dims += 1 # max_dims must be even
        p, i = np.meshgrid(np.arange(max_steps), np.arange(max_dims // 2))
        pos_emb = np.empty((1, max_steps, max_dims))
        pos_emb[0, :, ::2] = np.sin(p / 10000**(2 * i / max_dims)).T
        pos_emb[0, :, 1::2] = np.cos(p / 10000**(2 * i / max_dims)).T
        self.positional_embedding = tf.constant(pos_emb.astype(self.dtype))
    def call(self, inputs):
        shape = tf.shape(inputs)
        return inputs + self.positional_embedding[:, :shape[-2], :shape[-1]]
```

Then we can create the first layers of the Transformer:

```python
embed_size = 512; max_steps = 500; vocab_size = 10000
encoder_inputs = keras.layers.Input(shape=[None], dtype=np.int32)
decoder_inputs = keras.layers.Input(shape=[None], dtype=np.int32)
embeddings = keras.layers.Embedding(vocab_size, embed_size)
encoder_embeddings = embeddings(encoder_inputs)
decoder_embeddings = embeddings(decoder_inputs)
positional_encoding = PositionalEncoding(max_steps, max_dims=embed_size)
encoder_in = positional_encoding(encoder_embeddings)
decoder_in = positional_encoding(decoder_embeddings)
```

Now let's look deeper into the heart of the Transformer model: the Multi-Head Attention layer.

Multi-Head Attention

To understand how a Multi-Head Attention layer works, we must first understand the *Scaled Dot-Product Attention* layer, which it is based on. Let's suppose the encoder analyzed the input sentence "They played chess," and it managed to understand that the word "They" is the subject and the word "played" is the verb, so it encoded this information in the representations of these words. Now suppose the decoder has already translated the subject, and it thinks that it should translate the verb next. For this, it needs to fetch the verb from the input sentence. This is analog to a dictionary lookup: it's as if the encoder created a dictionary {"subject": "They", "verb": "played", ...} and the decoder wanted to look up the value that corresponds to the key "verb." However, the model does not have discrete tokens to represent the keys (like "subject" or "verb"); it has vectorized representations of these concepts (which it learned during training), so the key it will use for the lookup (called the *query*) will not perfectly match any key in the dictionary. The solution is to compute a similarity measure between the query and each key in the dictionary, and then use the softmax function to convert these similarity scores to weights that add up to 1. If the key that represents the verb is by far the most similar to the query, then that key's weight will be close to 1. Then the model can compute a weighted sum of the corresponding values, so if the weight of the "verb" key is close to 1, then the weighted sum will be very close to the representation of the word "played." In short, you can think of this whole process as a differentiable dictionary lookup. The similarity measure used by the Transformer is just the dot product, like in Luong attention. In fact, the equation is the same as for Luong attention, except for a scaling factor. The equation is shown in Equation 16-3, in a vectorized form.

Equation 16-3. Scaled Dot-Product Attention

$$\text{Attention}(\mathbf{Q}, \mathbf{K}, \mathbf{V}) = \text{softmax}\left(\frac{\mathbf{Q}\mathbf{K}^{\mathsf{T}}}{\sqrt{d_{keys}}}\right)\mathbf{V}$$

In this equation:

- **Q** is a matrix containing one row per query. Its shape is $[n_{queries}, d_{keys}]$, where $n_{queries}$ is the number of queries and d_{keys} is the number of dimensions of each query and each key.
- **K** is a matrix containing one row per key. Its shape is $[n_{keys}, d_{keys}]$, where n_{keys} is the number of keys and values.

- **V** is a matrix containing one row per value. Its shape is $[n_{keys}, d_{values}]$, where d_{values} is the number of each value.

- The shape of **Q K**$^\intercal$ is $[n_{queries}, n_{keys}]$: it contains one similarity score for each query/key pair. The output of the softmax function has the same shape, but all rows sum up to 1. The final output has a shape of $[n_{queries}, d_{values}]$: there is one row per query, where each row represents the query result (a weighted sum of the values).

- The scaling factor scales down the similarity scores to avoid saturating the softmax function, which would lead to tiny gradients.

- It is possible to mask out some key/value pairs by adding a very large negative value to the corresponding similarity scores, just before computing the softmax. This is useful in the Masked Multi-Head Attention layer.

In the encoder, this equation is applied to every input sentence in the batch, with **Q**, **K**, and **V** all equal to the list of words in the input sentence (so each word in the sentence will be compared to every word in the same sentence, including itself). Similarly, in the decoder's masked attention layer, the equation will be applied to every target sentence in the batch, with **Q**, **K**, and **V** all equal to the list of words in the target sentence, but this time using a mask to prevent any word from comparing itself to words located after it (at inference time the decoder will only have access to the words it already output, not to future words, so during training we must mask out future output tokens). In the upper attention layer of the decoder, the keys **K** and values **V** are simply the list of word encodings produced by the encoder, and the queries **Q** are the list of word encodings produced by the decoder.

The keras.layers.Attention layer implements Scaled Dot-Product Attention, efficiently applying Equation 16-3 to multiple sentences in a batch. Its inputs are just like **Q**, **K**, and **V**, except with an extra batch dimension (the first dimension).

> In TensorFlow, if A and B are tensors with more than two dimensions—say, of shape [2, 3, 4, 5] and [2, 3, 5, 6] respectively—then tf.matmul(A, B) will treat these tensors as 2 × 3 arrays where each cell contains a matrix, and it will multiply the corresponding matrices: the matrix at the ith row and j^{th} column in A will be multiplied by the matrix at the ith row and j^{th} column in B. Since the product of a 4 × 5 matrix with a 5 × 6 matrix is a 4 × 6 matrix, tf.matmul(A, B) will return an array of shape [2, 3, 4, 6].

If we ignore the skip connections, the layer normalization layers, the Feed Forward blocks, and the fact that this is Scaled Dot-Product Attention, not exactly Multi-Head Attention, then the rest of the Transformer model can be implemented like this:

```
Z = encoder_in
for N in range(6):
    Z = keras.layers.Attention(use_scale=True)([Z, Z])

encoder_outputs = Z
Z = decoder_in
for N in range(6):
    Z = keras.layers.Attention(use_scale=True, causal=True)([Z, Z])
    Z = keras.layers.Attention(use_scale=True)([Z, encoder_outputs])

outputs = keras.layers.TimeDistributed(
    keras.layers.Dense(vocab_size, activation="softmax"))(Z)
```

The use_scale=True argument creates an additional parameter that lets the layer learn how to properly downscale the similarity scores. This is a bit different from the Transformer model, which always downscales the similarity scores by the same factor ($\sqrt{d_{keys}}$). The causal=True argument when creating the second attention layer ensures that each output token only attends to previous output tokens, not future ones.

Now it's time to look at the final piece of the puzzle: what is a Multi-Head Attention layer? Its architecture is shown in Figure 16-10.

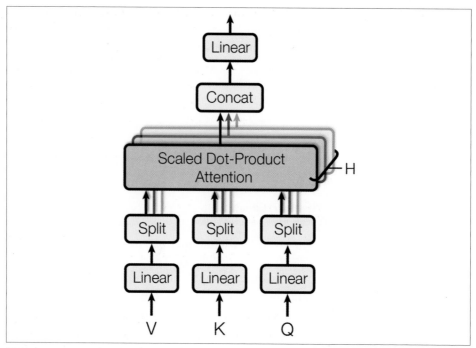

Figure 16-10. Multi-Head Attention layer architecture[23]

As you can see, it is just a bunch of Scaled Dot-Product Attention layers, each pre-ceded by a linear transformation of the values, keys, and queries (i.e., a time-distributed `Dense` layer with no activation function). All the outputs are simply concatenated, and they go through a final linear transformation (again, time-distributed). But why? What is the intuition behind this architecture? Well, consider the word "played" we discussed earlier (in the sentence "They played chess"). The encoder was smart enough to encode the fact that it is a verb. But the word represen-tation also includes its position in the text, thanks to the positional encodings, and it probably includes many other features that are useful for its translation, such as the fact that it is in the past tense. In short, the word representation encodes many differ-ent characteristics of the word. If we just used a single Scaled Dot-Product Attention layer, we would only be able to query all of these characteristics in one shot. This is why the Multi-Head Attention layer applies multiple different linear transformations of the values, keys, and queries: this allows the model to apply many different projec-tions of the word representation into different subspaces, each focusing on a subset of the word's characteristics. Perhaps one of the linear layers will project the word repre-sentation into a subspace where all that remains is the information that the word is a

---

23 This is the right part of figure 2 from the paper, reproduced with the kind authorization of the authors.

verb, another linear layer will extract just the fact that it is past tense, and so on. Then the Scaled Dot-Product Attention layers implement the lookup phase, and finally we concatenate all the results and project them back to the original space.

At the time of this writing, there is no `Transformer` class or `MultiHeadAttention` class available for TensorFlow 2. However, you can check out TensorFlow's great tutorial for building a Transformer model for language understanding (*https://homl.info/ transformertuto*). Moreover, the TF Hub team is currently porting several Transformer-based modules to TensorFlow 2, and they should be available soon. In the meantime, I hope I have demonstrated that it is not that hard to implement a Transformer yourself, and it is certainly a great exercise!

Recent Innovations in Language Models

The year 2018 has been called the "ImageNet moment for NLP": progress was astounding, with larger and larger LSTM and Transformer-based architectures trained on immense datasets. I highly recommend you check out the following papers, all published in 2018:

- The ELMo paper (*https://homl.info/elmo*)[24] by Matthew Peters introduced *Embeddings from Language Models* (ELMo): these are contextualized word embeddings learned from the internal states of a deep bidirectional language model. For example, the word "queen" will not have the same embedding in "Queen of the United Kingdom" and in "queen bee."

- The ULMFiT paper (*https://homl.info/ulmfit*)[25] by Jeremy Howard and Sebastian Ruder demonstrated the effectiveness of unsupervised pretraining for NLP tasks: the authors trained an LSTM language model using self-supervised learning (i.e., generating the labels automatically from the data) on a huge text corpus, then they fine-tuned it on various tasks. Their model outperformed the state of the art on six text classification tasks by a large margin (reducing the error rate by 18–24% in most cases). Moreover, they showed that by fine-tuning the pretrained model on just 100 labeled examples, they could achieve the same performance as a model trained from scratch on 10,000 examples.

- The GPT paper (*https://homl.info/gpt*)[26] by Alec Radford and other OpenAI researchers also demonstrated the effectiveness of unsupervised pretraining, but

---

24 Matthew Peters et al., "Deep Contextualized Word Representations," *Proceedings of the 2018 Conference of the North American Chapter of the Association for Computational Linguistics: Human Language Technologies* 1 (2018): 2227–2237.

25 Jeremy Howard and Sebastian Ruder, "Universal Language Model Fine-Tuning for Text Classification," *Proceedings of the 56th Annual Meeting of the Association for Computational Linguistics* 1 (2018): 328–339.

26 Alec Radford et al., "Improving Language Understanding by Generative Pre-Training" (2018).

this time using a Transformer-like architecture. The authors pretrained a large but fairly simple architecture composed of a stack of 12 Transformer modules (using only Masked Multi-Head Attention layers) on a large dataset, once again trained using self-supervised learning. Then they fine-tuned it on various language tasks, using only minor adaptations for each task. The tasks were quite diverse: they included text classification, *entailment* (whether sentence A entails sentence B),[27] similarity (e.g., "Nice weather today" is very similar to "It is sunny"), and question answering (given a few paragraphs of text giving some context, the model must answer some multiple-choice questions). Just a few months later, in February 2019, Alec Radford, Jeffrey Wu, and other OpenAI researchers published the GPT-2 paper, (*https://homl.info/gpt2*)[28] which proposed a very similar architecture, but larger still (with over 1.5 billion parameters!) and they showed that it could achieve good performance on many tasks without any fine-tuning. This is called *zero-shot learning* (ZSL). A smaller version of the GPT-2 model (with "just" 117 million parameters) is available at *https://github.com/openai/gpt-2*, along with its pretrained weights.

- The BERT paper (*https://homl.info/bert*)[29] by Jacob Devlin and other Google researchers also demonstrates the effectiveness of self-supervised pretraining on a large corpus, using a similar architecture to GPT but non-masked Multi-Head Attention layers (like in the Transformer's encoder). This means that the model is naturally bidirectional; hence the B in BERT (*Bidirectional Encoder Representations from Transformers*). Most importantly, the authors proposed two pretraining tasks that explain most of the model's strength:

Masked language model (MLM)

Each word in a sentence has a 15% probability of being masked, and the model is trained to predict the masked words. For example, if the original sentence is "She had fun at the birthday party," then the model may be given the sentence "She <mask> fun at the <mask> party" and it must predict the words "had" and "birthday" (the other outputs will be ignored). To be more precise, each selected word has an 80% chance of being masked, a 10% chance of being replaced by a random word (to reduce the discrepancy between pretraining and fine-tuning, since the model will not see <mask> tokens during fine-tuning), and a 10% chance of being left alone (to bias the model toward the correct answer).

---

27 For example, the sentence "Jane had a lot of fun at her friend's birthday party" entails "Jane enjoyed the party," but it is contradicted by "Everyone hated the party" and it is unrelated to "The Earth is flat."

28 Alec Radford et al., "Language Models Are Unsupervised Multitask Learners" (2019).

29 Jacob Devlin et al., "BERT: Pre-training of Deep Bidirectional Transformers for Language Understanding," *Proceedings of the 2018 Conference of the North American Chapter of the Association for Computational Linguistics: Human Language Technologies* 1 (2019).

Next sentence prediction (NSP)

The model is trained to predict whether two sentences are consecutive or not. For example, it should predict that "The dog sleeps" and "It snores loudly" are consecutive sentences, while "The dog sleeps" and "The Earth orbits the Sun" are not consecutive. This is a challenging task, and it significantly improves the performance of the model when it is fine-tuned on tasks such as question answering or entailment.

As you can see, the main innovations in 2018 and 2019 have been better subword tokenization, shifting from LSTMs to Transformers, and pretraining universal language models using self-supervised learning, then fine-tuning them with very few architectural changes (or none at all). Things are moving fast; no one can say what architectures will prevail next year. Today, it's clearly Transformers, but tomorrow it might be CNNs (e.g., check out the 2018 paper (*https://homl.info/pervasiveattention*)[30] by Maha Elbayad et al., where the researchers use masked 2D convolutional layers for sequence-to-sequence tasks). Or it might even be RNNs, if they make a surprise comeback (e.g., check out the 2018 paper (*https://homl.info/indrnn*)[31] by Shuai Li et al. that shows that by making neurons independent of each other in a given RNN layer, it is possible to train much deeper RNNs capable of learning much longer sequences).

In the next chapter we will discuss how to learn deep representations in an unsupervised way using autoencoders, and we will use generative adversarial networks (GANs) to produce images and more!

Exercises

1. What are the pros and cons of using a stateful RNN versus a stateless RNN?

2. Why do people use Encoder–Decoder RNNs rather than plain sequence-to-sequence RNNs for automatic translation?

3. How can you deal with variable-length input sequences? What about variable-length output sequences?

4. What is beam search and why would you use it? What tool can you use to implement it?

5. What is an attention mechanism? How does it help?

---

30 Maha Elbayad et al., "Pervasive Attention: 2D Convolutional Neural Networks for Sequence-to-Sequence Prediction," arXiv preprint arXiv:1808.03867 (2018).

31 Shuai Li et al., "Independently Recurrent Neural Network (IndRNN): Building a Longer and Deeper RNN," *Proceedings of the IEEE Conference on Computer Vision and Pattern Recognition* (2018): 5457–5466.

6. What is the most important layer in the Transformer architecture? What is its purpose?

7. When would you need to use sampled softmax?

8. *Embedded Reber grammars* were used by Hochreiter and Schmidhuber in their paper (*https://homl.info/93*) about LSTMs. They are artificial grammars that produce strings such as "BPBTSXXVPSEPE." Check out Jenny Orr's nice introduction (*https://homl.info/108*) to this topic. Choose a particular embedded Reber grammar (such as the one represented on Jenny Orr's page), then train an RNN to identify whether a string respects that grammar or not. You will first need to write a function capable of generating a training batch containing about 50% strings that respect the grammar, and 50% that don't.

9. Train an Encoder–Decoder model that can convert a date string from one format to another (e.g., from "April 22, 2019" to "2019-04-22").

10. Go through TensorFlow's Neural Machine Translation with Attention tutorial (*https://homl.info/nmttuto*).

11. Use one of the recent language models (e.g., BERT) to generate more convincing Shakespearean text.

Solutions to these exercises are available in Appendix A.

## CHAPTER 17

Representation Learning and Generative Learning Using Autoencoders and GANs

Autoencoders are artificial neural networks capable of learning dense representations of the input data, called *latent representations* or *codings*, without any supervision (i.e., the training set is unlabeled). These codings typically have a much lower dimensionality than the input data, making autoencoders useful for dimensionality reduction (see Chapter 8), especially for visualization purposes. Autoencoders also act as feature detectors, and they can be used for unsupervised pretraining of deep neural networks (as we discussed in Chapter 11). Lastly, some autoencoders are *generative models*: they are capable of randomly generating new data that looks very similar to the training data. For example, you could train an autoencoder on pictures of faces, and it would then be able to generate new faces. However, the generated images are usually fuzzy and not entirely realistic.

In contrast, faces generated by generative adversarial networks (GANs) are now so convincing that it is hard to believe that the people they represent do not exist. You can judge so for yourself by visiting *https://thispersondoesnotexist.com/*, a website that shows faces generated by a recent GAN architecture called *StyleGAN* (you can also check out *https://thisrentaldoesnotexist.com/* to see some generated Airbnb bedrooms). GANs are now widely used for super resolution (increasing the resolution of an image), colorization (*https://github.com/jantic/DeOldify*), powerful image editing (e.g., replacing photo bombers with realistic background), turning a simple sketch into a photorealistic image, predicting the next frames in a video, augmenting a dataset (to train other models), generating other types of data (such as text, audio, and time series), identifying the weaknesses in other models and strengthening them, and more.

Autoencoders and GANs are both unsupervised, they both learn dense representations, they can both be used as generative models, and they have many similar applications. However, they work very differently:

- Autoencoders simply learn to copy their inputs to their outputs. This may sound like a trivial task, but we will see that constraining the network in various ways can make it rather difficult. For example, you can limit the size of the latent representations, or you can add noise to the inputs and train the network to recover the original inputs. These constraints prevent the autoencoder from trivially copying the inputs directly to the outputs, which forces it to learn efficient ways of representing the data. In short, the codings are byproducts of the autoencoder learning the identity function under some constraints.

- GANs are composed of two neural networks: a *generator* that tries to generate data that looks similar to the training data, and a *discriminator* that tries to tell real data from fake data. This architecture is very original in Deep Learning in that the generator and the discriminator compete against each other during training: the generator is often compared to a criminal trying to make realistic counterfeit money, while the discriminator is like the police investigator trying to tell real money from fake. *Adversarial training* (training competing neural networks) is widely considered as one of the most important ideas in recent years. In 2016, Yann LeCun even said that it was "the most interesting idea in the last 10 years in Machine Learning."

In this chapter we will start by exploring in more depth how autoencoders work and how to use them for dimensionality reduction, feature extraction, unsupervised pretraining, or as generative models. This will naturally lead us to GANs. We will start by building a simple GAN to generate fake images, but we will see that training is often quite difficult. We will discuss the main difficulties you will encounter with adversarial training, as well as some of the main techniques to work around these difficulties. Let's start with autoencoders!

Efficient Data Representations

Which of the following number sequences do you find the easiest to memorize?

- 40, 27, 25, 36, 81, 57, 10, 73, 19, 68
- 50, 48, 46, 44, 42, 40, 38, 36, 34, 32, 30, 28, 26, 24, 22, 20, 18, 16, 14

At first glance, it would seem that the first sequence should be easier, since it is much shorter. However, if you look carefully at the second sequence, you will notice that it is just the list of even numbers from 50 down to 14. Once you notice this pattern, the second sequence becomes much easier to memorize than the first because you only need to remember the pattern (i.e., decreasing even numbers) and the starting and ending numbers (i.e., 50 and 14). Note that if you could quickly and easily memorize very long sequences, you would not care much about the existence of a pattern in the second sequence. You would just learn every number by heart, and that would be that. The fact that it is hard to memorize long sequences is what makes it useful to recognize patterns, and hopefully this clarifies why constraining an autoencoder during training pushes it to discover and exploit patterns in the data.

The relationship between memory, perception, and pattern matching was famously studied by William Chase and Herbert Simon in the early 1970s (*https://homl.info/111*).[1] They observed that expert chess players were able to memorize the positions of all the pieces in a game by looking at the board for just five seconds, a task that most people would find impossible. However, this was only the case when the pieces were placed in realistic positions (from actual games), not when the pieces were placed randomly. Chess experts don't have a much better memory than you and I; they just see chess patterns more easily, thanks to their experience with the game. Noticing patterns helps them store information efficiently.

Just like the chess players in this memory experiment, an autoencoder looks at the inputs, converts them to an efficient latent representation, and then spits out something that (hopefully) looks very close to the inputs. An autoencoder is always composed of two parts: an *encoder* (or *recognition network*) that converts the inputs to a latent representation, followed by a *decoder* (or *generative network*) that converts the internal representation to the outputs (see Figure 17-1).

---

1 William G. Chase and Herbert A. Simon, "Perception in Chess," *Cognitive Psychology* 4, no. 1 (1973): 55–81.

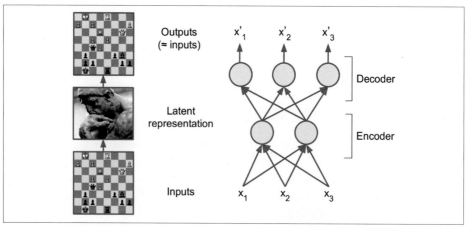

Figure 17-1. The chess memory experiment (left) and a simple autoencoder (right)

As you can see, an autoencoder typically has the same architecture as a Multi-Layer Perceptron (MLP; see Chapter 10), except that the number of neurons in the output layer must be equal to the number of inputs. In this example, there is just one hidden layer composed of two neurons (the encoder), and one output layer composed of three neurons (the decoder). The outputs are often called the *reconstructions* because the autoencoder tries to reconstruct the inputs, and the cost function contains a *reconstruction loss* that penalizes the model when the reconstructions are different from the inputs.

Because the internal representation has a lower dimensionality than the input data (it is 2D instead of 3D), the autoencoder is said to be *undercomplete*. An undercomplete autoencoder cannot trivially copy its inputs to the codings, yet it must find a way to output a copy of its inputs. It is forced to learn the most important features in the input data (and drop the unimportant ones).

Let's see how to implement a very simple undercomplete autoencoder for dimensionality reduction.

Performing PCA with an Undercomplete Linear Autoencoder

If the autoencoder uses only linear activations and the cost function is the mean squared error (MSE), then it ends up performing Principal Component Analysis (PCA; see Chapter 8).

The following code builds a simple linear autoencoder to perform PCA on a 3D dataset, projecting it to 2D:

```
from tensorflow import keras

encoder = keras.models.Sequential([keras.layers.Dense(2, input_shape=[3])])
decoder = keras.models.Sequential([keras.layers.Dense(3, input_shape=[2])])
autoencoder = keras.models.Sequential([encoder, decoder])

autoencoder.compile(loss="mse", optimizer=keras.optimizers.SGD(lr=0.1))
```

This code is really not very different from all the MLPs we built in past chapters, but there are a few things to note:

- We organized the autoencoder into two subcomponents: the encoder and the decoder. Both are regular Sequential models with a single Dense layer each, and the autoencoder is a Sequential model containing the encoder followed by the decoder (remember that a model can be used as a layer in another model).
- The autoencoder's number of outputs is equal to the number of inputs (i.e., 3).
- To perform simple PCA, we do not use any activation function (i.e., all neurons are linear), and the cost function is the MSE. We will see more complex autoencoders shortly.

Now let's train the model on a simple generated 3D dataset and use it to encode that same dataset (i.e., project it to 2D):

```
history = autoencoder.fit(X_train, X_train, epochs=20)
codings = encoder.predict(X_train)
```

Note that the same dataset, X_train, is used as both the inputs and the targets. Figure 17-2 shows the original 3D dataset (on the left) and the output of the autoencoder's hidden layer (i.e., the coding layer, on the right). As you can see, the autoencoder found the best 2D plane to project the data onto, preserving as much variance in the data as it could (just like PCA).

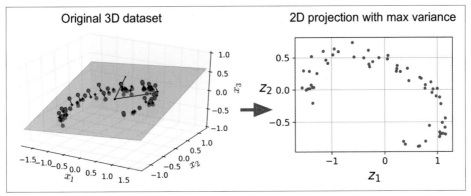

Figure 17-2. PCA performed by an undercomplete linear autoencoder

You can think of autoencoders as a form of self-supervised learning (i.e., using a supervised learning technique with automatically generated labels, in this case simply equal to the inputs).

Stacked Autoencoders

Just like other neural networks we have discussed, autoencoders can have multiple hidden layers. In this case they are called *stacked autoencoders* (or *deep autoencoders*). Adding more layers helps the autoencoder learn more complex codings. That said, one must be careful not to make the autoencoder too powerful. Imagine an encoder so powerful that it just learns to map each input to a single arbitrary number (and the decoder learns the reverse mapping). Obviously such an autoencoder will reconstruct the training data perfectly, but it will not have learned any useful data representation in the process (and it is unlikely to generalize well to new instances).

The architecture of a stacked autoencoder is typically symmetrical with regard to the central hidden layer (the coding layer). To put it simply, it looks like a sandwich. For example, an autoencoder for MNIST (introduced in Chapter 3) may have 784 inputs, followed by a hidden layer with 100 neurons, then a central hidden layer of 30 neurons, then another hidden layer with 100 neurons, and an output layer with 784 neurons. This stacked autoencoder is represented in Figure 17-3.

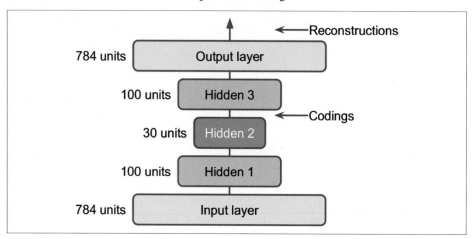

Figure 17-3. Stacked autoencoder

Implementing a Stacked Autoencoder Using Keras

You can implement a stacked autoencoder very much like a regular deep MLP. In particular, the same techniques we used in Chapter 11 for training deep nets can be applied. For example, the following code builds a stacked autoencoder for Fashion

MNIST (loaded and normalized as in Chapter 10), using the SELU activation function:

```
stacked_encoder = keras.models.Sequential([
    keras.layers.Flatten(input_shape=[28, 28]),
    keras.layers.Dense(100, activation="selu"),
    keras.layers.Dense(30, activation="selu"),
])
stacked_decoder = keras.models.Sequential([
    keras.layers.Dense(100, activation="selu", input_shape=[30]),
    keras.layers.Dense(28 * 28, activation="sigmoid"),
    keras.layers.Reshape([28, 28])
])
stacked_ae = keras.models.Sequential([stacked_encoder, stacked_decoder])
stacked_ae.compile(loss="binary_crossentropy",
                   optimizer=keras.optimizers.SGD(lr=1.5))
history = stacked_ae.fit(X_train, X_train, epochs=10,
                         validation_data=[X_valid, X_valid])
```

Let's go through this code:

- Just like earlier, we split the autoencoder model into two submodels: the encoder and the decoder.

- The encoder takes 28 × 28–pixel grayscale images, flattens them so that each image is represented as a vector of size 784, then processes these vectors through two Dense layers of diminishing sizes (100 units then 30 units), both using the SELU activation function (you may want to add LeCun normal initialization as well, but the network is not very deep so it won't make a big difference). For each input image, the encoder outputs a vector of size 30.

- The decoder takes codings of size 30 (output by the encoder) and processes them through two Dense layers of increasing sizes (100 units then 784 units), and it reshapes the final vectors into 28 × 28 arrays so the decoder's outputs have the same shape as the encoder's inputs.

- When compiling the stacked autoencoder, we use the binary cross-entropy loss instead of the mean squared error. We are treating the reconstruction task as a multilabel binary classification problem: each pixel intensity represents the probability that the pixel should be black. Framing it this way (rather than as a regression problem) tends to make the model converge faster.[2]

- Finally, we train the model using X_train as both the inputs and the targets (and similarly, we use X_valid as both the validation inputs and targets).

---

2 You might be tempted to use the accuracy metric, but it would not work properly, since this metric expects the labels to be either 0 or 1 for each pixel. You can easily work around this problem by creating a custom metric that computes the accuracy after rounding the targets and predictions to 0 or 1.

Visualizing the Reconstructions

One way to ensure that an autoencoder is properly trained is to compare the inputs and the outputs: the differences should not be too significant. Let's plot a few images from the validation set, as well as their reconstructions:

```
def plot_image(image):
    plt.imshow(image, cmap="binary")
    plt.axis("off")

def show_reconstructions(model, n_images=5):
    reconstructions = model.predict(X_valid[:n_images])
    fig = plt.figure(figsize=(n_images * 1.5, 3))
    for image_index in range(n_images):
        plt.subplot(2, n_images, 1 + image_index)
        plot_image(X_valid[image_index])
        plt.subplot(2, n_images, 1 + n_images + image_index)
        plot_image(reconstructions[image_index])

show_reconstructions(stacked_ae)
```

Figure 17-4 shows the resulting images.

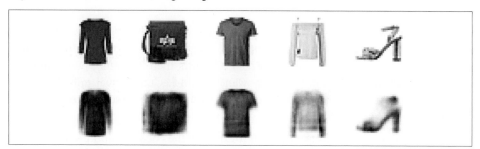

Figure 17-4. Original images (top) and their reconstructions (bottom)

The reconstructions are recognizable, but a bit too lossy. We may need to train the model for longer, or make the encoder and decoder deeper, or make the codings larger. But if we make the network too powerful, it will manage to make perfect reconstructions without having learned any useful patterns in the data. For now, let's go with this model.

Visualizing the Fashion MNIST Dataset

Now that we have trained a stacked autoencoder, we can use it to reduce the dataset's dimensionality. For visualization, this does not give great results compared to other dimensionality reduction algorithms (such as those we discussed in Chapter 8), but one big advantage of autoencoders is that they can handle large datasets, with many instances and many features. So one strategy is to use an autoencoder to reduce the dimensionality down to a reasonable level, then use another dimensionality reduction algorithm for visualization. Let's use this strategy to visualize Fashion MNIST. First, we use the encoder from our stacked autoencoder to reduce the dimensionality down to 30, then we use Scikit-Learn's implementation of the t-SNE algorithm to reduce the dimensionality down to 2 for visualization:

```
from sklearn.manifold import TSNE

X_valid_compressed = stacked_encoder.predict(X_valid)
tsne = TSNE()
X_valid_2D = tsne.fit_transform(X_valid_compressed)
```

Now we can plot the dataset:

```
plt.scatter(X_valid_2D[:, 0], X_valid_2D[:, 1], c=y_valid, s=10, cmap="tab10")
```

Figure 17-5 shows the resulting scatterplot (beautified a bit by displaying some of the images). The t-SNE algorithm identified several clusters which match the classes reasonably well (each class is represented with a different color).

Figure 17-5. Fashion MNIST visualization using an autoencoder followed by t-SNE

So, autoencoders can be used for dimensionality reduction. Another application is for unsupervised pretraining.

Unsupervised Pretraining Using Stacked Autoencoders

As we discussed in Chapter 11, if you are tackling a complex supervised task but you do not have a lot of labeled training data, one solution is to find a neural network that performs a similar task and reuse its lower layers. This makes it possible to train a high-performance model using little training data because your neural network won't have to learn all the low-level features; it will just reuse the feature detectors learned by the existing network.

Similarly, if you have a large dataset but most of it is unlabeled, you can first train a stacked autoencoder using all the data, then reuse the lower layers to create a neural network for your actual task and train it using the labeled data. For example, Figure 17-6 shows how to use a stacked autoencoder to perform unsupervised pretraining for a classification neural network. When training the classifier, if you really don't have much labeled training data, you may want to freeze the pretrained layers (at least the lower ones).

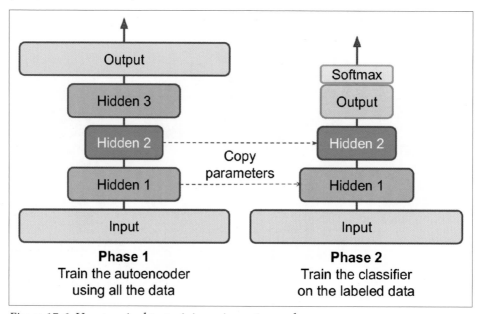

Figure 17-6. Unsupervised pretraining using autoencoders

Having plenty of unlabeled data and little labeled data is common. Building a large unlabeled dataset is often cheap (e.g., a simple script can download millions of images off the internet), but labeling those images (e.g., classifying them as cute or not) can usually be done reliably only by humans. Labeling instances is time-consuming and costly, so it's normal to have only a few thousand human-labeled instances.

There is nothing special about the implementation: just train an autoencoder using all the training data (labeled plus unlabeled), then reuse its encoder layers to create a new neural network (see the exercises at the end of this chapter for an example).

Next, let's look at a few techniques for training stacked autoencoders.

Tying Weights

When an autoencoder is neatly symmetrical, like the one we just built, a common technique is to *tie* the weights of the decoder layers to the weights of the encoder layers. This halves the number of weights in the model, speeding up training and limiting the risk of overfitting. Specifically, if the autoencoder has a total of N layers (not counting the input layer), and $\mathbf{W}_L$ represents the connection weights of the Lth layer (e.g., layer 1 is the first hidden layer, layer $N/2$ is the coding layer, and layer N is the output layer), then the decoder layer weights can be defined simply as: $\mathbf{W}_{N-L+1} = \mathbf{W}_L^\top$ (with $L = 1, 2, ..., N/2$).

To tie weights between layers using Keras, let's define a custom layer:

```
class DenseTranspose(keras.layers.Layer):
    def __init__(self, dense, activation=None, **kwargs):
        self.dense = dense
        self.activation = keras.activations.get(activation)
        super().__init__(**kwargs)
    def build(self, batch_input_shape):
        self.biases = self.add_weight(name="bias", initializer="zeros",
                                      shape=[self.dense.input_shape[-1]])
        super().build(batch_input_shape)
    def call(self, inputs):
        z = tf.matmul(inputs, self.dense.weights[0], transpose_b=True)
        return self.activation(z + self.biases)
```

This custom layer acts like a regular Dense layer, but it uses another Dense layer's weights, transposed (setting transpose_b=True is equivalent to transposing the second argument, but it's more efficient as it performs the transposition on the fly within the matmul() operation). However, it uses its own bias vector. Next, we can build a new stacked autoencoder, much like the previous one, but with the decoder's Dense layers tied to the encoder's Dense layers:

```
dense_1 = keras.layers.Dense(100, activation="selu")
dense_2 = keras.layers.Dense(30, activation="selu")

tied_encoder = keras.models.Sequential([
    keras.layers.Flatten(input_shape=[28, 28]),
    dense_1,
    dense_2
])
```

```
tied_decoder = keras.models.Sequential([
    DenseTranspose(dense_2, activation="selu"),
    DenseTranspose(dense_1, activation="sigmoid"),
    keras.layers.Reshape([28, 28])
])

tied_ae = keras.models.Sequential([tied_encoder, tied_decoder])
```

This model achieves a very slightly lower reconstruction error than the previous model, with almost half the number of parameters.

Training One Autoencoder at a Time

Rather than training the whole stacked autoencoder in one go like we just did, it is possible to train one shallow autoencoder at a time, then stack all of them into a single stacked autoencoder (hence the name), as shown in Figure 17-7. This technique is not used as much these days, but you may still run into papers that talk about "greedy layerwise training," so it's good to know what it means.

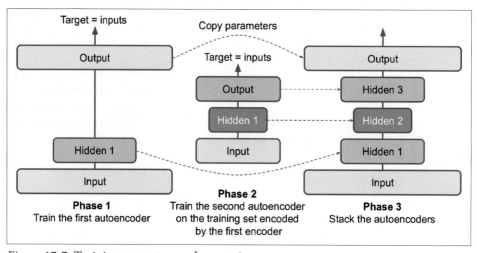

Figure 17-7. Training one autoencoder at a time

During the first phase of training, the first autoencoder learns to reconstruct the inputs. Then we encode the whole training set using this first autoencoder, and this gives us a new (compressed) training set. We then train a second autoencoder on this new dataset. This is the second phase of training. Finally, we build a big sandwich using all these autoencoders, as shown in Figure 17-7 (i.e., we first stack the hidden layers of each autoencoder, then the output layers in reverse order). This gives us the final stacked autoencoder (see the "Training One Autoencoder at a Time" section in the notebook for an implementation). We could easily train more autoencoders this way, building a very deep stacked autoencoder.

As we discussed earlier, one of the triggers of the current tsunami of interest in Deep Learning was the discovery in 2006 by Geoffrey Hinton et al. (*https://homl.info/136*) that deep neural networks can be pretrained in an unsupervised fashion, using this greedy layerwise approach. They used restricted Boltzmann machines (RBMs; see Appendix E) for this purpose, but in 2007 Yoshua Bengio et al. showed (*https://homl.info/112*)[3] that autoencoders worked just as well. For several years this was the only efficient way to train deep nets, until many of the techniques introduced in Chapter 11 made it possible to just train a deep net in one shot.

Autoencoders are not limited to dense networks: you can also build convolutional autoencoders, or even recurrent autoencoders. Let's look at these now.

Convolutional Autoencoders

If you are dealing with images, then the autoencoders we have seen so far will not work well (unless the images are very small): as we saw in Chapter 14, convolutional neural networks are far better suited than dense networks to work with images. So if you want to build an autoencoder for images (e.g., for unsupervised pretraining or dimensionality reduction), you will need to build a *convolutional autoencoder* (*https://homl.info/convae*).[4] The encoder is a regular CNN composed of convolutional layers and pooling layers. It typically reduces the spatial dimensionality of the inputs (i.e., height and width) while increasing the depth (i.e., the number of feature maps). The decoder must do the reverse (upscale the image and reduce its depth back to the original dimensions), and for this you can use transpose convolutional layers (alternatively, you could combine upsampling layers with convolutional layers). Here is a simple convolutional autoencoder for Fashion MNIST:

```
conv_encoder = keras.models.Sequential([
    keras.layers.Reshape([28, 28, 1], input_shape=[28, 28]),
    keras.layers.Conv2D(16, kernel_size=3, padding="same", activation="selu"),
    keras.layers.MaxPool2D(pool_size=2),
    keras.layers.Conv2D(32, kernel_size=3, padding="same", activation="selu"),
    keras.layers.MaxPool2D(pool_size=2),
    keras.layers.Conv2D(64, kernel_size=3, padding="same", activation="selu"),
    keras.layers.MaxPool2D(pool_size=2)
])
conv_decoder = keras.models.Sequential([
    keras.layers.Conv2DTranspose(32, kernel_size=3, strides=2, padding="valid",
                                 activation="selu",
                                 input_shape=[3, 3, 64]),
```

---

3 Yoshua Bengio et al., "Greedy Layer-Wise Training of Deep Networks," *Proceedings of the 19th International Conference on Neural Information Processing Systems* (2006): 153–160.

4 Jonathan Masci et al., "Stacked Convolutional Auto-Encoders for Hierarchical Feature Extraction," *Proceedings of the 21st International Conference on Artificial Neural Networks* 1 (2011): 52–59.

```
    keras.layers.Conv2DTranspose(16, kernel_size=3, strides=2, padding="same",
                                 activation="selu"),
    keras.layers.Conv2DTranspose(1, kernel_size=3, strides=2, padding="same",
                                 activation="sigmoid"),
    keras.layers.Reshape([28, 28])
])
conv_ae = keras.models.Sequential([conv_encoder, conv_decoder])
```

Recurrent Autoencoders

If you want to build an autoencoder for sequences, such as time series or text (e.g., for unsupervised learning or dimensionality reduction), then recurrent neural networks (see Chapter 15) may be better suited than dense networks. Building a *recurrent autoencoder* is straightforward: the encoder is typically a sequence-to-vector RNN which compresses the input sequence down to a single vector. The decoder is a vector-to-sequence RNN that does the reverse:

```
recurrent_encoder = keras.models.Sequential([
    keras.layers.LSTM(100, return_sequences=True, input_shape=[None, 28]),
    keras.layers.LSTM(30)
])
recurrent_decoder = keras.models.Sequential([
    keras.layers.RepeatVector(28, input_shape=[30]),
    keras.layers.LSTM(100, return_sequences=True),
    keras.layers.TimeDistributed(keras.layers.Dense(28, activation="sigmoid"))
])
recurrent_ae = keras.models.Sequential([recurrent_encoder, recurrent_decoder])
```

This recurrent autoencoder can process sequences of any length, with 28 dimensions per time step. Conveniently, this means it can process Fashion MNIST images by treating each image as a sequence of rows: at each time step, the RNN will process a single row of 28 pixels. Obviously, you could use a recurrent autoencoder for any kind of sequence. Note that we use a `RepeatVector` layer as the first layer of the decoder, to ensure that its input vector gets fed to the decoder at each time step.

OK, let's step back for a second. So far we have seen various kinds of autoencoders (basic, stacked, convolutional, and recurrent), and we have looked at how to train them (either in one shot or layer by layer). We also looked at a couple applications: data visualization and unsupervised pretraining.

Up to now, in order to force the autoencoder to learn interesting features, we have limited the size of the coding layer, making it undercomplete. There are actually many other kinds of constraints that can be used, including ones that allow the coding layer to be just as large as the inputs, or even larger, resulting in an *overcomplete autoencoder*. Let's look at some of those approaches now.

Denoising Autoencoders

Another way to force the autoencoder to learn useful features is to add noise to its inputs, training it to recover the original, noise-free inputs. This idea has been around since the 1980s (e.g., it is mentioned in Yann LeCun's 1987 master's thesis). In a 2008 paper (*https://homl.info/113*),[5] Pascal Vincent et al. showed that autoencoders could also be used for feature extraction. In a 2010 paper (*https://homl.info/114*),[6] Vincent et al. introduced *stacked denoising autoencoders*.

The noise can be pure Gaussian noise added to the inputs, or it can be randomly switched-off inputs, just like in dropout (introduced in Chapter 11). Figure 17-8 shows both options.

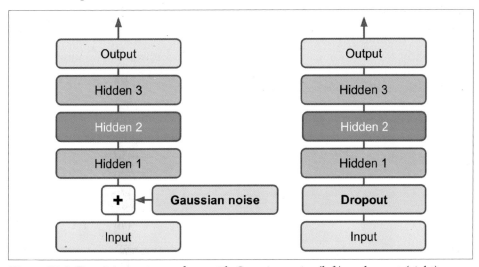

Figure 17-8. Denoising autoencoders, with Gaussian noise (left) or dropout (right)

The implementation is straightforward: it is a regular stacked autoencoder with an additional Dropout layer applied to the encoder's inputs (or you could use a Gaus sianNoise layer instead). Recall that the Dropout layer is only active during training (and so is the GaussianNoise layer):

5 Pascal Vincent et al., "Extracting and Composing Robust Features with Denoising Autoencoders," *Proceedings of the 25th International Conference on Machine Learning* (2008): 1096–1103.

6 Pascal Vincent et al., "Stacked Denoising Autoencoders: Learning Useful Representations in a Deep Network with a Local Denoising Criterion," *Journal of Machine Learning Research* 11 (2010): 3371–3408.

```
dropout_encoder = keras.models.Sequential([
    keras.layers.Flatten(input_shape=[28, 28]),
    keras.layers.Dropout(0.5),
    keras.layers.Dense(100, activation="selu"),
    keras.layers.Dense(30, activation="selu")
])
dropout_decoder = keras.models.Sequential([
    keras.layers.Dense(100, activation="selu", input_shape=[30]),
    keras.layers.Dense(28 * 28, activation="sigmoid"),
    keras.layers.Reshape([28, 28])
])
dropout_ae = keras.models.Sequential([dropout_encoder, dropout_decoder])
```

Figure 17-9 shows a few noisy images (with half the pixels turned off), and the images reconstructed by the dropout-based denoising autoencoder. Notice how the autoencoder guesses details that are actually not in the input, such as the top of the white shirt (bottom row, fourth image). As you can see, not only can denoising autoencoders be used for data visualization or unsupervised pretraining, like the other autoencoders we've discussed so far, but they can also be used quite simply and efficiently to remove noise from images.

Figure 17-9. Noisy images (top) and their reconstructions (bottom)

Sparse Autoencoders

Another kind of constraint that often leads to good feature extraction is *sparsity*: by adding an appropriate term to the cost function, the autoencoder is pushed to reduce the number of active neurons in the coding layer. For example, it may be pushed to have on average only 5% significantly active neurons in the coding layer. This forces the autoencoder to represent each input as a combination of a small number of activations. As a result, each neuron in the coding layer typically ends up representing a useful feature (if you could speak only a few words per month, you would probably try to make them worth listening to).

A simple approach is to use the sigmoid activation function in the coding layer (to constrain the codings to values between 0 and 1), use a large coding layer (e.g., with

300 units), and add some ℓ_1 regularization to the coding layer's activations (the decoder is just a regular decoder):

```
sparse_l1_encoder = keras.models.Sequential([
    keras.layers.Flatten(input_shape=[28, 28]),
    keras.layers.Dense(100, activation="selu"),
    keras.layers.Dense(300, activation="sigmoid"),
    keras.layers.ActivityRegularization(l1=1e-3)
])
sparse_l1_decoder = keras.models.Sequential([
    keras.layers.Dense(100, activation="selu", input_shape=[300]),
    keras.layers.Dense(28 * 28, activation="sigmoid"),
    keras.layers.Reshape([28, 28])
])
sparse_l1_ae = keras.models.Sequential([sparse_l1_encoder, sparse_l1_decoder])
```

This `ActivityRegularization` layer just returns its inputs, but as a side effect it adds a training loss equal to the sum of absolute values of its inputs (this layer only has an effect during training). Equivalently, you could remove the `ActivityRegularization` layer and set `activity_regularizer=keras.regularizers.l1(1e-3)` in the previous layer. This penalty will encourage the neural network to produce codings close to 0, but since it will also be penalized if it does not reconstruct the inputs correctly, it will have to output at least a few nonzero values. Using the ℓ_1 norm rather than the ℓ_2 norm will push the neural network to preserve the most important codings while eliminating the ones that are not needed for the input image (rather than just reducing all codings).

Another approach, which often yields better results, is to measure the actual sparsity of the coding layer at each training iteration, and penalize the model when the measured sparsity differs from a target sparsity. We do so by computing the average activation of each neuron in the coding layer, over the whole training batch. The batch size must not be too small, or else the mean will not be accurate.

Once we have the mean activation per neuron, we want to penalize the neurons that are too active, or not active enough, by adding a *sparsity loss* to the cost function. For example, if we measure that a neuron has an average activation of 0.3, but the target sparsity is 0.1, it must be penalized to activate less. One approach could be simply adding the squared error $(0.3 - 0.1)^2$ to the cost function, but in practice a better approach is to use the Kullback–Leibler (KL) divergence (briefly discussed in Chapter 4), which has much stronger gradients than the mean squared error, as you can see in Figure 17-10.

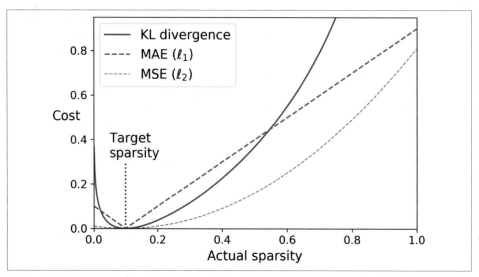

Figure 17-10. Sparsity loss

Given two discrete probability distributions P and Q, the KL divergence between these distributions, noted $D_{KL}(P \parallel Q)$, can be computed using Equation 17-1.

Equation 17-1. Kullback–Leibler divergence

$$D_{KL}(P \parallel Q) = \sum_i P(i) \log \frac{P(i)}{Q(i)}$$

In our case, we want to measure the divergence between the target probability p that a neuron in the coding layer will activate and the actual probability q (i.e., the mean activation over the training batch). So the KL divergence simplifies to Equation 17-2.

Equation 17-2. KL divergence between the target sparsity p and the actual sparsity q

$$D_{KL}(p \parallel q) = p \, \log \frac{p}{q} + (1 - p) \log \frac{1 - p}{1 - q}$$

Once we have computed the sparsity loss for each neuron in the coding layer, we sum up these losses and add the result to the cost function. In order to control the relative importance of the sparsity loss and the reconstruction loss, we can multiply the sparsity loss by a sparsity weight hyperparameter. If this weight is too high, the model will stick closely to the target sparsity, but it may not reconstruct the inputs properly, making the model useless. Conversely, if it is too low, the model will mostly ignore the sparsity objective and will not learn any interesting features.

We now have all we need to implement a sparse autoencoder based on the KL divergence. First, let's create a custom regularizer to apply KL divergence regularization:

```
K = keras.backend
kl_divergence = keras.losses.kullback_leibler_divergence

class KLDivergenceRegularizer(keras.regularizers.Regularizer):
    def __init__(self, weight, target=0.1):
        self.weight = weight
        self.target = target
    def __call__(self, inputs):
        mean_activities = K.mean(inputs, axis=0)
        return self.weight * (
            kl_divergence(self.target, mean_activities) +
            kl_divergence(1. - self.target, 1. - mean_activities))
```

Now we can build the sparse autoencoder, using the KLDivergenceRegularizer for the coding layer's activations:

```
kld_reg = KLDivergenceRegularizer(weight=0.05, target=0.1)
sparse_kl_encoder = keras.models.Sequential([
    keras.layers.Flatten(input_shape=[28, 28]),
    keras.layers.Dense(100, activation="selu"),
    keras.layers.Dense(300, activation="sigmoid", activity_regularizer=kld_reg)
])
sparse_kl_decoder = keras.models.Sequential([
    keras.layers.Dense(100, activation="selu", input_shape=[300]),
    keras.layers.Dense(28 * 28, activation="sigmoid"),
    keras.layers.Reshape([28, 28])
])
sparse_kl_ae = keras.models.Sequential([sparse_kl_encoder, sparse_kl_decoder])
```

After training this sparse autoencoder on Fashion MNIST, the activations of the neurons in the coding layer are mostly close to 0 (about 70% of all activations are lower than 0.1), and all neurons have a mean activation around 0.1 (about 90% of all neurons have a mean activation between 0.1 and 0.2), as shown in Figure 17-11.

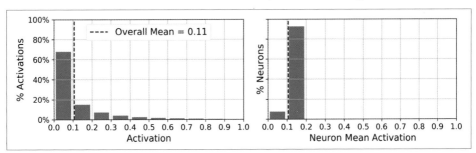

Figure 17-11. Distribution of all the activations in the coding layer (left) and distribution of the mean activation per neuron (right)

Variational Autoencoders

Another important category of autoencoders was introduced in 2013 (*https://homl.info/115*) by Diederik Kingma and Max Welling and quickly became one of the most popular types of autoencoders: *variational autoencoders*.[7]

They are quite different from all the autoencoders we have discussed so far, in these particular ways:

- They are *probabilistic autoencoders*, meaning that their outputs are partly determined by chance, even after training (as opposed to denoising autoencoders, which use randomness only during training).

- Most importantly, they are *generative autoencoders*, meaning that they can generate new instances that look like they were sampled from the training set.

Both these properties make them rather similar to RBMs, but they are easier to train, and the sampling process is much faster (with RBMs you need to wait for the network to stabilize into a "thermal equilibrium" before you can sample a new instance). Indeed, as their name suggests, variational autoencoders perform variational Bayesian inference (introduced in Chapter 9), which is an efficient way to perform approximate Bayesian inference.

Let's take a look at how they work. Figure 17-12 (left) shows a variational autoencoder. You can recognize the basic structure of all autoencoders, with an encoder followed by a decoder (in this example, they both have two hidden layers), but there is a twist: instead of directly producing a coding for a given input, the encoder produces a *mean coding* μ and a standard deviation σ. The actual coding is then sampled randomly from a Gaussian distribution with mean μ and standard deviation σ. After that the decoder decodes the sampled coding normally. The right part of the diagram shows a training instance going through this autoencoder. First, the encoder produces μ and σ, then a coding is sampled randomly (notice that it is not exactly located at μ), and finally this coding is decoded; the final output resembles the training instance.

---

7 Diederik Kingma and Max Welling, "Auto-Encoding Variational Bayes," arXiv preprint arXiv:1312.6114 (2013).

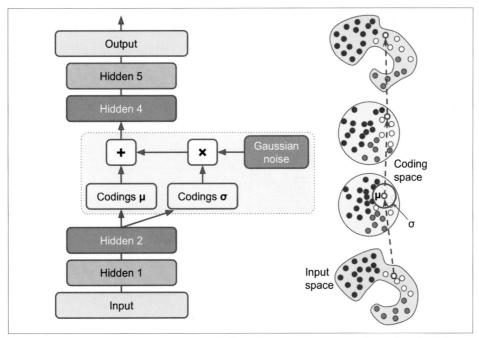

Figure 17-12. Variational autoencoder (left) and an instance going through it (right)

As you can see in the diagram, although the inputs may have a very convoluted distribution, a variational autoencoder tends to produce codings that look as though they were sampled from a simple Gaussian distribution:[8] during training, the cost function (discussed next) pushes the codings to gradually migrate within the coding space (also called the *latent space*) to end up looking like a cloud of Gaussian points. One great consequence is that after training a variational autoencoder, you can very easily generate a new instance: just sample a random coding from the Gaussian distribution, decode it, and voilà!

Now, let's look at the cost function. It is composed of two parts. The first is the usual reconstruction loss that pushes the autoencoder to reproduce its inputs (we can use cross entropy for this, as discussed earlier). The second is the *latent loss* that pushes the autoencoder to have codings that look as though they were sampled from a simple Gaussian distribution: it is the KL divergence between the target distribution (i.e., the Gaussian distribution) and the actual distribution of the codings. The math is a bit more complex than with the sparse autoencoder, in particular because of the Gaussian noise, which limits the amount of information that can be transmitted to the coding layer (thus pushing the autoencoder to learn useful features). Luckily, the

---

8 Variational autoencoders are actually more general; the codings are not limited to Gaussian distributions.

equations simplify, so the latent loss can be computed quite simply using Equation 17-3:[9]

Equation 17-3. Variational autoencoder's latent loss

$$\mathcal{L} = -\frac{1}{2} \sum_{i=1}^{K} 1 + \log\left(\sigma_i^2\right) - \sigma_i^2 - \mu_i^2$$

In this equation, $\mathcal{L}$ is the latent loss, n is the codings' dimensionality, and μ_i and σ_i are the mean and standard deviation of the ith component of the codings. The vectors $\mathbf{\mu}$ and $\mathbf{\sigma}$ (which contain all the μ_i and σ_i) are output by the encoder, as shown in Figure 17-12 (left).

A common tweak to the variational autoencoder's architecture is to make the encoder output $\mathbf{\gamma} = \log(\mathbf{\sigma}^2)$ rather than $\mathbf{\sigma}$. The latent loss can then be computed as shown in Equation 17-4. This approach is more numerically stable and speeds up training.

Equation 17-4. Variational autoencoder's latent loss, rewritten using $\mathbf{\gamma} = \log(\sigma^2)$

$$\mathcal{L} = -\frac{1}{2} \sum_{i=1}^{K} 1 + \gamma_i - \exp\left(\gamma_i\right) - \mu_i^2$$

Let's start building a variational autoencoder for Fashion MNIST (as shown in Figure 17-12, but using the $\mathbf{\gamma}$ tweak). First, we will need a custom layer to sample the codings, given $\mathbf{\mu}$ and $\mathbf{\gamma}$:

```python
class Sampling(keras.layers.Layer):
    def call(self, inputs):
        mean, log_var = inputs
        return K.random_normal(tf.shape(log_var)) * K.exp(log_var / 2) + mean
```

This `Sampling` layer takes two inputs: `mean` ($\mathbf{\mu}$) and `log_var` ($\mathbf{\gamma}$). It uses the function `K.random_normal()` to sample a random vector (of the same shape as $\mathbf{\gamma}$) from the Normal distribution, with mean 0 and standard deviation 1. Then it multiplies it by $\exp(\mathbf{\gamma} / 2)$ (which is equal to $\mathbf{\sigma}$, as you can verify), and finally it adds $\mathbf{\mu}$ and returns the result. This samples a codings vector from the Normal distribution with mean $\mathbf{\mu}$ and standard deviation $\mathbf{\sigma}$.

Next, we can create the encoder, using the Functional API because the model is not entirely sequential:

---

9 For more mathematical details, check out the original paper on variational autoencoders, or Carl Doersch's great tutorial (*https://homl.info/116*) (2016).

```
codings_size = 10

inputs = keras.layers.Input(shape=[28, 28])
z = keras.layers.Flatten()(inputs)
z = keras.layers.Dense(150, activation="selu")(z)
z = keras.layers.Dense(100, activation="selu")(z)
codings_mean = keras.layers.Dense(codings_size)(z)   # μ
codings_log_var = keras.layers.Dense(codings_size)(z)   # γ
codings = Sampling()([codings_mean, codings_log_var])
variational_encoder = keras.Model(
    inputs=[inputs], outputs=[codings_mean, codings_log_var, codings])
```

Note that the Dense layers that output codings_mean (**μ**) and codings_log_var (**γ**) have the same inputs (i.e., the outputs of the second Dense layer). We then pass both codings_mean and codings_log_var to the Sampling layer. Finally, the varia tional_encoder model has three outputs, in case you want to inspect the values of codings_mean and codings_log_var. The only output we will use is the last one (cod ings). Now let's build the decoder:

```
decoder_inputs = keras.layers.Input(shape=[codings_size])
x = keras.layers.Dense(100, activation="selu")(decoder_inputs)
x = keras.layers.Dense(150, activation="selu")(x)
x = keras.layers.Dense(28 * 28, activation="sigmoid")(x)
outputs = keras.layers.Reshape([28, 28])(x)
variational_decoder = keras.Model(inputs=[decoder_inputs], outputs=[outputs])
```

For this decoder, we could have used the Sequential API instead of the Functional API, since it is really just a simple stack of layers, virtually identical to many of the decoders we have built so far. Finally, let's build the variational autoencoder model:

```
_, _, codings = variational_encoder(inputs)
reconstructions = variational_decoder(codings)
variational_ae = keras.Model(inputs=[inputs], outputs=[reconstructions])
```

Note that we ignore the first two outputs of the encoder (we only want to feed the codings to the decoder). Lastly, we must add the latent loss and the reconstruction loss:

```
latent_loss = -0.5 * K.sum(
    1 + codings_log_var - K.exp(codings_log_var) - K.square(codings_mean),
    axis=-1)
variational_ae.add_loss(K.mean(latent_loss) / 784.)
variational_ae.compile(loss="binary_crossentropy", optimizer="rmsprop")
```

We first apply Equation 17-4 to compute the latent loss for each instance in the batch (we sum over the last axis). Then we compute the mean loss over all the instances in the batch, and we divide the result by 784 to ensure it has the appropriate scale com pared to the reconstruction loss. Indeed, the variational autoencoder's reconstruction loss is supposed to be the sum of the pixel reconstruction errors, but when Keras computes the "binary_crossentropy" loss, it computes the mean over all 784 pixels, rather than the sum. So, the reconstruction loss is 784 times smaller than we need it to be. We could define a custom loss to compute the sum rather than the mean, but it is simpler to divide the latent loss by 784 (the final loss will be 784 times smaller than it should be, but this just means that we should use a larger learning rate).

Note that we use the `RMSprop` optimizer, which works well in this case. And finally we can train the autoencoder!

```
history = variational_ae.fit(X_train, X_train, epochs=50, batch_size=128,
                             validation_data=[X_valid, X_valid])
```

Generating Fashion MNIST Images

Now let's use this variational autoencoder to generate images that look like fashion items. All we need to do is sample random codings from a Gaussian distribution and decode them:

```
codings = tf.random.normal(shape=[12, codings_size])
images = variational_decoder(codings).numpy()
```

Figure 17-13 shows the 12 generated images.

Figure 17-13. Fashion MNIST images generated by the variational autoencoder

The majority of these images look fairly convincing, if a bit too fuzzy. The rest are not great, but don't be too harsh on the autoencoder—it only had a few minutes to learn! Give it a bit more fine-tuning and training time, and those images should look better.

Variational autoencoders make it possible to perform *semantic interpolation*: instead of interpolating two images at the pixel level (which would look as if the two images were overlaid), we can interpolate at the codings level. We first run both images through the encoder, then we interpolate the two codings we get, and finally we decode the interpolated codings to get the final image. It will look like a regular Fashion MNIST image, but it will be an intermediate between the original images. In the following code example, we take the 12 codings we just generated, we organize them in a 3 × 4 grid, and we use TensorFlow's `tf.image.resize()` function to resize this grid to 5 × 7. By default, the `resize()` function will perform bilinear interpolation, so every other row and column will contain interpolated codings. We then use the decoder to produce all the images:

```
codings_grid = tf.reshape(codings, [1, 3, 4, codings_size])
larger_grid = tf.image.resize(codings_grid, size=[5, 7])
interpolated_codings = tf.reshape(larger_grid, [-1, codings_size])
images = variational_decoder(interpolated_codings).numpy()
```

Figure 17-14 shows the resulting images. The original images are framed, and the rest are the result of semantic interpolation between the nearby images. Notice, for example, how the shoe in the fourth row and fifth column is a nice interpolation between the two shoes located above and below it.

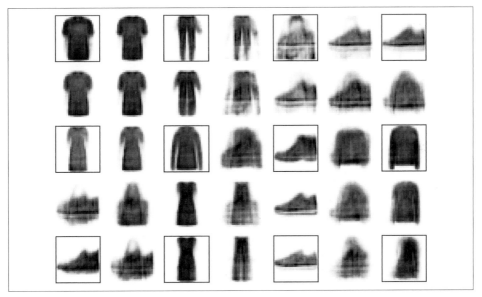

Figure 17-14. Semantic interpolation

For several years, variational autoencoders were quite popular, but GANs eventually took the lead, in particular because they are capable of generating much more realistic and crisp images. So let's turn our attention to GANs.

Generative Adversarial Networks

Generative adversarial networks were proposed in a 2014 paper (*https://homl.info/gan*)[10] by Ian Goodfellow et al., and although the idea got researchers excited almost instantly, it took a few years to overcome some of the difficulties of training GANs. Like many great ideas, it seems simple in hindsight: make neural networks compete against each other in the hope that this competition will push them to excel. As shown in Figure 17-15, a GAN is composed of two neural networks:

Generator

> Takes a random distribution as input (typically Gaussian) and outputs some data —typically, an image. You can think of the random inputs as the latent representations (i.e., codings) of the image to be generated. So, as you can see, the generator offers the same functionality as a decoder in a variational autoencoder, and it can be used in the same way to generate new images (just feed it some Gaussian noise, and it outputs a brand-new image). However, it is trained very differently, as we will soon see.

Discriminator

> Takes either a fake image from the generator or a real image from the training set as input, and must guess whether the input image is fake or real.

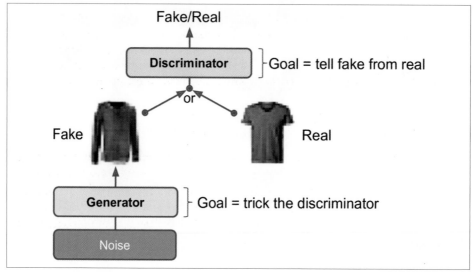

Figure 17-15. A generative adversarial network

---

10 Ian Goodfellow et al., "Generative Adversarial Nets," *Proceedings of the 27th International Conference on Neural Information Processing Systems* 2 (2014): 2672–2680.

During training, the generator and the discriminator have opposite goals: the discriminator tries to tell fake images from real images, while the generator tries to produce images that look real enough to trick the discriminator. Because the GAN is composed of two networks with different objectives, it cannot be trained like a regular neural network. Each training iteration is divided into two phases:

- In the first phase, we train the discriminator. A batch of real images is sampled from the training set and is completed with an equal number of fake images produced by the generator. The labels are set to 0 for fake images and 1 for real images, and the discriminator is trained on this labeled batch for one step, using the binary cross-entropy loss. Importantly, backpropagation only optimizes the weights of the discriminator during this phase.

- In the second phase, we train the generator. We first use it to produce another batch of fake images, and once again the discriminator is used to tell whether the images are fake or real. This time we do not add real images in the batch, and all the labels are set to 1 (real): in other words, we want the generator to produce images that the discriminator will (wrongly) believe to be real! Crucially, the weights of the discriminator are frozen during this step, so backpropagation only affects the weights of the generator.

 The generator never actually sees any real images, yet it gradually learns to produce convincing fake images! All it gets is the gradients flowing back through the discriminator. Fortunately, the better the discriminator gets, the more information about the real images is contained in these secondhand gradients, so the generator can make significant progress.

Let's go ahead and build a simple GAN for Fashion MNIST.

First, we need to build the generator and the discriminator. The generator is similar to an autoencoder's decoder, and the discriminator is a regular binary classifier (it takes an image as input and ends with a Dense layer containing a single unit and using the sigmoid activation function). For the second phase of each training iteration, we also need the full GAN model containing the generator followed by the discriminator:

```
codings_size = 30

generator = keras.models.Sequential([
    keras.layers.Dense(100, activation="selu", input_shape=[codings_size]),
    keras.layers.Dense(150, activation="selu"),
    keras.layers.Dense(28 * 28, activation="sigmoid"),
    keras.layers.Reshape([28, 28])
])
```

```
discriminator = keras.models.Sequential([
    keras.layers.Flatten(input_shape=[28, 28]),
    keras.layers.Dense(150, activation="selu"),
    keras.layers.Dense(100, activation="selu"),
    keras.layers.Dense(1, activation="sigmoid")
])
gan = keras.models.Sequential([generator, discriminator])
```

Next, we need to compile these models. As the discriminator is a binary classifier, we can naturally use the binary cross-entropy loss. The generator will only be trained through the gan model, so we do not need to compile it at all. The gan model is also a binary classifier, so it can use the binary cross-entropy loss. Importantly, the discriminator should not be trained during the second phase, so we make it non-trainable before compiling the gan model:

```
discriminator.compile(loss="binary_crossentropy", optimizer="rmsprop")
discriminator.trainable = False
gan.compile(loss="binary_crossentropy", optimizer="rmsprop")
```

> The trainable attribute is taken into account by Keras only when compiling a model, so after running this code, the discriminator *is* trainable if we call its fit() method or its train_on_batch() method (which we will be using), while it is *not* trainable when we call these methods on the gan model.

Since the training loop is unusual, we cannot use the regular fit() method. Instead, we will write a custom training loop. For this, we first need to create a Dataset to iterate through the images:

```
batch_size = 32
dataset = tf.data.Dataset.from_tensor_slices(X_train).shuffle(1000)
dataset = dataset.batch(batch_size, drop_remainder=True).prefetch(1)
```

We are now ready to write the training loop. Let's wrap it in a train_gan() function:

```
def train_gan(gan, dataset, batch_size, codings_size, n_epochs=50):
    generator, discriminator = gan.layers
    for epoch in range(n_epochs):
        for X_batch in dataset:
            # phase 1 - training the discriminator
            noise = tf.random.normal(shape=[batch_size, codings_size])
            generated_images = generator(noise)
            X_fake_and_real = tf.concat([generated_images, X_batch], axis=0)
            y1 = tf.constant([[0.]] * batch_size + [[1.]] * batch_size)
            discriminator.trainable = True
            discriminator.train_on_batch(X_fake_and_real, y1)
            # phase 2 - training the generator
            noise = tf.random.normal(shape=[batch_size, codings_size])
            y2 = tf.constant([[1.]] * batch_size)
            discriminator.trainable = False
            gan.train_on_batch(noise, y2)

train_gan(gan, dataset, batch_size, codings_size)
```

As discussed earlier, you can see the two phases at each iteration:

- In phase one we feed Gaussian noise to the generator to produce fake images, and we complete this batch by concatenating an equal number of real images. The targets y1 are set to 0 for fake images and 1 for real images. Then we train the discriminator on this batch. Note that we set the discriminator's trainable attribute to True: this is only to get rid of a warning that Keras displays when it notices that trainable is now False but was True when the model was compiled (or vice versa).

- In phase two, we feed the GAN some Gaussian noise. Its generator will start by producing fake images, then the discriminator will try to guess whether these images are fake or real. We want the discriminator to believe that the fake images are real, so the targets y2 are set to 1. Note that we set the trainable attribute to False, once again to avoid a warning.

That's it! If you display the generated images (see Figure 17-16), you will see that at the end of the first epoch, they already start to look like (very noisy) Fashion MNIST images.

Unfortunately, the images never really get much better than that, and you may even find epochs where the GAN seems to be forgetting what it learned. Why is that? Well, it turns out that training a GAN can be challenging. Let's see why.

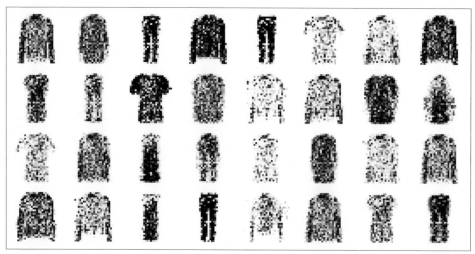

Figure 17-16. Images generated by the GAN after one epoch of training

The Difficulties of Training GANs

During training, the generator and the discriminator constantly try to outsmart each other, in a zero-sum game. As training advances, the game may end up in a state that game theorists call a *Nash equilibrium*, named after the mathematician John Nash: this is when no player would be better off changing their own strategy, assuming the other players do not change theirs. For example, a Nash equilibrium is reached when everyone drives on the left side of the road: no driver would be better off being the only one to switch sides. Of course, there is a second possible Nash equilibrium: when everyone drives on the *right* side of the road. Different initial states and dynamics may lead to one equilibrium or the other. In this example, there is a single optimal strategy once an equilibrium is reached (i.e., driving on the same side as everyone else), but a Nash equilibrium can involve multiple competing strategies (e.g., a predator chases its prey, the prey tries to escape, and neither would be better off changing their strategy).

So how does this apply to GANs? Well, the authors of the paper demonstrated that a GAN can only reach a single Nash equilibrium: that's when the generator produces perfectly realistic images, and the discriminator is forced to guess (50% real, 50% fake). This fact is very encouraging: it would seem that you just need to train the GAN for long enough, and it will eventually reach this equilibrium, giving you a perfect generator. Unfortunately, it's not that simple: nothing guarantees that the equilibrium will ever be reached.

The biggest difficulty is called *mode collapse*: this is when the generator's outputs gradually become less diverse. How can this happen? Suppose that the generator gets better at producing convincing shoes than any other class. It will fool the discriminator a bit more with shoes, and this will encourage it to produce even more images of shoes. Gradually, it will forget how to produce anything else. Meanwhile, the only fake images that the discriminator will see will be shoes, so it will also forget how to discriminate fake images of other classes. Eventually, when the discriminator manages to discriminate the fake shoes from the real ones, the generator will be forced to move to another class. It may then become good at shirts, forgetting about shoes, and the discriminator will follow. The GAN may gradually cycle across a few classes, never really becoming very good at any of them.

Moreover, because the generator and the discriminator are constantly pushing against each other, their parameters may end up oscillating and becoming unstable. Training may begin properly, then suddenly diverge for no apparent reason, due to these instabilities. And since many factors affect these complex dynamics, GANs are very sensitive to the hyperparameters: you may have to spend a lot of effort fine-tuning them.

These problems have kept researchers very busy since 2014: many papers were published on this topic, some proposing new cost functions[11] (though a 2018 paper (*https://homl.info/gansequal*)[12] by Google researchers questions their efficiency) or techniques to stabilize training or to avoid the mode collapse issue. For example, a popular technique called *experience replay* consists in storing the images produced by the generator at each iteration in a replay buffer (gradually dropping older generated images) and training the discriminator using real images plus fake images drawn from this buffer (rather than just fake images produced by the current generator). This reduces the chances that the discriminator will overfit the latest generator's outputs. Another common technique is called *mini-batch discrimination*: it measures how similar images are across the batch and provides this statistic to the discriminator, so it can easily reject a whole batch of fake images that lack diversity. This encourages the generator to produce a greater variety of images, reducing the chance of mode collapse. Other papers simply propose specific architectures that happen to perform well.

In short, this is still a very active field of research, and the dynamics of GANs are still not perfectly understood. But the good news is that great progress has been made, and some of the results are truly astounding! So let's look at some of the most successful architectures, starting with deep convolutional GANs, which were the state of the

---

11 For a nice comparison of the main GAN losses, check out this great GitHub project by Hwalsuk Lee (*https://homl.info/ganloss*).

12 Mario Lucic et al., "Are GANs Created Equal? A Large-Scale Study," *Proceedings of the 32nd International Conference on Neural Information Processing Systems* (2018): 698–707.

art just a few years ago. Then we will look at two more recent (and more complex) architectures.

Deep Convolutional GANs

The original GAN paper in 2014 experimented with convolutional layers, but only tried to generate small images. Soon after, many researchers tried to build GANs based on deeper convolutional nets for larger images. This proved to be tricky, as training was very unstable, but Alec Radford et al. finally succeeded in late 2015, after experimenting with many different architectures and hyperparameters. They called their architecture *deep convolutional GANs* (*https://homl.info/dcgan*) (DCGANs).[13] Here are the main guidelines they proposed for building stable convolutional GANs:

- Replace any pooling layers with strided convolutions (in the discriminator) and transposed convolutions (in the generator).
- Use Batch Normalization in both the generator and the discriminator, except in the generator's output layer and the discriminator's input layer.
- Remove fully connected hidden layers for deeper architectures.
- Use ReLU activation in the generator for all layers except the output layer, which should use tanh.
- Use leaky ReLU activation in the discriminator for all layers.

These guidelines will work in many cases, but not always, so you may still need to experiment with different hyperparameters (in fact, just changing the random seed and training the same model again will sometimes work). For example, here is a small DCGAN that works reasonably well with Fashion MNIST:

---

13 Alec Radford et al., "Unsupervised Representation Learning with Deep Convolutional Generative Adversarial Networks," arXiv preprint arXiv:1511.06434 (2015).

```
codings_size = 100

generator = keras.models.Sequential([
    keras.layers.Dense(7 * 7 * 128, input_shape=[codings_size]),
    keras.layers.Reshape([7, 7, 128]),
    keras.layers.BatchNormalization(),
    keras.layers.Conv2DTranspose(64, kernel_size=5, strides=2, padding="same",
                                 activation="selu"),
    keras.layers.BatchNormalization(),
    keras.layers.Conv2DTranspose(1, kernel_size=5, strides=2, padding="same",
                                 activation="tanh")
])
discriminator = keras.models.Sequential([
    keras.layers.Conv2D(64, kernel_size=5, strides=2, padding="same",
                        activation=keras.layers.LeakyReLU(0.2),
                        input_shape=[28, 28, 1]),
    keras.layers.Dropout(0.4),
    keras.layers.Conv2D(128, kernel_size=5, strides=2, padding="same",
                        activation=keras.layers.LeakyReLU(0.2)),
    keras.layers.Dropout(0.4),
    keras.layers.Flatten(),
    keras.layers.Dense(1, activation="sigmoid")
])
gan = keras.models.Sequential([generator, discriminator])
```

The generator takes codings of size 100, and it projects them to 6272 dimensions (7 *
7 * 128), and reshapes the result to get a $7 \times 7 \times 128$ tensor. This tensor is batch normalized and fed to a transposed convolutional layer with a stride of 2, which upsamples it from 7×7 to 14×14 and reduces its depth from 128 to 64. The result is batch
normalized again and fed to another transposed convolutional layer with a stride of 2,
which upsamples it from 14×14 to 28×28 and reduces the depth from 64 to 1. This
layer uses the tanh activation function, so the outputs will range from –1 to 1. For this
reason, before training the GAN, we need to rescale the training set to that same
range. We also need to reshape it to add the channel dimension:

```
X_train = X_train.reshape(-1, 28, 28, 1) * 2. - 1. # reshape and rescale
```

The discriminator looks much like a regular CNN for binary classification, except
instead of using max pooling layers to downsample the image, we use strided convolutions (strides=2). Also note that we use the leaky ReLU activation function.

Overall, we respected the DCGAN guidelines, except we replaced the BatchNormali
zation layers in the discriminator with Dropout layers (otherwise training was unstable in this case) and we replaced ReLU with SELU in the generator. Feel free to tweak
this architecture: you will see how sensitive it is to the hyperparameters (especially
the relative learning rates of the two networks).

Lastly, to build the dataset, then compile and train this model, we use the exact same
code as earlier. After 50 epochs of training, the generator produces images like those

shown in Figure 17-17. It's still not perfect, but many of these images are pretty convincing.

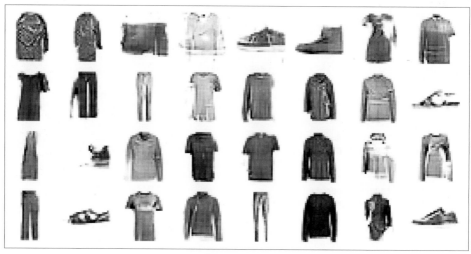

Figure 17-17. Images generated by the DCGAN after 50 epochs of training

If you scale up this architecture and train it on a large dataset of faces, you can get fairly realistic images. In fact, DCGANs can learn quite meaningful latent representations, as you can see in Figure 17-18: many images were generated, and nine of them were picked manually (top left), including three representing men with glasses, three men without glasses, and three women without glasses. For each of these categories, the codings that were used to generate the images were averaged, and an image was generated based on the resulting mean codings (lower left). In short, each of the three lower-left images represents the mean of the three images located above it. But this is not a simple mean computed at the pixel level (this would result in three overlapping faces), it is a mean computed in the latent space, so the images still look like normal faces. Amazingly, if you compute men with glasses, minus men without glasses, plus women without glasses—where each term corresponds to one of the mean codings—and you generate the image that corresponds to this coding, you get the image at the center of the 3 × 3 grid of faces on the right: a woman with glasses! The eight other images around it were generated based on the same vector plus a bit of noise, to illustrate the semantic interpolation capabilities of DCGANs. Being able to do arithmetic on faces feels like science fiction!

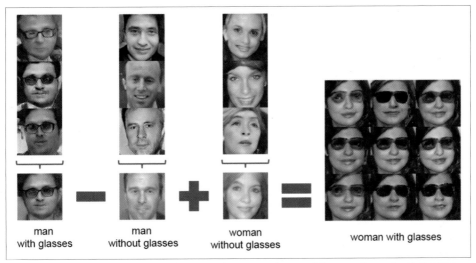

| man
with glasses | man
without glasses | woman
without glasses | woman with glasses |

Figure 17-18. Vector arithmetic for visual concepts (part of figure 7 from the DCGAN paper)[14]

 If you add each image's class as an extra input to both the generator and the discriminator, they will both learn what each class looks like, and thus you will be able to control the class of each image produced by the generator. This is called a *conditional GAN* (*https://homl.info/cgan*)[15] (CGAN).

DCGANs aren't perfect, though. For example, when you try to generate very large images using DCGANs, you often end up with locally convincing features but overall inconsistencies (such as shirts with one sleeve much longer than the other). How can you fix this?

Progressive Growing of GANs

An important technique was proposed in a 2018 paper (*https://homl.info/progan*)[16] by Nvidia researchers Tero Karras et al.: they suggested generating small images at the beginning of training, then gradually adding convolutional layers to both the generator and the discriminator to produce larger and larger images (4×4, 8×8, 16×16, …, 512×512, $1,024 \times 1,024$). This approach resembles greedy layer-wise training of

14 Reproduced with the kind authorization of the authors.

15 Mehdi Mirza and Simon Osindero, "Conditional Generative Adversarial Nets," arXiv preprint arXiv: 1411.1784 (2014).

16 Tero Karras et al., "Progressive Growing of GANs for Improved Quality, Stability, and Variation," *Proceedings of the International Conference on Learning Representations* (2018).

stacked autoencoders. The extra layers get added at the end of the generator and at the beginning of the discriminator, and previously trained layers remain trainable.

For example, when growing the generator's outputs from 4 × 4 to 8 × 8 (see Figure 17-19), an upsampling layer (using nearest neighbor filtering) is added to the existing convolutional layer, so it outputs 8 × 8 feature maps, which are then fed to the new convolutional layer (which uses "same" padding and strides of 1, so its outputs are also 8 × 8). This new layer is followed by a new output convolutional layer: this is a regular convolutional layer with kernel size 1 that projects the outputs down to the desired number of color channels (e.g., 3). To avoid breaking the trained weights of the first convolutional layer when the new convolutional layer is added, the final output is a weighted sum of the original output layer (which now outputs 8 × 8 feature maps) and the new output layer. The weight of the new outputs is α, while the weight of the original outputs is $1 - \alpha$, and α is slowly increased from 0 to 1. In other words, the new convolutional layers (represented with dashed lines in Figure 17-19) are gradually faded in, while the original output layer is gradually faded out. A similar fade-in/fade-out technique is used when a new convolutional layer is added to the discriminator (followed by an average pooling layer for downsampling).

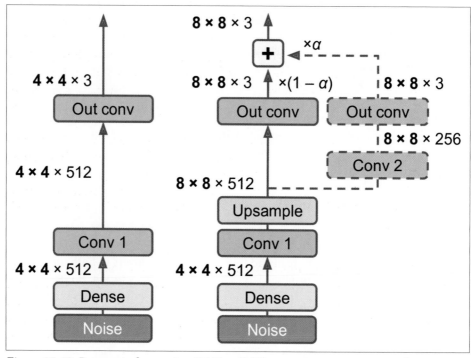

Figure 17-19. Progressively growing GAN: a GAN generator outputs 4 × 4 color images (left); we extend it to output 8 × 8 images (right)

The paper also introduced several other techniques aimed at increasing the diversity of the outputs (to avoid mode collapse) and making training more stable:

Minibatch standard deviation layer

Added near the end of the discriminator. For each position in the inputs, it computes the standard deviation across all channels and all instances in the batch (S = tf.math.reduce_std(inputs, axis=[0, -1])). These standard deviations are then averaged across all points to get a single value (v = tf.reduce_mean(S)). Finally, an extra feature map is added to each instance in the batch and filled with the computed value (tf.concat([inputs, tf.fill([batch_size, height, width, 1], v)], axis=-1)). How does this help? Well, if the generator produces images with little variety, then there will be a small standard deviation across feature maps in the discriminator. Thanks to this layer, the discriminator will have easy access to this statistic, making it less likely to be fooled by a generator that produces too little diversity. This will encourage the generator to produce more diverse outputs, reducing the risk of mode collapse.

Equalized learning rate

Initializes all weights using a simple Gaussian distribution with mean 0 and standard deviation 1 rather than using He initialization. However, the weights are scaled down at runtime (i.e., every time the layer is executed) by the same factor as in He initialization: they are divided by $\sqrt{2/n_{\text{inputs}}}$, where n_{inputs} is the number of inputs to the layer. The paper demonstrated that this technique significantly improved the GAN's performance when using RMSProp, Adam, or other adaptive gradient optimizers. Indeed, these optimizers normalize the gradient updates by their estimated standard deviation (see Chapter 11), so parameters that have a larger dynamic range[17] will take longer to train, while parameters with a small dynamic range may be updated too quickly, leading to instabilities. By rescaling the weights as part of the model itself rather than just rescaling them upon initialization, this approach ensures that the dynamic range is the same for all parameters, throughout training, so they all learn at the same speed. This both speeds up and stabilizes training.

Pixelwise normalization layer

Added after each convolutional layer in the generator. It normalizes each activation based on all the activations in the same image and at the same location, but across all channels (dividing by the square root of the mean squared activation). In TensorFlow code, this is inputs / tf.sqrt(tf.reduce_mean(tf.square(X), axis=-1, keepdims=True) + 1e-8) (the smoothing term 1e-8 is needed to

---

17 The dynamic range of a variable is the ratio between the highest and the lowest value it may take.

avoid division by zero). This technique avoids explosions in the activations due to excessive competition between the generator and the discriminator.

The combination of all these techniques allowed the authors to generate extremely convincing high-definition images of faces (*https://homl.info/progandemo*). But what exactly do we call "convincing"? Evaluation is one of the big challenges when working with GANs: although it is possible to automatically evaluate the diversity of the generated images, judging their quality is a much trickier and subjective task. One technique is to use human raters, but this is costly and time-consuming. So the authors proposed to measure the similarity between the local image structure of the generated images and the training images, considering every scale. This idea led them to another groundbreaking innovation: StyleGANs.

StyleGANs

The state of the art in high-resolution image generation was advanced once again by the same Nvidia team in a 2018 paper (*https://homl.info/stylegan*)[18] that introduced the popular StyleGAN architecture. The authors used *style transfer* techniques in the generator to ensure that the generated images have the same local structure as the training images, at every scale, greatly improving the quality of the generated images. The discriminator and the loss function were not modified, only the generator. Let's take a look at the StyleGAN. It is composed of two networks (see Figure 17-20):

Mapping network
> An eight-layer MLP that maps the latent representations **z** (i.e., the codings) to a vector **w**. This vector is then sent through multiple *affine transformations* (i.e., Dense layers with no activation functions, represented by the "A" boxes in Figure 17-20), which produces multiple vectors. These vectors control the style of the generated image at different levels, from fine-grained texture (e.g., hair color) to high-level features (e.g., adult or child). In short, the mapping network maps the codings to multiple style vectors.

Synthesis network
> Responsible for generating the images. It has a constant learned input (to be clear, this input will be constant *after* training, but *during* training it keeps getting tweaked by backpropagation). It processes this input through multiple convolutional and upsampling layers, as earlier, but there are two twists: first, some noise is added to the input and to all the outputs of the convolutional layers (before the activation function). Second, each noise layer is followed by an *Adaptive Instance Normalization* (AdaIN) layer: it standardizes each feature map independently (by

---

18 Tero Karras et al., "A Style-Based Generator Architecture for Generative Adversarial Networks," arXiv preprint arXiv:1812.04948 (2018).

subtracting the feature map's mean and dividing by its standard deviation), then it uses the style vector to determine the scale and offset of each feature map (the style vector contains one scale and one bias term for each feature map).

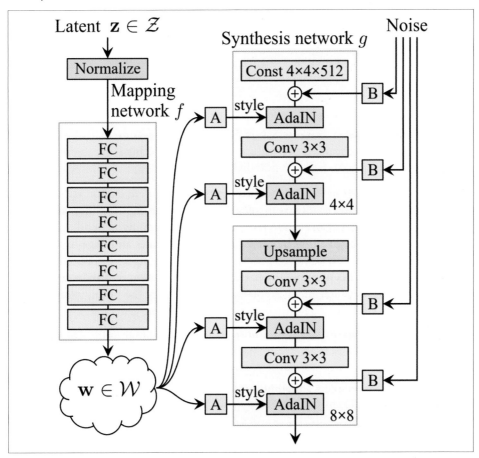

Figure 17-20. StyleGAN's generator architecture (part of figure 1 from the StyleGAN paper)[19]

The idea of adding noise independently from the codings is very important. Some parts of an image are quite random, such as the exact position of each freckle or hair. In earlier GANs, this randomness had to either come from the codings or be some pseudorandom noise produced by the generator itself. If it came from the codings, it meant that the generator had to dedicate a significant portion of the codings' representational power to store noise: this is quite wasteful. Moreover, the noise had to be

---

19 Reproduced with the kind authorization of the authors.

able to flow through the network and reach the final layers of the generator: this seems like an unnecessary constraint that probably slowed down training. And finally, some visual artifacts may appear because the same noise was used at different levels. If instead the generator tried to produce its own pseudorandom noise, this noise might not look very convincing, leading to more visual artifacts. Plus, part of the generator's weights would be dedicated to generating pseudorandom noise, which again seems wasteful. By adding extra noise inputs, all these issues are avoided; the GAN is able to use the provided noise to add the right amount of stochasticity to each part of the image.

The added noise is different for each level. Each noise input consists of a single feature map full of Gaussian noise, which is broadcast to all feature maps (of the given level) and scaled using learned per-feature scaling factors (this is represented by the "B" boxes in Figure 17-20) before it is added.

Finally, StyleGAN uses a technique called *mixing regularization* (or *style mixing*), where a percentage of the generated images are produced using two different codings. Specifically, the codings c_1 and c_2 are sent through the mapping network, giving two style vectors w_1 and w_2. Then the synthesis network generates an image based on the styles w_1 for the first levels and the styles w_2 for the remaining levels. The cutoff level is picked randomly. This prevents the network from assuming that styles at adjacent levels are correlated, which in turn encourages locality in the GAN, meaning that each style vector only affects a limited number of traits in the generated image.

There is such a wide variety of GANs out there that it would require a whole book to cover them all. Hopefully this introduction has given you the main ideas, and most importantly the desire to learn more. If you're struggling with a mathematical concept, there are probably blog posts out there that will help you understand it better. Then go ahead and implement your own GAN, and do not get discouraged if it has trouble learning at first: unfortunately, this is normal, and it will require quite a bit of patience before it works, but the result is worth it. If you're struggling with an implementation detail, there are plenty of Keras or TensorFlow implementations that you can look at. In fact, if all you want is to get some amazing results quickly, then you can just use a pretrained model (e.g., there are pretrained StyleGAN models available for Keras).

In the next chapter we will move to an entirely different branch of Deep Learning: Deep Reinforcement Learning.

Exercises

1. What are the main tasks that autoencoders are used for?

2. Suppose you want to train a classifier, and you have plenty of unlabeled training data but only a few thousand labeled instances. How can autoencoders help? How would you proceed?

3. If an autoencoder perfectly reconstructs the inputs, is it necessarily a good autoencoder? How can you evaluate the performance of an autoencoder?

4. What are undercomplete and overcomplete autoencoders? What is the main risk of an excessively undercomplete autoencoder? What about the main risk of an overcomplete autoencoder?

5. How do you tie weights in a stacked autoencoder? What is the point of doing so?

6. What is a generative model? Can you name a type of generative autoencoder?

7. What is a GAN? Can you name a few tasks where GANs can shine?

8. What are the main difficulties when training GANs?

9. Try using a denoising autoencoder to pretrain an image classifier. You can use MNIST (the simplest option), or a more complex image dataset such as CIFAR10 (*https://homl.info/122*) if you want a bigger challenge. Regardless of the dataset you're using, follow these steps:

   - Split the dataset into a training set and a test set. Train a deep denoising autoencoder on the full training set.

   - Check that the images are fairly well reconstructed. Visualize the images that most activate each neuron in the coding layer.

   - Build a classification DNN, reusing the lower layers of the autoencoder. Train it using only 500 images from the training set. Does it perform better with or without pretraining?

10. Train a variational autoencoder on the image dataset of your choice, and use it to generate images. Alternatively, you can try to find an unlabeled dataset that you are interested in and see if you can generate new samples.

11. Train a DCGAN to tackle the image dataset of your choice, and use it to generate images. Add experience replay and see if this helps. Turn it into a conditional GAN where you can control the generated class.

Solutions to these exercises are available in Appendix A.

## CHAPTER 18

Reinforcement Learning

Reinforcement Learning (RL) is one of the most exciting fields of Machine Learning today, and also one of the oldest. It has been around since the 1950s, producing many interesting applications over the years,[1] particularly in games (e.g., *TD-Gammon*, a Backgammon-playing program) and in machine control, but seldom making the headline news. But a revolution took place in 2013, when researchers from a British startup called DeepMind demonstrated a system that could learn to play just about any Atari game from scratch (*https://homl.info/dqn*),[2] eventually outperforming humans (*https://homl.info/dqn2*)[3] in most of them, using only raw pixels as inputs and without any prior knowledge of the rules of the games.[4] This was the first of a series of amazing feats, culminating in March 2016 with the victory of their system AlphaGo against Lee Sedol, a legendary professional player of the game of Go, and in May 2017 against Ke Jie, the world champion. No program had ever come close to beating a master of this game, let alone the world champion. Today the whole field of RL is boiling with new ideas, with a wide range of applications. DeepMind was bought by Google for over $500 million in 2014.

So how did DeepMind achieve all this? With hindsight it seems rather simple: they applied the power of Deep Learning to the field of Reinforcement Learning, and it worked beyond their wildest dreams. In this chapter we will first explain what

---

1 For more details, be sure to check out Richard Sutton and Andrew Barto's book on RL, *Reinforcement Learning: An Introduction* (*https://homl.info/126*)(MIT Press).

2 Volodymyr Mnih et al., "Playing Atari with Deep Reinforcement Learning," arXiv preprint arXiv:1312.5602 (2013).

3 Volodymyr Mnih et al., "Human-Level Control Through Deep Reinforcement Learning," *Nature* 518 (2015): 529–533.

4 Check out the videos of DeepMind's system learning to play *Space Invaders*, *Breakout*, and other video games at *https://homl.info/dqn3*.

Reinforcement Learning is and what it's good at, then present two of the most important techniques in Deep Reinforcement Learning: *policy gradients* and *deep Q-networks* (DQNs), including a discussion of *Markov decision processes* (MDPs). We will use these techniques to train models to balance a pole on a moving cart; then I'll introduce the TF-Agents library, which uses state-of-the-art algorithms that greatly simplify building powerful RL systems, and we will use the library to train an agent to play *Breakout*, the famous Atari game. I'll close the chapter by taking a look at some of the latest advances in the field.

Learning to Optimize Rewards

In Reinforcement Learning, a software *agent* makes *observations* and takes *actions* within an *environment*, and in return it receives *rewards*. Its objective is to learn to act in a way that will maximize its expected rewards over time. If you don't mind a bit of anthropomorphism, you can think of positive rewards as pleasure, and negative rewards as pain (the term "reward" is a bit misleading in this case). In short, the agent acts in the environment and learns by trial and error to maximize its pleasure and minimize its pain.

This is quite a broad setting, which can apply to a wide variety of tasks. Here are a few examples (see Figure 18-1):

a. The agent can be the program controlling a robot. In this case, the environment is the real world, the agent observes the environment through a set of *sensors* such as cameras and touch sensors, and its actions consist of sending signals to activate motors. It may be programmed to get positive rewards whenever it approaches the target destination, and negative rewards whenever it wastes time or goes in the wrong direction.

b. The agent can be the program controlling *Ms. Pac-Man*. In this case, the environment is a simulation of the Atari game, the actions are the nine possible joystick positions (upper left, down, center, and so on), the observations are screenshots, and the rewards are just the game points.

c. Similarly, the agent can be the program playing a board game such as Go.

d. The agent does not have to control a physically (or virtually) moving thing. For example, it can be a smart thermostat, getting positive rewards whenever it is close to the target temperature and saves energy, and negative rewards when humans need to tweak the temperature, so the agent must learn to anticipate human needs.

e. The agent can observe stock market prices and decide how much to buy or sell every second. Rewards are obviously the monetary gains and losses.

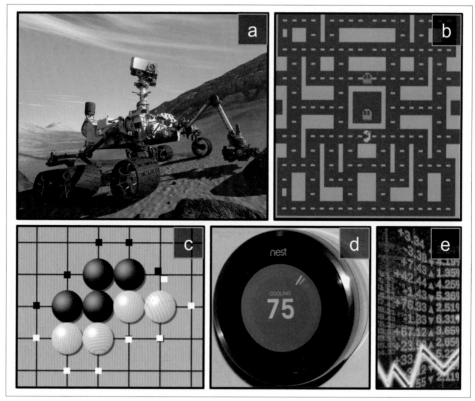

Figure 18-1. Reinforcement Learning examples: (a) robotics, (b) Ms. Pac-Man, (c) Go player, (d) thermostat, (e) automatic trader[5]

Note that there may not be any positive rewards at all; for example, the agent may move around in a maze, getting a negative reward at every time step, so it had better find the exit as quickly as possible! There are many other examples of tasks to which Reinforcement Learning is well suited, such as self-driving cars, recommender systems, placing ads on a web page, or controlling where an image classification system should focus its attention.

---

5 Image (a) is from NASA (public domain). (b) is a screenshot from the *Ms. Pac-Man* game, copyright Atari (fair use in this chapter). Images (c) and (d) are reproduced from Wikipedia. (c) was created by user Stevertigo and released under Creative Commons BY-SA 2.0 (*https://creativecommons.org/licenses/by-sa/2.0/*). (d) is in the public domain. (e) was reproduced from Pixabay, released under Creative Commons CC0 (*https://creativecommons.org/publicdomain/zero/1.0/*).

Policy Search

The algorithm a software agent uses to determine its actions is called its *policy*. The policy could be a neural network taking observations as inputs and outputting the action to take (see Figure 18-2).

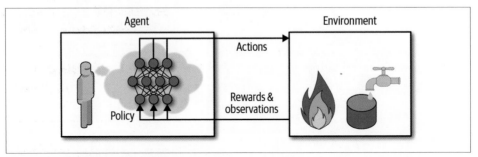

Figure 18-2. Reinforcement Learning using a neural network policy

The policy can be any algorithm you can think of, and it does not have to be deterministic. In fact, in some cases it does not even have to observe the environment! For example, consider a robotic vacuum cleaner whose reward is the amount of dust it picks up in 30 minutes. Its policy could be to move forward with some probability p every second, or randomly rotate left or right with probability $1 - p$. The rotation angle would be a random angle between $-r$ and $+r$. Since this policy involves some randomness, it is called a *stochastic policy*. The robot will have an erratic trajectory, which guarantees that it will eventually get to any place it can reach and pick up all the dust. The question is, how much dust will it pick up in 30 minutes?

How would you train such a robot? There are just two *policy parameters* you can tweak: the probability p and the angle range r. One possible learning algorithm could be to try out many different values for these parameters, and pick the combination that performs best (see Figure 18-3). This is an example of *policy search*, in this case using a brute force approach. When the *policy space* is too large (which is generally the case), finding a good set of parameters this way is like searching for a needle in a gigantic haystack.

Another way to explore the policy space is to use *genetic algorithms*. For example, you could randomly create a first generation of 100 policies and try them out, then "kill" the 80 worst policies[6] and make the 20 survivors produce 4 offspring each. An

---

6 It is often better to give the poor performers a slight chance of survival, to preserve some diversity in the "gene pool."

offspring is a copy of its parent[7] plus some random variation. The surviving policies plus their offspring together constitute the second generation. You can continue to iterate through generations this way until you find a good policy.[8]

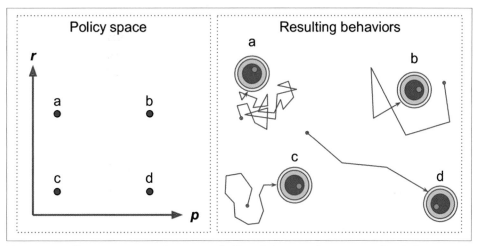

Figure 18-3. Four points in policy space (left) and the agent's corresponding behavior (right)

Yet another approach is to use optimization techniques, by evaluating the gradients of the rewards with regard to the policy parameters, then tweaking these parameters by following the gradients toward higher rewards.[9] We will discuss this approach, is called *policy gradients* (PG), in more detail later in this chapter. Going back to the vacuum cleaner robot, you could slightly increase p and evaluate whether doing so increases the amount of dust picked up by the robot in 30 minutes; if it does, then increase p some more, or else reduce p. We will implement a popular PG algorithm using TensorFlow, but before we do, we need to create an environment for the agent to live in—so it's time to introduce OpenAI Gym.

Introduction to OpenAI Gym

One of the challenges of Reinforcement Learning is that in order to train an agent, you first need to have a working environment. If you want to program an agent that

---

7 If there is a single parent, this is called *asexual reproduction*. With two (or more) parents, it is called *sexual reproduction*. An offspring's genome (in this case a set of policy parameters) is randomly composed of parts of its parents' genomes.

8 One interesting example of a genetic algorithm used for Reinforcement Learning is the *NeuroEvolution of Augmenting Topologies* (*https://homl.info/neat*) (NEAT) algorithm.

9 This is called *Gradient Ascent*. It's just like Gradient Descent but in the opposite direction: maximizing instead of minimizing.

will learn to play an Atari game, you will need an Atari game simulator. If you want to program a walking robot, then the environment is the real world, and you can directly train your robot in that environment, but this has its limits: if the robot falls off a cliff, you can't just click Undo. You can't speed up time either; adding more computing power won't make the robot move any faster. And it's generally too expensive to train 1,000 robots in parallel. In short, training is hard and slow in the real world, so you generally need a *simulated environment* at least for bootstrap training. For example, you may use a library like PyBullet (*https://pybullet.org/*) or MuJoCo (*http://www.mujoco.org/*) for 3D physics simulation.

OpenAI Gym (*https://gym.openai.com/*)[10] is a toolkit that provides a wide variety of simulated environments (Atari games, board games, 2D and 3D physical simulations, and so on), so you can train agents, compare them, or develop new RL algorithms.

Before installing the toolkit, if you created an isolated environment using virtualenv, you first need to activate it:

```
$ cd $ML_PATH                       # Your ML working directory (e.g., $HOME/ml)
$ source my_env/bin/activate # on Linux or MacOS
$ .\my_env\Scripts\activate  # on Windows
```

Next, install OpenAI Gym (if you are not using a virtual environment, you will need to add the --user option, or have administrator rights):

```
$ python3 -m pip install --upgrade gym
```

Depending on your system, you may also need to install the Mesa OpenGL Utility (GLU) library (e.g., on Ubuntu 18.04 you need to run apt install libglu1-mesa). This library will be needed to render the first environment. Next, open up a Python shell or a Jupyter notebook and create an environment with make():

```
>>> import gym
>>> env = gym.make("CartPole-v1")
>>> obs = env.reset()
>>> obs
array([-0.01258566, -0.00156614,  0.04207708, -0.00180545])
```

Here, we've created a CartPole environment. This is a 2D simulation in which a cart can be accelerated left or right in order to balance a pole placed on top of it (see Figure 18-4). You can get the list of all available environments by running gym.envs.registry.all(). After the environment is created, you must initialize it using the reset() method. This returns the first observation. Observations depend on the type of environment. For the CartPole environment, each observation is a 1D NumPy array containing four floats: these floats represent the cart's horizontal

---

10 OpenAI is an artificial intelligence research company, funded in part by Elon Musk. Its stated goal is to promote and develop friendly AIs that will benefit humanity (rather than exterminate it).

position (0.0 = center), its velocity (positive means right), the angle of the pole (0.0 = vertical), and its angular velocity (positive means clockwise).

Now let's display this environment by calling its render() method (see Figure 18-4). On Windows, this requires first installing an X Server, such as VcXsrv or Xming:

```
>>> env.render()
True
```

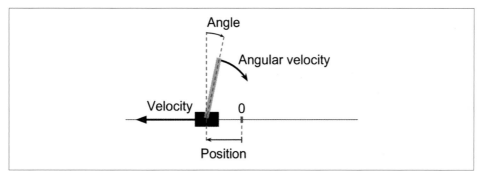

Figure 18-4. The CartPole environment

 If you are using a headless server (i.e., without a screen), such as a virtual machine on the cloud, rendering will fail. The only way to avoid this is to use a fake X server such as Xvfb or Xdummy. For example, you can install Xvfb (apt install xvfb on Ubuntu or Debian) and start Python using the following command: xvfb-run -s "-screen 0 1400x900x24" python3. Alternatively, install Xvfb and the pyvirtualdisplay library (*https://homl.info/pyvd*) (which wraps Xvfb) and run pyvirtualdisplay.Display(visible=0, size=(1400, 900)).start() at the beginning of your program.

If you want render() to return the rendered image as a NumPy array, you can set mode="rgb_array" (oddly, this environment will render the environment to screen as well):

```
>>> img = env.render(mode="rgb_array")
>>> img.shape  # height, width, channels (3 = Red, Green, Blue)
(800, 1200, 3)
```

Let's ask the environment what actions are possible:

```
>>> env.action_space
Discrete(2)
```

Discrete(2) means that the possible actions are integers 0 and 1, which represent accelerating left (0) or right (1). Other environments may have additional discrete actions, or other kinds of actions (e.g., continuous). Since the pole is leaning toward the right (obs[2] > 0), let's accelerate the cart toward the right:

```
>>> action = 1  # accelerate right
>>> obs, reward, done, info = env.step(action)
>>> obs
array([-0.01261699,  0.19292789,  0.04204097, -0.28092127])
>>> reward
1.0
>>> done
False
>>> info
{}
```

The step() method executes the given action and returns four values:

obs

    This is the new observation. The cart is now moving toward the right (obs[1] > 0). The pole is still tilted toward the right (obs[2] > 0), but its angular velocity is now negative (obs[3] < 0), so it will likely be tilted toward the left after the next step.

reward

    In this environment, you get a reward of 1.0 at every step, no matter what you do, so the goal is to keep the episode running as long as possible.

done

    This value will be True when the episode is over. This will happen when the pole tilts too much, or goes off the screen, or after 200 steps (in this last case, you have won). After that, the environment must be reset before it can be used again.

info

    This environment-specific dictionary can provide some extra information that you may find useful for debugging or for training. For example, in some games it may indicate how many lives the agent has.

 Once you have finished using an environment, you should call its close() method to free resources.

Let's hardcode a simple policy that accelerates left when the pole is leaning toward the left and accelerates right when the pole is leaning toward the right. We will run this policy to see the average rewards it gets over 500 episodes:

```
def basic_policy(obs):
    angle = obs[2]
    return 0 if angle < 0 else 1

totals = []
for episode in range(500):
    episode_rewards = 0
    obs = env.reset()
    for step in range(200):
        action = basic_policy(obs)
        obs, reward, done, info = env.step(action)
        episode_rewards += reward
        if done:
            break
    totals.append(episode_rewards)
```

This code is hopefully self-explanatory. Let's look at the result:

```
>>> import numpy as np
>>> np.mean(totals), np.std(totals), np.min(totals), np.max(totals)
(41.718, 8.858356280936096, 24.0, 68.0)
```

Even with 500 tries, this policy never managed to keep the pole upright for more than 68 consecutive steps. Not great. If you look at the simulation in the Jupyter notebooks (*https://github.com/ageron/handson-ml2*), you will see that the cart oscillates left and right more and more strongly until the pole tilts too much. Let's see if a neural network can come up with a better policy.

Neural Network Policies

Let's create a neural network policy. Just like with the policy we hardcoded earlier, this neural network will take an observation as input, and it will output the action to be executed. More precisely, it will estimate a probability for each action, and then we will select an action randomly, according to the estimated probabilities (see Figure 18-5). In the case of the CartPole environment, there are just two possible actions (left or right), so we only need one output neuron. It will output the probability p of action 0 (left), and of course the probability of action 1 (right) will be $1 - p$. For example, if it outputs 0.7, then we will pick action 0 with 70% probability, or action 1 with 30% probability.

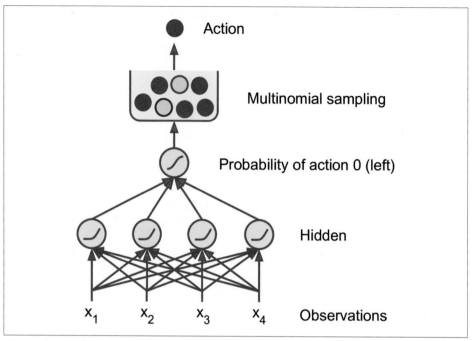

Figure 18-5. Neural network policy

You may wonder why we are picking a random action based on the probabilities given by the neural network, rather than just picking the action with the highest score. This approach lets the agent find the right balance between *exploring* new actions and *exploiting* the actions that are known to work well. Here's an analogy: suppose you go to a restaurant for the first time, and all the dishes look equally appealing, so you randomly pick one. If it turns out to be good, you can increase the probability that you'll order it next time, but you shouldn't increase that probability up to 100%, or else you will never try out the other dishes, some of which may be even better than the one you tried.

Also note that in this particular environment, the past actions and observations can safely be ignored, since each observation contains the environment's full state. If there were some hidden state, then you might need to consider past actions and observations as well. For example, if the environment only revealed the position of the cart but not its velocity, you would have to consider not only the current observation but also the previous observation in order to estimate the current velocity. Another example is when the observations are noisy; in that case, you generally want to use the past few observations to estimate the most likely current state. The CartPole problem is thus as simple as can be; the observations are noise-free, and they contain the environment's full state.

Here is the code to build this neural network policy using tf.keras:

```
import tensorflow as tf
from tensorflow import keras

n_inputs = 4 # == env.observation_space.shape[0]

model = keras.models.Sequential([
    keras.layers.Dense(5, activation="elu", input_shape=[n_inputs]),
    keras.layers.Dense(1, activation="sigmoid"),
])
```

After the imports, we use a simple `Sequential` model to define the policy network. The number of inputs is the size of the observation space (which in the case of Cart-Pole is 4), and we have just five hidden units because it's a simple problem. Finally, we want to output a single probability (the probability of going left), so we have a single output neuron using the sigmoid activation function. If there were more than two possible actions, there would be one output neuron per action, and we would use the softmax activation function instead.

OK, we now have a neural network policy that will take observations and output action probabilities. But how do we train it?

Evaluating Actions: The Credit Assignment Problem

If we knew what the best action was at each step, we could train the neural network as usual, by minimizing the cross entropy between the estimated probability distribution and the target probability distribution. It would just be regular supervised learning. However, in Reinforcement Learning the only guidance the agent gets is through rewards, and rewards are typically sparse and delayed. For example, if the agent manages to balance the pole for 100 steps, how can it know which of the 100 actions it took were good, and which of them were bad? All it knows is that the pole fell after the last action, but surely this last action is not entirely responsible. This is called the *credit assignment problem*: when the agent gets a reward, it is hard for it to know which actions should get credited (or blamed) for it. Think of a dog that gets rewarded hours after it behaved well; will it understand what it is being rewarded for?

To tackle this problem, a common strategy is to evaluate an action based on the sum of all the rewards that come after it, usually applying a *discount factor* γ (gamma) at each step. This sum of discounted rewards is called the action's *return*. Consider the example in Figure 18-6). If an agent decides to go right three times in a row and gets +10 reward after the first step, 0 after the second step, and finally –50 after the third step, then assuming we use a discount factor $\gamma = 0.8$, the first action will have a return of $10 + \gamma \times 0 + \gamma^2 \times (-50) = -22$. If the discount factor is close to 0, then future rewards won't count for much compared to immediate rewards. Conversely, if the discount factor is close to 1, then rewards far into the future will count almost as much as immediate rewards. Typical discount factors vary from 0.9 to 0.99. With a discount factor of 0.95, rewards 13 steps into the future count roughly for half as much as immediate rewards (since $0.95^{13} \approx 0.5$), while with a discount factor of 0.99, rewards 69 steps into the future count for half as much as immediate rewards. In the CartPole environment, actions have fairly short-term effects, so choosing a discount factor of 0.95 seems reasonable.

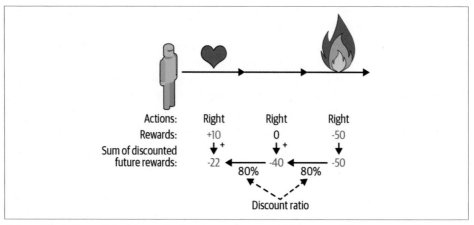

Figure 18-6. Computing an action's return: the sum of discounted future rewards

Of course, a good action may be followed by several bad actions that cause the pole to fall quickly, resulting in the good action getting a low return (similarly, a good actor may sometimes star in a terrible movie). However, if we play the game enough times, on average good actions will get a higher return than bad ones. We want to estimate how much better or worse an action is, compared to the other possible actions, on average. This is called the *action advantage*. For this, we must run many episodes and normalize all the action returns (by subtracting the mean and dividing by the standard deviation). After that, we can reasonably assume that actions with a negative advantage were bad while actions with a positive advantage were good. Perfect—now that we have a way to evaluate each action, we are ready to train our first agent using policy gradients. Let's see how.

Policy Gradients

As discussed earlier, PG algorithms optimize the parameters of a policy by following the gradients toward higher rewards. One popular class of PG algorithms, called

REINFORCE algorithms, was introduced back in 1992 (*https://homl.info/132*)[11] by Ronald Williams. Here is one common variant:

1. First, let the neural network policy play the game several times, and at each step, compute the gradients that would make the chosen action even more likely—but don't apply these gradients yet.

2. Once you have run several episodes, compute each action's advantage (using the method described in the previous section).

3. If an action's advantage is positive, it means that the action was probably good, and you want to apply the gradients computed earlier to make the action even more likely to be chosen in the future. However, if the action's advantage is negative, it means the action was probably bad, and you want to apply the opposite gradients to make this action slightly *less* likely in the future. The solution is simply to multiply each gradient vector by the corresponding action's advantage.

4. Finally, compute the mean of all the resulting gradient vectors, and use it to perform a Gradient Descent step.

Let's use tf.keras to implement this algorithm. We will train the neural network policy we built earlier so that it learns to balance the pole on the cart. First, we need a function that will play one step. We will pretend for now that whatever action it takes is the right one so that we can compute the loss and its gradients (these gradients will just be saved for a while, and we will modify them later depending on how good or bad the action turned out to be):

```
def play_one_step(env, obs, model, loss_fn):
    with tf.GradientTape() as tape:
        left_proba = model(obs[np.newaxis])
        action = (tf.random.uniform([1, 1]) > left_proba)
        y_target = tf.constant([[1.]]) - tf.cast(action, tf.float32)
        loss = tf.reduce_mean(loss_fn(y_target, left_proba))
    grads = tape.gradient(loss, model.trainable_variables)
    obs, reward, done, info = env.step(int(action[0, 0].numpy()))
    return obs, reward, done, grads
```

Let's walk though this function:

- Within the GradientTape block (see Chapter 12), we start by calling the model, giving it a single observation (we reshape the observation so it becomes a batch containing a single instance, as the model expects a batch). This outputs the probability of going left.

---

11 Ronald J. Williams, "Simple Statistical Gradient-Following Algorithms for Connectionist Reinforcement Leaning," *Machine Learning* 8 (1992) : 229–256.

- Next, we sample a random float between 0 and 1, and we check whether it is greater than `left_proba`. The `action` will be `False` with probability `left_proba`, or `True` with probability `1 - left_proba`. Once we cast this Boolean to a number, the action will be 0 (left) or 1 (right) with the appropriate probabilities.

- Next, we define the target probability of going left: it is 1 minus the action (cast to a float). If the action is 0 (left), then the target probability of going left will be 1. If the action is 1 (right), then the target probability will be 0.

- Then we compute the loss using the given loss function, and we use the tape to compute the gradient of the loss with regard to the model's trainable variables. Again, these gradients will be tweaked later, before we apply them, depending on how good or bad the action turned out to be.

- Finally, we play the selected action, and we return the new observation, the reward, whether the episode is ended or not, and of course the gradients that we just computed.

Now let's create another function that will rely on the `play_one_step()` function to play multiple episodes, returning all the rewards and gradients for each episode and each step:

```python
def play_multiple_episodes(env, n_episodes, n_max_steps, model, loss_fn):
    all_rewards = []
    all_grads = []
    for episode in range(n_episodes):
        current_rewards = []
        current_grads = []
        obs = env.reset()
        for step in range(n_max_steps):
            obs, reward, done, grads = play_one_step(env, obs, model, loss_fn)
            current_rewards.append(reward)
            current_grads.append(grads)
            if done:
                break
        all_rewards.append(current_rewards)
        all_grads.append(current_grads)
    return all_rewards, all_grads
```

This code returns a list of reward lists (one reward list per episode, containing one reward per step), as well as a list of gradient lists (one gradient list per episode, each containing one tuple of gradients per step and each tuple containing one gradient tensor per trainable variable).

The algorithm will use the `play_multiple_episodes()` function to play the game several times (e.g., 10 times), then it will go back and look at all the rewards, discount them, and normalize them. To do that, we need a couple more functions: the first will compute the sum of future discounted rewards at each step, and the second will normalize all these discounted rewards (returns) across many episodes by subtracting the mean and dividing by the standard deviation:

```
def discount_rewards(rewards, discount_factor):
    discounted = np.array(rewards)
    for step in range(len(rewards) - 2, -1, -1):
        discounted[step] += discounted[step + 1] * discount_factor
    return discounted

def discount_and_normalize_rewards(all_rewards, discount_factor):
    all_discounted_rewards = [discount_rewards(rewards, discount_factor)
                              for rewards in all_rewards]
    flat_rewards = np.concatenate(all_discounted_rewards)
    reward_mean = flat_rewards.mean()
    reward_std = flat_rewards.std()
    return [(discounted_rewards - reward_mean) / reward_std
            for discounted_rewards in all_discounted_rewards]
```

Let's check that this works:

```
>>> discount_rewards([10, 0, -50], discount_factor=0.8)
array([-22, -40, -50])
>>> discount_and_normalize_rewards([[10, 0, -50], [10, 20]],
...                                discount_factor=0.8)
...
[array([-0.28435071, -0.86597718, -1.18910299]),
 array([1.26665318, 1.0727777 ])]
```

The call to `discount_rewards()` returns exactly what we expect (see Figure 18-6). You can verify that the function `discount_and_normalize_rewards()` does indeed return the normalized action advantages for each action in both episodes. Notice that the first episode was much worse than the second, so its normalized advantages are all negative; all actions from the first episode would be considered bad, and conversely all actions from the second episode would be considered good.

We are almost ready to run the algorithm! Now let's define the hyperparameters. We will run 150 training iterations, playing 10 episodes per iteration, and each episode will last at most 200 steps. We will use a discount factor of 0.95:

```
n_iterations = 150
n_episodes_per_update = 10
n_max_steps = 200
discount_factor = 0.95
```

We also need an optimizer and the loss function. A regular Adam optimizer with learning rate 0.01 will do just fine, and we will use the binary cross-entropy loss function because we are training a binary classifier (there are two possible actions: left or right):

```
optimizer = keras.optimizers.Adam(lr=0.01)
loss_fn = keras.losses.binary_crossentropy
```

We are now ready to build and run the training loop!

```
for iteration in range(n_iterations):
    all_rewards, all_grads = play_multiple_episodes(
        env, n_episodes_per_update, n_max_steps, model, loss_fn)
    all_final_rewards = discount_and_normalize_rewards(all_rewards,
                                                       discount_factor)
    all_mean_grads = []
    for var_index in range(len(model.trainable_variables)):
        mean_grads = tf.reduce_mean(
            [final_reward * all_grads[episode_index][step][var_index]
             for episode_index, final_rewards in enumerate(all_final_rewards)
                 for step, final_reward in enumerate(final_rewards)], axis=0)
        all_mean_grads.append(mean_grads)
    optimizer.apply_gradients(zip(all_mean_grads, model.trainable_variables))
```

Let's walk through this code:

- At each training iteration, this loop calls the `play_multiple_episodes()` function, which plays the game 10 times and returns all the rewards and gradients for every episode and step.

- Then we call the `discount_and_normalize_rewards()` to compute each action's normalized advantage (which in the code we call the `final_reward`). This provides a measure of how good or bad each action actually was, in hindsight.

- Next, we go through each trainable variable, and for each of them we compute the weighted mean of the gradients for that variable over all episodes and all steps, weighted by the `final_reward`.

- Finally, we apply these mean gradients using the optimizer: the model's trainable variables will be tweaked, and hopefully the policy will be a bit better.

And we're done! This code will train the neural network policy, and it will successfully learn to balance the pole on the cart (you can try it out in the "Policy Gradients" section of the Jupyter notebook). The mean reward per episode will get very close to 200 (which is the maximum by default with this environment). Success!

 Researchers try to find algorithms that work well even when the agent initially knows nothing about the environment. However, unless you are writing a paper, you should not hesitate to inject prior knowledge into the agent, as it will speed up training dramatically. For example, since you know that the pole should be as vertical as possible, you could add negative rewards proportional to the pole's angle. This will make the rewards much less sparse and speed up training. Also, if you already have a reasonably good policy (e.g., hardcoded), you may want to train the neural network to imitate it before using policy gradients to improve it.

The simple policy gradients algorithm we just trained solved the CartPole task, but it would not scale well to larger and more complex tasks. Indeed, it is highly *sample inefficient*, meaning it needs to explore the game for a very long time before it can make significant progress. This is due to the fact that it must run multiple episodes to estimate the advantage of each action, as we have seen. However, it is the foundation of more powerful algorithms, such as *Actor-Critic* algorithms (which we will discuss briefly at the end of this chapter).

We will now look at another popular family of algorithms. Whereas PG algorithms directly try to optimize the policy to increase rewards, the algorithms we will look at now are less direct: the agent learns to estimate the expected return for each state, or for each action in each state, then it uses this knowledge to decide how to act. To understand these algorithms, we must first introduce *Markov decision processes*.

Markov Decision Processes

In the early 20th century, the mathematician Andrey Markov studied stochastic processes with no memory, called *Markov chains*. Such a process has a fixed number of states, and it randomly evolves from one state to another at each step. The probability for it to evolve from a state s to a state s' is fixed, and it depends only on the pair (s, s'), not on past states (this is why we say that the system has no memory).

Figure 18-7 shows an example of a Markov chain with four states.

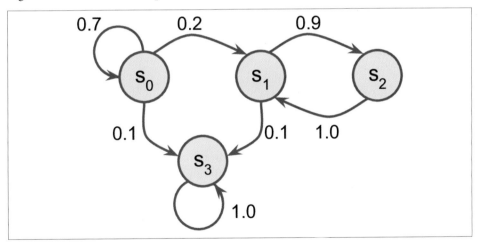

Figure 18-7. Example of a Markov chain

Suppose that the process starts in state s_0, and there is a 70% chance that it will remain in that state at the next step. Eventually it is bound to leave that state and never come back because no other state points back to s_0. If it goes to state s_1, it will then most likely go to state s_2 (90% probability), then immediately back to state s_1

(with 100% probability). It may alternate a number of times between these two states, but eventually it will fall into state s_3 and remain there forever (this is a *terminal state*). Markov chains can have very different dynamics, and they are heavily used in thermodynamics, chemistry, statistics, and much more.

Markov decision processes were first described in the 1950s by Richard Bellman (*https://homl.info/133*).[12] They resemble Markov chains but with a twist: at each step, an agent can choose one of several possible actions, and the transition probabilities depend on the chosen action. Moreover, some state transitions return some reward (positive or negative), and the agent's goal is to find a policy that will maximize reward over time.

For example, the MDP represented in Figure 18-8 has three states (represented by circles) and up to three possible discrete actions at each step (represented by diamonds).

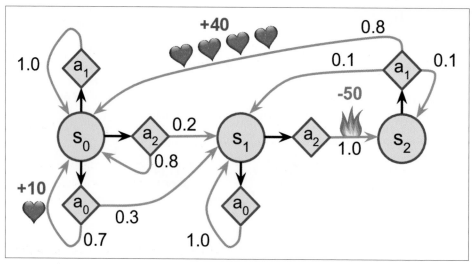

Figure 18-8. Example of a Markov decision process

If it starts in state s_0, the agent can choose between actions a_0, a_1, or a_2. If it chooses action a_1, it just remains in state s_0 with certainty, and without any reward. It can thus decide to stay there forever if it wants to. But if it chooses action a_0, it has a 70% probability of gaining a reward of +10 and remaining in state s_0. It can then try again and again to gain as much reward as possible, but at one point it is going to end up instead in state s_1. In state s_1 it has only two possible actions: a_0 or a_2. It can choose to stay put by repeatedly choosing action a_0, or it can choose to move on to state s_2 and get a negative reward of –50 (ouch). In state s_2 it has no other choice than to take

12 Richard Bellman, "A Markovian Decision Process," *Journal of Mathematics and Mechanics* 6, no. 5 (1957): 679–684.

action a_1, which will most likely lead it back to state s_0, gaining a reward of +40 on the way. You get the picture. By looking at this MDP, can you guess which strategy will gain the most reward over time? In state s_0 it is clear that action a_0 is the best option, and in state s_2 the agent has no choice but to take action a_1, but in state s_1 it is not obvious whether the agent should stay put (a_0) or go through the fire (a_2).

Bellman found a way to estimate the *optimal state value* of any state s, noted $V^*(s)$, which is the sum of all discounted future rewards the agent can expect on average after it reaches a state s, assuming it acts optimally. He showed that if the agent acts optimally, then the *Bellman Optimality Equation* applies (see Equation 18-1). This recursive equation says that if the agent acts optimally, then the optimal value of the current state is equal to the reward it will get on average after taking one optimal action, plus the expected optimal value of all possible next states that this action can lead to.

Equation 18-1. Bellman Optimality Equation

$$V^*(s) = \max_a \sum_{s'} T(s, a, s')[R(s, a, s') + \gamma \cdot V^*(s')] \quad \text{for all } s$$

In this equation:

- $T(s, a, s')$ is the transition probability from state s to state s', given that the agent chose action a. For example, in Figure 18-8, $T(s_2, a_1, s_0) = 0.8$.
- $R(s, a, s')$ is the reward that the agent gets when it goes from state s to state s', given that the agent chose action a. For example, in Figure 18-8, $R(s_2, a_1, s_0) = +40$.
- γ is the discount factor.

This equation leads directly to an algorithm that can precisely estimate the optimal state value of every possible state: you first initialize all the state value estimates to zero, and then you iteratively update them using the *Value Iteration* algorithm (see Equation 18-2). A remarkable result is that, given enough time, these estimates are guaranteed to converge to the optimal state values, corresponding to the optimal policy.

Equation 18-2. Value Iteration algorithm

$$V_{k+1}(s) \leftarrow \max_a \sum_{s'} T(s, a, s')[R(s, a, s') + \gamma \cdot V_k(s')] \quad \text{for all } s$$

In this equation, $V_k(s)$ is the estimated value of state s at the kth iteration of the algorithm.

This algorithm is an example of *Dynamic Programming*, which breaks down a complex problem into tractable subproblems that can be tackled iteratively.

Knowing the optimal state values can be useful, in particular to evaluate a policy, but it does not give us the optimal policy for the agent. Luckily, Bellman found a very similar algorithm to estimate the optimal *state-action values*, generally called *Q-Values* (Quality Values). The optimal Q-Value of the state-action pair (s, a), noted $Q^*(s, a)$, is the sum of discounted future rewards the agent can expect on average after it reaches the state s and chooses action a, but before it sees the outcome of this action, assuming it acts optimally after that action.

Here is how it works: once again, you start by initializing all the Q-Value estimates to zero, then you update them using the *Q-Value Iteration* algorithm (see Equation 18-3).

Equation 18-3. Q-Value Iteration algorithm

$$Q_{k+1}(s, a) \leftarrow \sum_{s'} T(s, a, s') \left[R(s, a, s') + \gamma \cdot \max_{a'} Q_k(s', a') \right] \quad \text{for all } (s'a)$$

Once you have the optimal Q-Values, defining the optimal policy, noted $\pi^*(s)$, is trivial: when the agent is in state s, it should choose the action with the highest Q-Value for that state: $\pi^*(s) = \underset{a}{\operatorname{argmax}} \ Q^*(s, a)$.

Let's apply this algorithm to the MDP represented in Figure 18-8. First, we need to define the MDP:

```
transition_probabilities = [ # shape=[s, a, s']
        [[0.7, 0.3, 0.0], [1.0, 0.0, 0.0], [0.8, 0.2, 0.0]],
        [[0.0, 1.0, 0.0], None, [0.0, 0.0, 1.0]],
        [None, [0.8, 0.1, 0.1], None]]
rewards = [ # shape=[s, a, s']
        [[+10, 0, 0], [0, 0, 0], [0, 0, 0]],
        [[0, 0, 0], [0, 0, 0], [0, 0, -50]],
        [[0, 0, 0], [+40, 0, 0], [0, 0, 0]]]
possible_actions = [[0, 1, 2], [0, 2], [1]]
```

For example, to know the transition probability from s_2 to s_0 after playing action a_1, we will look up `transition_probabilities[2][1][0]` (which is 0.8). Similarly, to get the corresponding reward, we will look up `rewards[2][1][0]` (which is +40). And to get the list of possible actions in s_2, we will look up `possible_actions[2]` (in this case, only action a_1 is possible). Next, we must initialize all the Q-Values to 0 (except for the the impossible actions, for which we set the Q-Values to $-\infty$):

```
Q_values = np.full((3, 3), -np.inf) # -np.inf for impossible actions
for state, actions in enumerate(possible_actions):
    Q_values[state, actions] = 0.0  # for all possible actions
```

Now let's run the Q-Value Iteration algorithm. It applies Equation 18-3 repeatedly, to all Q-Values, for every state and every possible action:

```
gamma = 0.90 # the discount factor

for iteration in range(50):
    Q_prev = Q_values.copy()
    for s in range(3):
        for a in possible_actions[s]:
            Q_values[s, a] = np.sum([
                    transition_probabilities[s][a][sp]
                    * (rewards[s][a][sp] + gamma * np.max(Q_prev[sp]))
                for sp in range(3)])
```

That's it! The resulting Q-Values look like this:

```
>>> Q_values
array([[18.91891892, 17.02702702, 13.62162162],
       [ 0.        ,        -inf, -4.87971488],
       [       -inf, 50.13365013,        -inf]])
```

For example, when the agent is in state s_0 and it chooses action a_1, the expected sum of discounted future rewards is approximately 17.0.

For each state, let's look at the action that has the highest Q-Value:

```
>>> np.argmax(Q_values, axis=1) # optimal action for each state
array([0, 0, 1])
```

This gives us the optimal policy for this MDP, when using a discount factor of 0.90: in state s_0 choose action a_0; in state s_1 choose action a_0 (i.e., stay put); and in state s_2 choose action a_1 (the only possible action). Interestingly, if we increase the discount factor to 0.95, the optimal policy changes: in state s_1 the best action becomes a_2 (go through the fire!). This makes sense because the more you value future rewards, the more you are willing to put up with some pain now for the promise of future bliss.

Temporal Difference Learning

Reinforcement Learning problems with discrete actions can often be modeled as Markov decision processes, but the agent initially has no idea what the transition probabilities are (it does not know $T(s, a, s')$), and it does not know what the rewards are going to be either (it does not know $R(s, a, s')$). It must experience each state and each transition at least once to know the rewards, and it must experience them multiple times if it is to have a reasonable estimate of the transition probabilities.

The *Temporal Difference Learning* (TD Learning) algorithm is very similar to the Value Iteration algorithm, but tweaked to take into account the fact that the agent has only partial knowledge of the MDP. In general we assume that the agent initially knows only the possible states and actions, and nothing more. The agent uses an *exploration policy*—for example, a purely random policy—to explore the MDP, and as it progresses, the TD Learning algorithm updates the estimates of the state values based on the transitions and rewards that are actually observed (see Equation 18-4).

Equation 18-4. TD Learning algorithm

$$V_{k+1}(s) \leftarrow (1 - \alpha)V_k(s) + \alpha\big(r + \gamma \cdot V_k(s')\big)$$

or, equivalently:

$$V_{k+1}(s) \leftarrow V_k(s) + \alpha \cdot \delta_k(s, r, s')$$
$$\text{with } \delta_k(s, r, s') = r + \gamma \cdot V_k(s') - V_k(s)$$

In this equation:

- α is the learning rate (e.g., 0.01).
- $r + \gamma \cdot V_k(s')$ is called the *TD target*.
- $\delta_k(s, r, s')$ is called the *TD error*.

A more concise way of writing the first form of this equation is to use the notation $a \underset{\alpha}{\leftarrow} b$, which means $a_{k+1} \leftarrow (1 - \alpha) \cdot a_k + \alpha \cdot b_k$. So, the first line of Equation 18-4 can be rewritten like this: $V(s) \underset{\alpha}{\leftarrow} r + \gamma \cdot V(s')$.

TD Learning has many similarities with Stochastic Gradient Descent, in particular the fact that it handles one sample at a time. Moreover, just like Stochastic GD, it can only truly converge if you gradually reduce the learning rate (otherwise it will keep bouncing around the optimum Q-Values).

For each state s, this algorithm simply keeps track of a running average of the immediate rewards the agent gets upon leaving that state, plus the rewards it expects to get later (assuming it acts optimally).

Q-Learning

Similarly, the Q-Learning algorithm is an adaptation of the Q-Value Iteration algorithm to the situation where the transition probabilities and the rewards are initially unknown (see Equation 18-5). Q-Learning works by watching an agent play (e.g., randomly) and gradually improving its estimates of the Q-Values. Once it has accurate Q-Value estimates (or close enough), then the optimal policy is choosing the action that has the highest Q-Value (i.e., the greedy policy).

Equation 18-5. Q-Learning algorithm

$$Q(s, a) \underset{\alpha}{\leftarrow} r + \gamma \cdot \max_{a'} \ Q(s', a')$$

For each state-action pair (s, a), this algorithm keeps track of a running average of the rewards r the agent gets upon leaving the state s with action a, plus the sum of discounted future rewards it expects to get. To estimate this sum, we take the maximum of the Q-Value estimates for the next state s', since we assume that the target policy would act optimally from then on.

Let's implement the Q-Learning algorithm. First, we will need to make an agent explore the environment. For this, we need a step function so that the agent can execute one action and get the resulting state and reward:

```
def step(state, action):
    probas = transition_probabilities[state][action]
    next_state = np.random.choice([0, 1, 2], p=probas)
    reward = rewards[state][action][next_state]
    return next_state, reward
```

Now let's implement the agent's exploration policy. Since the state space is pretty small, a simple random policy will be sufficient. If we run the algorithm for long enough, the agent will visit every state many times, and it will also try every possible action many times:

```
def exploration_policy(state):
    return np.random.choice(possible_actions[state])
```

Next, after we initialize the Q-Values just like earlier, we are ready to run the Q-Learning algorithm with learning rate decay (using power scheduling, introduced in Chapter 11):

```
alpha0 = 0.05 # initial learning rate
decay = 0.005 # learning rate decay
gamma = 0.90 # discount factor
state = 0 # initial state

for iteration in range(10000):
    action = exploration_policy(state)
    next_state, reward = step(state, action)
    next_value = np.max(Q_values[next_state])
    alpha = alpha0 / (1 + iteration * decay)
    Q_values[state, action] *= 1 - alpha
    Q_values[state, action] += alpha * (reward + gamma * next_value)
    state = next_state
```

This algorithm will converge to the optimal Q-Values, but it will take many iterations, and possibly quite a lot of hyperparameter tuning. As you can see in Figure 18-9, the Q-Value Iteration algorithm (left) converges very quickly, in fewer than 20 iterations, while the Q-Learning algorithm (right) takes about 8,000 iterations to converge. Obviously, not knowing the transition probabilities or the rewards makes finding the optimal policy significantly harder!

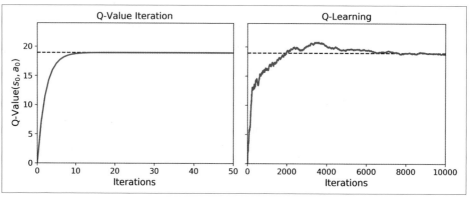

Figure 18-9. The Q-Value Iteration algorithm (left) versus the Q-Learning algorithm (right)

The Q-Learning algorithm is called an *off-policy* algorithm because the policy being trained is not necessarily the one being executed: in the previous code example, the policy being executed (the exploration policy) is completely random, while the policy being trained will always choose the actions with the highest Q-Values. Conversely, the Policy Gradients algorithm is an *on-policy* algorithm: it explores the world using the policy being trained. It is somewhat surprising that Q-Learning is capable of learning the optimal policy by just watching an agent act randomly (imagine learning to play golf when your teacher is a drunk monkey). Can we do better?

Exploration Policies

Of course, Q-Learning can work only if the exploration policy explores the MDP thoroughly enough. Although a purely random policy is guaranteed to eventually visit every state and every transition many times, it may take an extremely long time to do so. Therefore, a better option is to use the *ε-greedy policy* (ε is epsilon): at each step it acts randomly with probability ε, or greedily with probability 1–ε (i.e., choosing the action with the highest Q-Value). The advantage of the ε-greedy policy (compared to a completely random policy) is that it will spend more and more time exploring the interesting parts of the environment, as the Q-Value estimates get better and better, while still spending some time visiting unknown regions of the MDP. It is quite common to start with a high value for ε (e.g., 1.0) and then gradually reduce it (e.g., down to 0.05).

Alternatively, rather than relying only on chance for exploration, another approach is to encourage the exploration policy to try actions that it has not tried much before. This can be implemented as a bonus added to the Q-Value estimates, as shown in Equation 18-6.

Equation 18-6. Q-Learning using an exploration function

$$Q(s, a) \underset{\alpha}{\leftarrow} r + \gamma \cdot \max_{a'} \ f(Q(s', a'), N(s', a'))$$

In this equation:

- $N(s', a')$ counts the number of times the action a' was chosen in state s'.
- $f(Q, N)$ is an *exploration function*, such as $f(Q, N) = Q + \kappa/(1 + N)$, where κ is a curiosity hyperparameter that measures how much the agent is attracted to the unknown.

Approximate Q-Learning and Deep Q-Learning

The main problem with Q-Learning is that it does not scale well to large (or even medium) MDPs with many states and actions. For example, suppose you wanted to use Q-Learning to train an agent to play *Ms. Pac-Man* (see Figure 18-1). There are about 150 pellets that Ms. Pac-Man can eat, each of which can be present or absent (i.e., already eaten). So, the number of possible states is greater than $2^{150} \approx 10^{45}$. And if you add all the possible combinations of positions for all the ghosts and Ms. Pac-Man, the number of possible states becomes larger than the number of atoms in our planet, so there's absolutely no way you can keep track of an estimate for every single Q-Value.

The solution is to find a function $Q_\theta(s, a)$ that approximates the Q-Value of any state-action pair (s, a) using a manageable number of parameters (given by the parameter vector θ). This is called *Approximate Q-Learning*. For years it was recommended to use linear combinations of handcrafted features extracted from the state (e.g., distance of the closest ghosts, their directions, and so on) to estimate Q-Values, but in 2013, DeepMind (*https://homl.info/dqn*) showed that using deep neural networks can work much better, especially for complex problems, and it does not require any feature engineering. A DNN used to estimate Q-Values is called a *Deep Q-Network* (DQN), and using a DQN for Approximate Q-Learning is called *Deep Q-Learning*.

Now, how can we train a DQN? Well, consider the approximate Q-Value computed by the DQN for a given state-action pair (s, a). Thanks to Bellman, we know we want this approximate Q-Value to be as close as possible to the reward r that we actually observe after playing action a in state s, plus the discounted value of playing optimally from then on. To estimate this sum of future discounted rewards, we can simply execute the DQN on the next state s' and for all possible actions a'. We get an approximate future Q-Value for each possible action. We then pick the highest (since we assume we will be playing optimally) and discount it, and this gives us an estimate of the sum of future discounted rewards. By summing the reward r and the future discounted value estimate, we get a target Q-Value $y(s, a)$ for the state-action pair (s, a), as shown in Equation 18-7.

Equation 18-7. Target Q-Value

$$Q_{\text{target}}\left(s, a\right) = r + \gamma \cdot \max_{a'}\ Q_\theta\left(s', a'\right)$$

With this target Q-Value, we can run a training step using any Gradient Descent algorithm. Specifically, we generally try to minimize the squared error between the estimated Q-Value $Q(s, a)$ and the target Q-Value (or the Huber loss to reduce the algorithm's sensitivity to large errors). And that's all for the basic Deep Q-Learning algorithm! Let's see how to implement it to solve the CartPole environment.

Implementing Deep Q-Learning

The first thing we need is a Deep Q-Network. In theory, you need a neural net that takes a state-action pair and outputs an approximate Q-Value, but in practice it's much more efficient to use a neural net that takes a state and outputs one approximate Q-Value for each possible action. To solve the CartPole environment, we do not need a very complicated neural net; a couple of hidden layers will do:

```
env = gym.make("CartPole-v0")
input_shape = [4] # == env.observation_space.shape
n_outputs = 2 # == env.action_space.n

model = keras.models.Sequential([
    keras.layers.Dense(32, activation="elu", input_shape=input_shape),
    keras.layers.Dense(32, activation="elu"),
    keras.layers.Dense(n_outputs)
])
```

To select an action using this DQN, we pick the action with the largest predicted Q-Value. To ensure that the agent explores the environment, we will use an ε-greedy policy (i.e., we will choose a random action with probability ε):

```
def epsilon_greedy_policy(state, epsilon=0):
    if np.random.rand() < epsilon:
        return np.random.randint(2)
    else:
        Q_values = model.predict(state[np.newaxis])
        return np.argmax(Q_values[0])
```

Instead of training the DQN based only on the latest experiences, we will store all experiences in a *replay buffer* (or *replay memory*), and we will sample a random training batch from it at each training iteration. This helps reduce the correlations between the experiences in a training batch, which tremendously helps training. For this, we will just use a deque list:

```
from collections import deque

replay_buffer = deque(maxlen=2000)
```

 A *deque* is a linked list, where each element points to the next one and to the previous one. It makes inserting and deleting items very fast, but the longer the deque is, the slower random access will be. If you need a very large replay buffer, use a circular buffer; see the "Deque vs Rotating List" section of the notebook for an implementation.

Each experience will be composed of five elements: a state, the action the agent took, the resulting reward, the next state it reached, and finally a Boolean indicating whether the episode ended at that point (done). We will need a small function to sample a random batch of experiences from the replay buffer. It will return five NumPy arrays corresponding to the five experience elements:

```
def sample_experiences(batch_size):
    indices = np.random.randint(len(replay_buffer), size=batch_size)
    batch = [replay_buffer[index] for index in indices]
    states, actions, rewards, next_states, dones = [
        np.array([experience[field_index] for experience in batch])
        for field_index in range(5)]
    return states, actions, rewards, next_states, dones
```

Let's also create a function that will play a single step using the ε-greedy policy, then store the resulting experience in the replay buffer:

```
def play_one_step(env, state, epsilon):
    action = epsilon_greedy_policy(state, epsilon)
    next_state, reward, done, info = env.step(action)
    replay_buffer.append((state, action, reward, next_state, done))
    return next_state, reward, done, info
```

Finally, let's create one last function that will sample a batch of experiences from the replay buffer and train the DQN by performing a single Gradient Descent step on this batch:

```
batch_size = 32
discount_factor = 0.95
optimizer = keras.optimizers.Adam(lr=1e-3)
loss_fn = keras.losses.mean_squared_error
```

```
def training_step(batch_size):
    experiences = sample_experiences(batch_size)
    states, actions, rewards, next_states, dones = experiences
    next_Q_values = model.predict(next_states)
    max_next_Q_values = np.max(next_Q_values, axis=1)
    target_Q_values = (rewards +
                       (1 - dones) * discount_factor * max_next_Q_values)
    mask = tf.one_hot(actions, n_outputs)
    with tf.GradientTape() as tape:
        all_Q_values = model(states)
        Q_values = tf.reduce_sum(all_Q_values * mask, axis=1, keepdims=True)
        loss = tf.reduce_mean(loss_fn(target_Q_values, Q_values))
    grads = tape.gradient(loss, model.trainable_variables)
    optimizer.apply_gradients(zip(grads, model.trainable_variables))
```

Let's go through this code:

- First we define some hyperparameters, and we create the optimizer and the loss function.

- Then we create the training_step() function. It starts by sampling a batch of experiences, then it uses the DQN to predict the Q-Value for each possible action in each experience's next state. Since we assume that the agent will be playing optimally, we only keep the maximum Q-Value for each next state. Next, we use Equation 18-7 to compute the target Q-Value for each experience's state-action pair.

- Next, we want to use the DQN to compute the Q-Value for each experienced state-action pair. However, the DQN will also output the Q-Values for the other possible actions, not just for the action that was actually chosen by the agent. So we need to mask out all the Q-Values we do not need. The tf.one_hot() function makes it easy to convert an array of action indices into such a mask. For example, if the first three experiences contain actions 1, 1, 0, respectively, then the mask will start with [[0, 1], [0, 1], [1, 0],...]. We can then multiply the DQN's output with this mask, and this will zero out all the Q-Values we do not want. We then sum over axis 1 to get rid of all the zeros, keeping only the Q-Values of the experienced state-action pairs. This gives us the Q_values tensor, containing one predicted Q-Value for each experience in the batch.

- Then we compute the loss: it is the mean squared error between the target and predicted Q-Values for the experienced state-action pairs.

- Finally, we perform a Gradient Descent step to minimize the loss with regard to the model's trainable variables.

This was the hardest part. Now training the model is straightforward:

```
for episode in range(600):
    obs = env.reset()
    for step in range(200):
        epsilon = max(1 - episode / 500, 0.01)
        obs, reward, done, info = play_one_step(env, obs, epsilon)
        if done:
            break
    if episode > 50:
        training_step(batch_size)
```

We run 600 episodes, each for a maximum of 200 steps. At each step, we first compute the epsilon value for the ε-greedy policy: it will go from 1 down to 0.01, linearly, in a bit under 500 episodes. Then we call the play_one_step() function, which
will use the ε-greedy policy to pick an action, then execute it and record the experience in the replay buffer. If the episode is done, we exit the loop. Finally, if we are past
the 50th episode, we call the training_step() function to train the model on one
batch sampled from the replay buffer. The reason we play 50 episodes without training is to give the replay buffer some time to fill up (if we don't wait enough, then
there will not be enough diversity in the replay buffer). And that's it, we just implemented the Deep Q-Learning algorithm!

Figure 18-10 shows the total rewards the agent got during each episode.

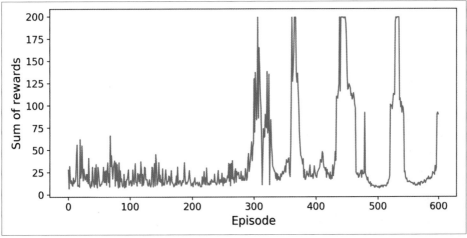

Figure 18-10. Learning curve of the Deep Q-Learning algorithm

As you can see, the algorithm made no apparent progress at all for almost 300 episodes (in part because ε was very high at the beginning), then its performance suddenly skyrocketed up to 200 (which is the maximum possible performance in this
environment). That's great news: the algorithm worked fine, and it actually ran much
faster than the Policy Gradient algorithm! But wait… just a few episodes later, it for-
got everything it knew, and its performance dropped below 25! This is called

catastrophic forgetting, and it is one of the big problems facing virtually all RL algorithms: as the agent explores the environment, it updates its policy, but what it learns in one part of the environment may break what it learned earlier in other parts of the environment. The experiences are quite correlated, and the learning environment keeps changing—this is not ideal for Gradient Descent! If you increase the size of the replay buffer, the algorithm will be less subject to this problem. Reducing the learning rate may also help. But the truth is, Reinforcement Learning is hard: training is often unstable, and you may need to try many hyperparameter values and random seeds before you find a combination that works well. For example, if you try changing the number of neurons per layer in the preceding from 32 to 30 or 34, the performance will never go above 100 (the DQN may be more stable with one hidden layer instead of two).

> Reinforcement Learning is notoriously difficult, largely because of the training instabilities and the huge sensitivity to the choice of hyperparameter values and random seeds.[13] As the researcher Andrej Karpathy put it: "[Supervised learning] wants to work. […] RL must be forced to work." You will need time, patience, perseverance, and perhaps a bit of luck too. This is a major reason RL is not as widely adopted as regular Deep Learning (e.g., convolutional nets). But there are a few real-world applications, beyond AlphaGo and Atari games: for example, Google uses RL to optimize its datacenter costs, and it is used in some robotics applications, for hyperparameter tuning, and in recommender systems.

You might wonder why we didn't plot the loss. It turns out that loss is a poor indicator of the model's performance. The loss might go down, yet the agent might perform worse (e.g., this can happen when the agent gets stuck in one small region of the environment, and the DQN starts overfitting this region). Conversely, the loss could go up, yet the agent might perform better (e.g., if the DQN was underestimating the Q-Values, and it starts correctly increasing its predictions, the agent will likely perform better, getting more rewards, but the loss might increase because the DQN also sets the targets, which will be larger too).

The basic Deep Q-Learning algorithm we've been using so far would be too unstable to learn to play Atari games. So how did DeepMind do it? Well, they tweaked the algorithm!

---

13 A great 2018 post (*https://homl.info/rlhard*) by Alex Irpan nicely lays out RL's biggest difficulties and limitations.

Deep Q-Learning Variants

Let's look at a few variants of the Deep Q-Learning algorithm that can stabilize and speed up training.

Fixed Q-Value Targets

In the basic Deep Q-Learning algorithm, the model is used both to make predictions and to set its own targets. This can lead to a situation analogous to a dog chasing its own tail. This feedback loop can make the network unstable: it can diverge, oscillate, freeze, and so on. To solve this problem, in their 2013 paper the DeepMind researchers used two DQNs instead of one: the first is the *online model*, which learns at each step and is used to move the agent around, and the other is the *target model* used only to define the targets. The target model is just a clone of the online model:

```
target = keras.models.clone_model(model)
target.set_weights(model.get_weights())
```

Then, in the `training_step()` function, we just need to change one line to use the target model instead of the online model when computing the Q-Values of the next states:

```
next_Q_values = target.predict(next_states)
```

Finally, in the training loop, we must copy the weights of the online model to the target model, at regular intervals (e.g., every 50 episodes):

```
if episode % 50 == 0:
    target.set_weights(model.get_weights())
```

Since the target model is updated much less often than the online model, the Q-Value targets are more stable, the feedback loop we discussed earlier is dampened, and its effects are less severe. This approach was one of the DeepMind researchers' main contributions in their 2013 paper, allowing agents to learn to play Atari games from raw pixels. To stabilize training, they used a tiny learning rate of 0.00025, they updated the target model only every 10,000 steps (instead of the 50 in the previous code example), and they used a very large replay buffer of 1 million experiences. They decreased `epsilon` very slowly, from 1 to 0.1 in 1 million steps, and they let the algorithm run for 50 million steps.

Later in this chapter, we will use the TF-Agents library to train a DQN agent to play *Breakout* using these hyperparameters, but before we get there, let's take a look at another DQN variant that managed to beat the state of the art once more.

Double DQN

In a 2015 paper (*https://homl.info/doubledqn*),[14] DeepMind researchers tweaked their DQN algorithm, increasing its performance and somewhat stabilizing training. They called this variant *Double DQN*. The update was based on the observation that the target network is prone to overestimating Q-Values. Indeed, suppose all actions are equally good: the Q-Values estimated by the target model should be identical, but since they are approximations, some may be slightly greater than others, by pure chance. The target model will always select the largest Q-Value, which will be slightly greater than the mean Q-Value, most likely overestimating the true Q-Value (a bit like counting the height of the tallest random wave when measuring the depth of a pool). To fix this, they proposed using the online model instead of the target model when selecting the best actions for the next states, and using the target model only to estimate the Q-Values for these best actions. Here is the updated `training_step()` function:

```
def training_step(batch_size):
    experiences = sample_experiences(batch_size)
    states, actions, rewards, next_states, dones = experiences
    next_Q_values = model.predict(next_states)
    best_next_actions = np.argmax(next_Q_values, axis=1)
    next_mask = tf.one_hot(best_next_actions, n_outputs).numpy()
    next_best_Q_values = (target.predict(next_states) * next_mask).sum(axis=1)
    target_Q_values = (rewards +
                          (1 - dones) * discount_factor * next_best_Q_values)
    mask = tf.one_hot(actions, n_outputs)
    [...] # the rest is the same as earlier
```

Just a few months later, another improvement to the DQN algorithm was proposed.

Prioritized Experience Replay

Instead of sampling experiences *uniformly* from the replay buffer, why not sample important experiences more frequently? This idea is called *importance sampling* (IS) or *prioritized experience replay* (PER), and it was introduced in a 2015 paper (*https://homl.info/prioreplay*)[15] by DeepMind researchers (once again!).

More specifically, experiences are considered "important" if they are likely to lead to fast learning progress. But how can we estimate this? One reasonable approach is to measure the magnitude of the TD error $\delta = r + \gamma \cdot V(s') - V(s)$. A large TD error indicates that a transition (s, r, s') is very surprising, and thus probably worth learning

---

14 Hado van Hasselt et al., "Deep Reinforcement Learning with Double Q-Learning," *Proceedings of the 30th AAAI Conference on Artificial Intelligence* (2015): 2094–2100.

15 Tom Schaul et al., "Prioritized Experience Replay," arXiv preprint arXiv:1511.05952 (2015).

from.[16] When an experience is recorded in the replay buffer, its priority is set to a very large value, to ensure that it gets sampled at least once. However, once it is sampled (and every time it is sampled), the TD error δ is computed, and this experience's priority is set to $p = |\delta|$ (plus a small constant to ensure that every experience has a non-zero probability of being sampled). The probability P of sampling an experience with priority p is proportional to p^ζ, where ζ is a hyperparameter that controls how greedy we want importance sampling to be: when $\zeta = 0$, we just get uniform sampling, and when $\zeta = 1$, we get full-blown importance sampling. In the paper, the authors used $\zeta = 0.6$, but the optimal value will depend on the task.

There's one catch, though: since the samples will be biased toward important experiences, we must compensate for this bias during training by downweighting the experiences according to their importance, or else the model will just overfit the important experiences. To be clear, we want important experiences to be sampled more often, but this also means we must give them a lower weight during training. To do this, we define each experience's training weight as $w = (n\ P)^{-\beta}$, where n is the number of experiences in the replay buffer, and β is a hyperparameter that controls how much we want to compensate for the importance sampling bias (0 means not at all, while 1 means entirely). In the paper, the authors used $\beta = 0.4$ at the beginning of training and linearly increased it to $\beta = 1$ by the end of training. Again, the optimal value will depend on the task, but if you increase one, you will usually want to increase the other as well.

Now let's look at one last important variant of the DQN algorithm.

Dueling DQN

The *Dueling DQN* algorithm (DDQN, not to be confused with Double DQN, although both techniques can easily be combined) was introduced in yet another 2015 paper (*https://homl.info/ddqn*)[17] by DeepMind researchers. To understand how it works, we must first note that the Q-Value of a state-action pair (s, a) can be expressed as $Q(s, a) = V(s) + A(s, a)$, where $V(s)$ is the value of state s and $A(s, a)$ is the *advantage* of taking the action a in state s, compared to all other possible actions in that state. Moreover, the value of a state is equal to the Q-Value of the best action a^* for that state (since we assume the optimal policy will pick the best action), so $V(s) = Q(s, a^*)$, which implies that $A(s, a^*) = 0$. In a Dueling DQN, the model estimates both the value of the state and the advantage of each possible action. Since the best action should have an advantage of 0, the model subtracts the maximum predicted

---

16 It could also just be that the rewards are noisy, in which case there are better methods for estimating an experience's importance (see the paper for some examples).

17 Ziyu Wang et al., "Dueling Network Architectures for Deep Reinforcement Learning," arXiv preprint arXiv: 1511.06581 (2015).

advantage from all predicted advantages. Here is a simple Dueling DQN model, implemented using the Functional API:

```
K = keras.backend
input_states = keras.layers.Input(shape=[4])
hidden1 = keras.layers.Dense(32, activation="elu")(input_states)
hidden2 = keras.layers.Dense(32, activation="elu")(hidden1)
state_values = keras.layers.Dense(1)(hidden2)
raw_advantages = keras.layers.Dense(n_outputs)(hidden2)
advantages = raw_advantages - K.max(raw_advantages, axis=1, keepdims=True)
Q_values = state_values + advantages
model = keras.Model(inputs=[input_states], outputs=[Q_values])
```

The rest of the algorithm is just the same as earlier. In fact, you can build a Double Dueling DQN and combine it with prioritized experience replay! More generally, many RL techniques can be combined, as DeepMind demonstrated in a 2017 paper (*https://homl.info/rainbow*).[18] The paper's authors combined six different techniques into an agent called *Rainbow*, which largely outperformed the state of the art.

Unfortunately, implementing all of these techniques, debugging them, fine-tuning them, and of course training the models can require a huge amount of work. So instead of reinventing the wheel, it is often best to reuse scalable and well-tested libraries, such as TF-Agents.

The TF-Agents Library

The TF-Agents library (*https://github.com/tensorflow/agents*) is a Reinforcement Learning library based on TensorFlow, developed at Google and open sourced in 2018. Just like OpenAI Gym, it provides many off-the-shelf environments (including wrappers for all OpenAI Gym environments), plus it supports the PyBullet library (for 3D physics simulation), DeepMind's DM Control library (based on MuJoCo's physics engine), and Unity's ML-Agents library (simulating many 3D environments). It also implements many RL algorithms, including REINFORCE, DQN, and DDQN, as well as various RL components such as efficient replay buffers and metrics. It is fast, scalable, easy to use, and customizable: you can create your own environments and neural nets, and you can customize pretty much any component. In this section we will use TF-Agents to train an agent to play *Breakout*, the famous Atari game (see Figure 18-11[19]), using the DQN algorithm (you can easily switch to another algorithm if you prefer).

---

18 Matteo Hessel et al., "Rainbow: Combining Improvements in Deep Reinforcement Learning," arXiv preprint arXiv:1710.02298 (2017): 3215–3222.

19 If you don't know this game, it's simple: a ball bounces around and breaks bricks when it touches them. You control a paddle near the bottom of the screen. The paddle can go left or right, and you must get the ball to break every brick, while preventing it from touching the bottom of the screen.

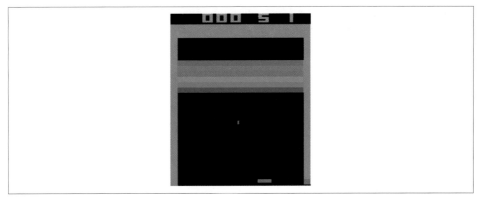

Figure 18-11. The famous Breakout game

Installing TF-Agents

Let's start by installing TF-Agents. This can be done using pip (as always, if you are using a virtual environment, make sure to activate it first; if not, you will need to use the --user option, or have administrator rights):

```
$ python3 -m pip install --upgrade tf-agents
```

At the time of this writing, TF-Agents is still quite new and improving every day, so the API may change a bit by the time you read this—but the big picture should remain the same, as well as most of the code. If anything breaks, I will update the Jupyter notebook accordingly, so make sure to check it out.

Next, let's create a TF-Agents environment that will just wrap OpenAI GGym's Breakout environment. For this, you must first install OpenAI Gym's Atari dependencies:

```
$ python3 -m pip install --upgrade 'gym[atari]'
```

Among other libraries, this command will install atari-py, which is a Python interface for the Arcade Learning Environment (ALE), a framework built on top of the Atari 2600 emulator Stella.

TF-Agents Environments

If everything went well, you should be able to import TF-Agents and create a Breakout environment:

```
>>> from tf_agents.environments import suite_gym
>>> env = suite_gym.load("Breakout-v4")
>>> env
<tf_agents.environments.wrappers.TimeLimit at 0x10c523c18>
```

This is just a wrapper around an OpenAI Gym environment, which you can access through the gym attribute:

```
>>> env.gym
<gym.envs.atari.atari_env.AtariEnv at 0x24dcab940>
```

TF-Agents environments are very similar to OpenAI Gym environments, but there are a few differences. First, the reset() method does not return an observation; instead it returns a TimeStep object that wraps the observation, as well as some extra information:

```
>>> env.reset()
TimeStep(step_type=array(0, dtype=int32),
        reward=array(0., dtype=float32),
        discount=array(1., dtype=float32),
        observation=array([[[0., 0., 0.], [0., 0., 0.],...]]], dtype=float32))
```

The step() method returns a TimeStep object as well:

```
>>> env.step(1) # Fire
TimeStep(step_type=array(1, dtype=int32),
        reward=array(0., dtype=float32),
        discount=array(1., dtype=float32),
        observation=array([[[0., 0., 0.], [0., 0., 0.],...]]], dtype=float32))
```

The reward and observation attributes are self-explanatory, and they are the same as for OpenAI Gym (except the reward is represented as a NumPy array). The step_type attribute is equal to 0 for the first time step in the episode, 1 for intermediate time steps, and 2 for the final time step. You can call the time step's is_last() method to check whether it is the final one or not. Lastly, the discount attribute indicates the discount factor to use at this time step. In this example it is equal to 1, so there will be no discount at all. You can define the discount factor by setting the discount parameter when loading the environment.

 At any time, you can access the environment's current time step by calling its current_time_step() method.

Environment Specifications

Conveniently, a TF-Agents environment provides the specifications of the observations, actions, and time steps, including their shapes, data types, and names, as well as their minimum and maximum values:

```
>>> env.observation_spec()
BoundedArraySpec(shape=(210, 160, 3), dtype=dtype('float32'), name=None,
                 minimum=[[[0. 0. 0.], [0. 0. 0.],...]],
                 maximum=[[[255., 255., 255.], [255., 255., 255.], ...]])
>>> env.action_spec()
BoundedArraySpec(shape=(), dtype=dtype('int64'), name=None,
                 minimum=0, maximum=3)
>>> env.time_step_spec()
TimeStep(step_type=ArraySpec(shape=(), dtype=dtype('int32'), name='step_type'),
         reward=ArraySpec(shape=(), dtype=dtype('float32'), name='reward'),
         discount=BoundedArraySpec(shape=(), ..., minimum=0.0, maximum=1.0),
         observation=BoundedArraySpec(shape=(210, 160, 3), ...))
```

As you can see, the observations are simply screenshots of the Atari screen, represented as NumPy arrays of shape [210, 160, 3]. To render an environment, you can
call env.render(mode="human"), and if you want to get back the image in the form of
a NumPy array, just call env.render(mode="rgb_array") (unlike in OpenAI Gym,
this is the default mode).

There are four actions available. Gym's Atari environments have an extra method that
you can call to know what each action corresponds to:

```
>>> env.gym.get_action_meanings()
['NOOP', 'FIRE', 'RIGHT', 'LEFT']
```

> Specs can be instances of a specification class, nested lists, or dic-
> tionaries of specs. If the specification is nested, then the specified
> object must match the specification's nested structure. For example,
> if the observation spec is {"sensors": ArraySpec(shape=[2]),
> "camera": ArraySpec(shape=[100, 100])}, then a valid observa-
> tion would be {"sensors": np.array([1.5, 3.5]), "camera":
> np.array(...)}. The tf.nest package provides tools to handle
> such nested structures (a.k.a. *nests*).

The observations are quite large, so we will downsample them and also convert them
to grayscale. This will speed up training and use less RAM. For this, we can use an
environment wrapper.

Environment Wrappers and Atari Preprocessing

TF-Agents provides several environment wrappers in the tf_agents.environ
ments.wrappers package. As their name suggests, they wrap an environment, for-
warding every call to it, but also adding some extra functionality. Here are some of
the available wrappers:

ActionClipWrapper
    Clips the actions to the action spec.

ActionDiscretizeWrapper

Quantizes a continuous action space to a discrete action space. For example, if the original environment's action space is the continuous range from –1.0 to +1.0, but you want to use an algorithm that only supports discrete action spaces, such as a DQN, then you can wrap the environment using discrete_env = ActionDiscretizeWrapper(env, num_actions=5), and the new discrete_env will have a discrete action space with five possible actions: 0, 1, 2, 3, 4. These actions correspond to the actions –1.0, –0.5, 0.0, 0.5, and 1.0 in the original environment.

ActionRepeat

Repeats each action over *n* steps, while accumulating the rewards. In many environments, this can speed up training significantly.

RunStats

Records environment statistics such as the number of steps and the number of episodes.

TimeLimit

Interrupts the environment if it runs for longer than a maximum number of steps.

VideoWrapper

Records a video of the environment.

To create a wrapped environment, you must create a wrapper, passing the wrapped environment to the constructor. That's all! For example, the following code will wrap our environment in an ActionRepeat wrapper so that every action is repeated four times:

```
from tf_agents.environments.wrappers import ActionRepeat

repeating_env = ActionRepeat(env, times=4)
```

OpenAI Gym has some environment wrappers of its own in the gym.wrappers package. They are meant to wrap Gym environments, though, not TF-Agents environments, so to use them you must first wrap the Gym environment with a Gym wrapper, then wrap the resulting environment with a TF-Agents wrapper. The suite_gym.wrap_env() function will do this for you, provided you give it a Gym environment and a list of Gym wrappers and/or a list of TF-Agents wrappers. Alternatively, the suite_gym.load() function will both create the Gym environment and wrap it for you, if you give it some wrappers. Each wrapper will be created without any arguments, so if you want to set some arguments, you must pass a lambda. For example, the following code creates a Breakout environment that will run for a maximum of 10,000 steps during each episode, and each action will be repeated four times:

```
from gym.wrappers import TimeLimit

limited_repeating_env = suite_gym.load(
    "Breakout-v4",
    gym_env_wrappers=[lambda env: TimeLimit(env, max_episode_steps=10000)],
    env_wrappers=[lambda env: ActionRepeat(env, times=4)])
```

For Atari environments, some standard preprocessing steps are applied in most papers that use them, so TF-Agents provides a handy AtariPreprocessing wrapper that implements them. Here is the list of preprocessing steps it supports:

Grayscale and downsampling
Observations are converted to grayscale and downsampled (by default to 84 × 84 pixels).

Max pooling
The last two frames of the game are max-pooled using a 1 × 1 filter. This is to remove the flickering that occurs in some Atari games due to the limited number of sprites that the Atari 2600 could display in each frame.

Frame skipping
The agent only gets to see every n frames of the game (by default $n = 4$), and its actions are repeated for each frame, collecting all the rewards. This effectively speeds up the game from the perspective of the agent, and it also speeds up training because rewards are less delayed.

End on life lost
In some games, the rewards are just based on the score, so the agent gets no immediate penalty for losing a life. One solution is to end the game immediately whenever a life is lost. There is some debate over the actual benefits of this strategy, so it is off by default.

Since the default Atari environment already applies random frame skipping and max pooling, we will need to load the raw, nonskipping variant called "BreakoutNoFrameskip-v4". Moreover, a single frame from the *Breakout* game is insufficient to know the direction and speed of the ball, which will make it very difficult for the agent to play the game properly (unless it is an RNN agent, which preserves some internal state between steps). One way to handle this is to use an environment wrapper that will output observations composed of multiple frames stacked on top of each other along the channels dimension. This strategy is implemented by the FrameStack4 wrapper, which returns stacks of four frames. Let's create the wrapped Atari environment!

```
from tf_agents.environments import suite_atari
from tf_agents.environments.atari_preprocessing import AtariPreprocessing
from tf_agents.environments.atari_wrappers import FrameStack4

max_episode_steps = 27000 # <=> 108k ALE frames since 1 step = 4 frames
environment_name = "BreakoutNoFrameskip-v4"

env = suite_atari.load(
    environment_name,
    max_episode_steps=max_episode_steps,
    gym_env_wrappers=[AtariPreprocessing, FrameStack4])
```

The result of all this preprocessing is shown in Figure 18-12. You can see that the resolution is much lower, but sufficient to play the game. Moreover, frames are stacked along the channels dimension, so red represents the frame from three steps ago, green is two steps ago, blue is the previous frame, and pink is the current frame.[20] From this single observation, the agent can see that the ball is going toward the lower-left corner, and that it should continue to move the paddle to the left (as it did in the previous steps).

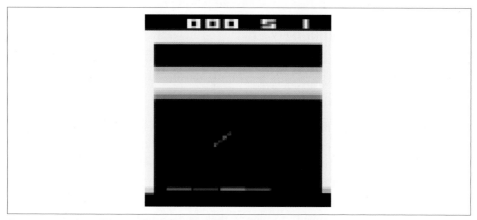

Figure 18-12. Preprocessed Breakout observation

Lastly, we can wrap the environment inside a TFPyEnvironment:

```
from tf_agents.environments.tf_py_environment import TFPyEnvironment

tf_env = TFPyEnvironment(env)
```

This will make the environment usable from within a TensorFlow graph (under the hood, this class relies on tf.py_function(), which allows a graph to call arbitrary

---

20 Since there are only three primary colors, you cannot just display an image with four color channels. For this reason, I combined the last channel with the first three to get the RGB image represented here. Pink is actually a mix of blue and red, but the agent sees four independent channels.

Python code). Thanks to the `TFPyEnvironment` class, TF-Agents supports both pure Python environments and TensorFlow-based environments. More generally, TF-Agents supports and provides both pure Python and TensorFlow-based components (agents, replay buffers, metrics, and so on).

Now that we have a nice Breakout environment, with all the appropriate preprocessing and TensorFlow support, we must create the DQN agent and the other components we will need to train it. Let's look at the architecture of the system we will build.

Training Architecture

A TF-Agents training program is usually split into two parts that run in parallel, as you can see in Figure 18-13: on the left, a *driver* explores the *environment* using a *collect policy* to choose actions, and it collects *trajectories* (i.e., experiences), sending them to an *observer*, which saves them to a *replay buffer*; on the right, an *agent* pulls batches of trajectories from the replay buffer and trains some *networks*, which the collect policy uses. In short, the left part explores the environment and collects trajectories, while the right part learns and updates the collect policy.

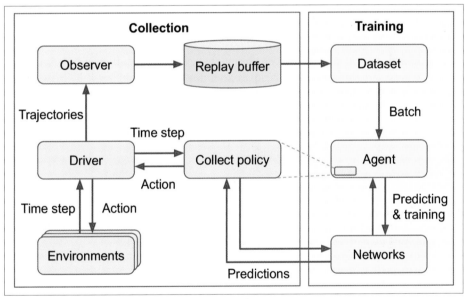

Figure 18-13. A typical TF-Agents training architecture

This figure begs a few questions, which I'll attempt to answer here:

- Why are there multiple environments? Instead of exploring a single environment, you generally want the driver to explore multiple copies of the environment in parallel, taking advantage of the power of all your CPU cores, keeping the training GPUs busy, and providing less-correlated trajectories to the training algorithm.

- What is a *trajectory*? It is a concise representation of a *transition* from one time step to the next, or a sequence of consecutive transitions from time step n to time step $n + t$. The trajectories collected by the driver are passed to the observer, which saves them in the replay buffer, and they are later sampled by the agent and used for training.

- Why do we need an observer? Can't the driver save the trajectories directly? Indeed, it could, but this would make the architecture less flexible. For example, what if you don't want to use a replay buffer? What if you want to use the trajectories for something else, like computing metrics? In fact, an observer is just any function that takes a trajectory as an argument. You can use an observer to save the trajectories to a replay buffer, or to save them to a TFRecord file (see Chapter 13), or to compute metrics, or for anything else. Moreover, you can pass multiple observers to the driver, and it will broadcast the trajectories to all of them.

Although this architecture is the most common, you can customize it as you please, and even replace some components with your own. In fact, unless you are researching new RL algorithms, you will most likely want to use a custom environment for your task. For this, you just need to create a custom class that inherits from the PyEnvironment class in the tf_agents.environments.py_environ ment package and overrides the appropriate methods, such as action_spec(), observation_spec(), _reset(), and _step() (see the "Creating a Custom TF_Agents Environment" section of the notebook for an example).

Now we will create all these components: first the Deep Q-Network, then the DQN agent (which will take care of creating the collect policy), then the replay buffer and the observer to write to it, then a few training metrics, then the driver, and finally the dataset. Once we have all the components in place, we will populate the replay buffer with some initial trajectories, then we will run the main training loop. So, let's start by creating the Deep Q-Network.

Creating the Deep Q-Network

The TF-Agents library provides many networks in the tf_agents.networks package and its subpackages. We will use the tf_agents.networks.q_network.QNetwork class:

```
from tf_agents.networks.q_network import QNetwork

preprocessing_layer = keras.layers.Lambda(
                          lambda obs: tf.cast(obs, np.float32) / 255.)
conv_layer_params=[(32, (8, 8), 4), (64, (4, 4), 2), (64, (3, 3), 1)]
fc_layer_params=[512]

q_net = QNetwork(
    tf_env.observation_spec(),
    tf_env.action_spec(),
    preprocessing_layers=preprocessing_layer,
    conv_layer_params=conv_layer_params,
    fc_layer_params=fc_layer_params)
```

This QNetwork takes an observation as input and outputs one Q-Value per action, so we must give it the specifications of the observations and the actions. It starts with a preprocessing layer: a simple Lambda layer that casts the observations to 32-bit floats and normalizes them (the values will range from 0.0 to 1.0). The observations contain unsigned bytes, which use 4 times less space than 32-bit floats, which is why we did not cast the observations to 32-bit floats earlier; we want to save RAM in the replay buffer. Next, the network applies three convolutional layers: the first has 32 8 × 8 filters and uses a stride of 4, the second has 64 4 × 4 filters and a stride of 2, and the third has 64 3 × 3 filters and a stride of 1. Lastly, it applies a dense layer with 512 units, followed by a dense output layer with 4 units, one per Q-Value to output (i.e., one per action). All convolutional layers and all dense layers except the output layer use the ReLU activation function by default (you can change this by setting the acti vation_fn argument). The output layer does not use any activation function.

Under the hood, a QNetwork is composed of two parts: an encoding network that processes the observations, followed by a dense output layer that outputs one Q-Value per action. TF-Agent's EncodingNetwork class implements a neural network architecture found in various agents (see Figure 18-14).

It may have one or more inputs. For example, if each observation is composed of some sensor data plus an image from a camera, you will have two inputs. Each input may require some preprocessing steps, in which case you can specify a list of Keras layers via the preprocessing_layers argument, with one preprocessing layer per input, and the network will apply each layer to the corresponding input (if an input requires multiple layers of preprocessing, you can pass a whole model, since a Keras model can always be used as a layer). If there are two inputs or more, you must also pass an extra layer via the preprocessing_combiner argument, to combine the outputs from the preprocessing layers into a single output.

Next, the encoding network will optionally apply a list of convolutions sequentially, provided you specify their parameters via the conv_layer_params argument. This must be a list composed of 3-tuples (one per convolutional layer) indicating the number of filters, the kernel size, and the stride. After these convolutional layers, the encoding network will optionally apply a sequence of dense layers, if you set the `fc_layer_params` argument: it must be a list containing the number of neurons for each dense layer. Optionally, you can also pass a list of dropout rates (one per dense layer) via the `dropout_layer_params` argument if you want to apply dropout after each dense layer. The `QNetwork` takes the output of this encoding network and passes it to the dense output layer (with one unit per action).

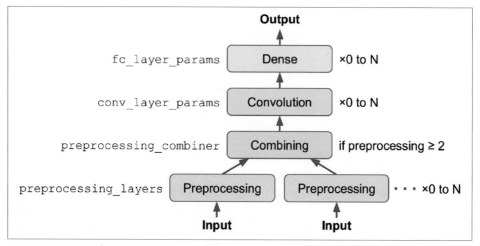

Figure 18-14. Architecture of an encoding network

 The `QNetwork` class is flexible enough to build many different architectures, but you can always build your own network class if you need extra flexibility: extend the `tf_agents.networks.Net work` class and implement it like a regular custom Keras layer. The `tf_agents.networks.Network` class is a subclass of the `keras.lay ers.Layer` class that adds some functionality required by some agents, such as the possibility to easily create shallow copies of the network (i.e., copying the network's architecture, but not its weights). For example, the `DQNAgent` uses this to create a copy of the online model.

Now that we have the DQN, we are ready to build the DQN agent.

Creating the DQN Agent

The TF-Agents library implements many types of agents, located in the `tf_agents .agents` package and its subpackages. We will use the `tf_agents.agents .dqn.dqn_agent.DqnAgent` class:

```
from tf_agents.agents.dqn.dqn_agent import DqnAgent

train_step = tf.Variable(0)
update_period = 4 # train the model every 4 steps
optimizer = keras.optimizers.RMSprop(lr=2.5e-4, rho=0.95, momentum=0.0,
                                     epsilon=0.00001, centered=True)
epsilon_fn = keras.optimizers.schedules.PolynomialDecay(
    initial_learning_rate=1.0, # initial ε
    decay_steps=250000 // update_period, # <=> 1,000,000 ALE frames
    end_learning_rate=0.01) # final ε
agent = DqnAgent(tf_env.time_step_spec(),
                 tf_env.action_spec(),
                 q_network=q_net,
                 optimizer=optimizer,
                 target_update_period=2000, # <=> 32,000 ALE frames
                 td_errors_loss_fn=keras.losses.Huber(reduction="none"),
                 gamma=0.99, # discount factor
                 train_step_counter=train_step,
                 epsilon_greedy=lambda: epsilon_fn(train_step))
agent.initialize()
```

Let's walk through this code:

- We first create a variable that will count the number of training steps.

- Then we build the optimizer, using the same hyperparameters as in the 2015 DQN paper.

- Next, we create a `PolynomialDecay` object that will compute the ε value for the ε-greedy collect policy, given the current training step (it is normally used to decay the learning rate, hence the names of the arguments, but it will work just fine to decay any other value). It will go from 1.0 down to 0.01 (the value used during in the 2015 DQN paper) in 1 million ALE frames, which corresponds to 250,000 steps, since we use frame skipping with a period of 4. Moreover, we will train the agent every 4 steps (i.e., 16 ALE frames), so ε will actually decay over 62,500 *training* steps.

- We then build the `DQNAgent`, passing it the time step and action specs, the `QNet` work to train, the optimizer, the number of training steps between target model updates, the loss function to use, the discount factor, the `train_step` variable, and a function that returns the ε value (it must take no argument, which is why we need a lambda to pass the `train_step`).

  Note that the loss function must return an error per instance, not the mean error, which is why we set `reduction="none"`.

- Lastly, we initialize the agent.

Next, let's build the replay buffer and the observer that will write to it.

Creating the Replay Buffer and the Corresponding Observer

The TF-Agents library provides various replay buffer implementations in the `tf_agents.replay_buffers` package. Some are purely written in Python (their module names start with py_), and others are written based on TensorFlow (their module names start with tf_). We will use the `TFUniformReplayBuffer` class in the `tf_agents.replay_buffers.tf_uniform_replay_buffer` package. It provides a high-performance implementation of a replay buffer with uniform sampling:[21]

```
from tf_agents.replay_buffers import tf_uniform_replay_buffer

replay_buffer = tf_uniform_replay_buffer.TFUniformReplayBuffer(
    data_spec=agent.collect_data_spec,
    batch_size=tf_env.batch_size,
    max_length=1000000)
```

Let's look at each of these arguments:

data_spec
> The specification of the data that will be saved in the replay buffer. The DQN agent knowns what the collected data will look like, and it makes the data spec available via its `collect_data_spec` attribute, so that's what we give the replay buffer.

batch_size
> The number of trajectories that will be added at each step. In our case, it will be one, since the driver will just execute one action per step and collect one trajectory. If the environment were a *batched environment*, meaning an environment that takes a batch of actions at each step and returns a batch of observations, then the driver would have to save a batch of trajectories at each step. Since we are using a TensorFlow replay buffer, it needs to know the size of the batches it will handle (to build the computation graph). An example of a batched environment is the `ParallelPyEnvironment` (from the `tf_agents.environments.paral lel_py_environment` package): it runs multiple environments in parallel in separate processes (they can be different as long as they have the same action and observation specs), and at each step it takes a batch of actions and executes them in the environments (one action per environment), then it returns all the resulting observations.

---

21 At the time of this writing, there is no prioritized experience replay buffer yet, but one will likely be open sourced soon.

max_length

The maximum size of the replay buffer. We created a large replay buffer that can store one million trajectories (as was done in the 2015 DQN paper). This will require a lot of RAM.

 When we store two consecutive trajectories, they contain two consecutive observations with four frames each (since we used the `Fra meStack4` wrapper), and unfortunately three of the four frames in the second observation are redundant (they are already present in the first observation). In other words, we are using about four times more RAM than necessary. To avoid this, you can instead use a `PyHashedReplayBuffer` from the `tf_agents.replay_buf fers.py_hashed_replay_buffer` package: it deduplicates data in the stored trajectories along the last axis of the observations.

Now we can create the observer that will write the trajectories to the replay buffer. An observer is just a function (or a callable object) that takes a trajectory argument, so we can directly use the `add_method()` method (bound to the `replay_buffer` object) as our observer:

```
replay_buffer_observer = replay_buffer.add_batch
```

If you wanted to create your own observer, you could write any function with a `trajectory` argument. If it must have a state, you can write a class with a `__call__(self, trajectory)` method. For example, here is a simple observer that will increment a counter every time it is called (except when the trajectory represents a boundary between two episodes, which does not count as a step), and every 100 increments it displays the progress up to a given total (the carriage return \r along with end="" ensures that the displayed counter remains on the same line):

```
class ShowProgress:
    def __init__(self, total):
        self.counter = 0
        self.total = total
    def __call__(self, trajectory):
        if not trajectory.is_boundary():
            self.counter += 1
        if self.counter % 100 == 0:
            print("\r{}/{}".format(self.counter, self.total), end="")
```

Now let's create a few training metrics.

Creating Training Metrics

TF-Agents implements several RL metrics in the `tf_agents.metrics` package, some purely in Python and some based on TensorFlow. Let's create a few of them in order to count the number of episodes, the number of steps taken, and most importantly the average return per episode and the average episode length:

```
from tf_agents.metrics import tf_metrics

train_metrics = [
    tf_metrics.NumberOfEpisodes(),
    tf_metrics.EnvironmentSteps(),
    tf_metrics.AverageReturnMetric(),
    tf_metrics.AverageEpisodeLengthMetric(),
]
```

> Discounting the rewards makes sense for training or to implement a policy, as it makes it possible to balance the importance of immediate rewards with future rewards. However, once an episode is over, we can evaluate how good it was overalls by summing the *undiscounted* rewards. For this reason, the AverageReturnMetric computes the sum of undiscounted rewards for each episode, and it keeps track of the streaming mean of these sums over all the episodes it encounters.

At any time, you can get the value of each of these metrics by calling its result() method (e.g., train_metrics[0].result()). Alternatively, you can log all metrics by calling log_metrics(train_metrics) (this function is located in the tf_agents.eval.metric_utils package):

```
>>> from tf_agents.eval.metric_utils import log_metrics
>>> import logging
>>> logging.get_logger().set_level(logging.INFO)
>>> log_metrics(train_metrics)
[...]
NumberOfEpisodes = 0
EnvironmentSteps = 0
AverageReturn = 0.0
AverageEpisodeLength = 0.0
```

Next, let's create the collect driver.

Creating the Collect Driver

As we explored in Figure 18-13, a driver is an object that explores an environment using a given policy, collects experiences, and broadcasts them to some observers. At each step, the following things happen:

- The driver passes the current time step to the collect policy, which uses this time step to choose an action and returns an *action step* object containing the action.

- The driver then passes the action to the environment, which returns the next time step.
- Finally, the driver creates a trajectory object to represent this transition and broadcasts it to all the observers.

Some policies, such as RNN policies, are stateful: they choose an action based on both the given time step and their own internal state. Stateful policies return their own state in the action step, along with the chosen action. The driver will then pass this state back to the policy at the next time step. Moreover, the driver saves the policy state to the trajectory (in the policy_info field), so it ends up in the replay buffer. This is essential when training a stateful policy: when the agent samples a trajectory, it must set the policy's state to the state it was in at the time of the sampled time step.

Also, as discussed earlier, the environment may be a batched environment, in which case the driver passes a *batched time step* to the policy (i.e., a time step object containing a batch of observations, a batch of step types, a batch of rewards, and a batch of discounts, all four batches of the same size). The driver also passes a batch of previous policy states. The policy then returns a *batched action step* containing a batch of actions and a batch of policy states. Finally, the driver creates a *batched trajectory* (i.e., a trajectory containing a batch of step types, a batch of observations, a batch of actions, a batch of rewards, and more generally a batch for each trajectory attribute, with all batches of the same size).

There are two main driver classes: DynamicStepDriver and DynamicEpisodeDriver. The first one collects experiences for a given number of steps, while the second collects experiences for a given number of episodes. We want to collect experiences for four steps for each training iteration (as was done in the 2015 DQN paper), so let's create a DynamicStepDriver:

```python
from tf_agents.drivers.dynamic_step_driver import DynamicStepDriver

collect_driver = DynamicStepDriver(
    tf_env,
    agent.collect_policy,
    observers=[replay_buffer_observer] + training_metrics,
    num_steps=update_period) # collect 4 steps for each training iteration
```

We give it the environment to play with, the agent's collect policy, a list of observers (including the replay buffer observer and the training metrics), and finally the number of steps to run (in this case, four). We could now run it by calling its run() method, but it's best to warm up the replay buffer with experiences collected using a purely random policy. For this, we can use the RandomTFPolicy class and create a second driver that will run this policy for 20,000 steps (which is equivalent to 80,000 simulator frames, as was done in the 2015 DQN paper). We can use our ShowProgress observer to display the progress:

```
from tf_agents.policies.random_tf_policy import RandomTFPolicy

initial_collect_policy = RandomTFPolicy(tf_env.time_step_spec(),
                                        tf_env.action_spec())
init_driver = DynamicStepDriver(
    tf_env,
    initial_collect_policy,
    observers=[replay_buffer.add_batch, ShowProgress(20000)],
    num_steps=20000) # <=> 80,000 ALE frames
final_time_step, final_policy_state = init_driver.run()
```

We're almost ready to run the training loop! We just need one last component: the dataset.

Creating the Dataset

To sample a batch of trajectories from the replay buffer, call its `get_next()` method. This returns the batch of trajectories plus a `BufferInfo` object that contains the sample identifiers and their sampling probabilities (this may be useful for some algorithms, such as PER). For example, the following code will sample a small batch of two trajectories (subepisodes), each containing three consecutive steps. These subepisodes are shown in Figure 18-15 (each row contains three consecutive steps from an episode):

```
>>> trajectories, buffer_info = replay_buffer.get_next(
...         sample_batch_size=2, num_steps=3)
...
>>> trajectories._fields
('step_type', 'observation', 'action', 'policy_info',
 'next_step_type', 'reward', 'discount')
>>> trajectories.observation.shape
TensorShape([2, 3, 84, 84, 4])
>>> trajectories.step_type.numpy()
array([[1, 1, 1],
       [1, 1, 1]], dtype=int32)
```

The `trajectories` object is a named tuple, with seven fields. Each field contains a tensor whose first two dimensions are 2 and 3 (since there are two trajectories, each with three steps). This explains why the shape of the `observation` field is [2, 3, 84, 84, 4]: that's two trajectories, each with three steps, and each step's observation is 84 × 84 × 4. Similarly, the `step_type` tensor has a shape of [2, 3]: in this example, both trajectories contain three consecutive steps in the middle on an episode (types 1, 1, 1). In the second trajectory, you can barely see the ball at the lower left of the first observation, and it disappears in the next two observations, so the agent is about to lose a life, but the episode will not end immediately because it still has several lives left.

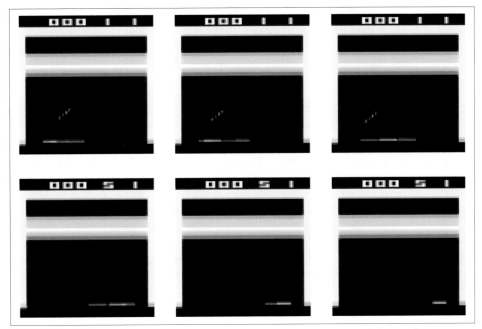

Figure 18-15. Two trajectories containing three consecutive steps each

Each trajectory is a concise representation of a sequence of consecutive time steps and action steps, designed to avoid redundancy. How so? Well, as you can see in Figure 18-16, transition n is composed of time step n, action step n, and time step $n + 1$, while transition $n + 1$ is composed of time step $n + 1$, action step $n + 1$, and time step $n + 2$. If we just stored these two transitions directly in the replay buffer, the time step $n + 1$ would be duplicated. To avoid this duplication, the nth trajectory step includes only the type and observation from time step n (not its reward and discount), and it does not contain the observation from time step $n + 1$ (however, it does contain a copy of the next time step's type; that's the only duplication).

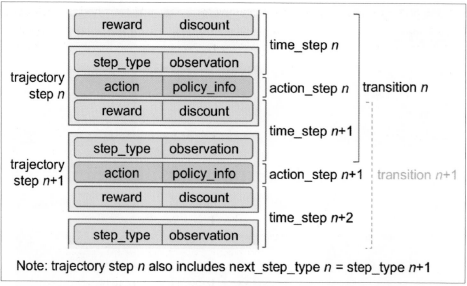

Note: trajectory step *n* also includes next_step_type *n* = step_type *n*+1

Figure 18-16. Trajectories, transitions, time steps, and action steps

So if you have a batch of trajectories where each trajectory has $t + 1$ steps (from time step *n* to time step $n + t$), then it contains all the data from time step *n* to time step $n + t$, except for the reward and discount from time step *n* (but it contains the reward and discount of time step $n + t + 1$). This represents *t* transitions (*n* to $n + 1$, $n + 1$ to $n + 2$, ..., $n + t - 1$ to $n + t$).

The to_transition() function in the tf_agents.trajectories.trajectory module converts a batched trajectory into a list containing a batched time_step, a batched action_step, and a batched next_time_step. Notice that the second dimension is 2 instead of 3, since there are *t* transitions between $t + 1$ time steps (don't worry if you're a bit confused; you'll get the hang of it):

```
>>> from tf_agents.trajectories.trajectory import to_transition
>>> time_steps, action_steps, next_time_steps = to_transition(trajectories)
>>> time_steps.observation.shape
TensorShape([2, 2, 84, 84, 4]) # 3 time steps = 2 transitions
```

A sampled trajectory may actually overlap two (or more) episodes! In this case, it will contain *boundary transitions*, meaning transitions with a step_type equal to 2 (end) and a next_step_type equal to 0 (start). Of course, TF-Agents properly handles such trajectories (e.g., by resetting the policy state when encountering a boundary). The trajectory's is_boundary() method returns a tensor indicating whether each step is a boundary or not.

For our main training loop, instead of calling the get_next() method, we will use a tf.data.Dataset. This way, we can benefit from the power of the Data API (e.g., parallelism and prefetching). For this, we call the replay buffer's as_dataset() method:

```
dataset = replay_buffer.as_dataset(
    sample_batch_size=64,
    num_steps=2,
    num_parallel_calls=3).prefetch(3)
```

We will sample batches of 64 trajectories at each training step (as in the 2015 DQN paper), each with 2 steps (i.e., 2 steps = 1 full transition, including the next step's observation). This dataset will process three elements in parallel, and prefetch three batches.

> For on-policy algorithms such as Policy Gradients, each experience should be sampled once, used from training, and then discarded. In this case, you can still use a replay buffer, but instead of using a Dataset, you would call the replay buffer's gather_all() method at each training iteration to get a tensor containing all the trajectories recorded so far, then use them to perform a training step, and finally clear the replay buffer by calling its clear() method.

Now that we have all the components in place, we are ready to train the model!

Creating the Training Loop

To speed up training, we will convert the main functions to TensorFlow Functions. For this we will use the tf_agents.utils.common.function() function, which wraps tf.function(), with some extra experimental options:

```
from tf_agents.utils.common import function

collect_driver.run = function(collect_driver.run)
agent.train = function(agent.train)
```

Let's create a small function that will run the main training loop for n_iterations:

```
def train_agent(n_iterations):
    time_step = None
    policy_state = agent.collect_policy.get_initial_state(tf_env.batch_size)
    iterator = iter(dataset)
    for iteration in range(n_iterations):
        time_step, policy_state = collect_driver.run(time_step, policy_state)
        trajectories, buffer_info = next(iterator)
        train_loss = agent.train(trajectories)
        print("\r{} loss:{:.5f}".format(
            iteration, train_loss.loss.numpy()), end="")
        if iteration % 1000 == 0:
            log_metrics(train_metrics)
```

The function first asks the collect policy for its initial state (given the environment batch size, which is 1 in this case). Since the policy is stateless, this returns an empty tuple (so we could have written `policy_state = ()`). Next, we create an iterator over the dataset, and we run the training loop. At each iteration, we call the driver's `run()` method, passing it the current time step (initially `None`) and the current policy state. It will run the collect policy and collect experience for four steps (as we configured earlier), broadcasting the collected trajectories to the replay buffer and the metrics. Next, we sample one batch of trajectories from the dataset, and we pass it to the agent's `train()` method. It returns a `train_loss` object which may vary depending on the type of agent. Next, we display the iteration number and the training loss, and every 1,000 iterations we log all the metrics. Now you can just call `train_agent()` for some number of iterations, and see the agent gradually learn to play *Breakout*!

```
train_agent(10000000)
```

This will take a lot of computing power and a lot of patience (it may take hours, or even days, depending on your hardware), plus you may need to run the algorithm several times with different random seeds to get good results, but once it's done, the agent will be superhuman (at least at *Breakout*). You can also try training this DQN agent on other Atari games: it can achieve superhuman skill at most action games, but it is not so good at games with long-running storylines.[22]

Overview of Some Popular RL Algorithms

Before we finish this chapter, let's take a quick look at a few popular RL algorithms:

Actor-Critic algorithms
> A family of RL algorithms that combine Policy Gradients with Deep Q-Networks. An Actor-Critic agent contains two neural networks: a policy net and a DQN. The DQN is trained normally, by learning from the agent's experiences. The policy net learns differently (and much faster) than in regular PG: instead of estimating the value of each action by going through multiple episodes, then summing the future discounted rewards for each action, and finally normalizing them, the agent (actor) relies on the action values estimated by the DQN (critic). It's a bit like an athlete (the agent) learning with the help of a coach (the DQN).

Asynchronous Advantage Actor-Critic (https://homl.info/a3c)[23] (A3C)
> An important Actor-Critic variant introduced by DeepMind researchers in 2016, where multiple agents learn in parallel, exploring different copies of the environ-

---

22 For a comparison of this algorithm's performance on various Atari games, see figure 3 in DeepMind's 2015 paper (*https://homl.info/dqn2*).

23 Volodymyr Mnih et al., "Asynchonous Methods for Deep Reinforcement Learning," *Proceedings of the 33rd International Conference on Machine Learning* (2016): 1928–1937.

ment. At regular intervals, but asynchronously (hence the name), each agent pushes some weight updates to a master network, then it pulls the latest weights from that network. Each agent thus contributes to improving the master network and benefits from what the other agents have learned. Moreover, instead of estimating the Q-Values, the DQN estimates the advantage of each action (hence the second A in the name), which stabilizes training.

Advantage Actor-Critic (https://homl.info/a2c) (A2C)

A variant of the A3C algorithm that removes the asynchronicity. All model updates are synchronous, so gradient updates are performed over larger batches, which allows the model to better utilize the power of the GPU.

Soft Actor-Critic (https://homl.info/sac)[24] (SAC)

An Actor-Critic variant proposed in 2018 by Tuomas Haarnoja and other UC Berkeley researchers. It learns not only rewards, but also to maximize the entropy of its actions. In other words, it tries to be as unpredictable as possible while still getting as many rewards as possible. This encourages the agent to explore the environment, which speeds up training, and makes it less likely to repeatedly execute the same action when the DQN produces imperfect estimates. This algorithm has demonstrated an amazing sample efficiency (contrary to all the previous algorithms, which learn very slowly). SAC is available in TF-Agents.

Proximal Policy Optimization (PPO) (https://homl.info/ppo)[25]

An algorithm based on A2C that clips the loss function to avoid excessively large weight updates (which often lead to training instabilities). PPO is a simplification of the previous *Trust Region Policy Optimization (https://homl.info/trpo)[26]* (TRPO) algorithm, also by John Schulman and other OpenAI researchers. OpenAI made the news in April 2019 with their AI called OpenAI Five, based on the PPO algorithm, which defeated the world champions at the multiplayer game *Dota 2*. PPO is also available in TF-Agents.

---

24 Tuomas Haarnoja et al., "Soft Actor-Critic: Off-Policy Maximum Entropy Deep Reinforcement Learning with a Stochastic Actor," *Proceedings of the 35th International Conference on Machine Learning* (2018): 1856–1865.

25 John Schulman et al., "Proximal Policy Optimization Algorithms," arXiv preprint arXiv:1707.06347 (2017).

26 John Schulman et al., "Trust Region Policy Optimization," *Proceedings of the 32nd International Conference on Machine Learning* (2015): 1889–1897.

Curiosity-based exploration (https://homl.info/curiosity)[27]

A recurring problem in RL is the sparsity of the rewards, which makes learning very slow and inefficient. Deepak Pathak and other UC Berkeley researchers have proposed an exciting way to tackle this issue: why not ignore the rewards, and just make the agent extremely curious to explore the environment? The rewards thus become intrinsic to the agent, rather than coming from the environment. Similarly, stimulating curiosity in a child is more likely to give good results than purely rewarding the child for getting good grades. How does this work? The agent continuously tries to predict the outcome of its actions, and it seeks situations where the outcome does not match its predictions. In other words, it wants to be surprised. If the outcome is predictable (boring), it goes elsewhere. However, if the outcome is unpredictable but the agent notices that it has no control over it, it also gets bored after a while. With only curiosity, the authors succeeded in training an agent at many video games: even though the agent gets no penalty for losing, the game starts over, which is boring so it learns to avoid it.

We covered many topics in this chapter: Policy Gradients, Markov chains, Markov decision processes, Q-Learning, Approximate Q-Learning, and Deep Q-Learning and its main variants (fixed Q-Value targets, Double DQN, Dueling DQN, and prioritized experience replay). We discussed how to use TF-Agents to train agents at scale, and finally we took a quick look at a few other popular algorithms. Reinforcement Learning is a huge and exciting field, with new ideas and algorithms popping out every day, so I hope this chapter sparked your curiosity: there is a whole world to explore!

Exercises

1. How would you define Reinforcement Learning? How is it different from regular supervised or unsupervised learning?

2. Can you think of three possible applications of RL that were not mentioned in this chapter? For each of them, what is the environment? What is the agent? What are some possible actions? What are the rewards?

3. What is the discount factor? Can the optimal policy change if you modify the discount factor?

4. How do you measure the performance of a Reinforcement Learning agent?

5. What is the credit assignment problem? When does it occur? How can you alleviate it?

6. What is the point of using a replay buffer?

---

27 Deepak Pathak et al., "Curiosity-Driven Exploration by Self-Supervised Prediction," *Proceedings of the 34th International Conference on Machine Learning* (2017): 2778–2787.

7. What is an off-policy RL algorithm?

8. Use policy gradients to solve OpenAI Gym's LunarLander-v2 environment. You will need to install the Box2D dependencies (`python3 -m pip install gym[box2d]`).

9. Use TF-Agents to train an agent that can achieve a superhuman level at SpaceInvaders-v4 using any of the available algorithms.

10. If you have about $100 to spare, you can purchase a Raspberry Pi 3 plus some cheap robotics components, install TensorFlow on the Pi, and go wild! For an example, check out this fun post (*https://homl.info/2*) by Lukas Biewald, or take a look at GoPiGo or BrickPi. Start with simple goals, like making the robot turn around to find the brightest angle (if it has a light sensor) or the closest object (if it has a sonar sensor), and move in that direction. Then you can start using Deep Learning: for example, if the robot has a camera, you can try to implement an object detection algorithm so it detects people and moves toward them. You can also try to use RL to make the agent learn on its own how to use the motors to achieve that goal. Have fun!

Solutions to these exercises are available in Appendix A.

## CHAPTER 19

Training and Deploying TensorFlow Models at Scale

Once you have a beautiful model that makes amazing predictions, what do you do with it? Well, you need to put it in production! This could be as simple as running the model on a batch of data and perhaps writing a script that runs this model every night. However, it is often much more involved. Various parts of your infrastructure may need to use this model on live data, in which case you probably want to wrap your model in a web service: this way, any part of your infrastructure can query your model at any time using a simple REST API (or some other protocol), as we discussed in Chapter 2. But as time passes, you need to regularly retrain your model on fresh data and push the updated version to production. You must handle model versioning, gracefully transition from one model to the next, possibly roll back to the previous model in case of problems, and perhaps run multiple different models in parallel to perform *A/B experiments*.[1] If your product becomes successful, your service may start to get plenty of *queries per second* (QPS), and it must scale up to support the load. A great solution to scale up your service, as we will see in this chapter, is to use TF Serving, either on your own hardware infrastructure or via a cloud service such as Google Cloud AI Platform. It will take care of efficiently serving your model, handle graceful model transitions, and more. If you use the cloud platform, you will also get many extra features, such as powerful monitoring tools.

Moreover, if you have a lot of training data, and compute-intensive models, then training time may be prohibitively long. If your product needs to adapt to changes quickly, then a long training time can be a showstopper (e.g., think of a news

---

1 An A/B experiment consists in testing two different versions of your product on different subsets of users in order to check which version works best and get other insights.

recommendation system promoting news from last week). Perhaps even more importantly, a long training time will prevent you from experimenting with new ideas. In Machine Learning (as in many other fields), it is hard to know in advance which ideas will work, so you should try out as many as possible, as fast as possible. One way to speed up training is to use hardware accelerators such as GPUs or TPUs. To go even faster, you can train a model across multiple machines, each equipped with multiple hardware accelerators. TensorFlow's simple yet powerful Distribution Strategies API makes this easy, as we will see.

In this chapter we will look at how to deploy models, first to TF Serving, then to Google Cloud AI Platform. We will also take a quick look at deploying models to mobile apps, embedded devices, and web apps. Lastly, we will discuss how to speed up computations using GPUs and how to train models across multiple devices and servers using the Distribution Strategies API. That's a lot of topics to discuss, so let's get started!

Serving a TensorFlow Model

Once you have trained a TensorFlow model, you can easily use it in any Python code: if it's a tf.keras model, just call its `predict()` method! But as your infrastructure grows, there comes a point where it is preferable to wrap your model in a small service whose sole role is to make predictions and have the rest of the infrastructure query it (e.g., via a REST or gRPC API).[2] This decouples your model from the rest of the infrastructure, making it possible to easily switch model versions or scale the service up as needed (independently from the rest of your infrastructure), perform A/B experiments, and ensure that all your software components rely on the same model versions. It also simplifies testing and development, and more. You could create your own microservice using any technology you want (e.g., using the Flask library), but why reinvent the wheel when you can just use TF Serving?

Using TensorFlow Serving

TF Serving is a very efficient, battle-tested model server that's written in C++. It can sustain a high load, serve multiple versions of your models and watch a model repository to automatically deploy the latest versions, and more (see Figure 19-1).

---

2 A REST (or RESTful) API is an API that uses standard HTTP verbs, such as GET, POST, PUT, and DELETE, and uses JSON inputs and outputs. The gRPC protocol is more complex but more efficient. Data is exchanged using Protocol Buffers (see Chapter 13).

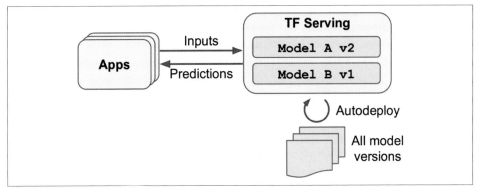

Figure 19-1. TF Serving can serve multiple models and automatically deploy the latest version of each model

So let's suppose you have trained an MNIST model using tf.keras, and you want to deploy it to TF Serving. The first thing you have to do is export this model to TensorFlow's *SavedModel format*.

Exporting SavedModels

TensorFlow provides a simple `tf.saved_model.save()` function to export models to the SavedModel format. All you need to do is give it the model, specifying its name and version number, and the function will save the model's computation graph and its weights:

```
model = keras.models.Sequential([...])
model.compile([...])
history = model.fit([...])

model_version = "0001"
model_name = "my_mnist_model"
model_path = os.path.join(model_name, model_version)
tf.saved_model.save(model, model_path)
```

It's usually a good idea to include all the preprocessing layers in the final model you export so that it can ingest data in its natural form once it is deployed to production. This avoids having to take care of preprocessing separately within the application that uses the model. Bundling the preprocessing steps within the model also makes it simpler to update them later on and limits the risk of mismatch between a model and the preprocessing steps it requires.

Since a SavedModel saves the computation graph, it can only be used with models that are based exclusively on TensorFlow operations, excluding the `tf.py_function()` operation (which wraps arbitrary Python code). It also excludes dynamic tf.keras models (see Appendix G), since these models cannot be converted to computation graphs. Dynamic models need to be served using other tools (e.g., Flask).

A SavedModel represents a version of your model. It is stored as a directory containing a *saved_model.pb* file, which defines the computation graph (represented as a serialized protocol buffer), and a *variables* subdirectory containing the variable values. For models containing a large number of weights, these variable values may be split across multiple files. A SavedModel also includes an *assets* subdirectory that may contain additional data, such as vocabulary files, class names, or some example instances for this model. The directory structure is as follows (in this example, we don't use assets):

```
my_mnist_model
└── 0001
    ├── assets
    ├── saved_model.pb
    └── variables
        ├── variables.data-00000-of-00001
        └── variables.index
```

As you might expect, you can load a SavedModel using the `tf.saved_model.load()` function. However, the returned object is not a Keras model: it represents the Saved-Model, including its computation graph and variable values. You can use it like a function, and it will make predictions (make sure to pass the inputs as tensors, and you must also set the `training` argument, generally to `False`):

```
saved_model = tf.saved_model.load(model_path)
y_pred = saved_model(X_new, training=False)
```

Alternatively, you can wrap this SavedModel's prediction function in a Keras model:

```
inputs = keras.layers.Input(shape=...)
outputs = saved_model(inputs, training=False)
model = keras.models.Model(inputs=[inputs], outputs=[outputs])
y_pred = model.predict(X_new)
```

TensorFlow also comes with a small `saved_model_cli` command-line tool to inspect SavedModels:

```
$ export ML_PATH="$HOME/ml" # point to this project, wherever it is
$ cd $ML_PATH
$ saved_model_cli show --dir my_mnist_model/0001 --all
MetaGraphDef with tag-set: 'serve' contains the following SignatureDefs:
signature_def['__saved_model_init_op']:
  [...]
```

```
signature_def['serving_default']:
  The given SavedModel SignatureDef contains the following input(s):
    inputs['flatten_input'] tensor_info:
        dtype: DT_FLOAT
        shape: (-1, 28, 28)
        name: serving_default_flatten_input:0
  The given SavedModel SignatureDef contains the following output(s):
    outputs['dense_1'] tensor_info:
        dtype: DT_FLOAT
        shape: (-1, 10)
        name: StatefulPartitionedCall:0
  Method name is: tensorflow/serving/predict
```

A SavedModel contains one or more *metagraphs*. A metagraph is a computation graph plus some function signature definitions (including their input and output names, types, and shapes). Each metagraph is identified by a set of tags. For example, you may want to have a metagraph containing the full computation graph, including the training operations (this one may be tagged "train", for example), and another metagraph containing a pruned computation graph with only the prediction operations, including some GPU-specific operations (this metagraph may be tagged "serve", "gpu"). However, when you pass a tf.keras model to the tf.saved_model.save() function, by default the function saves a much simpler SavedModel: it saves a single metagraph tagged "serve", which contains two signature definitions, an initialization function (called __saved_model_init_op, which you do not need to worry about) and a default serving function (called serving_default). When saving a tf.keras model, the default serving function corresponds to the model's call() function, which of course makes predictions.

The saved_model_cli tool can also be used to make predictions (for testing, not really for production). Suppose you have a NumPy array (X_new) containing three images of handwritten digits that you want to make predictions for. You first need to export them to NumPy's npy format:

```
np.save("my_mnist_tests.npy", X_new)
```

Next, use the saved_model_cli command like this:

```
$ saved_model_cli run --dir my_mnist_model/0001 --tag_set serve \
                      --signature_def serving_default \
                      --inputs flatten_input=my_mnist_tests.npy
[...] Result for output key dense_1:
[[1.1739199e-04 1.1239604e-07 6.0210604e-04 [...] 3.9471846e-04]
 [1.2294615e-03 2.9207937e-05 9.8599273e-01 [...] 1.1113169e-07]
 [6.4066830e-05 9.6359509e-01 9.0598064e-03 [...] 4.2495009e-04]]
```

The tool's output contains the 10 class probabilities of each of the 3 instances. Great! Now that you have a working SavedModel, the next step is to install TF Serving.

Installing TensorFlow Serving

There are many ways to install TF Serving: using a Docker image,[3] using the system's package manager, installing from source, and more. Let's use the Docker option, which is highly recommended by the TensorFlow team as it is simple to install, it will not mess with your system, and it offers high performance. You first need to install Docker (*https://docker.com/*). Then download the official TF Serving Docker image:

```
$ docker pull tensorflow/serving
```

Now you can create a Docker container to run this image:

```
$ docker run -it --rm -p 8500:8500 -p 8501:8501 \
           -v "$ML_PATH/my_mnist_model:/models/my_mnist_model" \
           -e MODEL_NAME=my_mnist_model \
           tensorflow/serving
[...]
2019-06-01 [...] loaded servable version {name: my_mnist_model version: 1}
2019-06-01 [...] Running gRPC ModelServer at 0.0.0.0:8500 ...
2019-06-01 [...] Exporting HTTP/REST API at:localhost:8501 ...
[evhttp_server.cc : 237] RAW: Entering the event loop ...
```

That's it! TF Serving is running. It loaded our MNIST model (version 1), and it is serving it through both gRPC (on port 8500) and REST (on port 8501). Here is what all the command-line options mean:

`-it`

Makes the container interactive (so you can press Ctrl-C to stop it) and displays the server's output.

`--rm`

Deletes the container when you stop it (no need to clutter your machine with interrupted containers). However, it does not delete the image.

`-p 8500:8500`

Makes the Docker engine forward the host's TCP port 8500 to the container's TCP port 8500. By default, TF Serving uses this port to serve the gRPC API.

`-p 8501:8501`

Forwards the host's TCP port 8501 to the container's TCP port 8501. By default, TF Serving uses this port to serve the REST API.

---

3 If you are not familiar with Docker, it allows you to easily download a set of applications packaged in a *Docker image* (including all their dependencies and usually some good default configuration) and then run them on your system using a *Docker engine*. When you run an image, the engine creates a *Docker container* that keeps the applications well isolated from your own system (but you can give it some limited access if you want). It is similar to a virtual machine, but much faster and more lightweight, as the container relies directly on the host's kernel. This means that the image does not need to include or run its own kernel.

```
-v "$ML_PATH/my_mnist_model:/models/my_mnist_model"
```
Makes the host's $ML_PATH/my_mnist_model directory available to the container at the path */models/mnist_model*. On Windows, you may need to replace / with \ in the host path (but not in the container path).

```
-e MODEL_NAME=my_mnist_model
```
Sets the container's MODEL_NAME environment variable, so TF Serving knows which model to serve. By default, it will look for models in the */models* directory, and it will automatically serve the latest version it finds.

```
tensorflow/serving
```
This is the name of the image to run.

Now let's go back to Python and query this server, first using the REST API, then the gRPC API.

Querying TF Serving through the REST API

Let's start by creating the query. It must contain the name of the function signature you want to call, and of course the input data:

```
import json

input_data_json = json.dumps({
    "signature_name": "serving_default",
    "instances": X_new.tolist(),
})
```

Note that the JSON format is 100% text-based, so the X_new NumPy array had to be converted to a Python list and then formatted as JSON:

```
>>> input_data_json
'{"signature_name": "serving_default", "instances": [[[0.0, 0.0, 0.0, [...]
0.3294117647058824, 0.725490196078431, [...very long], 0.0, 0.0, 0.0, 0.0]]]}'
```

Now let's send the input data to TF Serving by sending an HTTP POST request. This can be done easily using the requests library (it is not part of Python's standard library, so you will need to install it first, e.g., using pip):

```
import requests

SERVER_URL = 'http://localhost:8501/v1/models/my_mnist_model:predict'
response = requests.post(SERVER_URL, data=input_data_json)
response.raise_for_status() # raise an exception in case of error
response = response.json()
```

The response is a dictionary containing a single "predictions" key. The corresponding value is the list of predictions. This list is a Python list, so let's convert it to a NumPy array and round the floats it contains to the second decimal:

```
>>> y_proba = np.array(response["predictions"])
>>> y_proba.round(2)
array([[0.  , 0.  , 0.  , 0.  , 0.  , 0.  , 0.  , 1.  , 0.  , 0.  ],
       [0.  , 0.  , 0.99, 0.01, 0.  , 0.  , 0.  , 0.  , 0.  , 0.  ],
       [0.  , 0.96, 0.01, 0.  , 0.  , 0.  , 0.  , 0.01, 0.01, 0.  ]])
```

Hurray, we have the predictions! The model is close to 100% confident that the first image is a 7, 99% confident that the second image is a 2, and 96% confident that the third image is a 1.

The REST API is nice and simple, and it works well when the input and output data are not too large. Moreover, just about any client application can make REST queries without additional dependencies, whereas other protocols are not always so readily available. However, it is based on JSON, which is text-based and fairly verbose. For example, we had to convert the NumPy array to a Python list, and every float ended up represented as a string. This is very inefficient, both in terms of serialization/deserialization time (to convert all the floats to strings and back) and in terms of payload size: many floats end up being represented using over 15 characters, which translates to over 120 bits for 32-bit floats! This will result in high latency and bandwidth usage when transferring large NumPy arrays.[4] So let's use gRPC instead.

 When transferring large amounts of data, it is much better to use the gRPC API (if the client supports it), as it is based on a compact binary format and an efficient communication protocol (based on HTTP/2 framing).

Querying TF Serving through the gRPC API

The gRPC API expects a serialized `PredictRequest` protocol buffer as input, and it outputs a serialized `PredictResponse` protocol buffer. These protobufs are part of the `tensorflow-serving-api` library, which you must install (e.g., using pip). First, let's create the request:

```
from tensorflow_serving.apis.predict_pb2 import PredictRequest

request = PredictRequest()
request.model_spec.name = model_name
request.model_spec.signature_name = "serving_default"
input_name = model.input_names[0]
request.inputs[input_name].CopyFrom(tf.make_tensor_proto(X_new))
```

This code creates a `PredictRequest` protocol buffer and fills in the required fields, including the model name (defined earlier), the signature name of the function we

---

4 To be fair, this can be mitigated by serializing the data first and encoding it to Base64 before creating the REST request. Moreover, REST requests can be compressed using gzip, which reduces the payload size significantly.

want to call, and finally the input data, in the form of a `Tensor` protocol buffer. The `tf.make_tensor_proto()` function creates a `Tensor` protocol buffer based on the given tensor or NumPy array, in this case X_new.

Next, we'll send the request to the server and get its response (for this you will need the grpcio library, which you can install using pip):

```
import grpc
from tensorflow_serving.apis import prediction_service_pb2_grpc

channel = grpc.insecure_channel('localhost:8500')
predict_service = prediction_service_pb2_grpc.PredictionServiceStub(channel)
response = predict_service.Predict(request, timeout=10.0)
```

The code is quite straightforward: after the imports, we create a gRPC communication channel to *localhost* on TCP port 8500, then we create a gRPC service over this channel and use it to send a request, with a 10-second timeout (not that the call is synchronous: it will block until it receives the response or the timeout period expires). In this example the channel is insecure (no encryption, no authentication), but gRPC and TensorFlow Serving also support secure channels over SSL/TLS.

Next, let's convert the `PredictResponse` protocol buffer to a tensor:

```
output_name = model.output_names[0]
outputs_proto = response.outputs[output_name]
y_proba = tf.make_ndarray(outputs_proto)
```

If you run this code and print y_proba.numpy().round(2), you will get the exact same estimated class probabilities as earlier. And that's all there is to it: in just a few lines of code, you can now access your TensorFlow model remotely, using either REST or gRPC.

Deploying a new model version

Now let's create a new model version and export a SavedModel to the *my_mnist_model/0002* directory, just like earlier:

```
model = keras.models.Sequential([...])
model.compile([...])
history = model.fit([...])

model_version = "0002"
model_name = "my_mnist_model"
model_path = os.path.join(model_name, model_version)
tf.saved_model.save(model, model_path)
```

At regular intervals (the delay is configurable), TensorFlow Serving checks for new model versions. If it finds one, it will automatically handle the transition gracefully: by default, it will answer pending requests (if any) with the previous model version, while handling new requests with the new version.[5] As soon as every pending request has been answered, the previous model version is unloaded. You can see this at work in the TensorFlow Serving logs:

```
[...]
reserved resources to load servable {name: my_mnist_model version: 2}
[...]
Reading SavedModel from: /models/my_mnist_model/0002
Reading meta graph with tags { serve }
Successfully loaded servable version {name: my_mnist_model version: 2}
Quiescing servable version {name: my_mnist_model version: 1}
Done quiescing servable version {name: my_mnist_model version: 1}
Unloading servable version {name: my_mnist_model version: 1}
```

This approach offers a smooth transition, but it may use too much RAM (especially GPU RAM, which is generally the most limited). In this case, you can configure TF Serving so that it handles all pending requests with the previous model version and unloads it before loading and using the new model version. This configuration will avoid having two model versions loaded at the same time, but the service will be unavailable for a short period.

As you can see, TF Serving makes it quite simple to deploy new models. Moreover, if you discover that version 2 does not work as well as you expected, then rolling back to version 1 is as simple as removing the *my_mnist_model/0002* directory.

 Another great feature of TF Serving is its automatic batching capability, which you can activate using the --enable_batching option upon startup. When TF Serving receives multiple requests within a short period of time (the delay is configurable), it will automatically batch them together before using the model. This offers a significant performance boost by leveraging the power of the GPU. Once the model returns the predictions, TF Serving dispatches each prediction to the right client. You can trade a bit of latency for a greater throughput by increasing the batching delay (see the --batching_parameters_file option).

If you expect to get many queries per second, you will want to deploy TF Serving on multiple servers and load-balance the queries (see Figure 19-2). This will require deploying and managing many TF Serving containers across these servers. One way to handle that is to use a tool such as Kubernetes (*https://kubernetes.io/*), which is an open source system for simplifying container orchestration across many servers. If

---

5 If the SavedModel contains some example instances in the *assets/extra* directory, you can configure TF Serving to execute the model on these instances before starting to serve new requests with it. This is called *model warmup*: it will ensure that everything is properly loaded, avoiding long response times for the first requests.

you do not want to purchase, maintain, and upgrade all the hardware infrastructure, you will want to use virtual machines on a cloud platform such as Amazon AWS, Microsoft Azure, Google Cloud Platform, IBM Cloud, Alibaba Cloud, Oracle Cloud, or some other Platform-as-a-Service (PaaS). Managing all the virtual machines, handling container orchestration (even with the help of Kubernetes), taking care of TF Serving configuration, tuning and monitoring—all of this can be a full-time job. Fortunately, some service providers can take care of all this for you. In this chapter we will use Google Cloud AI Platform because it's the only platform with TPUs today, it supports TensorFlow 2, it offers a nice suite of AI services (e.g., AutoML, Vision API, Natural Language API), and it is the one I have the most experience with. But there are several other providers in this space, such as Amazon AWS SageMaker and Microsoft AI Platform, which are also capable of serving TensorFlow models.

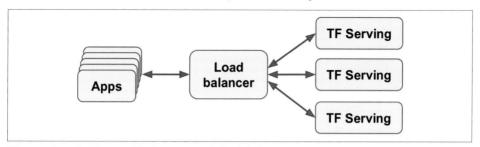

Figure 19-2. Scaling up TF Serving with load balancing

Now let's see how to serve our wonderful MNIST model on the cloud!

Creating a Prediction Service on GCP AI Platform

Before you can deploy a model, there's a little bit of setup to take care of:

1. Log in to your Google account, and then go to the Google Cloud Platform (GCP) console (*https://console.cloud.google.com/*) (see Figure 19-3). If you don't have a Google account, you'll have to create one.

2. If it is your first time using GCP, you will have to read and accept the terms and conditions. Click Tour Console if you want. At the time of this writing, new users are offered a free trial, including $300 worth of GCP credit that you can use over the course of 12 months. You will only need a small portion of that to pay for the services you will use in this chapter. Upon signing up for the free trial, you will still need to create a payment profile and enter your credit card number: it is used for verification purposes (probably to avoid people using the free trial multiple times), but you will not be billed. Activate and upgrade your account if requested.

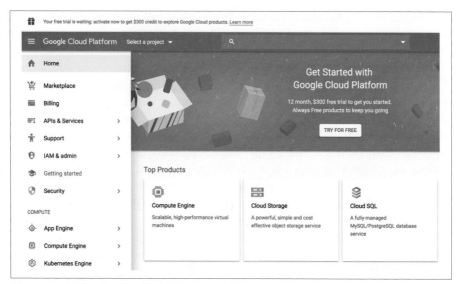

Figure 19-3. Google Cloud Platform console

3. If you have used GCP before and your free trial has expired, then the services you will use in this chapter will cost you some money. It should not be too much, especially if you remember to turn off the services when you do not need them anymore. Make sure you understand and agree to the pricing conditions before you run any service. I hereby decline any responsibility if services end up costing more than you expected! Also make sure your billing account is active. To check, open the navigation menu on the left and click Billing, and make sure you have set up a payment method and that the billing account is active.

4. Every resource in GCP belongs to a project. This includes all the virtual machines you may use, the files you store, and the training jobs you run. When you create an account, GCP automatically creates a project for you, called "My First Project." If you want, you can change its display name by going to the project settings: in the navigation menu (on the left of the screen), select IAM & admin → Settings, change the project's display name, and click Save. Note that the project also has a unique ID and number. You can choose the project ID when you create a project, but you cannot change it later. The project number is automatically generated and cannot be changed. If you want to create a new project, click the project name at the top of the page, then click New Project and enter the project ID. Make sure billing is active for this new project.

Always set an alarm to remind yourself to turn services off when you know you will only need them for a few hours, or else you might leave them running for days or months, incurring potentially significant costs.

5. Now that you have a GCP account with billing activated, you can start using the services. The first one you will need is Google Cloud Storage (GCS): this is where you will put the SavedModels, the training data, and more. In the navigation menu, scroll down to the Storage section, and click Storage → Browser. All your files will go in one or more *buckets*. Click Create Bucket and choose the bucket name (you may need to activate the Storage API first). GCS uses a single worldwide namespace for buckets, so simple names like "machine-learning" will most likely not be available. Make sure the bucket name conforms to DNS naming conventions, as it may be used in DNS records. Moreover, bucket names are public, so do not put anything private in there. It is common to use your domain name or your company name as a prefix to ensure uniqueness, or simply use a random number as part of the name. Choose the location where you want the bucket to be hosted, and the rest of the options should be fine by default. Then click Create.

6. Upload the *my_mnist_model* folder you created earlier (including one or more versions) to your bucket. To do this, just go to the GCS Browser, click the bucket, then drag and drop the *my_mnist_model* folder from your system to the bucket (see Figure 19-4). Alternatively, you can click "Upload folder" and select the *my_mnist_model* folder to upload. By default, the maximum size for a SavedModel is 250 MB, but it is possible to request a higher quota.

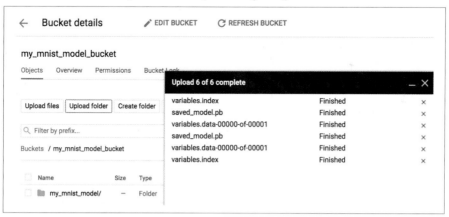

Figure 19-4. Uploading a SavedModel to Google Cloud Storage

7. Now you need to configure AI Platform (formerly known as ML Engine) so that it knows which models and versions you want to use. In the navigation menu, scroll down to the Artificial Intelligence section, and click AI Platform → Models. Click Activate API (it takes a few minutes), then click "Create model." Fill in the model details (see Figure 19-5) and click Create.

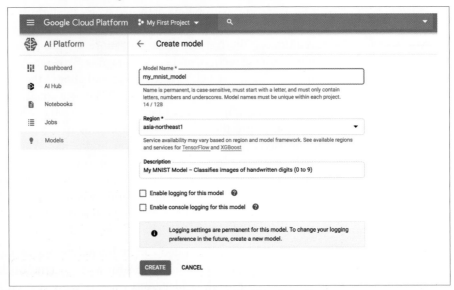

Figure 19-5. Creating a new model on Google Cloud AI Platform

8. Now that you have a model on AI Platform, you need to create a model version. In the list of models, click the model you just created, then click "Create version" and fill in the version details (see Figure 19-6): set the name, description, Python version (3.5 or above), framework (TensorFlow), framework version (2.0 if available, or 1.13),[6] ML runtime version (2.0, if available or 1.13), machine type (choose "Single core CPU" for now), model path on GCS (this is the full path to the actual version folder, e.g., *gs://my-mnist-model-bucket/my_mnist_model/0002/*), scaling (choose automatic), and minimum number of TF Serving containers to have running at all times (leave this field empty). Then click Save.

---

6 At the time of this writing, TensorFlow version 2 is not available yet on AI Platform, but that's OK: you can use 1.13, and it will run your TF 2 SavedModels just fine.

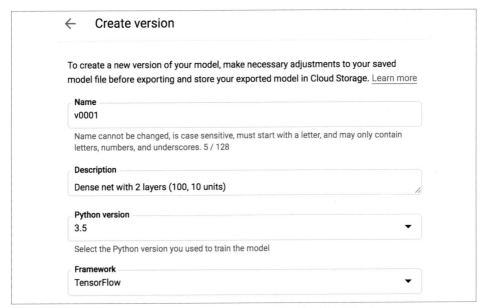

Figure 19-6. Creating a new model version on Google Cloud AI Platform

Congratulations, you have deployed your first model on the cloud! Because you selected automatic scaling, AI Platform will start more TF Serving containers when the number of queries per second increases, and it will load-balance the queries between them. If the QPS goes down, it will stop containers automatically. The cost is therefore directly linked to the QPS (as well as the type of machine you choose and the amount of data you store on GCS). This pricing model is particularly useful for occasional users and for services with important usage spikes, as well as for startups: the price remains low until the startup actually starts up.

If you do not use the prediction service, AI Platform will stop all containers. This means you will only pay for the amount of storage you use (a few cents per gigabyte per month). Note that when you query the service, AI Platform will need to start up a TF Serving container, which will take a few seconds. If this delay is unacceptable, you will have to set the minimum number of TF Serving containers to 1 when creating the model version. Of course, this means at least one machine will run constantly, so the monthly fee will be higher.

Now let's query this prediction service!

Using the Prediction Service

Under the hood, AI Platform just runs TF Serving, so in principle you could use the same code as earlier, if you knew which URL to query. There's just one problem: GCP also takes care of encryption and authentication. Encryption is based on SSL/TLS, and authentication is token-based: a secret authentication token must be sent to the server in every request. So before your code can use the prediction service (or any other GCP service), it must obtain a token. We will see how to do this shortly, but first you need to configure authentication and give your application the appropriate access rights on GCP. You have two options for authentication:

- Your application (i.e., the client code that will query the prediction service) could authenticate using user credentials with your own Google login and password. Using user credentials would give your application the exact same rights as on GCP, which is certainly way more than it needs. Moreover, you would have to deploy your credentials in your application, so anyone with access could steal your credentials and fully access your GCP account. In short, do not choose this option; it is only needed in very rare cases (e.g., when your application needs to access its user's GCP account).

- The client code can authenticate with a *service account*. This is an account that represents an application, not a user. It is generally given very restricted access rights: strictly what it needs, and no more. This is the recommended option.

So, let's create a service account for your application: in the navigation menu, go to IAM & admin → Service accounts, then click Create Service Account, fill in the form (service account name, ID, description), and click Create (see Figure 19-7). Next, you must give this account some access rights. Select the ML Engine Developer role: this will allow the service account to make predictions, and not much more. Optionally, you can grant some users access to the service account (this is useful when your GCP user account is part of an organization, and you wish to authorize other users in the organization to deploy applications that will be based on this service account or to manage the service account itself). Next, click Create Key to export the service account's private key, choose JSON, and click Create. This will download the private key in the form of a JSON file. Make sure to keep it private!

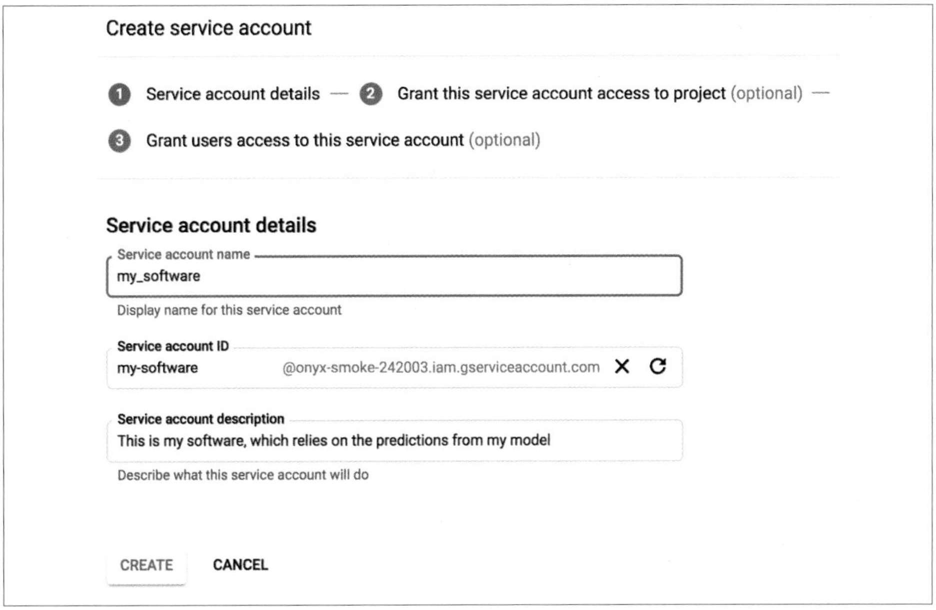

Figure 19-7. Creating a new service account in Google IAM

Great! Now let's write a small script that will query the prediction service. Google provides several libraries to simplify access to its services:

Google API Client Library
> This is a fairly thin layer on top of *OAuth 2.0* (*https://oauth.net/*) (for the authentication) and REST. You can use it with all GCP services, including AI Platform. You can install it using pip: the library is called `google-api-python-client`.

Google Cloud Client Libraries
> These are a bit more high-level: each one is dedicated to a particular service, such as GCS, Google BigQuery, Google Cloud Natural Language, and Google Cloud Vision. All these libraries can be installed using pip (e.g., the GCS Client Library is called `google-cloud-storage`). When a client library is available for a given service, it is recommended to use it rather than the Google API Client Library, as it implements all the best practices and will often use gRPC rather than REST, for better performance.

At the time of this writing there is no client library for AI Platform, so we will use the Google API Client Library. It will need to use the service account's private key; you can tell it where it is by setting the `GOOGLE_APPLICATION_CREDENTIALS` environment variable, either before starting the script or within the script like this:

```
import os

os.environ["GOOGLE_APPLICATION_CREDENTIALS"] = "my_service_account_key.json"
```

If you deploy your application to a virtual machine on Google Cloud Engine (GCE), or within a container using Google Cloud Kubernetes Engine, or as a web application on Google Cloud App Engine, or as a microservice on Google Cloud Functions, and if the GOOGLE_APPLICATION_CREDENTIALS environment variable is not set, then the library will use the default service account for the host service (e.g., the default GCE service account, if your application runs on GCE).

Next, you must create a resource object that wraps access to the prediction service:[7]

```
import googleapiclient.discovery

project_id = "onyx-smoke-242003" # change this to your project ID
model_id = "my_mnist_model"
model_path = "projects/{}/models/{}".format(project_id, model_id)
ml_resource = googleapiclient.discovery.build("ml", "v1").projects()
```

Note that you can append /versions/0001 (or any other version number) to the model_path to specify the version you want to query: this can be useful for A/B testing or for testing a new version on a small group of users before releasing it widely (this is called a *canary*). Next, let's write a small function that will use the resource object to call the prediction service and get the predictions back:

```
def predict(X):
    input_data_json = {"signature_name": "serving_default",
                       "instances": X.tolist()}
    request = ml_resource.predict(name=model_path, body=input_data_json)
    response = request.execute()
    if "error" in response:
        raise RuntimeError(response["error"])
    return np.array([pred[output_name] for pred in response["predictions"]])
```

The function takes a NumPy array containing the input images and prepares a dictionary that the client library will convert to the JSON format (as we did earlier). Then it prepares a prediction request, and executes it; it raises an exception if the response contains an error, or else it extracts the predictions for each instance and bundles them in a NumPy array. Let's see if it works:

```
>>> Y_probas = predict(X_new)
>>> np.round(Y_probas, 2)
array([[0. , 0. , 0. , 0. , 0. , 0. , 0. , 1. , 0. , 0. ],
       [0. , 0. , 0.99, 0.01, 0. , 0. , 0. , 0. , 0. , 0. ],
       [0. , 0.96, 0.01, 0. , 0. , 0. , 0. , 0.01, 0.01, 0. ]])
```

---

7 If you get an error saying that module google.appengine was not found, set cache_discovery=False in the call to the build() method; see *https://stackoverflow.com/q/55561354*.

Yes! You now have a nice prediction service running on the cloud that can automatically scale up to any number of QPS, plus you can query it from anywhere securely. Moreover, it costs you close to nothing when you don't use it: you'll pay just a few cents per month per gigabyte used on GCS. And you can also get detailed logs and metrics using Google Stackdriver (*https://cloud.google.com/stackdriver/*).

But what if you want to deploy your model to a mobile app? Or to an embedded device?

Deploying a Model to a Mobile or Embedded Device

If you need to deploy your model to a mobile or embedded device, a large model may simply take too long to download and use too much RAM and CPU, all of which will make your app unresponsive, heat the device, and drain its battery. To avoid this, you need to make a mobile-friendly, lightweight, and efficient model, without sacrificing too much of its accuracy. The TFLite (*https://tensorflow.org/lite*) library provides several tools[8] to help you deploy your models to mobile and embedded devices, with three main objectives:

- Reduce the model size, to shorten download time and reduce RAM usage.
- Reduce the amount of computations needed for each prediction, to reduce latency, battery usage, and heating.
- Adapt the model to device-specific constraints.

To reduce the model size, TFLite's model converter can take a SavedModel and compress it to a much lighter format based on FlatBuffers (*https://google.github.io/flatbuffers/*). This is an efficient cross-platform serialization library (a bit like Protocol Buffers) initially created by Google for gaming. It is designed so you can load FlatBuffers straight to RAM without any preprocessing: this reduces the loading time and memory footprint. Once the model is loaded into a mobile or embedded device, the TFLite interpreter will execute it to make predictions. Here is how you can convert a SavedModel to a FlatBuffer and save it to a *.tflite* file:

```
converter = tf.lite.TFLiteConverter.from_saved_model(saved_model_path)
tflite_model = converter.convert()
with open("converted_model.tflite", "wb") as f:
    f.write(tflite_model)
```

---

8 Also check out TensorFlow's Graph Transform Tools (*https://homl.info/tfgtt*) for modifying and optimizing computational graphs.

You can also save a tf.keras model directly to a FlatBuffer using `from_keras_model()`.

The converter also optimizes the model, both to shrink it and to reduce its latency. It prunes all the operations that are not needed to make predictions (such as training operations), and it optimizes computations whenever possible; for example, $3 \times a + 4 \times a + 5 \times a$ will be converted to $(3 + 4 + 5) \times a$. It also tries to fuse operations whenever possible. For example, Batch Normalization layers end up folded into the previous layer's addition and multiplication operations, whenever possible. To get a good idea of how much TFLite can optimize a model, download one of the pretrained TFLite models (*https://homl.info/litemodels*), unzip the archive, then open the excellent Netron graph visualization tool (*https://lutzroeder.github.io/netron/*) and upload the *.pb* file to view the original model. It's a big, complex graph, right? Next, open the optimized *.tflite* model and marvel at its beauty!

Another way you can reduce the model size (other than simply using smaller neural network architectures) is by using smaller bit-widths: for example, if you use half-floats (16 bits) rather than regular floats (32 bits), the model size will shrink by a factor of 2, at the cost of a (generally small) accuracy drop. Moreover, training will be faster, and you will use roughly half the amount of GPU RAM.

TFLite's converter can go further than that, by quantizing the model weights down to fixed-point, 8-bit integers! This leads to a fourfold size reduction compared to using 32-bit floats. The simplest approach is called *post-training quantization*: it just quantizes the weights after training, using a fairly basic but efficient symmetrical quantization technique. It finds the maximum absolute weight value, m, then it maps the floating-point range $-m$ to $+m$ to the fixed-point (integer) range -127 to $+127$. For example (see Figure 19-8), if the weights range from -1.5 to $+0.8$, then the bytes -127, 0, and $+127$ will correspond to the floats -1.5, 0.0, and $+1.5$, respectively. Note that 0.0 always maps to 0 when using symmetrical quantization (also note that the byte values $+68$ to $+127$ will not be used, since they map to floats greater than $+0.8$).

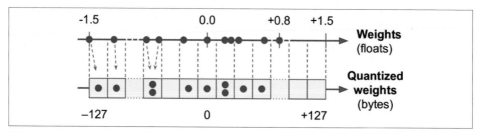

Figure 19-8. From 32-bit floats to 8-bit integers, using symmetrical quantization

To perform this post-training quantization, simply add `OPTIMIZE_FOR_SIZE` to the list of converter optimizations before calling the `convert()` method:

```
converter.optimizations = [tf.lite.Optimize.OPTIMIZE_FOR_SIZE]
```

This technique dramatically reduces the model's size, so it's much faster to download and store. However, at runtime the quantized weights get converted back to floats before they are used (these recovered floats are not perfectly identical to the original floats, but not too far off, so the accuracy loss is usually acceptable). To avoid recomputing them all the time, the recovered floats are cached, so there is no reduction of RAM usage. And there is no reduction either in compute speed.

The most effective way to reduce latency and power consumption is to also quantize the activations so that the computations can be done entirely with integers, without the need for any floating-point operations. Even when using the same bit-width (e.g., 32-bit integers instead of 32-bit floats), integer computations use less CPU cycles, consume less energy, and produce less heat. And if you also reduce the bit-width (e.g., down to 8-bit integers), you can get huge speedups. Moreover, some neural network accelerator devices (such as the Edge TPU) can only process integers, so full quantization of both weights and activations is compulsory. This can be done post-training; it requires a calibration step to find the maximum absolute value of the activations, so you need to provide a representative sample of training data to TFLite (it does not need to be huge), and it will process the data through the model and measure the activation statistics required for quantization (this step is typically fast).

The main problem with quantization is that it loses a bit of accuracy: it is equivalent to adding noise to the weights and activations. If the accuracy drop is too severe, then you may need to use *quantization-aware training*. This means adding fake quantization operations to the model so it can learn to ignore the quantization noise during training; the final weights will then be more robust to quantization. Moreover, the calibration step can be taken care of automatically during training, which simplifies the whole process.

I have explained the core concepts of TFLite, but going all the way to coding a mobile app or an embedded program would require a whole other book. Fortunately, one exists: if you want to learn more about building TensorFlow applications for mobile and embedded devices, check out the O'Reilly book *TinyML: Machine Learning with TensorFlow on Arduino and Ultra-Low Power Micro-Controllers* (*https://homl.info/tinyml*), by Pete Warden (who leads the TFLite team) and Daniel Situnayake.

TensorFlow in the Browser

What if you want to use your model in a website, running directly in the user's browser? This can be useful in many scenarios, such as:

- When your web application is often used in situations where the user's connectivity is intermittent or slow (e.g., a website for hikers), so running the model directly on the client side is the only way to make your website reliable.

- When you need the model's responses to be as fast as possible (e.g., for an online game). Removing the need to query the server to make predictions will definitely reduce the latency and make the website much more responsive.

- When your web service makes predictions based on some private user data, and you want to protect the user's privacy by making the predictions on the client side so that the private data never has to leave the user's machine.[9]

For all these scenarios, you can export your model to a special format that can be loaded by the TensorFlow.js JavaScript library (*https://tensorflow.org/js*). This library can then use your model to make predictions directly in the user's browser. The TensorFlow.js project includes a `tensorflowjs_converter` tool that can convert a TensorFlow SavedModel or a Keras model file to the *TensorFlow.js Layers* format: this is a directory containing a set of sharded weight files in binary format and a *model.json* file that describes the model's architecture and links to the weight files. This format is optimized to be downloaded efficiently on the web. Users can then download the model and run predictions in the browser using the TensorFlow.js library. Here is a code snippet to give you an idea of what the JavaScript API looks like:

```
import * as tf from '@tensorflow/tfjs';
const model = await tf.loadLayersModel('https://example.com/tfjs/model.json');
const image = tf.fromPixels(webcamElement);
const prediction = model.predict(image);
```

Once again, doing justice to this topic would require a whole book. If you want to learn more about TensorFlow.js, check out the O'Reilly book *Practical Deep Learning for Cloud, Mobile, and Edge* (*https://homl.info/tfjsbook*), by Anirudh Koul, Siddha Ganju, and Meher Kasam.

Next, we will see how to use GPUs to speed up computations!

---

9 If you're interested in this topic, check out *federated learning* (*https://tensorflow.org/federated*).

Using GPUs to Speed Up Computations

In Chapter 11 we discussed several techniques that can considerably speed up training: better weight initialization, Batch Normalization, sophisticated optimizers, and so on. But even with all of these techniques, training a large neural network on a single machine with a single CPU can take days or even weeks.

In this section we will look at how to speed up your models by using GPUs. We will also see how to split the computations across multiple devices, including the CPU and multiple GPU devices (see Figure 19-9). For now we will run everything on a single machine, but later in this chapter we will discuss how to distribute computations across multiple servers.

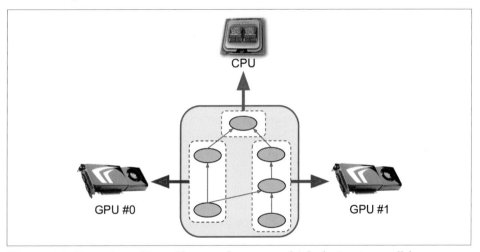

Figure 19-9. Executing a TensorFlow graph across multiple devices in parallel

Thanks to GPUs, instead of waiting for days or weeks for a training algorithm to complete, you may end up waiting for just a few minutes or hours. Not only does this save an enormous amount of time, but it also means that you can experiment with various models much more easily and frequently retrain your models on fresh data.

 You can often get a major performance boost simply by adding GPU cards to a single machine. In fact, in many cases this will suffice; you won't need to use multiple machines at all. For example, you can typically train a neural network just as fast using four GPUs on a single machine rather than eight GPUs across multiple machines, due to the extra delay imposed by network communications in a distributed setup. Similarly, using a single powerful GPU is often preferable to using multiple slower GPUs.

The first step is to get your hands on a GPU. There are two options for this: you can either purchase your own GPU(s), or you can use GPU-equipped virtual machines on the cloud. Let's start with the first option.

Getting Your Own GPU

If you choose to purchase a GPU card, then take some time to make the right choice. Tim Dettmers wrote an excellent blog post (*https://homl.info/66*) to help you choose, and he updates it regularly: I encourage you to read it carefully. At the time of this writing, TensorFlow only supports Nvidia cards with CUDA Compute Capability 3.5+ (*https://homl.info/cudagpus*) (as well as Google's TPUs, of course), but it may extend its support to other manufacturers. Moreover, although TPUs are currently only available on GCP, it is highly likely that TPU-like cards will be available for sale in the near future, and TensorFlow may support them. In short, make sure to check TensorFlow's documentation (*https://tensorflow.org/install*) to see what devices are supported at this point.

If you go for an Nvidia GPU card, you will need to install the appropriate Nvidia drivers and several Nvidia libraries.[10] These include the *Compute Unified Device Architecture* library (CUDA), which allows developers to use CUDA-enabled GPUs for all sorts of computations (not just graphics acceleration), and the *CUDA Deep Neural Network* library (cuDNN), a GPU-accelerated library of primitives for DNNs. cuDNN provides optimized implementations of common DNN computations such as activation layers, normalization, forward and backward convolutions, and pooling (see Chapter 14). It is part of Nvidia's Deep Learning SDK (note that you'll need to create an Nvidia developer account in order to download it). TensorFlow uses CUDA and cuDNN to control the GPU cards and accelerate computations (see Figure 19-10).

---

10 Please check the docs for detailed and up-to-date installation instructions, as they change quite often.

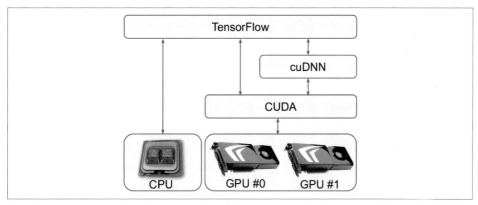

Figure 19-10. TensorFlow uses CUDA and cuDNN to control GPUs and boost DNNs

Once you have installed the GPU card(s) and all the required drivers and libraries, you can use the `nvidia-smi` command to check that CUDA is properly installed. It lists the available GPU cards, as well as processes running on each card:

```
$ nvidia-smi
Sun Jun  2 10:05:22 2019
+-----------------------------------------------------------------------------+
| NVIDIA-SMI 418.67       Driver Version: 410.79       CUDA Version: 10.0     |
|-------------------------------+----------------------+----------------------+
| GPU  Name        Persistence-M| Bus-Id        Disp.A | Volatile Uncorr. ECC |
| Fan  Temp  Perf  Pwr:Usage/Cap|         Memory-Usage | GPU-Util  Compute M. |
|===============================+======================+======================|
|   0  Tesla T4            Off  | 00000000:00:04.0 Off |                    0 |
| N/A   61C    P8    17W /  70W |      0MiB / 15079MiB |      0%      Default |
+-------------------------------+----------------------+----------------------+

+-----------------------------------------------------------------------------+
| Processes:                                                       GPU Memory |
|  GPU       PID   Type   Process name                             Usage      |
|=============================================================================|
|  No running processes found                                                 |
+-----------------------------------------------------------------------------+
```

At the time of this writing, you'll also need to install the GPU version of TensorFlow (i.e., the `tensorflow-gpu` library); however, there is ongoing work to have a unified installation procedure for both CPU-only and GPU machines, so please check the installation documentation to see which library you should install. In any case, since installing every required library correctly is a bit long and tricky (and all hell breaks loose if you do not install the correct library versions), TensorFlow provides a Docker image with everything you need inside. However, in order for the Docker container to have access to the GPU, you will still need to install the Nvidia drivers on the host machine.

To check that TensorFlow actually sees the GPUs, run the following tests:

```
>>> import tensorflow as tf
>>> tf.test.is_gpu_available()
True
>>> tf.test.gpu_device_name()
'/device:GPU:0'
>>> tf.config.experimental.list_physical_devices(device_type='GPU')
[PhysicalDevice(name='/physical_device:GPU:0', device_type='GPU')]
```

The is_gpu_available() function checks whether at least one GPU is available. The gpu_device_name() function gives the first GPU's name: by default, operations will run on this GPU. The list_physical_devices() function returns the list of all available GPU devices (just one in this example).[11]

Now, what if you don't want to invest time and money in getting your own GPU card? Just use a GPU VM on the cloud!

Using a GPU-Equipped Virtual Machine

All major cloud platforms now offer GPU VMs, some preconfigured with all the drivers and libraries you need (including TensorFlow). Google Cloud Platform enforces various GPU quotas, both worldwide and per region: you cannot just create thousands of GPU VMs without prior authorization from Google.[12] By default, the worldwide GPU quota is zero, so you cannot use any GPU VMs. Therefore, the very first thing you need to do is to request a higher worldwide quota. In the GCP console, open the navigation menu and go to IAM & admin → Quotas. Click Metric, click None to uncheck all locations, then search for "GPU" and select "GPUs (all regions)" to see the corresponding quota. If this quota's value is zero (or just insufficient for your needs), then check the box next to it (it should be the only selected one) and click "Edit quotas." Fill in the requested information, then click "Submit request." It may take a few hours (or up to a few days) for your quota request to be processed and (generally) accepted. By default, there is also a quota of one GPU per region and per GPU type. You can request to increase these quotas too: click Metric, select None to uncheck all metrics, search for "GPU," and select the type of GPU you want (e.g., NVIDIA P4 GPUs). Then click the Location drop-down menu, click None to uncheck all metrics, and click the location you want; check the boxes next to the quota(s) you want to change, and click "Edit quotas" to file a request.

---

11 Many code examples in this chapter use experimental APIs. They are very likely to be moved to the core API in future versions. So if an experimental function fails, try simply removing the word experimental, and hopefully it will work. If not, then perhaps the API has changed a bit; please check the Jupyter notebook, as I will ensure it contains the correct code.

12 Presumably, these quotas are meant to stop bad guys who might be tempted to use GCP with stolen credit cards to mine cryptocurrencies.

Once your GPU quota requests are approved, you can in no time create a VM equipped with one or more GPUs by using Google Cloud AI Platform's *Deep Learning VM Images*: go to *https://homl.info/dlvm*, click View Console, then click "Launch on Compute Engine" and fill in the VM configuration form. Note that some locations do not have all types of GPUs, and some have no GPUs at all (change the location to see the types of GPUs available, if any). Make sure to select TensorFlow 2.0 as the framework, and check "Install NVIDIA GPU driver automatically on first startup." It is also a good idea to check "Enable access to JupyterLab via URL instead of SSH": this will make it very easy to start a Jupyter notebook running on this GPU VM, powered by JupyterLab (this is an alternative web interface to run Jupyter notebooks). Once the VM is created, scroll down the navigation menu to the Artificial Intelligence section, then click AI Platform → Notebooks. Once the Notebook instance appears in the list (this may take a few minutes, so click Refresh once in a while until it appears), click its Open JupyterLab link. This will run JupyterLab on the VM and connect your browser to it. You can create notebooks and run any code you want on this VM, and benefit from its GPUs!

But if you just want to run some quick tests or easily share notebooks with your colleagues, then you should try Colaboratory.

Colaboratory

The simplest and cheapest way to access a GPU VM is to use *Colaboratory* (or *Colab*, for short). It's free! Just go to *https://colab.research.google.com/* and create a new Python 3 notebook: this will create a Jupyter notebook on your Google Drive (alternatively, you can open any notebook on GitHub, or on Google Drive, or you can even upload your own notebooks). Colab's user interface is similar to Jupyter's, except you can share and use the notebooks like regular Google Docs, and there are a few other minor differences (e.g., you can create handy widgets using special comments in your code).

When you open a Colab notebook, it runs on a free Google VM dedicated to you, called a *Colab Runtime*. By default the Runtime is CPU-only, but you can change this by going to Runtime → "Change runtime type," selecting GPU in the "Hardware accelerator" drop-down menu, then clicking Save. In fact, you could even select TPU! (Yes, you can actually use a TPU for free; we will talk about TPUs later in this chapter, though, so for now just select GPU.)

If you run multiple Colab notebooks using the same runtime type (see Figure 19-11), they will use the same Colab Runtime. So if one writes to a file, the others will be able to read that file. It's important to understand the security implications of this: if you run an untrusted Colab notebook written by a nasty hacker, it may read private data produced by the other notebooks and then leak this data back to the hacker. If this includes private access keys for some resources, the hacker will gain access to those resources. Moreover, if you install a library in the Colab Runtime, the other note-books will also have that library. Depending on what you want to do, this might be great or annoying (e.g., it means you cannot easily use different versions of the same library in different Colab notebooks).

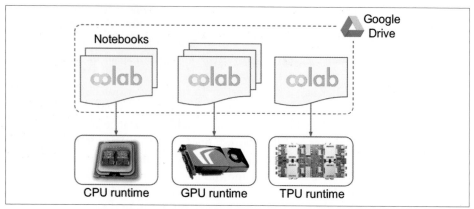

Figure 19-11. Colab Runtimes and notebooks

Colab does have some restrictions: as the FAQ states, "Colaboratory is intended for interactive use. Long-running background computations, particularly on GPUs, may be stopped. Please do not use Colaboratory for cryptocurrency mining." The web interface will automatically disconnect from the Colab Runtime if you leave it unat-tended for a while (~30 minutes). When you reconnect to the Colab Runtime, it may have been reset, so make sure you always download any data you care about. Even if you never disconnect, the Colab Runtime will automatically shut down after 12 hours, as it is not meant for long-running computations. Despite these limitations, it's a fantastic tool to run tests easily, get quick results, and collaborate with your colleagues.

Managing the GPU RAM

By default TensorFlow automatically grabs all the RAM in all available GPUs the first time you run a computation. It does this to limit GPU RAM fragmentation. This means that if you try to start a second TensorFlow program (or any program that requires the GPU), it will quickly run out of RAM. This does not happen as often as you might think, as you will most often have a single TensorFlow program running on a machine: usually a training script, a TF Serving node, or a Jupyter notebook. If you need to run multiple programs for some reason (e.g., to train two different mod-els in parallel on the same machine), then you will need to split the GPU RAM between these processes more evenly.

If you have multiple GPU cards on your machine, a simple solution is to assign each of them to a single process. To do this, you can set the CUDA_VISIBLE_DEVICES environment variable so that each process only sees the appropriate GPU card(s). Also set the `CUDA_DEVICE_ORDER` environment variable to `PCI_BUS_ID` to ensure that each ID always refers to the same GPU card. For example, if you have four GPU cards, you could start two programs, assigning two GPUs to each of them, by executing commands like the following in two separate terminal windows:

```
$ CUDA_DEVICE_ORDER=PCI_BUS_ID CUDA_VISIBLE_DEVICES=0,1 python3 program_1.py
# and in another terminal:
$ CUDA_DEVICE_ORDER=PCI_BUS_ID CUDA_VISIBLE_DEVICES=3,2 python3 program_2.py
```

Program 1 will then only see GPU cards 0 and 1, named /gpu:0 and /gpu:1 respectively, and program 2 will only see GPU cards 2 and 3, named /gpu:1 and /gpu:0 respectively (note the order). Everything will work fine (see Figure 19-12). Of course, you can also define these environment variables in Python by setting os.envi ron["CUDA_DEVICE_ORDER"] and os.environ["CUDA_VISIBLE_DEVICES"], as long as you do so before using TensorFlow.

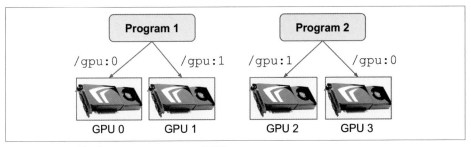

Figure 19-12. Each program gets two GPUs

Another option is to tell TensorFlow to grab only a specific amount of GPU RAM. This must be done immediately after importing TensorFlow. For example, to make TensorFlow grab only 2 GiB of RAM on each GPU, you must create a *virtual GPU device* (also called a *logical GPU device*) for each physical GPU device and set its memory limit to 2 GiB (i.e., 2,048 MiB):

```
for gpu in tf.config.experimental.list_physical_devices("GPU"):
    tf.config.experimental.set_virtual_device_configuration(
        gpu,
        [tf.config.experimental.VirtualDeviceConfiguration(memory_limit=2048)])
```

Now (supposing you have four GPUs, each with at least 4 GiB of RAM) two programs like this one can run in parallel, each using all four GPU cards (see Figure 19-13).

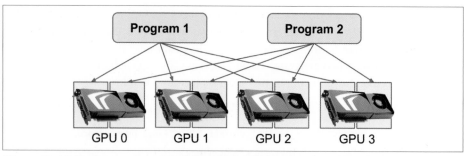

Figure 19-13. Each program gets all four GPUs, but with only 2 GiB of RAM on each GPU

If you run the `nvidia-smi` command while both programs are running, you should see that each process holds 2 GiB of RAM on each card:

```
$ nvidia-smi
[...]
+-----------------------------------------------------------------------+
| Processes:                                              GPU Memory |
|  GPU       PID   Type   Process name                    Usage      |
|=======================================================================|
|    0      2373     C   /usr/bin/python3                    2241MiB |
|    0      2533     C   /usr/bin/python3                    2241MiB |
|    1      2373     C   /usr/bin/python3                    2241MiB |
|    1      2533     C   /usr/bin/python3                    2241MiB |
[...]
```

Yet another option is to tell TensorFlow to grab memory only when it needs it (this also must be done immediately after importing TensorFlow):

```
for gpu in tf.config.experimental.list_physical_devices("GPU"):
    tf.config.experimental.set_memory_growth(gpu, True)
```

Another way to do this is to set the `TF_FORCE_GPU_ALLOW_GROWTH` environment variable to `true`. With this option, TensorFlow will never release memory once it has grabbed it (again, to avoid memory fragmentation), except of course when the program ends. It can be harder to guarantee deterministic behavior using this option (e.g., one program may crash because another program's memory usage went through the roof), so in production you'll probably want to stick with one of the previous options. However, there are some cases where it is very useful: for example, when you use a machine to run multiple Jupyter notebooks, several of which use TensorFlow. This is why the `TF_FORCE_GPU_ALLOW_GROWTH` environment variable is set to `true` in Colab Runtimes.

Lastly, in some cases you may want to split a GPU into two or more *virtual GPUs*—for example, if you want to test a distribution algorithm (this is a handy way to try out the code examples in the rest of this chapter even if you have a single GPU, such as in a Colab Runtime). The following code splits the first GPU into two virtual devices, with 2 GiB of RAM each (again, this must be done immediately after importing TensorFlow):

```
physical_gpus = tf.config.experimental.list_physical_devices("GPU")
tf.config.experimental.set_virtual_device_configuration(
    physical_gpus[0],
    [tf.config.experimental.VirtualDeviceConfiguration(memory_limit=2048),
     tf.config.experimental.VirtualDeviceConfiguration(memory_limit=2048)])
```

These two virtual devices will then be called `/gpu:0` and `/gpu:1`, and you can place operations and variables on each of them as if they were really two independent GPUs. Now let's see how TensorFlow decides which devices it should place variables and execute operations on.

Placing Operations and Variables on Devices

The TensorFlow whitepaper (*https://homl.info/67*)[13] presents a friendly *dynamic placer* algorithm that automagically distributes operations across all available devices, taking into account things like the measured computation time in previous runs of the graph, estimations of the size of the input and output tensors for each operation, the amount of RAM available in each device, communication delay when transferring data into and out of devices, and hints and constraints from the user. In practice this algorithm turned out to be less efficient than a small set of placement rules specified by the user, so the TensorFlow team ended up dropping the dynamic placer.

That said, tf.keras and tf.data generally do a good job of placing operations and variables where they belong (e.g., heavy computations on the GPU, and data preprocessing on the CPU). But you can also place operations and variables manually on each device, if you want more control:

- As just mentioned, you generally want to place the data preprocessing operations on the CPU, and place the neural network operations on the GPUs.

- GPUs usually have a fairly limited communication bandwidth, so it is important to avoid unnecessary data transfers in and out of the GPUs.

- Adding more CPU RAM to a machine is simple and fairly cheap, so there's usually plenty of it, whereas the GPU RAM is baked into the GPU: it is an expensive and thus limited resource, so if a variable is not needed in the next few training steps, it should probably be placed on the CPU (e.g., datasets generally belong on the CPU).

---

13 Martín Abadi et al., "TensorFlow: Large-Scale Machine Learning on Heterogeneous Distributed Systems" Google Research whitepaper (2015).

By default, all variables and all operations will be placed on the first GPU (named /gpu:0), except for variables and operations that don't have a GPU kernel:[14] these are placed on the CPU (named /cpu:0). A tensor or variable's device attribute tells you which device it was placed on:[15]

```
>>> a = tf.Variable(42.0)
>>> a.device
'/job:localhost/replica:0/task:0/device:GPU:0'
>>> b = tf.Variable(42)
>>> b.device
'/job:localhost/replica:0/task:0/device:CPU:0'
```

You can safely ignore the prefix /job:localhost/replica:0/task:0 for now (it allows you to place operations on other machines when using a TensorFlow cluster; we will talk about jobs, replicas, and tasks later in this chapter). As you can see, the first variable was placed on GPU 0, which is the default device. However, the second variable was placed on the CPU: this is because there are no GPU kernels for integer variables (or for operations involving integer tensors), so TensorFlow fell back to the CPU.

If you want to place an operation on a different device than the default one, use a tf.device() context:

```
>>> with tf.device("/cpu:0"):
...     c = tf.Variable(42.0)
...
>>> c.device
'/job:localhost/replica:0/task:0/device:CPU:0'
```

> The CPU is always treated as a single device (/cpu:0), even if your machine has multiple CPU cores. Any operation placed on the CPU may run in parallel across multiple cores if it has a multi-threaded kernel.

If you explicitly try to place an operation or variable on a device that does not exist or for which there is no kernel, then you will get an exception. However, in some cases you may prefer to fall back to the CPU; for example, if your program may run both on CPU-only machines and on GPU machines, you may want TensorFlow to ignore your tf.device("/gpu:*") on CPU-only machines. To do this, you can call tf.con fig.set_soft_device_placement(True) just after importing TensorFlow: when a

---

14 As we saw in Chapter 12, a kernel is a variable or operation's implementation for a specific data type and device type. For example, there is a GPU kernel for the float32 tf.matmul() operation, but there is no GPU kernel for int32 tf.matmul() (only a CPU kernel).

15 You can also use tf.debugging.set_log_device_placement(True) to log all device placements.

placement request fails, TensorFlow will fall back to its default placement rules (i.e., GPU 0 by default if it exists and there is a GPU kernel, and CPU 0 otherwise).

Now how exactly will TensorFlow execute all these operations across multiple devices?

Parallel Execution Across Multiple Devices

As we saw in Chapter 12, one of the benefits of using TF Functions is parallelism. Let's look at this a bit more closely. When TensorFlow runs a TF Function, it starts by analyzing its graph to find the list of operations that need to be evaluated, and it counts how many dependencies each of them has. TensorFlow then adds each operation with zero dependencies (i.e., each source operation) to the evaluation queue of this operation's device (see Figure 19-14). Once an operation has been evaluated, the dependency counter of each operation that depends on it is decremented. Once an operation's dependency counter reaches zero, it is pushed to the evaluation queue of its device. And once all the nodes that TensorFlow needs have been evaluated, it returns their outputs.

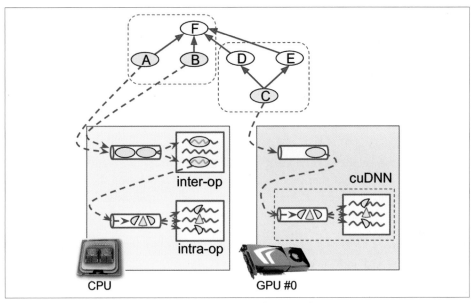

Figure 19-14. Parallelized execution of a TensorFlow graph

Operations in the CPU's evaluation queue are dispatched to a thread pool called the *inter-op thread pool*. If the CPU has multiple cores, then these operations will effectively be evaluated in parallel. Some operations have multithreaded CPU kernels: these kernels split their tasks into multiple suboperations, which are placed in another evaluation queue and dispatched to a second thread pool called the *intra-op thread pool* (shared by all multithreaded CPU kernels). In short, multiple operations and suboperations may be evaluated in parallel on different CPU cores.

For the GPU, things are a bit simpler. Operations in a GPU's evaluation queue are evaluated sequentially. However, most operations have multithreaded GPU kernels, typically implemented by libraries that TensorFlow depends on, such as CUDA and cuDNN. These implementations have their own thread pools, and they typically exploit as many GPU threads as they can (which is the reason why there is no need for an inter-op thread pool in GPUs: each operation already floods most GPU threads).

For example, in Figure 19-14, operations A, B, and C are source ops, so they can immediately be evaluated. Operations A and B are placed on the CPU, so they are sent to the CPU's evaluation queue, then they are dispatched to the inter-op thread pool and immediately evaluated in parallel. Operation A happens to have a multi-threaded kernel; its computations are split into three parts, which are executed in parallel by the intra-op thread pool. Operation C goes to GPU 0's evaluation queue, and in this example its GPU kernel happens to use cuDNN, which manages its own intra-op thread pool and runs the operation across many GPU threads in parallel. Suppose C finishes first. The dependency counters of D and E are decremented and they reach zero, so both operations are pushed to GPU 0's evaluation queue, and they are executed sequentially. Note that C only gets evaluated once, even though both D and E depend on it. Suppose B finishes next. Then F's dependency counter is decremented from 4 to 3, and since that's not 0, it does not run yet. Once A, D, and E are finished, then F's dependency counter reaches 0, and it is pushed to the CPU's evaluation queue and evaluated. Finally, TensorFlow returns the requested outputs.

An extra bit of magic that TensorFlow performs is when the TF Function modifies a stateful resource, such as a variable: it ensures that the order of execution matches the order in the code, even if there is no explicit dependency between the statements. For example, if your TF Function contains `v.assign_add(1)` followed by `v.assign(v * 2)`, TensorFlow will ensure that these operations are executed in that order.

You can control the number of threads in the inter-op thread pool by calling `tf.config.threading.set_inter_op_parallel ism_threads()`. To set the number of intra-op threads, use `tf.config.threading.set_intra_op_parallelism_threads()`. This is useful if you want do not want TensorFlow to use all the CPU cores or if you want it to be single-threaded.[16]

---

16 This can be useful if you want to guarantee perfect reproducibility, as I explain in this video (*https://homl.info/ repro*), based on TF 1.

With that, you have all you need to run any operation on any device, and exploit the power of your GPUs! Here are some of the things you could do:

- You could train several models in parallel, each on its own GPU: just write a training script for each model and run them in parallel, setting `CUDA_DEVICE_ORDER` and `CUDA_VISIBLE_DEVICES` so that each script only sees a single GPU device. This is great for hyperparameter tuning, as you can train in parallel multiple models with different hyperparameters. If you have a single machine with two GPUs, and it takes one hour to train one model on one GPU, then training two models in parallel, each on its own dedicated GPU, will take just one hour. Simple!

- You could train a model on a single GPU and perform all the preprocessing in parallel on the CPU, using the dataset's `prefetch()` method[17] to prepare the next few batches in advance so that they are ready when the GPU needs them (see Chapter 13).

- If your model takes two images as input and processes them using two CNNs before joining their outputs, then it will probably run much faster if you place each CNN on a different GPU.

- You can create an efficient ensemble: just place a different trained model on each GPU so that you can get all the predictions much faster to produce the ensemble's final prediction.

But what if you want to *train* a single model across multiple GPUs?

Training Models Across Multiple Devices

There are two main approaches to training a single model across multiple devices: *model parallelism*, where the model is split across the devices, and *data parallelism*, where the model is replicated across every device, and each replica is trained on a subset of the data. Let's look at these two options closely before we train a model on multiple GPUs.

Model Parallelism

So far we have trained each neural network on a single device. What if we want to train a single neural network across multiple devices? This requires chopping the model into separate chunks and running each chunk on a different device.

---

17 At the time of this writing it only prefetches the data to the CPU RAM, but you can use `tf.data.experimen tal.prefetch_to_device()` to make it prefetch the data and push it to the device of your choice so that the GPU does not waste time waiting for the data to be transferred.

Unfortunately, such model parallelism turns out to be pretty tricky, and it really depends on the architecture of your neural network. For fully connected networks, there is generally not much to be gained from this approach (see Figure 19-15). Intuitively, it may seem that an easy way to split the model is to place each layer on a different device, but this does not work because each layer needs to wait for the output of the previous layer before it can do anything. So perhaps you can slice it vertically— for example, with the left half of each layer on one device, and the right part on another device? This is slightly better, since both halves of each layer can indeed work in parallel, but the problem is that each half of the next layer requires the output of both halves, so there will be a lot of cross-device communication (represented by the dashed arrows). This is likely to completely cancel out the benefit of the parallel computation, since cross-device communication is slow (especially when the devices are located on different machines).

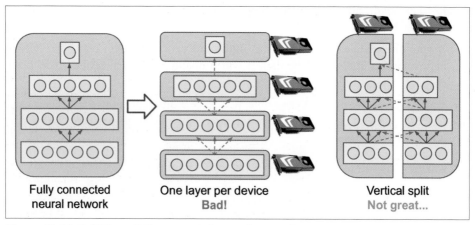

Figure 19-15. Splitting a fully connected neural network

Some neural network architectures, such as convolutional neural networks (see Chapter 14), contain layers that are only partially connected to the lower layers, so it is much easier to distribute chunks across devices in an efficient way (Figure 19-16).

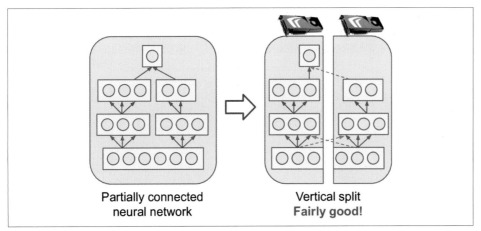

Figure 19-16. Splitting a partially connected neural network

Deep recurrent neural networks (see Chapter 15) can be split a bit more efficiently across multiple GPUs. If you split the network horizontally by placing each layer on a different device, and you feed the network with an input sequence to process, then at the first time step only one device will be active (working on the sequence's first value), at the second step two will be active (the second layer will be handling the output of the first layer for the first value, while the first layer will be handling the second value), and by the time the signal propagates to the output layer, all devices will be active simultaneously (Figure 19-17). There is still a lot of cross-device communication going on, but since each cell may be fairly complex, the benefit of running multiple cells in parallel may (in theory) outweigh the communication penalty. However, in practice a regular stack of LSTM layers running on a single GPU actually runs much faster.

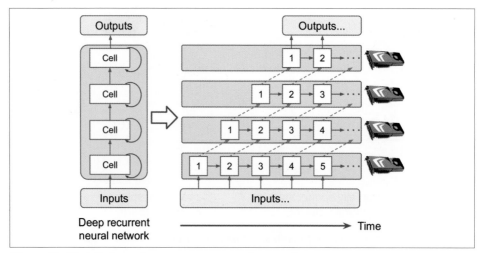

Figure 19-17. Splitting a deep recurrent neural network

In short, model parallelism may speed up running or training some types of neural networks, but not all, and it requires special care and tuning, such as making sure that devices that need to communicate the most run on the same machine.[18] Let's look at a much simpler and generally more efficient option: data parallelism.

Data Parallelism

Another way to parallelize the training of a neural network is to replicate it on every device and run each training step simultaneously on all replicas, using a different mini-batch for each. The gradients computed by each replica are then averaged, and the result is used to update the model parameters. This is called *data parallelism*. There are many variants of this idea, so let's look at the most important ones.

Data parallelism using the mirrored strategy

Arguably the simplest approach is to completely mirror all the model parameters across all the GPUs and always apply the exact same parameter updates on every GPU. This way, all replicas always remain perfectly identical. This is called the *mirrored strategy*, and it turns out to be quite efficient, especially when using a single machine (see Figure 19-18).

---

18 If you are interested in going further with model parallelism, check out Mesh TensorFlow (*https://github.com/ tensorflow/mesh*).

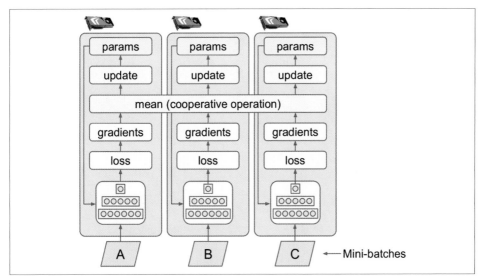

Figure 19-18. Data parallelism using the mirrored strategy

The tricky part when using this approach is to efficiently compute the mean of all the gradients from all the GPUs and distribute the result across all the GPUs. This can be done using an *AllReduce* algorithm, a class of algorithms where multiple nodes collaborate to efficiently perform a reduce operation (such as computing the mean, sum, and max), while ensuring that all nodes obtain the same final result. Fortunately, there are off-the-shelf implementations of such algorithms, as we will see.

Data parallelism with centralized parameters

Another approach is to store the model parameters outside of the GPU devices performing the computations (called *workers*), for example on the CPU (see Figure 19-19). In a distributed setup, you may place all the parameters on one or more CPU-only servers called *parameter servers*, whose only role is to host and update the parameters.

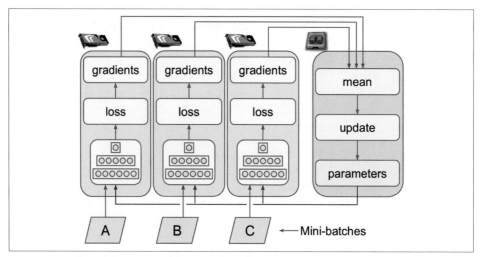

Figure 19-19. Data parallelism with centralized parameters

Whereas the mirrored strategy imposes synchronous weight updates across all GPUs, this centralized approach allows either synchronous or asynchronous updates. Let's see the pros and cons of both options.

Synchronous updates.   With *synchronous updates*, the aggregator waits until all gradients are available before it computes the average gradients and passes them to the optimizer, which will update the model parameters. Once a replica has finished computing its gradients, it must wait for the parameters to be updated before it can proceed to the next mini-batch. The downside is that some devices may be slower than others, so all other devices will have to wait for them at every step. Moreover, the parameters will be copied to every device almost at the same time (immediately after the gradients are applied), which may saturate the parameter servers' bandwidth.

To reduce the waiting time at each step, you could ignore the gradients from the slowest few replicas (typically ~10%). For example, you could run 20 replicas, but only aggregate the gradients from the fastest 18 replicas at each step, and just ignore the gradients from the last 2. As soon as the parameters are updated, the first 18 replicas can start working again immediately, without having to wait for the 2 slowest replicas. This setup is generally described as having 18 replicas plus 2 *spare replicas*.[19]

---

[19] This name is slightly confusing because it sounds like some replicas are special, doing nothing. In reality, all replicas are equivalent: they all work hard to be among the fastest at each training step, and the losers vary at every step (unless some devices are really slower than others). However, it does mean that if a server crashes, training will continue just fine.

Asynchronous updates. With asynchronous updates, whenever a replica has finished computing the gradients, it immediately uses them to update the model parameters. There is no aggregation (it removes the "mean" step in Figure 19-19) and no synchronization. Replicas work independently of the other replicas. Since there is no waiting for the other replicas, this approach runs more training steps per minute. Moreover, although the parameters still need to be copied to every device at every step, this happens at different times for each replica, so the risk of bandwidth saturation is reduced.

Data parallelism with asynchronous updates is an attractive choice because of its simplicity, the absence of synchronization delay, and a better use of the bandwidth. However, although it works reasonably well in practice, it is almost surprising that it works at all! Indeed, by the time a replica has finished computing the gradients based on some parameter values, these parameters will have been updated several times by other replicas (on average $N - 1$ times, if there are N replicas), and there is no guarantee that the computed gradients will still be pointing in the right direction (see Figure 19-20). When gradients are severely out-of-date, they are called *stale gradients*: they can slow down convergence, introducing noise and wobble effects (the learning curve may contain temporary oscillations), or they can even make the training algorithm diverge.

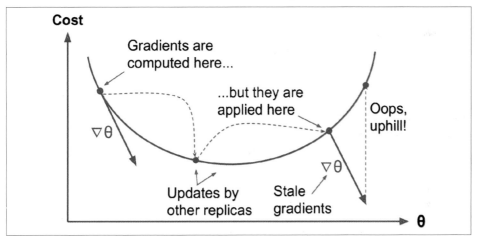

Figure 19-20. Stale gradients when using asynchronous updates

There are a few ways you can reduce the effect of stale gradients:

- Reduce the learning rate.
- Drop stale gradients or scale them down.
- Adjust the mini-batch size.

- Start the first few epochs using just one replica (this is called the *warmup phase*). Stale gradients tend to be more damaging at the beginning of training, when gradients are typically large and the parameters have not settled into a valley of the cost function yet, so different replicas may push the parameters in quite different directions.

A paper published by the Google Brain team in 2016 (*https://homl.info/68*)[20] benchmarked various approaches and found that using synchronous updates with a few spare replicas was more efficient than using asynchronous updates, not only converging faster but also producing a better model. However, this is still an active area of research, so you should not rule out asynchronous updates just yet.

Bandwidth saturation

Whether you use synchronous or asynchronous updates, data parallelism with centralized parameters still requires communicating the model parameters from the parameter servers to every replica at the beginning of each training step, and the gradients in the other direction at the end of each training step. Similarly, when using the mirrored strategy, the gradients produced by each GPU will need to be shared with every other GPU. Unfortunately, there always comes a point where adding an extra GPU will not improve performance at all because the time spent moving the data into and out of GPU RAM (and across the network in a distributed setup) will outweigh the speedup obtained by splitting the computation load. At that point, adding more GPUs will just worsen the bandwidth saturation and actually slow down training.

> For some models, typically relatively small and trained on a very large training set, you are often better off training the model on a single machine with a single powerful GPU with a large memory bandwidth.

Saturation is more severe for large dense models, since they have a lot of parameters and gradients to transfer. It is less severe for small models (but the parallelization gain is limited) and for large sparse models, where the gradients are typically mostly zeros and so can be communicated efficiently. Jeff Dean, initiator and lead of the Google Brain project, reported (*https://homl.info/69*) typical speedups of 25–40× when distributing computations across 50 GPUs for dense models, and a 300× speedup for sparser models trained across 500 GPUs. As you can see, sparse models really do scale better. Here are a few concrete examples:

---

20 Jianmin Chen et al., "Revisiting Distributed Synchronous SGD," arXiv preprint arXiv:1604.00981 (2016).

- Neural machine translation: 6× speedup on 8 GPUs
- Inception/ImageNet: 32× speedup on 50 GPUs
- RankBrain: 300× speedup on 500 GPUs

Beyond a few dozen GPUs for a dense model or few hundred GPUs for a sparse model, saturation kicks in and performance degrades. There is plenty of research going on to solve this problem (exploring peer-to-peer architectures rather than centralized parameter servers, using lossy model compression, optimizing when and what the replicas need to communicate, and so on), so there will likely be a lot of progress in parallelizing neural networks in the next few years.

In the meantime, to reduce the saturation problem, you probably want to use a few powerful GPUs rather than plenty of weak GPUs, and you should also group your GPUs on few and very well interconnected servers. You can also try dropping the float precision from 32 bits (`tf.float32`) to 16 bits (`tf.bfloat16`). This will cut in half the amount of data to transfer, often without much impact on the convergence rate or the model's performance. Lastly, if you are using centralized parameters, you can shard (split) the parameters across multiple parameter servers: adding more parameter servers will reduce the network load on each server and limit the risk of bandwidth saturation.

OK, now let's train a model across multiple GPUs!

Training at Scale Using the Distribution Strategies API

Many models can be trained quite well on a single GPU, or even on a CPU. But if training is too slow, you can try distributing it across multiple GPUs on the same machine. If that's still too slow, try using more powerful GPUs, or add more GPUs to the machine. If your model performs heavy computations (such as large matrix multiplications), then it will run much faster on powerful GPUs, and you could even try to use TPUs on Google Cloud AI Platform, which will usually run even faster for such models. But if you can't fit any more GPUs on the same machine, and if TPUs aren't for you (e.g., perhaps your model doesn't benefit much from TPUs, or perhaps you want to use your own hardware infrastructure), then you can try training it across several servers, each with multiple GPUs (if this is still not enough, as a last resort you can try adding some model parallelism, but this requires a lot more effort). In this section we will see how to train models at scale, starting with multiple GPUs on the same machine (or TPUs) and then moving on to multiple GPUs across multiple machines.

Luckily, TensorFlow comes with a very simple API that takes care of all the complexity for you: the *Distribution Strategies API*. To train a Keras model across all available GPUs (on a single machine, for now) using data parallelism with the mirrored strategy, create a `MirroredStrategy` object, call its `scope()` method to get a distribution context, and wrap the creation and compilation of your model inside that context. Then call the model's `fit()` method normally:

```
distribution = tf.distribute.MirroredStrategy()

with distribution.scope():
    mirrored_model = tf.keras.Sequential([...])
    mirrored_model.compile([...])

batch_size = 100 # must be divisible by the number of replicas
history = mirrored_model.fit(X_train, y_train, epochs=10)
```

Under the hood, tf.keras is distribution-aware, so in this `MirroredStrategy` context it knows that it must replicate all variables and operations across all available GPU devices. Note that the `fit()` method will automatically split each training batch across all the replicas, so it's important that the batch size be divisible by the number of replicas. And that's all! Training will generally be significantly faster than using a single device, and the code change was really minimal.

Once you have finished training your model, you can use it to make predictions efficiently: call the `predict()` method, and it will automatically split the batch across all replicas, making predictions in parallel (again, the batch size must be divisible by the number of replicas). If you call the model's `save()` method, it will be saved as a regular model, *not* as a mirrored model with multiple replicas. So when you load it, it will run like a regular model, on a single device (by default GPU 0, or the CPU if there are no GPUs). If you want to load a model and run it on all available devices, you must call `keras.models.load_model()` within a distribution context:

```
with distribution.scope():
    mirrored_model = keras.models.load_model("my_mnist_model.h5")
```

If you only want to use a subset of all the available GPU devices, you can pass the list to the `MirroredStrategy`'s constructor:

```
distribution = tf.distribute.MirroredStrategy(["/gpu:0", "/gpu:1"])
```

By default, the `MirroredStrategy` class uses the *NVIDIA Collective Communications Library* (NCCL) for the AllReduce mean operation, but you can change it by setting the `cross_device_ops` argument to an instance of the `tf.distribute.Hierarchical CopyAllReduce` class, or an instance of the `tf.distribute.ReductionToOneDevice` class. The default NCCL option is based on the `tf.distribute.NcclAllReduce` class, which is usually faster, but this depends on the number and types of GPUs, so you may want to give the alternatives a try.[21]

---

[21] For more details on AllReduce algorithms, read this great post (*https://homl.info/uenopost*) by Yuichiro Ueno, and this page on scaling with NCCL (*https://homl.info/ncclalgo*).

If you want to try using data parallelism with centralized parameters, replace the MirroredStrategy with the CentralStorageStrategy:

```
distribution = tf.distribute.experimental.CentralStorageStrategy()
```

You can optionally set the compute_devices argument to specify the list of devices you want to use as workers (by default it will use all available GPUs), and you can optionally set the parameter_device argument to specify the device you want to store the parameters on (by default it will use the CPU, or the GPU if there is just one).

Now let's see how to train a model across a cluster of TensorFlow servers!

Training a Model on a TensorFlow Cluster

A *TensorFlow cluster* is a group of TensorFlow processes running in parallel, usually on different machines, and talking to each other to complete some work—for example, training or executing a neural network. Each TF process in the cluster is called a *task*, or a *TF server*. It has an IP address, a port, and a type (also called its *role* or its *job*). The type can be either "worker", "chief", "ps" (parameter server), or "evaluator":

- Each *worker* performs computations, usually on a machine with one or more GPUs.

- The *chief* performs computations as well (it is a worker), but it also handles extra work such as writing TensorBoard logs or saving checkpoints. There is a single chief in a cluster. If no chief is specified, then the first worker is the chief.

- A *parameter server* only keeps track of variable values, and it is usually on a CPU-only machine. This type of task is only used with the ParameterServerStrategy.

- An *evaluator* obviously takes care of evaluation.

To start a TensorFlow cluster, you must first specify it. This means defining each task's IP address, TCP port, and type. For example, the following *cluster specification* defines a cluster with three tasks (two workers and one parameter server; see Figure 19-21). The cluster spec is a dictionary with one key per job, and the values are lists of task addresses (*IP:port*):

```
cluster_spec = {
    "worker": [
        "machine-a.example.com:2222",  # /job:worker/task:0
        "machine-b.example.com:2222"   # /job:worker/task:1
    ],
    "ps": ["machine-a.example.com:2221"] # /job:ps/task:0
}
```

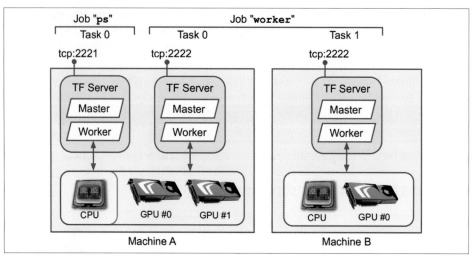

Figure 19-21. TensorFlow cluster

In general there will be a single task per machine, but as this example shows, you can configure multiple tasks on the same machine if you want (if they share the same GPUs, make sure the RAM is split appropriately, as discussed earlier).

> By default, every task in the cluster may communicate with every other task, so make sure to configure your firewall to authorize all communications between these machines on these ports (it's usually simpler if you use the same port on every machine).

When you start a task, you must give it the cluster spec, and you must also tell it what its type and index are (e.g., worker 0). The simplest way to specify everything at once (both the cluster spec and the current task's type and index) is to set the TF_CONFIG environment variable before starting TensorFlow. It must be a JSON-encoded dictionary containing a cluster specification (under the "cluster" key) and the type and index of the current task (under the "task" key). For example, the following TF_CON FIG environment variable uses the cluster we just defined and specifies that the task to start is the first worker:

```
import os
import json

os.environ["TF_CONFIG"] = json.dumps({
    "cluster": cluster_spec,
    "task": {"type": "worker", "index": 0}
})
```

In general you want to define the `TF_CONFIG` environment variable outside of Python, so the code does not need to include the current task's type and index (this makes it possible to use the same code across all workers).

Now let's train a model on a cluster! We will start with the mirrored strategy—it's surprisingly simple! First, you need to set the `TF_CONFIG` environment variable appropriately for each task. There should be no parameter server (remove the "ps" key in the cluster spec), and in general you will want a single worker per machine. Make extra sure you set a different task index for each task. Finally, run the following training code on every worker:

```
distribution = tf.distribute.experimental.MultiWorkerMirroredStrategy()

with distribution.scope():
    mirrored_model = tf.keras.Sequential([...])
    mirrored_model.compile([...])

batch_size = 100 # must be divisible by the number of replicas
history = mirrored_model.fit(X_train, y_train, epochs=10)
```

Yes, that's exactly the same code we used earlier, except this time we are using the `MultiWorkerMirroredStrategy` (in future versions, the `MirroredStrategy` will probably handle both the single machine and multimachine cases). When you start this script on the first workers, they will remain blocked at the AllReduce step, but as soon as the last worker starts up training will begin, and you will see them all advancing at exactly the same rate (since they synchronize at each step).

You can choose from two AllReduce implementations for this distribution strategy: a ring AllReduce algorithm based on gRPC for the network communications, and NCCL's implementation. The best algorithm to use depends on the number of workers, the number and types of GPUs, and the network. By default, TensorFlow will apply some heuristics to select the right algorithm for you, but if you want to force one algorithm, pass `CollectiveCommunication.RING` or `CollectiveCommunication.NCCL` (from `tf.distribute.experimental`) to the strategy's constructor.

If you prefer to implement asynchronous data parallelism with parameter servers, change the strategy to `ParameterServerStrategy`, add one or more parameter servers, and configure `TF_CONFIG` appropriately for each task. Note that although the workers will work asynchronously, the replicas on each worker will work synchronously.

Lastly, if you have access to TPUs on Google Cloud (*https://cloud.google.com/tpu/*), you can create a `TPUStrategy` like this (then use it like the other strategies):

```
resolver = tf.distribute.cluster_resolver.TPUClusterResolver()
tf.tpu.experimental.initialize_tpu_system(resolver)
tpu_strategy = tf.distribute.experimental.TPUStrategy(resolver)
```

 If you are a researcher, you may be eligible to use TPUs for free; see *https://tensorflow.org/tfrc* for more details.

You can now train models across multiple GPUs and multiple servers: give yourself a pat on the back! If you want to train a large model, you will need many GPUs, across many servers, which will require either buying a lot of hardware or managing a lot of cloud VMs. In many cases, it's going to be less hassle and less expensive to use a cloud service that takes care of provisioning and managing all this infrastructure for you, just when you need it. Let's see how to do that on GCP.

Running Large Training Jobs on Google Cloud AI Platform

If you decide to use Google AI Platform, you can deploy a training job with the same training code as you would run on your own TF cluster, and the platform will take care of provisioning and configuring as many GPU VMs as you desire (within your quotas).

To start the job, you will need the `gcloud` command-line tool, which is part of the Google Cloud SDK (*https://cloud.google.com/sdk/*). You can either install the SDK on your own machine, or just use the Google Cloud Shell on GCP. This is a terminal you can use directly in your web browser; it runs on a free Linux VM (Debian), with the SDK already installed and preconfigured for you. The Cloud Shell is available anywhere in GCP: just click the Activate Cloud Shell icon at the top right of the page (see Figure 19-22).

Figure 19-22. Activating the Google Cloud Shell

If you prefer to install the SDK on your machine, once you have installed it, you need to initialize it by running `gcloud init`: you will need to log in to GCP and grant access to your GCP resources, then select the GCP project you want to use (if you have more than one), as well as the region where you want the job to run. The `gcloud` command gives you access to every GCP feature, including the ones we used earlier.

You don't have to go through the web interface every time; you can write scripts that start or stop VMs for you, deploy models, or perform any other GCP action.

Before you can run the training job, you need to write the training code, exactly like you did earlier for a distributed setup (e.g., using the `ParameterServerStrategy`). AI Platform will take care of setting `TF_CONFIG` for you on each VM. Once that's done, you can deploy it and run it on a TF cluster with a command line like this:

```
$ gcloud ai-platform jobs submit training my_job_20190531_164700 \
    --region asia-southeast1 \
    --scale-tier PREMIUM_1 \
    --runtime-version 2.0 \
    --python-version 3.5 \
    --package-path /my_project/src/trainer \
    --module-name trainer.task \
    --staging-bucket gs://my-staging-bucket \
    --job-dir gs://my-mnist-model-bucket/trained_model \
    -- 
    --my-extra-argument1 foo --my-extra-argument2 bar
```

Let's go through these options. The command will start a training job named my_job_20190531_164700, in the `asia-southeast1` region, using a PREMIUM_1 *scale tier*: this corresponds to 20 workers (including a chief) and 11 parameter servers (check out the other available scale tiers (*https://homl.info/scaletiers*)). All these VMs will be based on AI Platform's 2.0 runtime (a VM configuration that includes TensorFlow 2.0 and many other packages)[22] and Python 3.5. The training code is located in the */my_project/src/trainer* directory, and the `gcloud` command will automatically bundle it into a pip package and upload it to GCS at *gs://my-staging-bucket*. Next, AI Platform will start several VMs, deploy the package to them, and run the `trainer.task` module. Lastly, the `--job-dir` argument and the extra arguments (i.e., all the arguments located after the `--` separator) will be passed to the training program: the chief task will usually use the `--job-dir` argument to find out where to save the final model on GCS, in this case at *gs://my-mnist-model-bucket/trained_model*. And that's it! In the GCP console, you can then open the navigation menu, scroll down to the Artificial Intelligence section, and open AI Platform → Jobs. You should see your job running, and if you click it you will see graphs showing the CPU, GPU, and RAM utilization for every task. You can click View Logs to access the detailed logs using Stackdriver.

---

22 At the time of this writing, the 2.0 runtime is not yet available, but it should be ready by the time you read this. Check out the list of available runtimes (*https://homl.info/runtimes*).

If you place the training data on GCS, you can create a `tf.data.TextLineDataset` or `tf.data.TFRecordDataset` to access it: just use the GCS paths as the filenames (e.g., *gs://my-data-bucket/my_data_001.csv*). These datasets rely on the `tf.io.gfile` package to access files: it supports both local files and GCS files (but make sure the service account you use has access to GCS).

If you want to explore a few hyperparameter values, you can simply run multiple jobs and specify the hyperparameter values using the extra arguments for your tasks. However, if you want to explore many hyperparameters efficiently, it's a good idea to use AI Platform's hyperparameter tuning service instead.

Black Box Hyperparameter Tuning on AI Platform

AI Platform provides a powerful Bayesian optimization hyperparameter tuning service called Google Vizier (*https://homl.info/vizier*).[23] To use it, you need to pass a YAML configuration file when creating the job (`--config tuning.yaml`). For example, it may look like this:

```
trainingInput:
  hyperparameters:
    goal: MAXIMIZE
    hyperparameterMetricTag: accuracy
    maxTrials: 10
    maxParallelTrials: 2
    params:
      - parameterName: n_layers
        type: INTEGER
        minValue: 10
        maxValue: 100
        scaleType: UNIT_LINEAR_SCALE
      - parameterName: momentum
        type: DOUBLE
        minValue: 0.1
        maxValue: 1.0
        scaleType: UNIT_LOG_SCALE
```

This tells AI Platform that we want to maximize the metric named `"accuracy"`, the job will run a maximum of 10 trials (each trial will run our training code to train the model from scratch), and it will run a maximum of 2 trials in parallel. We want it to tune two hyperparameters: the `n_layers` hyperparameter (an integer between 10 and 100) and the `momentum` hyperparameter (a float between 0.1 and 1.0). The `scaleType` argument specifies the prior for the hyperparameter value: `UNIT_LINEAR_SCALE`

---

23 Daniel Golovin et al., "Google Vizier: A Service for Black-Box Optimization," *Proceedings of the 23rd ACM SIGKDD International Conference on Knowledge Discovery and Data Mining* (2017): 1487–1495.

means a flat prior (i.e., no a priori preference), while UNIT_LOG_SCALE says we have a prior belief that the optimal value lies closer to the max value (the other possible prior is UNIT_REVERSE_LOG_SCALE, when we believe the optimal value to be close to the min value).

The n_layers and momentum arguments will be passed as command-line arguments to the training code, and of course it is expected to use them. The question is, how will the training code communicate the metric back to the AI Platform so that it can decide which hyperparameter values to use during the next trial? Well, AI Platform just monitors the output directory (specified via --job-dir) for any event file (introduced in Chapter 10) containing summaries for a metric named "accuracy" (or whatever metric name is specified as the hyperparameterMetricTag), and it reads those values. So your training code simply has to use the TensorBoard() callback (which you will want to do anyway for monitoring), and you're good to go!

Once the job is finished, all the hyperparameter values used in each trial and the resulting accuracy will be available in the job's output (available via the AI Platform → Jobs page).

> AI Platform jobs can also be used to efficiently execute your model on large amounts of data: each worker can read part of the data from GCS, make predictions, and save them to GCS.

Now you have all the tools and knowledge you need to create state-of-the-art neural net architectures and train them at scale using various distribution strategies, on your own infrastructure or on the cloud—and you can even perform powerful Bayesian optimization to fine-tune the hyperparameters!

Exercises

1. What does a SavedModel contain? How do you inspect its content?

2. When should you use TF Serving? What are its main features? What are some tools you can use to deploy it?

3. How do you deploy a model across multiple TF Serving instances?

4. When should you use the gRPC API rather than the REST API to query a model served by TF Serving?

5. What are the different ways TFLite reduces a model's size to make it run on a mobile or embedded device?

6. What is quantization-aware training, and why would you need it?

7. What are model parallelism and data parallelism? Why is the latter generally recommended?

8. When training a model across multiple servers, what distribution strategies can you use? How do you choose which one to use?

9. Train a model (any model you like) and deploy it to TF Serving or Google Cloud AI Platform. Write the client code to query it using the REST API or the gRPC API. Update the model and deploy the new version. Your client code will now query the new version. Roll back to the first version.

10. Train any model across multiple GPUs on the same machine using the `Mirrored Strategy` (if you do not have access to GPUs, you can use Colaboratory with a GPU Runtime and create two virtual GPUs). Train the model again using the `CentralStorageStrategy` and compare the training time.

11. Train a small model on Google Cloud AI Platform, using black box hyperparameter tuning.

Thank You!

Before we close the last chapter of this book, I would like to thank you for reading it up to the last paragraph. I truly hope that you had as much pleasure reading this book as I had writing it, and that it will be useful for your projects, big or small.

If you find errors, please send feedback. More generally, I would love to know what you think, so please don't hesitate to contact me via O'Reilly, through the *ageron/ handson-ml2* GitHub project, or on Twitter at @aureliengeron.

Going forward, my best advice to you is to practice and practice: try going through all the exercises (if you have not done so already), play with the Jupyter notebooks, join Kaggle.com or some other ML community, watch ML courses, read papers, attend conferences, and meet experts. It also helps tremendously to have a concrete project to work on, whether it is for work or for fun (ideally for both), so if there's anything you have always dreamt of building, give it a shot! Work incrementally; don't shoot for the moon right away, but stay focused on your project and build it piece by piece. It will require patience and perseverance, but when you have a walking robot, or a working chatbot, or whatever else you fancy to build, it will be immensely rewarding.

My greatest hope is that this book will inspire you to build a wonderful ML application that will benefit all of us! What will it be?

—Aurélien Géron, June 17, 2019

Exercise Solutions

 Solutions to the coding exercises are available in the online Jupyter notebooks at *https://github.com/ageron/handson-ml2*.

Chapter 1: The Machine Learning Landscape

1. Machine Learning is about building systems that can learn from data. Learning means getting better at some task, given some performance measure.

2. Machine Learning is great for complex problems for which we have no algorithmic solution, to replace long lists of hand-tuned rules, to build systems that adapt to fluctuating environments, and finally to help humans learn (e.g., data mining).

3. A labeled training set is a training set that contains the desired solution (a.k.a. a label) for each instance.

4. The two most common supervised tasks are regression and classification.

5. Common unsupervised tasks include clustering, visualization, dimensionality reduction, and association rule learning.

6. Reinforcement Learning is likely to perform best if we want a robot to learn to walk in various unknown terrains, since this is typically the type of problem that Reinforcement Learning tackles. It might be possible to express the problem as a supervised or semisupervised learning problem, but it would be less natural.

7. If you don't know how to define the groups, then you can use a clustering algorithm (unsupervised learning) to segment your customers into clusters of similar customers. However, if you know what groups you would like to have, then you

can feed many examples of each group to a classification algorithm (supervised learning), and it will classify all your customers into these groups.

8. Spam detection is a typical supervised learning problem: the algorithm is fed many emails along with their labels (spam or not spam).

9. An online learning system can learn incrementally, as opposed to a batch learning system. This makes it capable of adapting rapidly to both changing data and autonomous systems, and of training on very large quantities of data.

10. Out-of-core algorithms can handle vast quantities of data that cannot fit in a computer's main memory. An out-of-core learning algorithm chops the data into mini-batches and uses online learning techniques to learn from these mini-batches.

11. An instance-based learning system learns the training data by heart; then, when given a new instance, it uses a similarity measure to find the most similar learned instances and uses them to make predictions.

12. A model has one or more model parameters that determine what it will predict given a new instance (e.g., the slope of a linear model). A learning algorithm tries to find optimal values for these parameters such that the model generalizes well to new instances. A hyperparameter is a parameter of the learning algorithm itself, not of the model (e.g., the amount of regularization to apply).

13. Model-based learning algorithms search for an optimal value for the model parameters such that the model will generalize well to new instances. We usually train such systems by minimizing a cost function that measures how bad the system is at making predictions on the training data, plus a penalty for model complexity if the model is regularized. To make predictions, we feed the new instance's features into the model's prediction function, using the parameter values found by the learning algorithm.

14. Some of the main challenges in Machine Learning are the lack of data, poor data quality, nonrepresentative data, uninformative features, excessively simple models that underfit the training data, and excessively complex models that overfit the data.

15. If a model performs great on the training data but generalizes poorly to new instances, the model is likely overfitting the training data (or we got extremely lucky on the training data). Possible solutions to overfitting are getting more data, simplifying the model (selecting a simpler algorithm, reducing the number of parameters or features used, or regularizing the model), or reducing the noise in the training data.

16. A test set is used to estimate the generalization error that a model will make on new instances, before the model is launched in production.

17. A validation set is used to compare models. It makes it possible to select the best model and tune the hyperparameters.

18. The train-dev set is used when there is a risk of mismatch between the training data and the data used in the validation and test datasets (which should always be as close as possible to the data used once the model is in production). The train-dev set is a part of the training set that's held out (the model is not trained on it). The model is trained on the rest of the training set, and evaluated on both the train-dev set and the validation set. If the model performs well on the training set but not on the train-dev set, then the model is likely overfitting the training set. If it performs well on both the training set and the train-dev set, but not on the validation set, then there is probably a significant data mismatch between the training data and the validation + test data, and you should try to improve the training data to make it look more like the validation + test data.

19. If you tune hyperparameters using the test set, you risk overfitting the test set, and the generalization error you measure will be optimistic (you may launch a model that performs worse than you expect).

Chapter 2: End-to-End Machine Learning Project

See the Jupyter notebooks available at *https://github.com/ageron/handson-ml2*.

Chapter 3: Classification

See the Jupyter notebooks available at *https://github.com/ageron/handson-ml2*.

Chapter 4: Training Models

1. If you have a training set with millions of features you can use Stochastic Gradient Descent or Mini-batch Gradient Descent, and perhaps Batch Gradient Descent if the training set fits in memory. But you cannot use the Normal Equation or the SVD approach because the computational complexity grows quickly (more than quadratically) with the number of features.

2. If the features in your training set have very different scales, the cost function will have the shape of an elongated bowl, so the Gradient Descent algorithms will take a long time to converge. To solve this you should scale the data before training the model. Note that the Normal Equation or SVD approach will work just fine without scaling. Moreover, regularized models may converge to a suboptimal solution if the features are not scaled: since regularization penalizes large weights, features with smaller values will tend to be ignored compared to features with larger values.

3. Gradient Descent cannot get stuck in a local minimum when training a Logistic Regression model because the cost function is convex.[1]

4. If the optimization problem is convex (such as Linear Regression or Logistic Regression), and assuming the learning rate is not too high, then all Gradient Descent algorithms will approach the global optimum and end up producing fairly similar models. However, unless you gradually reduce the learning rate, Stochastic GD and Mini-batch GD will never truly converge; instead, they will keep jumping back and forth around the global optimum. This means that even if you let them run for a very long time, these Gradient Descent algorithms will produce slightly different models.

5. If the validation error consistently goes up after every epoch, then one possibility is that the learning rate is too high and the algorithm is diverging. If the training error also goes up, then this is clearly the problem and you should reduce the learning rate. However, if the training error is not going up, then your model is overfitting the training set and you should stop training.

6. Due to their random nature, neither Stochastic Gradient Descent nor Mini-batch Gradient Descent is guaranteed to make progress at every single training iteration. So if you immediately stop training when the validation error goes up, you may stop much too early, before the optimum is reached. A better option is to save the model at regular intervals; then, when it has not improved for a long time (meaning it will probably never beat the record), you can revert to the best saved model.

7. Stochastic Gradient Descent has the fastest training iteration since it considers only one training instance at a time, so it is generally the first to reach the vicinity of the global optimum (or Mini-batch GD with a very small mini-batch size). However, only Batch Gradient Descent will actually converge, given enough training time. As mentioned, Stochastic GD and Mini-batch GD will bounce around the optimum, unless you gradually reduce the learning rate.

8. If the validation error is much higher than the training error, this is likely because your model is overfitting the training set. One way to try to fix this is to reduce the polynomial degree: a model with fewer degrees of freedom is less likely to overfit. Another thing you can try is to regularize the model—for example, by adding an ℓ_2 penalty (Ridge) or an ℓ_1 penalty (Lasso) to the cost function. This will also reduce the degrees of freedom of the model. Lastly, you can try to increase the size of the training set.

---

1 If you draw a straight line between any two points on the curve, the line never crosses the curve.

9. If both the training error and the validation error are almost equal and fairly high, the model is likely underfitting the training set, which means it has a high bias. You should try reducing the regularization hyperparameter α.

10. Let's see:

   - A model with some regularization typically performs better than a model without any regularization, so you should generally prefer Ridge Regression over plain Linear Regression.

   - Lasso Regression uses an ℓ_1 penalty, which tends to push the weights down to exactly zero. This leads to sparse models, where all weights are zero except for the most important weights. This is a way to perform feature selection automatically, which is good if you suspect that only a few features actually matter. When you are not sure, you should prefer Ridge Regression.

   - Elastic Net is generally preferred over Lasso since Lasso may behave erratically in some cases (when several features are strongly correlated or when there are more features than training instances). However, it does add an extra hyperparameter to tune. If you want Lasso without the erratic behavior, you can just use Elastic Net with an `l1_ratio` close to 1.

11. If you want to classify pictures as outdoor/indoor and daytime/nighttime, since these are not exclusive classes (i.e., all four combinations are possible) you should train two Logistic Regression classifiers.

12. See the Jupyter notebooks available at *https://github.com/ageron/handson-ml2*.

Chapter 5: Support Vector Machines

1. The fundamental idea behind Support Vector Machines is to fit the widest possible "street" between the classes. In other words, the goal is to have the largest possible margin between the decision boundary that separates the two classes and the training instances. When performing soft margin classification, the SVM searches for a compromise between perfectly separating the two classes and having the widest possible street (i.e., a few instances may end up on the street). Another key idea is to use kernels when training on nonlinear datasets.

2. After training an SVM, a *support vector* is any instance located on the "street" (see the previous answer), including its border. The decision boundary is entirely determined by the support vectors. Any instance that is *not* a support vector (i.e., is off the street) has no influence whatsoever; you could remove them, add more instances, or move them around, and as long as they stay off the street they won't affect the decision boundary. Computing the predictions only involves the support vectors, not the whole training set.

3. SVMs try to fit the largest possible "street" between the classes (see the first answer), so if the training set is not scaled, the SVM will tend to neglect small features (see Figure 5-2).

4. An SVM classifier can output the distance between the test instance and the decision boundary, and you can use this as a confidence score. However, this score cannot be directly converted into an estimation of the class probability. If you set `probability=True` when creating an SVM in Scikit-Learn, then after training it will calibrate the probabilities using Logistic Regression on the SVM's scores (trained by an additional five-fold cross-validation on the training data). This will add the `predict_proba()` and `predict_log_proba()` methods to the SVM.

5. This question applies only to linear SVMs since kernelized SVMs can only use the dual form. The computational complexity of the primal form of the SVM problem is proportional to the number of training instances m, while the computational complexity of the dual form is proportional to a number between m^2 and m^3. So if there are millions of instances, you should definitely use the primal form, because the dual form will be much too slow.

6. If an SVM classifier trained with an RBF kernel underfits the training set, there might be too much regularization. To decrease it, you need to increase `gamma` or `C` (or both).

7. Let's call the QP parameters for the hard margin problem $\mathbf{H}'$, $\mathbf{f}'$, $\mathbf{A}'$, and $\mathbf{b}'$ (see "Quadratic Programming" on page 167). The QP parameters for the soft margin problem have m additional parameters ($n_p = n + 1 + m$) and m additional constraints ($n_c = 2m$). They can be defined like so:

   - $\mathbf{H}$ is equal to $\mathbf{H}'$, plus m columns of 0s on the right and m rows of 0s at the bottom: $\mathbf{H} = \begin{pmatrix} \mathbf{H}' & 0 & \cdots \\ 0 & 0 & \\ \vdots & & \ddots \end{pmatrix}$

   - $\mathbf{f}$ is equal to $\mathbf{f}'$ with m additional elements, all equal to the value of the hyperparameter C.

   - $\mathbf{b}$ is equal to $\mathbf{b}'$ with m additional elements, all equal to 0.

   - $\mathbf{A}$ is equal to $\mathbf{A}'$, with an extra $m \times m$ identity matrix $\mathbf{I}_m$ appended to the right, $-^*\mathbf{I}^*_m$ just below it, and the rest filled with 0s: $\mathbf{A} = \begin{pmatrix} \mathbf{A}' & \mathbf{I}_m \\ 0 & -\mathbf{I}_m \end{pmatrix}$

For the solutions to exercises 8, 9, and 10, please see the Jupyter notebooks available at *https://github.com/ageron/handson-ml2*.

Chapter 6: Decision Trees

1. The depth of a well-balanced binary tree containing m leaves is equal to $\log_2(m)$,[2] rounded up. A binary Decision Tree (one that makes only binary decisions, as is the case with all trees in Scikit-Learn) will end up more or less well balanced at the end of training, with one leaf per training instance if it is trained without restrictions. Thus, if the training set contains one million instances, the Decision Tree will have a depth of $\log_2(10^6) \approx 20$ (actually a bit more since the tree will generally not be perfectly well balanced).

2. A node's Gini impurity is generally lower than its parent's. This is due to the CART training algorithm's cost function, which splits each node in a way that minimizes the weighted sum of its children's Gini impurities. However, it is possible for a node to have a higher Gini impurity than its parent, as long as this increase is more than compensated for by a decrease in the other child's impurity. For example, consider a node containing four instances of class A and one of class B. Its Gini impurity is $1 - (1/5)^2 - (4/5)^2 = 0.32$. Now suppose the dataset is one-dimensional and the instances are lined up in the following order: A, B, A, A, A. You can verify that the algorithm will split this node after the second instance, producing one child node with instances A, B, and the other child node with instances A, A, A. The first child node's Gini impurity is $1 - (1/2)^2 - (1/2)^2 = 0.5$, which is higher than its parent's. This is compensated for by the fact that the other node is pure, so its overall weighted Gini impurity is $2/5 \times 0.5 + 3/5 \times 0 = 0.2$, which is lower than the parent's Gini impurity.

3. If a Decision Tree is overfitting the training set, it may be a good idea to decrease `max_depth`, since this will constrain the model, regularizing it.

4. Decision Trees don't care whether or not the training data is scaled or centered; that's one of the nice things about them. So if a Decision Tree underfits the training set, scaling the input features will just be a waste of time.

5. The computational complexity of training a Decision Tree is $O(n \times m \log(m))$. So if you multiply the training set size by 10, the training time will be multiplied by $K = (n \times 10m \times \log(10m)) / (n \times m \times \log(m)) = 10 \times \log(10m) / \log(m)$. If $m = 10^6$, then $K \approx 11.7$, so you can expect the training time to be roughly 11.7 hours.

6. Presorting the training set speeds up training only if the dataset is smaller than a few thousand instances. If it contains 100,000 instances, setting `presort=True` will considerably slow down training.

For the solutions to exercises 7 and 8, please see the Jupyter notebooks available at *https://github.com/ageron/handson-ml2*.

---

2 $\log_2$ is the binary log; $\log_2(m) = \log(m) / \log(2)$.

Chapter 7: Ensemble Learning and Random Forests

1. If you have trained five different models and they all achieve 95% precision, you can try combining them into a voting ensemble, which will often give you even better results. It works better if the models are very different (e.g., an SVM classifier, a Decision Tree classifier, a Logistic Regression classifier, and so on). It is even better if they are trained on different training instances (that's the whole point of bagging and pasting ensembles), but if not this will still be effective as long as the models are very different.

2. A hard voting classifier just counts the votes of each classifier in the ensemble and picks the class that gets the most votes. A soft voting classifier computes the average estimated class probability for each class and picks the class with the highest probability. This gives high-confidence votes more weight and often performs better, but it works only if every classifier is able to estimate class probabilities (e.g., for the SVM classifiers in Scikit-Learn you must set `probability=True`).

3. It is quite possible to speed up training of a bagging ensemble by distributing it across multiple servers, since each predictor in the ensemble is independent of the others. The same goes for pasting ensembles and Random Forests, for the same reason. However, each predictor in a boosting ensemble is built based on the previous predictor, so training is necessarily sequential, and you will not gain anything by distributing training across multiple servers. Regarding stacking ensembles, all the predictors in a given layer are independent of each other, so they can be trained in parallel on multiple servers. However, the predictors in one layer can only be trained after the predictors in the previous layer have all been trained.

4. With out-of-bag evaluation, each predictor in a bagging ensemble is evaluated using instances that it was not trained on (they were held out). This makes it possible to have a fairly unbiased evaluation of the ensemble without the need for an additional validation set. Thus, you have more instances available for training, and your ensemble can perform slightly better.

5. When you are growing a tree in a Random Forest, only a random subset of the features is considered for splitting at each node. This is true as well for Extra-Trees, but they go one step further: rather than searching for the best possible thresholds, like regular Decision Trees do, they use random thresholds for each feature. This extra randomness acts like a form of regularization: if a Random Forest overfits the training data, Extra-Trees might perform better. Moreover, since Extra-Trees don't search for the best possible thresholds, they are much faster to train than Random Forests. However, they are neither faster nor slower than Random Forests when making predictions.

6. If your AdaBoost ensemble underfits the training data, you can try increasing the number of estimators or reducing the regularization hyperparameters of the base estimator. You may also try slightly increasing the learning rate.

7. If your Gradient Boosting ensemble overfits the training set, you should try decreasing the learning rate. You could also use early stopping to find the right number of predictors (you probably have too many).

For the solutions to exercises 8 and 9, please see the Jupyter notebooks available at *https://github.com/ageron/handson-ml2*.

Chapter 8: Dimensionality Reduction

1. The main motivations for dimensionality reduction are:

   - To speed up a subsequent training algorithm (in some cases it may even remove noise and redundant features, making the training algorithm perform better)

   - To visualize the data and gain insights on the most important features

   - To save space (compression)

   The main drawbacks are:

   - Some information is lost, possibly degrading the performance of subsequent training algorithms.

   - It can be computationally intensive.

   - It adds some complexity to your Machine Learning pipelines.

   - Transformed features are often hard to interpret.

2. The curse of dimensionality refers to the fact that many problems that do not exist in low-dimensional space arise in high-dimensional space. In Machine Learning, one common manifestation is the fact that randomly sampled high-dimensional vectors are generally very sparse, increasing the risk of overfitting and making it very difficult to identify patterns in the data without having plenty of training data.

3. Once a dataset's dimensionality has been reduced using one of the algorithms we discussed, it is almost always impossible to perfectly reverse the operation, because some information gets lost during dimensionality reduction. Moreover, while some algorithms (such as PCA) have a simple reverse transformation procedure that can reconstruct a dataset relatively similar to the original, other algorithms (such as T-SNE) do not.

4. PCA can be used to significantly reduce the dimensionality of most datasets, even if they are highly nonlinear, because it can at least get rid of useless dimensions. However, if there are no useless dimensions—as in a Swiss roll dataset—then reducing dimensionality with PCA will lose too much information. You want to unroll the Swiss roll, not squash it.

5. That's a trick question: it depends on the dataset. Let's look at two extreme examples. First, suppose the dataset is composed of points that are almost perfectly aligned. In this case, PCA can reduce the dataset down to just one dimension while still preserving 95% of the variance. Now imagine that the dataset is composed of perfectly random points, scattered all around the 1,000 dimensions. In this case roughly 950 dimensions are required to preserve 95% of the variance. So the answer is, it depends on the dataset, and it could be any number between 1 and 950. Plotting the explained variance as a function of the number of dimensions is one way to get a rough idea of the dataset's intrinsic dimensionality.

6. Regular PCA is the default, but it works only if the dataset fits in memory. Incremental PCA is useful for large datasets that don't fit in memory, but it is slower than regular PCA, so if the dataset fits in memory you should prefer regular PCA. Incremental PCA is also useful for online tasks, when you need to apply PCA on the fly, every time a new instance arrives. Randomized PCA is useful when you want to considerably reduce dimensionality and the dataset fits in memory; in this case, it is much faster than regular PCA. Finally, Kernel PCA is useful for nonlinear datasets.

7. Intuitively, a dimensionality reduction algorithm performs well if it eliminates a lot of dimensions from the dataset without losing too much information. One way to measure this is to apply the reverse transformation and measure the reconstruction error. However, not all dimensionality reduction algorithms provide a reverse transformation. Alternatively, if you are using dimensionality reduction as a preprocessing step before another Machine Learning algorithm (e.g., a Random Forest classifier), then you can simply measure the performance of that second algorithm; if dimensionality reduction did not lose too much information, then the algorithm should perform just as well as when using the original dataset.

8. It can absolutely make sense to chain two different dimensionality reduction algorithms. A common example is using PCA to quickly get rid of a large number of useless dimensions, then applying another much slower dimensionality reduction algorithm, such as LLE. This two-step approach will likely yield the same performance as using LLE only, but in a fraction of the time.

For the solutions to exercises 9 and 10, please see the Jupyter notebooks available at *https://github.com/ageron/handson-ml2.*

Chapter 9: Unsupervised Learning Techniques

1. In Machine Learning, clustering is the unsupervised task of grouping similar instances together. The notion of similarity depends on the task at hand: for example, in some cases two nearby instances will be considered similar, while in others similar instances may be far apart as long as they belong to the same densely packed group. Popular clustering algorithms include K-Means, DBSCAN, agglomerative clustering, BIRCH, Mean-Shift, affinity propagation, and spectral clustering.

2. The main applications of clustering algorithms include data analysis, customer segmentation, recommender systems, search engines, image segmentation, semi-supervised learning, dimensionality reduction, anomaly detection, and novelty detection.

3. The elbow rule is a simple technique to select the number of clusters when using K-Means: just plot the inertia (the mean squared distance from each instance to its nearest centroid) as a function of the number of clusters, and find the point in the curve where the inertia stops dropping fast (the "elbow"). This is generally close to the optimal number of clusters. Another approach is to plot the silhouette score as a function of the number of clusters. There will often be a peak, and the optimal number of clusters is generally nearby. The silhouette score is the mean silhouette coefficient over all instances. This coefficient varies from +1 for instances that are well inside their cluster and far from other clusters, to –1 for instances that are very close to another cluster. You may also plot the silhouette diagrams and perform a more thorough analysis.

4. Labeling a dataset is costly and time-consuming. Therefore, it is common to have plenty of unlabeled instances, but few labeled instances. Label propagation is a technique that consists in copying some (or all) of the labels from the labeled instances to similar unlabeled instances. This can greatly extend the number of labeled instances, and thereby allow a supervised algorithm to reach better performance (this is a form of semi-supervised learning). One approach is to use a clustering algorithm such as K-Means on all the instances, then for each cluster find the most common label or the label of the most representative instance (i.e., the one closest to the centroid) and propagate it to the unlabeled instances in the same cluster.

5. K-Means and BIRCH scale well to large datasets. DBSCAN and Mean-Shift look for regions of high density.

6. Active learning is useful whenever you have plenty of unlabeled instances but labeling is costly. In this case (which is very common), rather than randomly selecting instances to label, it is often preferable to perform active learning, where human experts interact with the learning algorithm, providing labels for

specific instances when the algorithm requests them. A common approach is uncertainty sampling (see the description in "Active Learning" on page 255).

7. Many people use the terms *anomaly detection* and *novelty detection* interchangeably, but they are not exactly the same. In anomaly detection, the algorithm is trained on a dataset that may contain outliers, and the goal is typically to identify these outliers (within the training set), as well as outliers among new instances. In novelty detection, the algorithm is trained on a dataset that is presumed to be "clean," and the objective is to detect novelties strictly among new instances. Some algorithms work best for anomaly detection (e.g., Isolation Forest), while others are better suited for novelty detection (e.g., one-class SVM).

8. A Gaussian mixture model (GMM) is a probabilistic model that assumes that the instances were generated from a mixture of several Gaussian distributions whose parameters are unknown. In other words, the assumption is that the data is grouped into a finite number of clusters, each with an ellipsoidal shape (but the clusters may have different ellipsoidal shapes, sizes, orientations, and densities), and we don't know which cluster each instance belongs to. This model is useful for density estimation, clustering, and anomaly detection.

9. One way to find the right number of clusters when using a Gaussian mixture model is to plot the Bayesian information criterion (BIC) or the Akaike information criterion (AIC) as a function of the number of clusters, then choose the number of clusters that minimizes the BIC or AIC. Another technique is to use a Bayesian Gaussian mixture model, which automatically selects the number of clusters.

For the solutions to exercises 10 to 13, please see the Jupyter notebooks available at *https://github.com/ageron/handson-ml2*.

Chapter 10: Introduction to Artificial Neural Networks with Keras

1. Visit the TensorFlow Playground (*https://playground.tensorflow.org/*) and play around with it, as described in this exercise.

2. Here is a neural network based on the original artificial neurons that computes $A \oplus B$ (where $\oplus$ represents the exclusive OR), using the fact that $A \oplus B = (A \wedge \neg B) \vee (\neg A \wedge B)$. There are other solutions—for example, using the fact that $A \oplus B = (A \vee B) \wedge \neg(A \wedge B)$, or the fact that $A \oplus B = (A \vee B) \wedge (\neg A \vee \wedge B)$, and so on.

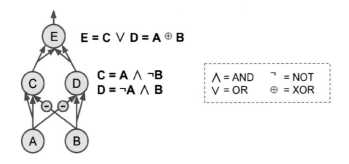

$$E = C \lor D = A \oplus B$$

$$C = A \land \lnot B$$
$$D = \lnot A \land B$$

$\land$ = AND $\lnot$ = NOT
$\lor$ = OR $\oplus$ = XOR

3. A classical Perceptron will converge only if the dataset is linearly separable, and it won't be able to estimate class probabilities. In contrast, a Logistic Regression classifier will converge to a good solution even if the dataset is not linearly separable, and it will output class probabilities. If you change the Perceptron's activation function to the logistic activation function (or the softmax activation function if there are multiple neurons), and if you train it using Gradient Descent (or some other optimization algorithm minimizing the cost function, typically cross entropy), then it becomes equivalent to a Logistic Regression classifier.

4. The logistic activation function was a key ingredient in training the first MLPs because its derivative is always nonzero, so Gradient Descent can always roll down the slope. When the activation function is a step function, Gradient Descent cannot move, as there is no slope at all.

5. Popular activation functions include the step function, the logistic (sigmoid) function, the hyperbolic tangent (tanh) function, and the Rectified Linear Unit (ReLU) function (see Figure 10-8). See Chapter 11 for other examples, such as ELU and variants of the ReLU function.

6. Considering the MLP described in the question, composed of one input layer with 10 passthrough neurons, followed by one hidden layer with 50 artificial neurons, and finally one output layer with 3 artificial neurons, where all artificial neurons use the ReLU activation function: ..The shape of the input matrix X is $m \times 10$, where m represents the training batch size.

   a. The shape of the hidden layer's weight vector W_h is 10×50, and the length of its bias vector b_h is 50.

   b. The shape of the output layer's weight vector W_o is 50×3, and the length of its bias vector b_o is 3.

   c. The shape of the network's output matrix Y is $m \times 3$.

   d. $Y^* = \text{ReLU}(\text{ReLU}(X W_h + b_h) W_o + b_o)$. Recall that the ReLU function just sets every negative number in the matrix to zero. Also note that when you are adding a bias vector to a matrix, it is added to every single row in the matrix, which is called *broadcasting*.

7. To classify email into spam or ham, you just need one neuron in the output layer of a neural network—for example, indicating the probability that the email is spam. You would typically use the logistic activation function in the output layer when estimating a probability. If instead you want to tackle MNIST, you need 10 neurons in the output layer, and you must replace the logistic function with the softmax activation function, which can handle multiple classes, outputting one probability per class. If you want your neural network to predict housing prices like in Chapter 2, then you need one output neuron, using no activation function at all in the output layer.[3]

8. Backpropagation is a technique used to train artificial neural networks. It first computes the gradients of the cost function with regard to every model parameter (all the weights and biases), then it performs a Gradient Descent step using these gradients. This backpropagation step is typically performed thousands or millions of times, using many training batches, until the model parameters converge to values that (hopefully) minimize the cost function. To compute the gradients, backpropagation uses reverse-mode autodiff (although it wasn't called that when backpropagation was invented, and it has been reinvented several times). Reverse-mode autodiff performs a forward pass through a computation graph, computing every node's value for the current training batch, and then it performs a reverse pass, computing all the gradients at once (see Appendix D for more details). So what's the difference? Well, backpropagation refers to the whole process of training an artificial neural network using multiple backpropagation steps, each of which computes gradients and uses them to perform a Gradient Descent step. In contrast, reverse-mode autodiff is just a technique to compute gradients efficiently, and it happens to be used by backpropagation.

9. Here is a list of all the hyperparameters you can tweak in a basic MLP: the number of hidden layers, the number of neurons in each hidden layer, and the activation function used in each hidden layer and in the output layer.[4] In general, the ReLU activation function (or one of its variants; see Chapter 11) is a good default for the hidden layers. For the output layer, in general you will want the logistic activation function for binary classification, the softmax activation function for multiclass classification, or no activation function for regression.

---

3 When the values to predict can vary by many orders of magnitude, you may want to predict the logarithm of the target value rather than the target value directly. Simply computing the exponential of the neural network's output will give you the estimated value (since $\exp(\log v) = v$).

4 In Chapter 11 we discuss many techniques that introduce additional hyperparameters: type of weight initialization, activation function hyperparameters (e.g., the amount of leak in leaky ReLU), Gradient Clipping threshold, type of optimizer and its hyperparameters (e.g., the momentum hyperparameter when using a MomentumOptimizer), type of regularization for each layer and regularization hyperparameters (e.g., dropout rate when using dropout), and so on.

If the MLP overfits the training data, you can try reducing the number of hidden layers and reducing the number of neurons per hidden layer.

10. See the Jupyter notebooks available at *https://github.com/ageron/handson-ml2*.

Chapter 11: Training Deep Neural Networks

1. No, all weights should be sampled independently; they should not all have the same initial value. One important goal of sampling weights randomly is to break symmetry: if all the weights have the same initial value, even if that value is not zero, then symmetry is not broken (i.e., all neurons in a given layer are equivalent), and backpropagation will be unable to break it. Concretely, this means that all the neurons in any given layer will always have the same weights. It's like having just one neuron per layer, and much slower. It is virtually impossible for such a configuration to converge to a good solution.

2. It is perfectly fine to initialize the bias terms to zero. Some people like to initialize them just like weights, and that's okay too; it does not make much difference.

3. A few advantages of the SELU function over the ReLU function are:

   - It can take on negative values, so the average output of the neurons in any given layer is typically closer to zero than when using the ReLU activation function (which never outputs negative values). This helps alleviate the vanishing gradients problem.

   - It always has a nonzero derivative, which avoids the dying units issue that can affect ReLU units.

   - When the conditions are right (i.e., if the model is sequential, and the weights are initialized using LeCun initialization, and the inputs are standardized, and there's no incompatible layer or regularization, such as dropout or ℓ_1 regularization), then the SELU activation function ensures the model is self-normalized, which solves the exploding/vanishing gradients problems.

4. The SELU activation function is a good default. If you need the neural network to be as fast as possible, you can use one of the leaky ReLU variants instead (e.g., a simple leaky ReLU using the default hyperparameter value). The simplicity of the ReLU activation function makes it many people's preferred option, despite the fact that it is generally outperformed by SELU and leaky ReLU. However, the ReLU activation function's ability to output precisely zero can be useful in some cases (e.g., see Chapter 17). Moreover, it can sometimes benefit from optimized implementation as well as from hardware acceleration. The hyperbolic tangent (tanh) can be useful in the output layer if you need to output a number between −1 and 1, but nowadays it is not used much in hidden layers (except in recurrent

nets). The logistic activation function is also useful in the output layer when you need to estimate a probability (e.g., for binary classification), but is rarely used in hidden layers (there are exceptions—for example, for the coding layer of variational autoencoders; see Chapter 17). Finally, the softmax activation function is useful in the output layer to output probabilities for mutually exclusive classes, but it is rarely (if ever) used in hidden layers.

5. If you set the momentum hyperparameter too close to 1 (e.g., 0.99999) when using an SGD optimizer, then the algorithm will likely pick up a lot of speed, hopefully moving roughly toward the global minimum, but its momentum will carry it right past the minimum. Then it will slow down and come back, accelerate again, overshoot again, and so on. It may oscillate this way many times before converging, so overall it will take much longer to converge than with a smaller momentum value.

6. One way to produce a sparse model (i.e., with most weights equal to zero) is to train the model normally, then zero out tiny weights. For more sparsity, you can apply ℓ_1 regularization during training, which pushes the optimizer toward sparsity. A third option is to use the TensorFlow Model Optimization Toolkit.

7. Yes, dropout does slow down training, in general roughly by a factor of two. However, it has no impact on inference speed since it is only turned on during training. MC Dropout is exactly like dropout during training, but it is still active during inference, so each inference is slowed down slightly. More importantly, when using MC Dropout you generally want to run inference 10 times or more to get better predictions. This means that making predictions is slowed down by a factor of 10 or more.

For the solutions to exercises 8, 9, and 10, please see the Jupyter notebooks available at *https://github.com/ageron/handson-ml2*.

Chapter 12: Custom Models and Training with TensorFlow

1. TensorFlow is an open-source library for numerical computation, particularly well suited and fine-tuned for large-scale Machine Learning. Its core is similar to NumPy, but it also features GPU support, support for distributed computing, computation graph analysis and optimization capabilities (with a portable graph format that allows you to train a TensorFlow model in one environment and run it in another), an optimization API based on reverse-mode autodiff, and several powerful APIs such as tf.keras, tf.data, tf.image, tf.signal, and more. Other popular Deep Learning libraries include PyTorch, MXNet, Microsoft Cognitive Toolkit, Theano, Caffe2, and Chainer.

2. Although TensorFlow offers most of the functionalities provided by NumPy, it is not a drop-in replacement, for a few reasons. First, the names of the functions are

not always the same (for example, `tf.reduce_sum()` versus `np.sum()`). Second, some functions do not behave in exactly the same way (for example, `tf.trans` `pose()` creates a transposed copy of a tensor, while NumPy's T attribute creates a transposed view, without actually copying any data). Lastly, NumPy arrays are mutable, while TensorFlow tensors are not (but you can use a `tf.Variable` if you need a mutable object).

3. Both `tf.range(10)` and `tf.constant(np.arange(10))` return a one-dimensional tensor containing the integers 0 to 9. However, the former uses 32-bit integers while the latter uses 64-bit integers. Indeed, TensorFlow defaults to 32 bits, while NumPy defaults to 64 bits.

4. Beyond regular tensors, TensorFlow offers several other data structures, including sparse tensors, tensor arrays, ragged tensors, queues, string tensors, and sets. The last two are actually represented as regular tensors, but TensorFlow provides special functions to manipulate them (in `tf.strings` and `tf.sets`).

5. When you want to define a custom loss function, in general you can just implement it as a regular Python function. However, if your custom loss function must support some hyperparameters (or any other state), then you should subclass the `keras.losses.Loss` class and implement the `__init__()` and `call()` methods. If you want the loss function's hyperparameters to be saved along with the model, then you must also implement the `get_config()` method.

6. Much like custom loss functions, most metrics can be defined as regular Python functions. But if you want your custom metric to support some hyperparameters (or any other state), then you should subclass the `keras.metrics.Metric` class. Moreover, if computing the metric over a whole epoch is not equivalent to computing the mean metric over all batches in that epoch (e.g., as for the precision and recall metrics), then you should subclass the `keras.metrics.Metric` class and implement the `__init__()`, `update_state()`, and `result()` methods to keep track of a running metric during each epoch. You should also implement the `reset_states()` method unless all it needs to do is reset all variables to 0.0. If you want the state to be saved along with the model, then you should implement the `get_config()` method as well.

7. You should distinguish the internal components of your model (i.e., layers or reusable blocks of layers) from the model itself (i.e., the object you will train). The former should subclass the `keras.layers.Layer` class, while the latter should subclass the `keras.models.Model` class.

8. Writing your own custom training loop is fairly advanced, so you should only do it if you really need to. Keras provides several tools to customize training without having to write a custom training loop: callbacks, custom regularizers, custom constraints, custom losses, and so on. You should use these instead of writing a custom training loop whenever possible: writing a custom training loop is more

error-prone, and it will be harder to reuse the custom code you write. However, in some cases writing a custom training loop is necessary—for example, if you want to use different optimizers for different parts of your neural network, like in the Wide & Deep paper (*https://homl.info/widedeep*). A custom training loop can also be useful when debugging, or when trying to understand exactly how training works.

9. Custom Keras components should be convertible to TF Functions, which means they should stick to TF operations as much as possible and respect all the rules listed in "TF Function Rules" on page 409. If you absolutely need to include arbitrary Python code in a custom component, you can either wrap it in a `tf.py_function()` operation (but this will reduce performance and limit your model's portability) or set `dynamic=True` when creating the custom layer or model (or set `run_eagerly=True` when calling the model's `compile()` method).

10. Please refer to "TF Function Rules" on page 409 for the list of rules to respect when creating a TF Function.

11. Creating a dynamic Keras model can be useful for debugging, as it will not compile any custom component to a TF Function, and you can use any Python debugger to debug your code. It can also be useful if you want to include arbitrary Python code in your model (or in your training code), including calls to external libraries. To make a model dynamic, you must set `dynamic=True` when creating it. Alternatively, you can set `run_eagerly=True` when calling the model's `compile()` method. Making a model dynamic prevents Keras from using any of TensorFlow's graph features, so it will slow down training and inference, and you will not have the possibility to export the computation graph, which will limit your model's portability.

For the solutions to exercises 12 and 13, please see the Jupyter notebooks available at *https://github.com/ageron/handson-ml2*.

Chapter 13: Loading and Preprocessing Data with TensorFlow

1. Ingesting a large dataset and preprocessing it efficiently can be a complex engineering challenge. The Data API makes it fairly simple. It offers many features, including loading data from various sources (such as text or binary files), reading data in parallel from multiple sources, transforming it, interleaving the records, shuffling the data, batching it, and prefetching it.

2. Splitting a large dataset into multiple files makes it possible to shuffle it at a coarse level before shuffling it at a finer level using a shuffling buffer. It also makes it possible to handle huge datasets that do not fit on a single machine. It's

also simpler to manipulate thousands of small files rather than one huge file; for example, it's easier to split the data into multiple subsets. Lastly, if the data is split across multiple files spread across multiple servers, it is possible to download several files from different servers simultaneously, which improves the bandwidth usage.

3. You can use TensorBoard to visualize profiling data: if the GPU is not fully utilized then your input pipeline is likely to be the bottleneck. You can fix it by making sure it reads and preprocesses the data in multiple threads in parallel, and ensuring it prefetches a few batches. If this is insufficient to get your GPU to 100% usage during training, make sure your preprocessing code is optimized. You can also try saving the dataset into multiple TFRecord files, and if necessary perform some of the preprocessing ahead of time so that it does not need to be done on the fly during training (TF Transform can help with this). If necessary, use a machine with more CPU and RAM, and ensure that the GPU bandwidth is large enough.

4. A TFRecord file is composed of a sequence of arbitrary binary records: you can store absolutely any binary data you want in each record. However, in practice most TFRecord files contain sequences of serialized protocol buffers. This makes it possible to benefit from the advantages of protocol buffers, such as the fact that they can be read easily across multiple platforms and languages and their definition can be updated later in a backward-compatible way.

5. The `Example` protobuf format has the advantage that TensorFlow provides some operations to parse it (the `tf.io.parse*example()` functions) without you having to define your own format. It is sufficiently flexible to represent instances in most datasets. However, if it does not cover your use case, you can define your own protocol buffer, compile it using `protoc` (setting the `--descriptor_set_out` and `--include_imports` arguments to export the protobuf descriptor), and use the `tf.io.decode_proto()` function to parse the serialized protobufs (see the "Custom protobuf" section of the notebook for an example). It's more complicated, and it requires deploying the descriptor along with the model, but it can be done.

6. When using TFRecords, you will generally want to activate compression if the TFRecord files will need to be downloaded by the training script, as compression will make files smaller and thus reduce download time. But if the files are located on the same machine as the training script, it's usually preferable to leave compression off, to avoid wasting CPU for decompression.

7. Let's look at the pros and cons of each preprocessing option:

   • If you preprocess the data when creating the data files, the training script will run faster, since it will not have to perform preprocessing on the fly. In some cases, the preprocessed data will also be much smaller than the original data, so you can save some space and speed up downloads. It may also be helpful to materialize the preprocessed data, for example to inspect it or archive it. However, this approach has a few cons. First, it's not easy to experiment with various preprocessing logics if you need to generate a preprocessed dataset for each variant. Second, if you want to perform data augmentation, you have to materialize many variants of your dataset, which will use a large amount of disk space and take a lot of time to generate. Lastly, the trained model will expect preprocessed data, so you will have to add preprocessing code in your application before it calls the model.

- If the data is preprocessed with the tf.data pipeline, it's much easier to tweak the preprocessing logic and apply data augmentation. Also, tf.data makes it easy to build highly efficient preprocessing pipelines (e.g., with multithreading and prefetching). However, preprocessing the data this way will slow down training. Moreover, each training instance will be preprocessed once per epoch rather than just once if the data was preprocessed when creating the data files. Lastly, the trained model will still expect preprocessed data.

- If you add preprocessing layers to your model, you will only have to write the preprocessing code once for both training and inference. If your model needs to be deployed to many different platforms, you will not need to write the preprocessing code multiple times. Plus, you will not run the risk of using the wrong preprocessing logic for your model, since it will be part of the model. On the downside, preprocessing the data will slow down training, and each training instance will be preprocessed once per epoch. Moreover, by default the preprocessing operations will run on the GPU for the current batch (you will not benefit from parallel preprocessing on the CPU, and prefetching). Fortunately, the upcoming Keras preprocessing layers should be able to lift the preprocessing operations from the preprocessing layers and run them as part of the tf.data pipeline, so you will benefit from multithreaded execution on the CPU and prefetching.

- Lastly, using TF Transform for preprocessing gives you many of the benefits from the previous options: the preprocessed data is materialized, each instance is preprocessed just once (speeding up training), and preprocessing layers get generated automatically so you only need to write the preprocessing code once. The main drawback is the fact that you need to learn how to use this tool.

8. Let's look at how to encode categorical features and text:

- To encode a categorical feature that has a natural order, such as a movie rating (e.g., "bad," "average," "good"), the simplest option is to use ordinal encoding: sort the categories in their natural order and map each category to its rank (e.g., "bad" maps to 0, "average" maps to 1, and "good" maps to 2). However, most categorical features don't have such a natural order. For example, there's no natural order for professions or countries. In this case, you can use one-hot encoding or, if there are many categories, embeddings.

- For text, one option is to use a bag-of-words representation: a sentence is represented by a vector counting the counts of each possible word. Since common words are usually not very important, you'll want to use TF-IDF to reduce their weight. Instead of counting words, it is also common to count n-grams, which are sequences of n consecutive words—nice and simple. Alternatively, you can encode each word using word embeddings, possibly pretrained. Rather than encoding words, it is also possible to encode each letter, or subword tokens (e.g., splitting "smartest" into "smart" and "est"). These last two options are discussed in Chapter 16.

For the solutions to exercises 9 and 10, please see the Jupyter notebooks available at *https://github.com/ageron/handson-ml2*.

Chapter 14: Deep Computer Vision Using Convolutional Neural Networks

1. These are the main advantages of a CNN over a fully connected DNN for image classification:

- Because consecutive layers are only partially connected and because it heavily reuses its weights, a CNN has many fewer parameters than a fully connected DNN, which makes it much faster to train, reduces the risk of overfitting, and requires much less training data.

- When a CNN has learned a kernel that can detect a particular feature, it can detect that feature anywhere in the image. In contrast, when a DNN learns a feature in one location, it can detect it only in that particular location. Since images typically have very repetitive features, CNNs are able to generalize much better than DNNs for image processing tasks such as classification, using fewer training examples.

- Finally, a DNN has no prior knowledge of how pixels are organized; it does not know that nearby pixels are close. A CNN's architecture embeds this prior knowledge. Lower layers typically identify features in small areas of the images, while higher layers combine the lower-level features into larger features. This works well with most natural images, giving CNNs a decisive head start compared to DNNs.

2. Let's compute how many parameters the CNN has. Since its first convolutional layer has 3 × 3 kernels, and the input has three channels (red, green, and blue),

each feature map has 3 × 3 × 3 weights, plus a bias term. That's 28 parameters per feature map. Since this first convolutional layer has 100 feature maps, it has a total of 2,800 parameters. The second convolutional layer has 3 × 3 kernels and its input is the set of 100 feature maps of the previous layer, so each feature map has 3 × 3 × 100 = 900 weights, plus a bias term. Since it has 200 feature maps, this layer has 901 × 200 = 180,200 parameters. Finally, the third and last convolutional layer also has 3 × 3 kernels, and its input is the set of 200 feature maps of the previous layers, so each feature map has 3 × 3 × 200 = 1,800 weights, plus a bias term. Since it has 400 feature maps, this layer has a total of 1,801 × 400 = 720,400 parameters. All in all, the CNN has 2,800 + 180,200 + 720,400 = 903,400 parameters.

Now let's compute how much RAM this neural network will require (at least) when making a prediction for a single instance. First let's compute the feature map size for each layer. Since we are using a stride of 2 and "same" padding, the horizontal and vertical dimensions of the feature maps are divided by 2 at each layer (rounding up if necessary). So, as the input channels are 200 × 300 pixels, the first layer's feature maps are 100 × 150, the second layer's feature maps are 50 × 75, and the third layer's feature maps are 25 × 38. Since 32 bits is 4 bytes and the first convolutional layer has 100 feature maps, this first layer takes up 4 × 100 × 150 × 100 = 6 million bytes (6 MB). The second layer takes up 4 × 50 × 75 × 200 = 3 million bytes (3 MB). Finally, the third layer takes up 4 × 25 × 38 × 400 = 1,520,000 bytes (about 1.5 MB). However, once a layer has been computed, the memory occupied by the previous layer can be released, so if everything is well optimized, only 6 + 3 = 9 million bytes (9 MB) of RAM will be required (when the second layer has just been computed, but the memory occupied by the first layer has not been released yet). But wait, you also need to add the memory occupied by the CNN's parameters! We computed earlier that it has 903,400 parameters, each using up 4 bytes, so this adds 3,613,600 bytes (about 3.6 MB). The total RAM required is therefore (at least) 12,613,600 bytes (about 12.6 MB).

Lastly, let's compute the minimum amount of RAM required when training the CNN on a mini-batch of 50 images. During training TensorFlow uses backpropagation, which requires keeping all values computed during the forward pass until the reverse pass begins. So we must compute the total RAM required by all layers for a single instance and multiply that by 50. At this point, let's start counting in megabytes rather than bytes. We computed before that the three layers require respectively 6, 3, and 1.5 MB for each instance. That's a total of 10.5 MB per instance, so for 50 instances the total RAM required is 525 MB. Add to that the RAM required by the input images, which is 50 × 4 × 200 × 300 × 3 = 36 million bytes (36 MB), plus the RAM required for the model parameters, which is about 3.6 MB (computed earlier), plus some RAM for the gradients (we will neglect this since it can be released gradually as backpropagation goes down the layers during the reverse pass). We are up to a total of roughly 525 + 36 + 3.6 = 564.6 MB, and that's really an optimistic bare minimum.

3. If your GPU runs out of memory while training a CNN, here are five things you could try to solve the problem (other than purchasing a GPU with more RAM):

- Reduce the mini-batch size.
- Reduce dimensionality using a larger stride in one or more layers.
- Remove one or more layers.
- Use 16-bit floats instead of 32-bit floats.
- Distribute the CNN across multiple devices.

4. A max pooling layer has no parameters at all, whereas a convolutional layer has quite a few (see the previous questions).

5. A local response normalization layer makes the neurons that most strongly activate inhibit neurons at the same location but in neighboring feature maps, which encourages different feature maps to specialize and pushes them apart, forcing them to explore a wider range of features. It is typically used in the lower layers to have a larger pool of low-level features that the upper layers can build upon.

6. The main innovations in AlexNet compared to LeNet-5 are that it is much larger and deeper, and it stacks convolutional layers directly on top of each other, instead of stacking a pooling layer on top of each convolutional layer. The main innovation in GoogLeNet is the introduction of *inception modules*, which make it possible to have a much deeper net than previous CNN architectures, with fewer parameters. ResNet's main innovation is the introduction of skip connections, which make it possible to go well beyond 100 layers. Arguably, its simplicity and consistency are also rather innovative. SENet's main innovation was the idea of using an SE block (a two-layer dense network) after every inception module in an inception network or every residual unit in a ResNet to recalibrate the relative importance of feature maps. Finally, Xception's main innovation was the use of depthwise separable convolutional layers, which look at spatial patterns and depthwise patterns separately.

7. Fully convolutional networks are neural networks composed exclusively of convolutional and pooling layers. FCNs can efficiently process images of any width and height (at least above the minimum size). They are most useful for object detection and semantic segmentation because they only need to look at the image once (instead of having to run a CNN multiple times on different parts of the image). If you have a CNN with some dense layers on top, you can convert these dense layers to convolutional layers to create an FCN: just replace the lowest dense layer with a convolutional layer with a kernel size equal to the layer's input size, with one filter per neuron in the dense layer, and using "valid" padding.

Generally the stride should be 1, but you can set it to a higher value if you want. The activation function should be the same as the dense layer's. The other dense layers should be converted the same way, but using 1 × 1 filters. It is actually possible to convert a trained CNN this way by appropriately reshaping the dense layers' weight matrices.

8. The main technical difficulty of semantic segmentation is the fact that a lot of the spatial information gets lost in a CNN as the signal flows through each layer, especially in pooling layers and layers with a stride greater than 1. This spatial information needs to be restored somehow to accurately predict the class of each pixel.

For the solutions to exercises 9 to 12, please see the Jupyter notebooks available at *https://github.com/ageron/handson-ml2*.

Chapter 15: Processing Sequences Using RNNs and CNNs

1. Here are a few RNN applications:

   - For a sequence-to-sequence RNN: predicting the weather (or any other time series), machine translation (using an Encoder–Decoder architecture), video captioning, speech to text, music generation (or other sequence generation), identifying the chords of a song

   - For a sequence-to-vector RNN: classifying music samples by music genre, analyzing the sentiment of a book review, predicting what word an aphasic patient is thinking of based on readings from brain implants, predicting the probability that a user will want to watch a movie based on their watch history (this is one of many possible implementations of *collaborative filtering* for a recommender system)

   - For a vector-to-sequence RNN: image captioning, creating a music playlist based on an embedding of the current artist, generating a melody based on a set of parameters, locating pedestrians in a picture (e.g., a video frame from a self-driving car's camera)

2. An RNN layer must have three-dimensional inputs: the first dimension is the batch dimension (its size is the batch size), the second dimension represents the time (its size is the number of time steps), and the third dimension holds the inputs at each time step (its size is the number of input features per time step). For example, if you want to process a batch containing 5 time series of 10 time steps each, with 2 values per time step (e.g., the temperature and the wind speed), the shape will be [5, 10, 2]. The outputs are also three-dimensional, with the same first two dimensions, but the last dimension is equal to the number of

neurons. For example, if an RNN layer with 32 neurons processes the batch we just discussed, the output will have a shape of [5, 10, 32].

3. To build a deep sequence-to-sequence RNN using Keras, you must set `return_sequences=True` for all RNN layers. To build a sequence-to-vector RNN, you must set `return_sequences=True` for all RNN layers except for the top RNN layer, which must have `return_sequences=False` (or do not set this argument at all, since `False` is the default).

4. If you have a daily univariate time series, and you want to forecast the next seven days, the simplest RNN architecture you can use is a stack of RNN layers (all with `return_sequences=True` except for the top RNN layer), using seven neurons in the output RNN layer. You can then train this model using random windows from the time series (e.g., sequences of 30 consecutive days as the inputs, and a vector containing the values of the next 7 days as the target). This is a sequence-to-vector RNN. Alternatively, you could set `return_sequences=True` for all RNN layers to create a sequence-to-sequence RNN. You can train this model using random windows from the time series, with sequences of the same length as the inputs as the targets. Each target sequence should have seven values per time step (e.g., for time step t, the target should be a vector containing the values at time steps $t + 1$ to $t + 7$).

5. The two main difficulties when training RNNs are unstable gradients (exploding or vanishing) and a very limited short-term memory. These problems both get worse when dealing with long sequences. To alleviate the unstable gradients problem, you can use a smaller learning rate, use a saturating activation function such as the hyperbolic tangent (which is the default), and possibly use gradient clipping, Layer Normalization, or dropout at each time step. To tackle the limited short-term memory problem, you can use LSTM or GRU layers (this also helps with the unstable gradients problem).

6. An LSTM cell's architecture looks complicated, but it's actually not too hard if you understand the underlying logic. The cell has a short-term state vector and a long-term state vector. At each time step, the inputs and the previous short-term state are fed to a simple RNN cell and three gates: the forget gate decides what to remove from the long-term state, the input gate decides which part of the output of the simple RNN cell should be added to the long-term state, and the output gate decides which part of the long-term state should be output at this time step (after going through the tanh activation function). The new short-term state is equal to the output of the cell. See Figure 15-9.

7. An RNN layer is fundamentally sequential: in order to compute the outputs at time step t, it has to first compute the outputs at all earlier time steps. This makes it impossible to parallelize. On the other hand, a 1D convolutional layer lends itself well to parallelization since it does not hold a state between time steps. In

other words, it has no memory: the output at any time step can be computed based only on a small window of values from the inputs without having to know all the past values. Moreover, since a 1D convolutional layer is not recurrent, it suffers less from unstable gradients. One or more 1D convolutional layers can be useful in an RNN to efficiently preprocess the inputs, for example to reduce their temporal resolution (downsampling) and thereby help the RNN layers detect long-term patterns. In fact, it is possible to use only convolutional layers, for example by building a WaveNet architecture.

8. To classify videos based on their visual content, one possible architecture could be to take (say) one frame per second, then run every frame through the same convolutional neural network (e.g., a pretrained Xception model, possibly frozen if your dataset is not large), feed the sequence of outputs from the CNN to a sequence-to-vector RNN, and finally run its output through a softmax layer, giving you all the class probabilities. For training you would use cross entropy as the cost function. If you wanted to use the audio for classification as well, you could use a stack of strided 1D convolutional layers to reduce the temporal resolution from thousands of audio frames per second to just one per second (to match the number of images per second), and concatenate the output sequence to the inputs of the sequence-to-vector RNN (along the last dimension).

For the solutions to exercises 9 and 10, please see the Jupyter notebooks available at *https://github.com/ageron/handson-ml2*.

Chapter 16: Natural Language Processing with RNNs and Attention

1. Stateless RNNs can only capture patterns whose length is less than, or equal to, the size of the windows the RNN is trained on. Conversely, stateful RNNs can capture longer-term patterns. However, implementing a stateful RNN is much harder—especially preparing the dataset properly. Moreover, stateful RNNs do not always work better, in part because consecutive batches are not independent and identically distributed (IID). Gradient Descent is not fond of non-IID datasets.

2. In general, if you translate a sentence one word at a time, the result will be terrible. For example, the French sentence "Je vous en prie" means "You are welcome," but if you translate it one word at a time, you get "I you in pray." Huh? It is much better to read the whole sentence first and then translate it. A plain sequence-to-sequence RNN would start translating a sentence immediately after reading the first word, while an Encoder–Decoder RNN will first read the whole sentence and then translate it. That said, one could imagine a plain sequence-to-sequence

RNN that would output silence whenever it is unsure about what to say next (just like human translators do when they must translate a live broadcast).

3. Variable-length input sequences can be handled by padding the shorter sequences so that all sequences in a batch have the same length, and using masking to ensure the RNN ignores the padding token. For better performance, you may also want to create batches containing sequences of similar sizes. Ragged tensors can hold sequences of variable lengths, and tf.keras will likely support them eventually, which will greatly simplify handling variable-length input sequences (at the time of this writing, it is not the case yet). Regarding variable-length output sequences, if the length of the output sequence is known in advance (e.g., if you know that it is the same as the input sequence), then you just need to configure the loss function so that it ignores tokens that come after the end of the sequence. Similarly, the code that will use the model should ignore tokens beyond the end of the sequence. But generally the length of the output sequence is not known ahead of time, so the solution is to train the model so that it outputs an end-of-sequence token at the end of each sequence.

4. Beam search is a technique used to improve the performance of a trained Encoder–Decoder model, for example in a neural machine translation system. The algorithm keeps track of a short list of the *k* most promising output sentences (say, the top three), and at each decoder step it tries to extend them by one word; then it keeps only the *k* most likely sentences. The parameter *k* is called the *beam width*: the larger it is, the more CPU and RAM will be used, but also the more accurate the system will be. Instead of greedily choosing the most likely next word at each step to extend a single sentence, this technique allows the system to explore several promising sentences simultaneously. Moreover, this technique lends itself well to parallelization. You can implement beam search fairly easily using TensorFlow Addons.

5. An attention mechanism is a technique initially used in Encoder–Decoder models to give the decoder more direct access to the input sequence, allowing it to deal with longer input sequences. At each decoder time step, the current decoder's state and the full output of the encoder are processed by an alignment model that outputs an alignment score for each input time step. This score indicates which part of the input is most relevant to the current decoder time step. The weighted sum of the encoder output (weighted by their alignment score) is then fed to the decoder, which produces the next decoder state and the output for this time step. The main benefit of using an attention mechanism is the fact that the Encoder–Decoder model can successfully process longer input sequences. Another benefit is that the alignment scores makes the model easier to debug and interpret: for example, if the model makes a mistake, you can look at which part of the input it was paying attention to, and this can help diagnose the issue. An

attention mechanism is also at the core of the Transformer architecture, in the Multi-Head Attention layers. See the next answer.

6. The most important layer in the Transformer architecture is the Multi-Head Attention layer (the original Transformer architecture contains 18 of them, including 6 Masked Multi-Head Attention layers). It is at the core of language models such as BERT and GPT-2. Its purpose is to allow the model to identify which words are most aligned with each other, and then improve each word's representation using these contextual clues.

7. Sampled softmax is used when training a classification model when there are many classes (e.g., thousands). It computes an approximation of the cross-entropy loss based on the logit predicted by the model for the correct class, and the predicted logits for a sample of incorrect words. This speeds up training considerably compared to computing the softmax over all logits and then estimating the cross-entropy loss. After training, the model can be used normally, using the regular softmax function to compute all the class probabilities based on all the logits.

For the solutions to exercises 8 to 11, please see the Jupyter notebooks available at *https://github.com/ageron/handson-ml2*.

Chapter 17: Representation Learning and Generative Learning Using Autoencoders and GANs

1. Here are some of the main tasks that autoencoders are used for:

   - Feature extraction

   - Unsupervised pretraining

   - Dimensionality reduction

   - Generative models

   - Anomaly detection (an autoencoder is generally bad at reconstructing outliers)

2. If you want to train a classifier and you have plenty of unlabeled training data but only a few thousand labeled instances, then you could first train a deep autoencoder on the full dataset (labeled + unlabeled), then reuse its lower half for the classifier (i.e., reuse the layers up to the codings layer, included) and train the classifier using the labeled data. If you have little labeled data, you probably want to freeze the reused layers when training the classifier.

3. The fact that an autoencoder perfectly reconstructs its inputs does not necessarily mean that it is a good autoencoder; perhaps it is simply an overcomplete autoencoder that learned to copy its inputs to the codings layer and then to the outputs.

In fact, even if the codings layer contained a single neuron, it would be possible for a very deep autoencoder to learn to map each training instance to a different coding (e.g., the first instance could be mapped to 0.001, the second to 0.002, the third to 0.003, and so on), and it could learn "by heart" to reconstruct the right training instance for each coding. It would perfectly reconstruct its inputs without really learning any useful pattern in the data. In practice such a mapping is unlikely to happen, but it illustrates the fact that perfect reconstructions are not a guarantee that the autoencoder learned anything useful. However, if it produces very bad reconstructions, then it is almost guaranteed to be a bad autoencoder. To evaluate the performance of an autoencoder, one option is to measure the reconstruction loss (e.g., compute the MSE, or the mean square of the outputs minus the inputs). Again, a high reconstruction loss is a good sign that the autoencoder is bad, but a low reconstruction loss is not a guarantee that it is good. You should also evaluate the autoencoder according to what it will be used for. For example, if you are using it for unsupervised pretraining of a classifier, then you should also evaluate the classifier's performance.

4. An undercomplete autoencoder is one whose codings layer is smaller than the input and output layers. If it is larger, then it is an overcomplete autoencoder. The main risk of an excessively undercomplete autoencoder is that it may fail to reconstruct the inputs. The main risk of an overcomplete autoencoder is that it may just copy the inputs to the outputs, without learning any useful features.

5. To tie the weights of an encoder layer and its corresponding decoder layer, you simply make the decoder weights equal to the transpose of the encoder weights. This reduces the number of parameters in the model by half, often making training converge faster with less training data and reducing the risk of overfitting the training set.

6. A generative model is a model capable of randomly generating outputs that resemble the training instances. For example, once trained successfully on the MNIST dataset, a generative model can be used to randomly generate realistic images of digits. The output distribution is typically similar to the training data. For example, since MNIST contains many images of each digit, the generative model would output roughly the same number of images of each digit. Some generative models can be parametrized—for example, to generate only some kinds of outputs. An example of a generative autoencoder is the variational autoencoder.

7. A generative adversarial network is a neural network architecture composed of two parts, the generator and the discriminator, which have opposing objectives. The generator's goal is to generate instances similar to those in the training set, to fool the discriminator. The discriminator must distinguish the real instances from the generated ones. At each training iteration, the discriminator is trained like a normal binary classifier, then the generator is trained to maximize the

discriminator's error. GANs are used for advanced image processing tasks such as super resolution, colorization, image editing (replacing objects with realistic background), turning a simple sketch into a photorealistic image, or predicting the next frames in a video. They are also used to augment a dataset (to train other models), to generate other types of data (such as text, audio, and time series), and to identify the weaknesses in other models and strengthen them.

8. Training GANs is notoriously difficult, because of the complex dynamics between the generator and the discriminator. The biggest difficulty is mode collapse, where the generator produces outputs with very little diversity. Moreover, training can be terribly unstable: it may start out fine and then suddenly start oscillating or diverging, without any apparent reason. GANs are also very sensitive to the choice of hyperparameters.

For the solutions to exercises 9, 10, and 11, please see the Jupyter notebooks available at *https://github.com/ageron/handson-ml2*.

Chapter 18: Reinforcement Learning

1. Reinforcement Learning is an area of Machine Learning aimed at creating agents capable of taking actions in an environment in a way that maximizes rewards over time. There are many differences between RL and regular supervised and unsupervised learning. Here are a few:

   - In supervised and unsupervised learning, the goal is generally to find patterns in the data and use them to make predictions. In Reinforcement Learning, the goal is to find a good policy.

   - Unlike in supervised learning, the agent is not explicitly given the "right" answer. It must learn by trial and error.

   - Unlike in unsupervised learning, there is a form of supervision, through rewards. We do not tell the agent how to perform the task, but we do tell it when it is making progress or when it is failing.

   - A Reinforcement Learning agent needs to find the right balance between exploring the environment, looking for new ways of getting rewards, and exploiting sources of rewards that it already knows. In contrast, supervised and unsupervised learning systems generally don't need to worry about exploration; they just feed on the training data they are given.

   - In supervised and unsupervised learning, training instances are typically independent (in fact, they are generally shuffled). In Reinforcement Learning, consecutive observations are generally *not* independent. An agent may remain in the same region of the environment for a while before it moves on, so consecutive observations will be very correlated. In some cases a replay memory (buffer) is used to ensure that the training algorithm gets fairly independent observations.

2. Here are a few possible applications of Reinforcement Learning, other than those mentioned in Chapter 18:

Music personalization

The environment is a user's personalized web radio. The agent is the software deciding what song to play next for that user. Its possible actions are to play any song in the catalog (it must try to choose a song the user will enjoy) or to play an advertisement (it must try to choose an ad that the user will be interested in). It gets a small reward every time the user listens to a song, a larger reward every time the user listens to an ad, a negative reward when the user skips a song or an ad, and a very negative reward if the user leaves.

Marketing

The environment is your company's marketing department. The agent is the software that defines which customers a mailing campaign should be sent to, given their profile and purchase history (for each customer it has two possible actions: send or don't send). It gets a negative reward for the cost of the mailing campaign, and a positive reward for estimated revenue generated from this campaign.

Product delivery

Let the agent control a fleet of delivery trucks, deciding what they should pick up at the depots, where they should go, what they should drop off, and so on. It will get positive rewards for each product delivered on time, and negative rewards for late deliveries.

3. When estimating the value of an action, Reinforcement Learning algorithms typically sum all the rewards that this action led to, giving more weight to immediate rewards and less weight to later rewards (considering that an action has more influence on the near future than on the distant future). To model this, a discount factor is typically applied at each time step. For example, with a discount factor of 0.9, a reward of 100 that is received two time steps later is counted as only $0.9^2 \times 100 = 81$ when you are estimating the value of the action. You can think of the discount factor as a measure of how much the future is valued relative to the present: if it is very close to 1, then the future is valued almost as much as the present; if it is close to 0, then only immediate rewards matter. Of course, this impacts the optimal policy tremendously: if you value the future, you may be willing to put up with a lot of immediate pain for the prospect of eventual rewards, while if you don't value the future, you will just grab any immediate reward you can find, never investing in the future.

4. To measure the performance of a Reinforcement Learning agent, you can simply sum up the rewards it gets. In a simulated environment, you can run many episodes and look at the total rewards it gets on average (and possibly look at the min, max, standard deviation, and so on).

5. The credit assignment problem is the fact that when a Reinforcement Learning agent receives a reward, it has no direct way of knowing which of its previous actions contributed to this reward. It typically occurs when there is a large delay between an action and the resulting reward (e.g., during a game of Atari's *Pong*, there may be a few dozen time steps between the moment the agent hits the ball and the moment it wins the point). One way to alleviate it is to provide the agent with shorter-term rewards, when possible. This usually requires prior knowledge about the task. For example, if we want to build an agent that will learn to play chess, instead of giving it a reward only when it wins the game, we could give it a reward every time it captures one of the opponent's pieces.

6. An agent can often remain in the same region of its environment for a while, so all of its experiences will be very similar for that period of time. This can introduce some bias in the learning algorithm. It may tune its policy for this region of the environment, but it will not perform well as soon as it moves out of this region. To solve this problem, you can use a replay memory; instead of using only the most immediate experiences for learning, the agent will learn based on a buffer of its past experiences, recent and not so recent (perhaps this is why we dream at night: to replay our experiences of the day and better learn from them?).

7. An off-policy RL algorithm learns the value of the optimal policy (i.e., the sum of discounted rewards that can be expected for each state if the agent acts optimally) while the agent follows a different policy. Q-Learning is a good example of such an algorithm. In contrast, an on-policy algorithm learns the value of the policy that the agent actually executes, including both exploration and exploitation.

For the solutions to exercises 8, 9, and 10, please see the Jupyter notebooks available at *https://github.com/ageron/handson-ml2*.

Chapter 19: Training and Deploying TensorFlow Models at Scale

1. A SavedModel contains a TensorFlow model, including its architecture (a computation graph) and its weights. It is stored as a directory containing a *saved_model.pb* file, which defines the computation graph (represented as a serialized protocol buffer), and a *variables* subdirectory containing the variable values. For models containing a large number of weights, these variable values may be split across multiple files. A SavedModel also includes an *assets* subdirectory that may contain additional data, such as vocabulary files, class names, or some

example instances for this model. To be more accurate, a SavedModel can contain one or more *metagraphs*. A metagraph is a computation graph plus some function signature definitions (including their input and output names, types, and shapes). Each metagraph is identified by a set of tags. To inspect a SavedModel, you can use the command-line tool `saved_model_cli` or just load it using `tf.saved_model.load()` and inspect it in Python.

2. TF Serving allows you to deploy multiple TensorFlow models (or multiple versions of the same model) and make them accessible to all your applications easily via a REST API or a gRPC API. Using your models directly in your applications would make it harder to deploy a new version of a model across all applications. Implementing your own microservice to wrap a TF model would require extra work, and it would be hard to match TF Serving's features. TF Serving has many features: it can monitor a directory and autodeploy the models that are placed there, and you won't have to change or even restart any of your applications to benefit from the new model versions; it's fast, well tested, and scales very well; and it supports A/B testing of experimental models and deploying a new model version to just a subset of your users (in this case the model is called a *canary*). TF Serving is also capable of grouping individual requests into batches to run them jointly on the GPU. To deploy TF Serving, you can install it from source, but it is much simpler to install it using a Docker image. To deploy a cluster of TF Serving Docker images, you can use an orchestration tool such as Kubernetes, or use a fully hosted solution such as Google Cloud AI Platform.

3. To deploy a model across multiple TF Serving instances, all you need to do is configure these TF Serving instances to monitor the same *models* directory, and then export your new model as a SavedModel into a subdirectory.

4. The gRPC API is more efficient than the REST API. However, its client libraries are not as widely available, and if you activate compression when using the REST API, you can get almost the same performance. So, the gRPC API is most useful when you need the highest possible performance and the clients are not limited to the REST API.

5. To reduce a model's size so it can run on a mobile or embedded device, TFLite uses several techniques:

   - It provides a converter which can optimize a SavedModel: it shrinks the model and reduces its latency. To do this, it prunes all the operations that are not needed to make predictions (such as training operations), and it optimizes and fuses operations whenever possible.

   - The converter can also perform post-training quantization: this technique dramatically reduces the model's size, so it's much faster to download and store.

- It saves the optimized model using the FlatBuffer format, which can be loaded to RAM directly, without parsing. This reduces the loading time and memory footprint.

6. Quantization-aware training consists in adding fake quantization operations to the model during training. This allows the model to learn to ignore the quantization noise; the final weights will be more robust to quantization.

7. Model parallelism means chopping your model into multiple parts and running them in parallel across multiple devices, hopefully speeding up the model during training or inference. Data parallelism means creating multiple exact replicas of your model and deploying them across multiple devices. At each iteration during training, each replica is given a different batch of data, and it computes the gradients of the loss with regard to the model parameters. In synchronous data parallelism, the gradients from all replicas are then aggregated and the optimizer performs a Gradient Descent step. The parameters may be centralized (e.g., on parameter servers) or replicated across all replicas and kept in sync using AllReduce. In asynchronous data parallelism, the parameters are centralized and the replicas run independently from each other, each updating the central parameters directly at the end of each training iteration, without having to wait for the other replicas. To speed up training, data parallelism turns out to work better than model parallelism, in general. This is mostly because it requires less communication across devices. Moreover, it is much easier to implement, and it works the same way for any model, whereas model parallelism requires analyzing the model to determine the best way to chop it into pieces.

8. When training a model across multiple servers, you can use the following distribution strategies:

- The `MultiWorkerMirroredStrategy` performs mirrored data parallelism. The model is replicated across all available servers and devices, and each replica gets a different batch of data at each training iteration and computes its own gradients. The mean of the gradients is computed and shared across all replicas using a distributed AllReduce implementation (NCCL by default), and all replicas perform the same Gradient Descent step. This strategy is the simplest to use since all servers and devices are treated in exactly the same way, and it performs fairly well. In general, you should use this strategy. Its main limitation is that it requires the model to fit in RAM on every replica.

- The `ParameterServerStrategy` performs asynchronous data parallelism. The model is replicated across all devices on all workers, and the parameters are sharded across all parameter servers. Each worker has its own training loop, running asynchronously with the other workers; at each training iteration, each worker gets its own batch of data and fetches the latest version of the model parameters from the parameter servers, then it computes the gradients of the loss with regard to these parameters, and it sends them to the parameter servers. Lastly, the parameter servers perform a Gradient Descent step using these gradients. This strategy is generally slower than the previous strategy, and a bit harder to deploy, since it requires managing parameter servers. However, it is useful to train huge models that don't fit in GPU RAM.

For the solutions to exercises 9, 10, and 11, please see the Jupyter notebooks available at *https://github.com/ageron/handson-ml2*.

Machine Learning Project Checklist

This checklist can guide you through your Machine Learning projects. There are eight main steps:

1. Frame the problem and look at the big picture.
2. Get the data.
3. Explore the data to gain insights.
4. Prepare the data to better expose the underlying data patterns to Machine Learning algorithms.
5. Explore many different models and shortlist the best ones.
6. Fine-tune your models and combine them into a great solution.
7. Present your solution.
8. Launch, monitor, and maintain your system.

Obviously, you should feel free to adapt this checklist to your needs.

Frame the Problem and Look at the Big Picture

1. Define the objective in business terms.
2. How will your solution be used?
3. What are the current solutions/workarounds (if any)?
4. How should you frame this problem (supervised/unsupervised, online/offline, etc.)?
5. How should performance be measured?
6. Is the performance measure aligned with the business objective?

7. What would be the minimum performance needed to reach the business objective?

8. What are comparable problems? Can you reuse experience or tools?

9. Is human expertise available?

10. How would you solve the problem manually?

11. List the assumptions you (or others) have made so far.

12. Verify assumptions if possible.

Get the Data

Note: automate as much as possible so you can easily get fresh data.

1. List the data you need and how much you need.

2. Find and document where you can get that data.

3. Check how much space it will take.

4. Check legal obligations, and get authorization if necessary.

5. Get access authorizations.

6. Create a workspace (with enough storage space).

7. Get the data.

8. Convert the data to a format you can easily manipulate (without changing the data itself).

9. Ensure sensitive information is deleted or protected (e.g., anonymized).

10. Check the size and type of data (time series, sample, geographical, etc.).

11. Sample a test set, put it aside, and never look at it (no data snooping!).

Explore the Data

Note: try to get insights from a field expert for these steps.

1. Create a copy of the data for exploration (sampling it down to a manageable size if necessary).

2. Create a Jupyter notebook to keep a record of your data exploration.

3. Study each attribute and its characteristics:

   - Name

   - Type (categorical, int/float, bounded/unbounded, text, structured, etc.)

- % of missing values
- Noisiness and type of noise (stochastic, outliers, rounding errors, etc.)
- Usefulness for the task
- Type of distribution (Gaussian, uniform, logarithmic, etc.)

4. For supervised learning tasks, identify the target attribute(s).
5. Visualize the data.
6. Study the correlations between attributes.
7. Study how you would solve the problem manually.
8. Identify the promising transformations you may want to apply.
9. Identify extra data that would be useful (go back to "Get the Data" on page 756).
10. Document what you have learned.

Prepare the Data

Notes:

- Work on copies of the data (keep the original dataset intact).
- Write functions for all data transformations you apply, for five reasons:
  - So you can easily prepare the data the next time you get a fresh dataset
  - So you can apply these transformations in future projects
  - To clean and prepare the test set
  - To clean and prepare new data instances once your solution is live
  - To make it easy to treat your preparation choices as hyperparameters

1. Data cleaning:
   - Fix or remove outliers (optional).
   - Fill in missing values (e.g., with zero, mean, median…) or drop their rows (or columns).

2. Feature selection (optional):
   - Drop the attributes that provide no useful information for the task.

3. Feature engineering, where appropriate:
   - Discretize continuous features.

- Decompose features (e.g., categorical, date/time, etc.).
- Add promising transformations of features (e.g., $\log(x)$, sqrt(x), x^2, etc.).
- Aggregate features into promising new features.

4. Feature scaling:

- Standardize or normalize features.

Shortlist Promising Models

Notes:

- If the data is huge, you may want to sample smaller training sets so you can train many different models in a reasonable time (be aware that this penalizes complex models such as large neural nets or Random Forests).
- Once again, try to automate these steps as much as possible.

1. Train many quick-and-dirty models from different categories (e.g., linear, naive Bayes, SVM, Random Forest, neural net, etc.) using standard parameters.
2. Measure and compare their performance.

- For each model, use N-fold cross-validation and compute the mean and standard deviation of the performance measure on the N folds.

3. Analyze the most significant variables for each algorithm.
4. Analyze the types of errors the models make.

- What data would a human have used to avoid these errors?

5. Perform a quick round of feature selection and engineering.
6. Perform one or two more quick iterations of the five previous steps.
7. Shortlist the top three to five most promising models, preferring models that make different types of errors.

Fine-Tune the System

Notes:

- You will want to use as much data as possible for this step, especially as you move toward the end of fine-tuning.

- As always, automate what you can.

1. Fine-tune the hyperparameters using cross-validation:

   - Treat your data transformation choices as hyperparameters, especially when you are not sure about them (e.g., if you're not sure whether to replace missing values with zeros or with the median value, or to just drop the rows).

   - Unless there are very few hyperparameter values to explore, prefer random search over grid search. If training is very long, you may prefer a Bayesian optimization approach (e.g., using Gaussian process priors, as described by Jasper Snoek et al. (*https://homl.info/134*)).[1]

2. Try Ensemble methods. Combining your best models will often produce better performance than running them individually.

3. Once you are confident about your final model, measure its performance on the test set to estimate the generalization error.

 Don't tweak your model after measuring the generalization error: you would just start overfitting the test set.

Present Your Solution

1. Document what you have done.
2. Create a nice presentation.

   - Make sure you highlight the big picture first.

3. Explain why your solution achieves the business objective.
4. Don't forget to present interesting points you noticed along the way.

   - Describe what worked and what did not.
   - List your assumptions and your system's limitations.

---

1 Jasper Snoek et al., "Practical Bayesian Optimization of Machine Learning Algorithms," *Proceedings of the 25th International Conference on Neural Information Processing Systems* 2 (2012): 2951–2959.

5. Ensure your key findings are communicated through beautiful visualizations or easy-to-remember statements (e.g., "the median income is the number-one predictor of housing prices").

Launch!

1. Get your solution ready for production (plug into production data inputs, write unit tests, etc.).

2. Write monitoring code to check your system's live performance at regular intervals and trigger alerts when it drops.

   - Beware of slow degradation: models tend to "rot" as data evolves.

   - Measuring performance may require a human pipeline (e.g., via a crowdsourcing service).

   - Also monitor your inputs' quality (e.g., a malfunctioning sensor sending random values, or another team's output becoming stale). This is particularly important for online learning systems.

3. Retrain your models on a regular basis on fresh data (automate as much as possible).

SVM Dual Problem

To understand *duality*, you first need to understand the *Lagrange multipliers* method. The general idea is to transform a constrained optimization objective into an unconstrained one, by moving the constraints into the objective function. Let's look at a simple example. Suppose you want to find the values of x and y that minimize the function $f(x, y) = x^2 + 2y$, subject to an *equality constraint*: $3x + 2y + 1 = 0$. Using the Lagrange multipliers method, we start by defining a new function called the *Lagrangian* (or *Lagrange function*): $g(x, y, \alpha) = f(x, y) - \alpha(3x + 2y + 1)$. Each constraint (in this case just one) is subtracted from the original objective, multiplied by a new variable called a Lagrange multiplier.

Joseph-Louis Lagrange showed that if $(\hat{x}, \hat{y})$ is a solution to the constrained optimization problem, then there must exist an $\hat{\alpha}$ such that $(\hat{x}, \hat{y}, \hat{\alpha})$ is a *stationary point* of the Lagrangian (a stationary point is a point where all partial derivatives are equal to zero). In other words, we can compute the partial derivatives of $g(x, y, \alpha)$ with regard to x, y, and α; we can find the points where these derivatives are all equal to zero; and the solutions to the constrained optimization problem (if they exist) must be among these stationary points.

In this example the partial derivatives are:
$$\begin{cases} \frac{\partial}{\partial x}g(x, y, \alpha) = 2x - 3\alpha \\ \frac{\partial}{\partial y}g(x, y, \alpha) = 2 - 2\alpha \\ \frac{\partial}{\partial \alpha}g(x, y, \alpha) = -3x - 2y - 1 \end{cases}$$

When all these partial derivatives are equal to 0, we find that $2\hat{x} - 3\hat{\alpha} = 2 - 2\hat{\alpha} = -3\hat{x} - 2\hat{y} - 1 = 0$, from which we can easily find that $\hat{x} = \frac{3}{2}$, $\hat{y} = -\frac{11}{4}$, and $\hat{\alpha} = 1$. This is the only stationary point, and as it respects the constraint, it must be the solution to the constrained optimization problem.

However, this method applies only to equality constraints. Fortunately, under some regularity conditions (which are respected by the SVM objectives), this method can be generalized to *inequality constraints* as well (e.g., $3x + 2y + 1 \geq 0$). The *generalized Lagrangian* for the hard margin problem is given by Equation C-1, where the $\alpha^{(i)}$ variables are called the *Karush–Kuhn–Tucker* (KKT) multipliers, and they must be greater or equal to zero.

Equation C-1. Generalized Lagrangian for the hard margin problem

$$\mathscr{L}(\mathbf{w}, b, \alpha) = \frac{1}{2}\mathbf{w}^\mathsf{T}\mathbf{w} - \sum_{i=1}^{m} \alpha^{(i)}\left(t^{(i)}\left(\mathbf{w}^\mathsf{T}\mathbf{x}^{(i)} + b\right) - 1\right)$$

$$\text{with} \quad \alpha^{(i)} \geq 0 \quad \text{for } i = 1, 2, \cdots, m$$

Just like with the Lagrange multipliers method, you can compute the partial derivatives and locate the stationary points. If there is a solution, it will necessarily be among the stationary points $\left(\widehat{\mathbf{w}}, \hat{b}, \hat{\alpha}\right)$ that respect the *KKT conditions*:

- Respect the problem's constraints: $t^{(i)}\left(\widehat{\mathbf{w}}^\mathsf{T}\mathbf{x}^{(i)} + \hat{b}\right) \geq 1 \quad \text{for } i = 1, 2, \dots, m$.
- Verify $\hat{\alpha}^{(i)} \geq 0 \quad \text{for } i = 1, 2, \cdots, m$.
- Either $\hat{\alpha}^{(i)} = 0$ or the ith constraint must be an *active constraint*, meaning it must hold by equality: $t^{(i)}\left(\widehat{\mathbf{w}}^\mathsf{T}\mathbf{x}^{(i)} + \hat{b}\right) = 1$. This condition is called the *complementary slackness* condition. It implies that either $\hat{\alpha}^{(i)} = 0$ or the ith instance lies on the boundary (it is a support vector).

Note that the KKT conditions are necessary conditions for a stationary point to be a solution of the constrained optimization problem. Under some conditions, they are also sufficient conditions. Luckily, the SVM optimization problem happens to meet these conditions, so any stationary point that meets the KKT conditions is guaranteed to be a solution to the constrained optimization problem.

We can compute the partial derivatives of the generalized Lagrangian with regard to **w** and *b* with Equation C-2.

Equation C-2. Partial derivatives of the generalized Lagrangian

$$\nabla_\mathbf{w}\mathscr{L}(\mathbf{w}, b, \alpha) = \mathbf{w} - \sum_{i=1}^{m} \alpha^{(i)}t^{(i)}\mathbf{x}^{(i)}$$

$$\frac{\partial}{\partial b}\mathscr{L}(\mathbf{w}, b, \alpha) = -\sum_{i=1}^{m} \alpha^{(i)}t^{(i)}$$

When these partial derivatives are equal to zero, we have Equation C-3.

Equation C-3. Properties of the stationary points

$$\widehat{\mathbf{w}} = \sum_{i=1}^{m} \widehat{\alpha}^{(i)} t^{(i)} \mathbf{x}^{(i)}$$

$$\sum_{i=1}^{m} \widehat{\alpha}^{(i)} t^{(i)} = 0$$

If we plug these results into the definition of the generalized Lagrangian, some terms disappear and we find Equation C-4.

Equation C-4. Dual form of the SVM problem

$$\mathscr{L}\left(\widehat{\mathbf{w}}, \widehat{b}, \alpha\right) = \frac{1}{2} \sum_{i=1}^{m} \sum_{j=1}^{m} \alpha^{(i)} \alpha^{(j)} t^{(i)} t^{(j)} \mathbf{x}^{(i)\top} \mathbf{x}^{(j)} \quad - \quad \sum_{i=1}^{m} \alpha^{(i)}$$

$$\text{with} \quad \alpha^{(i)} \geq 0 \quad \text{for } i = 1, 2, \cdots, m$$

The goal is now to find the vector $\widehat{\alpha}$ that minimizes this function, with $\widehat{\alpha}^{(i)} \geq 0$ for all instances. This constrained optimization problem is the dual problem we were looking for.

Once you find the optimal $\widehat{\alpha}$, you can compute $\widehat{\mathbf{w}}$ using the first line of Equation C-3. To compute $\widehat{b}$, you can use the fact that a support vector must verify $t^{(i)}(\widehat{\mathbf{w}}^\top \mathbf{x}^{(i)} + \widehat{b}) = 1$, so if the kth instance is a support vector (i.e., $\widehat{\alpha}^{(k)} > 0$), you can use it to compute $\widehat{b} = t^{(k)} - \widehat{\mathbf{w}}^\top \mathbf{x}^{(k)}$. However, it is often preferred to compute the average over all support vectors to get a more stable and precise value, as in Equation C-5.

Equation C-5. Bias term estimation using the dual form

$$\widehat{b} = \frac{1}{n_s} \sum_{\substack{i=1 \\ \widehat{\alpha}^{(i)} > 0}}^{m} \left[t^{(i)} - \widehat{\mathbf{w}}^\top \mathbf{x}^{(i)} \right]$$

Autodiff

This appendix explains how TensorFlow's autodifferentiation (autodiff) feature works, and how it compares to other solutions.

Suppose you define a function $f(x, y) = x^2y + y + 2$, and you need its partial derivatives $\partial f/\partial x$ and $\partial f/\partial y$, typically to perform Gradient Descent (or some other optimization algorithm). Your main options are manual differentiation, finite difference approximation, forward-mode autodiff, and reverse-mode autodiff. TensorFlow implements reverse-mode autodiff, but to understand it, it's useful to look at the other options first. So let's go through each of them, starting with manual differentiation.

Manual Differentiation

The first approach to compute derivatives is to pick up a pencil and a piece of paper and use your calculus knowledge to derive the appropriate equation. For the function $f(x, y)$ just defined, it is not too hard; you just need to use five rules:

- The derivative of a constant is 0.
- The derivative of λx is λ (where λ is a constant).
- The derivative of x^λ is $\lambda x^{\lambda - 1}$, so the derivative of x^2 is $2x$.
- The derivative of a sum of functions is the sum of these functions' derivatives.
- The derivative of λ times a function is λ times its derivative.

From these rules, you can derive Equation D-1.

Equation D-1. Partial derivatives of f(x, y)

$$\frac{\partial f}{\partial x} = \frac{\partial\left(x^2 y\right)}{\partial x} + \frac{\partial y}{\partial x} + \frac{\partial 2}{\partial x} = y\frac{\partial\left(x^2\right)}{\partial x} + 0 + 0 = 2xy$$

$$\frac{\partial f}{\partial y} = \frac{\partial\left(x^2 y\right)}{\partial y} + \frac{\partial y}{\partial y} + \frac{\partial 2}{\partial y} = x^2 + 1 + 0 = x^2 + 1$$

This approach can become very tedious for more complex functions, and you run the risk of making mistakes. Fortunately, there are other options. Let's look at finite difference approximation now.

Finite Difference Approximation

Recall that the derivative $h'(x_0)$ of a function $h(x)$ at a point x_0 is the slope of the function at that point. More precisely, the derivative is defined as the limit of the slope of a straight line going through this point x_0 and another point x on the function, as x gets infinitely close to x_0 (see Equation D-2).

Equation D-2. Definition of the derivative of a function h(x) at point x_0

$$h'(x_0) = \lim_{x \to x_0} \frac{h(x) - h(x_0)}{x - x_0}$$

$$= \lim_{\varepsilon \to 0} \frac{h(x_0 + \varepsilon) - h(x_0)}{\varepsilon}$$

So, if we wanted to calculate the partial derivative of $f(x, y)$ with regard to x at $x = 3$ and $y = 4$, we could compute $f(3 + \varepsilon, 4) - f(3, 4)$ and divide the result by ε, using a very small value for ε. This type of numerical approximation of the derivative is called a *finite difference approximation*, and this specific equation is called *Newton's difference quotient*. That's exactly what the following code does:

```
def f(x, y):
    return x**2*y + y + 2

def derivative(f, x, y, x_eps, y_eps):
    return (f(x + x_eps, y + y_eps) - f(x, y)) / (x_eps + y_eps)

df_dx = derivative(f, 3, 4, 0.00001, 0)
df_dy = derivative(f, 3, 4, 0, 0.00001)
```

Unfortunately, the result is imprecise (and it gets worse for more complicated functions). The correct results are respectively 24 and 10, but instead we get:

```
>>> print(df_dx)
24.000039999805264
>>> print(df_dy)
10.000000000331966
```

Notice that to compute both partial derivatives, we have to call f() at least three times (we called it four times in the preceding code, but it could be optimized). If there were 1,000 parameters, we would need to call f() at least 1,001 times. When you are dealing with large neural networks, this makes finite difference approximation way too inefficient.

However, this method is so simple to implement that it is a great tool to check that the other methods are implemented correctly. For example, if it disagrees with your manually derived function, then your function probably contains a mistake.

So far, we have considered two ways to compute gradients: using manual differentiation and using finite difference approximation. Unfortunately, both were fatally flawed to train a large-scale neural network. So let's turn to autodiff, starting with forward mode.

Forward-Mode Autodiff

Figure D-1 shows how forward-mode autodiff works on an even simpler function, $g(x, y) = 5 + xy$. The graph for that function is represented on the left. After forward-mode autodiff, we get the graph on the right, which represents the partial derivative $\partial g/\partial x = 0 + (0 \times x + y \times 1) = y$ (we could similarly obtain the partial derivative with regard to y).

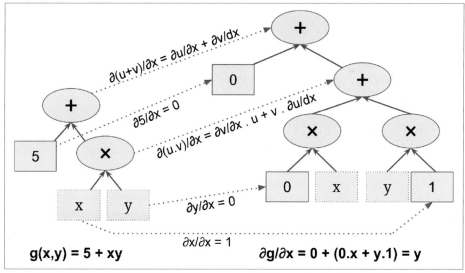

Figure D-1. Forward-mode autodiff

The algorithm will go through the computation graph from the inputs to the outputs (hence the name "forward mode"). It starts by getting the partial derivatives of the leaf nodes. The constant node (5) returns the constant 0, since the derivative of a constant is always 0. The variable x returns the constant 1 since $\partial x/\partial x = 1$, and the variable y returns the constant 0 since $\partial y/\partial x = 0$ (if we were looking for the partial derivative with regard to y, it would be the reverse).

Now we have all we need to move up the graph to the multiplication node in function g. Calculus tells us that the derivative of the product of two functions u and v is $\partial(u \times v)/\partial x = \partial v/\partial x \times u + v \times \partial u/\partial x$. We can therefore construct a large part of the graph on the right, representing $0 \times x + y \times 1$.

Finally, we can go up to the addition node in function g. As mentioned, the derivative of a sum of functions is the sum of these functions' derivatives. So we just need to create an addition node and connect it to the parts of the graph we have already computed. We get the correct partial derivative: $\partial g/\partial x = 0 + (0 \times x + y \times 1)$.

However, this equation can be simplified (a lot). A few pruning steps can be applied to the computation graph to get rid of all unnecessary operations, and we get a much smaller graph with just one node: $\partial g/\partial x = y$. In this case simplification is fairly easy, but for a more complex function forward-mode autodiff can produce a huge graph that may be tough to simplify and lead to suboptimal performance.

Note that we started with a computation graph, and forward-mode autodiff produced another computation graph. This is called *symbolic differentiation*, and it has two nice features: first, once the computation graph of the derivative has been produced, we can use it as many times as we want to compute the derivatives of the given function for any value of x and y; second, we can run forward-mode autodiff again on the resulting graph to get second-order derivatives if we ever need to (i.e., derivatives of derivatives). We could even compute third-order derivatives, and so on.

But it is also possible to run forward-mode autodiff without constructing a graph (i.e., numerically, not symbolically), just by computing intermediate results on the fly. One way to do this is to use *dual numbers*, which are weird but fascinating numbers of the form $a + b\varepsilon$, where a and b are real numbers and ε is an infinitesimal number such that $\varepsilon^2 = 0$ (but $\varepsilon \neq 0$). You can think of the dual number $42 + 24\varepsilon$ as something akin to $42.0000\cdots000024$ with an infinite number of 0s (but of course this is simplified just to give you some idea of what dual numbers are). A dual number is represented in memory as a pair of floats. For example, $42 + 24\varepsilon$ is represented by the pair (42.0, 24.0).

Dual numbers can be added, multiplied, and so on, as shown in Equation D-3.

Equation D-3. A few operations with dual numbers

$$\lambda(a + b\varepsilon) = \lambda a + \lambda b\varepsilon$$
$$(a + b\varepsilon) + (c + d\varepsilon) = (a + c) + (b + d)\varepsilon$$
$$(a + b\varepsilon) \times (c + d\varepsilon) = ac + (ad + bc)\varepsilon + (bd)\varepsilon^2 = ac + (ad + bc)\varepsilon$$

Most importantly, it can be shown that $h(a + b\varepsilon) = h(a) + b \times h'(a)\varepsilon$, so computing $h(a + \varepsilon)$ gives you both $h(a)$ and the derivative $h'(a)$ in just one shot. Figure D-2 shows that the partial derivative of $f(x, y)$ with regard to x at $x = 3$ and $y = 4$ (which we will write $\partial f/\partial x$ (3, 4)) can be computed using dual numbers. All we need to do is compute $f(3 + \varepsilon, 4)$; this will output a dual number whose first component is equal to $f(3, 4)$ and whose second component is equal to $\partial f/\partial x$ (3, 4).

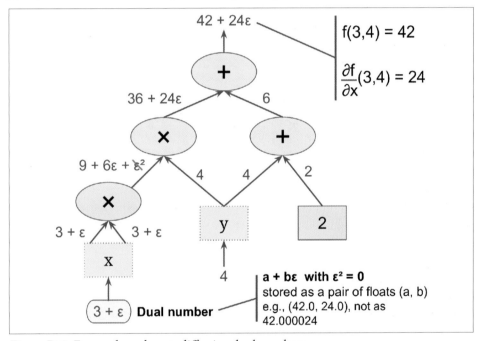

Figure D-2. Forward-mode autodiff using dual numbers

To compute $\partial f/\partial x$ (3, 4) we would have to go through the graph again, but this time with $x = 3$ and $y = 4 + \varepsilon$.

So forward-mode autodiff is much more accurate than finite difference approximation, but it suffers from the same major flaw, at least when there are many inputs and few outputs (as is the case when dealing with neural networks): if there were 1,000 parameters, it would require 1,000 passes through the graph to compute all the partial derivatives. This is where reverse-mode autodiff shines: it can compute all of them in just two passes through the graph. Let's see how.

Reverse-Mode Autodiff

Reverse-mode autodiff is the solution implemented by TensorFlow. It first goes through the graph in the forward direction (i.e., from the inputs to the output) to compute the value of each node. Then it does a second pass, this time in the reverse direction (i.e., from the output to the inputs), to compute all the partial derivatives. The name "reverse mode" comes from this second pass through the graph, where gradients flow in the reverse direction. Figure D-3 represents the second pass. During the first pass, all the node values were computed, starting from $x = 3$ and $y = 4$. You can see those values at the bottom right of each node (e.g., $x \times x = 9$). The nodes are labeled n_1 to n_7 for clarity. The output node is n_7: $f(3, 4) = n_7 = 42$.

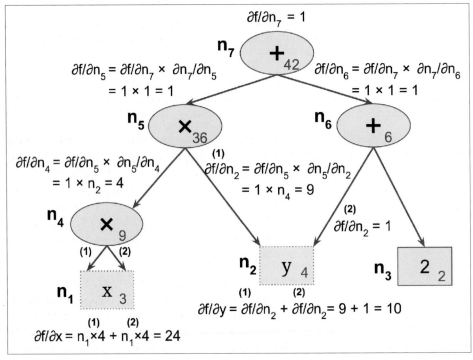

Figure D-3. Reverse-mode autodiff

The idea is to gradually go down the graph, computing the partial derivative of $f(x, y)$ with regard to each consecutive node, until we reach the variable nodes. For this, reverse-mode autodiff relies heavily on the *chain rule*, shown in Equation D-4.

Equation D-4. Chain rule

$$\frac{\partial f}{\partial x} = \frac{\partial f}{\partial n_i} \times \frac{\partial n_i}{\partial x}$$

Since n_7 is the output node, $f = n_7$ so $\partial f / \partial n_7 = 1$.

Let's continue down the graph to n_5: how much does f vary when n_5 varies? The answer is $\partial f / \partial n_5 = \partial f / \partial n_7 \times \partial n_7 / \partial n_5$. We already know that $\partial f / \partial n_7 = 1$, so all we need is $\partial n_7 / \partial n_5$. Since n_7 simply performs the sum $n_5 + n_6$, we find that $\partial n_7 / \partial n_5 = 1$, so $\partial f / \partial n_5 = 1 \times 1 = 1$.

Now we can proceed to node n_4: how much does f vary when n_4 varies? The answer is $\partial f / \partial n_4 = \partial f / \partial n_5 \times \partial n_5 / \partial n_4$. Since $n_5 = n_4 \times n_2$, we find that $\partial n_5 / \partial n_4 = n_2$, so $\partial f / \partial n_4 = 1 \times n_2 = 4$.

The process continues until we reach the bottom of the graph. At that point we will have calculated all the partial derivatives of $f(x, y)$ at the point $x = 3$ and $y = 4$. In this example, we find $\partial f / \partial x = 24$ and $\partial f / \partial y = 10$. Sounds about right!

Reverse-mode autodiff is a very powerful and accurate technique, especially when there are many inputs and few outputs, since it requires only one forward pass plus one reverse pass per output to compute all the partial derivatives for all outputs with regard to all the inputs. When training neural networks, we generally want to minimize the loss, so there is a single output (the loss), and hence only two passes through the graph are needed to compute the gradients. Reverse-mode autodiff can also handle functions that are not entirely differentiable, as long as you ask it to compute the partial derivatives at points that are differentiable.

In Figure D-3, the numerical results are computed on the fly, at each node. However, that's not exactly what TensorFlow does: instead, it creates a new computation graph. In other words, it implements *symbolic* reverse-mode autodiff. This way, the computation graph to compute the gradients of the loss with regard to all the parameters in the neural network only needs to be generated once, and then it can be executed over and over again, whenever the optimizer needs to compute the gradients. Moreover, this makes it possible to compute higher-order derivatives if needed.

If you ever want to implement a new type of low-level TensorFlow operation in C++, and you want to make it compatible with autodiff, then you will need to provide a function that returns the partial derivatives of the function's outputs with regard to its inputs. For example, suppose you implement a function that computes the square of its input: $f(x) = x^2$. In that case you would need to provide the corresponding derivative function: $f'(x) = 2x$.

Other Popular ANN Architectures

In this appendix I will give a quick overview of a few historically important neural network architectures that are much less used today than deep Multilayer Perceptrons (Chapter 10), convolutional neural networks (Chapter 14), recurrent neural networks (Chapter 15), or autoencoders (Chapter 17). They are often mentioned in the literature, and some are still used in a range of applications, so it is worth knowing about them. Additionally, we will discuss *deep belief nets*, which were the state of the art in Deep Learning until the early 2010s. They are still the subject of very active research, so they may well come back with a vengeance in the future.

Hopfield Networks

Hopfield networks were first introduced by W. A. Little in 1974, then popularized by J. Hopfield in 1982. They are *associative memory* networks: you first teach them some patterns, and then when they see a new pattern they (hopefully) output the closest learned pattern. This made them useful for character recognition, in particular, before they were outperformed by other approaches: you first train the network by showing it examples of character images (each binary pixel maps to one neuron), and then when you show it a new character image, after a few iterations it outputs the closest learned character.

Hopfield networks are fully connected graphs (see Figure E-1); that is, every neuron is connected to every other neuron. Note that in the diagram the images are 6 × 6 pixels, so the neural network on the left should contain 36 neurons (and 630 connections), but for visual clarity a much smaller network is represented.

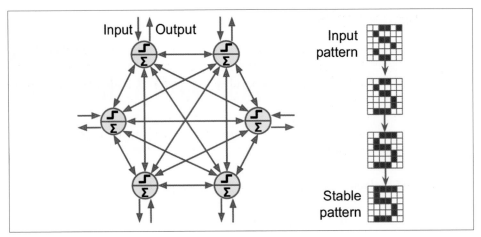

Figure E-1. Hopfield network

The training algorithm works by using Hebb's rule (see "The Perceptron" on page 284): for each training image, the weight between two neurons is increased if the corresponding pixels are both on or both off, but decreased if one pixel is on and the other is off.

To show a new image to the network, you just activate the neurons that correspond to active pixels. The network then computes the output of every neuron, and this gives you a new image. You can then take this new image and repeat the whole process. After a while, the network reaches a stable state. Generally, this corresponds to the training image that most resembles the input image.

A so-called *energy function* is associated with Hopfield nets. At each iteration, the energy decreases, so the network is guaranteed to eventually stabilize to a low-energy state. The training algorithm tweaks the weights in a way that decreases the energy level of the training patterns, so the network is likely to stabilize in one of these low-energy configurations. Unfortunately, some patterns that were not in the training set also end up with low energy, so the network sometimes stabilizes in a configuration that was not learned. These are called *spurious patterns*.

Another major flaw with Hopfield nets is that they don't scale very well—their memory capacity is roughly equal to 14% of the number of neurons. For example, to classify 28 × 28–pixel images, you would need a Hopfield net with 784 fully connected neurons and 306,936 weights. Such a network would only be able to learn about 110 different characters (14% of 784). That's a lot of parameters for such a small memory.

Boltzmann Machines

Boltzmann machines were invented in 1985 by Geoffrey Hinton and Terrence Sejnow-ski. Just like Hopfield nets, they are fully connected ANNs, but they are based on *stochastic neurons*: instead of using a deterministic step function to decide what value to output, these neurons output 1 with some probability, and 0 otherwise. The probability function that these ANNs use is based on the Boltzmann distribution (used in statistical mechanics), hence their name. Equation E-1 gives the probability that a particular neuron will output 1.

Equation E-1. Probability that the ith neuron will output 1

$$p\left(s_i^{(\text{next step})} = 1\right) = \sigma\left(\frac{\sum_{j=1}^{N} w_{i,j} s_j + b_i}{T}\right)$$

- s_j is the jth neuron's state (0 or 1).
- $w_{i,j}$ is the connection weight between the ith and j^{th} neurons. Note that $w_{i,i} = 0$.
- b_i is the ith neuron's bias term. We can implement this term by adding a bias neuron to the network.
- N is the number of neurons in the network.
- T is a number called the network's *temperature*; the higher the temperature, the more random the output is (i.e., the more the probability approaches 50%).
- σ is the logistic function.

Neurons in Boltzmann machines are separated into two groups: *visible units* and *hidden units* (see Figure E-2). All neurons work in the same stochastic way, but the visible units are the ones that receive the inputs and from which outputs are read.

Because of its stochastic nature, a Boltzmann machine will never stabilize into a fixed configuration; instead, it will keep switching between many configurations. If it is left running for a sufficiently long time, the probability of observing a particular configuration will only be a function of the connection weights and bias terms, not of the original configuration (similarly, after you shuffle a deck of cards for long enough, the configuration of the deck does not depend on the initial state). When the network reaches this state where the original configuration is "forgotten," it is said to be in *thermal equilibrium* (although its configuration keeps changing all the time). By setting the network parameters appropriately, letting the network reach thermal equilibrium, and then observing its state, we can simulate a wide range of probability distributions. This is called a *generative model*.

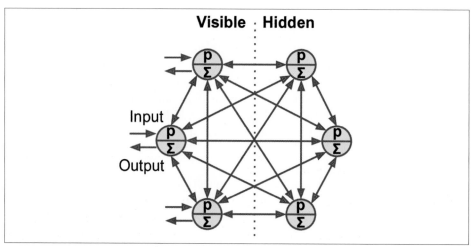

Figure E-2. Boltzmann machine

Training a Boltzmann machine means finding the parameters that will make the network approximate the training set's probability distribution. For example, if there are three visible neurons and the training set contains 75% (0, 1, 1) triplets, 10% (0, 0, 1) triplets, and 15% (1, 1, 1) triplets, then after training a Boltzmann machine, you could use it to generate random binary triplets with about the same probability distribution. For example, about 75% of the time it would output the (0, 1, 1) triplet.

Such a generative model can be used in a variety of ways. For example, if it is trained on images, and you provide an incomplete or noisy image to the network, it will automatically "repair" the image in a reasonable way. You can also use a generative model for classification. Just add a few visible neurons to encode the training image's class (e.g., add 10 visible neurons and turn on only the fifth neuron when the training image represents a 5). Then, when given a new image, the network will automatically turn on the appropriate visible neurons, indicating the image's class (e.g., it will turn on the fifth visible neuron if the image represents a 5).

Unfortunately, there is no efficient technique to train Boltzmann machines. However, fairly efficient algorithms have been developed to train *restricted Boltzmann machines* (RBMs).

Restricted Boltzmann Machines

An RBM is simply a Boltzmann machine in which there are no connections between visible units or between hidden units, only between visible and hidden units. For example, Figure E-3 represents an RBM with three visible units and four hidden units.

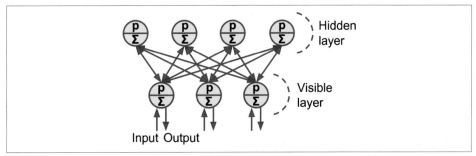

Figure E-3. Restricted Boltzmann machine

A very efficient training algorithm called *Contrastive Divergence* was introduced in 2005 by Miguel Á. Carreira-Perpiñán and Geoffrey Hinton (*https://homl.info/135*).[1] Here is how it works: for each training instance **x**, the algorithm starts by feeding it to the network by setting the state of the visible units to $x_1, x_2, \cdots, x_n$. Then you compute the state of the hidden units by applying the stochastic equation described before (Equation E-1). This gives you a hidden vector **h** (where h_i is equal to the state of the ith unit). Next you compute the state of the visible units, by applying the same stochastic equation. This gives you a vector **x'**. Then once again you compute the state of the hidden units, which gives you a vector **h'**. Now you can update each connection weight by applying the rule in Equation E-2, where η is the learning rate.

Equation E-2. Contrastive divergence weight update

$$w_{i,j} \leftarrow w_{i,j} + \eta\left(\mathbf{x}\mathbf{h}^\mathsf{T} - \mathbf{x'}\mathbf{h'}^\mathsf{T}\right)$$

The great benefit of this algorithm is that it does not require waiting for the network to reach thermal equilibrium: it just goes forward, backward, and forward again, and that's it. This makes it incomparably more efficient than previous algorithms, and it was a key ingredient to the first success of Deep Learning based on multiple stacked RBMs.

Deep Belief Nets

Several layers of RBMs can be stacked; the hidden units of the first-level RBM serve as the visible units for the second-layer RBM, and so on. Such an RBM stack is called a *deep belief net* (DBN).

---

1 Miguel Á. Carreira-Perpiñán and Geoffrey E. Hinton, "On Contrastive Divergence Learning," *Proceedings of the 10th International Workshop on Artificial Intelligence and Statistics* (2005): 59–66.

Yee-Whye Teh, one of Geoffrey Hinton's students, observed that it was possible to train DBNs one layer at a time using Contrastive Divergence, starting with the lower layers and then gradually moving up to the top layers. This led to the groundbreaking article that kickstarted the Deep Learning tsunami in 2006 (*https://homl.info/136*).[2]

Just like RBMs, DBNs learn to reproduce the probability distribution of their inputs, without any supervision. However, they are much better at it, for the same reason that deep neural networks are more powerful than shallow ones: real-world data is often organized in hierarchical patterns, and DBNs take advantage of that. Their lower layers learn low-level features in the input data, while higher layers learn high-level features.

Just like RBMs, DBNs are fundamentally unsupervised, but you can also train them in a supervised manner by adding some visible units to represent the labels. Moreover, one great feature of DBNs is that they can be trained in a semisupervised fashion. Figure E-4 represents such a DBN configured for semisupervised learning.

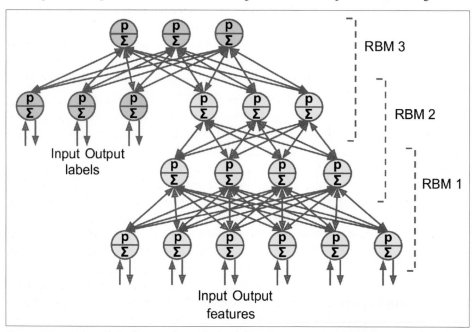

Figure E-4. A deep belief network configured for semisupervised learning

First, RBM 1 is trained without supervision. It learns low-level features in the training data. Then RBM 2 is trained with RBM 1's hidden units as inputs, again without

---

2 Geoffrey E. Hinton et al., "A Fast Learning Algorithm for Deep Belief Nets," *Neural Computation* 18 (2006): 1527–1554.

supervision: it learns higher-level features (note that RBM 2's hidden units include only the three rightmost units, not the label units). Several more RBMs could be stacked this way, but you get the idea. So far, training was 100% unsupervised. Lastly, RBM 3 is trained using RBM 2's hidden units as inputs, as well as extra visible units used to represent the target labels (e.g., a one-hot vector representing the instance class). It learns to associate high-level features with training labels. This is the supervised step.

At the end of training, if you feed RBM 1 a new instance, the signal will propagate up to RBM 2, then up to the top of RBM 3, and then back down to the label units; hopefully, the appropriate label will light up. This is how a DBN can be used for classification.

One great benefit of this semisupervised approach is that you don't need much labeled training data. If the unsupervised RBMs do a good enough job, then only a small amount of labeled training instances per class will be necessary. Similarly, a baby learns to recognize objects without supervision, so when you point to a chair and say "chair," the baby can associate the word "chair" with the class of objects it has already learned to recognize on its own. You don't need to point to every single chair and say "chair"; only a few examples will suffice (just enough so the baby can be sure that you are indeed referring to the chair, not to its color or one of the chair's parts).

Quite amazingly, DBNs can also work in reverse. If you activate one of the label units, the signal will propagate up to the hidden units of RBM 3, then down to RBM 2, and then RBM 1, and a new instance will be output by the visible units of RBM 1. This new instance will usually look like a regular instance of the class whose label unit you activated. This generative capability of DBNs is quite powerful. For example, it has been used to automatically generate captions for images, and vice versa: first a DBN is trained (without supervision) to learn features in images, and another DBN is trained (again without supervision) to learn features in sets of captions (e.g., "car" often comes with "automobile"). Then an RBM is stacked on top of both DBNs and trained with a set of images along with their captions; it learns to associate high-level features in images with high-level features in captions. Next, if you feed the image DBN an image of a car, the signal will propagate through the network, up to the top-level RBM, and back down to the bottom of the caption DBN, producing a caption. Due to the stochastic nature of RBMs and DBNs, the caption will keep changing randomly, but it will generally be appropriate for the image. If you generate a few hundred captions, the most frequently generated ones will likely be a good description of the image.[3]

---

3 See this video by Geoffrey Hinton for more details and a demo: *https://homl.info/137*.

Self-Organizing Maps

Self-organizing maps (SOMs) are quite different from all the other types of neural networks we have discussed so far. They are used to produce a low-dimensional representation of a high-dimensional dataset, generally for visualization, clustering, or classification. The neurons are spread across a map (typically 2D for visualization, but it can be any number of dimensions you want), as shown in Figure E-5, and each neuron has a weighted connection to every input (note that the diagram shows just two inputs, but there are typically a very large number, since the whole point of SOMs is to reduce dimensionality).

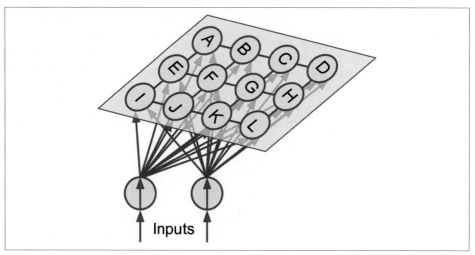

Figure E-5. Self-organizing map

Once the network is trained, you can feed it a new instance and this will activate only one neuron (i.e., one point on the map): the neuron whose weight vector is closest to the input vector. In general, instances that are nearby in the original input space will activate neurons that are nearby on the map. This makes SOMs useful not only for visualization (in particular, you can easily identify clusters on the map), but also for applications like speech recognition. For example, if each instance represents an audio recording of a person pronouncing a vowel, then different pronunciations of the vowel "a" will activate neurons in the same area of the map, while instances of the vowel "e" will activate neurons in another area, and intermediate sounds will generally activate intermediate neurons on the map.

One important difference from the other dimensionality reduction techniques discussed in Chapter 8 is that all instances get mapped to a discrete number of points in the low-dimensional space (one point per neuron). When there are very few neurons, this technique is better described as clustering rather than dimensionality reduction.

The training algorithm is unsupervised. It works by having all the neurons compete against each other. First, all the weights are initialized randomly. Then a training instance is picked randomly and fed to the network. All neurons compute the distance between their weight vector and the input vector (this is very different from the artificial neurons we have seen so far). The neuron that measures the smallest distance wins and tweaks its weight vector to be slightly closer to the input vector, making it more likely to win future competitions for other inputs similar to this one. It also recruits its neighboring neurons, and they too update their weight vectors to be slightly closer to the input vector (but they don't update their weights as much as the winning neuron). Then the algorithm picks another training instance and repeats the process, again and again. This algorithm tends to make nearby neurons gradually specialize in similar inputs.[4]

---

4 You can imagine a class of young children with roughly similar skills. One child happens to be slightly better at basketball. This motivates them to practice more, especially with their friends. After a while, this group of friends gets so good at basketball that other kids cannot compete. But that's okay, because the other kids specialize in other areas. After a while, the class is full of little specialized groups.

Special Data Structures

In this appendix we will take a very quick look at the data structures supported by TensorFlow, beyond regular float or integer tensors. This includes strings, ragged tensors, sparse tensors, tensor arrays, sets, and queues.

Strings

Tensors can hold byte strings, which is useful in particular for natural language processing (see Chapter 16):

```
>>> tf.constant(b"hello world")
<tf.Tensor: id=149, shape=(), dtype=string, numpy=b'hello world'>
```

If you try to build a tensor with a Unicode string, TensorFlow automatically encodes it to UTF-8:

```
>>> tf.constant("café")
<tf.Tensor: id=138, shape=(), dtype=string, numpy=b'caf\xc3\xa9'>
```

It is also possible to create tensors representing Unicode strings. Just create an array of 32-bit integers, each representing a single Unicode code point:[1]

```
>>> tf.constant([ord(c) for c in "café"])
<tf.Tensor: id=211, shape=(4,), dtype=int32,
          numpy=array([ 99,  97, 102, 233], dtype=int32)>
```

---

1 If you are not familiar with Unicode code points, please check out *https://homl.info/unicode*.

In tensors of type tf.string, the string length is not part of the tensor's shape. In other words, strings are considered as atomic values. However, in a Unicode string tensor (i.e., an int32 tensor), the length of the string *is* part of the tensor's shape.

The `tf.strings` package contains several functions to manipulate string tensors, such as `length()` to count the number of bytes in a byte string (or the number of code points if you set `unit="UTF8_CHAR"`), `unicode_encode()` to convert a Unicode string tensor (i.e., int32 tensor) to a byte string tensor, and `unicode_decode()` to do the reverse:

```
>>> b = tf.strings.unicode_encode(u, "UTF-8")
>>> tf.strings.length(b, unit="UTF8_CHAR")
<tf.Tensor: id=386, shape=(), dtype=int32, numpy=4>
>>> tf.strings.unicode_decode(b, "UTF-8")
<tf.Tensor: id=393, shape=(4,), dtype=int32,
        numpy=array([ 99,  97, 102, 233], dtype=int32)>
```

You can also manipulate tensors containing multiple strings:

```
>>> p = tf.constant(["Café", "Coffee", "caffè", "咖啡"])
>>> tf.strings.length(p, unit="UTF8_CHAR")
<tf.Tensor: id=299, shape=(4,), dtype=int32,
        numpy=array([4, 6, 5, 2], dtype=int32)>
>>> r = tf.strings.unicode_decode(p, "UTF8")
>>> r
tf.RaggedTensor(values=tf.Tensor(
[   67    97   102   233    67   111   102   102   101   101    99    97
   102   102   232 21654 21857], shape=(17,), dtype=int32),
   row_splits=tf.Tensor([ 0  4 10 15 17], shape=(5,), dtype=int64))
>>> print(r)
<tf.RaggedTensor [[67, 97, 102, 233], [67, 111, 102, 102, 101, 101],
                [99, 97, 102, 102, 232], [21654, 21857]]>
```

Notice that the decoded strings are stored in a `RaggedTensor`. What is that?

Ragged Tensors

A ragged tensor is a special kind of tensor that represents a list of arrays of different sizes. More generally, it is a tensor with one or more *ragged dimensions*, meaning dimensions whose slices may have different lengths. In the ragged tensor r, the second dimension is a ragged dimension. In all ragged tensors, the first dimension is always a regular dimension (also called a *uniform dimension*).

All the elements of the ragged tensor r are regular tensors. For example, let's look at the second element of the ragged tensor:

```
>>> print(r[1])
tf.Tensor([ 67 111 102 102 101 101], shape=(6,), dtype=int32)
```

The tf.ragged package contains several functions to create and manipulate ragged tensors. Let's create a second ragged tensor using tf.ragged.constant() and concatenate it with the first ragged tensor, along axis 0:

```
>>> r2 = tf.ragged.constant([[65, 66], [], [67]])
>>> print(tf.concat([r, r2], axis=0))
<tf.RaggedTensor [[67, 97, 102, 233], [67, 111, 102, 102, 101, 101], [99, 97,
102, 102, 232], [21654, 21857], [65, 66], [], [67]]>
```

The result is not too surprising: the tensors in r2 were appended after the tensors in r along axis 0. But what if we concatenate r and another ragged tensor along axis 1?

```
>>> r3 = tf.ragged.constant([[68, 69, 70], [71], [], [72, 73]])
>>> print(tf.concat([r, r3], axis=1))
<tf.RaggedTensor [[67, 97, 102, 233, 68, 69, 70], [67, 111, 102, 102, 101, 101,
71], [99, 97, 102, 102, 232], [21654, 21857, 72, 73]]>
```

This time, notice that the ith tensor in r and the ith tensor in r3 were concatenated. Now that's more unusual, since all of these tensors can have different lengths.

If you call the to_tensor() method, it gets converted to a regular tensor, padding shorter tensors with zeros to get tensors of equal lengths (you can change the default value by setting the default_value argument):

```
>>> r.to_tensor()
<tf.Tensor: id=1056, shape=(4, 6), dtype=int32, numpy=
array([[   67,    97,   102,   233,     0,     0],
       [   67,   111,   102,   102,   101,   101],
       [   99,    97,   102,   102,   232,     0],
       [21654, 21857,     0,     0,     0,     0]], dtype=int32)>
```

Many TF operations support ragged tensors. For the full list, see the documentation of the tf.RaggedTensor class.

Sparse Tensors

TensorFlow can also efficiently represent *sparse tensors* (i.e., tensors containing mostly zeros). Just create a tf.SparseTensor, specifying the indices and values of the nonzero elements and the tensor's shape. The indices must be listed in "reading order" (from left to right, and top to bottom). If you are unsure, just use tf.sparse.reorder(). You can convert a sparse tensor to a dense tensor (i.e., a regular tensor) using tf.sparse.to_dense():

```
>>> s = tf.SparseTensor(indices=[[0, 1], [1, 0], [2, 3]],
                        values=[1., 2., 3.],
                        dense_shape=[3, 4])
>>> tf.sparse.to_dense(s)
<tf.Tensor: id=1074, shape=(3, 4), dtype=float32, numpy=
array([[0., 1., 0., 0.],
       [2., 0., 0., 0.],
       [0., 0., 0., 3.]], dtype=float32)>
```

Note that sparse tensors do not support as many operations as dense tensors. For example, you can multiply a sparse tensor by any scalar value, and you get a new sparse tensor, but you cannot add a scalar value to a sparse tensor, as this would not return a sparse tensor:

```
>>> s * 3.14
<tensorflow.python.framework.sparse_tensor.SparseTensor at 0x13205d470>
>>> s + 42.0
[...] TypeError: unsupported operand type(s) for +: 'SparseTensor' and 'float'
```

Tensor Arrays

A tf.TensorArray represents a list of tensors. This can be handy in dynamic models containing loops, to accumulate results and later compute some statistics. You can read or write tensors at any location in the array:

```
array = tf.TensorArray(dtype=tf.float32, size=3)
array = array.write(0, tf.constant([1., 2.]))
array = array.write(1, tf.constant([3., 10.]))
array = array.write(2, tf.constant([5., 7.]))
tensor1 = array.read(1) # => returns (and pops!) tf.constant([3., 10.])
```

Notice that reading an item pops it from the array, replacing it with a tensor of the same shape, full of zeros.

 When you write to the array, you must assign the output back to the array, as shown in this code example. If you don't, although your code will work fine in eager mode, it will break in graph mode (these modes were presented in Chapter 12).

When creating a TensorArray, you must provide its size, except in graph mode. Alternatively, you can leave the size unset and instead set dynamic_size=True, but this will hinder performance, so if you know the size in advance, you should set it. You must also specify the dtype, and all elements must have the same shape as the first one written to the array.

You can stack all the items into a regular tensor by calling the stack() method:

```
>>> array.stack()
<tf.Tensor: id=2110875, shape=(3, 2), dtype=float32, numpy=
array([[1., 2.],
       [0., 0.],
       [5., 7.]], dtype=float32)>
```

Sets

TensorFlow supports sets of integers or strings (but not floats). It represents them using regular tensors. For example, the set {1, 5, 9} is just represented as the tensor [[1, 5, 9]]. Note that the tensor must have at least two dimensions, and the sets must be in the last dimension. For example, [[1, 5, 9], [2, 5, 11]] is a tensor holding two independent sets: {1, 5, 9} and {2, 5, 11}. If some sets are shorter than others, you must pad them with a padding value (0 by default, but you can use any other value you prefer).

The tf.sets package contains several functions to manipulate sets. For example, let's create two sets and compute their union (the result is a sparse tensor, so we call to_dense() to display it):

```
>>> a = tf.constant([[1, 5, 9]])
>>> b = tf.constant([[5, 6, 9, 11]])
>>> u = tf.sets.union(a, b)
>>> u
<tensorflow.python.framework.sparse_tensor.SparseTensor at 0x132b60d30>
>>> tf.sparse.to_dense(u)
<tf.Tensor: [...] numpy=array([[ 1,  5,  6,  9, 11]], dtype=int32)>
```

You can also compute the union of multiple pairs of sets simultaneously:

```
>>> a = tf.constant([[1, 5, 9], [10, 0, 0]])
>>> b = tf.constant([[5, 6, 9, 11], [13, 0, 0, 0, 0]])
>>> u = tf.sets.union(a, b)
>>> tf.sparse.to_dense(u)
<tf.Tensor: [...] numpy=array([[ 1,  5,  6,  9, 11],
                               [ 0, 10, 13,  0,  0]], dtype=int32)>
```

If you prefer to use a different padding value, you must set default_value when calling to_dense():

```
>>> tf.sparse.to_dense(u, default_value=-1)
<tf.Tensor: [...] numpy=array([[ 1,  5,  6,  9, 11],
                               [ 0, 10, 13, -1, -1]], dtype=int32)>
```

> The default default_value is 0, so when dealing with string sets, you must set the default_value (e.g., to an empty string).

Other functions available in tf.sets include difference(), intersection(), and size(), which are self-explanatory. If you want to check whether or not a set contains some given values, you can compute the intersection of that set and the values. If you want to add some values to a set, you can compute the union of the set and the values.

Queues

A queue is a data structure to which you can push data records, and later pull them out. TensorFlow implements several types of queues in the tf.queue package. They used to be very important when implementing efficient data loading and preprocessing pipelines, but the tf.data API has essentially rendered them useless (except perhaps in some rare cases) because it is much simpler to use and provides all the tools you need to build efficient pipelines. For the sake of completeness, though, let's take a quick look at them.

The simplest kind of queue is the first-in, first-out (FIFO) queue. To build it, you need to specify the maximum number of records it can contain. Moreover, each record is a tuple of tensors, so you must specify the type of each tensor, and optionally their shapes. For example, the following code example creates a FIFO queue with maximum three records, each containing a tuple with a 32-bit integer and a string. Then it pushes two records to it, looks at the size (which is 2 at this point), and pulls a record out:

```
>>> q = tf.queue.FIFOQueue(3, [tf.int32, tf.string], shapes=[(), ()])
>>> q.enqueue([10, b"windy"])
>>> q.enqueue([15, b"sunny"])
>>> q.size()
<tf.Tensor: id=62, shape=(), dtype=int32, numpy=2>
>>> q.dequeue()
[<tf.Tensor: id=6, shape=(), dtype=int32, numpy=10>,
 <tf.Tensor: id=7, shape=(), dtype=string, numpy=b'windy'>]
```

It is also possible to enqueue and dequeue multiple records at once (the latter requires specifying the shapes when creating the queue):

```
>>> q.enqueue_many([[13, 16], [b'cloudy', b'rainy']])
>>> q.dequeue_many(3)
[<tf.Tensor: [...] numpy=array([15, 13, 16], dtype=int32)>,
 <tf.Tensor: [...] numpy=array([b'sunny', b'cloudy', b'rainy'], dtype=object)>]
```

Other queue types include:

PaddingFIFOQueue

> Same as FIFOQueue, but its dequeue_many() method supports dequeueing multiple records of different shapes. It automatically pads the shortest records to ensure all the records in the batch have the same shape.

`PriorityQueue`

A queue that dequeues records in a prioritized order. The priority must be a 64-bit integer included as the first element of each record. Surprisingly, records with a lower priority will be dequeued first. Records with the same priority will be dequeued in FIFO order.

`RandomShuffleQueue`

A queue whose records are dequeued in random order. This was useful to implement a shuffle buffer before tf.data existed.

If a queue is already full and you try to enqueue another record, the `enqueue*()` method will freeze until a record is dequeued by another thread. Similarly, if a queue is empty and you try to dequeue a record, the `dequeue*()` method will freeze until records are pushed to the queue by another thread.

TensorFlow Graphs

In this appendix, we will explore the graphs generated by TF Functions (see Chapter 12).

TF Functions and Concrete Functions

TF Functions are polymorphic, meaning they support inputs of different types (and shapes). For example, consider the following `tf_cube()` function:

```
@tf.function
def tf_cube(x):
    return x ** 3
```

Every time you call a TF Function with a new combination of input types or shapes, it generates a new *concrete function*, with its own graph specialized for this particular combination. Such a combination of argument types and shapes is called an *input signature*. If you call the TF Function with an input signature it has already seen before, it will reuse the concrete function it generated earlier. For example, if you call `tf_cube(tf.constant(3.0))`, the TF Function will reuse the same concrete function it used for `tf_cube(tf.constant(2.0))` (for float32 scalar tensors). But it will generate a new concrete function if you call `tf_cube(tf.constant([2.0]))` or `tf_cube(tf.constant([3.0]))` (for float32 tensors of shape [1]), and yet another for `tf_cube(tf.constant([[1.0, 2.0], [3.0, 4.0]]))` (for float32 tensors of shape [2, 2]). You can get the concrete function for a particular combination of inputs by calling the TF Function's `get_concrete_function()` method. It can then be called like a regular function, but it will only support one input signature (in this example, float32 scalar tensors):

```
>>> concrete_function = tf_cube.get_concrete_function(tf.constant(2.0))
>>> concrete_function
<tensorflow.python.eager.function.ConcreteFunction at 0x155c29240>
>>> concrete_function(tf.constant(2.0))
<tf.Tensor: id=19068249, shape=(), dtype=float32, numpy=8.0>
```

Figure G-1 shows the tf_cube() TF Function, after we called tf_cube(2) and tf_cube(tf.constant(2.0)): two concrete functions were generated, one for each signature, each with its own optimized *function graph* (FuncGraph), and its own *function definition* (FunctionDef). A function definition points to the parts of the graph that correspond to the function's inputs and outputs. In each FuncGraph, the nodes (ovals) represent operations (e.g., power, constants, or placeholders for arguments like x), while the edges (the solid arrows between the operations) represent the tensors that will flow through the graph. The concrete function on the left is specialized for x = 2, so TensorFlow managed to simplify it to just output 8 all the time (note that the function definition does not even have an input). The concrete function on the right is specialized for float32 scalar tensors, and it could not be simplified. If we call tf_cube(tf.constant(5.0)), the second concrete function will be called, the placeholder operation for x will output 5.0, then the power operation will compute 5.0 ** 3, so the output will be 125.0.

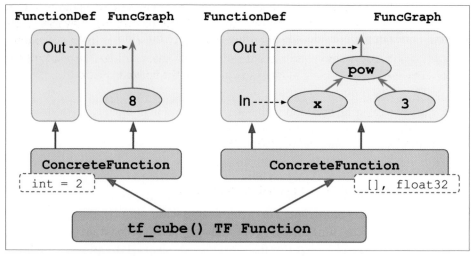

Figure G-1. The tf_cube() TF Function, with its ConcreteFunctions and their Function-Graphs

The tensors in these graphs are *symbolic tensors*, meaning they don't have an actual value, just a data type, a shape, and a name. They represent the future tensors that will flow through the graph once an actual value is fed to the placeholder x and the graph is executed. Symbolic tensors make it possible to specify ahead of time how to connect operations, and they also allow TensorFlow to recursively infer the data types and shapes of all tensors, given the data types and shapes of their inputs.

Now let's continue to peek under the hood, and see how to access function definitions and function graphs and how to explore a graph's operations and tensors.

Exploring Function Definitions and Graphs

You can access a concrete function's computation graph using the `graph` attribute, and get the list of its operations by calling the graph's `get_operations()` method:

```
>>> concrete_function.graph
<tensorflow.python.framework.func_graph.FuncGraph at 0x14db5ef98>
>>> ops = concrete_function.graph.get_operations()
>>> ops
[<tf.Operation 'x' type=Placeholder>,
 <tf.Operation 'pow/y' type=Const>,
 <tf.Operation 'pow' type=Pow>,
 <tf.Operation 'Identity' type=Identity>]
```

In this example, the first operation represents the input argument x (it is called a *placeholder*), the second "operation" represents the constant 3, the third operation represents the power operation (`**`), and the final operation represents the output of this function (it is an identity operation, meaning it will do nothing more than copy the output of the addition operation[1]). Each operation has a list of input and output tensors that you can easily access using the operation's `inputs` and `outputs` attributes. For example, let's get the list of inputs and outputs of the power operation:

```
>>> pow_op = ops[2]
>>> list(pow_op.inputs)
[<tf.Tensor 'x:0' shape=() dtype=float32>,
 <tf.Tensor 'pow/y:0' shape=() dtype=float32>]
>>> pow_op.outputs
[<tf.Tensor 'pow:0' shape=() dtype=float32>]
```

This computation graph is represented in Figure G-2.

---

1 You can safely ignore it—it is only here for technical reasons, to ensure that TF Functions don't leak internal structures.

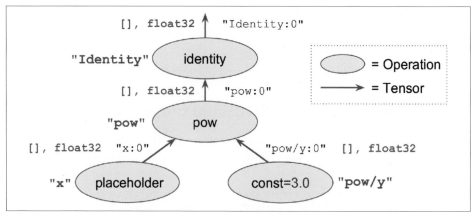

Figure G-2. Example of a computation graph

Note that each operation has a name. It defaults to the name of the operation (e.g., "pow"), but you can define it manually when calling the operation (e.g., tf.pow(x, 3, name="other_name")). If a name already exists, TensorFlow automatically adds a unique index (e.g., "pow_1", "pow_2", etc.). Each tensor also has a unique name: it is always the name of the operation that outputs this tensor, plus :0 if it is the operation's first output, or :1 if it is the second output, and so on. You can fetch an operation or a tensor by name using the graph's get_operation_by_name() or get_tensor_by_name() methods:

```
>>> concrete_function.graph.get_operation_by_name('x')
<tf.Operation 'x' type=Placeholder>
>>> concrete_function.graph.get_tensor_by_name('Identity:0')
<tf.Tensor 'Identity:0' shape=() dtype=float32>
```

The concrete function also contains the function definition (represented as a protocol buffer[2]), which includes the function's signature. This signature allows the concrete function to know which placeholders to feed with the input values, and which tensors to return:

```
>>> concrete_function.function_def.signature
name: "__inference_cube_19068241"
input_arg {
  name: "x"
  type: DT_FLOAT
}
output_arg {
  name: "identity"
  type: DT_FLOAT
}
```

---

2 A popular binary format discussed in Chapter 13.

Now let's look more closely at tracing.

A Closer Look at Tracing

Let's tweak the tf_cube() function to print its input:

```
@tf.function
def tf_cube(x):
    print("x =", x)
    return x ** 3
```

Now let's call it:

```
>>> result = tf_cube(tf.constant(2.0))
x = Tensor("x:0", shape=(), dtype=float32)
>>> result
<tf.Tensor: id=19068290, shape=(), dtype=float32, numpy=8.0>
```

The result looks good, but look at what was printed: x is a symbolic tensor! It has a shape and a data type, but no value. Plus it has a name ("x:0"). This is because the print() function is not a TensorFlow operation, so it will only run when the Python function is traced, which happens in graph mode, with arguments replaced with symbolic tensors (same type and shape, but no value). Since the print() function was not captured into the graph, the next times we call tf_cube() with float32 scalar tensors, nothing is printed:

```
>>> result = tf_cube(tf.constant(3.0))
>>> result = tf_cube(tf.constant(4.0))
```

But if we call tf_cube() with a tensor of a different type or shape, or with a new Python value, the function will be traced again, so the print() function will be called:

```
>>> result = tf_cube(2) # new Python value: trace!
x = 2
>>> result = tf_cube(3) # new Python value: trace!
x = 3
>>> result = tf_cube(tf.constant([[1., 2.]])) # New shape: trace!
x = Tensor("x:0", shape=(1, 2), dtype=float32)
>>> result = tf_cube(tf.constant([[3., 4.], [5., 6.]])) # New shape: trace!
x = Tensor("x:0", shape=(None, 2), dtype=float32)
>>> result = tf_cube(tf.constant([[7., 8.], [9., 10.]])) # Same shape: no trace
```

If your function has Python side effects (e.g., it saves some logs to disk), be aware that this code will only run when the function is traced (i.e., every time the TF Function is called with a new input signature). It best to assume that the function may be traced (or not) any time the TF Function is called.

In some cases, you may want to restrict a TF Function to a specific input signature. For example, suppose you know that you will only ever call a TF Function with batches of 28 × 28–pixel images, but the batches will have very different sizes. You may not want TensorFlow to generate a different concrete function for each batch size, or count on it to figure out on its own when to use None. In this case, you can specify the input signature like this:

```
@tf.function(input_signature=[tf.TensorSpec([None, 28, 28], tf.float32)])
def shrink(images):
    return images[:, ::2, ::2] # drop half the rows and columns
```

This TF Function will accept any float32 tensor of shape [*, 28, 28], and it will reuse the same concrete function every time:

```
img_batch_1 = tf.random.uniform(shape=[100, 28, 28])
img_batch_2 = tf.random.uniform(shape=[50, 28, 28])
preprocessed_images = shrink(img_batch_1) # Works fine. Traces the function.
preprocessed_images = shrink(img_batch_2) # Works fine. Same concrete function.
```

However, if you try to call this TF Function with a Python value, or a tensor of an unexpected data type or shape, you will get an exception:

```
img_batch_3 = tf.random.uniform(shape=[2, 2, 2])
preprocessed_images = shrink(img_batch_3)  # ValueError! Unexpected signature.
```

Using AutoGraph to Capture Control Flow

If your function contains a simple for loop, what do you expect will happen? For example, let's write a function that will add 10 to its input, by just adding 1 10 times:

```
@tf.function
def add_10(x):
    for i in range(10):
        x += 1
    return x
```

It works fine, but when we look at its graph, we find that it does not contain a loop: it just contains 10 addition operations!

```
>>> add_10(tf.constant(0))
<tf.Tensor: id=19280066, shape=(), dtype=int32, numpy=10>
>>> add_10.get_concrete_function(tf.constant(0)).graph.get_operations()
[<tf.Operation 'x' type=Placeholder>, [...],
 <tf.Operation 'add' type=Add>, [...],
 <tf.Operation 'add_1' type=Add>, [...],
 <tf.Operation 'add_2' type=Add>, [...],
 [...]
 <tf.Operation 'add_9' type=Add>, [...],
 <tf.Operation 'Identity' type=Identity>]
```

This actually makes sense: when the function got traced, the loop ran 10 times, so the x += 1 operation was run 10 times, and since it was in graph mode, it recorded this operation 10 times in the graph. You can think of this for loop as a "static" loop that gets unrolled when the graph is created.

If you want the graph to contain a "dynamic" loop instead (i.e., one that runs when the graph is executed), you can create one manually using the tf.while_loop() operation, but it is not very intuitive (see the "Using AutoGraph to Capture Control Flow" section of the Chapter 12 notebook for an example). Instead, it is much simpler to use TensorFlow's *AutoGraph* feature, discussed in Chapter 12. AutoGraph is actually activated by default (if you ever need to turn it off, you can pass autograph=False to tf.function()). So if it is on, why didn't it capture the for loop in the add_10() function? Well, it only captures for loops that iterate over tf.range(), not range(). This is to give you the choice:

- If you use range(), the for loop will be static, meaning it will only be executed when the function is traced. The loop will be "unrolled" into a set of operations for each iteration, as we saw.

- If you use tf.range(), the loop will be dynamic, meaning that it will be included in the graph itself (but it will not run during tracing).

Let's look at the graph that gets generated if you just replace range() with tf.range() in the add_10() function:

```
>>> add_10.get_concrete_function(tf.constant(0)).graph.get_operations()
[<tf.Operation 'x' type=Placeholder>, [...],
 <tf.Operation 'range' type=Range>, [...],
 <tf.Operation 'while' type=While>, [...],
 <tf.Operation 'Identity' type=Identity>]
```

As you can see, the graph now contains a While loop operation, as if you had called the tf.while_loop() function.

Handling Variables and Other Resources in TF Functions

In TensorFlow, variables and other stateful objects, such as queues or datasets, are called *resources*. TF Functions treat them with special care: any operation that reads or updates a resource is considered stateful, and TF Functions ensure that stateful operations are executed in the order they appear (as opposed to stateless operations, which may be run in parallel, so their order of execution is not guaranteed). Moreover, when you pass a resource as an argument to a TF Function, it gets passed by reference, so the function may modify it. For example:

```
counter = tf.Variable(0)

@tf.function
def increment(counter, c=1):
    return counter.assign_add(c)

increment(counter) # counter is now equal to 1
increment(counter) # counter is now equal to 2
```

If you peek at the function definition, the first argument is marked as a resource:

```
>>> function_def = increment.get_concrete_function(counter).function_def
>>> function_def.signature.input_arg[0]
name: "counter"
type: DT_RESOURCE
```

It is also possible to use a `tf.Variable` defined outside of the function, without explicitly passing it as an argument:

```
counter = tf.Variable(0)

@tf.function
def increment(c=1):
    return counter.assign_add(c)
```

The TF Function will treat this as an implicit first argument, so it will actually end up with the same signature (except for the name of the argument). However, using global variables can quickly become messy, so you should generally wrap variables (and other resources) inside classes. The good news is `@tf.function` works fine with methods too:

```
class Counter:
    def __init__(self):
        self.counter = tf.Variable(0)

    @tf.function
    def increment(self, c=1):
        return self.counter.assign_add(c)
```

Do not use =, +=, -=, or any other Python assignment operator with TF variables. Instead, you must use the `assign()`, `assign_add()`, or `assign_sub()` methods. If you try to use a Python assignment operator, you will get an exception when you call the method.

A good example of this object-oriented approach is, of course, tf.keras. Let's see how to use TF Functions with tf.keras.

Using TF Functions with tf.keras (or Not)

By default, any custom function, layer, or model you use with tf.keras will automatically be converted to a TF Function; you do not need to do anything at all! However, in some cases you may want to deactivate this automatic conversion—for example, if your custom code cannot be turned into a TF Function, or if you just want to debug your code, which is much easier in eager mode. To do this, you can simply pass dynamic=True when creating the model or any of its layers:

```
model = MyModel(dynamic=True)
```

If your custom model or layer will always be dynamic, you can instead call the base class's constructor with dynamic=True:

```
class MyLayer(keras.layers.Layer):
    def __init__(self, units, **kwargs):
        super().__init__(dynamic=True, **kwargs)
        [...]
```

Alternatively, you can pass run_eagerly=True when calling the compile() method:

```
model.compile(loss=my_mse, optimizer="nadam", metrics=[my_mae],
              run_eagerly=True)
```

Now you know how TF Functions handle polymorphism (with multiple concrete functions), how graphs are automatically generated using AutoGraph and tracing, what graphs look like, how to explore their symbolic operations and tensors, how to handle variables and resources, and how to use TF Functions with tf.keras.

Index

Symbols

1cycle scheduling, 361
1D convolutional layers, 520

A

A/B experiments, 667
accelerated K-Means, 244
accuracy
    defined, 89
    example of, 2
    measuring using cross-validation, 89
action advantage, 620
action step, 656
actions
    evaluating, 619
    exploiting versus exploring, 618
activation functions
    exponential linear unit (ELU), 336-338
    hyperbolic tangent (tanh), 291
    Logistic (sigmoid), 143, 293, 302, 332
    nonsaturating, 335
    Rectified Linear Unit function (ReLU),
        292-293
    Scaled Exponential Linear Unit (SELU), 334,
        337-338, 368
    softmax, 294, 299, 470, 482, 488, 543
    softplus, 293
active constraint, 762
active learning, 255
Actor-Critic algorithms, 625, 662
AdaBoost, 200
AdaGrad, 354
Adam and Nadam optimization, 356
Adaptive Boosting, 200

adaptive instance normalization (AdaIN), 604
adaptive learning rate, 355
adaptive moment estimation, 356
additive attention, 550
Advantage Actor-Critic (A2C), 663
adversarial learning, 495, 568
affine transformations, 604
affinity, 237
affinity propagation, 259
agents, 14
agglomerative clustering, 258
AI Platform, 680
Akaike information criterion (AIC), 267
AlexNet, 464
algorithms
    Actor-Critic algorithms, 625, 662
    Advantage Actor-Critic (A2C), 663
    AllReduce algorithm, 705
    Asynchronous Advantage Actor-Critic
        (A3C), 662
    BIRCH algorithm, 259
    CART training algorithm, 177, 179
    clustering algorithms, 10
    Dueling DQN algorithm, 641
    dynamic placer algorithm, 697
    Expectation-Maximization (EM) algorithm,
        262
    for anomaly detection, 274
    genetic algorithms, 612
    greedy algorithms, 180
    hierarchical clustering algorithms, 10
    importance of data over, 24
    Isolation Forest algorithm, 274
    isomap algorithm, 233

K-Means algorithm, 238
Lloyd–Forgy algorithm, 238
Mean-Shift algorithm, 259
off-policy algorithms, 632
on-policy algorithms, 632
one-class SVM algorithm, 275
Proximal Policy Optimization (PPO), 663
Randomized PCA algorithm, 225
REINFORCE algorithms, 620
Soft Actor-Critic algorithm, 663
supervised learning, 8
unsupervised learning, 9
Value Iteration algorithm, 627
visualization algorithms, 11
AllReduce algorithm, 705
alpha channels, 250
anchor boxes, 490
anomaly detection
additional algorithms for, 274
examples of, 12
goal of, 236
using clustering, 237
using Gaussian Mixtures, 266
Approximate Q-Learning, 633
area under the curve (AUC), 98
argmax operator, 149
artificial neural networks (ANNs)
Boltzmann machines, 775
fine-tuning hyperparameters for, 320-327
from biological to artificial neurons,
280-295
Hopfield networks, 773
implementing MLPs with Keras, 295-320
overview of, 279
restricted Boltzmann machines (RBMs), 776
self-organizing maps (SOMs), 780
artificial neurons, 283
association rule learning, 12
associative memory networks, 773
Asynchronous Advantage Actor-Critic (A3C),
662
asynchronous updates, 707
Atari preprocessing, 645
attention mechanisms
defined, 526
explainability and, 553
overview of, 549
Transformer architecture, 554
visual attention, 552

attributes, 8
autoencoders
convolutional, 579
denoising, 581
efficient data representations, 569
generative, 586
versus Generative Adversarial Networks
(GANs), 568
overview of, 567
parts of, 569
PCA with undercomplete linear autoencoders, 570
probabilistic, 586
recurrent, 580
sparse, 582
stacked, 572-575
undercomplete, 570
unsupervised pretraining using stacked,
576-579
variational, 586-591
AutoGraphs, 407
automatic differentiation (autodiff), 290, 399,
765-772
AutoML, 323
autonomous driving systems, 497
autoregressive integrated moving average
(ARIMA) models, 506
average absolute deviation, 41
average pooling layer, 459
Average Precision (AP), 491

B
backpropagation, 289-292
backpropagation through time (BPTT), 502
bag of words, 438
bagging and pasting
out-of-bag evaluation, 195
overview of, 192
in Scikit-Learn, 194
Bahdanau attention, 550
bandwidth saturation, 708
basic cells, 500
Batch Gradient Descent, 121
batch learning, 15
Batch Normalization (BN), 339
batch size, 325
batched action step, 657
batched time step, 657
batched trajectory, 657

Bayesian Gaussian Mixture models, 270
Bayesian inference, 586
Bayesian information criterion (BIC), 267
beam search, 547
beam width, 547
Bellman Optimality Equation, 627
Better Life Index, 19
bias neurons, 285
bias terms, 112
bias/variance trade-off, 134
bidirectional recurrent layers, 546
bidirectional RNNs, 546
binary classifiers, 88
binary trees, 177
biological neural networks (BNN), 282
biological neurons, 280
BIRCH algorithm, 259
black box models, 178
black box stochastic variational inference
    (BBSVI), 273
blenders, 208
Boltzmann machines, 775
boosting
    AdaBoost, 200
    Gradient Boosting, 203
    overview of, 199
bottleneck layers, 468
boundary transitions, 660
bounding box priors, 490
break the symmetry, 291
Byte-Pair Encoding, 536

C

calculus, 112
California Housing Prices dataset, 36
callbacks, 315
canary testing, 684
CART training algorithm, 177, 179
catastrophic forgetting, 637
categorical distribution, 261
categorical features
    encoding using embeddings, 433
    encoding using one-hot vectors, 431
causal models, 510
centroids, 238
chain rule, 290
chaining transformations, 415
character RNNs (Char-RNNs)
    building and training, 530

chopping sequential datasets, 528
generating Shakespearean text, 531
overview of, 526
splitting sequential datasets, 527
stateful RNNs and, 532
training dataset creation, 527
using, 531
chatbots, 525
chi-squared test, 182
Classification and Regression Tree (CART),
    177, 179
classification problems
    AdaBoost classifiers, 200
    binary classifiers, 88
    classification and localization, 483
    classification MLPs, 294
    error analysis, 102
    example of, 8
    Extra-Trees classifier, 198
    hard margin classification, 154
    image classifiers using Sequential APIs,
        297-307
    large margin classification, 153
    linear SVM classification, 153
    MNIST dataset, 85
    multiclass classification, 100
    multilabel classification, 106
    multioutput classification, 107
    multitask classification, 311
    nonlinear SVM classification, 157-162
    performance measures, 88-100
    soft margin classification, 154
    voting classifiers, 189
closed-form solution, 114
cluster specification, 711
clustering algorithms
    additional algorithms, 258
    applications for, 10, 237
    DBSCAN, 255
    goal of, 236
    for image segmentation, 238, 249
    K-Means, 238-249
    overview of, 236
    for preprocessing, 251
    for semi-supervised learning, 253
code examples, obtaining and using, xxi
codings, 567
Colab Runtime, 693
Colaboratory (Colab), 693

collect policy, 649
color channels, 451
color segmentation, 249
column vectors, 113
comments and questions, xxiii, 718
complementary slackness, 762
components, 38
compression, 224
computation graphs, 376
Compute Unified Device Architecture library (CUDA), 690
concatenative attention, 550
concrete functions, 791
conditional probability, 547
confusion matrix, 90
connectionism, 280
constrained optimization, 166
Contrastive Divergence, 777
convergence, 118
convex function, 120
convolution kernels, 450
convolutional autoencoders, 579
convolutional layer
    filters, 450
    memory requirements, 456
    overview of, 448
    stacking multiple feature maps, 451
    TensorFlow implementation, 453
Convolutional Neural Networks (CNNs)
    architecture of visual cortex, 446
    classification and localization, 483
    CNN architectures, 460-478
    convolutional layer, 448-456
    object detection, 485-492
    overview of, 445
    pooling layer, 456
    pretrained models for transfer learning, 481
    pretrained models from Keras, 479
    ResNet-34 using Keras, 478
    semantic segmentation, 492
core instances, 255
corpus development, 24
correlation coefficient, 58
cost functions
    cross-entropy loss (log loss), 149
    hinge loss, 155, 173
    mean absolute error (MAE), 41, 293
    mean squared error, 120, 293, 308, 384, 570, 573, 583, 636

role of, 20
credit assignment problem, 619
cross-entropy loss (log loss), 149, 295
cross-validation, 31, 73, 89
CUDA Deep Neural Network library (cuDNN), 690
curiosity-based exploration, 664
curse of dimensionality, 214
custom models
    about, 375
    activation functions, initializers, regularizers, and constraints, 387
    computing gradients using Autodiff, 399, 765-772
    layers, 391
    loss functions, 384
    losses and metrics, 397
    metrics, 388
    models, 394
    saving and loading, 385
    training loops, 402
customer segmentation, 237

D

data (see also data preparation; data visualization; training data)
    analyzing through clustering, 237
    California Housing Prices dataset, 36
    chopping sequential datasets, 528
    compressing, 224
    data mismatch, 32
    decompressing, 224
    downloading, 46
    efficient data representations, 569
    Fashion MNIST dataset, 297, 574, 590
    flat datasets, 529
    geographical data, 56
    Google News 7B corpus, 541
    helper function creation, 420
    importance of over algorithms, 24
    Internet Movie Database, 534
    iris dataset, 145
    loading and preprocessing with TensorFlow, 413-442
    MNIST dataset, 85
    nested datasets, 529
    noisy data, 19
    prefetching, 421
    preprocessing, 251, 419, 430-439

reconstruction error, 224
reducing dimensionality of, 222
shuffling, 416
skewed datasets, 89
sources for, 35
splitting sequential datasets, 527
training dataset creation, 527
training sparse models, 359
using datasets with tf.Keras, 423
Data API (TensorFlow)
  chaining transformations, 415
  helper function creation, 420
  overview of, 414
  prefetching data, 421
  preprocessing data, 419
  shuffling data, 416
  using datasets with tf.keras, 423
data augmentation, 464
data parallelism, 701, 704
data preparation
  benefits of functions for, 62
  custom transformers, 68
  data cleaning, 63
  feature scaling, 69
  handling text and categorical attributes, 65
  transformation pipelines, 70
data snooping bias, 51
data visualization
  attribute combinations, 61
  computing correlations, 58
  dimensionality reduction, 213
  geographical data, 56
  test, training, and exploration sets, 56
  using TensorBoard for, 317
  visualizing Fashion MNIST Dataset, 574
  visualizing reconstructions, 574
datasets, defined, 414
DataViz (see data visualization)
DBSCAN (density-based spatial clustering of
  applications with noise), 255
decision boundaries, 145
decision function, 93
Decision Stumps, 203
Decision Trees
  benefits of, 175
  CART training algorithm, 179
  computational complexity, 180
  estimating class probabilities, 178
  evaluating, 73

Gini impurity versus entropy, 180
instability drawbacks, 185
making predictions, 176
regression tasks, 183
regularization hyperparameters, 181
training and visualizing, 175
decoders, 501, 569
decompression, 224
deep autoencoders, 572
deep belief networks (DBNs), 13, 777
deep computer vision (see Convolutional Neural Networks (CNNs))
deep convolutional GANs, 598
Deep Learning VM Images, 693
deep neural networks (DNNs)
  avoiding overfitting, 364-371
  default configuration, 371
  defined, xv, 289
  faster optimizers, 351-364
  overview of, 331
  reusing pretrained layers, 345-351
  vanishing/exploding gradients problems,
    332-345
Deep Neuroevolution, 323
Deep Q-Learning
  Double DQN, 640
  Dueling DQN, 641
  fixed Q-Value targets, 639
  implementing, 634
  overview of, 633
  prioritized experience replay, 640
  variants of, 639
deep Q-networks (DQNs), 633, 650, 650
denoising autoencoders, 581
dense layer, 285
dense vectors, 556
density estimation, 236, 264
depth concat layer, 467
depth radius, 466
depthwise separable convolution, 474
deques, 635
development sets (dev sets), 31
differencing, 506
dimensionality reduction
  additional techniques, 232
  approaches for, 215-218
  using clustering, 237
  curse of dimensionality, 214
  goal of, 12

LLE (Locally Linear Embedding), 230
    overview of, 213
    PCA (Principal Component Analysis),
      219-230
discount factors, 619
discriminators, 568
Distribution Strategies API, 668, 709
dot product, 551
Double DQN, 640
Double Dueling DQN, 642
DQN agents, 652
dropout, 365
dual numbers, 768
dual problem, 168, 761
duck typing, 68
Dueling DQN algorithm, 641
dummy attributes, 67
dying ReLUs problem, 335
dynamic models, 313
dynamic placer algorithm, 697
Dynamic Programming, 628

E

eager execution/eager mode, 408
early stopping, 141
Elastic Net, 140
ELU (exponential linear unit), 336-338
embedded devices, 685
Embedded Reber grammars, 566
embedding, 68, 413, 433
embedding matrix, 435
encoders, 501, 569
Encoder–Decoder model, 501, 542-548
end-of-sequence (EoS) token, 542, 556
energy function, 774
Ensemble Learning
    bagging and pasting, 192-196
    benefits of, 74
    best uses of, 191
    boosting, 199-208
    defined, 189
    examples of, 189
    Random Forests, 189, 197
    random patches and random subspaces, 196
    stacking, 208
    voting classifiers, 189
Ensemble methods, 189
ensembles, 189
entailment, 564

entropy impurity measure, 180
epochs, 125, 290
equalized learning rates, 603
equivariance, 458
error analysis, 102
estimators, 64
Euclidean norm, 41
event files, 317
evidence lower bound (ELBO), 272
example project
    data downloading, 42-55, 756
    data preparation, 62-72, 757
    data visualization, 56-62, 756
    framing the problem, 37, 755
    launching, monitoring, and maintaining, 80,
      760
    Machine Learning project checklist, 37, 755
    model fine-tuning, 75-80, 759
    model selection and training, 72, 758
    overview of, 35
    project goals, 37
    real-world data for, 35
    selecting performance measure, 39
    verifying assumptions, 42
Exclusive OR (XOR) classification problem,
    288
exercise solutions, 719-753
expectation step, 262
Expectation-Maximization (EM) algorithm,
    262
experience replay, 597
explainability, 553
explained variance ratio, 222
exploding gradients problem, 332
exploration policy, 630, 632
exploration sets, 56
exponential linear unit (ELU), 336-338
exponential scheduling, 360
Extra-Trees classifier, 198
Extremely Randomized Trees ensemble, 198

F

F1 score, 92
fake quantization, 687
false positive rate (FPR), 97
fan-in/fan-out numbers, 333
Fashion MNIST dataset, 297, 574, 590
Fast-MCD (minimum covariance determi-
    nant), 274

feature engineering, 27
feature extraction, 12, 27
feature maps, 228, 450
feature scaling, 69
feature selection, 27
feature space, 226
feature vector, 113
features, 8
feedforward neural networks (FNNs), 289
filters, 450
final trained models, 20
finite difference approximation, 766
First In, First Out (FIFO) queues, 383
first-order partial derivatives (Jacobians), 358
fitness functions, 20
fixed Q-Value targets, 639
flat datasets, 529
folds, 73, 89
forecasting, 503
forget gate, 516
forward pass, 290
forward-mode autodiff, 767
fraud detection, 237
Full Gradient Descent, 122
fully connected layer, 285
fully convolutional networks (FCNs), 487
fully-specified model architecture, 20
function definitions, 792
function graphs, 792
Functional API, 308-313

G

gate controllers, 516
Gated Recurrent Unit (GRU) cell, 518
Gaussian mixture model (GMM)
    additional algorithms for anomaly and nov-
        elty detection, 274
    anomaly detection using, 266
    Bayesian Gaussian Mixture models, 270
    graphical model of, 260
    overview of, 260
    selecting cluster number, 267
    variants, 260
Gaussian Radial Basis Function (RBF), 159
generalization error, 30
generalized Lagrangian, 762
Generative Adversarial Networks (GANs)
    versus autoencoders, 568
    deep convolutional GANs (DCGANs), 598

difficulties of training, 596
    overview of, 592
    progressive growing of, 601
    StyleGANs, 604
    uses for, 567
generative autoencoders, 586
generative models, 263, 567, 775 (see also
    autencoders; Generative Adversarial Net-
    works (GANs))
generative network, 569
generators, 568
genetic algorithms, 612
Gini impurity measure, 180
global average pooling layer, 460
global minimum, 119
Glorot and He initialization, 333
Google Cloud Platform (GCP)
    prediction service creation, 677-681
    prediction service use, 682-685
Google Cloud Storage (GCS), 679
Google News 7B corpus, 541
GoogLeNet, 466
GPUs (graphics processing units)
    adding to single machines, 689
    Colaboratory (Colab), 693
    GPU-equipped virtual machines, 692
    managing GPU RAM, 694
    parallel execution across multiple devices,
        699
    placing operations and variables on devices,
        697
    selecting, 690
    speeding computations with, 689
Gradient Boosted Regression Trees (GBRT),
    203
Gradient Boosting, 203
gradient clipping, 345
Gradient Descent (GD)
    Batch Gradient Descent, 121
    Mini-batch Gradient Descent, 127
    overview of, 111, 118
    Stochastic Gradient Descent, 124
Gradient Tree Boosting, 203
graph mode, 408
greedy algorithms, 180
greedy layer-wise pretraining, 349
greedy layer-wise training, 578

H

hard clustering, 240
hard margin classification, 154
hard voting classifiers, 190
harmonic mean, 92
HDF5 format, 314
He initialization, 333
Heaviside step function, 285
Hebb's rule, 286
Hebbian learning, 286
helper functions, 420
hidden layers
    in MLPs, 289
    neurons per hidden layer, 325
    number of, 323
hidden units, 775
hierarchical clustering algorithms, 10
Hierarchical DBSCAN (HDBSCAN), 258
high-dimensional training sets, 213
hinge loss function, 155, 173
Hinton, Geoffrey, xv
histograms, 50
hold outs, 31
holdout validation, 31
Hopfield networks, 773
Huber loss, 293, 384
Hyperas, 322
Hyperband, 323
hyperbolic tangent function (tanh), 291
Hyperopt, 322
hyperparameters
    defined, 29
    fine-tuning for neural networks, 320-327
    hyperparameter tuning, 31, 75
    learning rate, 118
    Python libraries for optimization, 322
    regularization hyperparameters, 181
hyperplanes, 165
hypothesis boosting, 199

I

identity matrix, 137
image classification
    multitask classification, 311
    using Sequential API, 297-307
image generation, 495
image segmentation, 238, 249
importance sampling (IS), 640
impurity, 177, 180
imputation, 503
incremental learning, 16
Incremental PCA (IPCA), 225
independent and identically distributed (IID), 126
inequality constraints, 762
inertia, 243
inference, 23
information theory, 180
initialization
    centroid initialization methods, 243
    Glorot and He initialization, 333
    LeCun initialization, 334
    random initialization, 118
    Xavier initialization, 333
inliers, 266
input and output sequences, 501
input gate, 516
input layers, 289
input neurons, 285
input signatures, 791
instability, 185
instance segmentation, 249, 495
instance-based learning, 17, 22
inter-op thread pool, 699
intercept terms, 112
Internet Movie Database, 534
intra-op thread pool, 699
invariance, 457
inverse transformation, 225
iris dataset, 145
isolated environments, 43
Isolation Forest algorithm, 274
isomap algorithm, 233

J

JupyterLab, 693
just-in-time (JIT) compiler, 376

K

K-fold cross-validation, 73, 89
K-Means
    accelerated and mini-batch, 244
    centroid initialization methods, 243
    hard and soft clustering, 240
    image segmentation, 249
    K-Means algorithm, 241
    limits of, 248
    optimal cluster number, 245

overview of, 238
preprocessing with, 251
proposed improvement to, 243
scaling input features, 249
for semi-supervised learning, 253
k-Nearest Neighbors regression, 22
Karush–Kuhn–Tucker (KKT) multipliers, 762
keep probability, 367
Keras
benefits of, xvi
complex architectures, 314
gradient clipping in, 345
implementing Batch Normalization with, 341
implementing dropout using, 367
implementing MLPs with, 295-320
implementing ResNet-34 with, 478
keras.callbacks package, 316
loading datasets with, 297
low-level API, 381
multibackend Keras, 295
preprocessing layers, 437
saving and restoring models in, 314
stacked autoencoders using, 572
transfer learning with, 347
using code examples from keras.io, 300
using pretrained models from, 479
Keras Tuner, 322
Kernel PCA (kPCA), 226-230
kernel trick, 158, 228
kernelized SVM, 169
kernels, 170, 226, 377
kopt library, 322
Kullback–Leibler divergence, 150

L

label propagation, 254
labels, 8, 39, 239
Lagrange multipliers, 761
landmarks, 159
language models, 563 (see also natural language processing (NLP))
large margin classification, 153
Lasso Regression, 137
latent loss, 587
latent representations, 567
latent variables, 262
law of large numbers, 191
Layer Normalization, 512

layers
1D convolutional layer, 520
adaptive instance normalization (AdaIN), 604
bidirectional recurrent layer, 546
convolutional layer, 448-456
dense (fully connected) layer, 285
hidden layer, 289
input layer, 289
Masked Multi-Head Attention layer, 556
minibatch standard deviation layer, 603
Multi-Head Attention layer, 556, 559
output layer, 289
pooling layer, 456
recurrent, 498-502
reusing pretrained, 345-351
Scaled Dot-Product Attention layer, 559
leaf nodes, 176
leaky ReLU function, 335
learning curves, 130-134
learning rate, 16, 118, 325, 603
learning rate scheduling, 359
learning schedules, 125, 360
LeCun initialization, 334
LeNet-5, 463
Levenshtein distance, 161
liblinear library, 162
libsvm library, 162
likelihood function, 267
linear algebra, 112
linear autoencoders, 570
Linear Discriminant Analysis (LDA), 233
linear models, 19
Linear Regression model
approaches to training, 111, 113
computational complexity, 117
Normal Equation, 114
overview of, 112
linear SVM classification, 153
lists of lists, using SequenceExample Protobuf, 429
LLE (Locally Linear Embedding), 230
Lloyd-Forgy algorithm, 238
local minimum, 119
Local Outlier Factor (LOF), 274
local response normalization, 465
localization, 483
log loss, 144
log-odds, 144

logical computations, 283
logical GPU devices, 695
Logistic (sigmoid) function, 143, 293-294, 302, 332
Logistic Regression
    classification with, 8
    decision boundaries, 145
    estimating probabilities, 143
    overview of, 142
    Softmax Regression, 148
    training and cost function, 144
logit, 144
Logit Regression (see Logistic Regression)
long sequences
    overview of, 511
    short-term memory problems, 514-523
    unstable gradients problem, 512
Long Short-Term Memory (LSTM) cell, 514
loss functions (see cost functions)
Luong attention, 551

M

Machine Learning (ML)
    additional resources, xix
    applications for, xv, 5
    approach to learning, xvi
    benefits of, 2
    challenges of, 23-30
    defined, 1
    history of, xv
    locating papers on, 378
    notations for, 40, 164
    overview of, 30
    prerequisites to learning, xvii
    testing and validating, 30-33
    topics covered, xvii
    types of, 7-23
Machine Learning project checklist, 37, 755
majority-vote classifiers, 190
majority-vote predictions, 187
Manhattan norm, 41
manifold assumption, 218
manifold hypothesis, 218
Manifold Learning, 218
manual differentiation, 765
margin violations, 155
Markov chains, 625
Markov Decision Processes (MDP), 625-629
Mask R-CNN, 495

mask tensors, 539
masked language model (MLM), 564
Masked Multi-Head Attention layer, 556
masking, 538
max pooling layer, 457
max-norm regularization, 370
maximization step, 262
maximum a-posteriori (MAP) estimation, 269
maximum likelihood estimate (MLE), 269
mean absolute error (MAE), 41
mean Average Precision (mAP), 491
mean coding, 586
mean field variational inference, 273
Mean-Shift algorithm, 259
measure of similarity, 18
memory bandwidth, 422
memory cells, 500
Mercer's conditions, 171
Mercer's theorem, 171
meta learners, 208
metagraphs, 671
metrics
    accuracy, 388
    area under the curve (AUC), 98
    confusion matrix, 90, 90
    F1 score, 92
    mean absolute error (MAE), 41, 293
    mean average precision, 491
    mean squared error, 183, 505
    precision, 91-97
    recall, 91-97
    RMSE, 39
    ROC curve, 97
Microsoft Cognitive Toolkit (CNTK), 295
min-max scaling, 69
Mini-batch Gradient Descent, 127
mini-batch K-Means, 244
mini-batches, 15, 127
minibatch discrimination, 597
minibatch standard deviation layer, 603
mirrored strategy, 704
mixing regularization, 606
ML Engine, 680
MNIST dataset, 85
mobile devices, 685
mode collapse, 597
model parallelism, 701
model parameters, 20
model selection, 19, 31, 72

model-based learning, 18
models (see also custom models)
    causal models, 510
    complex using Functional API, 308-313
    custom with TensorFlow, 384-405
    defined, 20
    dynamic using Subclassing API, 313
    fine-tuning, 75-80
    parametric versus nonparametric, 181
    pretrained models for transfer learning, 481
    pretrained models from Keras, 479
    saving and restoring, 314
    sequence-to-sequence models, 510
    training, 20, 72 (see also training models)
    training across multiple devices, 701-717
    training sparse models, 359
    using callbacks, 315
    using TensorBoard for visualization, 317
    white versus black box, 178
modules, 540
momentum optimization, 351
momentum vector, 352
Monte Carlo (MC) dropout, 368
Multi-Head Attention layer, 556, 559
multibackend Keras, 295
multiclass classification, 100
Multidimensional Scaling (MDS), 232
multilabel classification, 106
Multilayer Perceptrons (MLPs)
    backpropagation and, 289-292
    classification MLPs, 294
    regression MLPs, 292
multinomial classifiers, 100
Multinomial Logistic Regression, 148
multioutput classification, 107
multiple outputs, 311
multiple regression problems, 39
multiplicative attention, 551
multitask classification, 311
multivariate regression problems, 39
multivariate time series, 503

N

naive forecasting, 505
Nash equilibrium, 596
natural language processing (NLP)
    attention mechanisms, 549-563
    CNNs for, 445
    Encoder–Decoder network for, 542-548

    generating text using character RNNs, 526-534
    overview of, 525
    recent innovations in, 563
    RNNS for, 497
    sentiment analysis, 534-542
    uses for, 351
nested datasets, 529
Nesterov Accelerated Gradient (NAG), 353
Nesterov momentum optimization, 353
neural machine translation (NMT), 542-563
    (see also natural language processing (NLP))
neurons
    bias neurons, 285
    fan-in/fan-out numbers, 333
    from biological to artificial, 280-295
    input neurons, 285
    logical computations with, 283
    per hidden layer, 325
    recurrent neurons, 498-502
    stochastic neurons, 775
Newton's difference quotient, 766
next sentence prediction (NSP), 565
No Free Lunch (NFL) theorem, 33
noisy data, 19
non-max suppression, 486
nonlinear dimensionality reduction (NLDR), 230
nonlinear SVM classification, 157-162
nonparametric models, 181
nonsaturating activation functions, 335
nonsequential neural networks, 308
Normal Equation, 114
normalization, 69, 339, 603
normalized exponential, 148
novelty detection, 12, 267, 274
NP-Complete problem, 180
null hypothesis, 182
NumPy
    array_split() function, 226
    dense arrays, 67
    installing, 42
    inv() function, 115
    memmap class, 226
    randint() function, 107
    serializing large arrays, 75
    svd() function, 221
    using TensorFlow like, 379-384

NVIDIA Collective Communications Library (NCCL), 710
Nvidia GPU cards, 690

O

object detection
    fully convolutional networks (FCNs), 487
    overview of, 485
    You Only Look Once (YOLO), 489
objectness output, 486
observed variables, 262
observers, 654
off-policy algorithms, 632
offline learning, 15
on-policy algorithms, 632
one-class SVM algorithm, 275
one-hot encoding, 67
one-hot vectors, 431
one-versus-all (OvA) strategy, 100
one-versus-one (OvO) strategy, 100
one-versus-the-rest (OvR) strategy, 100
online learning, 15, 88
online model, 639
online SVMs, 172
OpenAI Gym, 613-617
Optical Character Recognition (OCR), 1
optimal state value, 627
optimizers
    AdaGrad, 354
    Adam and Nadam optimization, 356
    creating faster, 351
    first- and second-order partial derivatives, 358
    learning rate scheduling, 359
    momentum optimization, 351
    Nesterov Accelerated Gradient (NAG), 353
    RMSProp, 355
    Stochastic Gradient Descent (SGD), 88, 124
original space, 226
out-of-core learning, 16
out-of-sample error, 30
out-of-vocabulary (oov) buckets, 432
outlier detection, 237, 266
output gate, 516
output layers, 289
overcomplete autoencoders, 580, 580
overfitting
    avoiding through regularization, 364-371
    defined, 27

limiting risk of, 457

P

p (posterior) distribution, 272
p (prior) distribution, 271
p-value, 182
parameter efficiency, 323
parameter matrix, 148
parameter servers, 705
parameter space, 121
parameter vector, 113
parametric leaky ReLU (PReLU), 335
parametric models, 181
partial derivatives, 121
pasting (see bagging and pasting)
pattern matching, 569
PCA (Principal Component Analysis)
    anomaly and novelty detection using, 274
    choosing dimension number, 223
    for compression, 224
    explained variance ratio, 222
    incremental, 225
    Kernel PCA (kPCA), 226-230
    overview of, 219
    preserving variance, 219
    principal component axis, 220
    projecting down to d dimensions, 221
    randomized, 225
    using Scikit-Learn, 222
    undercomplete linear autoencoders for, 570
Pearson's r, 58
peephole connections, 518
penalties, 14
Perceptron, 284-288
Perceptron convergence theorem, 287
performance measures (see metrics)
performance scheduling, 361
piecewise constant scheduling, 361
pipelines, 38, 424
pixelwise normalization layers, 603
policies, 14, 612
policy gradients (PG), 613, 620-625
policy parameters, 612
policy search, 612
policy space, 612
polynomial features, 158
polynomial kernels, 170
Polynomial Regression, 112, 128
pooling kernel, 457

pooling layer, 456
positional embeddings, 556
post-training quantization, 686
power scheduling, 360
pre-images, 228
precision, 91-97
prediction problems, 8, 17, 189
prediction service
    creating on GCP AI, 677-681
    using, 682-685
predictors, 65
preprocessing, 251, 430-439
pretraining
    for transfer learning, 481
    greedy layer-wise pretraining, 349
    models from Keras, 479
    on auxiliary tasks, 350
    reusing pretrained embeddings, 540
    reusing pretrained layers, 345-351
    unsupervised pretraining, 349
    using stacked autoencoders, 576-579
primal problem, 168
prioritized experience replay (PER), 640
probabilistic autoencoders, 586
probability density function (PDF), 236, 264
projection, 215
propositional logic, 280
protocol buffers (protobufs), 425
Proximal Policy Optimization (PPO), 663
pruning, 182
PyTorch library, 296

Q

Q-Learning
    Approximate Q-Learning and Deep Q-
      Learning, 633
    exploration policy, 632
    implementing, 631
    overview of, 630
Q-Value Iteration, 628
Q-Values, 628
Quadratic Programming (QP) problems, 167
quantization-aware training, 687
queries per second (QPS), 667
questions and comments, xxiii, 718
queues, 383, 788

R

Radial Basis Function (RBF), 159

ragged tensors, 383, 784
Rainbow agent, 642
Random Forests
    benefits of, 189
    Extra-Trees, 198
    feature importance, 198
    overview of, 197
random initialization, 118
random patches and random subspaces, 196
random projections, 232
randomized leaky ReLU (RReLU), 335
Randomized PCA, 225
recall, 91-97
receiver operating characteristic (ROC) curve,
    97
recognition network, 569
recommender systems, 237
reconstruction error, 224
reconstruction loss, 397, 570
reconstruction pre-images, 228
reconstructions, 570
Rectified Linear Unit function (ReLU), 292-293
recurrent autoencoders, 580
recurrent neural networks (RNNs)
    bidirectional RNNs, 546
    forecasting time series, 503-511
    generating text using character RNNS,
      526-534
    handling long sequences, 511-523
    overview of, 497
    recurrent neurons and layers, 498-502
    stateless and stateful, 525, 532
    training, 502
recurrent neurons, 498
Region Proposal Network (RPN), 492
regression problems
    Decision Trees, 183
    defined, 8
    k-Nearest Neighbors regression, 22
    Lasso Regression, 137
    Linear Regression, 112-117
    Logistic Regression, 142-151
    multiple regression problems, 39
    multivariate regression problems, 39
    Polynomial Regression, 128
    regression MLPs, 292
    regression MLPs using Sequential API, 307
    Ridge Regression, 135
    Softmax Regression, 148-151

SVM regression, 162
univariate regression problems, 39
regular expressions, 536
regularization
avoiding overfitting through, 364-371
defined, 28
hyperparameters for Decision Trees, 181
multiple outputs for, 311
shrinkage technique, 205
regularization terms, 135
regularized linear models
Elastic Net, 140
Lasso Regression, 137
overview of, 134
Ridge Regression, 135
REINFORCE algorithms, 620
Reinforcement Learning (RL)
algorithms for, 662
Deep Q-Learning, 633-638
evaluating actions, 619
Markov Decision Processes (MDP), 625-629
neural network policies, 617
OpenAI Gym, 613-617
optimizing rewards, 610
overview of, 14, 609
policy gradients, 620-625
policy search, 612
Q-Learning, 630-634
Temporal Difference Learning, 629
TF-Agents library, 642-662
ReLU (Rectified Linear Unit function), 292-293
replay buffers, 635, 649, 654
replay memory, 635
representation learning, 68, 434 (see also autoencoders)
residual blocks, 395
residual errors, 203
residual learning, 471
residual units, 471
ResNet (Residual Network), 471
ResNet-34 CNN, 478
responsibilities (clustering), 262
restoring models, 314
restricted Boltzmann machines (RBMs), 13, 349, 776
reverse-mode autodiff, 290, 770
rewards, 14
Ridge Regression, 135
RMSProp, 355

Root Mean Square Error (RMSE), 39, 120
root nodes, 176

S

SAMME (Stagewise Additive Modeling using a Multiclass Exponential loss function), 203
sample inefficiency, 625
sampled softmax technique, 544
sampling bias, 25
sampling noise, 25
SavedModel format, 669
saving and restoring models, 314
Scaled Dot-Product Attention layer, 559
Scaled Exponential Linear Unit (SELU) function, 334, 337-338, 368
Scikit-Learn
AdaBoost version used in, 203
anomaly and novelty detection, 274
automatic reconstruction with, 229
bagging and pasting in, 194
benefits of, xvi
CART training algorithm, 177, 179
clustering algorithms in, 258
computing classifier metrics, 92-107
converting text to numbers, 66
cross_val_score() function, 89
data centering in, 221
dataset dictionary structure, 85
DecisionTreeRegressor class, 183
design principles, 64
dimensionality reduction in, 232
ExtraTreesClassifier class, 198
feature importance scoring, 198
feature scaling, 154
full SVD approach, 225
GBRT ensemble training in, 204
GridSearchCV, 76
incremental training in, 207
IncrementalPCA class, 226
installing, 42
K-fold cross-validation feature, 73
KernelPCA class, 227
launching, monitoring, and maintaining your system, 80
linear model using, 21
linear regression using, 116
LLE (Locally Linear Embedding), 230, 232
max_depth hyperparameter, 181
mean_squared_error function, 72

missing value handling, 63
one-hot vectors, 67
out-of-bag evaluation, 195
PCA using, 222
Perceptron class, 287
presorting data with, 180
Randomized PCA algorithm, 225
random_state hyperparameter, 185
saving models, 75
SGDClassifier class, 88
splitting datasets into subsets, 53
stratified sampling using, 54
SVM classification classes, 162
SVM models, 155
tolerance hyperparameter, 162
transformation sequences, 70
transformers and, 68
voting classifiers in, 191
Scikit-Optimize, 322
SE block, 476
SE-Inception, 476
SE-ResNet, 476
search engines, 238
second-order partial derivatives (Hessians), 358
self-attention mechanism, 556
self-normalization, 337
self-organizing maps (SOMs), 780
self-supervised learning, 351
SELU (Scaled Exponential Linear Unit) func-
    tion (see Scaled Exponential Linear Unit
    (SELU) function)
semantic interpolation, 590
semantic segmentation, 249, 458, 492
semi-supervised learning
    clustering algorithms for, 237, 253
    defined, 13
    examples of, 13
SENet (Squeeze-and-Excitation Network), 476
sensitivity, 91
sentence encoders, 541
sentiment analysis
    defined, 526
    masking, 538
    overview of, 534
    reusing pretrained embeddings, 540
separable convolution, 474
sequence-to-sequence models, 510
sequence-to-vector networks, 501
SequenceExample protobuf (TensorFlow), 429

sequences
    forecasting time series, 503-511
    handling long, 511-523
    input and output, 501
    RNNS for, 497
Sequential API
    image classifiers using, 297-307
    regression MLP using, 307
service account, 682
sets, 383, 787
Shannon's information theory, 180
short-term memory problems, 514-523
shortcut connections, 471
shrinkage, 205
shuffling-buffer approach, 417
sigmoid (Logistic) activation function, 143,
    293-294, 302, 332
sigmoid kernel, 171
silhouette coefficient, 246
silhouette diagram, 247
silhouette score, 246
similarity functions, 159
simulated annealing, 125
simulated environments, 614
single-shot learning, 495
Singular Value Decomposition (SVD), 117, 221
skewed datasets, 89
skip connections, 337, 471
Sklearn-Deap, 323
slack variables, 167
smoothing term, 340
Soft Actor-Critic algorithm, 663
soft clustering, 240
soft margin classification, 154
soft voting, 192
softmax function, 148, 294, 299, 470, 482, 488,
    543
Softmax Regression, 148
softplus activation function, 293
spam filters, 1, 2
spare replicas, 706
sparse autoencoders, 582
sparse matrix, 67
sparse models, 359
sparse tensors, 383, 785
sparsity, 582
sparsity loss, 583
Spearmint library, 322
spectral clustering, 259

spurious patterns, 774
stacked autoencoders
    overview of, 572
        stacked denoising autoencoders, 581
        unsupervised pretraining using, 576-579
        using Keras, 572
        visualizing Fashion MNIST Dataset, 574
        visualizing reconstructions, 574
stacked denoising autoencoders, 581
stacked generalization, 208
stacking, 208
stale gradients, 707
standard correlation coefficient, 58
standardization, 69
start of sequence (SoS) token, 535
state-action values, 628
stateful metrics, 389
stationary point, 761
statistical mode, 193
statistical significance, 182
step function, 284
Stochastic Gradient Boosting, 207
Stochastic Gradient Descent (SGD), 88, 124
stochastic neurons, 775
stochastic policy, 612
stratified sampling, 53
streaming metrics, 389
stride, 449
string kernels, 161
string subsequence kernel, 161
string tensors, 383, 783
strong learners, 190
style mixing, 606
style transfer, 604
StyleGANs, 567, 604
Subclassing API, 313
subderivatives, 173
subgradient vector, 140
subsampling, 456
subspace, 215
summaries (TensorFlow), 317
supervised learning
    algorithms covered, 9
    common tasks, 8
    defined, 8
Support Vector Machines (SVMs)
    benefits of, 153
    decision function and prediction, 165
    dual problem, 168, 761

kernelized SVM, 169
linear SVM classification, 153
nonlinear SVM classification, 157-162
online SVMs, 172
SVM regression, 162
training objective, 166
support vectors, 154
symbolic differentiation, 768
symbolic tensors, 408, 792
symmetry, breaking in backpropagation, 291
synchronous updates, 706

T

t-Distributed Stochastic Neighbor Embedding
    (t-SNE), 233
tail-heavy histograms, 51
Talos library, 322
target model, 639
TD error, 630
TD target, 630
temperature
    in Boltzmann machines, 775
    in text generation, 531
Temporal Difference Learning (TD Learning),
    629
tensor arrays, 383, 786
TensorBoard, 317
TensorFlow Addons, 545
TensorFlow cluster, 711
TensorFlow Extended (TFX), 440
TensorFlow Hub, 378, 540
TensorFlow Lite, 378
TensorFlow Model Optimization Toolkit (TF-
    MOT), 359
TensorFlow Playground, 295
TensorFlow, basics of
    architecture, 377
    benefits, xvi, 376
    community support, 379
    features, 376
    getting help, 379
    installing, 296
    library ecosystem, 378
    operating system compatibility, 378
    PyTorch library and, 296
    versions covered, 375
TensorFlow, CNNs
    convolution operations, 494
    convolutional layers, 453

pooling layer, 458
TensorFlow, custom models and training
  about, 375
  activation functions, initializers, regulariz-
    ers, and constraints, 387
  computing gradients using Autodiff, 399,
    765-772
  implementing learning rate scheduling, 363
  layers, 391
  loss functions, 384
  losses and metrics, 397
  metrics, 388
  models, 394
  saving and loading, 385
  special data structures, 783-789
  training loops, 402
TensorFlow, data loading and preprocessing
  Data API, 414-424
  overview of, 413
  preprocessing input features, 430-439
  TensorFlow Datasets (TFDS) Project, 441,
    441
  TF Transform, 439
  TFRecord format, 424-430
TensorFlow, functions and graphs
  AutoGraph and tracing, 407, 791-799
  overview of, 405
  TF Function rules, 409
TensorFlow, model deployment at scale
  deploying on AI platforms, 81
  deploying to mobile and embedded devices,
    685-688
  overview of, 667
  serving TensorFlow models, 668-685
  training models across multiple devices,
    701-717
  using GPUs to speed computations, 689-701
TensorFlow, NumPy-like operations
  other data structures, 383
  tensors and NumPy, 381
  tensors and operations, 379
  type conversions, 381
  variables, 382
TensorFlow.js, 378
tensors, 379
Term-Frequency × Inverse-Document-
  Frequency (TF-IDF), 439
terminal state, 626
test sets, 30, 51

testing and validation
  data mismatch, 32
  hyperparameter tuning, 31
  model selection, 31
text generation
  building and training models for, 530
  chopping sequential datasets, 528
  generating Shakespearean text, 531
  overview of, 526
  splitting sequential datasets, 527
  stateful RNNs and, 532
  training dataset creation, 527
  using models for, 531
TF Datasets (TFDS), 414, 441
TF Functions
  graphs generated by, 791-799
  rules, 409
TF Transform (tf.Transform), 414, 439
TF-Agents library
  collect driver, 656
  datasets, 658
  deep Q-networks (DQNs), 650
  DQN agents, 652
  environment specifications, 644
  environment wrappers, 645
  environments, 643
  installing, 643
  overview of, 642
  replay buffer and observer, 654
  training architecture, 649
  training loops, 661
  training metrics, 655
tf.keras, 295, 363, 363, 423
tf.summary package, 319
TF.Text library, 536
TFRecord format
  compressed TFRecord files, 425
  lists of lists using SequenceExample Proto-
    buf, 429
  loading and parsing examples, 428
  overview of, 424
  protocol buffers (protobufs), 425
  TensorFlow protobufs, 427
Theano, 295
theoretical information criterion, 267
thermal equilibrium, 775
threshold logic unit (TLU), 284
Tikhonov regularization, 135
time series data

additional models for, 506
baseline metrics, 505
deep RNNS, 506
forecasting several steps ahead, 508
overview of, 503
RNNS for, 497
simple RNNs, 505
time step, 498
tokenization, 536
tolerance, 123
TPUs (tensor processing units), 377
train-dev sets, 32
training data
defined, 2
hold outs, 31
insufficient quantity of, 23
irrelevant features, 27
nonrepresentative, 25
overfitting, 27
poor quality, 26
training dataset creation, 527
underfitting, 29
training instances, 2, 215
training models
defined, 20
example project, 72
Gradient Descent, 118-128
learning curves, 130-134
Linear Regression, 112-117
Logistic Regression, 142-151
overview of, 111
Polynomial Regression, 128-130
regularized linear models, 134-142
training samples, 2
training set rotation, 185
training sets, 2, 30, 213
training/serving skew, 440
trajectories, 649
trajectory, 650
transfer learning, 324, 346, 481
transformations
affine transformations, 604
chaining, 415
custom, 68
inverse transformation, 225
purpose of, 64
transformation pipelines, 70
Transformer architecture, 554
transposed convolutional layer, 493

true negative rate (TNR), 97
true positive rate (TPR), 91
truncated backpropagation through time, 529
Turing test, 525
tying weights, 577
type conversions, 381

U

uncertainty sampling, 255
undercomplete autoencoders, 570
underfitting, 29
undiscounted rewards, 656
univariate regression problems, 39
univariate time series, 503
unrolling the network through time, 498
unstable gradients problem, 512
unsupervised learning
algorithms covered, 10
clustering, 236-260
common tasks, 10
defined, 9
Gaussian mixtures model (GMM), 260-275
overview of, 235
pretraining using stacked autoencoders,
576-579
unsupervised pretraining, 349
upsampling layer, 493
utility functions, 20

V

validation sets, 31
Value Iteration algorithm, 627
vanishing/exploding gradients problems,
332-345
variables, 382
variance
explained variance ratio, 222
preserving, 219
variational autoencoders, 586-591
variational inference, 272
variational parameters, 272
vector-to-sequence networks, 501
vectors
column vectors, 113
feature vectors, 113
momentum vector, 352
parameter vectors, 113
subgradient vectors, 140
VGGNet, 470

virtual GPU devices, 695
visible units, 775
visual attention, 552
visualization algorithms, 11
vocabulary, 432
voice recognition, 445

W

wall time, 341
warmup phase, 708
WaveNet, 498, 521
weak learners, 190
weighted moving average model, 506
white box models, 178
Wide & Deep neural networks, 308
wisdom of the crowd, 189
word embeddings, 434

word tokenization, 536
WordTrees, 490
workspace creation, 42

X

Xavier initialization, 333
Xception (Extreme Inception), 474
XGBoost, 208

Y

You Only Look Once (YOLO), 489

Z

zero padding, 449
zero-shot learning (ZSL), 564
ZF Net, 466

About the Author

Aurélien Géron is a Machine Learning consultant and lecturer. A former Googler, he led YouTube's video classification team from 2013 to 2016. He's been a founder of and CTO at a few different companies: Wifirst, a leading wireless ISP in France; Polyconseil, a consulting firm focused on telecoms, media, and strategy; and Kiwisoft, a consulting firm focused on Machine Learning and data privacy.

Before all that he worked as an engineer in a variety of domains: finance (JP Morgan and Société Générale), defense (Canada's DOD), and healthcare (blood transfusion). He also published a few technical books (on C++, WiFi, and internet architectures) and lectured about computer science at a French engineering school.

A few fun facts: he taught his three children to count in binary with their fingers (up to 1,023), he studied microbiology and evolutionary genetics before going into software engineering, and his parachute didn't open on the second jump.

Colophon

The animal on the cover of *Hands-On Machine Learning with Scikit-Learn, Keras, and TensorFlow* is the fire salamander (*Salamandra salamandra*), an amphibian found across most of Europe. Its black, glossy skin features large yellow spots on the head and back, signaling the presence of alkaloid toxins. This is a possible source of this amphibian's common name: contact with these toxins (which they can also spray short distances) causes convulsions and hyperventilation. Either the painful poisons or the moistness of the salamander's skin (or both) led to a misguided belief that these creatures not only could survive being placed in fire but could extinguish it as well.

Fire salamanders live in shaded forests, hiding in moist crevices and under logs near the pools or other freshwater bodies that facilitate their breeding. Though they spend most of their lives on land, they give birth to their young in water. They subsist mostly on a diet of insects, spiders, slugs, and worms. Fire salamanders can grow up to a foot in length, and in captivity may live as long as 50 years.

The fire salamander's numbers have been reduced by destruction of their forest habitat and capture for the pet trade, but the greatest threat they face is the susceptibility of their moisture-permeable skin to pollutants and microbes. Since 2014, they have become extinct in parts of the Netherlands and Belgium due to an introduced fungus.

Many of the animals on O'Reilly covers are endangered; all of them are important to the world. The cover illustration is by Karen Montgomery, based on an engraving from *Wood's Illustrated Natural History*. The cover fonts are URW Typewriter and Guardian Sans. The text font is Adobe Minion Pro; the heading font is Adobe Myriad Condensed; and the code font is Dalton Maag's Ubuntu Mono.

O'REILLY®

There's much more
where this came from.

Experience books, videos, live online
training courses, and more from O'Reilly
and our 200+ partners—all in one place.

Learn more at oreilly.com/online-learning